D0931580

Advanced Asymmetric Synthesis

Advanced Asymmetric Synthesis

Edited by

G. R. STEPHENSON
School of Chemical Sciences
University of East Anglia
Norwich

CHAPMAN & HALL
London · Weinheim · New York · Tokyo · Melbourne · Madras

Published by
Blackie Academic & Professional, an imprint of Chapman & Hall,
2–6 Boundary Row, London SE1 8HN

Chapman & Hall, 2–6 Boundary Row, London SE1 8HN, UK

Chapman & Hall GmbH, Pappelallee 3, 69469 Weinheim, Germany

Chapman & Hall USA, 115 Fifth Avenue, Fourth Floor, New York, NY 10003, USA

Chapman & Hall Japan, ITP-Japan, Kyowa Building, 3F, 2-2-1 Hirakawacho, Chiyoda-ku, Tokyo 102, Japan

DA Book (Aust.) Pty Ltd, 648 Whitehorse Road, Mitcham 3132, Victoria, Australia

Chapman & Hall India, R. Seshadri, 32 Second Main Road, CIT East, Madras 600 035, India

First edition 1996

Typeset in 10/12pt Times by AFS Image Setters Ltd, Glasgow
Printed in Great Britain by Hartnolls Limited, Bodmin, Cornwall

ISBN 0 7514 0049 1

A catalogue record for this book is available from the British Library

Library of Congress Catalog Card Number: 95-78252

Printed on acid-free text paper, manufactured in accordance with ANSI/NISO Z39.48-1992 (Permanence of Paper)

Preface

This book has its origins in a series of annual short courses for industrial chemists, given in Norwich at the UEA since 1989. In these courses, the University invites visiting faculty to present review lectures and tutorial exercises on topics in the field of asymmetric synthesis. Speakers are asked to give a personal overview of their own research, placed in the context of the main contributions of other workers in the field. Most of the chapter authors for this book have participated in teaching on at least one of the UEA short courses, and the personal perspective has been carried over into this series of review articles. There is no better way to pass on insight and expertise, whether it be to course participants visiting the UEA for a few days, or to readers of a monograph.

The *Asymmetric Synthesis Short Course* makes no attempt at a comprehensive coverage of the subject in any one year, and though, over the period since the course began, almost all the topics of major importance have been reviewed, a comprehensive survey is similarly not the objective of this compilation. Areas have been chosen which are either of major industrial importance, or where particularly rapid progress in research has been made over the last few years. (These, after all, may be the major industrial topics of the future.) In this way, in a relatively short volume, we can draw attention to the main procedures that researchers must choose from when planning asymmetric target molecule synthesis, and offer a commentary on how these choices can be made. As it is for the courses, that is the purpose of this book.

Richard Stephenson
The University of East Anglia
Norwich

To Diana

Contributors

Dr Alexandre Alexakis	Laboratoire de Chimie des Organo-éléments, Tour 45, Université P. et M. Curie, 4, Place Jussieu, F-75252, Paris, Cedex 05, France.
Professor Carsten Bolm	Fachbereich Chemie, Philipps-Universität Marburg, D-35032 Marburg, Germany.
Professor Kevin Burgess	Department of Chemistry, Texas A & M University, College Station, Texas 77843, USA.
Dr Eva Campi	Department of Chemistry, Monash University, Clayton, Victoria 3168, Australia.
Dr Rudolf O. Duthaler	Ciba-Geigy AG, Materials Research, CH-1723 Marly 1, Switzerland.
Professor Jean-Pierre Genet	Laboratoire de Synthèse Organique, Ecole Nationale Supérieure de Chimie de Paris, Rue Pierre et Marie Curie, 75231 Paris, 11 Cedex 05, France.
Dr Andreas Hafner	Ciba-Geigy AG, Materials Research, CH-1723 Marly 1, Switzerland.
Dr Karl J. Hale	Department of Chemistry, Christopher Ingold Laboratories, University College London, 20 Gordon Street, London WC10 0AJ, UK.
Dr Soraya Manaviazar	Department of Chemistry, Christopher Ingold Laboratories, University College London, 20 Gordon Street, London WC10 0AJ, UK.
Dr Pierre Mangeney	Laboratoire de Chimie des Organo-éléments, Tour 45, Université P. et M. Curie, 4, Place Jussieu, F-75252, Paris, Cedex 05, France.
Dr Patrick Perlmutter	Department of Chemistry, Monash University, Clayton, Victoria 3168, Australia.
Dr Nigel S. Simpkins	Department of Chemistry, The University of Nottingham, Nottingham Park, Nottingham NG7 2RD, UK.

Professor Guy Solladié Ecole Européenne des Hautes Etudes des Industries Chimiques, Université Louis Pasteur, F-67008, Strasbourg, France.

Dr G. Richard Stephenson School of Chemical Sciences, University of East Anglia, Norwich NR4 7TJ, UK.

Dr Nicholas J. Turner Department of Chemistry, University of Edinburgh, King's Buildings, West Mains Road, Edinburgh, EH9 3JJ, UK.

Dr Wilfred A. van der Donk Department of Chemistry, 18-390, Massachusetts Institute of Technology, 77 Massachusetts Avenue, Cambridge, MA 02139, USA.

Dr Andrew Whiting Department of Chemistry, U.M.I.S.T., PO Box 88, Manchester M69 1QD, UK.

Dr Jonathan M.J. Williams Department of Chemistry, Loughborough University of Technology, Loughborough, LE11 3TU, UK.

Contents

1 Design of asymmetric synthesis **1**
G.R. STEPHENSON and J.-P. GENET

1.1 Introduction 1
1.2 Asymmetric and enantioselective synthesis 1
1.3 Design of asymmetric synthesis 1
1.3.1 *De novo* asymmetric synthesis 2
1.3.2 Asymmetric induction 2
1.3.3 Chirality relay 2
1.3.4 The design challenge 3
1.4 Beyond the fourth generation? 4
1.5 Strategies to improve the chances of success 4
1.6 Importance of chirality and asymmetric synthesis 7
1.7 Conclusion 8
References 8

2 Non-linear effects in enantioselective synthesis: asymmetric amplification **9**
C. BOLM

2.1 Enantioselective alkylation of aldehydes: use of zinc reagents 11
2.2 Enantioselective oxidations and C–C bond formations: use of titanium reagents 15
2.3 Asymmetric conjugate additions to enones: use of copper and zinc reagents, nickel catalysis 19
2.4 Enantioselective reductions and C–C bond formations: use of lanthanide catalysts 21
2.5 Asymmetric C–C bond formations: use of amino acid and peptide catalysts 23
2.6 Conclusion 24
References 25

3 Chiral enolates **27**
K.J. HALE and S. MANAVIAZAR

3.1 Introduction 27
3.2 The aldol reactions of chiral enolates 27
3.2.1 Chiral auxiliary-mediated aldol-type additions 27
3.2.2 Asymmetric aldol reactions mediated by external chiral complexing agents under stoichiometric conditions 33
3.2.3 Catalytic asymmetric aldol reactions 39

3.3 Diastereoselective alkylation of chiral metal enolates 46
3.4 Asymmetric oxidation, amination and thioalkylation of chiral enolates 53
Acknowledgements 56
References 56

4 Optically active β-ketosulfoxides in asymmetric synthesis 60
G. SOLLADIÉ

4.1 Introduction 60
4.2 Optically active sulfoxide preparations 60
4.3 Stereoselective reduction of β-ketosulfoxides 64
4.4 Application of the β-ketosulfoxide reduction to total synthesis 73
4.5 Conclusions 89
References 90

5 Chiral aminals in asymmetric synthesis 93
A. ALEXAKIS and P. MANGENEY

5.1 Introduction 93
5.2 Chiral aminals as resolving agents 94
5.3 Conjugate addition reactions 96
5.4 Reactions with C=O and C=N 98
5.5 Chiral dihydropyridines 102
5.5.1 Chiral 4-substituted 1,4-dihydropyridines 102
5.5.2 Chiral 6-substituted 1,6-dihydropyridines 105
5.6 Cycloaddition reactions 105
5.6.1 Asymmetric dipolar additions with *N*-metallated azomethines 105
5.6.2 Cycloaddition of nitrile oxides 106
5.6.3 Inverse Diels–Alder reaction 106
5.7 Asymmetric tricarbonylchromium complexes 108
5.8 Cleavage of aminals 108
5.9 Conclusions 109
References 109

6 Asymmetric deprotonation reactions using enantiopure lithium amide bases 111
N.S. SIMPKINS

6.1 Introduction 111
6.2 Asymmetric deprotonation reactions 111
6.2.1 Prochiral ketones 111
6.2.2 Salt effects and the *in situ* quench protocol 117
6.2.3 Rationalisation of stereochemical results 119
6.2.4 Unsymmetrical ketone substrates 120
6.2.5 Non-ketone substrates 122
6.3 Conclusions 124
References 124

7 Asymmetric Diels–Alder reactions **126**
A. WHITING

7.1 Introduction 126
7.2 Diels–Alder reactions of C=C 126
7.2.1 Chiral dienophiles 126
7.2.2 Chiral dienes 131
7.2.3 Chiral catalysts 133
7.3 Diels–Alder reactions of C=O 137
7.3.1 Chiral dienophiles 137
7.3.2 Chiral catalysts 138
7.4 Diels–Alder reactions of C=N 139
7.4.1 Chiral dienophiles 139
7.4.2 Chiral dienes 140
7.4.3 Chiral catalysts 141
7.5 Diels–Alder reactions of N=O 141
7.5.1 Chiral dienophiles 141
7.6 Conclusions 142
References 142

8 Transition metal catalysts for asymmetric reduction **146**
J.-P. GENET

8.1 Transition metal complexes for asymmetric hydrogenation reactions 146
8.2 Chiral ligands 146
8.3 Preparation of chiral catalysts 148
8.3.1 Chiral rhodium catalysts 148
8.3.2 Chiral ruthenium catalysts 148
8.3.3 Chiral iridium and cobalt catalysts 150
8.4 Asymmetric hydrogenation of alkenes 151
8.4.1 Hydrogenation of dehydroamino acids and related substrates 151
8.4.2 Simple alkenes 162
8.5 Asymmetric hydrogenation of keto groups 163
8.5.1 Ketones 163
8.5.2 Alpha-keto esters, acids, amides, lactones 165
8.5.3 Hydrogenation of β-keto derivatives 166
8.6 Asymmetric hydrogenation of C=N 174
References 176

9 Asymmetric hydroboration **181**
K. BURGESS and W.A. VAN DER DONK

9.1 Introduction 181
9.2 Reagent-controlled diastereoselectivity in hydroborations 182
9.2.1 Scope and limitations 182
9.2.2 Stereodiscriminating hydroborating reagents from optically active alkenes 183
9.2.3 Resolution of a chiral borane 192
9.3 Enantioselective hydroborations 194
9.3.1 Mechanism of rhodium-catalysed hydroboration 196
9.3.2 Scope and limitations 196
9.4 Substrate-controlled diastereoselectivity in hydroborations 200

9.4.1 Intermolecular reactions 200
9.4.2 Intramolecular reactions 205
9.5 Conclusions 209
References 210

10 Asymmetric hydrosilylation and hydroformylation reactions 212
E. CAMPI and P. PERLMUTTER

10.1 Introduction 212
10.2 Asymmetric hydrosilylation 212
10.2.1 Asymmetric hydrosilylation of ketones 212
10.2.2 Asymmetric hydrosilylation of alkenes 213
10.3 Asymmetric hydroformylation of alkenes 217
10.4 Conclusions 220
References 220

11 Asymmetric conjugate addition reactions 222
P. PERLMUTTER

11.1 Introduction 222
11.2 Stereoselective conjugate addition reactions 222
11.3 Organocuprate reagents 224
11.3.1 Lithium heterocuprates 224
11.3.2 Lithium homocuprates 226
11.4 Organomagnesium reagents 227
11.5 Organozinc reagents 228
11.6 Conclusions 229
References 229

12 Stereoselective transformations mediated by chiral titanium complexes 231
A. HAFNER and R.O. DUTHALER

12.1 Introduction 231
12.2 Synthesis and structures of chiral complexes 232
12.2.1 Titanium complexes with chiral monodentate ligands 233
12.2.2 Titanium complexes with bidentate ligands 234
12.3 Addition of d^3-nucleophiles to carbonyl compounds 239
12.3.1 Enantioselective aldol reaction with titanium enolates 239
12.3.2 Enantioselective allytitanation of aldehydes 243
12.4 Addition of d^1-nucleophiles to carbonyl compounds 251
12.4.1 Stoichiometric use of chiral alkyltitanium complexes 251
12.4.2 Chiral titanium complexes as catalysts for enantioselective additions using dialkylzinc compounds 253
References 256

13 Asymmetric synthesis using enzymes and whole cells 260
N.J. TURNER

13.1 Introduction 260
13.2 Hydrolysis 260

13.3 Reduction 265
13.4 Oxidation 268
13.5 Formation of C–C bonds 270
13.6 Conclusions 272
References 273

14 Asymmetric palladium-catalysed coupling reactions 274
G.R. STEPHENSON

14.1 Introduction 274
14.1.1 Mechanism of Heck coupling 276
14.1.2 Mechanism of cross coupling 277
14.2 Enantioselective modification of palladium-catalysed coupling reactions 277
14.2.1 Role of chiral auxiliaries in Heck coupling reactions 278
14.3 Strategy 1: enantioface differentiation 279
14.3.1 The intermolecular case 280
14.3.2 The intramolecular case 281
14.3.3 Combining asymmetric induction and kinetic resolution 285
14.4 Strategy 2: enantiotopos differentiation 288
14.4.1 Enantiotopic ends of a symmetrical alkene 288
14.4.2 Enantiotopos differentiation with two alkenes 289
14.5 Strategy 3: combined enantioface and enantiotopos recognition 293
14.5.1 Use of a chiral nucleophile in the differentiation of prochiral alkenes 294
14.6 Strategy 4: asymmetric transformation 295
14.6.1 Kinetic resolution at the coupling step 295
14.6.2 Kinetic resolution selecting between two interconverting precursors 295
14.6.3 Differentiation at the reductive elimination step 297
14.7 Conclusions 297
References 297

15 Palladium allyl π-complexes in asymmetric synthesis 299
J.M.J. WILLIAMS

15.1 Introduction 299
15.2 Palladium allyl π-complexes 299
15.2.1 Simple palladium-catalysed allylic substitution 299
15.2.2 Palladium-catalysed reactions involving diastereomerically pure substrates 301
15.2.3 Palladium-catalysed reactions involving enantiomerically pure substrates 302
15.3 Diastereoselectivity with allylpalladium complexes 303
15.4 Enantiocontrol of palladium-catalysed allylic substitution by the use of enantiomerically pure ligands 305
15.4.1 Enantiocontrol via *meso*-complexes 305
15.4.2 Enantioselective palladium-catalysed allylic substitution reactions which do not proceed via *meso*-complexes 309
15.5 Conclusions 310
References 311

16 Stoichiometric π-complexes in asymmetric synthesis **313**
G.R. STEPHENSON

16.1 Introduction 313
16.1.1 Stoichiometric metal complexes in synthesis design 313
16.1.2 Chirality properties of transition metal π-complexes 314
16.1.3 Potential symmetry planes in transition metal π-complexes 315
16.1.4 Stereochemistry of reactions of transition metal π-complexes 318
16.2 Working ligands and their role in synthesis design 319
16.3 Stoichiometric metal complexes in skeletal bond-formation steps 320
16.3.1 η^2-Complexes 321
16.3.2 η^3-Complexes 323
16.3.3 η^4-Complexes 323
16.3.4 η^5-Complexes 327
16.3.5 η^6-Complexes 337
16.4 Reaching out beyond η^n 340
16.4.1 Metal-controlled reactions of metal-stabilised cations and conventional functionality, adjacent to the η^n unit 340
16.4.2 Metal-derived activation derived from the η^n unit 351
16.5 Electrophilicity and nucleophilicity of transition metal π-complexes 354
16.6 Skeletal bond formation during the introduction and detachment of the metal/ligand fragment 354
16.6.1 Strategies to improve the efficiency of reaction sequences that make multiple use of the metal 354
16.6.2 Reactions that form metal complexes with concomitant skeletal bond formation 355
16.6.3 Decomplexation with concomitant carbon–carbon bond formation 359
16.7 Conclusions 363
References 363

17 Asymmetric oxidation **367**
G.R. STEPHENSON

17.1 Introduction 367
17.2 Importance of the Sharpless asymmetric epoxidation 368
17.3 Mechanism of the Sharpless asymmetric epoxidation 371
17.4 Jacobsen–Katsuki epoxidation 373
17.5 Other methods of asymmetric epoxidation 377
17.6 Sharpless asymmetric *cis*-dihydroxylation 378
17.7 Choice of auxiliary for Sharpless asymmetric dihydroxylation 381
17.8 Limitations and alternative dihydroxylation procedures 384
17.9 Asymmetric oxidation of thioethers 386
17.10 Other asymmetric oxidations 387
17.11 Conclusions 389
Acknowledgement 389
References 389

18 Worked examples in asymmetric synthesis design **392**
G.R. STEPHENSON

18.1 Symmetry and synthesis design 392
18.1.1 C_2 axes in synthetic planning: simplification of synthetic routes 392
18.1.2 *S* axes: prochiral intermediates in synthetic planning 392

18.2 How to use symmetrical intermediates in enantioselective synthesis 395
18.2.1 Enantioface differentiation 395
18.2.2 Enantiotopic group differentiation 397
18.2.3 Deleting a chiral centre to induce asymmetry 399
18.3 Comparison of enantioface and enantiotopos differentiation 400
18.4 Diastereoselectivity 401
18.5 Planning for stereocontrol 403
18.6 Conclusions 405
References 405

Key to abbreviations for chiral auxiliaries 406

Index 415

1 Design of asymmetric synthesis

G.R. STEPHENSON and J.-P. GENET

1.1 Introduction

There is a sense in which asymmetric synthesis is the last frontier for organic chemistry, at least as far as its application to the construction of target molecules is concerned. Specialists in the field, however, need not fear for their jobs, because there is still a great deal to achieve to perfect the science of the enantioselective preparation for organic compounds. While the challenges of chemo- and regioselective bond formation are well understood, perhaps even conquered, stereoselective reactions still pose problems and constitute a formidable task if required on a plant-scale, as, for example, in the manufacture of enantiopure pharmaceutical reagents. This book surveys the methods of asymmetric synthesis and illustrates their use, particularly in routes to natural products.

1.2 Asymmetric and enantioselective synthesis

There are many definitions of asymmetric synthesis but most are cumbersome to state, because of technical concerns about the differences between asymmetric induction and asymmetric reactions. For the purposes of this book, an asymmetric synthesis is one which creates new stereogenic units in a controlled way. Such processes may be enantioselective or diastereoselective (or both), and so they may give rise to enantiopure products or racemic products, with control of relative stereochemistry. Readers should take care. It is not strictly correct to refer to a synthesis as chiral, and, though starting materials, reagents, catalysts, auxiliaries and reaction products may be chiral, to say this indicates nothing about enantiomeric purity. However, nowadays, when we refer to asymmetric synthesis, we generally mean one which is capable of giving rise to enantiopure products. Certainly it is this type of asymmetric synthesis which is the most important and poses the greatest challenge.

1.3 Design of asymmetric synthesis

Three types of reaction, or sequence of reactions, are relevant to the design of asymmetric synthesis:

1. *de novo* asymmetric synthesis
2. asymmetric induction
3. chirality relay.

1.3.1 De novo *asymmetric synthesis*

The first type of process is rare. Achiral starting materials can in certain circumstances become converted into chiral non-racemic products, and some process of this type must have been responsible for the original biasing of natural products to favour one enantiomeric series. The related concept of enantioselective autocatalysis is in the vanguard of new developments in asymmetric synthesis. Here, a chiral product catalyses its own enantioselective preparation. *De novo* methods will be of growing practical importance as laboratory procedures but are not at present in routine use in enantioselective synthetic routes.

1.3.2 Asymmetric induction

With asymmetric induction, we are in the heartland of modern asymmetric synthesis. A prochiral substrate or functional group is converted into an enantiopure product in a reaction mediated by a chiral auxiliary, either in a stoichiometric or catalytic fashion. When a chiral catalyst is involved, a tiny amount of the chiral auxiliary can produce a large amount of enantiopure product, so this can be a highly cost-efficient approach. Reactions of this type are enantioselective. Since the chiral auxiliary is not incorporated into the final product, the success of the stereoselectivity is determined by measuring the excess of one enantiomer of the product over the other.

1.3.3 Chirality relay

In some enantiomer syntheses of target molecules, optically pure starting materials are incorporated directly. These chiral centres are 'bought-in', and, of course, if they are simply taken through into the product, no real *asymmetric* synthesis has been performed. The supply of enantiopure starting materials for this purpose is sometimes referred to as the 'chiral pool'. Normally, however, the bought-in chirality must be modified, or new chiral centres introduced (substrate-controlled diastereoselectivity), a situation where diastereoselectivity is the measure of success. The efficiency of stereocontrol is determined by measuring the excess of the desired diastereoisomer over the others. In other cases, the original chirality of the building block is destroyed in the reaction that forms the new chiral centre. Now it is the ratio of enantiomers that is measured.

1.3.4 *The design challenge*

In planning asymmetric synthesis, a combination of asymmetric induction and chirality relay is normally employed. It can be induced chirality, not bought-in chirality, that is used in diastereoselective reactions to form new chiral centres as the intermediates are elaborated and the target structure takes shape. In the planning stage, the relationship between stereogenic features and the functional groups of the target molecule will determine which centres are best made by asymmetric induction and which by diastereoselective later steps. Furthermore, there can be short cuts in the design process, since several chiral centres can be formed in a single reaction step, so the features of the target structure should not be considered in isolation. In practice, a few highly versatile, well understood, classes of reaction are called into play in the vast majority of enantioselective synthetic routes. It is sensible in the design process to look to see if these popular 'classic' reactions can be employed before opting for a route that relies on a more novel (and inevitably more speculative) step. Approaches to asymmetric synthesis are sometimes grouped by 'generations', reflecting increasing sophistication of approach, and this can be helpful in choosing between the 'classic' and the 'speculative'.

(a) *First-generation asymmetric synthesis.* Diastereoselective reactions in which the formation of a new chiral centre is under the control of an existing centre in the same molecule. This is often referred to as substrate-controlled diastereoselectivity. Chirality relay is based on first-generation asymmetric synthesis.

(b) *Second-generation asymmetric synthesis.* A stoichiometric chiral auxiliary is covalently attached to the substrate before chirality relay is performed. Chirality in the auxiliary controls the asymmetric induction, and the auxiliary is removed for reuse once the new chiral centre is built.

(c) *Third-generation asymmetric synthesis.* When a stoichiometric chiral reagent effects asymmetric induction, the inducing chirality is not part of the substrate and so is not determined directly, in the planning process, by the structure of the target molecule. Efficient asymmetric induction is the only concern. The use of a chiral reagent is referred to as third-generation asymmetric synthesis or reagent-controlled diastereoselectivity.

(d) *Fourth-generation asymmetric synthesis.* Catalytic modifications of second- and third-generation methods tend to be considered together as a fourth-generation approach. The chiral auxiliary (and other catalyst components, e.g. a transition metal) can become covalently attached to the substrate in an intermediate in the catalytic cycle (cf. second-generation methods) or might act in an intermolecular fashion, inducing asymmetry in a single step.

1.4 Beyond the fourth generation?

When several chiral auxiliaries operate together in the same process, unexpected effects can be observed. Products can be formed in greater optical purity than the auxiliaries used to make them (asymmetric amplification) and reaction products can catalyse their own enantioselective formation (enantioselective autocatalysis). Systematic investigation of these effects is still in its infancy, but since the goal of asymmetric synthesis design is complete control of enantiomeric purity, new generation methods that are unrestricted by the availability (or expense) of optically pure starting materials, reagents and auxiliaries will have a profound effect on the way asymmetric synthesis is planned.

Before turning to a survey of some of the main methods of asymmetric synthesis, this book examines the new frontier of asymmetric amplification (Chapter 2).

1.5 Strategies to improve the chances of success

Whether the method chosen is 'classic' or 'speculative', old or new, it is unlikely to work perfectly in every situation! Fortunately, there are rational approaches to enhance incomplete stereoselectivity.

The first is double stereodifferentiation, which is illustrated by the comparison of Schemes 1.1 and 1.2. For double stereodifferentiation, an additional stereogenic unit is introduced into the reaction to bias selectivity between transition states. In Scheme 1.2, the additional feature is the chirality of R*, and this becomes covalently incorporated into the product. The principle, however, would apply equally well to third- or fourth-generation modifications, in which the extra chirality is in a reagent or catalyst.

Although completely *erythro* selective, the aldol reaction in Scheme 1.1 was only partially selective relative to the chiral centre in the optically active aldehyde (**1**). A 4:1 mixture of diastereomers was produced. This ratio could be improved[1] by the use of a modified enolate in which a second chiral centre *reinforces* the ability to recognise the diastereofaces of the aldehyde. In this situation, the stereocentres are referred to as 'matched'. If the enantiomer of (**1**) was used with the enolate derived from (**2**) the centres would be 'mismatched' and stereoselectivity would be worse.

A second approach to improve stereoselectivity combines enantiotopic group and diastereoface selection — simultaneous asymmetric induction and kinetic resolution. Such reactions have been the subject of a theoretical treatment by Schreiber, who reached the startling conclusion that an arbitrarily high enantiomeric excess can be obtained in such circumstances by allowing increasing amounts of double addition products to be formed (Table 1.1). Effective kinetic resolution can be relied on to enhance the initial asymmetric induction in such circumstances.

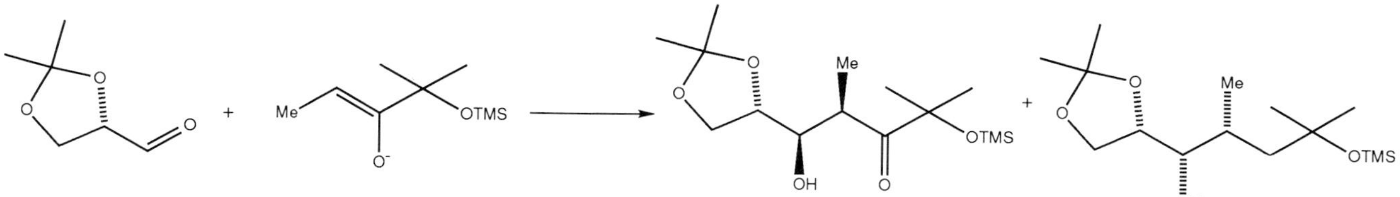

1

Me OTMS O⁻ O

OH Me O OTMS

4 : 1

Scheme 1.1

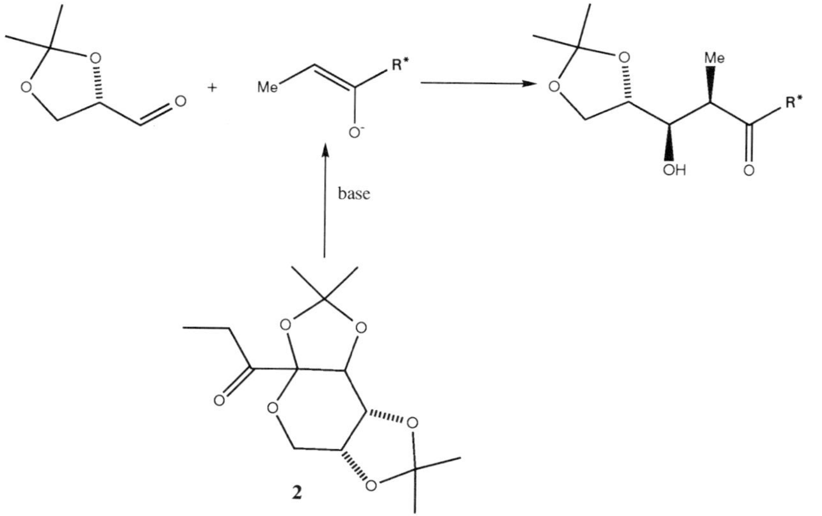

Me R* O⁻

base

2

OH Me R* O

Scheme 1.2

Table 1.1 Calculated values for combined enantiotopic group and diastereoface selectivity

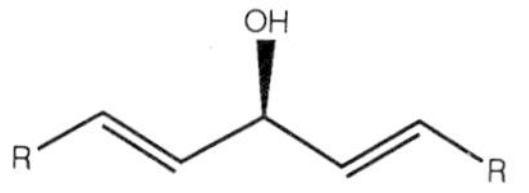

$[S]/[S]_{init}$	% yield	% ee
1	0	–
0.5	0.48	99.4
10^{-2}	0.93 (max)	99.96
10^{-3}	0.91	99.994
10^{-6}	0.85	99.99999

Initial concentration, $[S]_{init}$

This effect has been demonstrated[2] in Sharpless epoxidations, but should be general for reactions under kinetic control. The key is to allow some of

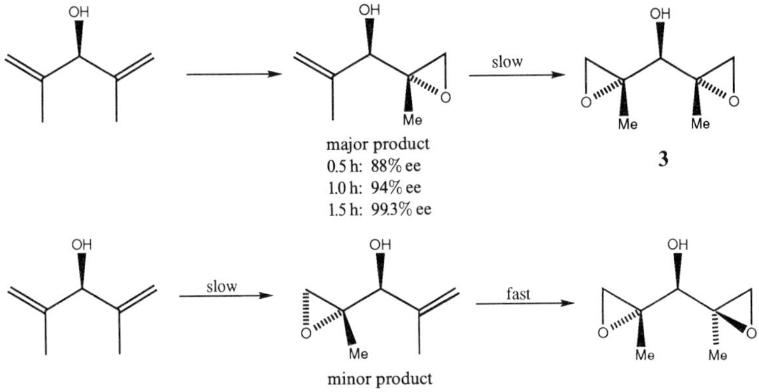

major product
0.5 h: 88% ee
1.0 h: 94% ee
1.5 h: 99.3% ee

minor product

Scheme 1.3

the product to progress on to (**3**). The minor product from the first reaction will react quicker because the remaining alkene is the form that is favoured in interaction with the catalyst.

It is clear from Table 1.1, that exceptionally high ee is achieved in vanishingly low yield. None-the-less, if the objective is to improve a moderately good 88% ee to >99% (Scheme 1.3), it can be seen that controlled double epoxidation provides a workable solution, though one which inevitably requires separation of reaction products.

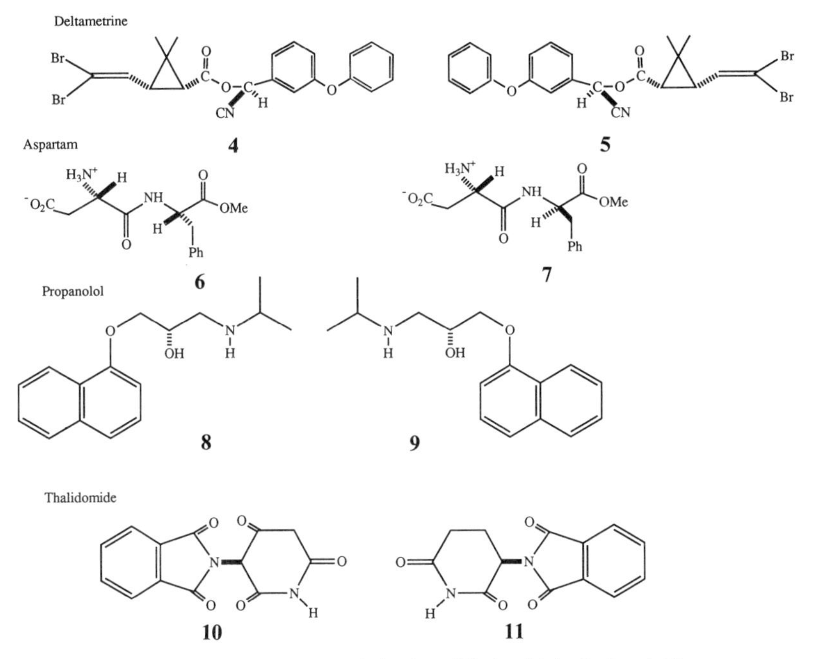

Figure 1.1 Chiral compounds having differing biological activities.

1.6 Importance of chirality and asymmetric synthesis

The importance of enantiomerically pure compounds comes from the central role of enantiomer recognition in biological activity.[3] There are many examples of pharmaceutical drugs,[4] agrochemicals[5] and other chemical compounds where the desired biological property is related to the absolute configuration. For example, in the series of eight possible stereoisomers of dibromovinyl-2,2-dimethylcyclopropane carboxylic esters (pyrethroid insecticides) (Figure 1.1), deltamethrin ((*R*,*R*,*S*)-**4**) is the most powerful insecticide whereas the stereoisomer ((*S*,*S*,*R*)-**5**) is inactive.[6] In food additives, aspartame ((*S*,*S*)-**6**) is used as an artificial sweetener whilst the (*S*,*R*) form (**7**) (Figure 1.1) tastes bitter and must be avoided in the manufacturing process.

The three main classes of biopolymer are all chiral. Nucleic acids and polysaccharides contain sugar subunits. Common ones such as ribose (in RNA) and glucose are chiral. Proteins contain chiral amino acids. Furthermore, these subunits are normally at high optical purity. The recognition events in biology, and the action of drugs that intervene in these events, will, therefore, almost always involve the molecular recognition of a biologically active molecule by a chiral non-racemic receptor structure. The two enantiomers of a drug molecule cannot be expected to bind equally well to the receptor and so should cause different biological responses.

In drugs tested for therapeutical use, it is often found that only one enantiomer possesses the desired biological activity, whereas the other enantiomer is inactive or possesses a different activity and causes toxic side effects. In the case of propanolol (Figure 1.1), the (*S*) enantiomer (**8**) is a β-blocker whilst the (*R*) form (**9**) possesses contraceptive activity. The most dramatic example is the well-known example of thalidomide (Figure 1.1) since the (*S*) enantiomer (**10**) exhibits teratogenic activity and the (*R*) enantiomer (**11**) was used as a sedative. Therefore, racemic mixtures had to be withdrawn from the market. Even when the other enantiomers are inert, it may be desirable to synthesise and use the active one in its pure form. The (*S*) enantiomer of ibuprophen is active as pain reliever whereas the (*R*) enantiomer is inactive. For drug delivery, the potency of an active enantiomer compared with a racemic of active and inactive enantiomers is such that the dose can be reduced in half. The second reason for the preparation of the active isomer is economic, the production of the inert isomer represents a waste of starting materials and resources. There are also less obvious, but no less significant, problems with optically impure pharmaceuticals, even when employed at high ee. Enantiomers may be competitive antagonists. This is the case for (+)- and (−)-isopropylnoradrenaline acting on the α_1-adrenergic receptors in rats.[7] A further possibility is that the non-beneficial enantiomer may be preferentially involved in biotoxication. An example is Deprenyl, an antidepressant and anti-Parkinson's disease drug, for which the less active (*S*) isomer is converted into (*S*)-(+)-methamphetamine and (*S*)-(+)-amphetamine, which cause undesired CNS stimulation.[8] It follows, in principle, that selective biotransformation of the 'inactive' enantiomer could also produce a far more potent antagonist for the receptor, reducing efficacy of even a substantially optically pure drug.

1.7 Conclusions

Economic, environmental and pharmacodynamic considerations all point to the need for enantiomerically pure drugs and agrochemicals. Enantioselective synthesis is of growing importance, particularly in the pharmaceutical industry, and is currently the subject of intense research in academic and industrial laboratories around the world.

References

1. C.H. Heathcock, C.T. White, J.J. Morrison and D. VanDerveer, *J. Org. Chem.*, 1981, **46**, 1296.
2. S.L. Schreiber, T.S. Schreiber and D.B. Smith, *J. Am. Chem. Soc.*, 1987, **109**, 1525.
3. R. Bentley, in *Stereochemistry*, C.H. Tamm (ed.), Elsevier, Amsterdam, 1982.
4. (a) S.C. Stinson (1992) Chiral Drugs *Chem. Eng. News*, 1992 (Sept.) **28**, 46; (b) A.S.C. Chan, *Chem. Tech.*, 1993, 46.
5. R.G.M. Tombo and D. Bellus, *Angew. Chem. Int. Ed. Engl.*, 1991, **30**(10), 1193.
6. J. Tessier and J.J. Hervé, *Deltametrine*, Roussel Uclaf Ed., 1982, p. 25.
7. E.J. Ariens, *Proc. 1st Int. Pharmacol. Meet.*, Vol. 7, Pergamon Press, Oxford, 1963, pp. 247–264.
8. G.P. Reynolds, J.D. Elsworth, K. Blau, M. Sandler, A.J. Lees and G.M. Stern, *Br. Clin. Pharmacol.*, 1978, **6**, 542.

2 Non-linear effects in enantioselective synthesis: asymmetric amplification

C. BOLM

In 1986 Kagan, Agami *et al.*[1] published a remarkable paper in the *Journal of the American Chemical Society*. The French groups had studied asymmetric oxidation and aldol reactions* and they reported unique *non-linear* relationships between the enantiomeric excess (ee) of a chiral auxiliary and the ee of the product obtained in these reactions.[1] In the titanium-mediated Sharpless epoxidation of geraniol (**1**) using diethyl tartrate (DET) as the chiral ligand (Scheme 2.1), the optical purities of the epoxide (**2**) were significantly higher than the values calculated by assuming a linear relationship between tartrate and epoxide enantiomeric excesses (Figure 2.1, curve b). In the asymmetric oxidation of methyl-*p*-tolyl sulfide (**3**) and in the proline-catalysed Robinson annulation of triketone (**5**) (Hajos–Parrish reaction) (Scheme 2.1), negative deviations from linearity were observed (Figure 2.1, curve c).

Two years later, Oguni *et al.* introduced the term 'asymmetric amplification' to describe the phenomena of obtaining products with very high enantiomeric excess using chiral auxiliaries of low ee.[3] They have found that the addition of diethylzinc to benzaldehyde (**7**) was catalysed by sterically constrained β-amino alcohols (Scheme 2.2), and that even with chiral catalysts of only 10–20% ee the optical purity of the product (**8**) was in the range of 80–90% ee.

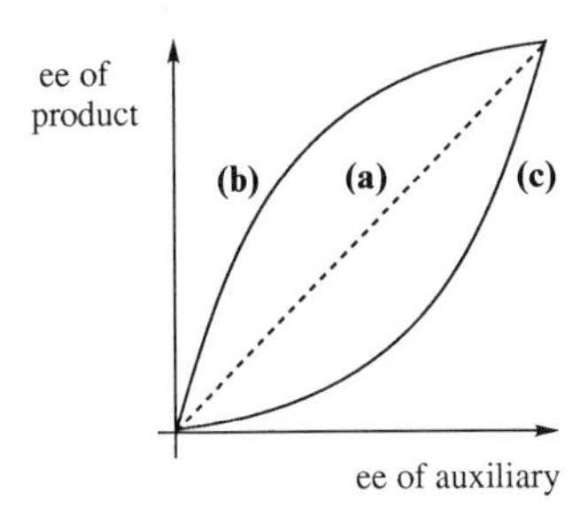

Figure 2.1 Schematic representations of correlations between the ee of the product and the ee of the chiral auxiliary.

* Recently, after completion of this manuscript, an extension of Kagan's earlier study was published. Two models were introduced which allow analysis of the experimental data. Various catalyst systems have been considered including those in which reactive species with more than two chiral ligands operate.[2]

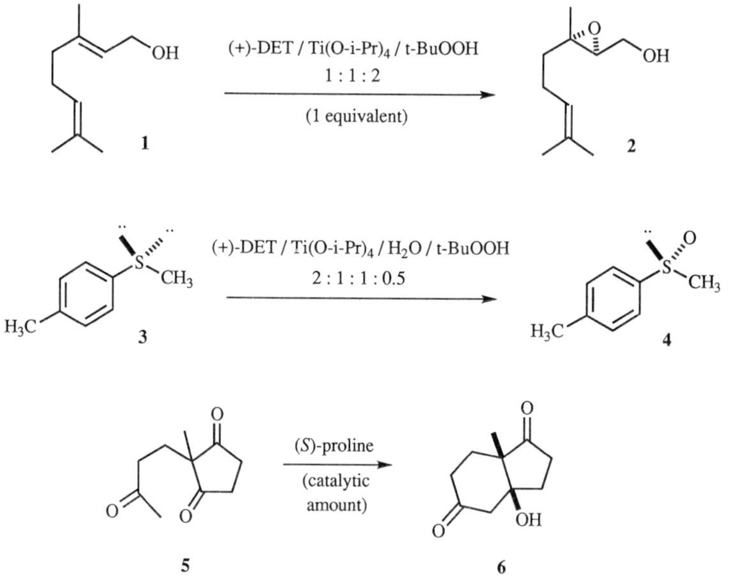

1 (+)-DET / Ti(O-i-Pr)$_4$ / t-BuOOH 1 : 1 : 2 (1 equivalent) → 2

3 (+)-DET / Ti(O-i-Pr)$_4$ / H_2O / t-BuOOH 2 : 1 : 1 : 0.5 → 4

5 (*S*)-proline (catalytic amount) → 6

Scheme 2.1

The mechanism of this catalysed asymmetric aldehyde alkylation was subsequently investigated by Noyori *et al.* and a comprehensive explanation of the origin of this phenomenon and the enantioselection was presented.[4]

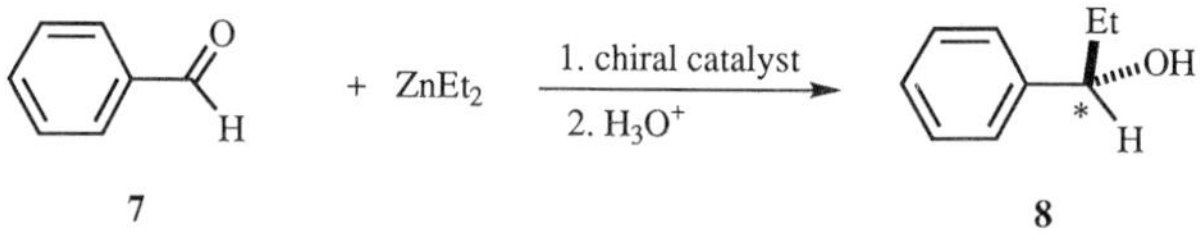

7 + $ZnEt_2$ 1. chiral catalyst 2. H_3O^+ → 8

Scheme 2.2

Non-linear effects have been observed in *stoichiometric* asymmetric reactions and in enantioselective *catalyses*, the latter being more difficult to interpret because of the kinetic factors of the overall catalysed process. All four reactions mentioned above, titanium-promoted alkene and sulfide oxidations, the Hajos–Parrish reaction and the asymmetric alkylation, have in common that species bearing more than one chiral ligand play a dominant role for enantioselection. Deviations from linearity are a result of diastereomeric interactions between aggregated complexes, or between chiral molecules in the transition state of the enantiodifferentiating step.

A simple scheme should illustrate the basic concept (Scheme 2.3).[1,2] If a chiral complex of the type $[M]L_2^*$ (where L* stands for a chiral ligand and [M] for the rest of the molecule) is used to mediate an asymmetric reaction, three different complexes can be expected if L* is *not* enantiomerically pure.

[M] + 2 L* ⟶ L_S-[M]-L_R *meso*-promoter

where L* is **not** enantiomerically pure

L_R-[M]-L_R } optically active promoters
L_S-[M]-L_S

---

case **A**: if *meso*-promoter is **ACTIVE**, and optically active promoter is **inactive**:

⟹ racemic product

case **B**: if *meso*-promoter is **inactive**, and optically active promoter is **ACTIVE**:

⟹ ASYMMETRIC AMPLIFICATION

Scheme 2.3

Two enantiomeric complexes, L_S-[M]-L_S and L_R-[M]-L_R, each having ligands with the same absolute configuration (homochiral ligands), and one diastereomeric '*meso*-type' complex, L_S-[M]-L_R, bearing two ligands of opposite chirality can be formed. If the last *achiral* complex is more active than L_S-[M]-L_S or L_R-[M]-L_R (Case A), the product of the asymmetric transformation will have a lower enantiomeric excess than expected from a linear correlation between ligand and product ee. If, however, L_S-[M]-L_S or L_R-[M]-L_R is more active than L_S-[M]-L_R (Case B), an *asymmetric amplification* can be expected, and the product ee will exceed the optical purity of the chiral auxiliary L*.

2.1 Enantioselective alkylation of aldehydes: use of zinc reagents

The basic model illustrated above has been used by Noyori *et al.* in their excellent study of enantioselective alkyl transfer from dialkylzincs to aldehydes to explain the origin of non-linear effects.[4] The addition of zinc reagents to aldehydes is markedly accelerated by the presence of catalytic amounts of β-amino alcohols. The corresponding alkylated products are obtained with very high optical purity.[5] For example, ethyl transfer from diethylzinc to benzaldehyde (**7**), catalysed by 8 mol% of (−)-DAIB (**9**), gives (*S*)-1-phenylpropanol (**8**) with 98% ee (Scheme 2.2)[4]

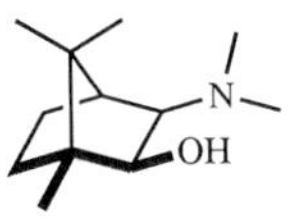

N OH OH N N OH

9 (DAIB) 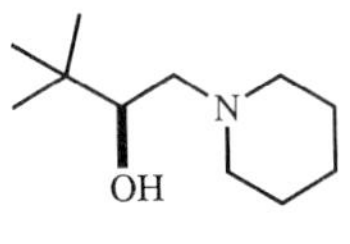 **10** 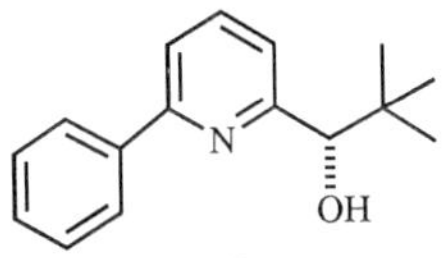 **11**

Table 2.1 Asymmetric amplification in enantioselective addition of diethylzinc to benzaldehyde

β-amino alcohol	% ee of catalyst	% ee of product	Reference
9	10.0	90.0	4a
9	15.0	95.0	4a
9	24.0	96.0	4a
9	53.0	97.0	4a
9	81.0	97.0	4a
9	100.0	98.0	4a
10	3.1	36.0	3
10	6.5	74.0	3
10	10.7	82.0	3
10	20.5	88.0	3
10	59.8	92.0	3
10	77.1	94.0	3
10	100.0	98.0	4c
11	2.1	39.0	6
11	9.6	74.7	6
11	14.0	86.6	6
11	30.8	87.8	6
11	33.1	87.4	6
11	38.5	87.7	6
11	57.8	86.5	6
11	78.9	85.7	6
11	95.6	86.5	6
11	>99.0	85.7	6

Determination of the correlation between product and catalyst ee showed a significant positive deviation from linearity. When (−)-DAIB of only 15% ee was used, ((*S*)-**8**) was obtained in 92% yield with 95% ee! Strong amplifications have also been found by Oguni,[3] Bolm[6] and others[7] using different catalyst precursors. Table 2.1 summarizes the results of amplification studies with catalytic amounts of β-amino alcohols (**9**), (**10**) and pyridyl alcohol (**11**)

The reaction mechanism has been elucidated by Noyori *et al.* on the basis of spectroscopic studies, kinetic measurements and isolation of key intermediates.[4] Accordingly, treatment of a β-amino alcohol with dialkylzinc affords a zinc alkoxide (**A**) which is in equilibrium with its more stable dinuclear dimer (**B**). In the alkyl transfer, unsaturated monomeric zinc complex (**A**) is the actual catalyst. Both reagents, aldehyde and dialkylzinc, are activated by coordination to **A**, and an appropriate three-dimensional arrangement of associate (**C**) leads to high enantioselection in the alkylation (Scheme 2.4).

If a β-amino alcohol of low optical purity is used, two monomeric zinc alkoxides of opposite chirality ($\mathbf{A}_R$ and $\mathbf{A}_S$) are formed. Dimerization leads to diastereomeric associates $\mathbf{B}_{S,R}$ and $\mathbf{B}_{S,S}/\mathbf{B}_{R,R}$. These hetero- and homochiral dinuclear zinc complexes differ in stability, and formation of the achiral *meso*-complex $\mathbf{B}_{S,R}$ is significantly favoured thermodynamically. Consequently, in a mixture of unequal amounts of zinc alkoxides, $\mathbf{A}_S$ and $\mathbf{A}_R$, the

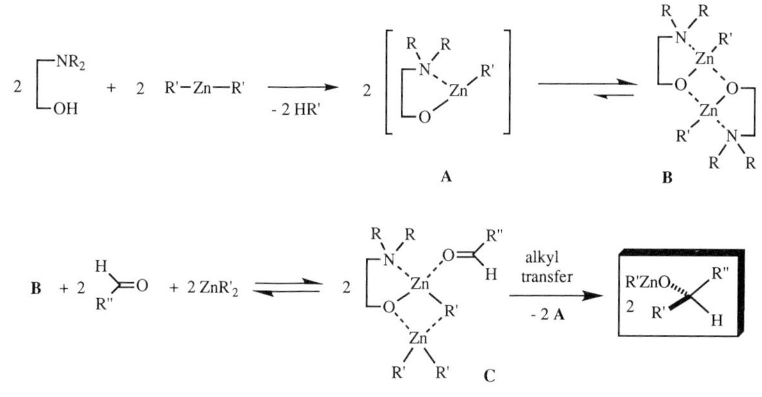

Scheme 2.4

minor enantiomer will be 'trapped' in the stable *meso*-complex. The enantiomer in excess forms a more labile dinuclear dimer which has a high tendency to dissociate, thereby forming the monomeric zinc species which serves as the alkylation catalyst (Scheme 2.5).

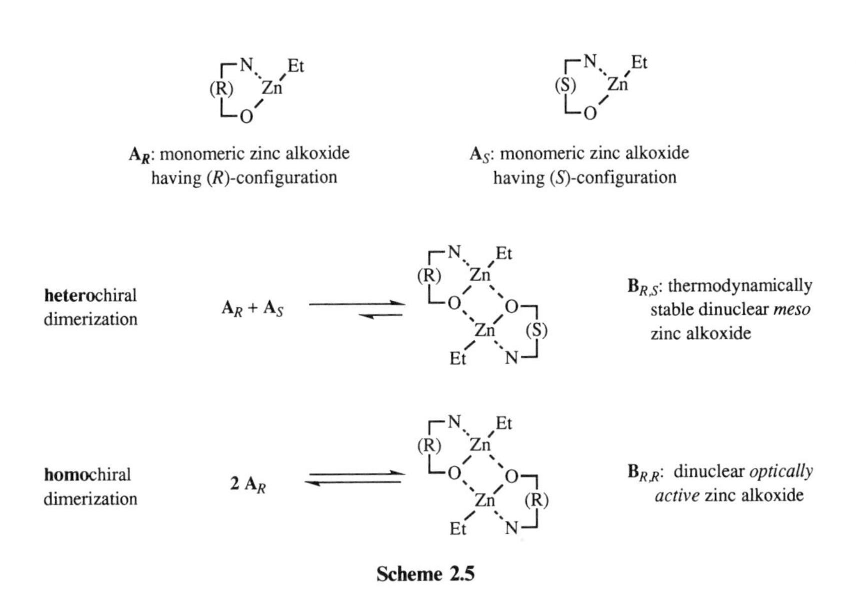

Scheme 2.5

Dinuclear zinc alkoxides derived from DAIB (**9**) and dimethylzinc have been characterized by X-ray crystallography.[4] Enantiomerically pure and racemic DAIB gave C_2-symmetric associate (**12**) and *meso*-complex (**13**) respectively. Both dimers have *endo*-fused central 5/4/5 tricyclic structures bridged via Zn_2O_2 four-membered rings. The *syn*-geometry of dimer (**12**) leads to a complex structure which is more congested than the one of *anti*-configurated *meso*-complex (**13**). NMR studies and molecular weight determinations show that zinc alkoxides (**12**) and (**13**) retain their dimeric nature in solution.

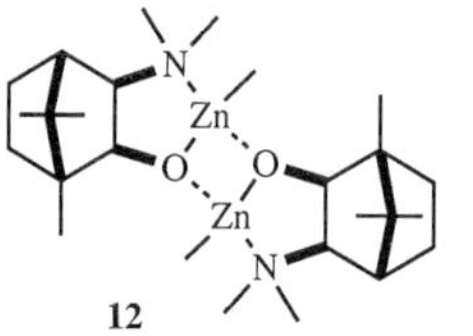

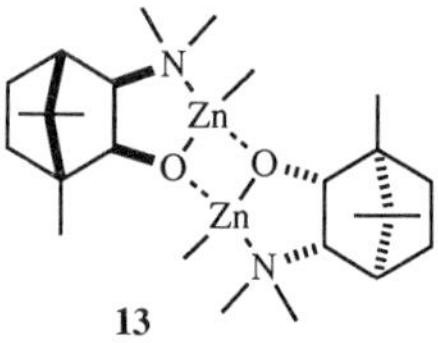

Solid-state molecular structures of ethylzinc-containing complexes have also been determined crystallographically.[6,8] Reaction of (rac-**11**) and (rac-**15**) with diethylzinc resulted in the formation of chelated alkoxy zinc dimers (**14**) and (**16**), respectively. Both possess an achiral *meso* structure similar to methylzinc complex (**13**).

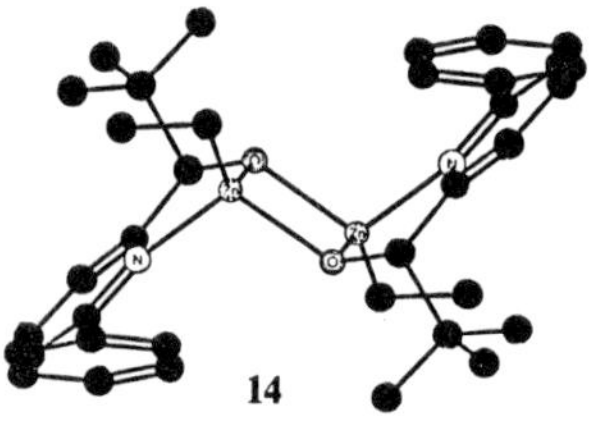

NMR-spectra of chiral dimers in the presence of dialkylzinc and benzaldehyde revealed that dynamic mixtures of aggregates were formed. In contrast, *meso*-complexes appear to be more inert under identical conditions.[4] These observations support the mechanistic scheme in which the minor enantiomer is sequestered into a catalytically inactive species. Upon addition of the zinc reagent to the aldehyde, product (**17**) is formed. This new chiral zinc alkoxide slowly converts into a stable cubic tetramer (**18**) (Scheme 2.6).

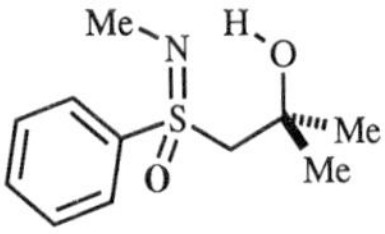

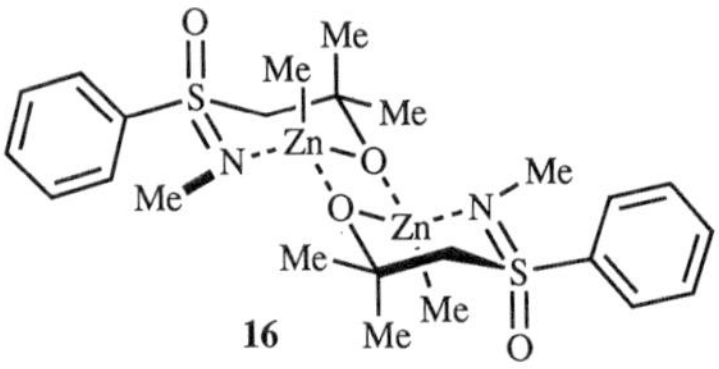

NMR studies indicated an influence of the absolute configuration of (**17**) on the overall aggregate composition.[9] Only when (**17**) had the appropriate stereochemistry at the chiral centre was an interaction with the aggregated catalyst precursor observed. A possible cleavage of higher associates promoted by the product could then result in improved catalyst activity. This effect

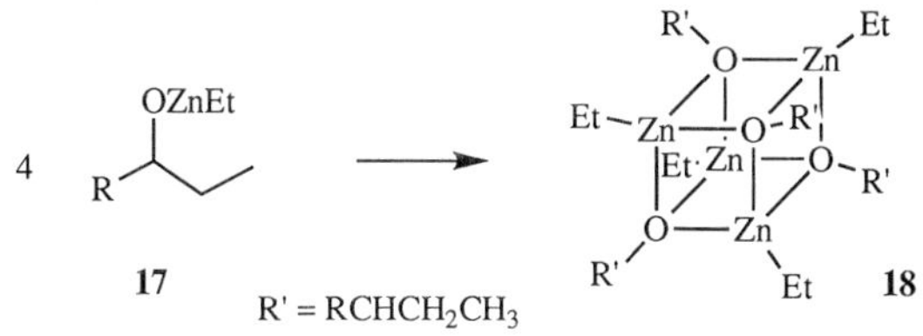

4 17 R' = $RCHCH_2CH_3$ 18

Scheme 2.6

would be a very special case of an *enantioselective autocatalysis*,[10,11] a transformation in which an optically active product catalyses its own formation.

2.2 Enantioselective oxidations and C–C bond formations: use of titanium reagents

In the enantioselective titanium-catalysed epoxidation of allylic alcohols developed by Sharpless *et al.*, the dinuclear titanium complex $[Ti(tartrate\ ester)(OR)_2]_2$ was identified as the active catalyst. From a number of mechanistic studies and structural investigations, Finn and Sharpless concluded that Ti–O-bridged dimer (**19**) is the major and most active species involved in the asymmetric oxygen transfer.[12] Different aggregates involving more than one chiral ligand have also been proposed.[13] Titanium/tartrate-catalysed epoxidation of allylic alcohols has been recently reviewed.[14]

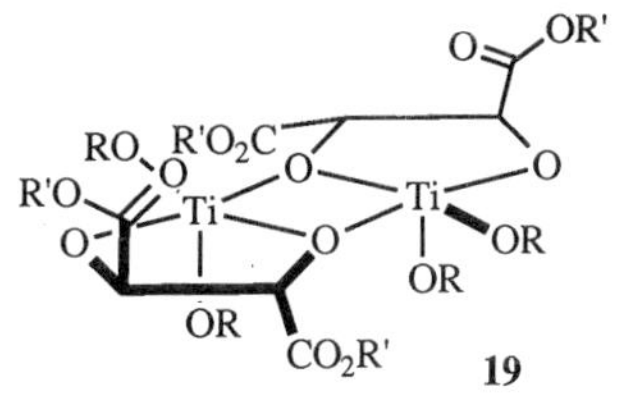

19

Although in solution rapid ligand exchange leads to a complicated mixture of equilibrating titanium alkoxides, only the one(s) bearing the chiral ligand catalyse(s) the alkene epoxidation efficiently. This accelerating effect of the chiral ligand is very beneficial and may even be essential for the formation of products with high optical purity ('Ligand Accelerated Catalysis' (LAC)).[15] For the explanation of the observed asymmetric amplification, dimeric complexes such as (**19**) or species containing more than one tartrate ligand are of major importance. The asymmetric amplification indicates that homochiral associates must have a higher efficiency in the enantioselective oxygen transfer than the corresponding heterochiral ones.

In Kagan's sulfide oxidation of (**3**), negative deviations from linearity have been observed.[1,16] The oxidizing reagent, which is prepared from diethyl tartrate, titanium tetraisopropoxide and water, is believed to be a μ-oxotitanium dimer having tartrate ligands on titanium. Although the precise nature of the active complex formed upon addition of the oxidant, t-butyl hydroperoxide, is unknown, the non-linear correlation between ligand and product ee shows that species having at least two chiral ligands must be involved in the enantiodifferentiating step.

A related *catalytic* system developed by Uemura *et al.* uses binaphthol (BINOL, (**20**)) as chiral auxiliary. Over a wide range, the ee values of the produced methyl-*p*-tolyl sulfoxide (**4**) *exceed* the optical purities of (**20**).[17] This observation has been interpreted as evidence of fundamental composition differences between Kagan's and Uemura's catalyst systems.

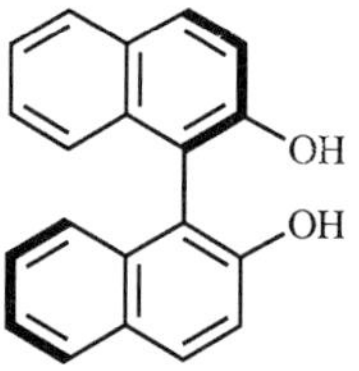

OH
OH

20
(*R*)-BINOL

Oguni *et al.* used a modified Sharpless reagent for the preparation of optically active cyanohydrins.[18] They found an asymmetric amplification in the enantioselective trimethylsilylcyanation of benzaldehyde (**7**). Using equimolar amounts of the titanium/tartrate reagent, the optical purity of product (**22**) was significantly higher than the ee of the chiral ligand (Scheme 2.7).

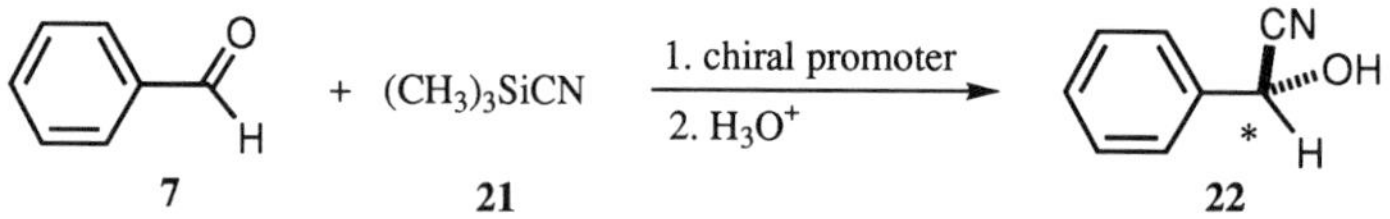

7 + $(CH_3)_3SiCN$ **21** → 1. chiral promoter, 2. H_3O^+ → **22**

chiral promoter: $Ti(O\text{-}i\text{-}Pr)_4$ / tartrate ester / i-PrOH

tartrate ester ee [%]	product ee [%]
15	27
25	50
50	70

Scheme 2.7

The chiral titanium reagent (**24**), derived from tartrate derivative (**23**) and $TiCl_2(O\text{-}i\text{-}Pr)_2$, has been used by Iwasawa *et al.* in enantioselective Diels–Alder reactions (Scheme 2.8).[19] Using partially resolved diol (**23**), with only 25% ee, as chiral auxiliary gave the cycloddition product (**27**) with 83% ee. This

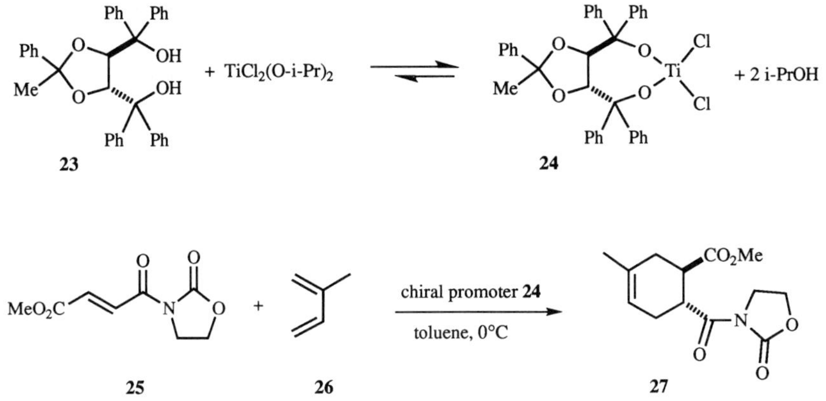

Scheme 2.8

strong amplification of chirality has been explained by the formation of a non-productive insoluble titanium species, which contains (*R*)- and (*S*)-tartrates in a 1:1 ratio. The rest of the major enantiomer remains in solution and catalyses the asymmetric cycloaddition giving product (**27**) with very high optical purity.

Mikami and Terada reported asymmetric amplifications in carbonyl-ene reactions using BINOL/titanium catalysts.[20] Treatment of enantiomerically pure BINOL (**20**) with diisopropoxytitanium dibromide in the presence of molecular sieves affords a titanium alkoxide which efficiently catalyses the formation of the ene-product (**30**) with 95% ee (Scheme 2.9). The same complex derived from (**20**) with 33% ee gives the product with 91% ee (Table 2.2).

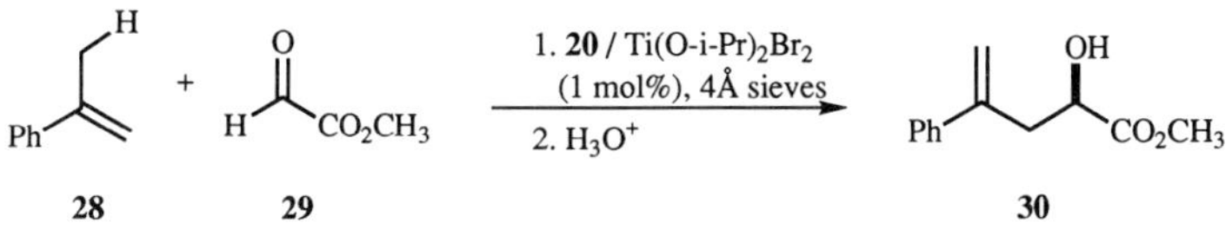

Scheme 2.9

Table 2.2 Catalysed enantioselective glyoxylate-ene reaction[20]

% ee of BINOL	% ee of product	% yield of ene product
13.0	59.9	94
33.0	91.4	92
46.8	92.9	88
66.8	94.4	96
100.0	94.6	98

Table 2.3 Effect of catalyst[a] concentration on enantioselective amplification[20]

Concentration (mM)	% ee of product	% yield of ene product
1.7	87.6	97
8.4	83.9	93
17.0	81.3	99
34.0	80.6	96
170.0	74.9	88

[a] BINOL/$TiCl_2(O\text{-}i\text{-}Pr)_2$ (31% ee) was used as chiral catalyst.

Molecular weight (MW) measurements indicated the dimeric nature of both the homo- and heterochiral titanium complexes obtained from enantiopure and racemic BINOL, respectively. However, the dimers differ in stability. Whereas the homochiral one shows a concentration-dependent molecular weight, indicating its lability and high tendency to dissociate into catalytically active monomers, the heterochiral dimer exhibits no MW change over a wide concentration range. As a consequence, the degree of asymmetric amplification increases at lower catalyst concentrations (Table 2.3).

Kinetic studies revealed that a catalyst derived from enantiopure (**20**) was about 35 times more active than the corresponding complex prepared from racemic BINOL. The departure of the observed ee correlation from a calculated curve based on this rate difference was interpreted as an indication for the involvement of *trinuclear* associates. A MW measurement of a complex derived from BINOL (**20**) with 33% ee supported this assumption.[20a]

Mikami *et al.* also used the titanium/BINOL system in the asymmetric catalysis of (hetero-)Diels–Alder reactions* (Scheme 2.10).[21] In these

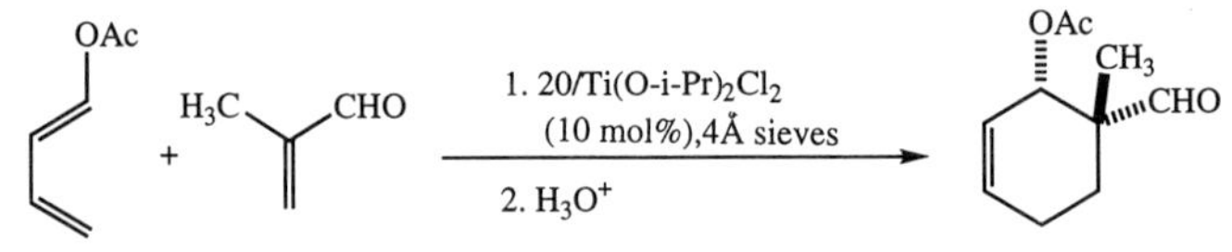

OAc

H_3C CHO

+

1. 20/Ti(O-i-Pr)$_2$Cl$_2$ (10 mol%),4Å sieves

2. H_3O^+

OAc CH_3 CHO

31 **32** **33**

Scheme 2.10

cycloadditions, the asymmetric amplification depends on the mixing manner of the catalyst. A positive non-linear effect was only observed in reactions catalysed by $[(R)\text{-BINOL/Ti}]_2$ and *meso*-$[\text{BINOL/Ti}]_2$ mixtures. When the catalyst was prepared by mixing the two enantiomeric complexes $[(R)$-BINOL/Ti$]_2$ and $[(S)$-BINOL/Ti$]_2$ a *linear* relationship between ligand ee and product ee was found. These results indicate that under these conditions

* A related catalyst system prepared from BINOL and $Ti(O\text{-}i\text{-}Pr)_4$ has been used in the catalysed enantioselective methallylation of aldehydes. A positive non-linear effect was observed.[21b]

(and in the absence of molecular sieves) the formation of the catalytically less active *meso*-complex via ligand exchange between $[(R)\text{-BINOL/Ti}]_2$ and $[(S)\text{-BINOL/Ti}]_2$ is slow. The catalysts derived from these homochiral associates operate independently and simultaneously.

2.3 Asymmetric conjugate additions* to enones: use of copper and zinc reagents, nickel catalysis

Rossiter *et al.* observed an asymmetric amplification in the conjugate addition of a chirally modified cuprate to cyclic enone (**35**) (Scheme 2.11).[22] The

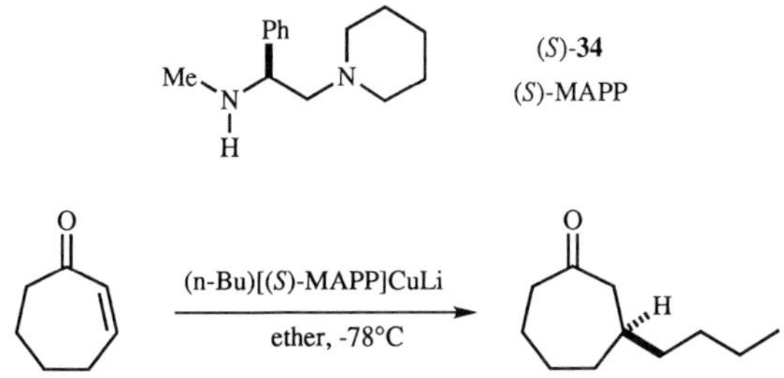

(*S*)-**34**
(*S*)-MAPP

(n-Bu)[(*S*)-MAPP]CuLi
ether, -78°C

35 **36**

Scheme 2.11

enantiopure amidocuprate, derived from (*S*)-MAPP (**34**), CuI and n-butyllithium, reacts enantioselectively with 2-cycloheptenone (**35**) to give 3-n-butylheptanone (**36**) with 96% ee. When the optical purity of (**34**) was reduced to 56% ee, (**36**) was obtained with 82% ee. These experimental results, and their comparison with computed ee values, suggest that the copper reagent reacts as a dimer. A statistical mixture of (*S,S*), (*R,R*) and (*S,R*) dimeric complexes is formed, and the *meso*-complex is the least reactive one.

Non-linear effects in conjugate addition reactions have also been observed in the muscone synthesis described by Tanaka *et al.*[23] The copper reagent obtained by sequential treatment of amino alcohol (**37**), with MeLi, CuI and MeLi in a 2:2:1:2 stoichiometric ratio enantioselectively adds to (*E*)-cyclopentadec-2-enone (**38**), affording optically pure (*R*)-muscone (**39**) in 89% yield (Scheme 2.12).

With cuprates obtained from ligands with lower optical purity, positive *and* negative deviations from linearity were observed (Table 2.4). Asymmetric amplification was only achieved with chiral ligands of relatively high optical purity. As a result of negative deviations, use of (**37**) with low ee gave almost racemic muscone.

* See also Chapter 11.

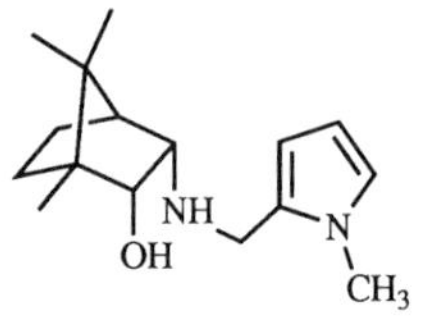

(1*R*,2*R*,3*S*,4*S*)-**37**

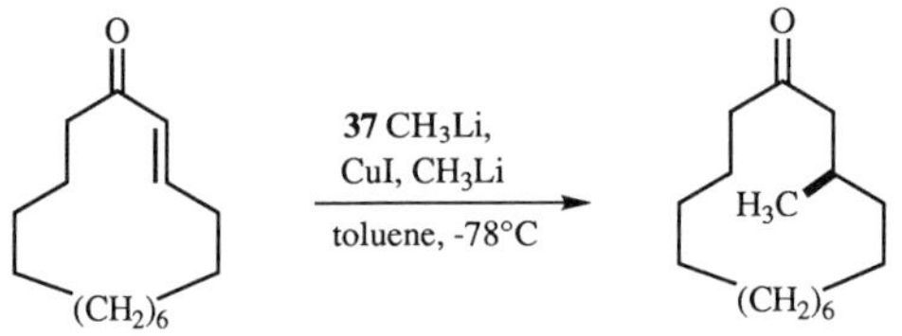

37 CH_3Li, CuI, CH_3Li

toluene, -78°C

38

39 (*R*)-muscone

Scheme 2.12

Table 2.4 Non-linear effects in the muscone synthesis[23]

% ee of chiral ligand	% ee of muscone	% yield of muscone
100	100	89
80	93	75
60	76	82
50	33	77
40	26	96
20	6	93

The non-linear effects were suggested to arise from differences in chemical behaviour of diastereomeric dinuclear copper complexes. The homochiral dimeric associate (**40**) was proposed to play a dominant role in chirality transfer.[23] Non-linear effects have also been observed in conjugate addition reactions using chiral copper(I) arenethiolate catalysts.[24]

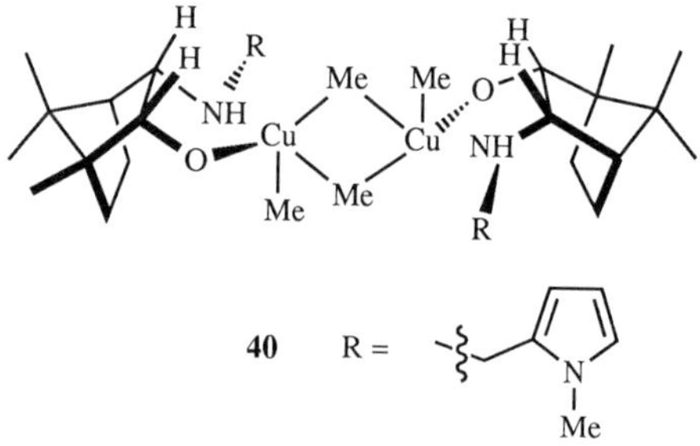

40 R =

In contrast to most reactions with organocopper reagents, where stoichiometric amounts of valuable and expensive chiral auxiliaries are required,

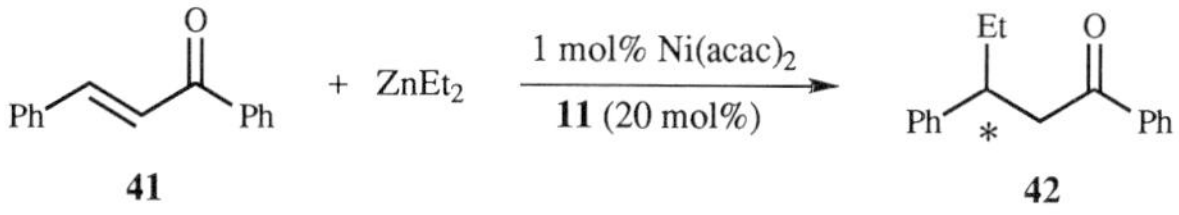

41 + $ZnEt_2$ $\xrightarrow[\textbf{11}\ (20\ mol\%)]{1\ mol\%\ Ni(acac)_2}$ **42**

Scheme 2.13

conjugate enone additions of dialkylzincs can be stereochemically directed by the use of nickel *catalysts*. An asymmetric amplification has been observed by Bolm in the nickel-catalysed diethylzinc addition to chalcone (**41**) in the presence of catalytic amounts of pyridyl alcohol (**11**) (Scheme 2.13).[25]

Variation of the optical purity of (**11**) revealed a positive non-linear relationship between product and ligand ee over the whole range of enantiopurity (Figure 2.2). The degree of asymmetric amplification increased with decrease of the enantiopurity of (**11**).

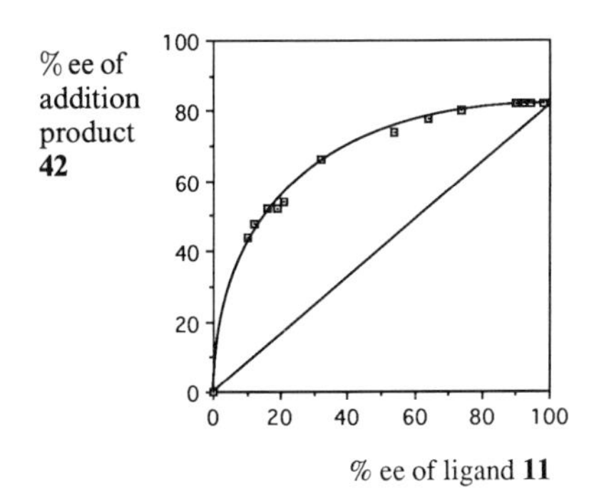

Figure 2.2 Asymmetric amplification in nickel-catalysed conjugate addition of diethylzinc to chalcone.[25]

As demonstrated by Jansen and Feringa, the ee of the diethylzinc addition product (**42**) also significantly exceeds the enantiopurity of the ligand when (**9**) is used instead of (**11**).[26] Asymmetric amplification also occurs in a β-hydroxysulfoximine/Ni catalysis.[27]

2.4 Enantioselective reductions and C–C bond formations: use of lanthanide catalysts

Evans *et al.* reported an enantioselective Meerwein–Ponndorf–Verley reduction of aryl alkyl ketones using a chiral samarium catalyst (**44**).[28] Double deprotonation of (**43**) with n-butyllithium, followed by treatment with SmI_3, gives soluble complex (**44**), which catalyses the ketone reduction by 2-propanol, giving secondary alcohols in high yields and up to 97% ee. When (**43**) with 80% ee was used as ligand in the reduction of *o*-chloroacetophenone

(**45**, R = *o*-Cl), the corresponding alcohol (**46**) was formed in 95% ee (Scheme 2.14). This optical yield was identical to that obtained with enantiomerically pure (**43**).

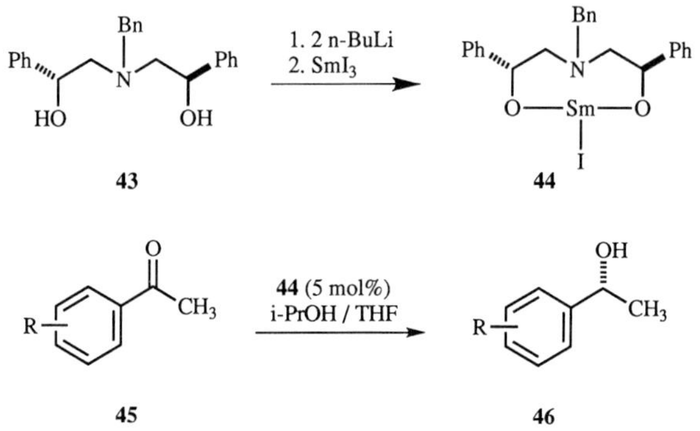

Scheme 2.14

A different lanthanide catalyst was developed by Shibasaki *et al.*[29] Treatment of a suspension of $LaCl_3 \cdot 7H_2O$ with THF solutions of dilithium binaphthoxide (**47**), NaO-t-Bu and water gave a catalytically active La complex (**48**) for asymmetric nitro aldol reactions (Scheme 2.15). The

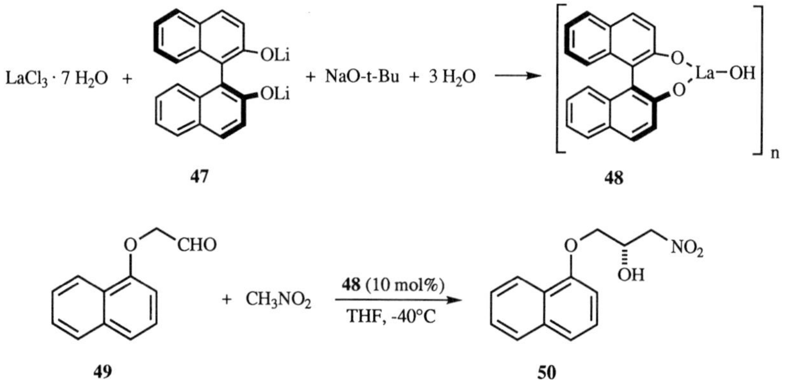

Scheme 2.15

asymmetric amplification, which was discovered in the reaction of nitromethane with α-naphthoxyacetaldehyde (**49**) (ligand ee: 56%; product ee: 68%), indicated the non-monomeric character of (**48**). The La catalyst structure was investigated.[29b]

Kobayashi *et al.* used chiral ytterbium and scandium catalysts in enantioselective Diels–Alder reactions.[30] The non-linear effects depend on the metal and the presence of achiral additives. Both positive and negative deviations from linearity have been observed.

2.5 Asymmetric C–C bond formations: use of amino acid and peptide catalysts

Although enantiomeric interactions between organic molecules have been subjects of study for a long time,[31] investigations of non-linear effects in enantioselective transformations mediated (or catalysed) by purely organic compounds are still rare. The proline-catalysed intramolecular aldolization of triketone (**5**) (Hajos–Parrish reaction), studied by Agami *et al.*[32] is an excellent example of the importance of diastereomeric interactions in hetero- and homochiral associates on the magnitude of asymmetric induction. Kinetic experiments[32b] revealed that *two* proline molecules are involved in the transition state of the enantiodifferentiating step. Formation of enamine intermediate (**51**) was proposed to be followed by internal proton transfer, catalysed by the second proline molecule. When non-enantiomerically pure proline is used, the dual role of the amino acid leads to diastereomeric intermediates. The observed deviation from the linear relationship between the enantiomeric excesses of proline and ketol then is attributed to the reactivity differences of these intermediates (Scheme 2.16).

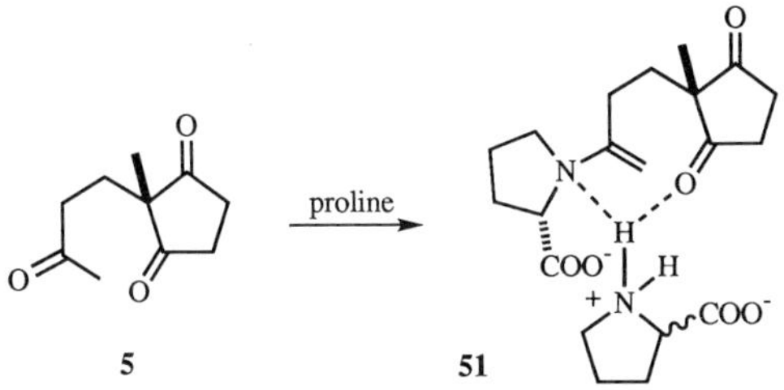

5 —proline→ **51**

Scheme 2.16

Whereas a *negative* non-linear effect has been observed in the proline-catalysed Hajos–Parrish reaction, *positive* deviations from linearity were found in an enantioselective aldehyde hydrocyanation. Cyclic dipeptide (**52**) catalyses the HCN addition to 3-phenoxybenzaldehyde (**53**). The corresponding cyanohydrin (**54**) is formed in good yield with high enantiomeric excess (Scheme 2.17).[11]

Interestingly, product (**54**) has a significant influence on its own formation. Therefore, a much more efficient catalysis is achieved when small amounts of cyanohydrin (**54**) (with appropriate absolute configuration) are present initially. This effect was used in the amplification study. A catalyst with only 12% ee showed low catalytic activity, and (**54**) was obtained with 11% ee.

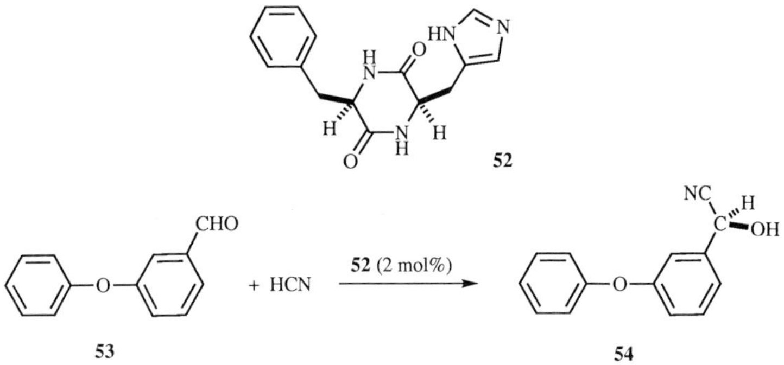

52

CHO

53 + HCN —52 (2 mol%)→ 54

NC H C OH

Scheme 2.17

However, the initial presence of small amounts of optically active product led to catalyst activation and chirality amplification. Dipeptide (**52**) with 12% ee now gave the cyanohydrin (**54**) with an ee of 87%. These observations suggest that a complex composed of both dipeptide and product is the actual catalyst, and that this species of unknown composition is catalytically much more active than (**52**) alone. Thus, the product* accelerates its own catalysed asymmetric formation.[33]

2.6 Conclusion

The study of the correlation between ligand and product ee is a useful tool for mechanistic analysis. A non-linear relationship indicates that the active species, which controls the product stereochemistry, has at least two chiral ligands in its vicinity. In contrast, and with great caution, a linear ee correlation can be interpreted as evidence for a process in which enantiodifferentiation is achieved by a species bearing only one single chiral ligand.[34]

In asymmetric synthesis, non-linear effects become particularly interesting when high enantioselectivities are achieved using only partially resolved reagents or ligands. The practical consequences are obvious. Tedious and troublesome enantiomer separation, or absolute configurational control in ligand synthesis, becomes unnecessary when a small enantiomeric excess in the ligand is sufficient for high enantiocontrol of the asymmetric transformation.

In the near future, new mechanistic insights will lead to a deeper understanding of the underlying principles of *chirality-amplifying* phenomena. Further investigations and process developments will demonstrate possible applications of non-linear effects in asymmetric synthesis.

* For studies on the stereochemical effects of the product in asymmetric C–C bond formations, 'enantioselective antoinduction', see ref. 33.

References

1. C. Puchot, O. Samuel, E. Duñach, S. Zhao, C. Agami and H.B. Kagan, *J. Am. Chem. Soc.*, 1986, **108**, 2353.
2. D. Guillaneux, S.-H. Zhao, O. Samuel, D. Rainford and H.B. Kagan, *J. Am. Chem. Soc.*, 1994, **116**, 9430.
3. N. Oguni, Y. Matsuda and T. Kaneko, *J. Am. Chem. Soc.*, 1988, **110**, 7877.
4. (a) M. Kitamura, S. Okada, S. Suga and R. Noyori, *J. Am. Chem. Soc.*, 1989, **111**, 4028; (b) R. Noyori and M. Kitamura, *Angew. Chem.*, 1991, **103**, 34; *Angew. Chem. Int. Ed. Engl.*, 1991, **30**, 49; (c) R. Noyori, S. Suga, K. Kawai, S. Okada, M. Kitamura, M. Oguni, M. Hayashi, T. Kaneko and Y. Matsuda, *J. Organomet. Chem.*, 1990, **382**, 19.
5. K. Soai and S. Niwa, *Chem. Rev.*, 1992, **92**, 833 (review).
6. C. Bolm, G. Schlingloff and K. Harms, *Chem. Ber.*, 1992, **125**, 1191.
7. F. Biesemeier, PhD thesis, University of Marburg, 1993.
8. C. Bolm, J. Müller, G. Schlingloff, M. Zehnder and M. Neuburger, *J. Chem. Soc., Chem. Commun.*, 1993, 182.
9. C. Bolm and J. Müller, *Tetrahedron*, 1994, **50**, 4355.
10. (a) H. Wynberg, *Chimia*, 1989, **43**, 150; (b) H. Wynberg, *J. Macromol. Sci., Chem., Sect. A*, 1989, **26**, 1033; (c) K. Soai, S. Niwa and H. Hori, *J. Chem. Soc., Chem. Commun.*, 1990, 982.
11. H. Danda, H. Nishikawa and K. Otaka, *J. Org. Chem.*, 1991, **56**, 6740.
12. (a) M.G. Finn and K.B. Sharpless, *J. Am. Chem. Soc.*, 1991, **113**, 113; (b) S. Woodard, M.G. Finn and K.B. Sharpless, *J. Am. Chem. Soc.*, 1991, **113**, 106; (c) I.D. Williams, S.F. Petersen, K.B. Sharpless and S.J. Lippard, *J. Am. Chem. Soc.*, 1984, **106**, 6430; (d) S.F. Petersen, J.C. Dewan, R.R. Eckman and K.B. Sharpless, *J. Am. Chem. Soc.*, 1987, **109**, 1279.
13. (a) P.G. Potvin, P.C.C. Kwong and M.A. Brook, *J. Chem. Soc., Chem. Commun.*, 1988, 773; (b) E.J. Corey, *J. Org. Chem.*, 1990, **55**, 1693.
14. R.A. Johnson and K.B. Sharpless, in *Catalytic Asymmetric Synthesis*, I. Ojima (ed.), VCH, New York, 1993, p. 103.
15. D.J. Berrisford, C. Bolm and K.B. Sharpless, *Angew. Chem.*, 1995, **107**, 1159; *Angew. Chem. Int. Ed. Engl.*, 1995, **34**, 1059.
16. (a) H.B. Kagan and F. Rebiere, *Synlett*, 1990, 643; (b) H.B. Kagan, in ref. 14, p. 203.
17. (a) N. Komatsu, M. Hashizume, T. Sugita and S. Uemura, *J. Org. Chem.*, 1993, **58**, 4529; (b) N. Komatsu, M. Hashizume, T. Sugita and S. Uemura, *J. Org. Chem.*, 1993, **58**, 7624.
18. (a) M. Hayashi, T. Matsuda and N. Oguni, *J. Chem. Soc., Chem. Commun.*, 1990, 1364; (b) M. Hayashi, T. Matsuda and N. Oguni, *J. Chem. Soc., Perkin Trans. 1*, 1992, 3135.
19. N. Iwasawa, Y. Hayashi, H. Sakurai and K. Narasaka, *Chem. Lett.*, 1989, 1581.
20. (a) K. Mikami and M. Terada, *Tetrahedron*, 1992, **48**, 5671; (b) M. Terada, K. Mikami and T. Nakai, *J. Chem. Soc., Chem. Commun.*, 1990, 1623; (c) M. Terada and K. Mikami, *J. Chem. Soc., Chem. Commun.*, 1994, 833; (d) K. Mikami and M. Shimizu, *Chem. Rev.*, 1992, **92**, 1021 (review on asymmetric ene reactions); (e) K. Mikami, M. Terada, S. Narisawa and T. Nakai, *Synlett*, 1992, 255 (review).
21. (a) K. Mikami, Y. Motoyama and M. Terada, *J. Am. Chem. Soc.*, 1994, **116**, 2812; (b) G.E. Keck, D. Krishnamurthy and M.C. Grier, *J. Org. Chem.*, 1993, **58**, 6543.
22. (a) B.E. Rossiter, M. Eguchi, A.E. Hernández and D. Vickers, *Tetrahedron Lett.*, 1991, **32**, 3973; (b) B.E. Rossiter, G. Miao, N.M. Swingle, M. Eguchi, A.E. Hernández and R.G. Patterson, *Tetrahedron: Asymmetry*, 1992, **3**, 231; (c) B.E. Rossiter, M. Eguchi, G. Miao, N.M. Swingle, A.E. Hernández, D. Vickers, E. Fluckiger, R.G. Patterson and K.V. Reddy, *Tetrahedron*, 1993, **49**, 965; (d) B.E. Rossiter and N.M. Swingle, *Chem. Rev.*, 1992, **92**, 771 (review).
23. K. Tanaka, J. Matsui, Y. Kawabata, H. Suzuki and A. Watanabe, *J. Chem. Soc., Chem. Commun.*, 1991, 1632.
24. (a) Q.-L. Zhou and A. Pfaltz, *Tetrahedron*, 1994, **50**, 4467; (b) G. van Koten, *Pure Appl. Chem.*, 1994, **66**, 1455.
25. C. Bolm, M. Ewald and M. Felder, *Chem. Ber.*, 1992, **125**, 1205; (b) C. Bolm, *Tetrahedron: Asymmetry*, 1991, **2**, 701.
26. (a) J.F.G.A. Jansen and B.L. Feringa, *Tetrahedron: Asymmetry*, 1992, **3**, 581; (b) A.H.M. de Vries, J.F.G.A. Jansen and B.L. Feringa, *Tetrahedron*, 1994, **50**, 4479.
27. C. Bolm, M. Felder and J. Müller, *Synlett*, 1992, 439.
28. D.A. Evans, S.G. Nelson, M.R. Gagné and A.R. Muci, *J. Am. Chem. Soc.*, 1993, **115**, 9800.

29. (a) H. Sasai, T. Suzuki, N. Itoh and M. Shibasaki, *Tetrahedron Lett.*, 1993, **34**, 851; (b) H. Sasai, T. Suzuki, N. Itoh, K. Tanaka, T. Date, K. Okamura and M. Shibasaki, *J. Am. Chem. Soc.*, 1993, **115**, 10372.
30. (a) S. Kobayashi, H. Ishitani, M. Araki and I. Hachiya, *Tetrahedron Lett.*, 1994, 35, 6325; (b) S. Kobayashi, M. Araki and I. Hachiya, *J. Org. Chem.*, 1994, **59**, 3758; (c) S. Kobayashi, H. Ishitani, I. Hachiya and M. Araki, *Tetrahedron*, 1994, **50**, 11623; (d) S. Kobayashi, *Synlett*, 1994, 689 (review).
31. For examples see references 1–11 in ref. 1.
32. (a) C. Agami, *Bull. Soc. Chim. Fr.*, 1988, 499 (review); (b) C. Agami and C. Puchot, *J. Mol. Cat.*, 1986, **38**, 341.
33. (a) A.H. Alberts and H. Wynberg, *J. Am. Chem. Soc.*, 1989, **111**, 7265; (b) A.H. Alberts and H. Wynberg, *J. Chem. Soc., Chem. Commun.*, 1990, 453.
34. (a) Z. Li, K.R. Conser and E.N. Jacobsen, *J. Am. Chem. Soc.*, 1993, **115**, 5326; (b) B. Schmidt and D. Seebach, *Angew. Chem.*, 1991, **103**, 1383; *Angew. Chem. Int. Ed. Engl.*, 1991, **30**, 1321.

3 Chiral enolates

K.J. HALE and S. MANAVIAZAR

3.1 Introduction

For more than half a century, organic chemists have sought to control the stereochemistry of reactions involving chiral enolates for the purpose of constructing single diastereoisomers of molecules in enantiomerically pure form; yet, it is only since the early 1980s that significant progress has been made on this topic. In this short chapter, we attempt to summarise some of the more important developments that have occurred in chiral enolate chemistry during this period.

3.2 The aldol reactions of chiral enolates

The asymmetric aldol reaction of pre-prepared, stereodefined, chiral enolates with achiral aldehydes is one of the fundamental carbon–carbon bond-forming reactions of modern-day organic synthesis.[1] A vast array of methods now exists for accomplishing this synthetic transformation, some of which exploit enolates covalently bound to a removable chiral auxiliary, others which utilise enolates coordinated to an external chiral ligand present in stoichiometric or sub-stoichiometric quantity.

3.2.1 Chiral auxiliary-mediated aldol-type additions

The first noteworthy attempt at using a chiral auxiliary to dictate the stereochemical outcome of an asymmetric aldol reaction was by Mitsui *et al.* in 1964.[2] They condensed the bromomagnesium enolate of (−)-menthyl acetate with acetophenone to obtain the (*S*) aldol adduct in 58% ee and 53% yield. Although this methodology did not prove generally applicable, it did, nevertheless, herald the birth of the chiral auxiliary-directed asymmetric aldol reaction. The next significant advance in this technology came in the early 1980s, when research groups headed by Evans[3] and Masamune[4] introduced a range of chiral auxiliaries which, when bound to dialkylboron enolates, initiated highly enantioselective aldol reactions with a host of aldehydes (Scheme 3.1).

The chiral boron enolates obtained from *N*-acyl oxazolidinones have proven the most popular in this capacity, mainly because of their ease of

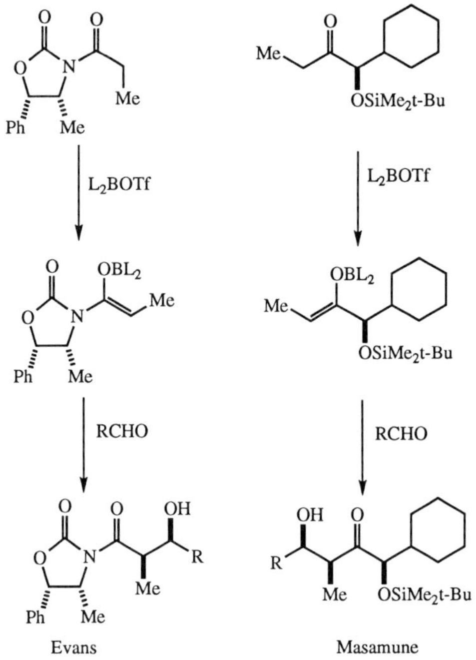
L2BOTf
L2BOTf
RCHO
RCHO
Evans
Masamune

Scheme 3.1

preparation, their excellent diastereofacial selectivity and the ease with which the auxiliary can be removed and recycled. Ordinarily, *N*-acyl oxazolidinones enolise rapidly at −78°C with di-n-butylboron triflate (1.1 equiv.) and triethylamine (1.3 equiv.) in dichloromethane to provide the (*Z*)-boron enolates almost exclusively.[3-5] The latter react readily with aldehydes at this temperature to give *syn*-aldol adducts[3-5] in excellent yield after oxidative work-up (Schemes 3.1 and 3.2). As a rule, the quantities of *anti*-aldol adducts formed under the standard Evans conditions never usually exceed more than

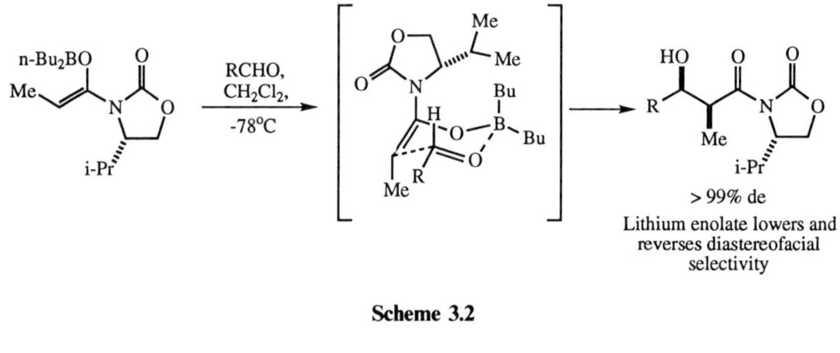
RCHO,
CH_2Cl_2,
-78°C
> 99% de
Lithium enolate lowers and reverses diastereofacial selectivity

Scheme 3.2

0.9% of the total reaction products.[6] Changing the counterion from boron to lithium generally leads to a drop in diastereoselectivity along with the formation of enantiomeric products. Reverse and high *syn*-diastereofacial selectivity is also observed when tri-isopropoxytitanium enolates of *N*-acyl oxazolidinones are used in the aldol reaction.[7] The stereoselectivity of both these processes is presumably the result of bidentate metal chelation to the oxazolidinone carbonyl and the enolate oxygen.[7]

The Evans group have extended their methodology to encompass β-ketoimides (Scheme 3.3).[8] Depending on the precise enolisation conditions

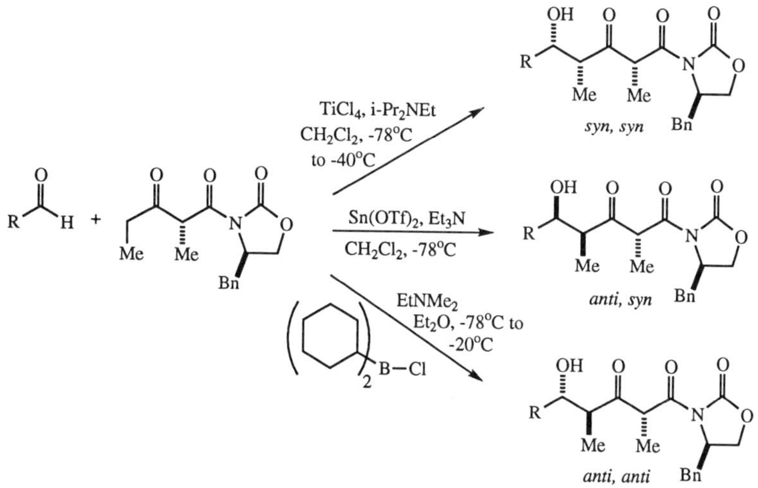

Scheme 3.3

chosen, aldols can be generated with either a *syn–syn*, or an *anti–syn*, or even an *anti–anti* stereochemical relationship between the adjacent chiral regions. A spectacular example of the *syn–syn* variant of this aldol tactic being applied on a complex aldehyde substrate is to be found in Evans' synthesis of (+)-calyculin A (Scheme 3.4).[9]

An interesting departure from the standard Evans aldol reaction has been Kende's use of *N*-acylated oxazolidinone tin(II) enolates for asymmetric 'Reformatsky-type' reactions.[10] This methodology appears to be particularly useful for the preparation of enantiomerically enriched α,α-dimethyl-β-hydroxy acids (Scheme 3.5) and recently featured in an asymmetric synthesis of neooxazolomycin by Kende *et al.*[11]

Oppolzer has discovered that (*Z*)-di-*n*-butylboron enolates derived from optically active bornane-10,2-sultams are excellent participants in the *syn*-selective asymmetric aldol process.[12] Typically, the diastereomeric excesses of these reactions lie in the 88–98% range (Scheme 3.6). Significantly, when

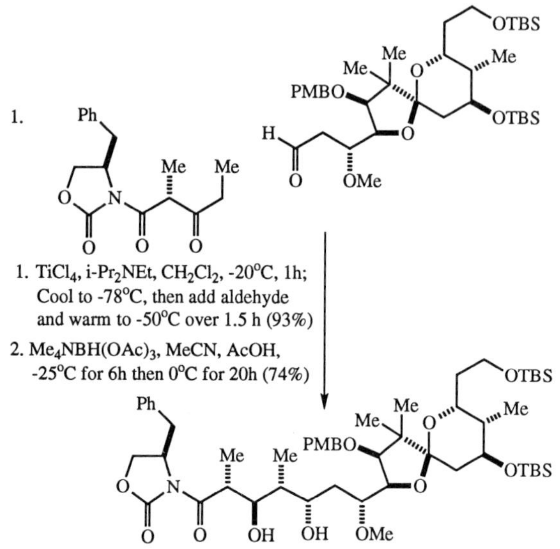

1. $TiCl_4$, i-Pr_2NEt, CH_2Cl_2, -20°C, 1h; Cool to -78°C, then add aldehyde and warm to -50°C over 1.5 h (93%)
2. $Me_4NBH(OAc)_3$, MeCN, AcOH, -25°C for 6h then 0°C for 20h (74%)

Scheme 3.4

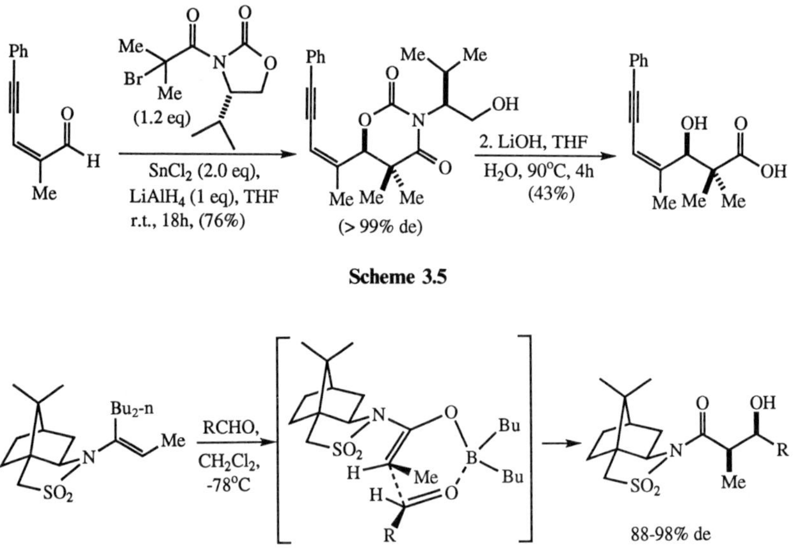

(1.2 eq)
$SnCl_2$ (2.0 eq), $LiAlH_4$ (1 eq), THF r.t., 18h, (76%)
(> 99% de)
2. LiOH, THF H_2O, 90°C, 4h (43%)

Scheme 3.5

RCHO, CH_2Cl_2, -78°C

88-98% de

Tri-*n*-butylstannyl enolate reverses diastereofacial selectivity

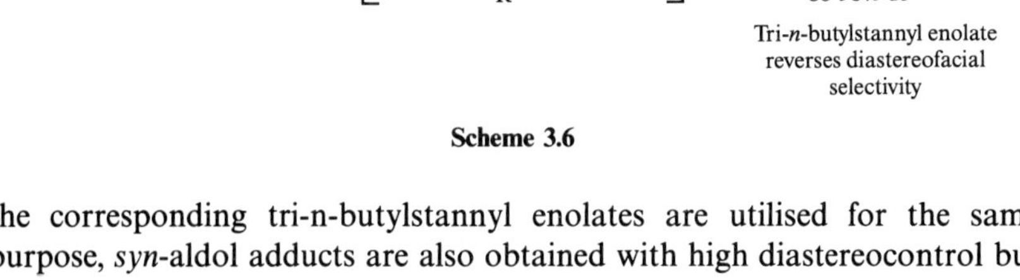

Scheme 3.6

the corresponding tri-n-butylstannyl enolates are utilised for the same purpose, *syn*-aldol adducts are also obtained with high diastereocontrol but with an opposite sense of asymmetric induction, presumably because of the occurrence of internal chelation between the sultam oxygen and the tin moiety during the aldol transition state; a possibility denied to the boron enolates.

Excellent *syn*-stereoselectivity has been observed in the reactions of chlorozirconocene (*Z*)-enolates obtained from chiral *N*-propionyl-2,5-disubstituted pyrrolidines with aldehydes (Scheme 3.7).[13] As with the Evans

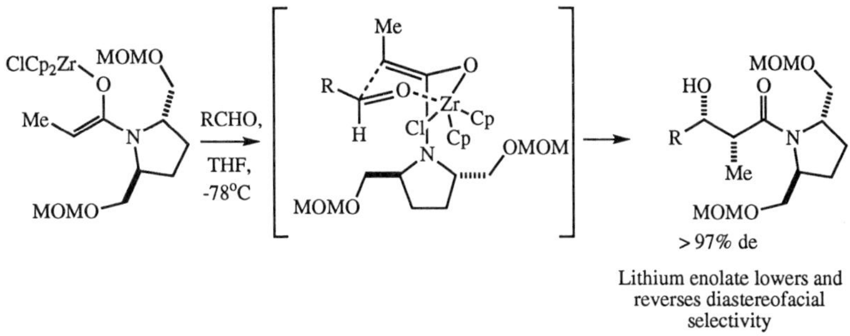

ClCp$_2$Zr MOMO O Me N MOMO RCHO, THF, -78°C Me R O O Zr Cp Cp Cl H N OMOM MOMO HO O MOMO R N Me MOMO

>97% de

Lithium enolate lowers and reverses diastereofacial selectivity

Scheme 3.7

N-acyl oxazolidinone enolates, the diastereofacial preference is reversed and lowered when the corresponding lithium enolates are used for these reactions.

Several noteworthy procedures are now available for accomplishing the *anti*-asymmetric aldol reaction. One of these, devised by Myers and Widdowson, is based on the reactions of a silicon bridged (*Z*)-*O*-silyl ketene *N*,*O*-acetal derived from L-prolinol (Scheme 3.8) with aldehydes at ambient temperature.[14] Intriguingly, these additions proceed without Lewis acid additives and yield silicon-bridged *anti*-aldol adducts with diastereoselectivities frequently in excess of 94%. The enantioselectivity is thought to arise from a boat-like pericyclic transition state proceeding through hypervalent silicon. Conditions for removing these chiral auxiliaries have already been developed by Evans and McGee.[15]

An especially powerful version of the *anti*-selective aldol reaction has been developed by Oppolzer's group that utilises a chiral bornyl-10,2-sultam *O*-silyl *O*,*N*-ketene acetal as the nucleophilic coupling partner for the aldehyde (Scheme 3.8).[16] Under normal circumstances, these reactions are carried out at low temperature in the presence of a Lewis acid promoter. An open transition state has been invoked to rationalise the product stereochemistry. Diastereoselectivity appears to be maximal (>99%) when titanium tetrachloride is used as the Lewis acid promoter, although very good results are also obtained when zinc chloride or t-butyldimethylsilyl triflate are employed. The aldol adducts are typically liberated from the auxiliary without loss of optical integrity by base hydrolysis in the presence of hydrogen peroxide. An attractive aspect of this procedure is that the auxiliary can frequently be

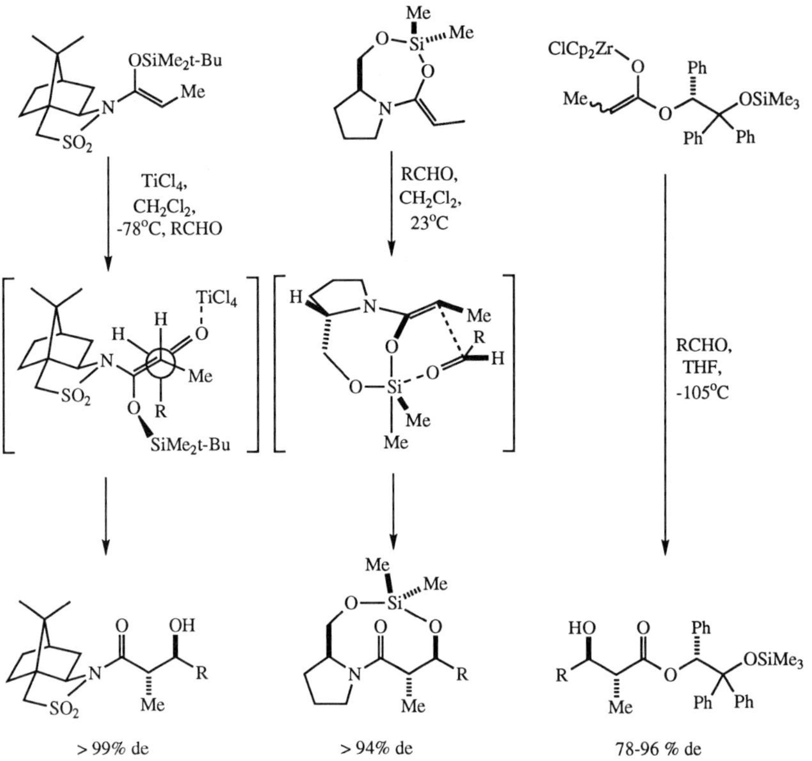

> 99% de

> 94% de

78-96 % de

Scheme 3.8

recovered in yields exceeding 82%, which is clearly of significance for large-scale industrial work.

Braun has discovered that the lithium enolate acquired from deprotonation of (*R*)-2-trimethylsiloxy-1,2,2-triphenylethyl propionate with lithium cyclohexylisopropylamide will transmetallate to the corresponding zirconium enolate after adding zirconocene dichloride.[17] This enolate will then attack the *Si*-face of simple unfunctionalised aldehydes at −105°C to produce *anti*-aldols, although with sterically hindered systems one sometimes observes incomplete reactions at this temperature (Scheme 3.8).[18] Corey has found that this reactivity problem can be overcome simply by conducting the reactions at higher temperatures (−20°C) for longer time periods.[18]

A choice of auxiliary-based procedures now exists for carrying out the 'acetate' asymmetric aldol reaction. Among the more convenient methods is Oppolzer's titanium tetrachloride-promoted addition of an *O*-silyl-*N*,*O*-ketene acetal to aldehydes (Scheme 3.9). This affords almost enantiomerically pure α-unsubstituted β-hydroxy carboxylic acids or esters in good yield after removal of the chiral auxiliary.[19]

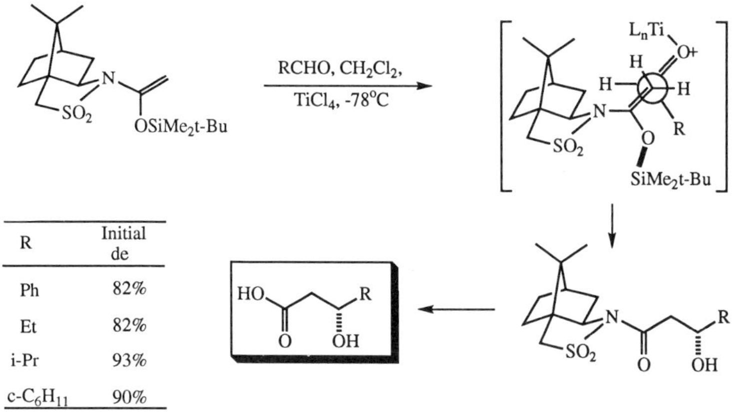

R	Initial de
Ph	82%
Et	82%
i-Pr	93%
c-C_6H_{11}	90%

Scheme 3.9

Braun has also introduced a chiral acetate synthon that shows excellent diastereofacial discriminatory properties in its asymmetric aldol reactions (Scheme 3.10). For this particular chiral 'acetate', dianion formation is

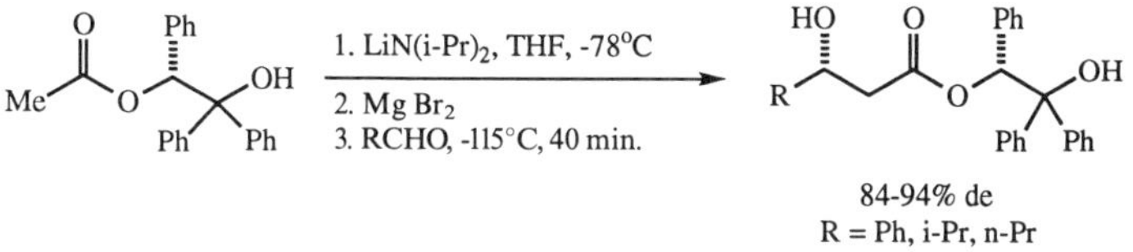

84-94% de
R = Ph, i-Pr, n-Pr

Scheme 3.10

accomplished with 2 equivalents of lithium diisopropylamide in THF; the dianion is then transmetallated with magnesium bromide or iodide, and the resulting magnesium enolate reacted with the aldehyde at − 115°C in the presence of dimethyl ester or isopentane (Scheme 3.10).[20] Base hydrolysis allows the (*R*)-triphenylethane diol auxiliary to be removed in excellent yield without loss of optical purity in the *β*-hydroxy acid.

3.2.2 *Asymmetric aldol reactions mediated by external chiral complexing agents under stoichiometric conditions*

In 1981, Meyers and Yamamoto published their seminal paper on the use of (+)-diisopinylcampheylboron triflate as an external chiral Lewis acid for mediating the *anti*-selective asymmetric aza-aldol reaction of achiral oxazolines with achiral aldehydes.[21] Meyers and Yamamoto showed that the asymmetric aldol products produced after acid hydrolysis of the oxazoline moiety were considerably enriched in the *anti*-diastereoisomers (*anti:syn* ratios, 9 to 19:1), and that these products were formed in significant enantiomeric excess (77–85%) (Scheme 3.11).[21] Despite the rather obvious synthetic advantages

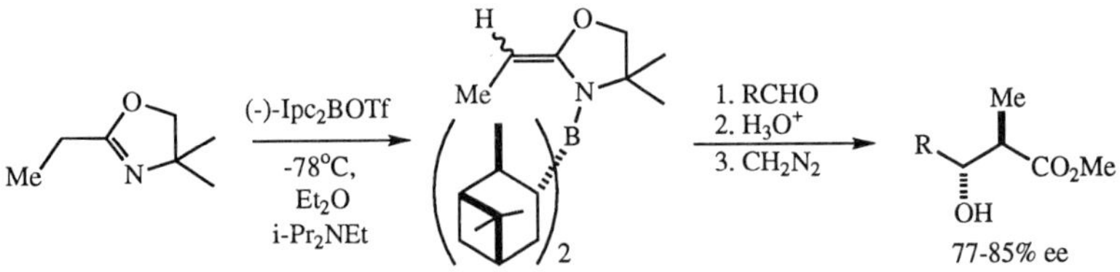

(-)-Ipc_2BOTf, -78°C, Et_2O, i-Pr_2NEt; 1. RCHO 2. H_3O^+ 3. CH_2N_2; 77-85% ee

Scheme 3.11

that this type of approach offers when compared with chiral auxiliary-based methods, organic chemists were rather slow to embrace these concepts, a fact that in hindsight now seems rather astonishing!

The next major advance in chiral reagent design for the externally directed asymmetric aldol reaction was Mukaiyama's introduction of optically active L-proline-derived diamines in conjunction with tin(II) triflate.[22] These chiral complexing agents enable highly reactive chiral tin(II) enolates to be generated from a variety of *N*-acylated thiazolidines. Mukaiyama utilised these intermediates in a range of aldol reactions with aldehydes and ketones to produce adducts in 85–95% optical yield (Scheme 3.12).[22a]

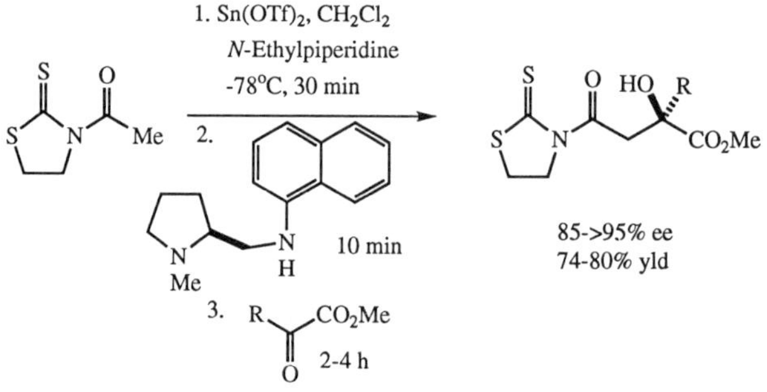

1. $Sn(OTf)_2$, CH_2Cl_2, *N*-Ethylpiperidine, -78°C, 30 min; 2. (diamine) 10 min; 3. $RCOCO_2Me$ 2-4 h; 85->95% ee, 74-80% yld

Scheme 3.12

Another ingenious development was Masamune's use of chiral *trans*-2,5-dimethylborolane triflates for the Lewis acid-mediated aldol reaction.[23] In the case of (2*S*,5*S*)-(*E*)-boron enolates formed from bulky thioesters, these react with aldehydes at −78°C to form products with 2,3-*anti*-stereochemistry with excellent stereocontrol (Scheme 3.13 and Table 3.1). Significantly, these external chiral controllers are recoverable at the end of the aldol reaction by complexation with 2,2-dimethylaminoethanol.

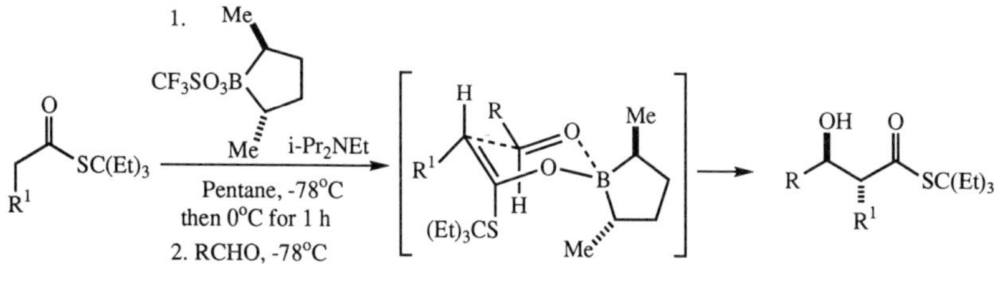

Scheme 3.13

Table 3.1 Diastereo- and enantioselectivity in asymmetric thioester enolate addition

R	R^1	*anti*:*syn*	% ee
n-Pr	H	–	87
n-Pr	Me	33:1	93
i-Pr	H	–	87
i-Pr	Me	30:1	95
c-C_6H_{11}	H	–	86
c-C_6H_{11}	H	32:1	93
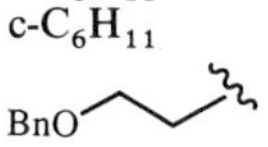 BnO	H	–	85
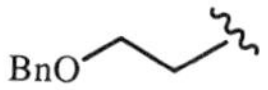 BnO	Me	30:1	92

R and R^1 in Scheme 3.13.

A more accessible family of chiral Lewis acids having C_2 symmetry are the bromides shown in Scheme 3.14.[24] These are recommended by Corey *et al.* for the highly stereocontrolled assembly of both *syn*- and *anti*-propionate aldols and their acetate counterparts (Scheme 3.14). The breadth of target compounds accessible via these reagents make their preparation particularly worthwhile.

The use of 1,2:5,6-diacetone-D-glucose as a chiral complexing agent for titanium enolates of t-butyl acetate has been advocated by Duthaler *et al.* These enolates show an overwhelming preference to add to the *Re*-face of aldehydes and grant a convenient synthetic pathway to chiral α-unsubstituted β-hydroxy esters (Scheme 3.15).[25] Extension of this methodology to an (*E*)-propionyl enolate (Scheme 3.15)[26] and a *N*-bis-silylated glycine enolate (Scheme 3.16)[27] afforded *syn*-aldol adducts in ee and de values exceeding 95%, respectively. Interestingly, the (*E*)-propionyl enolate of Scheme 3.15 is readily isomerised to the (*Z*)-enolate after warming to −30°C, allowing *anti*-aldols also to be prepared via this chemistry (Scheme 3.17).[26] Contrary to popular belief, L-glucose is a readily accessible monosaccharide, and so it should prove relatively straightforward to prepare the optical antipodes of the products shown in Schemes 3.15, 3.16 and 3.17 utilising the enantiomeric sugar complex.

Ph Ph
$ArSO_2N$ B NSO_2Ar
Br

1. Ar = p-MeC_6H_4
2. Ar = p-$NO_2C_6H_4$
3. Ar = 3,5-$(CF_3)_2C_6H_3$

PhS(C=O)Me

1. (*R,R*)-**1**, Et_3N, CH_2Cl_2, -78°C to 23°C, 1h;
2. RCHO, -90°C, 2h (82-84%)

PhS(C=O) OH R

R	% ee
i-Pr	83
Ph	91

PhS(C=O) Me

1. (*R,R*)-**2**, *i*-Pr_2NEt (2 eq), CH_2Cl_2, -45°C;
2. RCHO, (70-72%)

PhS(C=O) OH R Me

R	% ee	*syn:anti*
i-Pr	97	94.5:5.5
Ph	95	98:2

t-BuO(C=O) R^1

1. (*R,R*)-**3**, Et_3N, PhMe or PhMe-Hexane, -78°C;
2. RCHO, (90-93%)

t-BuO(C=O) OH R R^1

R	R^1	% ee	*syn:anti*
i-Pr	Br	92	98:2
Ph	Me	94	98:2

Scheme 3.14

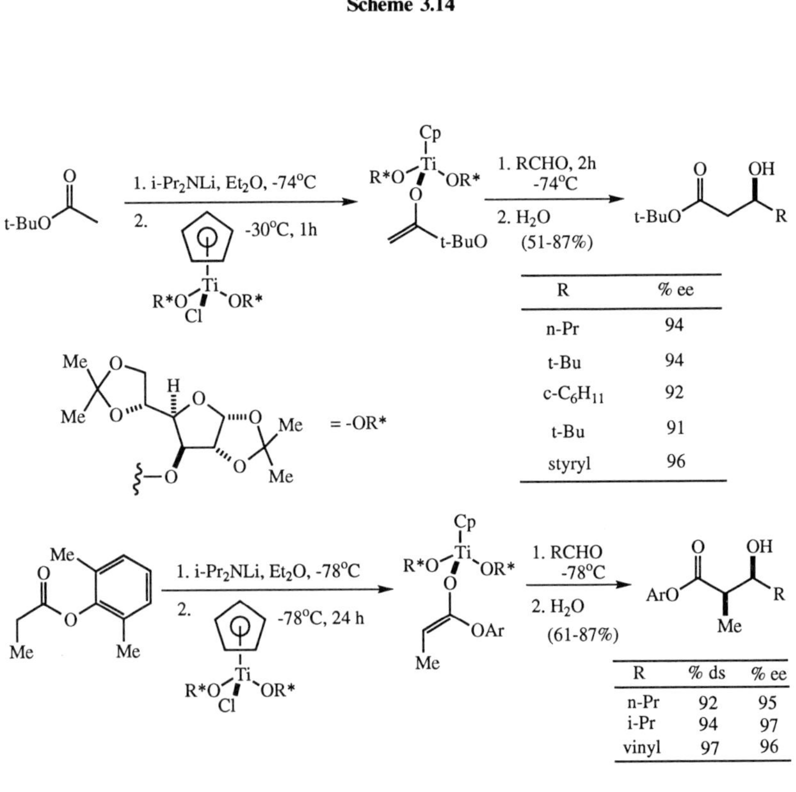

R	% ee
n-Pr	94
t-Bu	94
c-C_6H_{11}	92
t-Bu	91
styryl	96

R	% ds	% ee
n-Pr	92	95
i-Pr	94	97
vinyl	97	96

Scheme 3.15

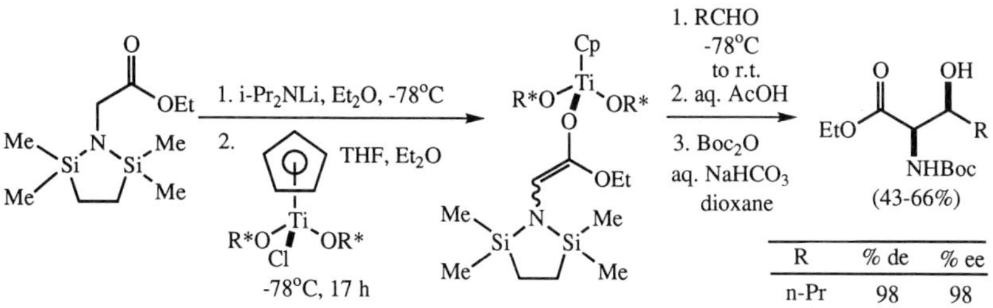

O, OEt, Me, N, Si, Me, Me, Si, Me

1. i-Pr_2NLi, Et_2O, -78°C
2. CpTi(OR*)$_2$Cl (R*O–Ti(Cl)–OR*), THF, Et_2O, -78°C, 17 h

Cp, R*O–Ti–OR*, O, OEt, Me, N, Si, Me, Me, Si, Me

1. RCHO, -78°C to r.t.
2. aq. AcOH
3. Boc_2O, aq. $NaHCO_3$, dioxane

EtO, O, OH, R, NHBoc (43-66%)

R	% de	% ee
n-Pr	98	98
t-Bu	96	96
vinyl	97	97
Ph	96	97

R*O = 1,2:5,6-Di-*O*-isopropylidene-α-D-glucofuranos-3-yl

Scheme 3.16

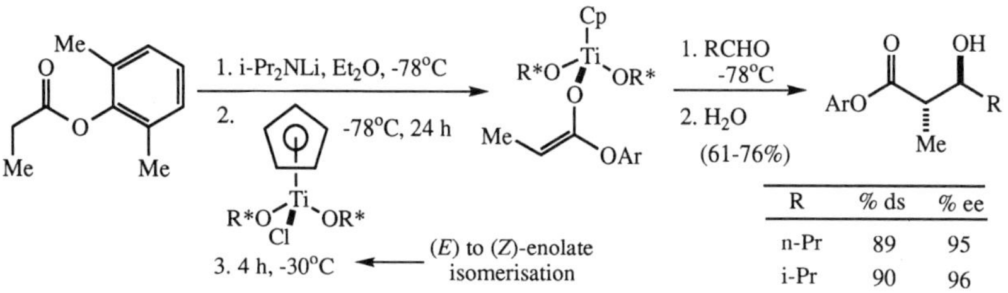

Me, O, O, Me, Me

1. i-Pr_2NLi, Et_2O, -78°C
2. CpTi(OR*)$_2$Cl (R*O–Ti(Cl)–OR*), -78°C, 24 h
3. 4 h, -30°C ← (*E*) to (*Z*)-enolate isomerisation

Cp, R*O–Ti–OR*, O, Me, OAr

1. RCHO, -78°C
2. H_2O

(61-76%)

ArO, O, OH, R, Me

R	% ds	% ee
n-Pr	89	95
i-Pr	90	96
vinyl	81	98

R*O = 1,2:5,6-di-*O*-isopropylidene-α-D-glucofuranos-3-yl

Scheme 3.17

A markedly enantioselective pathway to differentially protected *syn*- or *anti*-1,2-dihydroxy thioesters can be procured by the addition of *O*-silyl-*S,O*-ketene acetals to aldehydes in the presence of a proline-derived chiral amine, tin(II) triflate, and tri-n-butyltin acetate (Scheme 3.18).[28]

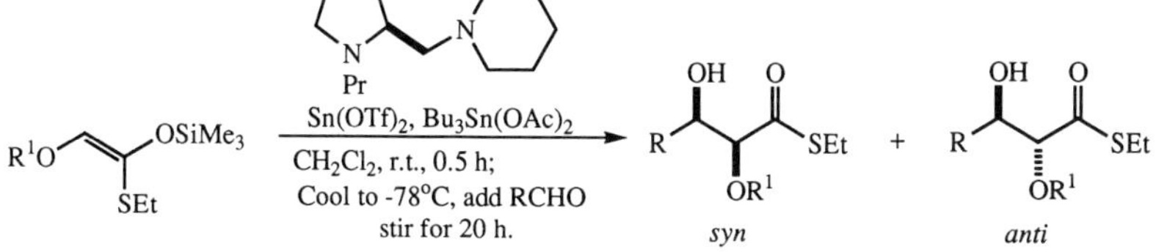

R^1O, $OSiMe_3$, SEt

N, Pr, N

$Sn(OTf)_2$, $Bu_3Sn(OAc)_2$
CH_2Cl_2, r.t., 0.5 h;
Cool to -78°C, add RCHO
stir for 20 h.

OH, O, R, SEt, OR^1 — *syn* + OH, O, R, SEt, OR^1 — *anti*

R^1 = $SiMe_2$t-Bu, 46-93% yld, *syn:anti* = 88: 12-97: 3, *syn* aldol = 82-94% ee
R^1 = CH_2Ph, 59-88% yld, *syn:anti* = 9: 91-1: 99, *anti* aldol = 95-98% ee

Scheme 3.18

In recent years, there has been considerable interest in the stoichiometric use of external chiral Lewis acids for mediating the asymmetric aldol reactions of ketones. Paterson's group at Cambridge have been particularly active in this area, and it is because of their efforts that the stereoselective enolisation of 3-pentanone has been shown to be possible with either (+)- or (−)-Ipc_2BOTf and i-Pr_2NEt in dichloromethane at −78°C.[29] Under these conditions, the (*Z*)-boron enolate is formed almost exclusively; it undergoes a series of highly diastereoselective (⩾95%) aldol reactions with aldehydes to produce *syn*-α-methyl-β-hydroxy ketones with good enantioselectivity, after oxidative work-up (Scheme 3.19 and Table 3.2). The method appears best suited to

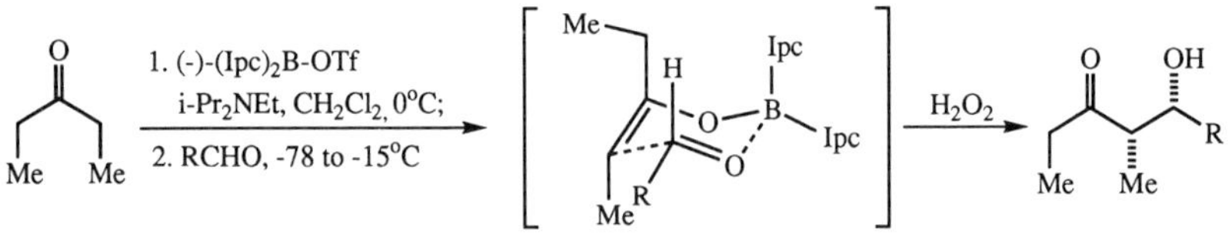

1. (-)-$(Ipc)_2B$-OTf, i-Pr_2NEt, CH_2Cl_2, 0°C; 2. RCHO, -78 to -15°C; H_2O_2

Scheme 3.19

Table 3.2 Diastereo- and entantioselectivity in asymmetric enolate addition

	R	Yield (%)	*syn*:*anti*	% ee
= (-)-$(Ipc)_2B$-OTf	Me	91	97:3	82
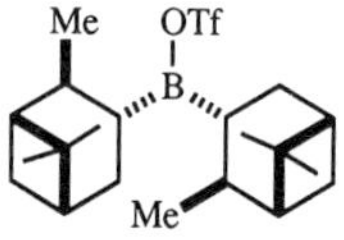	n-Pr	92	97:3	80
	i-Pr	45	96:4	66
	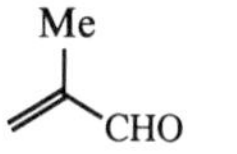 $CH_2=C(Me)$–CHO	78	98:2	91
	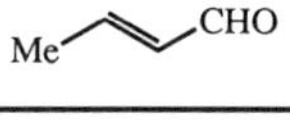 Me–CH=CH–CHO	75	98:2	86

R from Scheme 3.19.

reactions with sterically undemanding aldehydes with little or no branching at the α-position. The same workers have also investigated the enantioselective aldol reactions of methyl ketones with these chiral boron reagents and have noted that product enantioselectives are lower (53–78%) and opposite to those with the corresponding 3-pentanone boron enolates (Scheme 3.20).[29]

Corey's group later examined the aldol reactions of 3-pentanone with their chiral boron bromide shown in Scheme 3.21.[24a] Extremely high enantiofacial and diastereofacial selectivity was noted for a wide range of aldehydes including sterically demanding substrates such as isobutyraldehyde (Scheme 3.21).

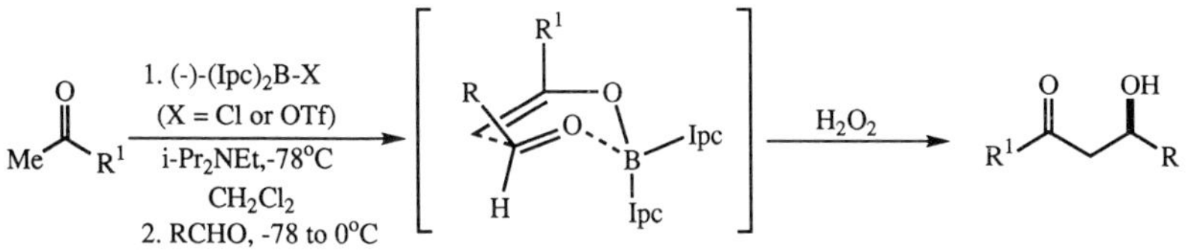

1. (-)-$(Ipc)_2B$-X (X = Cl or OTf), i-Pr_2NEt, -78°C, CH_2Cl_2; 2. RCHO, -78 to 0°C; H_2O_2

R^1	R	Yield (%)	% ee
Me	n-Pr	68	78
Me	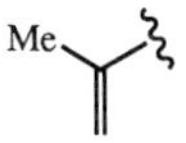 Me (isopropenyl)	62	67
i-$PrCH_2$	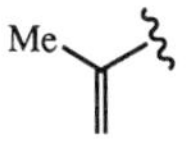 Me (isopropenyl)	56	65

Scheme 3.20

1. TsN–B(Br)–NTs (Ph, Ph diazaborolidine); i-Pr_2NEt (2 equiv), CH_2Cl_2, -78°C, 1h; 2. RCHO, -78°C, 2h

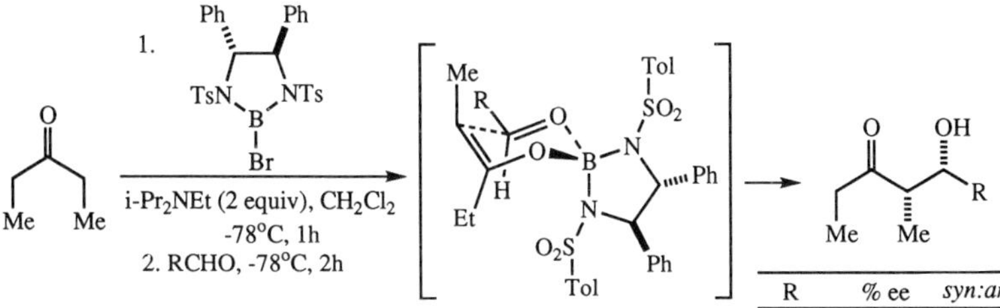

R	% ee	*syn:anti*
Et	>98	>98:2
i-Pr	95	98:2
Ph	97	94:6

Scheme 3.21

3.2.3 Catalytic asymmetric aldol reactions

In terms of synthetic efficiency, the catalytic variant of the asymmetric aldol reaction probably represents its apotheosis, and in recent times it has evolved considerably. The first successful manoeuvres in this direction appeared in 1986, when Ito *et al.* reported that sub-stoichiometric quantities of gold(I) complexes, generated from chiral ferrocenylphosphine ligands, were effective at catalysing the asymmetric aldol reactions of methyl isocyanoacetate with aldehydes. Generally speaking, *trans*-oxazolines are obtained in high optical purity from these reactions (Scheme 3.22 and Table 3.3) provided they are conducted with aldehydes bearing non-electron-withdrawing substituents.[30] The *trans*-oxazolines are believed to arise from the addition of the ammonium enolate onto the *Si*-face of the aldehyde after it has become involved in an ion-pair similar to that shown in Scheme 3.22.

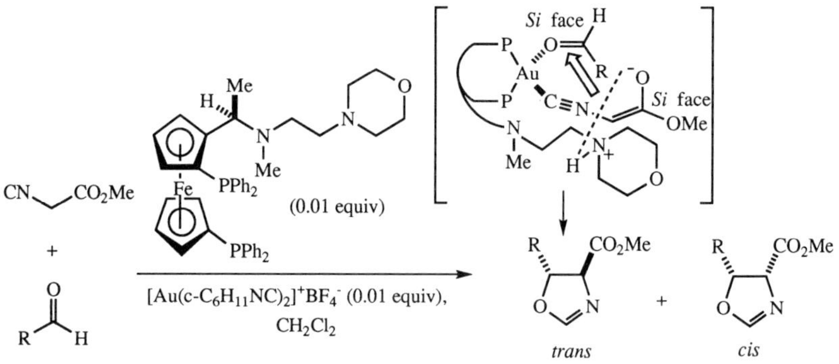

Scheme 3.22

Table 3.3 Use of ferrocene-based chiral auxiliary

R	Temperature (°C)	Time (h)	Yield (%)	*trans*:*cis*	% ee (trans)
Me	25	10–40	99	89:11	89
i-Pr	25	10–40	100	99:1	92
n-Pr–CH=CH–	25	10–40	85	87:13	92
Ph	25	20–40	93	95:5	95
4-F-C_6H_4	0	70	92	94:6	94
C_6F_5	0	100	92	37:63	36

R from Scheme 3.22.

In the presence of the chiral catalyst derived from (*S*)-1-methyl-2-[(*N*-naphthylamino)methyl]pyrollidine and tin(II) triflate, Mukaiyama *et al.* have discovered that different aldehydes will react with the silyl enol ether of (*S*)-ethyl propanethioate to produce aldol adducts with excellent relative and absolute stereochemical control (Scheme 3.23 and Table 3.4).[31] Intriguingly, the enantiomeric products are available from a procedure that utilises the same chiral amine, tin(II) oxide, and trimethylsilyltriflate; the precise reasons

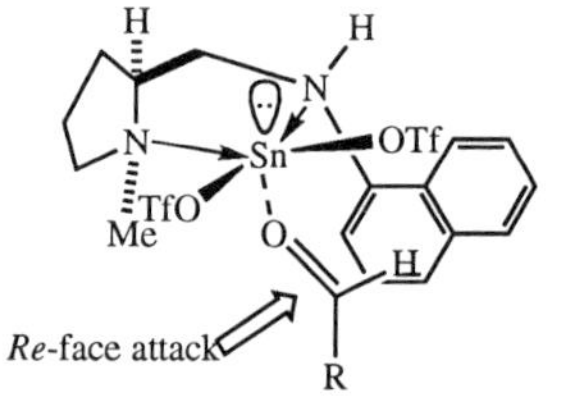

Aldehyde *Si*-face sterically shielded by napthylamine unit

Re-face attack

for this reversal of stereoselectivity are unclear and await mechanistic elucidation.[32]

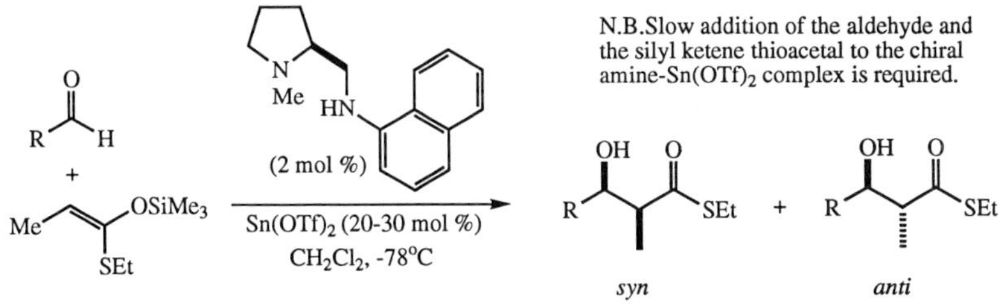

N.B.Slow addition of the aldehyde and the silyl ketene thioacetal to the chiral amine-$Sn(OTf)_2$ complex is required.

(2 mol %)

$Sn(OTf)_2$ (20-30 mol %)

CH_2Cl_2, -78°C

syn *anti*

Scheme 3.23

Table 3.4 Diastereo- and enantioselectivity in asymmetric tin enolate addition

R	Yield (%)	*syn*:*anti*	% ee
Ph	86	93:7	91
$Me(CH_2)_6$	76	100:0	>98
cyclohexyl	64	>99:1	92
p-ClC_6H_4	80	93:7	93
Me_3Si—≡	73	95:5	91

R from Scheme 3.23.

Mukaiyama has also shown that the stereochemistry of asymmetric aldol reactions between 1-ethylthio-1-trimethylsiloxyethene and chiral aldehydes is governed primarily by the structure of the chiral tin(II) catalyst rather than the inherent chirality of the substrate (Scheme 3.24).[33]

Employing catalytic amounts of the chiral complex formed from a tartaric

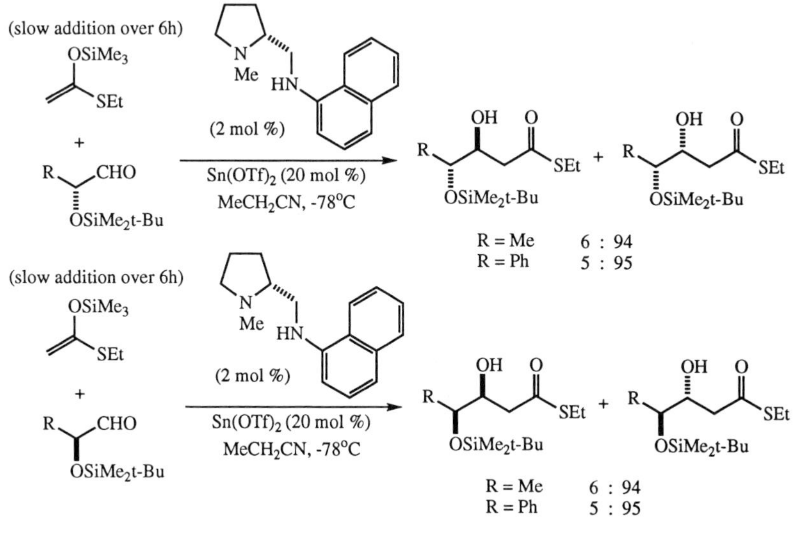

(slow addition over 6h)

(2 mol %)

$Sn(OTf)_2$ (20 mol %)

$MeCH_2CN$, -78°C

R = Me 6 : 94
R = Ph 5 : 95

(slow addition over 6h)

(2 mol %)

$Sn(OTf)_2$ (20 mol %)

$MeCH_2CN$, -78°C

R = Me 6 : 94
R = Ph 5 : 95

Scheme 3.24

acid derivative and borane-THF, Yamamoto *et al.* have been able to induce exceptionally good stereoselectivity in aldol-type reactions between aldehydes and *O*-silyl enol ethers (Scheme 3.25) or *O*-silyl ketene acetals.[34] In both

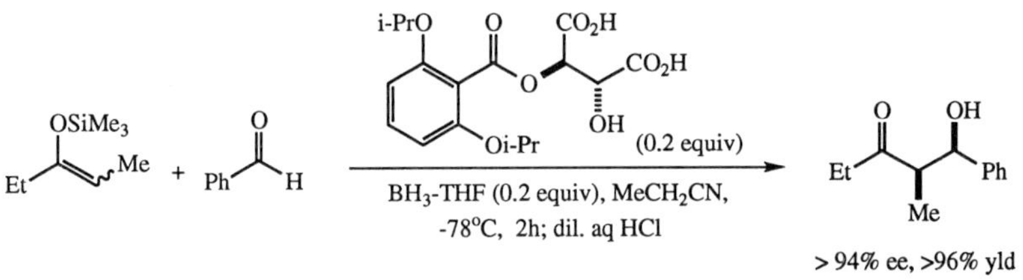

i-PrO O CO_2H CO_2H O OH Oi-Pr (0.2 equiv)

$OSiMe_3$ Et Me + Ph O H → BH_3-THF (0.2 equiv), $MeCH_2CN$, -78°C, 2h; dil. aq HCl → Et O OH Ph Me

> 94% ee, >96% yld

Scheme 3.25

cases, the *erythro*-β-hydroxyl carbonyl adducts are obtained in very high ee. For the *O*-silyl enol ethers of ethyl ketones, it was observed that the *erythro*-selectivity was independent of the alkene geometry in the starting enol ether, which suggests that these reactions proceed through an extended acyclic transition state.

Another breakthrough in chiral catalyst design has been Masamune's discovery[35] that certain α,α-disubstituted sulfonamido acyloxyboranes can function as excellent chiral Lewis acid catalysts for the asymmetric aldol reaction; this development arose out of earlier work by Kiyooka *et al.*[36] Utilising this approach, Masamune was able to prepare α,α-dimethyl-β-hydroxy esters in 84–96% ee (Scheme 3.26 and Table 3.5) from the addition

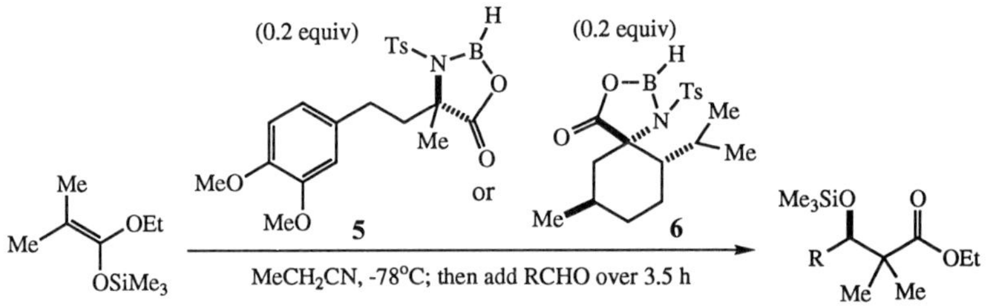

(0.2 equiv) Ts N B H O Me O MeO MeO **5** or (0.2 equiv) O B H Ts N Me Me O Me **6**

Me Me OEt $OSiMe_3$ → $MeCH_2CN$, -78°C; then add RCHO over 3.5 h → Me_3SiO O R OEt Me Me

Scheme 3.26

Table 3.5 Use of sulfonamido acyloxy boranes

Chiral catalyst	R	Yield (%)	% ee
5	Ph	80	84
5	c-C_6H_{11}	68	91
6	Ph	83	91
6	c-C_6H_{11}	59	96

Chiral catalyst and R as in Scheme 3.26.

of 1-trimethylsilyloxy-1-ethoxy-2-methyl-2-propene to primary or secondary aldehydes.[35] The method beautifully complements the earlier auxiliary-based approach of Kende[10,11] as a source of α,α-dimethyl-β-hydroxy acids in high optical purity (Scheme 3.5).

Masamune has extended the methodology to unsubstituted and mono-substituted ketene acetals.[37] With the former nucleophiles, 'acetate'-type aldols are synthesised with high enantioselectivity (Scheme 3.27 and

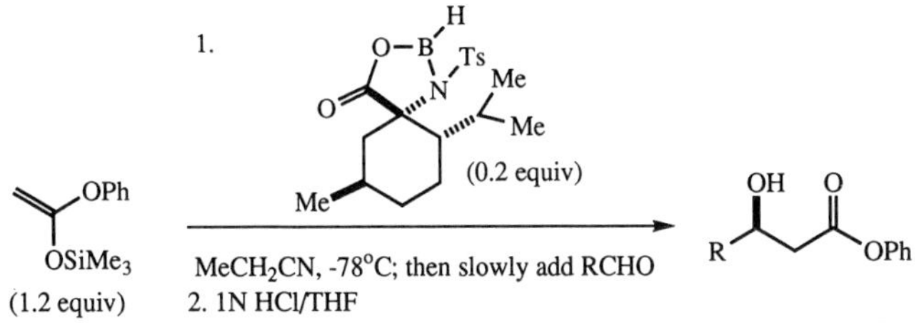

Scheme 3.27

Table 3.6 Alternative approach to α,α-dimethyl-β-hydroxy acids

R	Yield (%)	% ee
Ph	77	93
c-C_6H_{11}	87	84
$Ph(CH_2)_2CHO$	78	85

R as in Scheme 3.27.

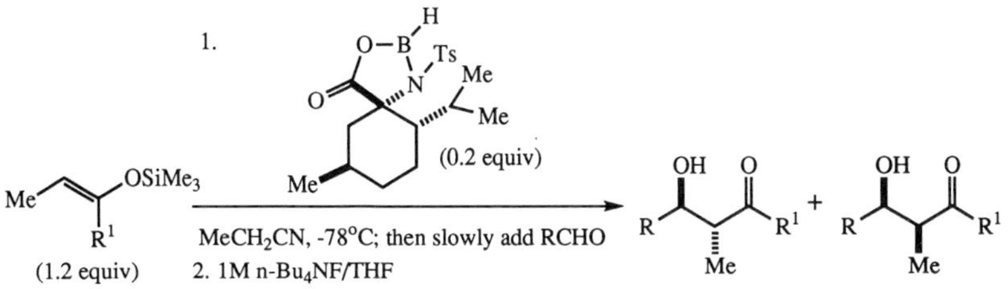

Scheme 3.28

Table 3.7 *Anti*-selectivity

R	R^1	*anti*:*syn*	Yield (%)	% ee *anti*
Ph	SBu^t	94:6	78	82
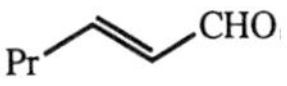 Pr⁄⁄CHO	SEt	80:20	80	60
$4\text{-}MeOC_6H_4CHO$	SEt	89:11	78	75

R and R^1 from Scheme 3.28.

Table 3.6), while *anti*-adducts are available with reasonable diastereoselectivity from the latter (Scheme 3.28 and Table 3.7).

A more accessible Lewis acid catalyst is the toluenesulfonamido (*S*)-tryptophan oxazaborolidinone of Corey *et al.* (Scheme 3.29 and Table 3.8).[38] It allows silyl enol ethers of methyl ketones to be added to aldehydes in

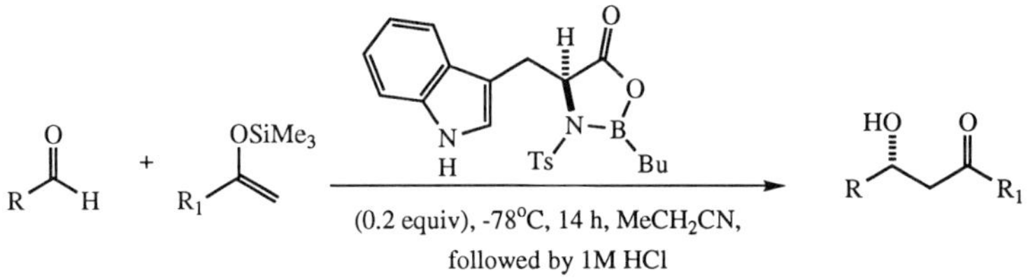

(0.2 equiv), -78°C, 14 h, $MeCH_2CN$, followed by 1M HCl

Scheme 3.29

Table 3.8 Use of a chiral Lewis acid catalyst derived from (*S*)-tryptophan

R	R^1	Yield (%)	% ee
Ph	Ph	82	89
c-C_6H_{11}	Ph	67	93
n-Pr	Ph	94	89
2-furyl	Ph	100	92
Ph	n-Bu	100	90
c-C_6H_{11}	n-Bu	56	86

R and R^1 from Scheme 3.29.

86–93% ee when employed in propionitrile at −78°C. Curiously, when used to mediate aldol additions between *O*-silyl ketene acetals and aldehydes, this borane catalyst leads to disappointing product ee values.

A powerful approach to thioacetate aldols has been devised by Mikami *et al.* which capitalises on the use of a chiral binaphthol (BINOL) titanium dichloride catalyst (Scheme 3.30 and Table 3.9).[39] It mediates remarkably enantioselective aldol reactions with aldehydes of great structural diversity, when present in only small quantities (0.05 equiv.). Another attractive feature of this method is that it leads to good optical yields of products even when the reactions are performed at 0°C.

A related chiral binaphthyl catalyst has been introduced by Carreira *et al.* that leads to even better product ee values in the 'acetate' aldol reaction with various aldehydes (Scheme 3.31).[40] Like the former method, low temperatures are not required for maximal ee and so again this method should be commercially significant.

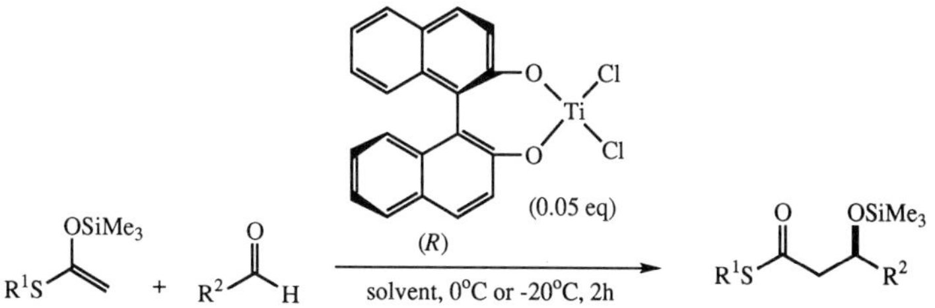

(0.05 eq)

(*R*)

solvent, 0°C or -20°C, 2h

Scheme 3.30

Table 3.9 Use of BINOL as a chiral auxiliary

R^1	R^2	Solvent	Yield (%)	% ee
Et	$BnOCH_2$	PhMe	81	94
t-Bu	$ClCH_2$	PhMe	61	91
Et	$BocNHCH_2$	CH_2Cl_2	64	88
t-Bu	C_8H_{17}	PhMe	60	91
Et	i-Pr	PhMe	61	85
Et	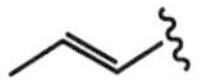 (E)-propenyl	PhMe	60	81
Et	$n\text{-}BuO_2C$	PhMe	84	95

R^1 and R^2 as in Scheme 3.30.

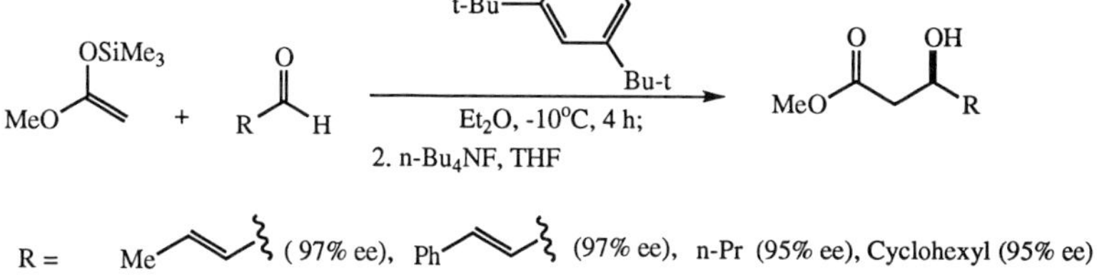

1. (2 mol %)

Et_2O, -10°C, 4 h;

2. $n\text{-}Bu_4NF$, THF

R = Me–CH=CH– (97% ee), Ph–CH=CH– (97% ee), n-Pr (95% ee), Cyclohexyl (95% ee)

Scheme 3.31

3.3 Diastereoselective alkylation of chiral metal enolates[1b,41]

In comparison to the asymmetric aldol reaction, the number of chiral auxiliaries available for the diastereoselective α-alkylation of carboxylic acid derivatives is far less extensive.[1b,41] Outstanding in this regard are the (*S*)-prolinol amides of Evans;[42] these form chelated (*Z*)-enolates upon treatment with 2 equiv. of lithium diisopropylamide in THF at −78°C. They readily C-alkylate *β*-branched primary alkyl iodides and activated alkyl bromides from their less hindered *Si*-face, to provide substitution products in 92–96% de (Scheme 3.32).[42] Significantly, the (*S*)-prolinol amide enolates

1. LDA, THF, -78°C
2. 5% HMPA;
3. R^1X, -100°C
Si face
(78-98% yld)

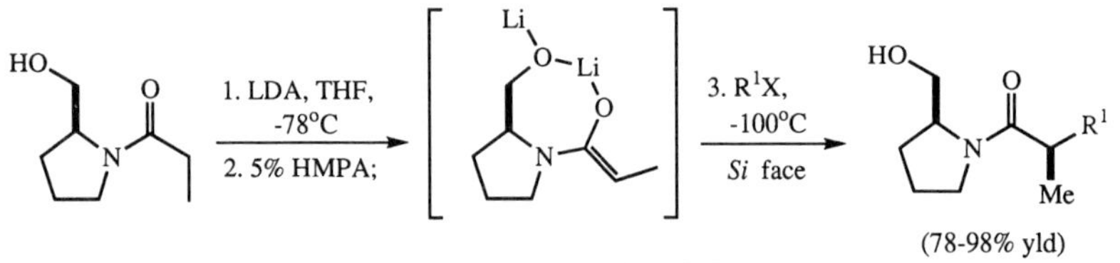

R^1X	Ratio
Allyl-Br	96:4
i-Bu-I	98:2
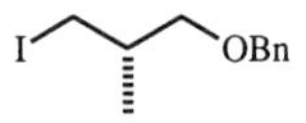 I, OBn	98:2

Scheme 3.32

are sufficiently nucleophilic to attack terminal epoxides, but an opposite facial preference is noted to that with haloalkanes (Scheme 3.33).[43] Presumably,

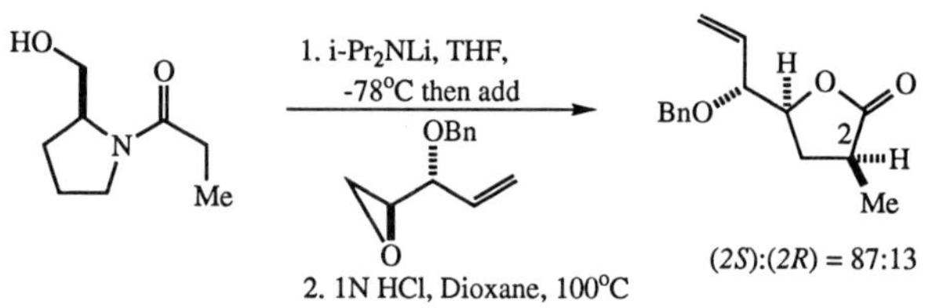

1. i-Pr_2NLi, THF, -78°C then add
2. 1N HCl, Dioxane, 100°C
(2*S*):(2*R*) = 87:13

Scheme 3.33

this reversal of enantioselectivity reflects intermolecular chelation of the epoxide oxygen with the lithio-alkoxymethyl group, an event which would favour attack on the seemingly more hindered *Re*-face of the enolate.

Enolates derived from *N*-acyl 2,5-*trans*-disubstituted pyrrolidines also react with alkyl halides, but give products with an opposite sense of chirality to those observed with the (*S*)-prolinol amide enolates (Scheme 3.34).[44] It appears that with this class of chiral controller, π-facial selectivity is governed

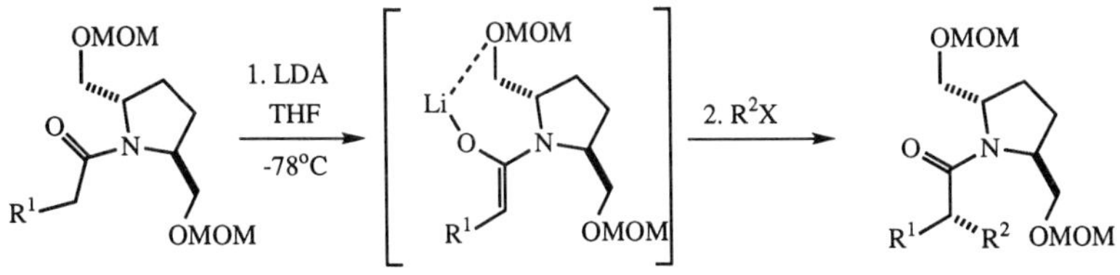

R^1	R^2X	Ratio
Me	BnBr	65:1
BnO	i-PrOTf	98.5:1.5
$(MeS)_2C{=}N$	$HC{\equiv}CCH_2Br$	99:1

Scheme 3.34

primarily by the sterically hindered environment created around the enolate by the C_2 symmetrically disposed alkoxymethyl groups adorning the pyroliidine ring.

Trans-2,5-bis(methoxymethyloxymethyl)pyrrolidine auxiliaries are proving particularly versatile for the stereoselective introduction of quarternary asymmetric centres in cyanoacetic acid derivatives (Scheme 3.35).[45] As can

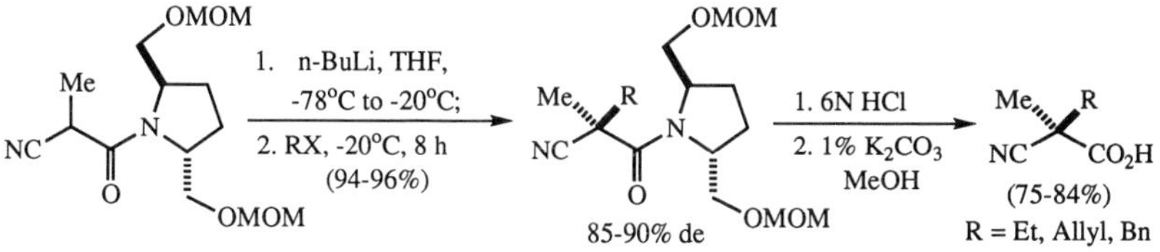

Scheme 3.35

be seen from the scheme, α,α-dialkylated products can be prepared in 80–90% ee when primary alkyl iodides and activated alkyl bromides are employed as electrophiles.

A considerably less nucleophilic family of chiral metal enolates are those derived from *N*-acylated oxazolidinones.[46] These only undergo rapid alkylation with allylic and benzylic halides or α-halo ethers and tend to be rather unreactive towards many unactivated primary alkyl iodides. A recent application of this technology in a natural product synthesis can be found in Smith's synthesis of (+)-phyllanthocin (Scheme 3.36).[47]

Helmchen *et al.* have reported that (*E*)-lithium enolates generated from the chiral propionate in Scheme 3.37 attack alkyl halides mainly from the front face of the enolate double bond, to provide products with (*R*) stereochemistry at the newly created chiral centre. As one would expect, opposite enantioselectivity results when (*Z*)-enolates are employed for the same reactions (Scheme 3.37).[48] The (*E*)-enolates are usually available by

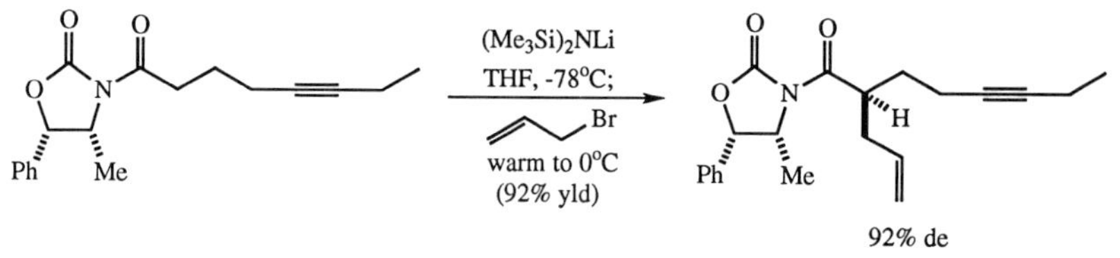
(Me3Si)2NLi
THF, -78°C;
Br
warm to 0°C
(92% yld)
Ph
Me
H
92% de

Scheme 3.36

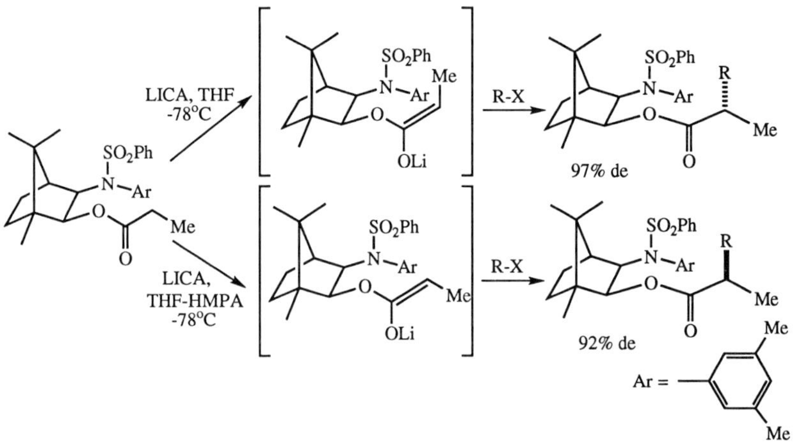
LICA, THF
-78°C
SO2Ph
N
Ar
Me
OLi
R-X
97% de
LICA,
THF-HMPA
-78°C
92% de
Ar =

Scheme 3.37

treatment of the starting ester with LICA and HMPA in THF at low temperature.

Oppolzer *et al.* have demonstrated that *N*-acylated bornane-10,2-sultams undergo deprotonation when treated with either n-BuLi or sodium hexamethyldisilazide in THF, to give chelated lithium- and sodium (*Z*)-enolates.[49] The latter react preferentially on their *Re*-face with α-halo esters, benzylic, allylic and propargylic halides, and with non-activated primary alkyl iodides (Scheme 3.38). The scope of this method has been further expanded to provide a chiral glycine equivalent (Scheme 3.39), which

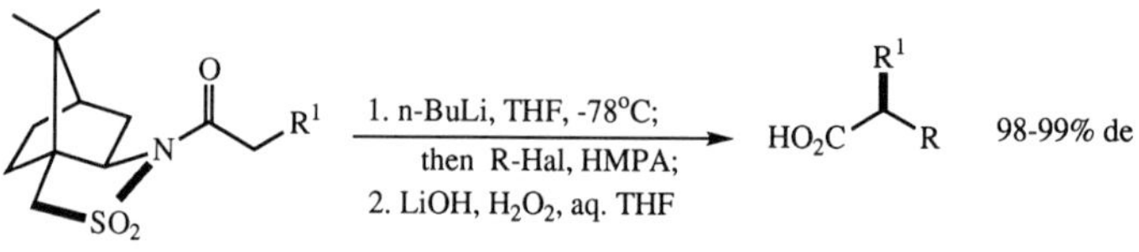
1. n-BuLi, THF, -78°C;
then R-Hal, HMPA;
2. LiOH, H2O2, aq. THF
HO2C
R
98-99% de

Scheme 3.38

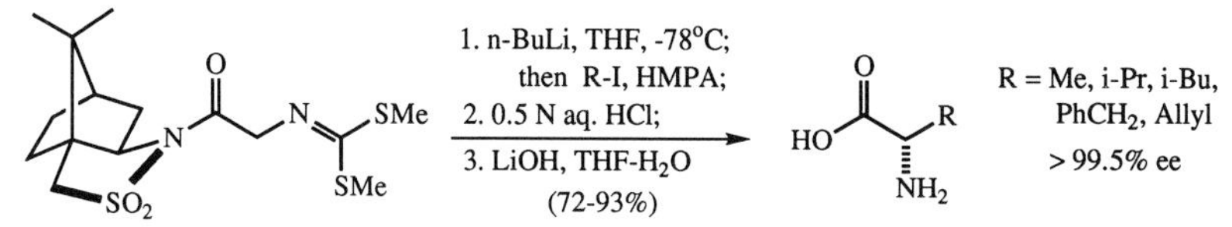

1. n-BuLi, THF, -78°C; then R-I, HMPA; 2. 0.5 N aq. HCl; 3. LiOH, THF-H_2O (72-93%)

R = Me, i-Pr, i-Bu, $PhCH_2$, Allyl
> 99.5% ee

Scheme 3.39

allows access to a wide variety of α-amino acids in high enantiomeric purity after alkylation and removal of the chiral auxiliary. Remarkably, these glycine enolates are of sufficient nucleophilicity to even readily displace secondary alkyl iodides.[50,51] The *N*-deprotection step is generally accomplished by mild acid hydrolysis, the bornyl-10,2-sultam being detached afterwards by the standard saponification method.

Oppolzer has also found that deprotonation and alkylation of this chiral glycine bornyl-10,2-sultam can be achieved at 0°C under the auspices of phase-transfer catalysis ($LiOH/Bu_4NSO_4$, CH_2Cl_2/H_2O, $PhCH_2Cl$, ultrasound).[51]

An effective class of chiral enolates for alkylation reactions are those produced from acyl complexes of $[Fe(\eta\text{-}C_5H_5)(CO)(PPh_3)]$.[52-55] As a rule, (*E*)-enolates are formed almost exclusively (>200:1) from these systems so long as the deprotonation step is carried out at −78°C with *n*-butyllithium. Alkylation generally proceeds with excellent levels of stereocontrol from the less hindered side of the enolate, opposite to the Ph_3P group when it is in a conformation where the acyl-oxygen is *anti* to the CO ligand (Scheme 3.40).[52] Although primary alkyl halides and tosylates are most frequently cited as the electrophiles for these alkylations, epoxides can also be cajoled

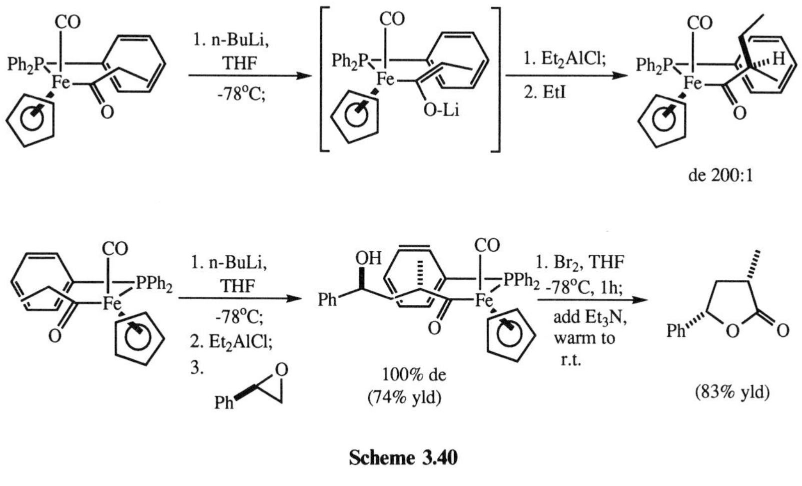

1. n-BuLi, THF, -78°C; 1. Et_2AlCl; 2. EtI

de 200:1

1. n-BuLi, THF, -78°C; 2. Et_2AlCl; 3. (phenyl epoxide)

100% de (74% yld)

1. Br_2, THF -78°C, 1h; add Et_3N, warm to r.t.

(83% yld)

Scheme 3.40

into reacting if diethylaluminium chloride is incorporated as an additive.[53] Presumably, the latter functions as a Lewis acid, coordinating to the epoxide oxygen and thus increasing its susceptibility to nucleophilic attack by the enolate (Scheme 3.40). Significantly, Davies has developed the latter reaction into an effective method for the kinetic resolution of racemic monosubstituted epoxides.[54] According to Davies,[55] the iron auxiliary is best detached from the alkylated products by treatment with a one-electron oxidant such as bromine, *N*-bromosuccinimide (NBS), ceric ion or ferric ion. Bromine is claimed to be the reagent of choice, but for acid-labile substrates the use of NBS is also recommended. The main advantage of the brominative method of removal is that it frequently allows the chiral auxiliary to be readily recovered as $[Fe(\eta\text{-}C_5H_5)(CO)(PPh_3)Br]$, a feature that makes it attractive for large-scale applications.

Myers *et al.* have recently examined the asymmetric alkylation of carboxylic acid derivatives bound to the chiral auxiliary pseudoephedrine and found them to be highly diastereoselective.[56] As a rule, these alkylations require dianion formation with lithium diisopropylamide (2.25 equiv.) at −78°C in tetrahydrofuran in the presence of lithium chloride (6 equiv.); activated or non-hindered primary alkyl halides are the electrophiles that react most effectively with these systems (Scheme 3.41).[56] A noteworthy feature of this method is

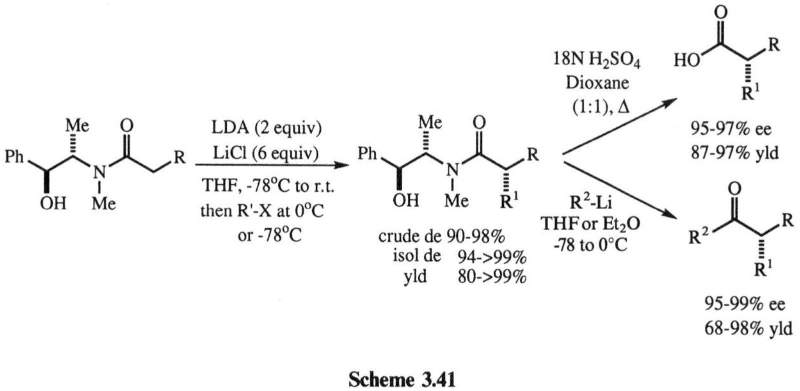

Ph, Me, OH, N, Me, O, R
LDA (2 equiv)
LiCl (6 equiv)
THF, -78°C to r.t.
then R'-X at 0°C
or -78°C
Ph, Me, OH, N, Me, O, R, R^1
crude de 90-98%
isol de 94->99%
yld 80->99%
18N H_2SO_4
Dioxane
(1:1), Δ
HO, O, R, R^1
95-97% ee
87-97% yld
R^2-Li
THF or Et_2O
-78 to 0°C
R^2, O, R, R^1
95-99% ee
68-98% yld

Scheme 3.41

the ease with which the alkylated products can be converted directly into chiral aldehydes, ketones and acids of high enantiomeric purity. Thus, with lithium triethoxyaluminium hydride, aldehydes are obtained in 90–98% ee and 75–82% yield. Ketones are also available in 95–99% ee simply by exposing the alkylated product to an alkyllithium reagent; chiral acids are accessible by acid hydrolysis (Scheme 3.41).[56]

A new method for the stereoselective introduction of quaternary asymmetric carbon atoms has been developed by Meyers *et al.* that is based upon iterative lithiation and alkylation of chiral bicyclic lactams derived from (*S*)-valinol.[57]

As a rule, the initial alkylation step proceeds with poor diastereocontrol, while the second alkylation proceeds with excellent *endo*-selectivity (Scheme 3.42). For the lactam shown in Scheme 3.42, the dialkylated product was

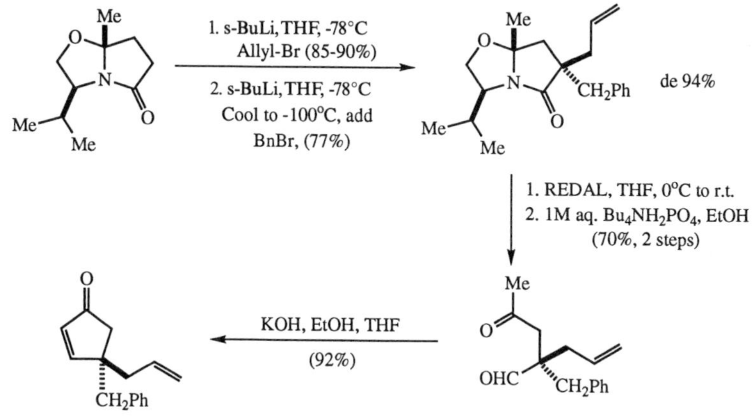

Scheme 3.42

obtained in 94% de. Subsequent Red-Al reduction and acid hydrolysis furnished a keto-aldehyde that was converted to the 4,4-dialkylated cyclopentenone by intramolecular aldol ring-closure with base.[57] Utilising variations of this technology, the Colorado group have synthesised a significant number of natural products. Their exploits have been collated in a review article by Romo and Meyers.[58]

An interesting device for installing quaternary asymmetric carbon centres involves the stereoselective alkylation of enolate dianions from the chiral half-esters of monosubstituted malonic acid derivatives (Scheme 3.43).[59] In

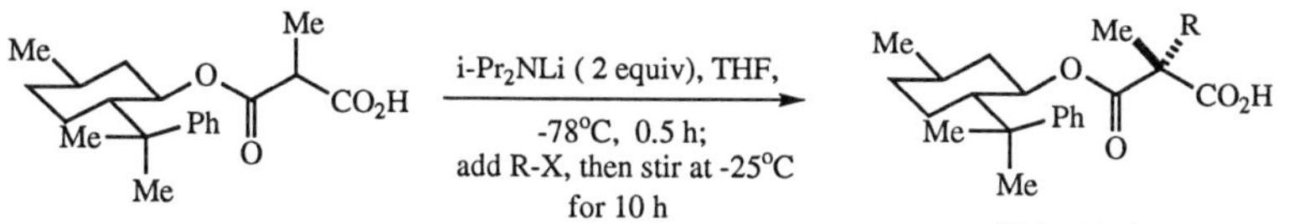

R-X	Yield (%)	Ratio
Et-I	83	4:1
n-Pr-I	72	4:1
allyl-I	77	7:1
Bn-Br	72	12:1
2-$MeOC_6H_4CH_2Br$	73	16:1

Scheme 3.43

most cases, the diastereoselectivity varies from modest, (4:1) with unactivated alkyl iodides, to good (12–16:1) with benzylic bromides.

A remarkable chiral phase-transfer catalyst has been described by Grabowski that originates from cinchonide (Scheme 3.44).[60] When used in conjunction

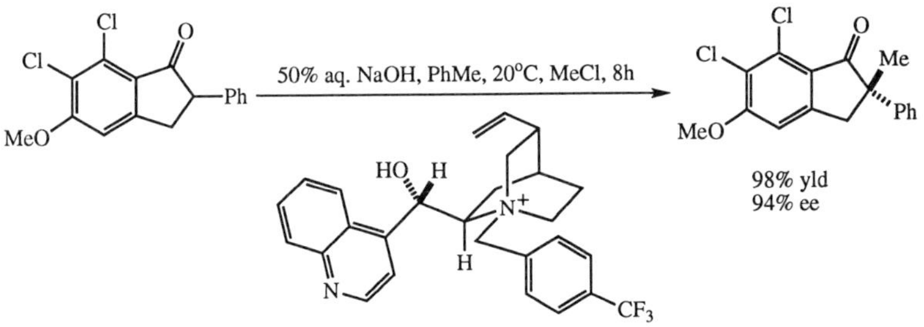

Scheme 3.44

with aqueous sodium hydroxide, it transiently generates chiral enolates from racemic α-phenyl indanone derivatives, which then undergo highly stereoselective alkylation reactions from their less hindered side. Although these reactions work well with unactivated alkyl halides, they do suffer the disadvantage that they are not applicable to a wide range of starting ketones. Similar phase-transfer catalysts have also been used successfully for the asymmetric alkylation of glycine derivatives by O'Donnell *et al.*[61] Both methods represent important milestones in catalytic enantioselective enolate alkylation, and we anticipate that further progress will be made along these lines in the coming years.

Before departing from the topic of asymmetric alkylation, we would like to draw the reader's attention to an interesting paper that has appeared, in which a chiral naphthyl ketone has been deprotonated at its sole asymmetric carbon atom to generate an enolate with fleeting axial chirality; the latter was then reacted with a variety of alkylating agents to give products that were also chiral (Scheme 3.45).[62] These workers termed this phenomenon

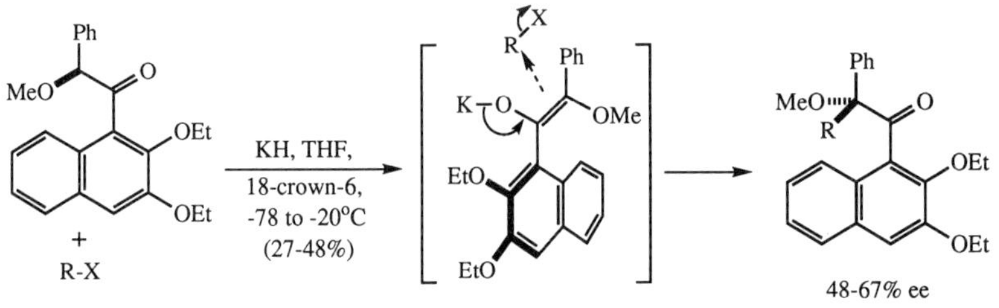

Scheme 3.45

'Memory of Chirality' and suggested that the concept might lead to new developments in asymmetric synthesis in the near future.

3.4 Asymmetric oxidation, amination and thioalkylation of chiral enolates

Several noteworthy papers have now appeared on the asymmetric oxidation of chiral enolates. Tamm *et al.* have discovered that excellent diastereoselectivity can be attained in the oxidation of certain chiral ester potassium enolates using the MoOPH complex (Scheme 3.46).[63]

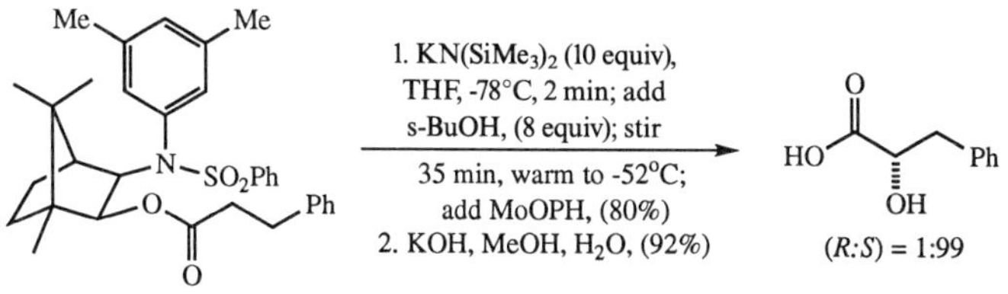

1. $KN(SiMe_3)_2$ (10 equiv), THF, -78°C, 2 min; add s-BuOH, (8 equiv); stir 35 min, warm to -52°C; add MoOPH, (80%)
2. KOH, MeOH, H_2O, (92%)

(*R*:*S*) = 1:99

Scheme 3.46

Evans has also successfully oxidised enolates of *N*-acyl oxazolidinones utilising the Davis oxaziridine reagents.[64] Generally speaking, the levels of chirality transfer improve as the degree of steric congestion around the enolate is raised. Significantly, however, this increase in steric crowding around the reaction site does not lead to marked diminution of the product yield (Scheme 3.47).[9]

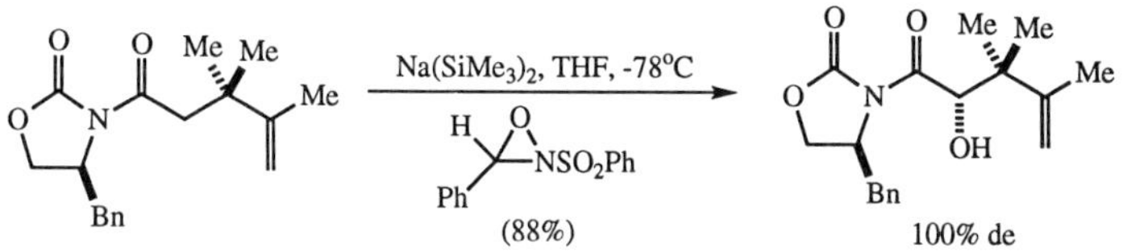

$Na(SiMe_3)_2$, THF, -78°C; (88%)

100% de

Scheme 3.47

A number of good procedures have recently been introduced for the electrophilic amination of chiral enolates. The protocols of Evans[65] and Vederas[66] are now proving to be the methods of choice for many such applications. These entail trapping the lithium enolates of *N*-acylated oxazolidinones with di-t-butyl azodicarboxylate at −78°C. They produce α-hydrazino acid derivatives in exceptionally high de and in good yield. The free hydrazines can then be converted to the α-amino acids by hydrogenolysis of the N–N bond using a Raney nickel catalyst (Scheme 3.48). An interesting

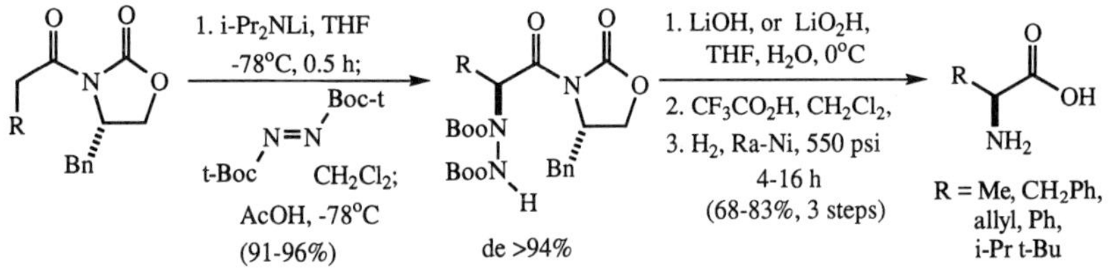

Scheme 3.48

twist on this chemistry that leads to the synthesis of enantiomerically pure piperazic acid utilises the highly stereoselective *tandem* conjugate addition/intramolecular S_N2 cyclisation of chiral bromovaleryl enolates with di-*t*-butylazodicarboxylate (Scheme 3.49).[67]

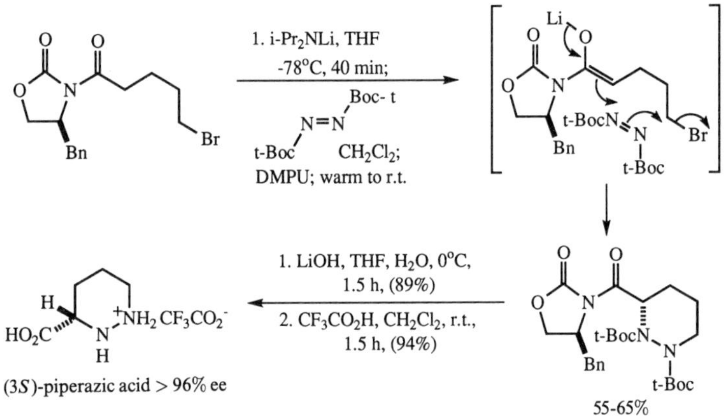

Scheme 3.49

Oppolzer has now provided an elegant solution to the problem of chiral α-hydroxamic acid synthesis by exploiting the reaction of chelated (*Z*)-sodium enolates from *N*-acylated bornane-10,2-sultams with 1-chloro-1-nitroso-cyclohexane. In this approach, the enolate is attacked preferentially from its C(α)-*Re*-face by the $[NHOH]^+$ synthon to generate a nitrone intermediate that is converted to the α-hydroxamic acid by treatment with acid followed by base (Scheme 3.50).[68] Oppolzer has also developed conditions (Zn/H^+) for cleaving the N–O bond in these products to give access to chiral α-amino acids.

A particularly expeditious synthesis of α-azido acids has been invented by Evans *et al.* that is based on enolisation of chiral *N*-acyl oxazolidinones with potassium hexamethyldisilazide, trapping with tri-isopropylbenzenesulfonyl azide at −78°C, and quenching of the resulting triazene intermediates with acetic acid (Scheme 3.51).[69] Other azide-transfer agents such as tosyl azide generally proved less satisfactory in this regard. An alternative but less direct

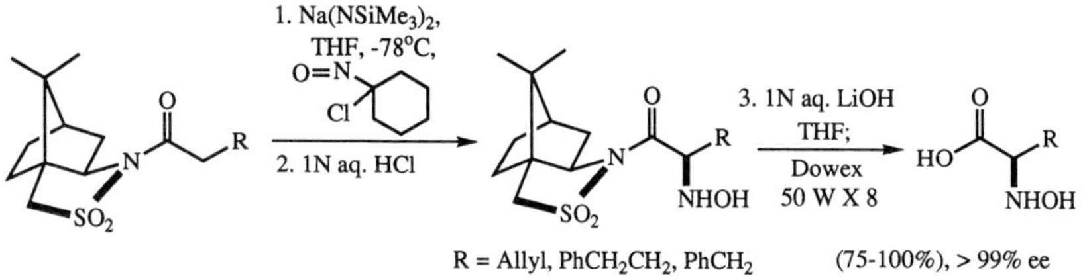

1. $Na(NSiMe_3)_2$, THF, -78°C, O=N–(1-chlorocyclohexyl)
2. 1N aq. HCl

3. 1N aq. LiOH THF; Dowex 50 W X 8

NHOH NHOH SO$_2$ HO

R = Allyl, $PhCH_2CH_2$, $PhCH_2$ (75-100%), > 99% ee

Scheme 3.50

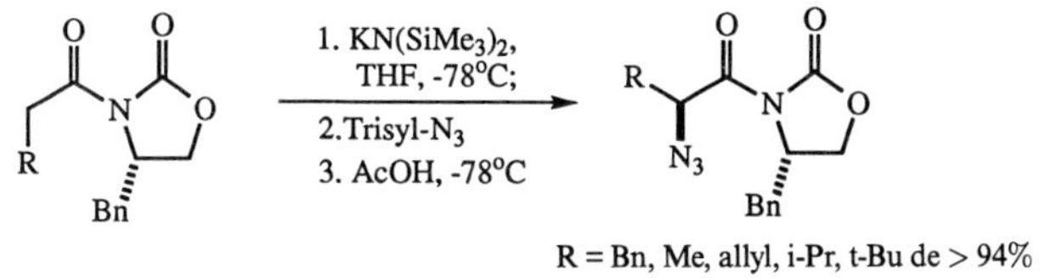

1. $KN(SiMe_3)_2$, THF, -78°C;
2. Trisyl-N_3
3. AcOH, -78°C

R Bn N_3 Bn

R = Bn, Me, allyl, i-Pr, t-Bu de > 94%

Scheme 3.51

method for obtaining chiral α-azido acids is shown in Scheme 3.52; it features an asymmetric α-bromination of a *N*-acyl oxazolidinone followed by nucleophilic displacement with azide ion.[69,70]

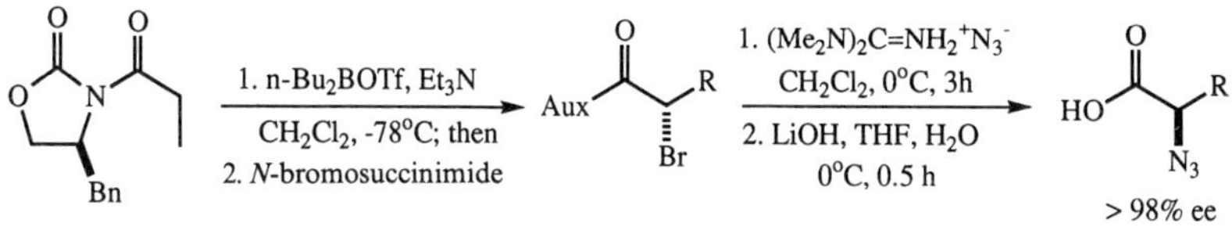

1. n-Bu_2BOTf, Et_3N CH_2Cl_2, -78°C; then
2. *N*-bromosuccinimide

Aux R Br

1. $(Me_2N)_2C{=}NH_2^+N_3^-$ CH_2Cl_2, 0°C, 3h
2. LiOH, THF, H_2O 0°C, 0.5 h

HO R N_3

Bn

> 98% ee

Scheme 3.52

Davies *et al.* have recently shown that the enolate derived from a chiral iron–acyl propionyl complex can be intercepted stereoselectively (16:1) with phenyl disulfide (Scheme 3.53).[71] Subsequent separation of these diastereoisomers

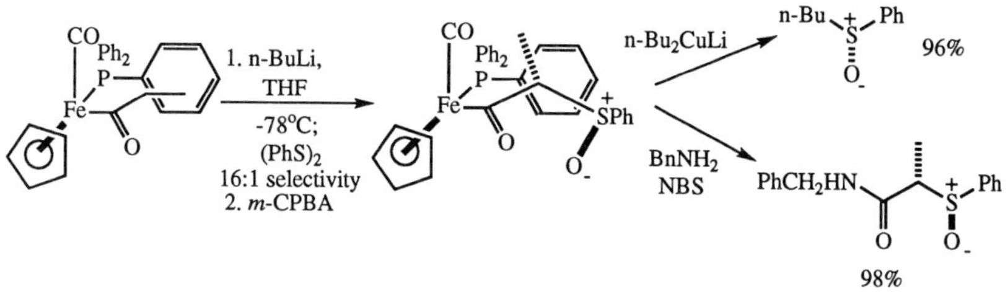

CO Ph_2 P Fe O

1. n-BuLi, THF -78°C; $(PhS)_2$ 16:1 selectivity
2. *m*-CPBA

CO Ph_2 P Fe O $\overset{+}{S}Ph$ O^-

n-Bu_2CuLi → n-Bu–$\overset{+}{S}$(O$^-$)–Ph 96%

$BnNH_2$ NBS → $PhCH_2HN$–C(O)–CH(CH$_3$)–$\overset{+}{S}$(O$^-$)–Ph 98%

Scheme 3.53

and oxidation of the major component with *m*-CPBA yielded the phenylsulfoxide as a single product. Treatment of this compound with lithium di-*n*-butylcuprate cleanly detached the iron auxiliary and simultaneously inverted the chirality of the sulfoxide. An oxidative method of decomplexation was also investigated with *N*-bromosuccinimide and benzylamine; this gave a chiral β-amido sulfoxide (Scheme 3.53).

Chibale and Warren have described a valuable asymmetric synthesis of α-thiophenyl aldehydes that capitalises on the highly diastereoselective capture of chelated lithium enolates derived from *N*-acyl oxazolidinones with phenyl disulfide (Scheme 3.54).[72] The auxiliary is best excised from the chiral

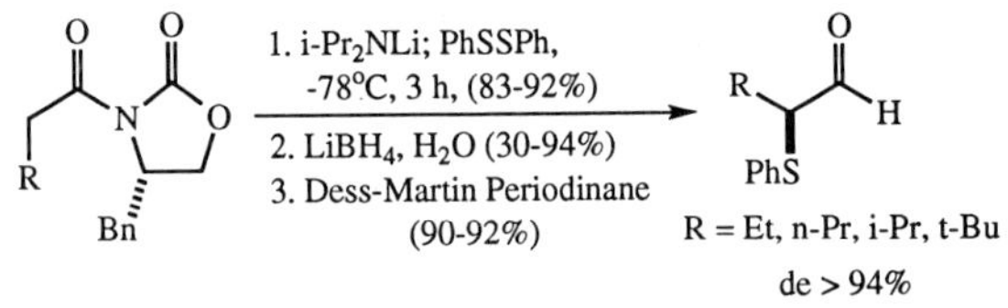

1. i-Pr_2NLi; PhSSPh, -78°C, 3 h, (83-92%)
2. $LiBH_4$, H_2O (30-94%)
3. Dess-Martin Periodinane (90-92%)

R = Et, n-Pr, i-Pr, t-Bu

de > 94%

Scheme 3.54

products by reduction with lithium borohydride in water. The resulting alcohols are then oxidised to the chiral aldehydes with the Dess–Martin periodinane. These conditions avoid racemisation at the newly created asymmetric carbon, which tends to be problematic in these systems.

Acknowledgements

We thank Drs Ian Paterson and Alison Franklin of Cambridge University for kindly supplying us with a preprint of their article on the asymmetric aldol reaction.

References

1. (a) A.S. Franklin and I. Paterson, *Contemporary Organic Synthesis*, 1994, p. 317; (b) C.H. Heathcock, in *Modern Synthetic Methods*, Vol. VI, R. Scheffold (ed.), Verlag Chemie, Basel, 1992, p. 1; (c) C.H. Heathcock in *Comprehensive Organic Synthesis*, Vol. 2, B.M. Trost and I. Fleming (eds), Pergamon Press, Oxford, 1991, p. 181; (d) B. Moon Kim, S.F. Williams and S. Masamune in *Comprehensive Organic Synthesis*, Vol. 2, B.M. Trost and I. Fleming (eds), Pergamon Press, Oxford, 1991, p. 239; (e) I. Paterson in *Comprehensive Organic Synthesis*, Vol. 2, B.M. Trost and I. Fleming (eds), Pergamon Press, Oxford, 1991, p. 301; (f) C. Gennari in *Comprehensive Organic Synthesis*, Vol. 2, B.M. Trost and I. Fleming, (eds), Pergamon Press, Oxford, 1991, p. 629; (g) D.A. Evans in *Asymmetric Synthesis*, Vol. 3, J.C. Morrison (ed.), Academic Press, New York, 1984, p. 213; (h) M. Braun, *Angew. Chem., Int. Edn. Engl.*, 1987, **26**, 24; (i) C.H. Heathcock in *Asymmetric Synthesis*, Vol. 3, J.C. Morrison

(ed.), Academic Press, New York, 1984, p. 111; (j) D.A. Evans, J.V. Nelson and T.R. Taber, *Topics in Stereochemistry*, Vol. 13, 1982 p. 1; (k) D.A. Evans, *Aldrichima Acta*, 1982, **15**, 23.
2. S. Mitsui, K. Konno, I. Onuma and K. Shimizu, *J. Chem. Soc. Jap.*, 1964, **85**, 437; E.B. Dongala, D.L. Hull, C. Mioskowski and G. Solladie, *Tetrahedron Lett.*, 1973, 4983.
3. D.A. Evans, J. Bartroli and T.L. Shih, *J. Am. Chem. Soc.*, 1981, **103**, 2127.
4. S. Masamune, W. Choy, F.A.J. Kerdesky and B. Imperiali, *J. Am. Chem. Soc.*, 1981, **103**, 1566.
5. R. Baker and J.L. Castro, *J. Chem. Soc., Perkin Trans. 1*, 1990, 47.
6. H. Danda, M.M. Hansen and C.H. Heathcock, *J. Org. Chem.*, 1990, **55**, 173.
7. M. Nerz-Stormes and E.R. Thornton, *Tetrahedron Lett.*, 1986, **27**, 897; *J. Org. Chem.*, 1991, **56**, 2489; C. Siegel and E.R. Thornton, *Tetrahedron Lett.*, 1986, **27**, 457.
8. D.A. Evans, J.S. Clark, R. Metternich, V.J. Novack and G.S. Sheppard, *J. Am. Chem. Soc.*, 1990, **112**, 866; D.A. Evans, H.P. Ng, J.S. Clark and D.L. Rieger, *Tetrahedron*, 1992, **48**, 2127.
9. D.A. Evans, J.R. Gage and J.L. Leighton, *J. Am. Chem. Soc.*, 1992, **114**, 9434.
10. A.S. Kende, K. Kawamura and M.J. Orwat, *Tetrahedron Lett.*, 1989, **30**, 5821.
11. A.S. Kende, K. Kawamura and R.J. DeVita, *J. Am. Chem. Soc.*, 1990, **112**, 4070.
12. W. Oppolzer, J. Blagg, I. Rodriguez and E. Walther, *J. Am. Chem. Soc.*, 1990, **112**, 2767.
13. T. Katsuki and M. Yamaguchi, *Tetrahedron Lett.*, 1985, **26**, 5807.
14. A.G. Myers and K.L. Widdowson, *J. Am. Chem. Soc.*, 1990, **112**, 9672.
15. D.A. Evans and L.R. McGee, *J. Am. Chem. Soc.*, 1981, **103**, 2876.
16. W. Oppolzer, C. Starkemann, I. Rodriguez and G. Bernadinelli, *Tetrahedron Lett.*, 1991, **32**, 61.
17. M. Braun and H. Sacha, *Angew. Chem., Int. Edn. Engl.*, 1991, **30**, 1318; H. Sacha, D. Waldmuller and M. Braun, *Chem. Ber.*, 1994, **127**, 1959.
18. E.J. Corey, G.A. Reichard and R. Kania, *Tetrahedron Lett.*, 1993, **34**, 6977.
19. W. Oppolzer and C. Starkemann, *Tetrahedron Lett.*, 1992, **33**, 2439.
20. M. Braun and R. Devant, *Tetrahedron Lett.*, 1984, **25**, 5031.
21. (a) A.I. Meyers and Y. Yamamoto, *J. Am. Chem. Soc.*, 1981, **103**, 4278; (b) A.I. Meyers and Y. Yamamoto, *Tetrahedron*, 1984, **40**, 2309.
22. (a) R.W. Stevens and T. Mukaiyama, *Chem. Lett.*, 1983, 1799; N. Iwasawa and T. Mukaiyama, *Chem. Lett.*, 1982, 1441; (b) N. Iwasawa and T. Mukaiyama, *Chem. Lett.*, 1983, 297; (c) T. Mukaiyama, N. Iwasawa, R.W. Stevens and T. Haga, *Tetrahedron*, 1984, **40**, 1381; (d) S. Kobayashi, H. Uchiro, Y. Fujishita, I. Shiina and T. Mukaiyama, *J. Am. Chem. Soc.*, 1991, **113**, 4247; (e) T. Mukaiyama, S. Kobayashi and T. Sano, *Tetrahedron*, 1990, **46**, 4653.
23. (a) S. Masamune, T. Sato, B.M. Kim and T.A. Wollmann, *J. Am. Chem. Soc.*, 1986, **108**, 8279; (b) M.A. Blanchette, M.S. Malamas, M.H. Nantz, J.C. Roberts, P. Somfai, D.C. Whritenour, S. Masamune, M. Kageyama and T. Tamura, *J. Org. Chem.*, 1989, **54**, 2817; (c) N. Tanimoto, S.W. Gerritz, A. Sawabe, T. Noda, S.A. Filla and S. Masamune, *Angew. Chem., Int. Edn. Engl.*, 1994, **33**, 673.
24. (a) E.J. Corey, R. Imwinkelreid, S. Pikul and Y.B. Xiang, *J. Am. Chem. Soc.*, 1989, **111**, 5493; (b) E.J. Corey and S.S. Kim, *J. Am. Chem. Soc.*, 1990, **112**, 4976; (c) E.J. Corey, D.-H. Lee and S. Choi, *Tetrahedron Lett.*, 1992, **33**, 6735.
25. R.O. Duthaler, P. Herold, W. Lottenbach, K. Oertle and M. Riediker, *Angew. Chem., Int. Edn. Engl.*, 1989, **28**, 495.
26. R.O. Duthaler, P. Herold, S. Wyler-Helfer and M. Reidiker, *Helv. Chim. Acta*, 1990, **73**, 659.
27. G. Bold, R.O. Duthaler and M. Reidiker, *Angew. Chem., Int. Edn. Engl.*, 1989, **28**, 497.
28. T. Mukaiyama, H. Uchiro, I. Shiina and S. Kobayashi, *Chem. Lett.*, 1990, 1019; T. Mukaiyama, I. Shiina and S. Kobayashi, *Chem. Lett.*, 1991, 1901; S. Kobayashi and T. Kawasuji, *Tetrahedron Lett.*, 1994, **35**, 3329.
29. I. Paterson, M.A. Lister and C.K. McClure, *Tetrahedron Lett.*, 1986, **27**, 4787; I. Paterson, *Chem. Ind. (Lond.)*, 1988, 390; I. Paterson and J.M. Goodman, *Tetrahedron Lett.*, 1989, **30**, 997; I. Paterson, J.M. Goodman, M.A. Lister, R.C. Schumann, C.K. McClure and R. Norcross, *Tetrahedron*, 1990, **46**, 4663.
30. Y. Ito, M. Sawamura and T. Hayashi, *J. Am. Chem. Soc.*, 1986, **108**, 6405; Y. Ito, M. Sawamura and T. Hayashi, *Tetrahedron Lett.*, 1987, **28**, 6215; V.A. Soloshonok and T. Hayashi, *Tetrahedron Lett.*, 1994, **35**, 2713; for a similar Ag(I) catalysed asymmetric aldol process see: V.A. Soloshonok and T. Hayashi, *Tetrahedron Lett.*, 1994, **35**, 1055.
31. T. Mukaiyama, M. Furuya, A. Ohtsubo and S. Kobayashi, *Chem. Lett.*, 1991, 989;

T. Mukaiyama, S. Kobayashi, H. Uchiro and I. Shiina, *Chem. Lett.*, 1990, 129; S. Kobayashi, Y. Fujishita and T. Mukaiyama, *Chem. Lett.*, 1990, 1455.
32. T. Mukaiyama, H. Uchiro and S. Kobayashi, *Chem. Lett.*, 1990, 1147.
33. S. Kobayashi, A. Ohtsubo and T. Mukaiyama, *Chem. Lett.*, 1991, 831.
34. K. Furuta, T. Maruyama and H. Yamamoto, *J. Am. Chem. Soc.*, 1991, **113**, 1041; K. Furuta, T. Maruyama and H. Yamamoto, *Synlett*, 1991, 439.
35. E.R. Parmee, O. Tempkin, S. Masamune and A. Abiko, *J. Am. Chem. Soc.*, 1991, **113**, 9365.
36. S. Kiyooka, Y. Kaneko, M. Komura, H. Matsuo and M. Nakano, *J. Org. Chem.*, 1991, **56**, 2276; S. Kiyooka, Y. Kaneko and K. Kume, *Tetrahedron Lett.*, 1992, **33**, 4927.
37. E.R. Parmee, Y. Hong, O. Tempkin and S. Masamune, *Tetrahedron Lett.*, 1992, **33**, 1729.
38. E.J. Corey, C.L. Cywin and T.D. Roper, *Tetrahedron Lett.*, 1992, **33**, 6907.
39. K. Mikami and S. Matsukawa, *J. Am. Chem. Soc.*, 1994, **116**, 4077; K. Mikami and S. Matsukawa, *J. Am. Chem. Soc.*, 1993, **115**, 7039.
40. E.M. Carreira, R.A. Singer and W. Lee, *J. Am. Chem. Soc.*, 1994, **116**, 8837.
41. D. Caine in *Comprehensive Organic Synthesis*, Vol. 3, B.M. Trost and I. Fleming (eds), Pergamon Press, Oxford, 1991, p. 1.
42. D.A. Evans and J.M. Takacs, *Tetrahedron Lett.*, 1980, **21**, 4233; D.A. Evans, R.L. Dow, T.L. Shih, J.M. Takacs and R. Zahler, *J. Am. Chem. Soc.*, 1990, **112**, 5290; P.E. Sonnet and R.R. Heath, *J. Org. Chem.*, 1980, **45**, 3137.
43. D. Askin, R.P. Volante, K.M. Ryan, R.A. Reamer and I. Shinkai, *Tetrahedron Lett.*, 1988, **29**, 4245.
44. Y. Kawenami, Y. Ito, T. Kitagawa, Y. Taniguchi, T. Katsuki, M. Yamaguchi, *Tetrahedron Lett.*, 1984, **25**, 857; M. Enomoto, Y. Ito, T. Katsuki and M. Yamaguchi, *Tetrahedron Lett.*, 1985, **26**, 1343.
45. T. Hanamoto, T. Katsuki and M. Yamaguchi, *Tetrahedron Lett.*, 1986, **27**, 2463.
46. D.A. Evans, M.D. Ennis and D.J. Mathre, *J. Am. Chem. Soc.*, 1982, **104**, 1737.
47. A.B. Smith, III, M. Fukui, H.A. Vaccaro and J.R. Empfield, *J. Am. Chem. Soc.*, 1991, 113, 2071.
48. R. Schmierer, G. Grotemeier, G. Helmchen and A. Selim, *Angew. Chem., Int. Ed. Engl.*, 1981, **20**, 207; G. Helmchen, A. Selim, D. Dorsch and I. Taufer, *Tetrahedron Lett.*, 1983, **24**, 3213.
49. W. Oppolzer, R. Moretti and S. Thomi, *Tetrahedron Lett.*, 1989, **30**, 5603.
50. W. Oppolzer, R. Moretti and S. Thomi, *Tetrahedron Lett.*, 1989, **30**, 6009.
51. W. Oppolzer, *Pure Appl. Chem.*, 1990, **62**, 1241; W. Oppolzer, H. Bienayme and A. Genevois-Borella, *J. Am. Chem. Soc.*, 1991, **113**, 9660.
52. G. J. Baird and S.G. Davies, *J. Organomet., Chem.*, 1983, **248**, C1; G.J. Baird, J.A. Bandy, S.G. Davies and K. Prout, *J. Chem. Soc., Chem. Comm.*, 1983, 1202; S.G. Davies and J.I. Seeman, *Tetrahedron Lett.*, 1984, **25**, 1845; S.L. Brown, S.G. Davies, D.F. Foster, J.I. Seeman and P. Warner, *Tetrahedron Lett.*, 1986, **27**, 623; G. Bashiardes and S.G. Davies, *Tetrahedron Lett.*, 1988, **29**, 6509; S.P. Collingwood, S.G. Davies and S.C. Preston, *Tetrahedron Lett.*, 1990, **31**, 4067; G.J. Bodwell and S.G. Davies, *Tetrahedron: Asymmetry*, 1993, **2**, 1075; see also: L.S. Liebeskind, M.E. Welker and R.W. Fengl, *J. Am. Chem. Soc.*, 1986, **108**, 6328 and the references cited therein.
53. S.G. Davies, R. Polywka and P. Warner, *Tetrahedron*, 1990, **46**, 4847.
54. S.L. Brown, S.G. Davies, P. Warner, R.H. Jones and K. Prout, *J. Chem. Soc., Chem. Comm.*, 1985, 1446.
55. S.G. Davies, *Aldrichimica Acta*, 1990, **23**, 31.
56. A.G. Myers, B.H. Yang, H. Chen and J.L. Gleason, *J. Am. Chem. Soc.*, 1994, **116**, 9361.
57. A.I. Meyers and K.T. Wanner, *Tetrahedron Lett.*, 1985, **26**, 863.
58. D.A. Romo and A.I. Meyers, *Tetrahedron*, 1991, **47**, 9503.
59. M. Ihara, M. Takahashi, N. Taniguchi, K. Yasui, H. Niitsuma and K. Fukomoto, *J. Chem. Soc., Perkin Trans. 1*, 1991, 525.
60. D.L. Hughes, U.-H. Dolling, K.M. Ryan, E.F. Schoenewaldt and E.J.J. Grabowski, *J. Org. Chem.*, 1987, **52**, 4745.
61. M.J. O'Donnell, W.D. Bennett and S. Wu, *J. Am. Chem. Soc.*, 1989, **111**, 2353; M.J. O'Donnell and S. Wu, *Tetrahedron: Asymmetry*, 1992, **3**, 591.
62. T. Kawabata, K. Yahiro and K. Fuji, *J. Am. Chem. Soc.*, 1991, **113**, 9694.
63. R. Gamboni and C. Tamm, *Helv. Chim. Acta*, 1986, **69**, 615.
64. D.A. Evans, M.M. Morrissey and R.L. Dorow, *J. Am. Chem. Soc.*, 1985, **107**, 4346.
65. D.A. Evans, T.C. Britton, R.L. Dorow and J.F. Dellaria, *J. Am. Chem. Soc.*, 1986, **108**, 6395; D.A. Evans, T.C. Britton, R.L. Dorow and J.F. Dellaria, *Tetrahedron*, 1988, **44**, 5525.

66. L.A. Trimble and J.C. Vederas, *J. Am. Chem. Soc.*, 1986, **108**, 6397.
67. K.J. Hale, V.M. Delisser and S. Manaviazar, *Tetrahedron Lett.*, 1992, **33**, 7613.
68. W. Oppolzer and O. Tamura, *Tetrahedron Lett.*, 1990, **31**, 991; W. Oppolzer, O. Tamura and J. Deerberg, *Helv. Chim. Acta*, 1992, **75**, 1965.
69. D.A. Evans, T.C. Britton, J.A. Ellman and R.L. Dorow, *J. Am. Chem. Soc.*, 1990, **112**, 4011.
70. D.A. Evans, J.A. Ellman and R.L. Dorow, *Tetrahedron Lett.*, 1987, **28**, 1123.
71. S.G. Davies and G.L. Gravatt, *J. Chem. Soc., Chem. Comm.*, 1988, 780.
72. K. Chibale and S. Warren, *Tetrahedron Lett.*, 1994, **35**, 3991.

4 Optically active β-ketosulfoxides in asymmetric synthesis

G. SOLLADIÉ

4.1 Introduction

During the 1980s, organic sulfur compounds have become increasingly useful and important in organic synthesis. Sulfur, incorporated into an organic molecule, stabilizes a negative charge on an adjacent carbon atom, a property which has been especially important in the development of new ways to form carbon–carbon bonds. With respect to sulfides and sulfones, the sulfoxide group is of special interest because of its chirality and the presence of three different kinds of substituent, from a steric and stereoelectronic point of view: the lone pair of electrons, the oxygen atom and two aryl or alkyl groups give a special efficiency to sulfoxides in asymmetric synthesis.

The purpose of this chapter will be limited to asymmetric syntheses from β-ketosulfoxides. After a general overview on the different ways to introduce an optically active sulfoxide group on a target molecule, the asymmetric reduction of β-ketosulfoxides will be described. Finally, several applications in total synthesis of natural products will be presented.

4.2 Optically active sulfoxide preparations

Until now, optimally active sulfoxides have been obtained in many different ways: optical resolution, asymmetric synthesis, kinetic resolution and stereospecific synthesis.

Optical resolution could be achieved, following the pioneering work of Harrison,[1] by means of an acidic or basic group present in the molecule. The resolution of ethyl *p*-tolyl sulfoxide was also achieved in 1966, through the formation and separation of the diastereoisomeric complexes with *trans*-dichloroethylene platinum(II) containing optically active α-phenylethylamine as a ligand.[2] The more recent work on optical resolution was reviewed in detail by Mikolajczyk *et al.*[3]

Asymmetric oxidation of sulfides with optically active peracids was first reported by Montanari[4] and Balenovic,[5] though their products were of low enantiomeric purity, generally not higher than 10%. More recently Kagan[6] reported that high enantioselectivities could be obtained with a modified Sharpless reagent [$Ti(O\text{-}i\text{-}Pr)_4$/DET/t-BuOOH/$H_2O$]; ee values in the range of 80 to 90% were obtained (see Chapter 17) in the case of simple alkyl aryl sulfides. Enzymatic oxidation of sulfides also gave very good results in a few cases.[3,7]

New approaches starting from cyclic disulfides[8] or oxazolidinones[9] were also reported.

However, all these methods, which give good results with specific substrates, are not yet general enough. A great achievement in the stereochemistry of organosulfur compounds was the stereoselective synthesis of optically active sulfoxides, originally proposed by Gilman[10a] and developed later by Andersen[10b]. This approach to sulfoxides of high optical purity, still most important and widely used, is based on the reaction of the diastereoisomerically pure (−)-(*S*)-menthyl *p*-toluene sulfinate (**1**)* with Grignard reagents. (+)-(*R*)-Ethyl *p*-tolyl sulfoxide (**2**) was the first optically active sulfoxide obtained by this method (Scheme 4.1).[10b] The reaction proceeds with complete inversion

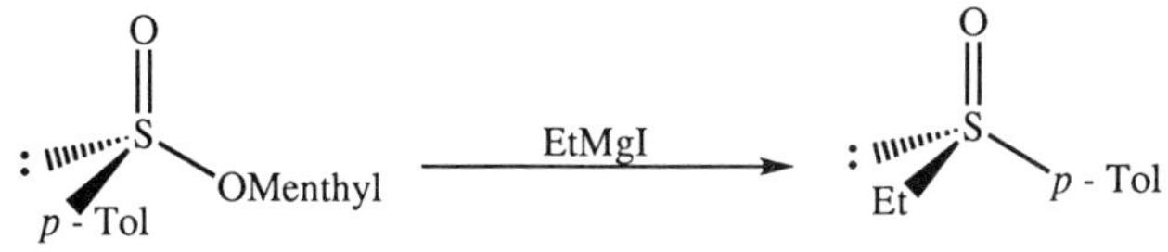

O
S
: OMenthyl
p - Tol
EtMgI
O
S
: Et *p* - Tol

(-)-(*S*)-1 (+)-(*R*)-2

Scheme 4.1

of configuration at sulfur. This was demonstrated by chemical correlation[11–13] and ORD studies.[11, 13–16] A Cotton effect was observed between 235 and 255 nm for alkyl aryl sulfoxides and near 200 nm for dialkyl sulfoxides, characteristic of the absolute configuration at sulfur. The absolute configuration of (−)-menthyl *p*-toluene sulfinate was previously established[14] by correlation with that of (−)-menthyl (−)-*p*-iodobenzene sulfinate determined by X-ray diffraction.[16]

The Andersen sulfoxide synthesis is general in scope and its application to the synthesis of complex optically active sulfoxides will be shown later. However, a major drawback in this reaction is the requirement for optically pure (−)-(*S*)-menthyl *p*-tolyl sulfinate (**1**). In the numerous examples reported by Andersen,[10b,15,17] Mislow[14,16,18] and others,[11,12,19] ((−)-(*S*)-**1**) was obtained from the reaction of 1-menthol with *p*-toluene sulfinyl chloride followed by fractional crystallization of the mixture of the two diastereoisomers. This esterification reaction showed no particular stereoselectivity, giving a 1:1 diastereomeric mixture. The process can be improved and the fractional crystallization of the diastereoisomers avoided by using the acid-catalysed

* An ambiguity arises in the configurational designation of sulfinate esters because the prefixes (*R*) and (*S*) are reversed according to whether the S–O bond is regarded as a single or as a double bond. We followed a previously established custom and considered for nomenclatural purposes the S–O bond as a single bond.

epimerization of sulfinates. Philipps reported in 1925 that 1-menthyl 1-*p*-toluene sulfinate underwent mutarotation very slowly.[20] It was shown later that this was the result of the catalysis by *p*-toluene sulfinic acid[21] and indeed that this epimerization could be catalysed by HCl.[22] In 1964, it was shown that sulfoxides are also rapidly and cleanly racemized at room temperature by HCl in organic solvents such as benzene, dioxan or THF.[18] Kinetic studies and ^{18}O-labelling experiments on sulfoxides[22] and sulfinate esters[23] confirmed the mechanism proposed for this sulfur epimerization (Scheme 4.2).

Therefore, starting from the mixture of (*R*)- and (*S*)-menthyl *p*-toluene sulfinate, the two diastereoisomers can be equilibrated in acidic medium and, within a few days, the equilibrium is displaced towards the less-soluble isomer. In this way, a 90% yield was obtained.[24] This epimerization process can be performed on large quantities of product (Scheme 4.3).[25]

The reaction of arene sulfinates with Grignard reagents is usually carried out in ethyl ether solution. However, in this solvent, chiral sulfoxides are formed in moderate or low yields depending on the structure of both the sulfinic ester and the Grignard reagent. Harpp carried out detailed studies on this reaction and reported that the reaction conditions must be carefully selected, otherwise considerable quantities of impurities, which are difficult to separate, are formed.[26] They also found that the use of lithium organocuprates instead of Grignard reagents gives a cleaner conversion of sulfinates to sulfoxides, but with moderate yields (16–59%). Mikolajczyk reported later that chiral sulfoxides of greater chemical and optical purity and in higher yields are obtained when the reactions of menthyl sulfinate with Grignard reagents are carried out in a benzene solution.[27] However, the application of this reaction to large-scale experiments was not straightforward because of the difficult separation of menthol from the resulting sulfoxide, usually requiring a purification by chromatography. The separation of menthol can be performed by an appropriate change of solvent, and this enabled a large-scale procedure for the preparation of optically pure methyl *p*-tolyl sulfoxide to be achieved.[25] A great variety of sulfoxides have been prepared by this method (Scheme 4.4).[10b,18,25,28–36]

Besides alkyl and aryl *p*-tolyl sulfoxides, sulfinyl esters of type (**3**) can be easily obtained by condensation of magnesium enolate of esters[29] and sulfinyl amides (**4**) from lithiated tertiary amides.[30] Lithiated anions α to sulfides also react cleanly with the menthyl sulfinate (**1**) to give the corresponding optically pure sulfinyl sulfides (**5**).[31] Enantiopure vinylic sulfoxides (**6**) in the (*E*)-configuration were also prepared from vinylic Grignard reagents and the menthyl sulfinate (**1**).[31,33a] Both ((*E*)-**6**) and ((*Z*)-**7**) were readily obtained in a stereocontrolled manner and in two steps from alkynic Grignard reagents followed by hydride or catalytic reduction of the triple bond.[35,36] Finally the (*E*)-vinylic sulfoxides (**6**) can also be obtained in two steps using the Wittig–Horner type condensation of optically active sulfinyl phosphonates on aldehydes.[36] A recent method gives (*E*)-1,3-butadienyl sulfoxides by a

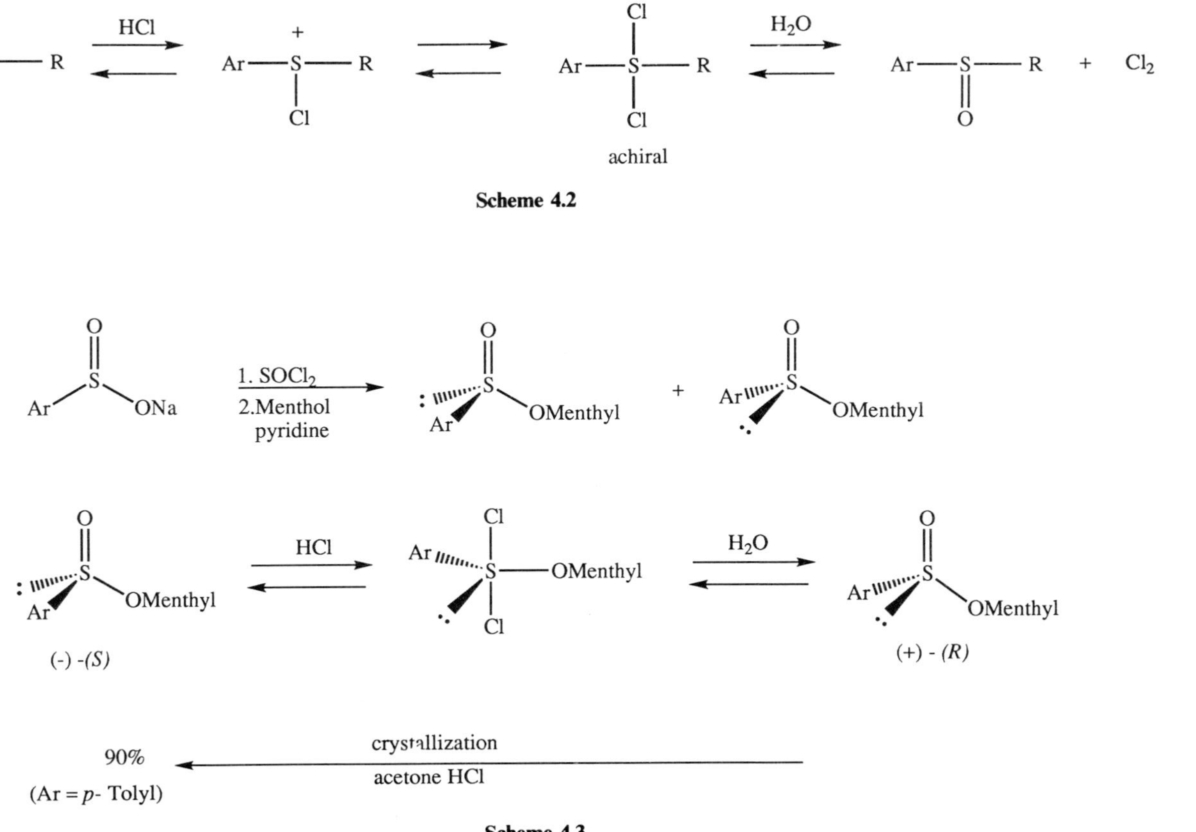

Scheme 4.2

Scheme 4.3

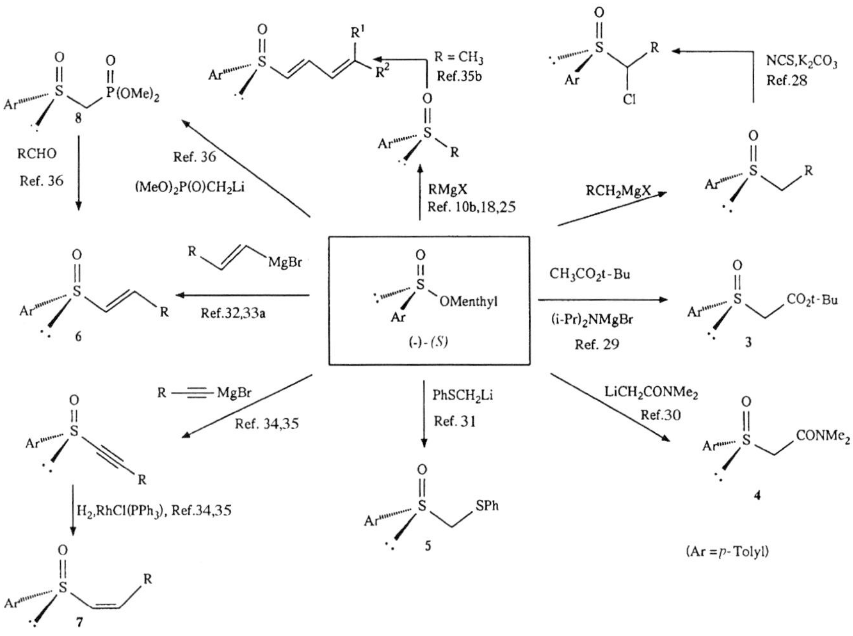

Scheme 4.4

condensation–elimination sequence from lithiated methyl *p*-tolyl sulfoxide and α,β-unsaturated aldehydes.[33b]

Finally, dianions of 1,3-diketones will react also with menthyl *p*-tolyl sulfinate (**1**) to give in high yield the corresponding sulfinyl-2,4-diketone (**9**) (Scheme 4.5).[37]

4.3 Stereoselective reduction of β-ketosulfoxides

Since the pioneering work of Corey, who was the first to prepare racemic β-ketosulfoxides from the anion of dimethyl sulfoxide and esters,[38] numerous racemic β-ketosulfoxides were synthesized and widely used in organic synthesis. However, Kunieda was the first to prepare (+)-(*R*)-α-(*p*-tolylsulfinyl)-acetophenone from (+)-(*R*)-methyl *p*-tolyl sulfoxide and ethyl benzoate and to report its reaction with Grignard reagents, leading to a mixture of diastereomeric alcohols in a 7:3 ratio.[39] Using this general synthetic procedure (Scheme 4.6), Annunziata and Cinquini prepared several β-ketosulfoxides and studied the stereoselectivity of the ketone reduction.[40]

Schneider reported another synthesis of chiral *p*-tolyl sulfinylmethyl ketone based on decarboxylation of optically active sulfinyl esters, obtained from

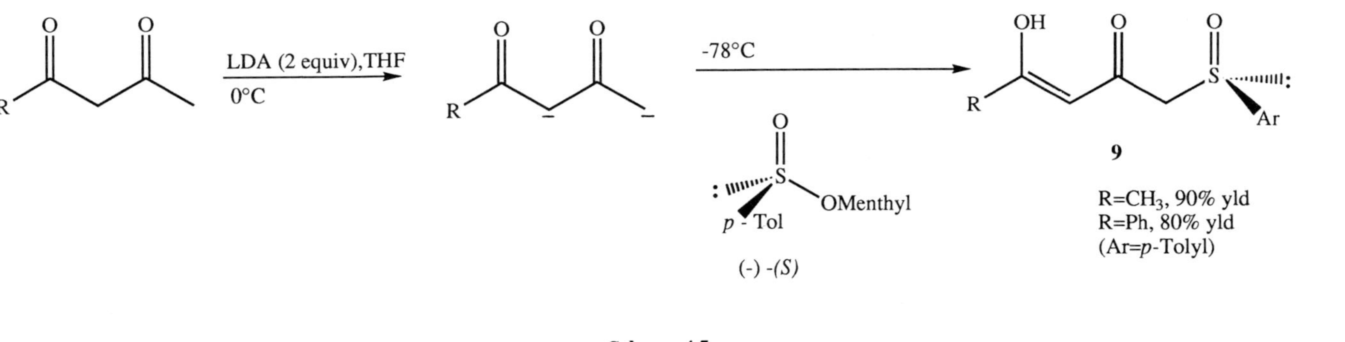
O
O
R
LDA (2 equiv),THF
0°C
-78°C
p - Tol
OMenthyl
(-) -(S)
OH
Ar
9
R=CH3, 90% yld
R=Ph, 80% yld
(Ar=p-Tolyl)

Scheme 4.5

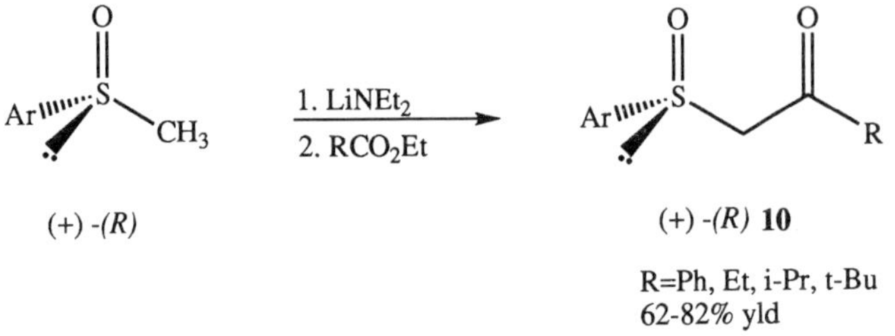

(+) -(*R*) → 1. $LiNEt_2$ 2. RCO_2Et → (+) -(*R*) **10**

R=Ph, Et, i-Pr, t-Bu
62-82% yld

Scheme 4.6

menthyl *p*-toluene sulfinate (**1**) and the dianion of methyl acetoacetate (Scheme 4.7).[41]

The stereoselctivity of the reduction of *β*-ketosulfoxides of type (**10**) was first investigated by Annunziata and Cinquini with $NaBH_4$ and $LiAlH_4$ at $-70°C$.[40] They determined the de% by ^1H NMR without the identification of the main diastereoisomer. The results show that the extent of asymmetric induction was in the range 60–70% with $LiAlH_4$ and lower with $NaBH_4$.

The author and co-workers reinvestigated this reduction process with many different reducing agents.[42] The diastereoselectivity was determined by NMR from the AB pattern displayed by the methylene protons α to the sulfoxide group. The absolute configuration of the main diastereoisomer was determined by chemical correlation with known methyl carbinols after desulfurization (Scheme 4.8).

As shown in Table 4.1, the author and co-workers obtained the same results as Annunziata[40] with $LiAlH_4$ and $NaBH_4$. However, with diborane and DIBAL, a reverse asymmetric induction with respect to

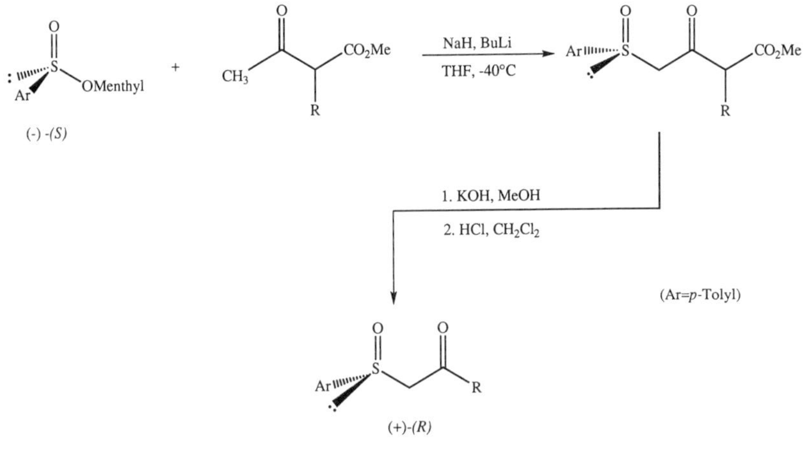

(-) -(*S*) OMenthyl + CO_2Me → NaH, BuLi / THF, -40°C → CO_2Me → 1. KOH, MeOH 2. HCl, CH_2Cl_2 → (+)-(*R*)

(Ar=*p*-Tolyl)

Scheme 4.7

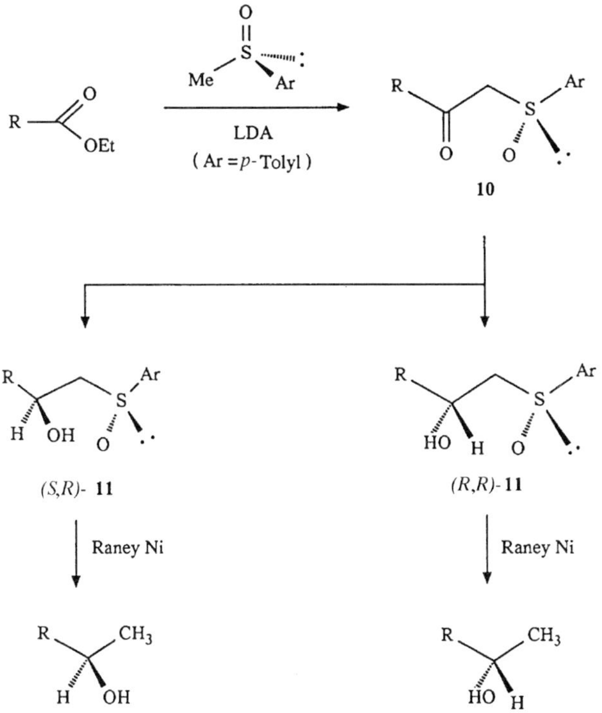

Scheme 4.8

$LiAlH_4$ occurred. Sodium, lithium, tetrabutylammonium borohydride as well as $LiAlH_4$ gave mainly the (*R*,*R*) diastereoisomer (about 60% de), which indicated that the cation was not playing an important role in the control

Table 4.1 Reduction of β-ketosulfoxides (**10**) at −78°C

Reducing agent	Solvent	((*R*,*R*)-**11**):((*S*,*R*)-**11**)
$NaBH_4$	Et_2O/THF	69:31
$NaBH_4$	EtOH	80:20
$LiBH_4$	Et_2O/THF	81:19
n-Bu_4NBH_4	Et_2O/THF	85:15
$LiAlH_4$	Et_2O/THF	84:16
$LiEt_3BH$	THF	80:20
Li-s-Bu_3BH	THF	66:34
$Zn(BH_4)_2$	Et_2O/THF	66:34
$Zn(BH_4)_2$	EtOH	60:40
Me_2S, BH_3	THF	53:47
B_2H_6, THF	THF	30:70
i-Bu_2Al	THF	22:78

of the stereochemistry. However, diborane and DIBAL gave mainly the (*S*,*R*) diastereomer (60% de with DIBAL). Other reducing agents, such as lithium tri-*s*-butyl borohydride, zinc borohydride and borane–methyl sulfide complex showed a lower stereoselectivity.

In many cases, the diastereoselectivity was significantly increased by adding DIBAL at $-78°C$ to the β-ketosulfoxide solution (Table 4.2, method B) instead adding the β-ketosulfoxide solution to DIBAL (Table 4.2, method A). Moreover, the addition of DIBAL to a β-ketosulfoxide THF solution containing 1 equiv. anhydrous zinc chloride at $-78°C$ gave a reverse stereoselectivity with a very high de (Table 4.2).[43,44]

These results show that it is possible to reduce β-ketosulfoxides with the appropriate reducing agent (DIBAL or $ZnCl_2$/DIBAL), with a very high diastereoselectivity, into the corresponding diastereoisomeric (*R*,*R*) or (*R*,*S*) β-hydroxy sulfoxides, which are extremely useful synthons in organic synthesis. Both enantiomers of methylcarbinols,[42] allylic methyl carbinols,[45] epoxides[43,44] and lactones[44] were prepared following this methodology (Scheme 4.9). Desulfurization of β-hydroxy sulfoxides was easily carried out with Raney nickel. However in the presence of an ethylenic linkage, the desulfurization has to be done with lithium in ethylamine thus allowing a nice synthesis of chiral allylic alcohols. For the epoxide preparation, the sulfoxide was reduced to sulfide either with $LiAlH_4$[43] or Zn-Me_3SiCl[44] and ring-closure was carried out in the presence of a base from the corresponding sulfonium salt.

Optically active 4-substituted butenolides (**12**) were also obtained from β-hydroxy sulfoxides.[46] The synthesis of one enantiomer is described in

Table 4.2 Reduction of β-ketosulfoxides (**10**) at $-78°C$

R	Method[a]	Reducing agent	((*R*,*R*)-**11**):((*R*,*S*)-**11**)	Yield (%)	Ref.
Ph	A	DIBAL	20:80	95	43
	B	DIBAL	>5:95	95	43
		$LiAlH_4$	80:20	80	43
		DIBAL, $ZnCl_2$	>95:5	90	43
		DIBAL, $ZnCl_2$	>99:1	80	44
$Ph(CH_2)_2$	A	DIBAL	13:87	98	43
	B	DIBAL	7:93	95	43
		$LiAlH_4$	88:12	90	43
		DIBAL, $ZnCl_2$	>95:5	95	43
n-C_8H_{17}	B	DIBAL	5:95	95	43
		DIBAL, $ZnCl_2$	>95:5	92	43
n-$C_{13}H_{27}$	B	DIBAL	5:95	95	43
		DIBAL, $ZnCl_2$	>95:5	95	43
t-$BuO_2C(CH_2)_3$		DIBAL, $ZnCl_2$	99:1	78	44
t-$BuO_2C(CH_2)_2$		DIBAL, $ZnCl_2$	97:3	93	44
C_2H_5		DIBAL, $ZnCl_2$	>99:1	80	44

[a] Method A, addition of the β-ketosulfoxide solution to DIBAL; method B, addition of DIBAL to a β-ketosulfoxide solution.

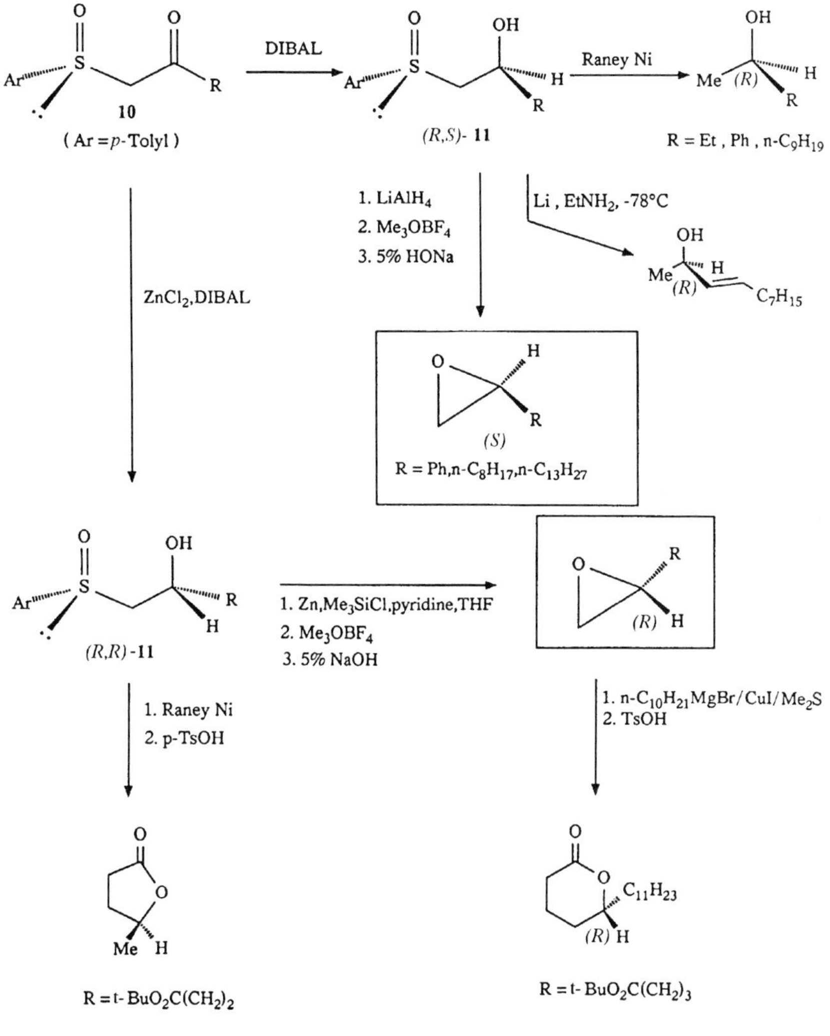

DIBAL
Raney Ni
10
(Ar = p-Tolyl)
(R,S)- 11
R = Et , Ph , n-C_9H_{19}
1. $LiAlH_4$
2. Me_3OBF_4
3. 5% HONa
Li , $EtNH_2$, -78°C
$ZnCl_2$,DIBAL
(S)
R = Ph,n-C_8H_{17},n-$C_{13}H_{27}$
(R,R)-11
1. Zn,Me_3SiCl,pyridine,THF
2. Me_3OBF_4
3. 5% NaOH
1. Raney Ni
2. p-TsOH
1. n-$C_{10}H_{21}MgBr/CuI/Me_2S$
2. TsOH
R = t-$BuO_2C(CH_2)_2$
R = t-$BuO_2C(CH_2)_3$

Scheme 4.9

Scheme 4.10. After the reduction step, the alkylation was carried out on the dianion of the hydroxysulfone with sodium iodoacetate. Then the molecule was lactonized with a catalytic amount of p-toluene sulfonic acid and desulfonylated in the presence of triethylamine.

The stereoselectivity of these reductions was first explained[43,47] by an intramolecular hydride transfer in the case of DIBAL, and the corresponding intermolecular reaction from a $ZnCl_2$-chelated β-ketosulfoxide in the case of $ZnCl_2$/DIBAL reduction, both controlled by steric and stereoelectronic effects.

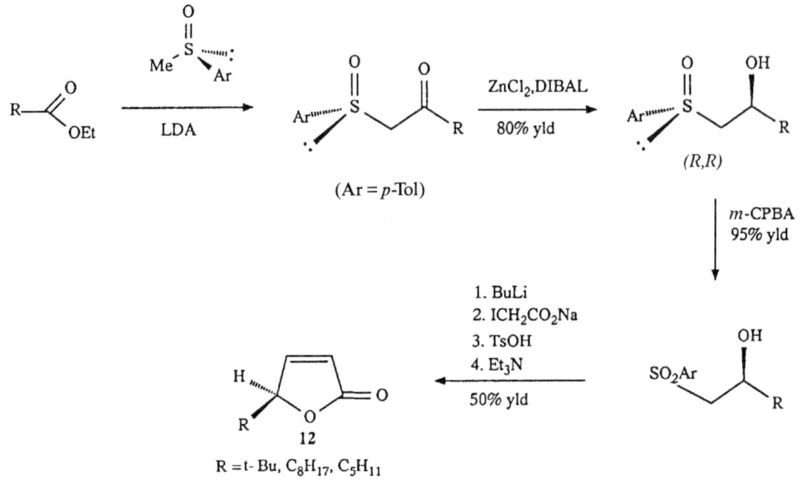

Scheme 4.10

However, recent results[48] afforded important information about the reaction mechanism of the $ZnCl_2$/DIBAL reduction. We found that only a catalytic amount (0.05 to 0.1 equiv.) of $ZnCl_2$ was necessary for the reaction. This suggested an intramolecular hydride transfer, not an intermolecular one, assuming that the DIBAL approach was Cl directed as shown in Figure 4.1. The $ZnCl_2$-chelated β-ketosulfoxide adopts the favoured twisted conformation $\mathbf{C_1}$ where the *p*-tolyl group is pseudo-equatorial, the absolute configuration at sulfur being (*R*). In the early stage of the reaction, the approach of $HAl(i\text{-}Bu)_2$ is then directed by complexation with the geometrically well-located Cl atom leading to a bimetallic bridged species where aluminium is dsp^3 hybridized. In this model of approach, $\mathbf{M_1}$, the hydride is in just the right position to be transferred intramolecularly from the top, leading to the (*R*) configuration at C-2 as observed. In conformation $\mathbf{C_2}$, where the *p*-tolyl group has an unfavourable pseudo-axial position, the Cl-directed approach of $HAl(Bu^i)_2$ is now greatly hindered by the *p*-tolyl group, which explains the small contribution of the corresponding approach $\mathbf{M_2}$ to the stereoselectivity. The high asymmetric induction obtained with non-stoichiometric amounts of $ZnCl_2$ suggest that after the hydride transfer $ZnCl_2$ is displaced from the resulting aluminium alkoxide and used to chelate another molecule of β-ketosulfoxide.

Considering again the Lewis acid character of Al^{3+} in $HAl(i\text{-}Bu)_2$, it seems likely that DIBAL reduction of β-ketosulfoxides involves also, in an early stage of the reaction, a chelated dsp^3 hybridized aluminium as shown in model $\mathbf{M_3}$ (where the *p*-tolyl group has a favourable equatorial orientation). Model $\mathbf{M_3}$ leads through an intramolecular hydride transfer to the (*S*) configuration at C-2, as observed (Figure 4.2).

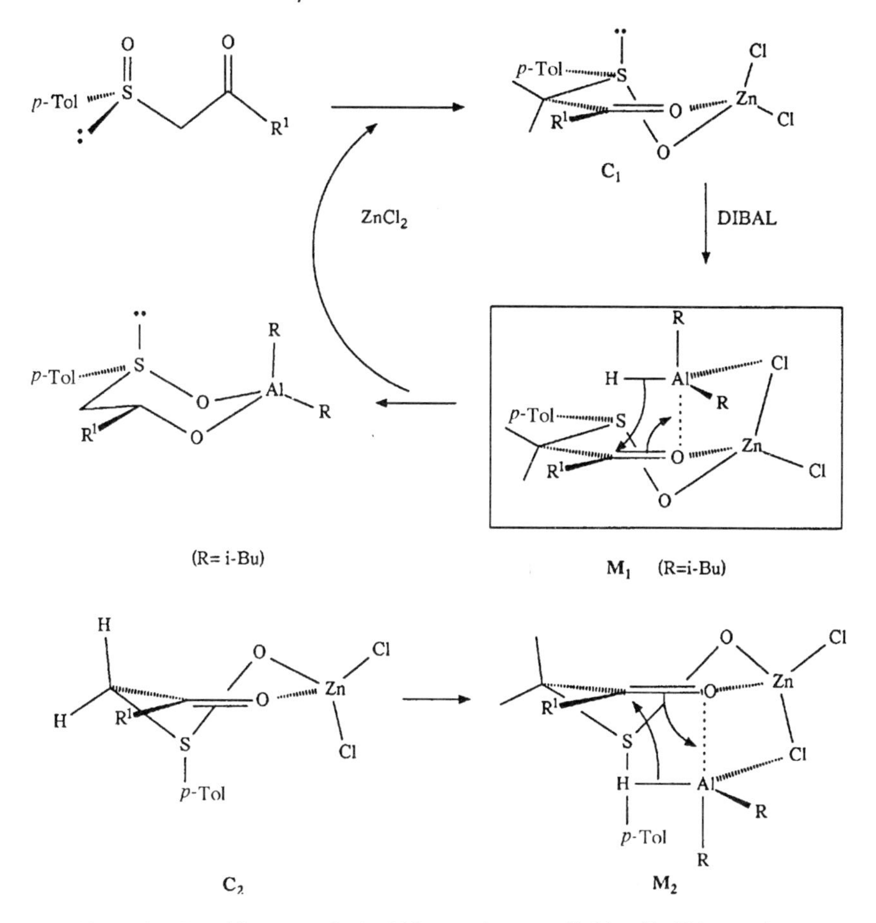

Figure 4.1 Transition states for hydride transfer controlled by Al–Cl interactions.

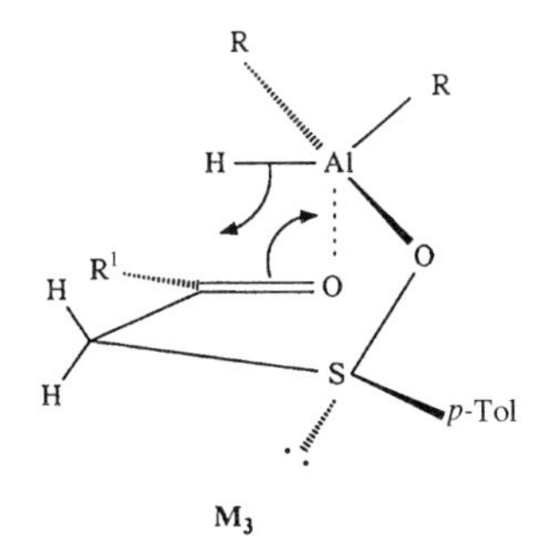

Figure 4.2 Transition state for hydride transfer without Al–Cl interactions.

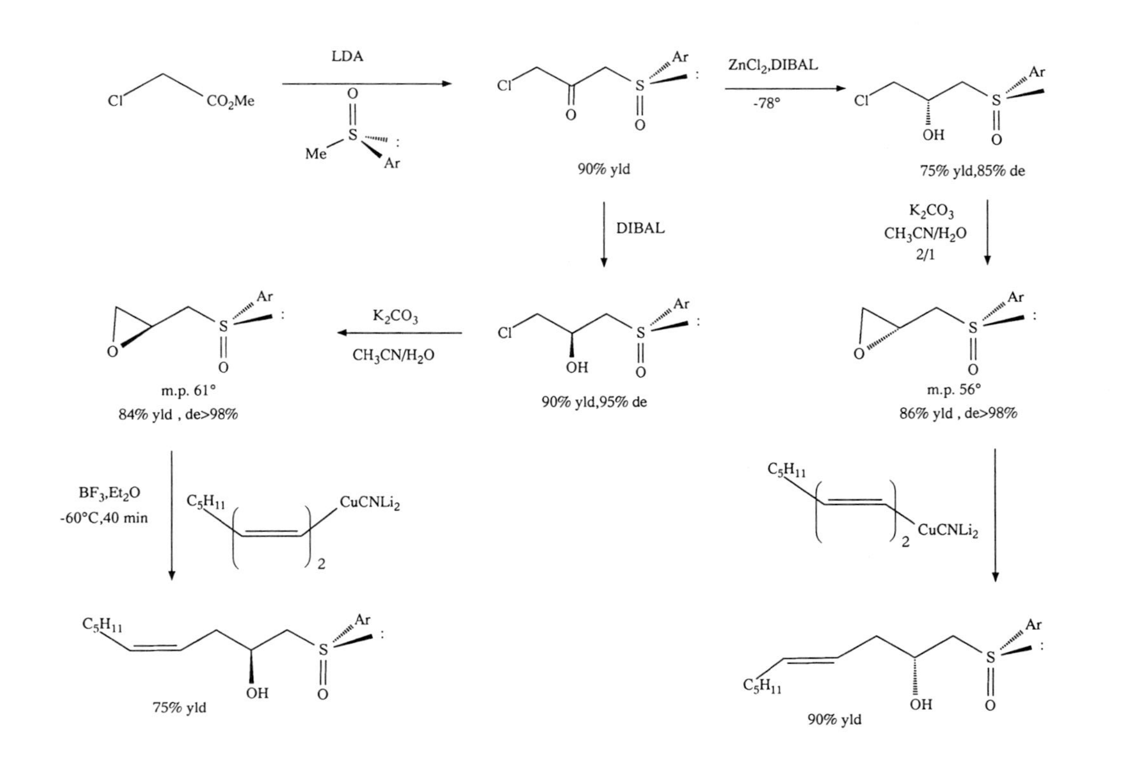
LDA
ZnCl2,DIBAL
-78°
90% yld
75% yld,85% de
DIBAL
K2CO3
CH3CN/H2O
2/1
K2CO3
CH3CN/H2O
90% yld,95% de
m.p. 61°
84% yld , de>98%
m.p. 56°
86% yld , de>98%
BF3,Et2O
-60°C,40 min
CuCNLi2
75% yld
90% yld

Scheme 4.11

The γ-chloro β-ketosulfoxides, readily prepared from methyl chloroacetate and (+)-(*R*)-methyl *p*-tolyl sulfoxide, can be reduced to the corresponding β-hydroxy sulfoxides in the (*R*,*R*) or (*S*,*S*) configuration with $ZnCl_2$/DIBAL or DIBAL alone (Scheme 4.11).[49] The γ-chloro-β-hydroxy sulfoxides can be easily transformed into optically pure α-sulfinyl epoxides, precursors of chiral homoallylic β-hydroxy sulfoxides, by reaction with cyanocuprates.[49,50]

The β,γ-diketosulfoxides are also reduced with very high diastereoselectivity with DIBAL to the corresponding γ-keto β-hydroxy sulfoxides.[51] The γ-ketone group being totally enolized, 2 equiv. of DIBAL must be used. The absolute configuration of the reduced product was determined by correlation with the corresponding *anti*-diol obtained by the well established Evans' procedure (Scheme 4.12). (*R*)-β,γ-diketosulfoxides and (*R*)-β-ketosulfoxides are both reduced with DIBAL to (*R*,*S*)-β-hydroxy sulfoxides.

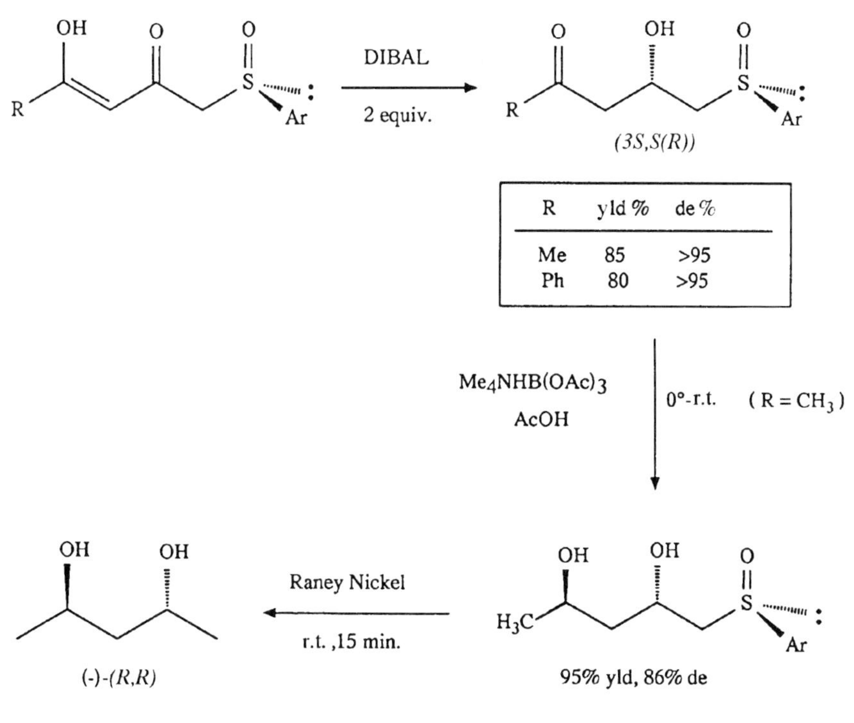

R	yld %	de %
Me	85	>95
Ph	80	>95

Scheme 4.12

4.4 Application of the β-ketosulfoxide reduction to total synthesis

One of the great advantages of chiral sulfoxides in total synthesis is to allow the asymmetric induction step to be performed in the very last part of the synthesis, via the stereoselective β-ketosulfoxide reduction. Furthermore, both configurations of the chiral hydroxylic centre can be prepared according to the reduction conditions (DIBAL or $ZnCl_2$/DIBAL), allowing access to both enantiomers of the target molecule.

This is shown in the asymmetric synthesis of the two macrolides, lasiodiplodin and zearalenone. The total synthesis of lasiodiplodin dimethylether (**13**) was divided in two parts:[52] first the synthesis of the achiral diester (**14**) and then the introduction of the chiral carbinol part via a β-ketosulfoxide functionality in the very last steps of the synthesis, allowing the preparation of both configurations of the macrolide (Scheme 4.13). Reduction of the β-ketosulfoxide (**15**) with DIBAL in the presence of $ZnCl_2$ yielded the (*R,R*)-β-hydroxy sulfoxide, while the reduction with DIBAL alone afforded the (*R,S*) isomer. After desulfurization, both enantiomers of the seco-acid were cyclized using the known Gerlach's method.

Zearalenone dimethyl ether (**16**) was prepared by first making the chiral part (**17**) using a chiral sulfoxide auxiliary and the achiral sulfone ester (**18**).[53] The hydroxyester (**17**) was obtained from the β-ketosulfoxide, which was readily prepared from glutaric anhydride (Scheme 4.14). After coupling the sulfonyl anion (**18**) with the ester (**17**), desulfurization and carbonyl protection, the cyclization to zearalenone dimethyl ether was carried out following the Masamune method.

The α-sulfinyl epoxide (**19**), which is prepared by reduction of γ-chloro β-ketosulfoxide (Scheme 4.11), is a very important precursor of functionalized

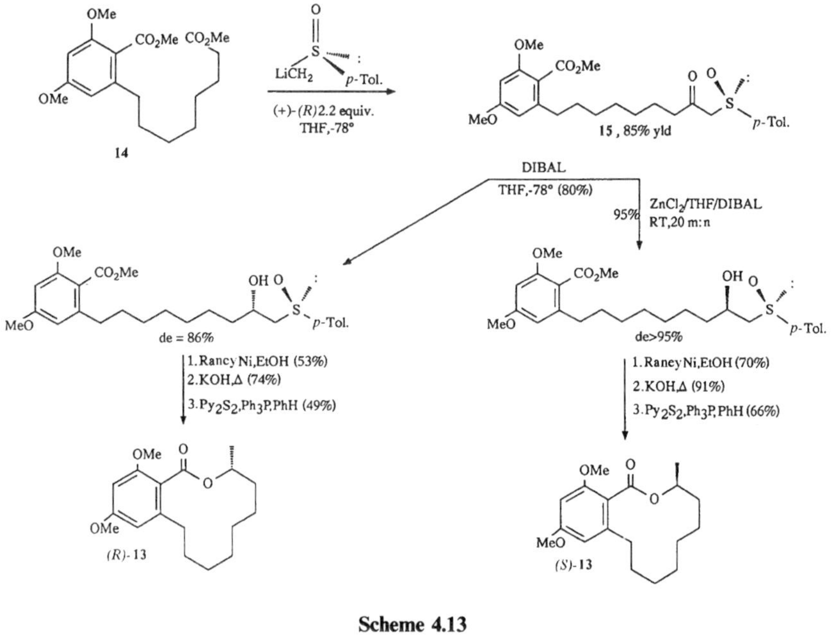

OMe
CO_2Me CO_2Me
OMe
14
O
S
$LiCH_2$ *p*-Tol.
(+)-(*R*)2.2 equiv.
THF,-78°
OMe
CO_2Me
O O
S
MeO
p-Tol.
15 , 85% yld
DIBAL
THF,-78° (80%)
95%
$ZnCl_2$/THF/DIBAL
RT,20 m:n
OMe
CO_2Me
OH O
S
p-Tol.
MeO
de = 86%
1. Raney Ni,EtOH (53%)
2. KOH,Δ (74%)
3. Py_2S_2,Ph_3P,PhH (49%)
OMe
CO_2Me
OH O
S
p-Tol.
MeO
de>95%
1. Raney Ni,EtOH (70%)
2. KOH,Δ (91%)
3. Py_2S_2,Ph_3P,PhH (66%)
OMe O
O
OMe
(*R*)-13
OMe O
O
MeO
(*S*)-13

Scheme 4.13

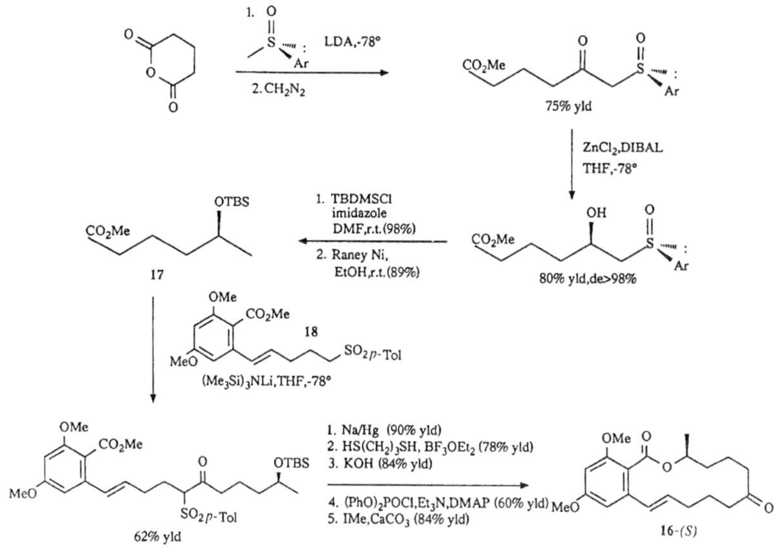

Scheme 4.14

chiral homoallylic carbinols. It was applied to the synthesis of the C-11–C-20 fragment of leukotriene B_4 (Scheme 4.15),[50] the sulfoxide group being easily transformed into an aldehyde by a Pummerer rearrangement. The epoxide ((*R*,*R*)-**19**) was reacted with (*E*)-cyanocuprate to give the homoallylic β-hydroxy sulfoxide ((*R*,*R*)-**20**) in 90% yield. After protecting the OH group

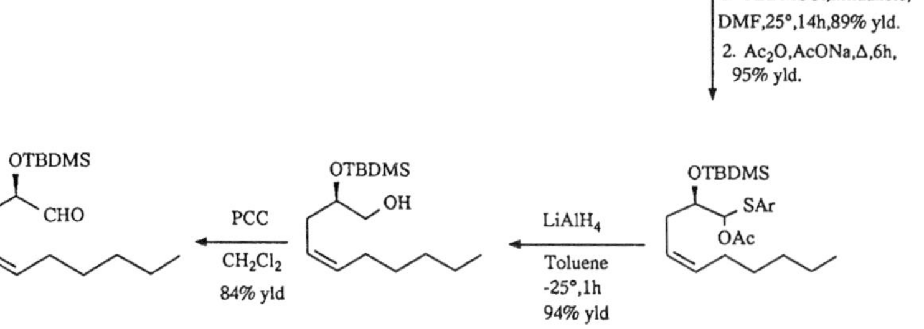

Scheme 4.15

with a TBDMS, the molecule was submitted to a Pummerer rearrangement in acetic anhydride and the resulting acetate reduced with $LiAlH_4$ in toluene. Finally oxidation of the primary alcohol gave the target, the (*R*) homoallylic hydroxy aldehyde corresponding to the C-11–C-20 fragment of leukotriene B_4.

Scheme 4.9 showed that β-hydroxy sulfoxides can easily be transformed into chiral epoxides. That result was applied to the synthesis of chiral *syn*- and *anti*-1,3-diols present in the C-1–C-12 unit of amphotericin B (Scheme 4.16).[54] The β-ketosulfoxide ((*R*)-**21**) was reduced with $ZnCl_2$/DIBAL and transformed into the epoxide ((*S*)-**22**) by the method already described. Epoxide opening with dithiane followed by protection of the hydroxyl led to the aldehyde ((*S*)-**23**). Condensation of 2-bromomagnesium-1,3-dithiane to aldehyde (**23**) gave in 70% yield only the (*S*,*S*) diastereoisomer (**24**), because of chelation control in the 1,3-asymmetric induction. The intermediate (**24**) was then easily transformed into the aldehyde (**25**) with an *anti*-diol part, which can be completely isomerized in basic medium under thermodynamic control to the aldehyde (**26**) with a *syn*-diol moiety. Finally, condensation of 2-bromomagnesium-1,3-dithiane with the aldehyde (**26**) gave only the *syn*-adduct (**27**), as expected from chelation control.

Another application of chiral epoxides obtained from β-hydroxy sulfoxides was the total synthesis of yashabushiketol having a chiral aldol functionality.[55] In this rather short asymmetric synthesis, the optically active epoxide ((*S*)-**28**) was opened by reaction with a substituted dithiane anion to give the corresponding chiral alcohol, which is easily transformed into (−)-yashabushiketol (Scheme 4.17). Both enantiomers of this natural product can be easily obtained from the epoxide enantiomers.

Natural gingerols[55] were also prepared in a very similar way from the β-ketosulfoxide (**29**): DIBAL reduction, transformation of the resulting β-hydroxy sulfoxide into the epoxide (**30**) reaction with a substituted dithiane anion, formation of the primary tosylate (**31**) and finally reduction with $LiAlH_4$ into (+)-(6*S*)-gingerol or reaction with cyanocuprates to give (+)-(8*S*)- and (+)-(10*S*)-gingerol (Schemes 4.18 and 4.19).

Two different ways to prepare optically active *syn*- and *anti*-1,3-diols have been described above (Schemes 4.12 and 4.16). It is also possible to transform commercially available α-hydroxy esters into *syn*- and *anti*-1,2-diols via the corresponding γ-hydroxy β-ketosulfoxides, as shown in Scheme 4.20.[56]

Protected (*S*)-methyl lactate reacted cleanly with (*R*)-methyl-*p*-tolyl sulfoxide to give in high yield the corresponding γ-alkoxy-β-ketosulfoxide. The reduction with DIBAL and $ZnBr_2$/DIBAL gave in high yields the expected *syn*- and *anti*-dihydroxy sulfoxides. The diastereoselectivity was very high in both cases when the protecting group was TBDMS (only one diastereomer was detected by 200 MHz in ^1H NMR). When the protecting group was MEM, the diastereoselectivity was only 84% with $ZnBr_2$/DIBAL, probably because of the presence of several oxygen atoms in the protecting group competing in the chelated transition states. Therefore, this method

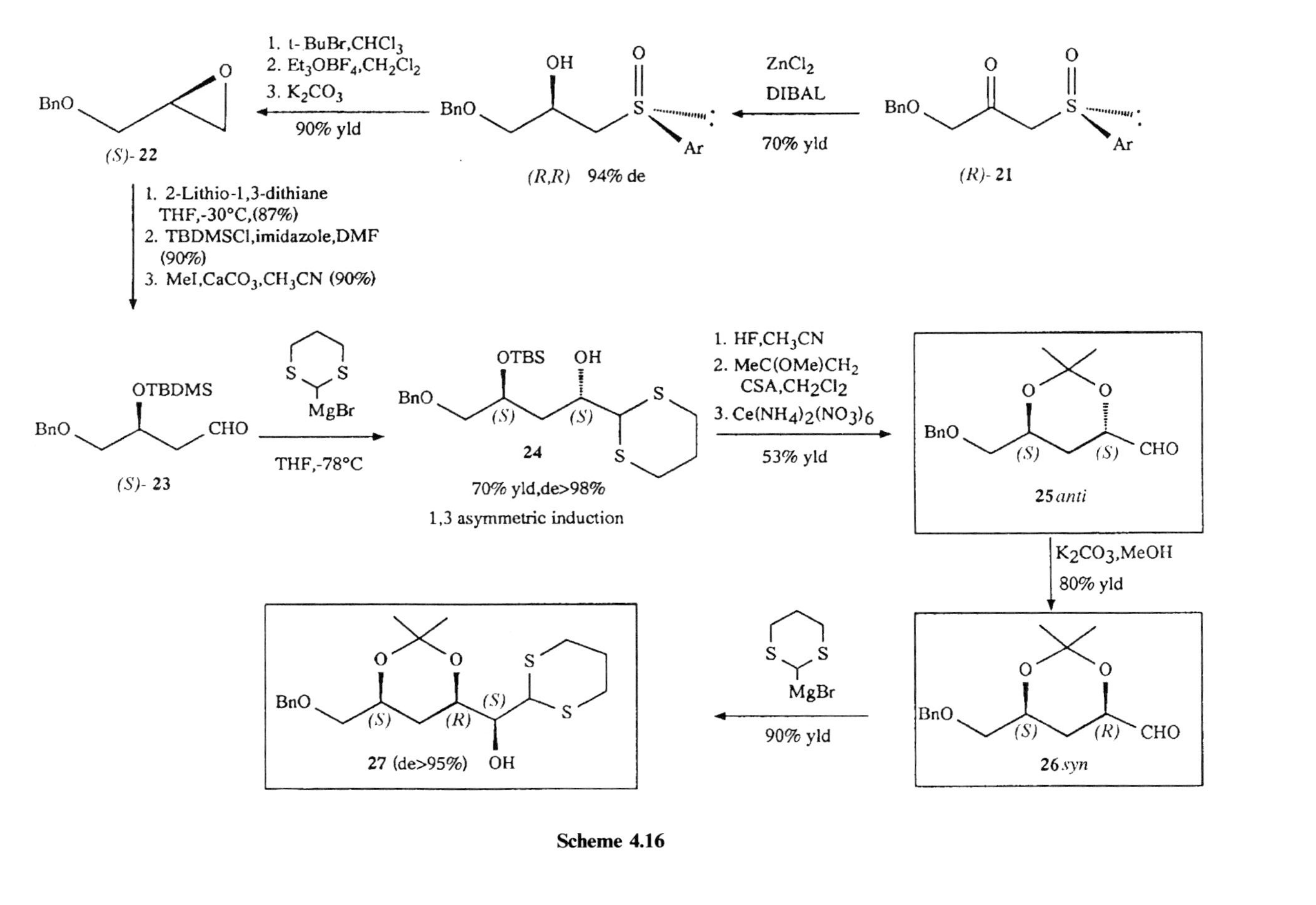
ZnCl2
DIBAL
70% yld
(R)- 21
(R,R) 94% de
1. t-BuBr,CHCl3
2. Et3OBF4,CH2Cl2
3. K2CO3
90% yld
(S)- 22
1. 2-Lithio-1,3-dithiane
THF,-30°C,(87%)
2. TBDMSCl,imidazole,DMF
(90%)
3. MeI,CaCO3,CH3CN (90%)
OTBDMS
CHO
(S)- 23
MgBr
THF,-78°C
OTBS
OH
24
70% yld,de>98%
1,3 asymmetric induction
1. HF,CH3CN
2. MeC(OMe)CH2
CSA,CH2Cl2
3. Ce(NH4)2(NO3)6
53% yld
25 anti
K2CO3,MeOH
80% yld
26 syn
90% yld
27 (de>95%)

Scheme 4.16

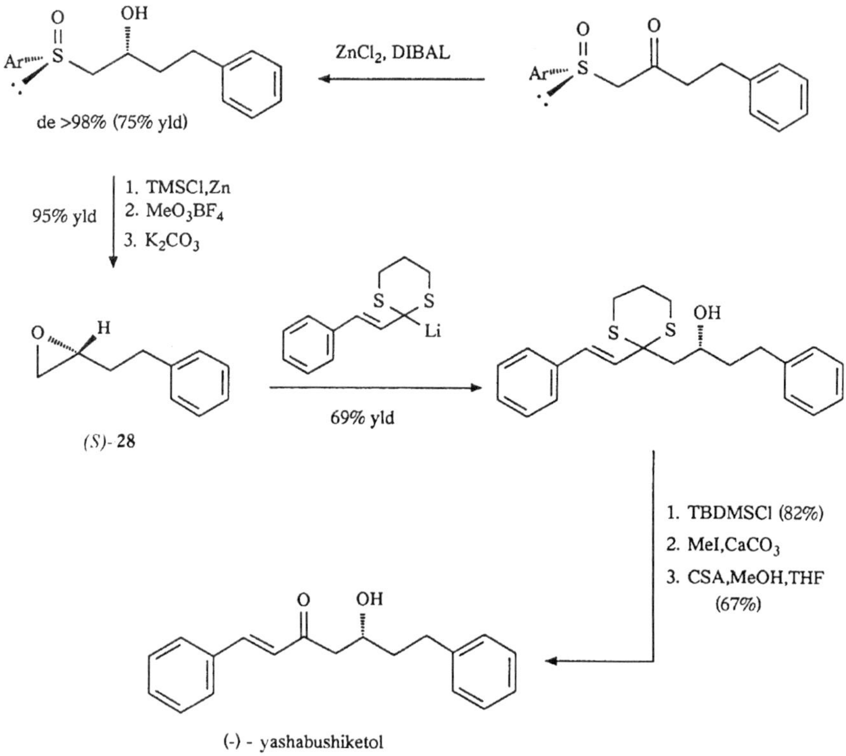

$ZnCl_2$, DIBAL

de >98% (75% yld)

95% yld
1. TMSCl,Zn
2. MeO_3BF_4
3. K_2CO_3

(*S*)-28

69% yld

1. TBDMSCl (82%)
2. MeI,$CaCO_3$
3. CSA,MeOH,THF
(67%)

(-) - yashabushiketol

Scheme 4.17

allowed very efficient transformation of α-hydroxy esters into enantiomerically pure *syn*- and *anti*-1,2-diols, as long as the protecting group on the γ-hydroxyl is TBDMS. Finally the β,γ-dihydroxy sulfoxides can be either desulfurized or transformed, as shown, via a Pummerer rearrangement affording triols.

If there is an oxygen-containing protecting group such as a MEM group in the γ-position, the enantiomerically pure *syn*-1,2-diols could be obtained by changing the sulfoxide configuration and using DIBAL reduction as shown in Scheme 4.21 for malic ester. The resulting highly functionalized diols from malic ester can be easily transformed via a Pummerer rearrangement into 2-deoxy-L-xylose or 2-deoxy-D-ribose derivatives.[57]

The synthesis of the macrolide cladospolide A is another example of this methodology.[58] The retrosynthesis is shown in Scheme 4.22. The secoacid can be made from the γ-alkoxy β-ketosulfoxide (**32**), followed by a Pummerer rearrangement of the dihydroxy sulfoxide and a Wittig reaction. The product (**32**) can be obtained from the α-alkoxy ester (**33**) prepared by a Wittig reaction between (**34**) and (**35**), respectively, made from the β-ketosulfoxides (**36**) and (**37**).

The β-ketosulfoxide (**36**) was prepared from the corresponding γ-hydroxy

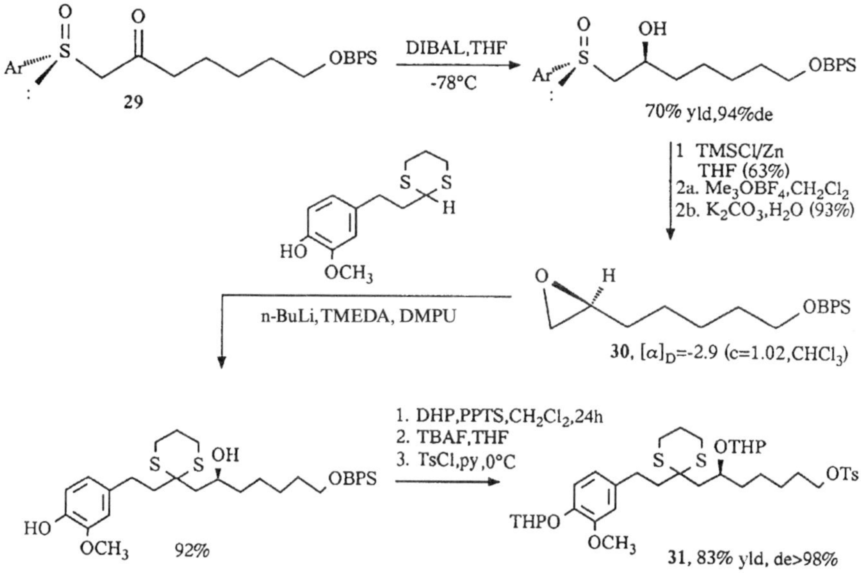

Scheme 4.18

ester, reduced with DIBAL (de > 95%) and transformed after desulfurization with Raney nickel into the phosphonium salt (**34**) by standard methods. The molecule (**37**) was obtained by reaction of the *t*-butyl acetoacetate dianion with (−)-(*S*)-menthyl sulfinate, reduced with DIBAL (de > 95%)[58,70] and easily transformed into (**35**) via a Pummerer rearrangement. The Wittig reaction gave only the (*Z*)-isomer, which was reduced with Pd–C. The CH_2O-TBS group was directly oxidized to the corresponding carboxylic acid with PDC in DMF, esterified with CH_2N_2 and reacted with (*R*)-methyl *p*-tolyl sulfoxide to give the desired γ-alkoxy-β-ketosulfoxide (**32**) (Scheme 4.23).

DIBAL reduction gave only the *anti*-diol (**38**), which, after protection, was submitted to a Pummerer rearrangement and reduced to the primary alcohol (**39**). Swern oxidation followed by a Wittig reaction gave the enantiomerically pure secoester (Scheme 4.24). The cyclization was carried out using Yamaguchi's methodology.

Spiroacetals are part of many natural products, with a broad spectrum of biological activities. The two enantiomers, (2*R*,6*S*,8*R*) and (2*S*,6*R*,8*S*) of 2,8-dimethyl-1,7-dioxaspiro-[5,5]-undecane (**40**), have been synthesized. As shown in the retrosynthetic Scheme 4.25 for the enantiomer ((+)-(2*R*,6*S*,8*R*)-**40**), the spiroacetal structure can be easily made by cyclization of the (2*R*,10*R*)-6-keto-2,10-undecanediol. The C_2 symmetry of the desired diol suggested that the diketodisulfoxide (**41**) was a very simple intermediate for making both chiral hydroxylic centres in the desired configurations.[59]

The diketodisulfoxide (**41**) (Scheme 4.26) was prepared in high yield by

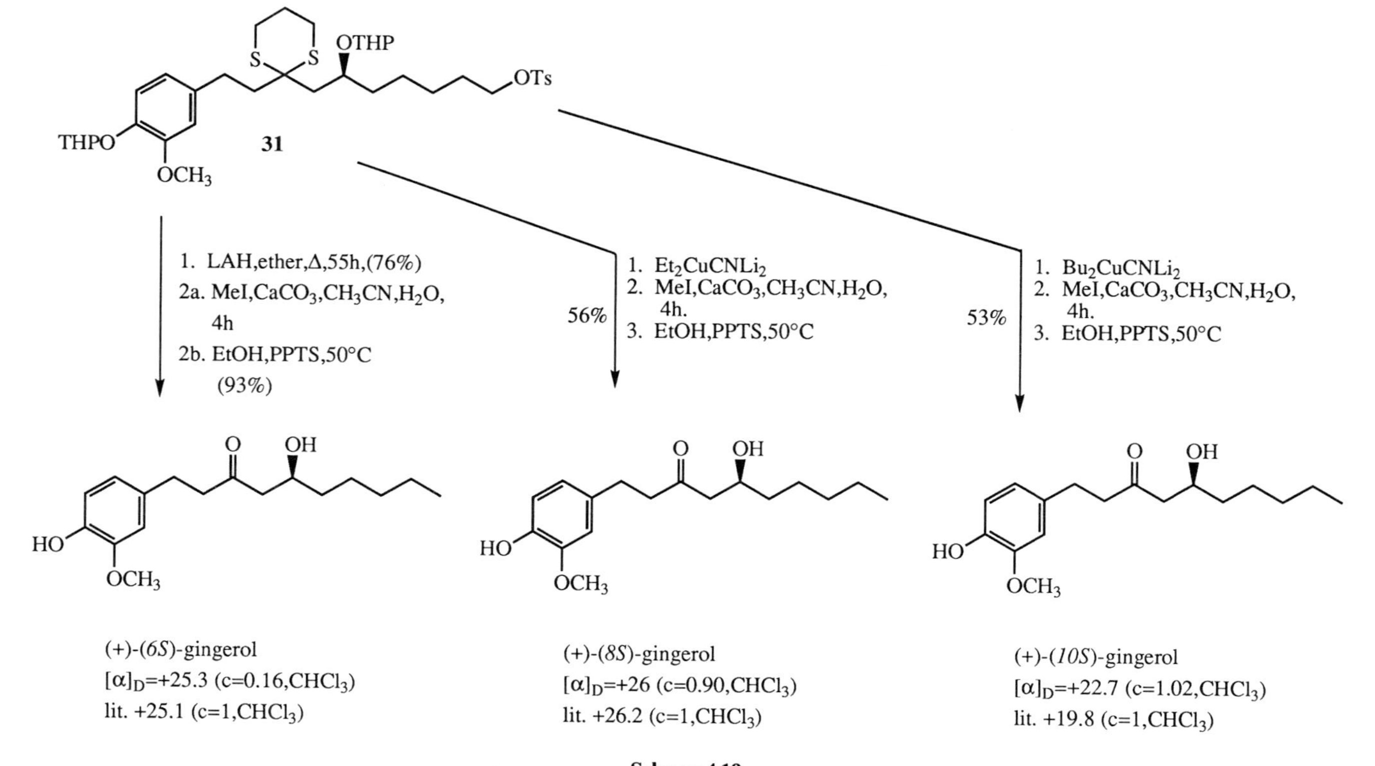
THPO
OCH3
S
S
OTHP
OTs
31
1. LAH,ether,Δ,55h,(76%)
2a. MeI,CaCO3,CH3CN,H2O,
4h
2b. EtOH,PPTS,50°C
(93%)
56%
1. Et2CuCNLi2
2. MeI,CaCO3,CH3CN,H2O,
4h.
3. EtOH,PPTS,50°C
53%
1. Bu2CuCNLi2
2. MeI,CaCO3,CH3CN,H2O,
4h.
3. EtOH,PPTS,50°C
O
OH
HO
OCH3
(+)-(6S)-gingerol
[α]D=+25.3 (c=0.16,CHCl3)
lit. +25.1 (c=1,CHCl3)
(+)-(8S)-gingerol
[α]D=+26 (c=0.90,CHCl3)
lit. +26.2 (c=1,CHCl3)
(+)-(10S)-gingerol
[α]D=+22.7 (c=1.02,CHCl3)
lit. +19.8 (c=1,CHCl3)

Scheme 4.19

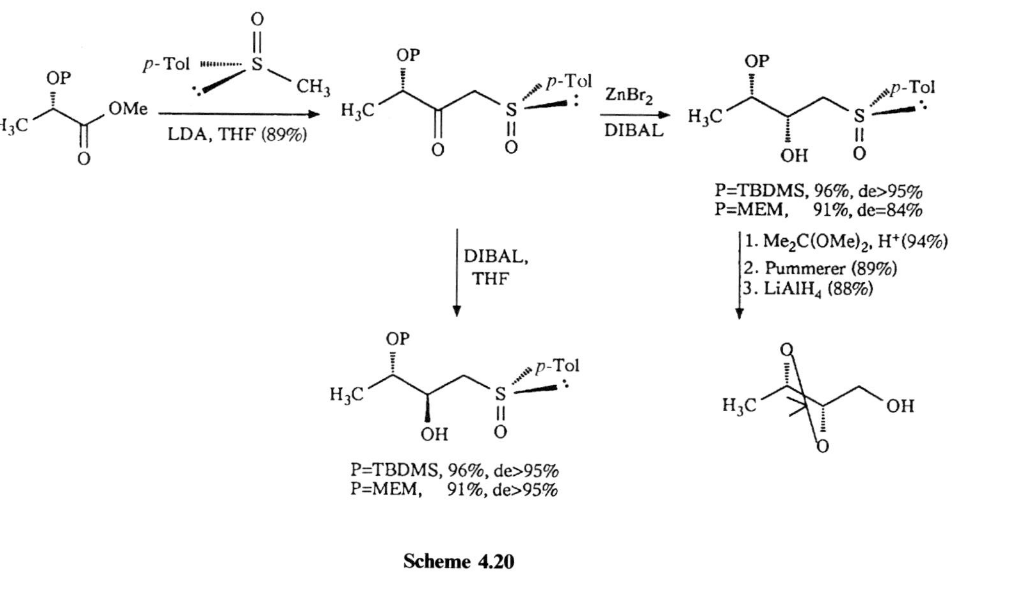
OP
H3C
OMe
O
p-Tol
S
CH3
LDA, THF (89%)
ZnBr2
DIBAL
DIBAL,
THF
OH
P=TBDMS, 96%, de>95%
P=MEM, 91%, de=84%
1. Me2C(OMe)2, H+(94%)
2. Pummerer (89%)
3. LiAlH4 (88%)
P=TBDMS, 96%, de>95%
P=MEM, 91%, de>95%

Scheme 4.20

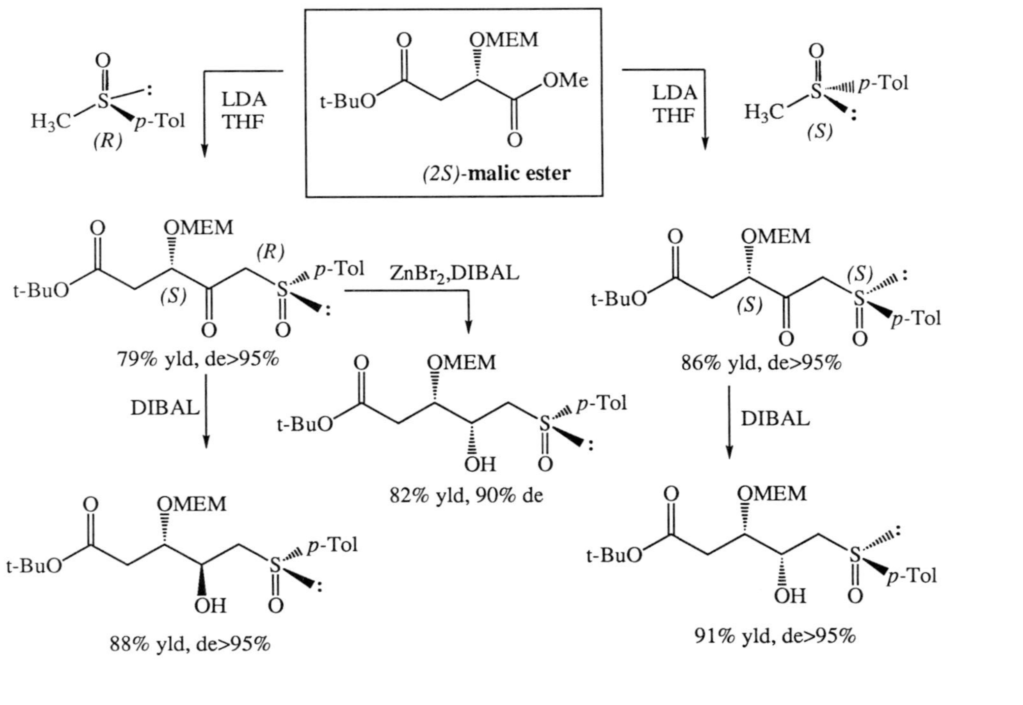
(R)
LDA
THF
(2S)-malic ester
(S)
79% yld, de>95%
DIBAL
ZnBr₂,DIBAL
82% yld, 90% de
86% yld, de>95%
88% yld, de>95%
91% yld, de>95%

Scheme 4.21

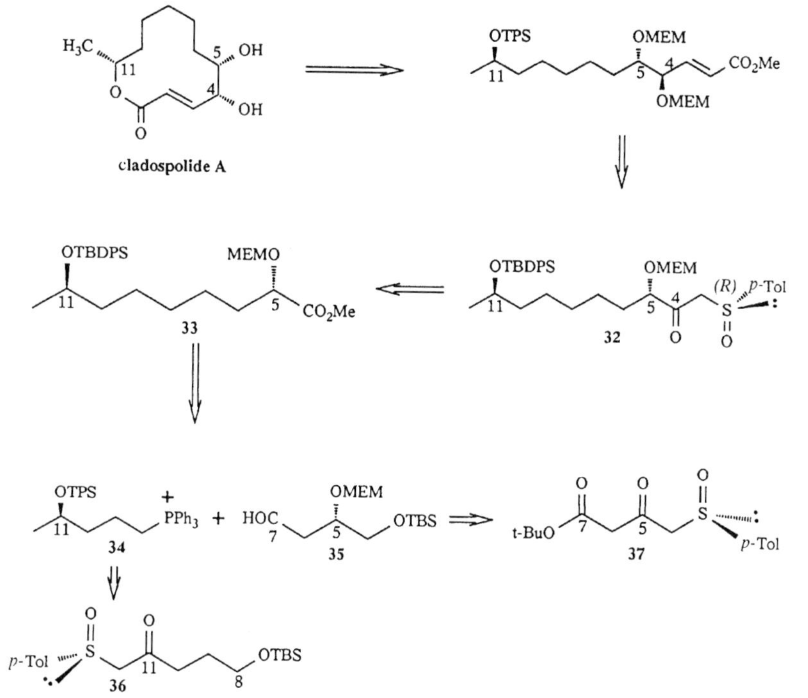

H_3C 11 5 OH 4 OH O O

cladospolide A

OTPS OMEM 11 5 4 CO_2Me OMEM

OTBDPS MEMO 11 5 CO_2Me 33

OTBDPS OMEM (R) p-Tol 11 5 4 S O 32 O

OTPS + PPh_3 11 34 + HOC 7 OMEM 5 OTBS 35

O O O t-BuO 7 5 S p-Tol 37

O O p-Tol S 11 8 OTBS 36

Scheme 4.22

O O p-Tol S OTBS 36

1. DIBAL, 67%, de>95%
2. TPSCl, imid., DMF, 96%

O OTPS p-Tol S OTBS

1. p-TsOH, EtOH, 95%
2a. Raney Ni
2b. $P(Ph)_3$, imid., I_2
2c. $P(Ph)_3$, toluene
81% overall

OTPS $\overset{+}{P}Ph_3\ I^-$

34

81%, $[\alpha]_D = -4.4$

1. n-Buli, THF, 0°C
2.

O OMEM H OTBS 35

OTPS OTBS OMEM

67% overall
1. H_2, Pd-C
2. PDC / DMF
3. CH_2N_2, ether

OTPS OMEM OMe 33 O

p-Tol S CH_3 O

LDA, -78°C
92%

OTPS OMEM p-Tol S 32 O O

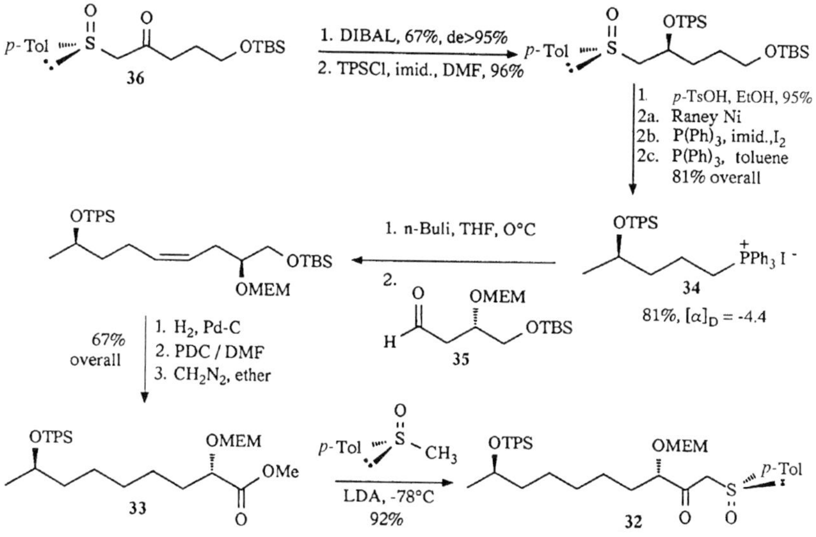

Scheme 4.23

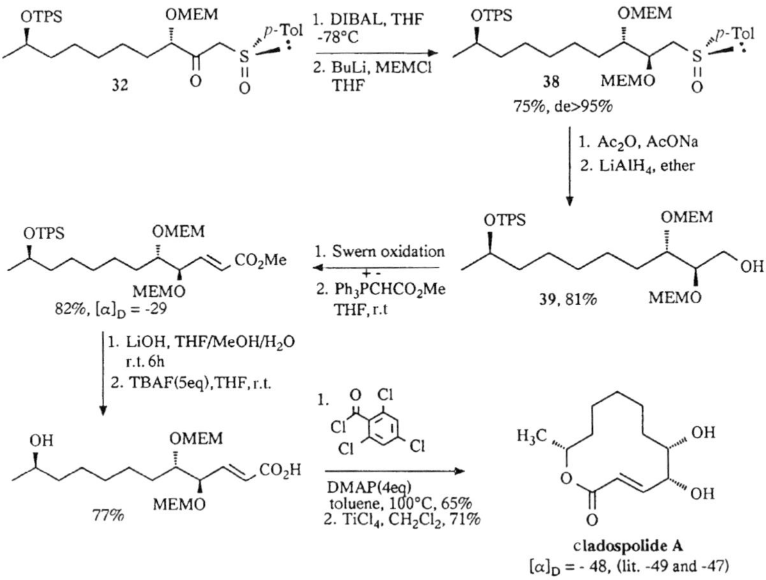

OTPS OMEM 32 O O p-Tol S 1. DIBAL, THF -78°C 2. BuLi, MEMCl THF OTPS OMEM 38 MEMO 75%, de>95% 1. Ac_2O, AcONa 2. $LiAlH_4$, ether OTPS OMEM 39, 81% MEMO OH 1. Swern oxidation 2. Ph_3PCHCO_2Me THF, r.t OTPS OMEM CO_2Me MEMO 82%, $[\alpha]_D = -29$ 1. LiOH, THF/MeOH/H_2O r.t. 6h 2. TBAF(5eq),THF, r.t. OH OMEM CO_2H MEMO 77% 1. DMAP(4eq) toluene, 100°C, 65% 2. $TiCl_4$, CH_2Cl_2, 71% H_3C OH OH O O cladospolide A $[\alpha]_D = -48$, (lit. -49 and -47)

Scheme 4.24

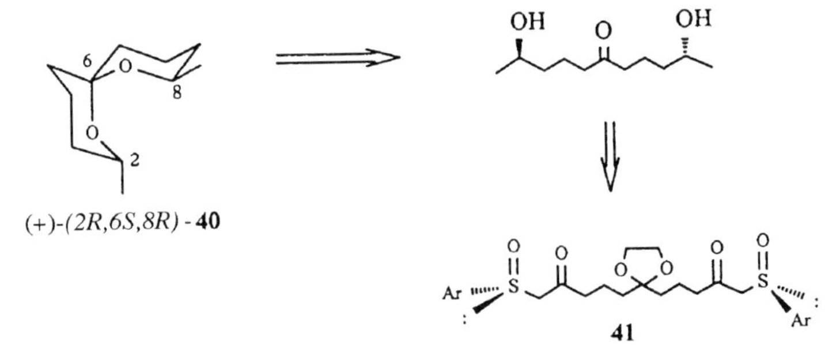

(+)-(2R,6S,8R)-**40** OH O OH Ar S O O O O O O S Ar **41**

Scheme 4.25

reaction of the carbanion of (*R*)-(+)-methyl-*p*-tolyl sulfoxide with the acetal of diethyl 5-ketoazelate. DIBAL reduction yielded, as expected, the dihydroxy disulfoxide (**42**) with the (*S*)-configuration at both hydroxylic centres. Similarly $ZnCl_2$/DIBAL reduction of (**41**) led to the dihydroxy disulfoxide (**43**) with the (*R*) configuration at the hydroxylic centres in 80% yield and de > 95%. Deprotection of the acetal with PPTS in wet acetone allowed the immediate spiroacetalization of the product in almost a quantitative yield. Compound (**42**) led to the spiroacetal ([2(*S*),6(*S*),8(*S*),S(*R*)]-**44**) and (**43**) to ([2(*R*),6(*R*),8(*R*),S(*R*)]-**45**). The stereochemistry of the created chiral spirocarbon was totally controlled by the anomeric effect. Finally desulfurization with

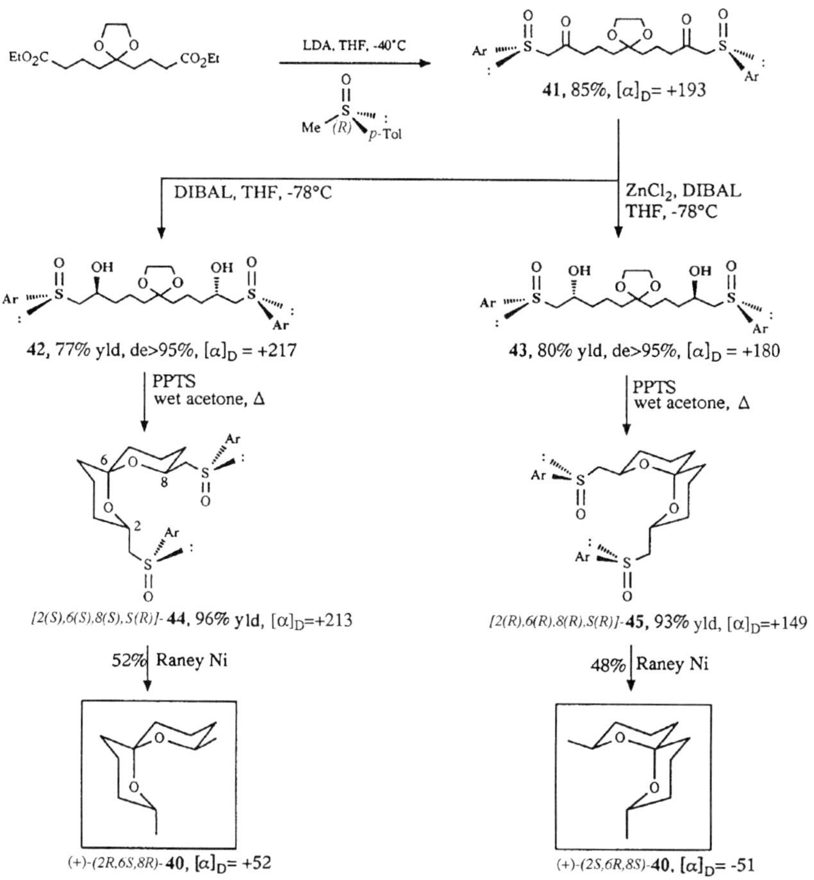

LDA, THF, -40°C

41, 85%, $[\alpha]_D$= +193

DIBAL, THF, -78°C

$ZnCl_2$, DIBAL THF, -78°C

42, 77% yld, de>95%, $[\alpha]_D$ = +217

43, 80% yld, de>95%, $[\alpha]_D$ = +180

PPTS wet acetone, Δ

[2(S),6(S),8(S),S(R)]-44, 96% yld, $[\alpha]_D$=+213

[2(R),6(R),8(R),S(R)]-45, 93% yld, $[\alpha]_D$=+149

52% Raney Ni

48% Raney Ni

(+)-(2R,6S,8R)-40, $[\alpha]_D$= +52

(+)-(2S,6R,8S)-40, $[\alpha]_D$= -51

Scheme 4.26

Raney nickel in methanol afforded the corresponding enantiomeric spiroacetals (**40**).

This synthesis of spiroacetals demonstrated that enantiomerically pure ω-diols of C_2 symmetry could be prepared by reduction of the corresponding diketodisulfoxides. This method was applied to the synthesis of enantiomerically pure *syn*-1,4-diols (Scheme 4.27).[61] The required diketodisulfoxide (**46**) was prepared from methyl succinate. Reduction with DIBAL or $ZnBr_2$/DIBAL gave the corresponding dihydroxy disulfoxides (**47**) and (**48**), which after desulfurization afforded enantiomerically pure 1,4-hexanediol, respectively, in the (*R*,*R*) and (*S*,*S*) configuration.

For the preparation of enantiomerically pure 1,5-diols, the synthesis of the diketodisulfoxide, derived from the homologue methyl glutarate and (*R*)-methyl-*p*-tolyl sulfoxide, gave mainly secondary cyclic products. Therefore,

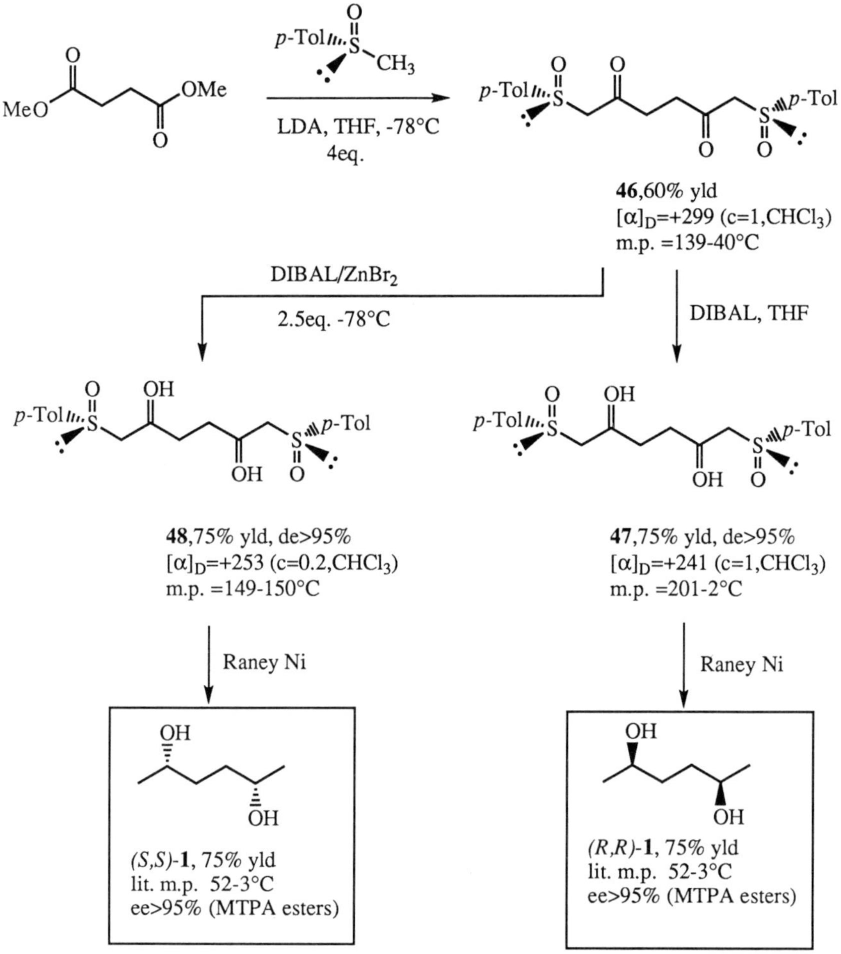

p-Tol S(O) CH_3

LDA, THF, -78°C
4eq.

46,60% yld
$[\alpha]_D$=+299 (c=1,$CHCl_3$)
m.p. =139-40°C

DIBAL/$ZnBr_2$
2.5eq. -78°C

DIBAL, THF

48,75% yld, de>95%
$[\alpha]_D$=+253 (c=0.2,$CHCl_3$)
m.p. =149-150°C

47,75% yld, de>95%
$[\alpha]_D$=+241 (c=1,$CHCl_3$)
m.p. =201-2°C

Raney Ni

Raney Ni

(S,S)-**1**, 75% yld
lit. m.p. 52-3°C
ee>95% (MTPA esters)

(R,R)-**1**, 75% yld
lit. m.p. 52-3°C
ee>95% (MTPA esters)

Scheme 4.27

the asymmetric synthesis of 1,5-heptane diol had to be carried out from a stepwise process based on the reduction of *β*-ketosulfoxides,[60] as shown in Scheme 4.28.

The *β*,*δ*-diketosulfoxides (Scheme 4.12) are also excellent precursors of enantiomerically pure 1,3-diols. This method was applied to the asymmetric synthesis of (+)-nonactic acid and (−)-8-*epi*-nonactic acid.[58,61] As shown in the retrosynthetic Scheme 4.29, the approach was based on the asymmetric reduction of the (*R*)-*β*,*δ*-diketosulfoxide (**49**).

The synthesis of the t-butyl ester (**52**) of the *β*,*δ*-diketosulfoxide (**49**) was carried out following this methodology[37,51] from the diester (**51**) and the

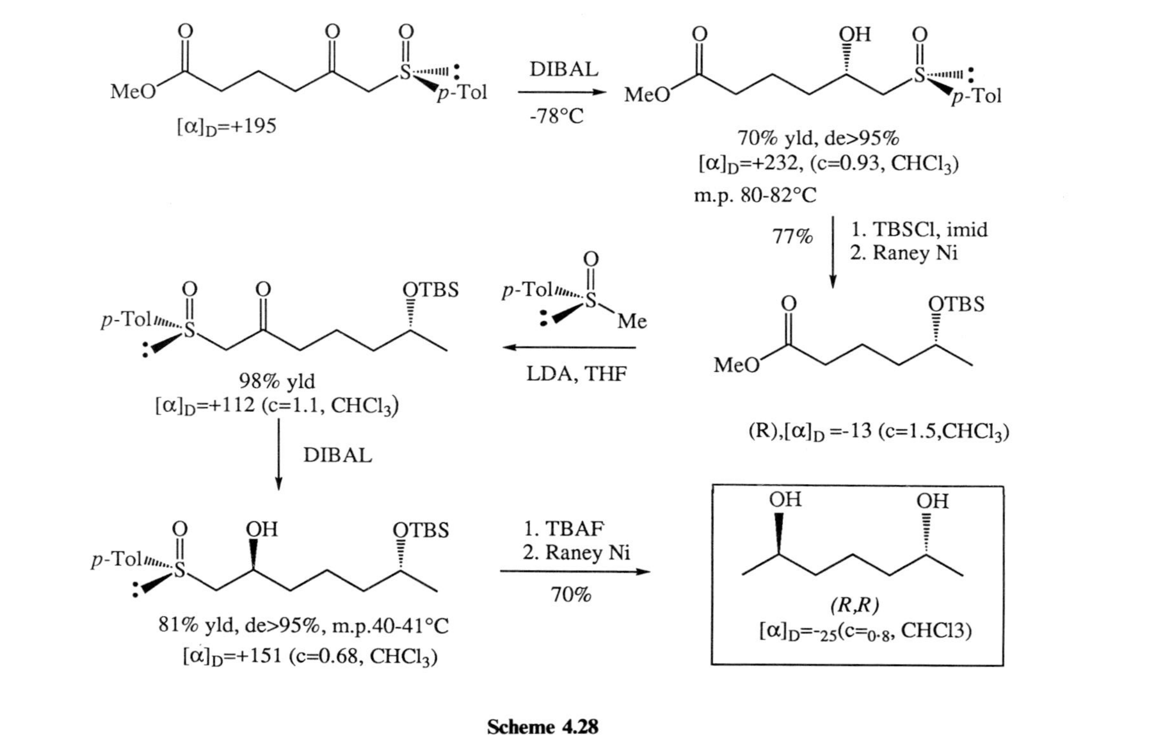
MeO
p-Tol
[α]D=+195
DIBAL
-78°C
MeO
p-Tol
70% yld, de>95%
[α]D=+232, (c=0.93, CHCl3)
m.p. 80-82°C
77%
1. TBSCl, imid
2. Raney Ni
OTBS
MeO
(R),[α]D =-13 (c=1.5,CHCl3)
p-Tol
Me
LDA, THF
p-Tol
OTBS
98% yld
[α]D=+112 (c=1.1, CHCl3)
DIBAL
p-Tol
OH
OTBS
81% yld, de>95%, m.p.40-41°C
[α]D=+151 (c=0.68, CHCl3)
1. TBAF
2. Raney Ni
70%
OH
OH
(R,R)
[α]D=-25(c=0.8, CHCl3)

Scheme 4.28

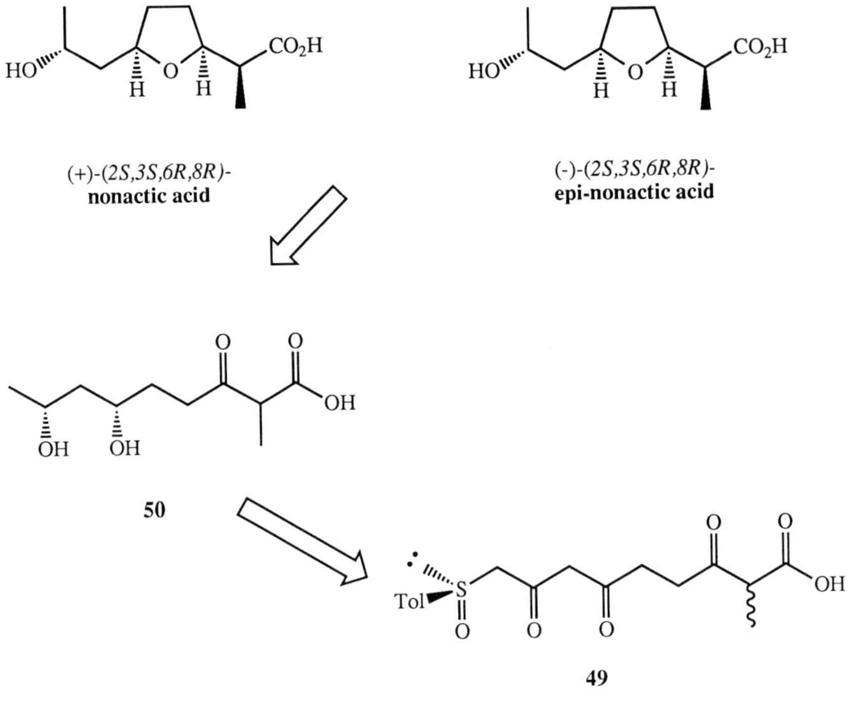

(+)-(2*S*,3*S*,6*R*,8*R*)-
nonactic acid

(-)-(2*S*,3*S*,6*R*,8*R*)-
epi-nonactic acid

50

49

Scheme 4.29

dianion of (*R*)-(+)-1-(*p*-toylsulfinyl)-2-propanone in 75% yield (Scheme 4.30).[61]

As expected from the preceding results, the δ-carbonyl in (**52**) was totally enolized and the reduction required 2 equiv. DIBAL: one to quench the δ-enolate and one to reduce the β-carbonyl. As shown before, this reduction involves an intramolecular hydride transfer from an intermediate in which DIBAL is chelated on the sulfoxide oxygen. As expected, the reduced product was the [S(*R*),8(*S*)]-hydroxy sulfoxide (**53**) (Scheme 4.30). The intramolecular hydride transfer allowed the β-carbonyl to be reduced without protecting the other carbonyl groups. In the next step, the δ-keto group was reduced to the *syn*-diol using diethylmethoxyborane/$NaBH_4$.[62] However, the resulting *syn*-diol (**54**) could not be isolated. It spontaneously cyclized to the lactol (**55**). PPTS-catalysed dehydration of the lactol gave the (+)-enantiomer of [6(*S*),8(*S*),S(*R*)]-(2*E*)-2,3-dehydro-*epi*-nonactate (**56**) in 95% yield, which was desulfurized with Raney nickel and transformed into (−)-(2*R*,3*R*,6*S*,8*R*)-*epi*-nonactic acid by known procedures.[63]

For the synthesis of (+)-t-butyl nonactate (Scheme 4.31), the hydroxy sulfoxide (**53**) was reduced with the Evans reagent,[64] tetramethylammonium triacetoxyborohydride, into the *anti*-diol (**57**), which cyclized spontaneously into the bicyclic acetal (**58**). Acidic treatment of (**58**) yielded

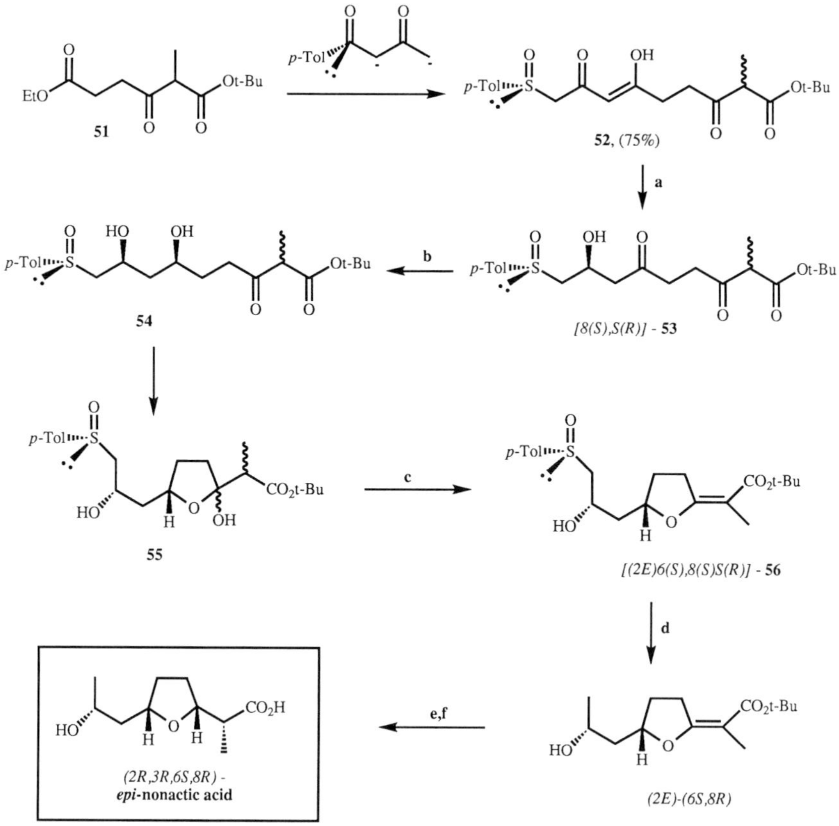

Scheme 4.30 a: DIBAL, THF, −78°C, 15 min., 60%; b: Et_2BOMe, $NaBH_4$, THF, MeOH, −78°C, 2.5 h, 80%; c: PPTS, CH_2Cl_2, r.t., 1 h, 95%; d: Raney Ni, MeOH, r.t., 30 min., 98%; e: Rh/Al_2O_3, H_2, MeOH, 3 days, 98%; f: 2N KOH, H_2O, r.t., 48%.

[6(*R*),8(*S*),S(*R*)]-dehydrononactate (**59**) which was transformed as before into (+)-(2*S*,3*S*,6*R*,8*R*)-nonactic acid. This synthesis is the shortest asymmetric synthesis of *epi*-nonactic acid and (+)-nonactic acid, requiring only six steps (no protecting groups were used in any step).

4.5 Conclusions

It was not possible to review in a single chapter all the literature dealing with optically active β-ketosulfoxides used in asymmetric synthesis. This report is based mainly on the results obtained at Strasbourg. Several reviews, published in recent years, reported complementary results.[65–69]

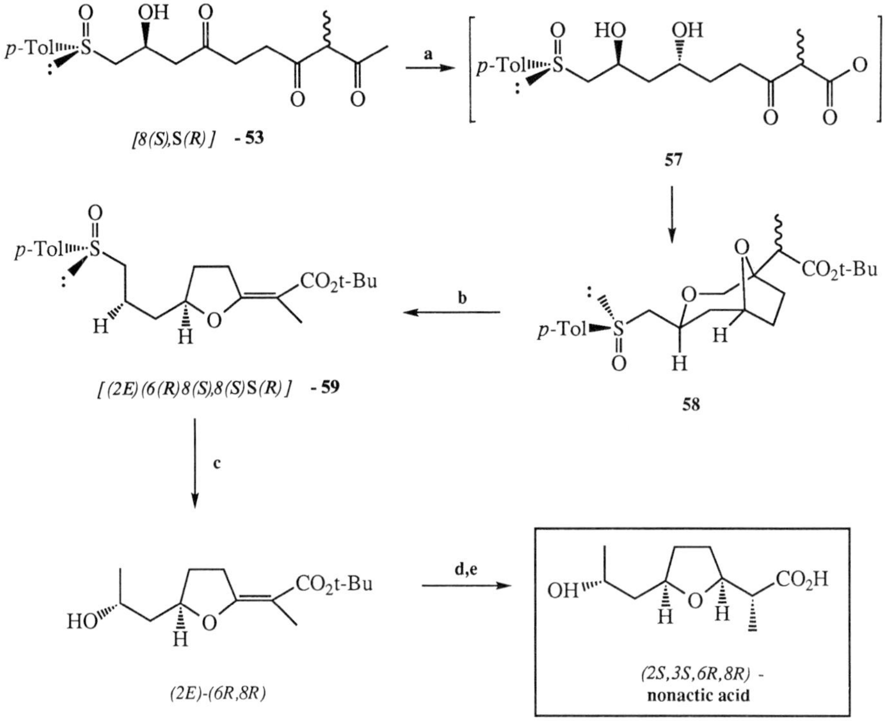

[8(S),S(R)] - **53**

57

58

[(2E)(6(R)8(S),8(S)S(R)] - **59**

(2E)-(6R,8R)

(2S,3S,6R,8R) - nonactic acid

Scheme 4.31 **a**: $(Me_4N)BH(OAc)_3$, AcOH, r.t., 2.5 h, 78%; **b**: HCL, r.t., 3 h, 50%; **c**: Raney Ni, MeOH, r.t., 30 min., 97%; **d**: Rh/Al_2O_3, H_2, 3 days, 97%; **e**: 2N KOH, H_2O, r.t., 58%.

References

1. P.W.B. Harrison, J. Kenyon and H. Phillips, *J. Chem. Soc.*, 1926, **128**, 2079.
2. A.C. Cope and E. Caress, *J. Am. Chem. Soc.*, 1966, **88**, 1711.
3. J. Drabowicz, P. Kielbasinski and M. Mikolajczyk, in *Chemistry of Sulfones and Sulfoxides*, S. Patai, Z. Rappoport and C.J.M. Stirling (eds), Wiley, New York, 1988, p. 233.
4. A. Macconi, F. Montanari, M. Secci and M. Tramontini, *Tetrahedron Lett.*, 1961, 607; U. Folli, D. Iarossi, F. Montanari and G. Torre, *J. Chem. Soc. C*, 1968, 1317.
5. K. Balenovic, M. Bregant and D. Francetic, *Tetrahedron Lett.*, 1960, 20.
6. (a) P. Pitchen, F. Dunach, M.N. Deshmukh and H.B. Kagan, *J. Am. Chem. Soc.*, 1984, **106**, 8188; (b) E. Dunach and H.B. Kagan, *Nouv. J. Chim.*, 1985, **9**, 1.
7. (a) B.J. Auret, D.R. Boyd, H.B. Henbest and S. Ross, *J. Chem. Soc. C*, 1968, 2371; (b) E. Abushanab, D. Reed, F. Suzuki and C.J. Sih, *Tetrahedron Lett.*, 1978, 3415; (c) S. Colonna, N. Gaggero, L. Casella, G. Cerrea and T. Pasta, *Tetrahedron Asymmetry*, 1992, **3**, 95.
8. H.B. Kagan and F. Rebière, *Synlett*, 1990, 643.
9. D.A. Evans, M.M. Faul, L. Colombo, J.J. Bisaha, J. Clardy and D. Cherry, *J. Am. Chem. Soc.*, 1992, **114**, 5977.
10. (a) H. Gilman, J. Robinson and N.H. Beaber, *J. Am. Chem. Soc.*, 1926, **48**, 2715; (b) K.K. Andersen, *Tetrahedron Lett.*, 1962, 93.

11. M. Axelrod, P. Bickart, J. Jacobus, M.M. Green and K. Mislow, *J. Am. Chem. Soc.*, 1968, **90**, 4835.
12. M. Nishio and K. Nishihata, *J. Chem. Soc., Chem. Commun.*, 1970, 1485.
13. S. Juge and H.B. Kagan, *Tetrahedron Lett.*, 1975, 2733; for a general method of determining the absolute configuration of sulfoxides.
14. K. Mislow, M.M. Green, P. Laur, J.P. Melillo, T. Simmons and A.L. Ternay, *J. Am. Chem. Soc.*, 1965, **87**, 1958.
15. K.K. Andersen, W. Gaffield, N.E. Papanikolaou, J.W. Foley and R.I. Perkins, *J. Am. Chem. Soc.*, 1964, **86**, 5637.
16. K. Mislow, A. Ternay and J.T. Melillo, *J. Am. Chem. Soc.*, 1963, **85**, 2329.
17. K.K. Andersen, *J. Org. Chem.*, 1964, **29**, 1953.
18. K. Mislow, T. Simmons, J.T. Melillo and A.L. Ternay, *J. Am. Chem. Soc.*, 1964, **86**, 1452.
19. C.J.M. Stirling, *J. Chem. Soc.*, 1963, 5741.
20. H. Philipps, *J. Chem. Soc.*, 1925, **127**, 2552.
21. K. Ziegler and A. Wenz, *Justus Liebigs Ann. Chem.*, 1934, **511**, 109.
22. M. Cioni and E. Ciuffarin, *J. Chem. Res. (S)*, 1978, 270, 272, 274.
23. J. Drabowicz and S. Oae, *Tetrahedron*, 1978, **34**, 63.
24. C. Mioskowski and G. Solladié, *Tetrahedron*, 1980, **36**, 227.
25. G. Solladié, J. Hutt and A. Girardin, *Synthesis*, 1987, 173.
26. D.N. Harpp, S.M. Vines, J.P. Montillier and T.H. Chan, *J. Org. Chem.*, 1976, **41**, 3987.
27. J. Drabowicz, B. Bujnicki and M. Mikolajczyk, *J. Org. Chem.*, 1982, **47**, 3325.
28. T. Satoh, T. Oohara, Y. Ueda and K. Yamakawa, *Tetrahedron Lett.*, 1988, **29**, 313; the reaction proceeds with inversion of configuration at sulfur.
29. C. Mioskowski and G. Solladié, *Tetrahedron Lett.*, 1975, 3341.
30. R. Annunziata, M. Cinquini, S. Colonna and F. Cozzi, *J. Chem. Soc., Perkin Trans. 1*, 1981, 614.
31. L. Colombo, G. Gennari and E. Narisamo, *Tetrahedron Lett.*, 1978, 3861.
32. D.J. Abott, S. Colonna and C.J.M. Stirling, *J. Chem. Soc., Chem. Commun.*, 1971, 471.
33. (a) G.H. Posner and P.W. Tang, *J. Org. Chem.*, 1978, **43**, 4131; (b) G. Solladié, P. Ruiz, F. Colobert, C. Hamdouchi, M.C. Carreño and J.L. Garcia Ruano, *Synthesis*, 1991, 1011.
34. H. Kosugi, M. Kitaoka, K. Tagami and H. Uda, *Chem. Lett.*, 1985, 805.
35. H. Kosugi, M. Kitaoka, K. Tagami, A. Takahashi and H. Uda, *J. Org. Chem.*, 1987, **52**, 1078.
36. M. Mikolajczyk, W. Midura, S. Grejszczak, A. Zatorski and A. Chefczynska, *J. Org. Chem.*, 1973, **43**, 473.
37. G. Solladié and N. Ghiatou, *Tetrahedron: Asymmetry*, 1992, **3**, 33.
38. E.J. Corey and M. Chaykowski, *J. Am. Chem. Soc.*, 1962, **84**, 866; *J. Am. Chem. Soc.*, 1965, **87**, 1345.
39. N. Kunieda, J. Nokami and M. Kinoshita, *Chem. Lett.*, 1974, 369.
40. R. Annunziata, M. Cinquini and F. Cozzi, *J. Chem. Soc. Perkin Trans. 1*, 1979, 1687.
41. F. Schneider and R. Simon, *Synthesis*, 1986, 582.
42. G. Solladié, C. Greck, G. Demailly and A. Solladié-Cavallo, *Tetrahedron Lett.*, 1982, **23**, 5047.
43. G. Solladié, G. Demailly and C. Greck, *Tetrahedron Lett.*, 1985, **26**, 435.
44. H. Kosugi, H. Konta and H. Uda, *J. Chem. Soc., Chem. Commun.*, 1985, 211.
45. G. Solladié, G. Demailly and C. Greck, *J. Org. Chem.*, 1985, **50**, 1552.
46. G. Solladié, C. Fréchou, G. Demailly and C. Greck, *J. Org. Chem.*, 1986, **51**, 1912.
47. M.C. Carreño, J.L. Garcia Ruano, A.M. Martin, C. Pedregal, J.H. Rodriguez, A. Rubio, J. Sanchez and G. Solladié, *J. Org. Chem.*, 1990, **55**, 2120.
48. A. Solladié-Cavallo, J. Suffert, A. Adib and G. Solladié, *Tetrahedron Lett.*, 1990, **31**, 6649.
49. G. Solladié, C. Hamdouchi and M. Vicente, *Tetrahedron Lett.*, 1988, **29**, 5929.
50. G. Solladié, C. Hamdouchi and C. Ziani-Cherif, *Tetrahedron: Asymmetry*, 1991, **2**, 457.
51. G. Solladié and N. Ghiatou, *Tetrahedron Lett.*, 1991, **33**, 1605.
52. G. Solladié, A. Rubio, M.C. Carreño and J.L. Garcia Ruano, *Tetrahedron: Asymmetry*, 1990, **1**, 187.
53. G. Solladié, M.C. Maestro, A. Rubio, C. Pedregal, M.C. Carreño and J.L. Garcia Ruano, *J. Org. Chem.*, 1991, **56**, 2317.
54. G. Solladié and J. Hutt, *Tetrahedron Lett.*, 1987, **28**, 797.
55. G. Solladié and C. Ziani-Cherif, *J. Org. Chem.*, 1993, **58**, 2181.
56. G. Solladié and A. Almario, *Tetrahedron Lett.*, 1994, **35**, 1937.
57. G. Solladié and A. Almario, *Tetrahedron; Asymmetry.*, 1994, **5**, 1717.
58. G. Solladié, A. Almario and C. Dominguez, *Pure and Appl. Chem.*, 1994, **66**, 2159.
59. G. Solladié and N. Huser, *Tetrahedron; Asymmetry*, 1994, **5**, 255.

60. G. Solladié, N. Huser, J.L. Garcia-Ruano, J. Adrio, C. Carreño and A. Tito, *Tetrahedron Lett.*, 1994, **35**, 5297.
61. G. Solladié and C. Dominguez, *J. Org. Chem.*, 1994, **59**, 3898.
62. K. Chen, G. Hardtmann, K. Prasad, O. Repic and M.J. Shapiro, *Tetrahedron Lett.*, 1987, **28**, 155.
63. P.A. Bartlett, J.D. Meadows and E. Ottow, *J. Am. Chem. Soc.*, 1984, **105**, 5304.
64. D.A. Evans, K.T. Chapman and E.M. Carreira, *J. Am. Chem. Soc.*, 1988, **110**, 3560.
65. (a) G. Solladié, *Synthesis*, 1981, 185; (b) S. Colonna, R. Annunziata and M. Cinquini, *Phosphorus Sulfur*, 1981, **10**, 197.
66. G. Solladié, in *Asymmetric Synthesis*, Vol. 2, J.D. Morison (ed.), Academic Press, New York, 1983, p. 184.
67. (a) M. Cinquini, F. Cozzi and F. Montanari, in *Organic Sulfur Chemistry*, F. Bernardi, I.G. Csizmadia and A. Mangini (eds), Elsevier, Amsterdam, 1985, pp. 305–407; (b) G. Solladié, in *Perspectives in the Organic Chemistry of Sulfur*, B. Zwanenburg and A.J.H. Klunder (eds), Elsevier, Amsterdam, 1987, pp. 293–314; (c) G.H. Posner, in *Perspectives in the Organic Chemistry of Sulfur*, B. Zwanenburg and A.J.H. Klunder (eds), Elsevier, Amsterdam, 1987, pp. 145–152.
68. (a) G. Posner, in *Chemistry of Sulfones and Sulfoxides*, S. Patai, Z. Rappoport and C.J.M. Stirling (eds), Wiley, New York, 1988, pp. 823–849; (b) G.H. Posner, in *Asymmetric Synthesis*, Vol. 2, J.D. Morrison (ed.), Academic Press, New York, 1983, p. 225.
69. (a) G. Solladié, in *Comprehensive Organic Synthesis*, Vol. 6, B. Trost (ed.), Pergamon, Oxford, 1991, pp. 133–170; (b) G. Solladié and C. Carreño, in *Organic Sulfur Chemistry*, P. Page (ed.), Academic Press, 1995, Vol. 1, pp. 1–47; (c) G. Solladié, in *Stereoselective Synthesis*, Houben-Weyl (ed.), Georg Thieme Verlag, 1995, Vol. E21a, pp. 1056–1075 and 1995, Vol. E21b, pp. 1793–1816.
70. G. Solladié and A. Almario, *Tetrahedron Lett.*, 1992, **33**, 2477.

5 Chiral aminals in asymmetric synthesis

A. ALEXAKIS and P. MANGENEY

5.1 Introduction

Aminals are the nitrogen equivalents of acetals.[1] Acyclic aminals are not usually very stable. In contrast, the cyclic ones are stable compounds in neutral and basic media. Particularly attractive as reagents are the 5-membered ring systems (imidazolidine ring), because of the availability of chiral 1,2-diamines and the ease of hydrolysis back to the aldehyde, which can be done under very mild acidic conditions.

Chiral aminals were first introduced by Mukaiyama using diamines derived from proline.[2] More recently, chiral aminals have been prepared from diamines possessing a C_2 axis of symmetry. These are particularly attractive because the formation of aminals does not create any new stereogenic centre. The use of such aminals in asymmetric synthesis is briefly highlighted in this chapter. Chiral diamines may also be used for analytical purposes for the measurement of enantiomeric excess (ee).

Diamines (**1**)–(**4**) have been generally used. Diamine (**1**) is commercially

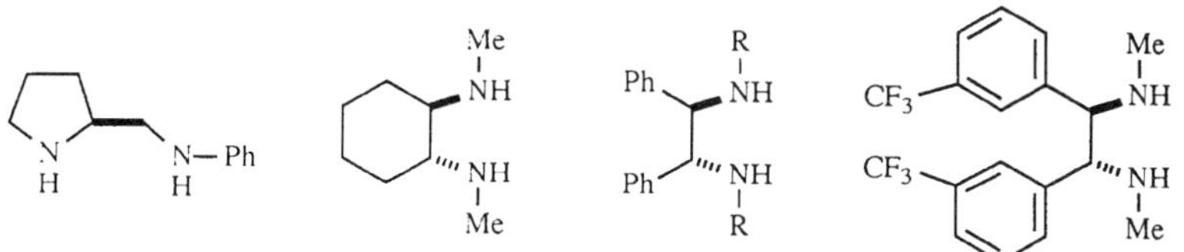

1 2 3a R=Me 4
3b R=Ph
3c R=i-Pr

available or may be prepared in four steps from commercially available (*S*)-proline.[2] Diamine (**2**) was obtained by *N*-methylation of commercial (*R*,*R*)-diaminocyclohexane.[3] Diamines (**3a**) and (**4**) are commercially available. They may also be prepared according to known procedures.[3-7]

The aminals are formed very easily by stirring equimolecular amounts of the *N*,*N*′-disubstituted diamine with an aldehyde, generally without any catalyst, in various solvents (even in aqueous media), with or without molecular sieves. Ketones do not react. The reaction is usually quantitative, and, depending on its structure, the obtained aminal is purified by chromatography or used in crude form. Aminals formed with diamine (**2**) are very sensitive to hydrolysis, because of ring strain, and care should be

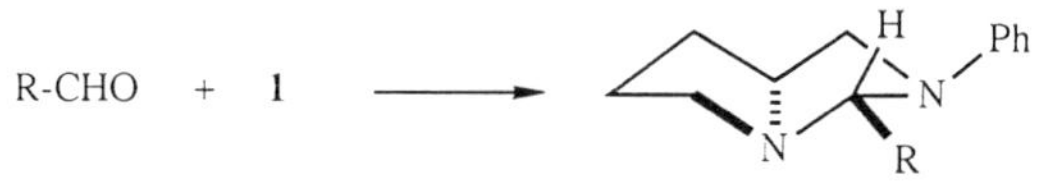

R-CHO + 1 ⟶

Scheme 5.1

taken during chromatographic purifications (NEt_3 should be used as coeluent). The hydrolysis back to the aldehyde is usually done in slightly acidic media, such as 2% HCl. Alternatively, wet acidic (with oxalic acid) silica gel may be used.

Aminals, prepared as shown in Scheme 5.1, with diamine (**1**) form a *cis* fused 5-membered ring structure as a single diastereomer at the aminalic centre. This particular conformation creates the asymmetric environment which allows the further diastereoselective reactions.

Aminals prepared with C_2 symmetrical diamines (**2**–**4**) do not create a new stereogenic centre. The diastereoselective reactions are controlled by the specific conformation of such imidazolidine rings. Therefore, according to the X-ray structures known for such aminals, each substituent on the nitrogen atoms is located *trans* to the substituent of the adjacent carbons, one being pseudoequatorial and the other pseudoaxial. The new stereogenic centres are, in fact, the nitrogen atoms, and they may exert a dual control: (i) steric control by the size of the *N*-substituent or (ii) chelation control by the lone pair of one of the nitrogen atoms (Scheme 5.2).

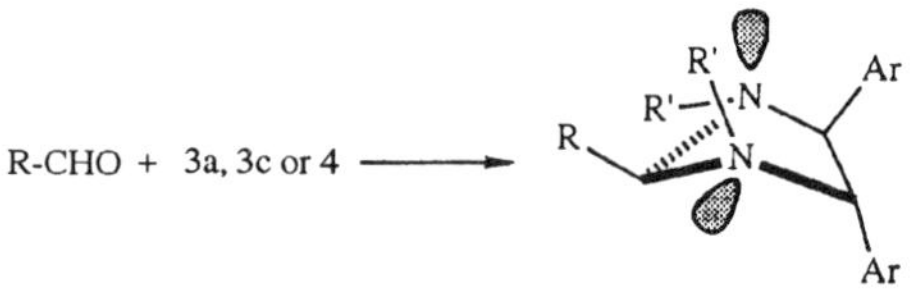

R-CHO + 3a, 3c or 4 ⟶

Scheme 5.2

5.2 Chiral aminals as resolving agents

The diastereomeric aminals formed with a chiral racemic aldehyde and an optically pure C_2 symmetrical diamine are usually very well separated by chromatography. One of the simplest synthetic applications of chiral aminals is the resolution of such aldehydes. Once the diastereomeric aminals are formed, they are separated by flash column chromatography on silica gel (the reaction could be run on 5–10 g scale). Each diastereomerically pure aminal is then hydrolysed back to the optically pure aldehyde, without racemization. A large variety of aldehydes have been resolved in this way using diamine (**2**)[8a] or, more often, diamine (**3a**)[9,10] (a few examples are shown in Scheme 5.3).

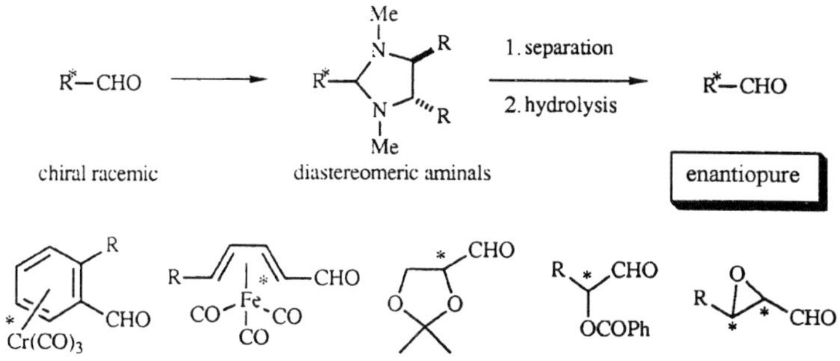

Scheme 5.3

In addition to their easy separation, the diastereomeric aminals can be distinguished analytically. Therefore, the method is suitable as an analytical tool for the determination of the enantiomeric excess of a chiral aldehyde (Scheme 5.4).[9] For this particular purpose, diamine (**4**) allows an accurate measurement of ee by ^{19}F NMR, in addition to ^{1}H and ^{13}C NMR and high performance liquid chromatography (HPLC).[7]

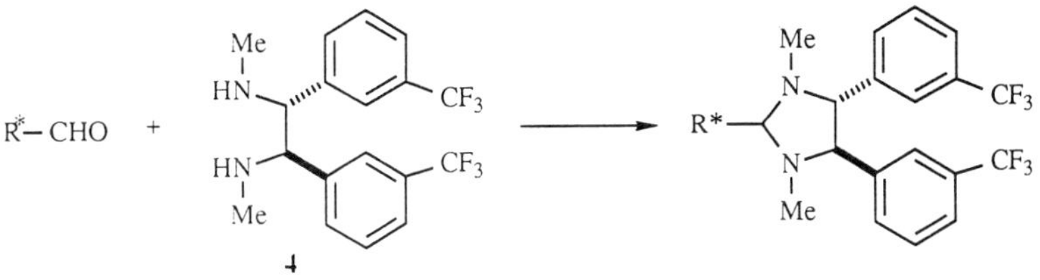

Scheme 5.4

The non-*N*-substituted diamine (**5**) is able to form cyclic aminals with 5-, 6- and 7-membered cyclic ketones. This property allows the determination of the ee of such ketones bearing a stereogenic centre by ^{13}C NMR (Scheme 5.5).[11] The method is as efficient but faster than the usual formation of diastereomeric ketals.[12]

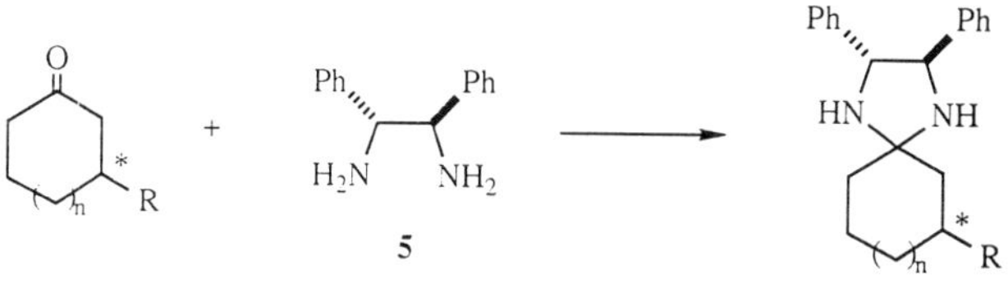

Scheme 5.5

5.3 Conjugate addition reactions*

The chiral aminal moiety can be placed on the *ortho*-position of the aromatic ring of a cinnamate system. The aminal (**6**) (formed with diamine (**2**)) is the most efficient, affording high yields and excellent enantioselectivity in the final conjugate adduct (Scheme 5.6).[13] However, the aminal (**7**) (formed with diamine (**3a**) and having the same absolute configuration as aminal (**6**) gave, after hydrolysis of the aminal moiety, the *opposite* enantiomer, albeit in lower ee (Scheme 5.6). This contrasting behaviour could be explained by the different type of stereocontrol exerted by the chiral auxiliary. Aminal (**6**) possesses a rigid *trans*-fused bicyclic structure which does not allow any conformational flexibility. It is clear on models that one *N*-methyl group masks one face of the π-alkenic system. In aminal (**7**), the flexibility of the imidazolidine ring

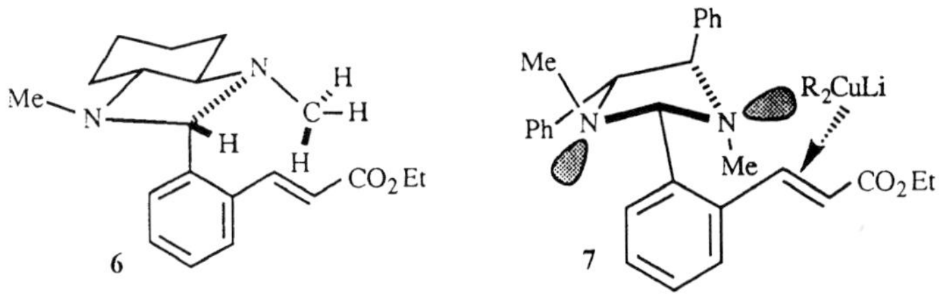

allows an efficient chelation of the organometallic reagent by one nitrogen, delivering the R group on the side of this N atom. Experiments performed in the presence of trimethylsilyl chloride (TMSCl) gave substantial support to this hypothesis. In conjugate additions reactions, it is known that TMSCl favours a steric control and disfavours a possible chelation control. In the above aminals, only aminal (**7**) showed a complete reversal of diastereoselectivity.[14]

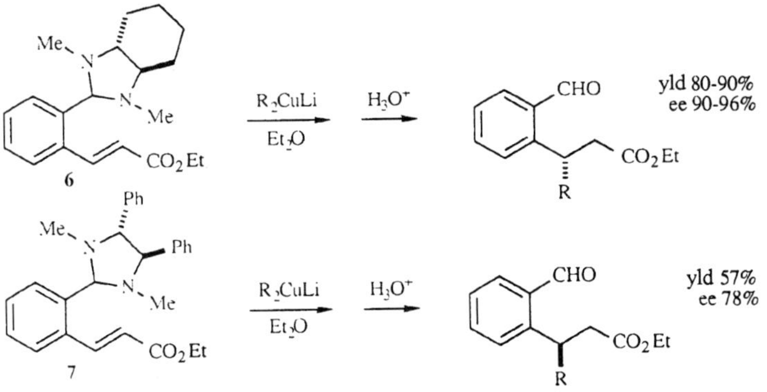

Scheme 5.6

* See also Chapter 11.

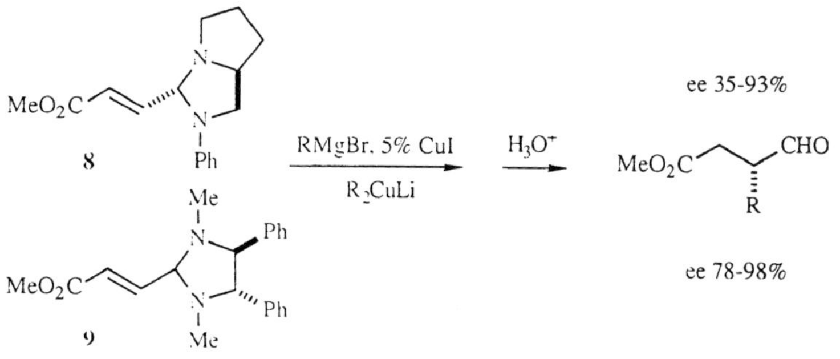

Scheme 5.7

Substrates bearing the chiral aminal controller in different position have also been tested (Scheme 5.7). Thus the fumaric ester derivatives (**8**)[15] and (**9**)[16] react with high levels of stereocontrol, affording, after hydrolysis of the aminal, the corresponding substituted fumaric derivatives. The stereochemical control of these two examples is contrasting. Aminal (**9**) exerts a steric control (reactions run in the presence of TMSCl afforded the same diastereomer), whereas in aminal (**8**) a chelation control was claimed on the basis that the Lewis acidic Mg atom chelates to the more basic N atom (not the Ph substituted one).

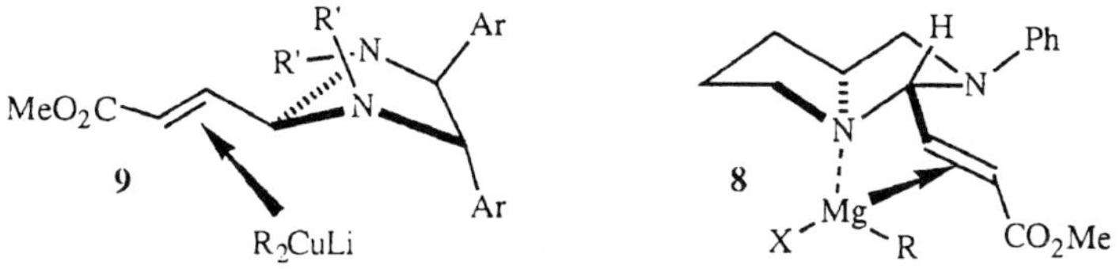

However it is possible to place the aminal auxiliary on the nucleophilic partner, viz. the organometallic reagent. This approach was also tested (Scheme 5.8) with arylcuprate reagents of type (**10**).[17] These organometallic

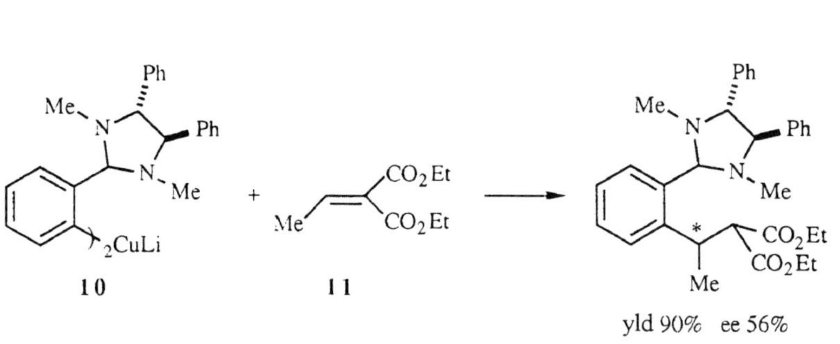

Scheme 5.8

reagents are poorly reactive with α,β-ethylenic esters and ketones but react quite well with the more reactive alkylidene malonate (**11**), affording the conjugate adduct with a modest diastereoselectivity.

5.4 Reactions with C=O and C=N

Aryl organometallics such as (**10**) bearing an *ortho*-aminal moiety have been more successfully used in reaction with aldehydes (Scheme 5.9).[18,19] After hydrolysis of the aminal, the lactol (**12**) was either reduced and acetylated, or oxidized into the lactone. Both aminal systems gave high enantioselectivity (ee 52–96%).

The same intermediate lactol (**12**) could be obtained by swopping the location of the electrophilic and nucleophilic centres. Thus, phthalaldehyde reacts only once with diamine (**2**) or (**3a**) to give monoaminal (**13**) or (**14**). Upon reaction with various organometallic reagents, it was found that aminal (**14**) gave the best selectivities (ee 72–98%) with organocuprate reagents (Scheme 5.10).[16,20]

In a series of papers Mukaiyama *et al.* explored the reaction of aminals of glyoxal derivatives. Phenyl glyoxal was condensed with diamine (**1**) to form the aminal (**15**). Reaction of (**15**) with a Grignard reagent afforded α-hydroxy α-phenyl aldehydes (Scheme 5.11).[21]

A more general approach to the other glyoxal derivatives was disclosed starting from glyoxylate aminal (**16**) (Scheme 5.12). Upon addition of a first Grignard reagent, the intermediate α-keto aminal (**17**) was reacted with a second Grignard reagent to afford, after hydrolysis, the disubstituted α-hydroxy aldehydes (**18**) in high ee (78–99%).[22] This synthetic approach was successfully used for the asymmetric synthesis of both enantiomers of frontalin (**19**)[23] and for the synthesis of malyngolide (**20**), a new antibiotic isolated from marine algae.[24] Alternatively, the second Grignard reagent could be replaced by the lithium enolate of ethyl acetate, giving access to β-hydroxy esters of type (**21**).[25]

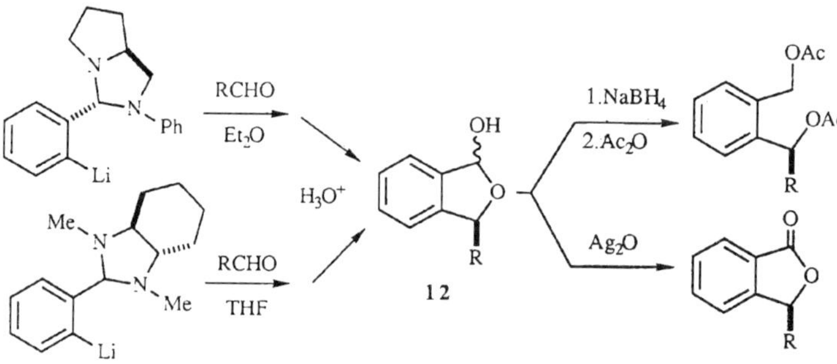

Ph Li RCHO Et_2O Me Me Li RCHO THF H_3O^+ OH O R 12 1.NaBH$_4$ 2.Ac$_2$O OAc OAc R Ag_2O O O R

Scheme 5.9

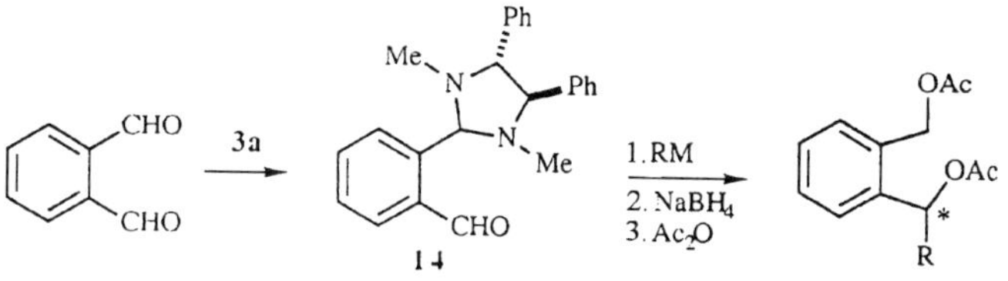

RM	Yield (%)	ee (%)	*S*/*R*
BuLi	76	40	*S*
BuMgCl	72	68	*S*
BuMnBr	69	99	*R*
R_2CuLi	74–93	72–99	*S*

Scheme 5.10

The same methodology was also applied in an indirect preparation of the monoaminal of glyoxal itself (**22**).[26] In this case, the first organometallic is an aluminium hydride reagent. This monoaminal could be isolated in crude form but could not be purified. In a second step, a Grignard reagent was added, followed by benzylation of the hydroxy group and reduction of the free aldehyde to afford the monoprotected diol (**23**) in high ee (83–97%). This sequence was applied to the synthesis of *exo*-(+)-brevicomin (**24**).

An alternative route to the monoaminals of glyoxal starts from the commercial 40% aqueous glyoxal solution (Scheme 5.13).[27] Thus, momoaminals (**25**) and (**26**) were obtained from diamines (**3a**) and (**3c**) and used as crude products. Aminal (**26**) was stable enough to be isolated pure upon silica gel chromatography. These monoaminals react diastereoselectively with organolithium reagents, aminal (**26**) being the best one, affording a single diastereomer.[28] After acetylation, followed by mild acidic hydrolysis, the α-acetoxy aldehyde could be obtained (Scheme 5.14).

Aminal (**25**) is also the starting point for several other chiral synthons (Scheme 5.15).[28] Wittig–Horner olefination afforded the synthon (**9**) with which conjugate additions have been performed (see Scheme 5.7). Alternatively, the aldehyde functionality of aminal (**25**) could be transformed into an imine funtionality by reaction with primary amines. Imine (**27**) reacts diastereoselectively (de > 95%) with organolithium or Grignard reagents. The trityl-protecting group was selectively removed under the mild conditions shown

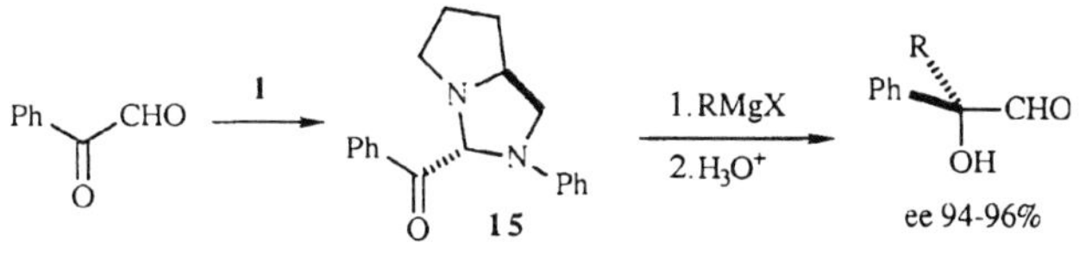

Scheme 5.11

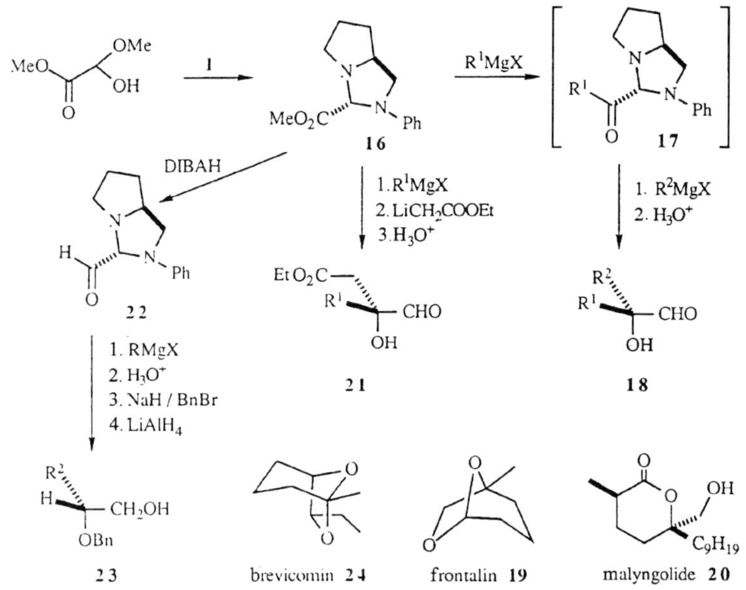

Scheme 5.12

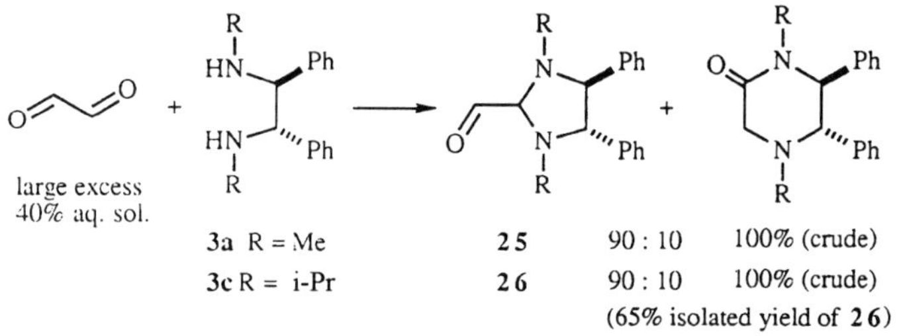

Scheme 5.13

in Scheme 5.15 without touching the aminal functionality. Alternatively, imine (**28**), bearing the *p*-anisyl protecting group, could be reacted with an enolate to yield a β-lactam (Scheme 5.16).[28]

The monohydrazone of glyoxal (**29**) was a starting point for a general methodology directed toward the synthesis of α-amino aldehydes. Such

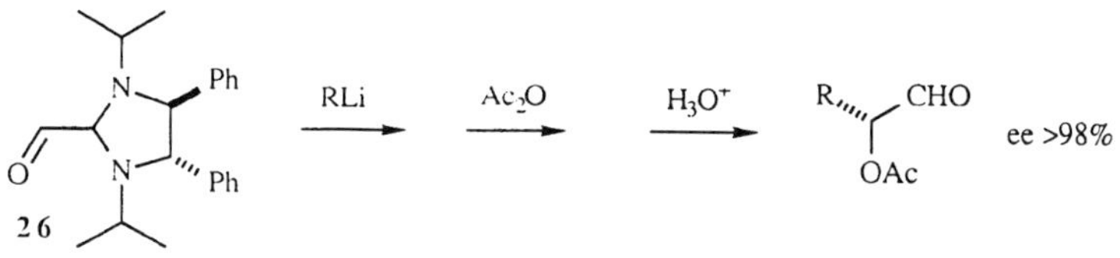

26 — RLi → Ac_2O → H_3O^+ → R–CH(OAc)–CHO ee >98%

Scheme 5.14

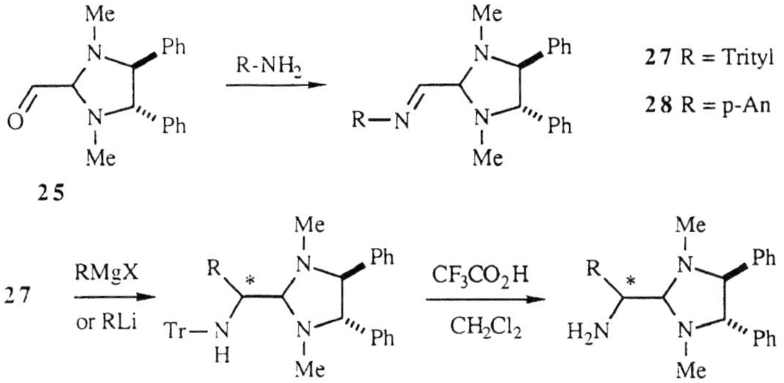

25 — R-NH_2 → 27 R = Trityl; 28 R = p-An

27 — RMgX or RLi → (Tr–NH) — CF_3CO_2H, CH_2Cl_2 → (H_2N)

Scheme 5.15

28 + (OLi, OMe) → (p-An) yld 92% de 71%

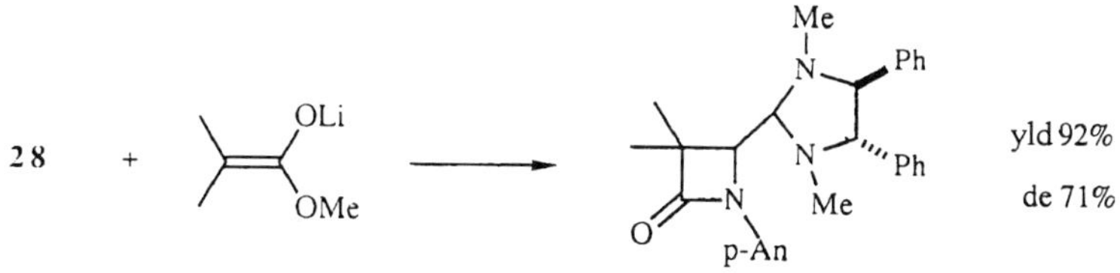

Scheme 5.16

aldehydes are ideal precursors of α-amino acids or α-amino alcohols. Thus, compound (**29**) was treated with diamine (**3a**) to afford the crystalline aminal (**30**) (Scheme 5.17). This aminal gave a single diastereomer upon reaction with an organolithium reagent in THF, through steric control.[29] By contrast, with Grignard reagents in toluene as solvent, (**30**) gave the adduct of opposite stereochemistry, through a chelation-controlled reaction (de 88–99%) (Scheme 5.17).[30]

The cleavage of the hydrazine N–N bond was best achieved with Raney nickel under ultrasonic conditions.[31] Protection of the primary amine functionality, as t-Boc, followed by slightly acidic hydrolysis of the aminal-protecting group, gave the desired enantiomerically pure α-amino aldehyde (Scheme 5.18).[29]

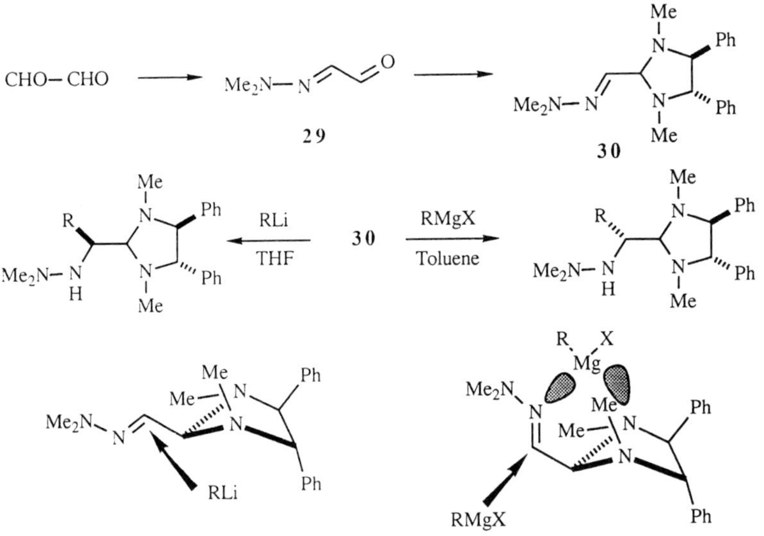

Scheme 5.17

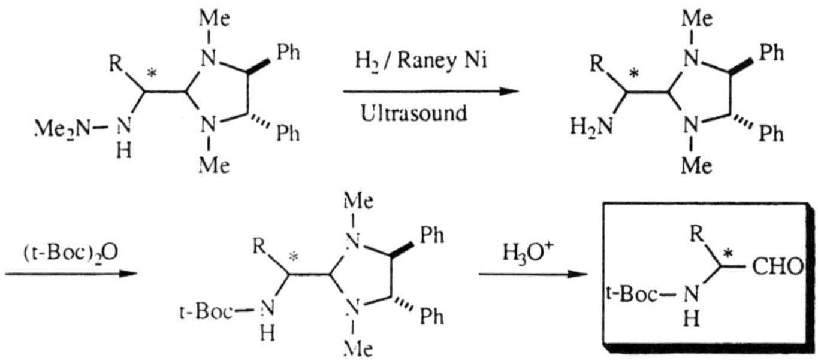

Scheme 5.18

5.5 Chiral dihydropyridines

5.5.1 *Chiral 4-substituted 1,4-dihydropyridines*

An asymmetric synthesis of 3-formyl-4-substituted 1,4-dihydropyridines (**31**), important chiral synthons, has been described via the aminal (**32**), which is easily prepared by condensation of diamine (**3a**) with 3-formylpyridine.[32] This synthesis involves the regio- and diastereoselective addition of organocopper reagents on the *in situ* prepared acylpyridinium salt (Scheme 5.19). Aminals

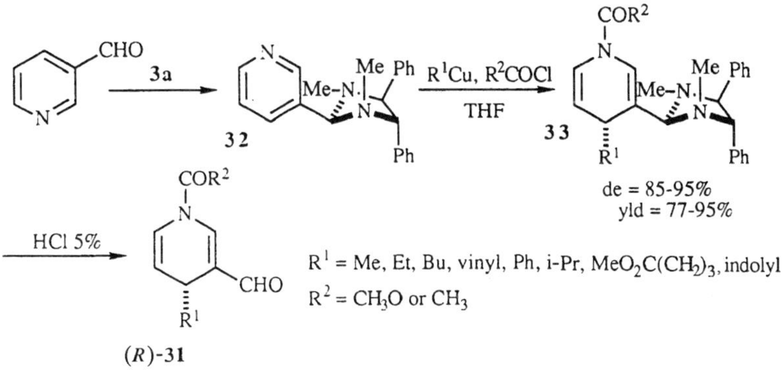

3a; R^1Cu, R^2COCl, THF; HCl 5%

de = 85-95%
yld = 77-95%

R^1 = Me, Et, Bu, vinyl, Ph, i-Pr, $MeO_2C(CH_2)_3$, indolyl
R^2 = CH_3O or CH_3

(*R*)-**31**

Scheme 5.19

(**33**) were hydrolysed to the corresponding aldehydes (**31**) by acidic hydrolysis. The configuration of the newly formed stereogenic centre was established by correlation and shown to be (*R*) starting from an (*S*,*S*) diamine (**3a**). The same reaction, performed on an alkylpyridinium salt, afforded the 1,4-dihydropyridine in good yield but with a poor diastereoselectivity (Scheme 5.20). In order to explain the stereochemistry observed for (**33**), a transition state involving a chelation of the organocopper reagent with the lone pair of the nitrogen atom N-2 *and* the oxygen of the carbamate (or amide) function was postulated (**34**).

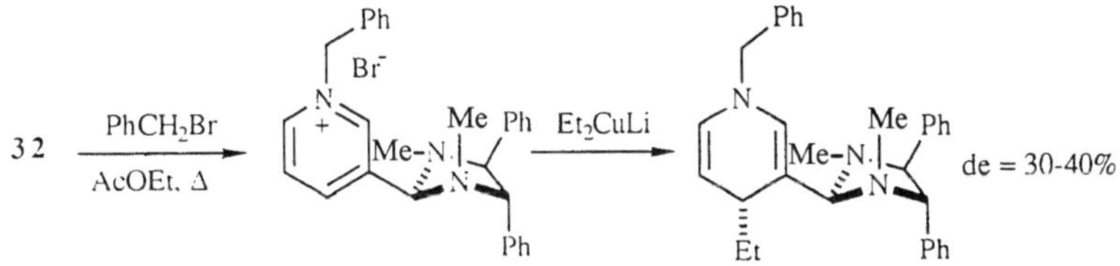

32; $PhCH_2Br$, AcOEt, Δ; Et_2CuLi; de = 30-40%

Scheme 5.20

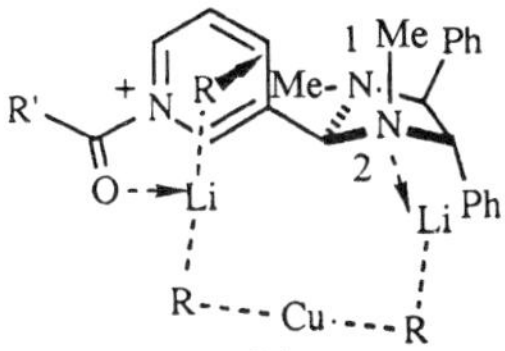

34

Organocopper reagents, known to undergo 1,2 addition rather than 1,4 (conjugate) addition, afford the 6-substituted 1,6-dihydropyridines shown in

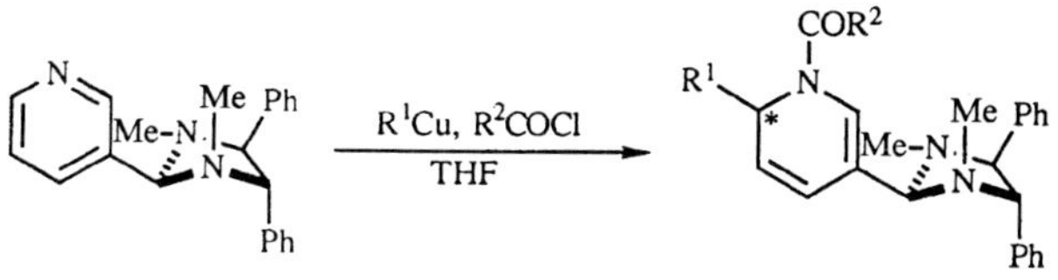

COR^2 R^1 * Me Ph Me–N N Ph

R^1Cu, R^2COCl

THF

R^1 = Me_3Si–C≡C–, CH_2=CH–CH_2–, EtO_2C–CH_2–, $(CH_3O_2C)_2CH$–

Scheme 5.21

Scheme 5.21 in various yields and with poor to good diastereoselectivity (10–75%).[33] These chiral dihydropyridines were the starting point for the synthesis of several classes of alkaloid. Thus, the possibility of using functionalized acyl chlorides was exploited in the enantioselective syntheses of indoloquinolizine[33,34] and benzoquinolizine frameworks.[33] This approach (Schemes 5.22 and 5.23) allows an exceedingly short asymmetric construction of such skeletons.

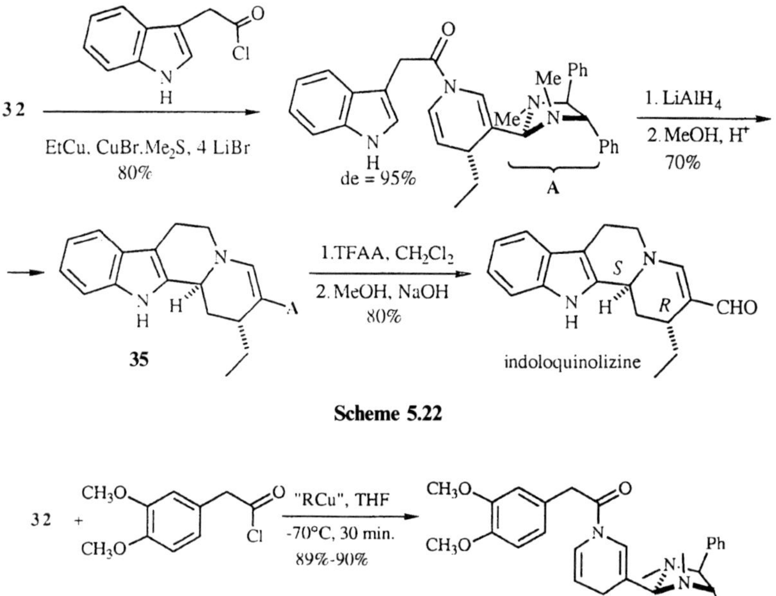

32 → EtCu, $CuBr.Me_2S$, 4 LiBr, 80%; de = 95%; A

1. $LiAlH_4$ 2. MeOH, H^+ 70%

35 → 1. TFAA, CH_2Cl_2 2. MeOH, NaOH 80% → indoloquinolizine (S, R, CHO)

Scheme 5.22

32 + (CH_3O)$_2$-aryl acetyl chloride → "RCu", THF, -70°C, 30 min. 89%-90%; de = 92-95%; A

$LiAlH_4$, Et_2O → 36 → 1. TFA, 0°C 2. TFAA, CH_2Cl_2 3. MeOH, NaOH → Benzoquinolizine (11b, 2, CHO, R)

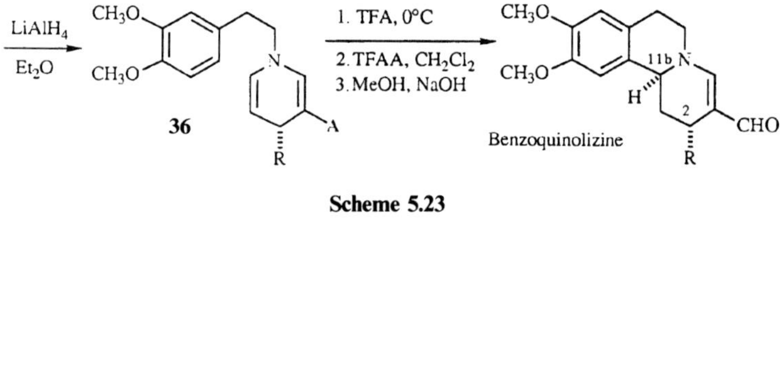

Scheme 5.23

Aminals (**35**) and (**36**), bearing a basic β-nitrogen are exceptionally stable towards acidic hydrolysis. The deprotection could be performed with trifluoroacetic anhydride followed by an alkaline methanolysis.

5.5.2 *Chiral 6-substituted 1,6-dihydropyridines*

Synthesis of chiral 1,6-dihydropyridines was performed by addition of organocopper reagents to the pyridines (**37**) and (**38**) bearing the aminal in position 4.[35] The best selectivity was attained with aminal (**38**), where the bulky N-i-Pr group exerts a better steric control (Scheme 5.24).

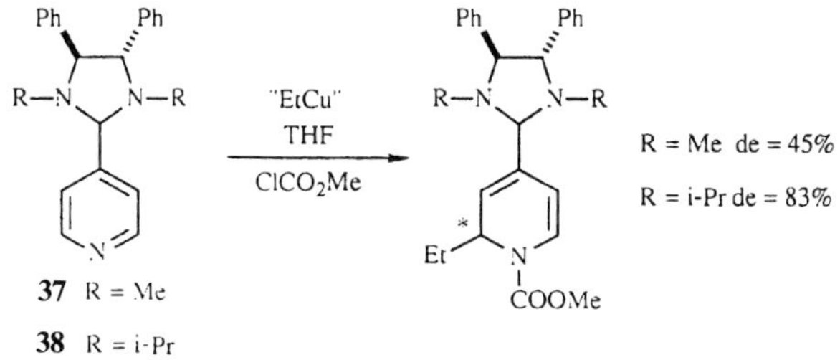

"EtCu"
THF
$ClCO_2Me$

37 R = Me
38 R = i-Pr

R = Me de = 45%
R = i-Pr de = 83%

Scheme 5.24

5.6 Cycloaddition reactions

5.6.1 *Asymmetric dipolar additions with* N*-metallated azomethines*

Syntheses of chiral pyrrolidines by asymmetric dipolar addition of *N*-metallated azomethines with fumaric aminals (**8**), (**9**) and (**39**) were reported by Kanemasa *et al.*[5,36,37] When the aminals (**8**) and (**9**) were allowed to react with azomethines (**40**) (R′ = Me or t-Bu), the cycloadducts (**41a**) were obtained as single diastereomers. A slightly lower diastereoselectivity was observed with the cyano-stabilized *N*-lithiated azomethine ylide (**42**) and aminal (**8**). By contrast, *N*-phenyl aminal (**39**) gave the opposite diastereomer (**41b**), albeit with a lower selectivity. All the cycloadducts (**41a,b**) were transformed into acetals (**43a,b**).

The stereochemistry of the cycloadditions with aminals (**8**) and (**9**) was postulated to arise from the exclusive participation of the thermodynamically less favoured *syn*-periplanar conformer, the attack of the ylide occurring on the less hindered side. It has been shown by X-ray analysis that the conformation of the imidazolidine ring of (**39**) was relatively flat compared with that of the *N*-methyl-substituted aminal (**9**). Therefore, the steric differentiation of the two faces of the enoate is effected by C–Ph groups rather than the N–Me groups.

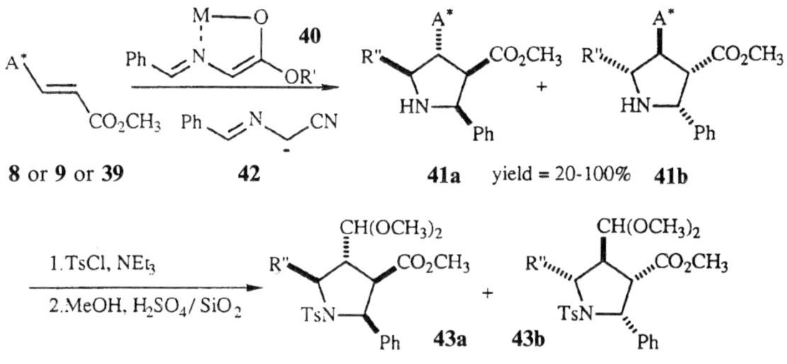

8 or 9 or 39 + 42 (via 40) → 41a + 41b, yield = 20-100%

1. TsCl, NEt_3; 2. MeOH, H_2SO_4/ SiO_2 → 43a + 43b

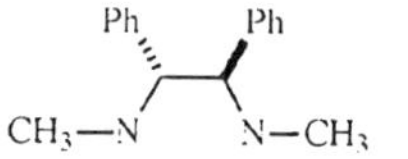

Aminal	A*	R″	(41a)/(41b)
(*S*)-**8**		CO_2CH_3	100/0
		CO_2t-Bu	100/0
		CN	93/7
(*R*,*R*)-**9**		CO_2CH_3	100/0
		CO_2t-Bu	100/0
(*R*,*R*)-**39**		CO_2CH_3	9/91
		CO_2t-Bu	55/45

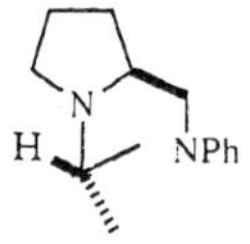

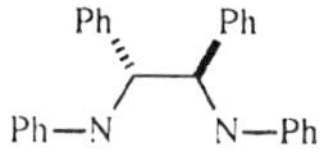

Scheme 5.25

5.6.2 Cycloaddition of nitrile oxides

Nitrile oxides were reacted with aminals (**8**), (**9**) and (**39**)[38] giving the cycloadducts (**44a,b**), which were converted into the acetals (**45a,b**) (Scheme 5.26). The relative stereochemistry of each of the acetals was determined by X-ray analysis of (**44**).

A poor regioselectivity was obtained with aminal (**8**) and only the minor regioisomer was found to be optically pure. Aminal (**9**) behaved similarly but each regioisomer was found to be optically pure. By contrast, the aminal (**39**) gave a single regioisomer, albeit with a poor diastereoselectivity.

5.6.3 Inverse Diels–Alder reaction

Enantioselective Diels–Alder reactions with inverse electron demand were attempted with aminals of type (**46**) with or without C_2 symmetry, used as chiral ketene equivalents.[39,40] The reactions gave aminals (**47**) which rearrange to (**48**) and (**49**) (Scheme 5.27). This mixture was transformed into the

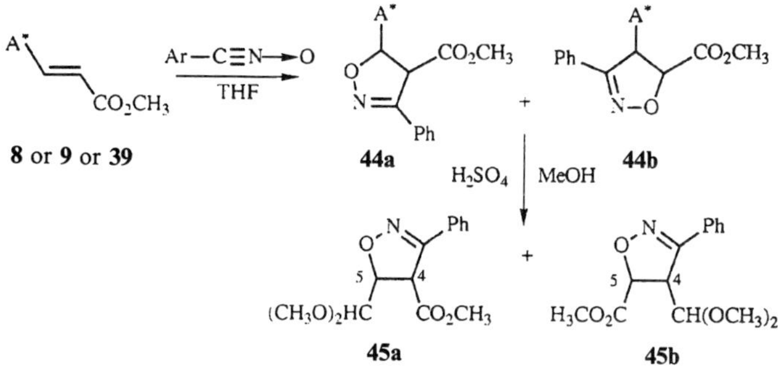

A* CO₂CH₃ → Ar—C≡N→O, THF → 44a + 44b → H₂SO₄ | MeOH → 45a + 45b

8 or **9** or **39**

A* CO$_2$CH$_3$ O N Ph **44a**

A* Ph CO$_2$CH$_3$ N–O **44b**

O N Ph 5 4 (CH$_3$O)$_2$HC CO$_2$CH$_3$ **45a**

O N Ph 5 4 H$_3$CO$_2$C CH(OCH$_3$)$_2$ **45b**

Aminal	Cycloadduct	Yield (%)	ee (%)
8	(4*S*,5*S*)-**45a**	38	40
	(4*S*,5*S*)-**45b**	14	100
9	(4*R*,5*R*)-**45a**	29	100
	(4*R*,5*R*)-**45b**	10	100
39	(4*R*,5*R*)-**45a**	71	43

Scheme 5.26

semicarbazone (**50**) in order to determine the optical purity and absolute configuration. It was found that the optical yields are kinetically controlled and that the cycloaddition was stereospecific with regards to dienophile. The best optical yield was obtained for $R^1 = R^2 = CH_3$, $R^3 = Ph$, $R^4 = H$ ((*S*)-configuration). In this case, the semicarbazone obtained with methyl hexadienoate (2*E*,4*Z*) was of (*R*)-configuration (ee = 46%).

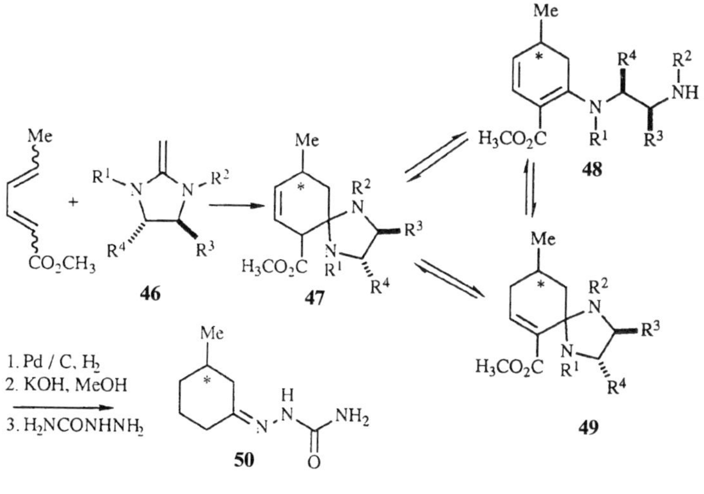

Me CO$_2$CH$_3$ + R^1 N N R^2 R^4 R^3 **46** → Me * R^2 N R^3 N H$_3$CO$_2$C R^1 R^4 **47**

Me * H$_3$CO$_2$C N R^1 R^4 R^2 NH R^3 **48**

Me * R^2 N R^3 N H$_3$CO$_2$C R^1 R^4 **49**

1. Pd / C, H$_2$
2. KOH, MeOH
3. H$_2$NCONHNH$_2$

Me * N H N NH$_2$ O **50**

Scheme 5.27

5.7 Asymmetric tricarbonylchromium complexes

Enantiopure *o*-benzaldehyde tricarbonylchromium complexes (**51a,b**) are available by resolution of the corresponding aminals (see Scheme 5.3). The diastereoselective introduction of the tricarbonyl chromium moiety on aminals (**52**) was also reported (Scheme 5.28).[8] The reaction, performed with $Cr(CO)_6$ at 140°C, gave selectively the corresponding complexes (**53a**) (de = 76–82%). Tricarbonylchromium was also introduced under mild conditions, at room temperature, by exchange with tricarbonyl(naphthalene)chromium. The opposite diastereomers (**53b**) were obtained with very high selectivity (de 94–96%). This inversion of selectivity was ascribed to a thermodynamic control with $Cr(CO)_6$ and to a kinetic control with tricarbonyl(naphthalene)chromium. Indeed, the kinetic products (**53b**), when heated at 140°C for 20 h, isomerize to form thermodynamically preferred (**53a**) (Scheme 5.28).

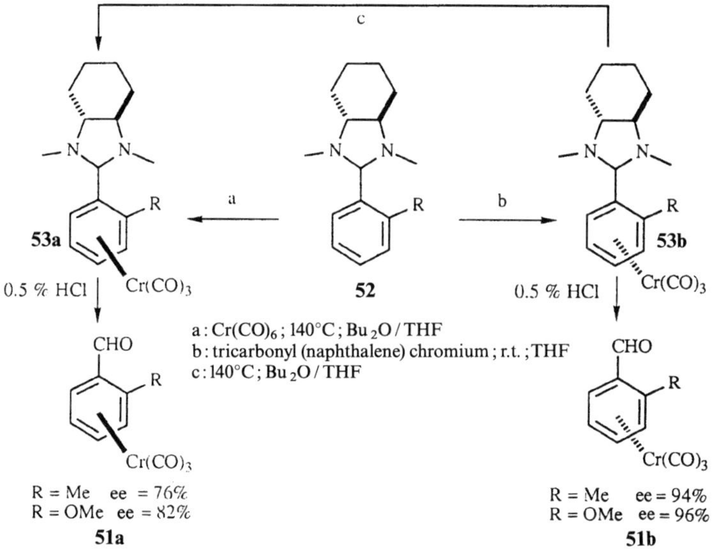

a: $Cr(CO)_6$; 140°C; Bu_2O/THF
b: tricarbonyl (naphthalene) chromium; r.t.; THF
c: 140°C; Bu_2O/THF

53a: 0.5 % HCl → 51a: R = Me ee = 76%; R = OMe ee = 82%

53b: 0.5 % HCl → 51b: R = Me ee = 94%; R = OMe ee = 96%

Scheme 5.28

5.8 Cleavage of aminals

Asymmetric synthesis of chiral aldehydes of type (**54**) was attempted from α,β-ethylenic aminals (**55**) and (**56**) by using organocopper reagents in the presence of methyl chloroformate.[41] By analogy with chiral acetals,[42] the imidazolidine ring is diastereoselectively opened by the chloroformate with concomitant introduction of the R group at the γ-position. Because of the ring strain in aminal (**55**), the reaction is faster and more selective than with aminal (**56**). After hydrolysis of the resulting enamines, chiral aldehydes (**54**) are obtained with a low to moderate optical purity (Scheme 5.29). The same reaction, performed on the aminal (**55**) with an acyl chloride (CH_3COCl instead of CH_3OCOCl) gave the aldehyde with a poor optical purity (ee = 25%).

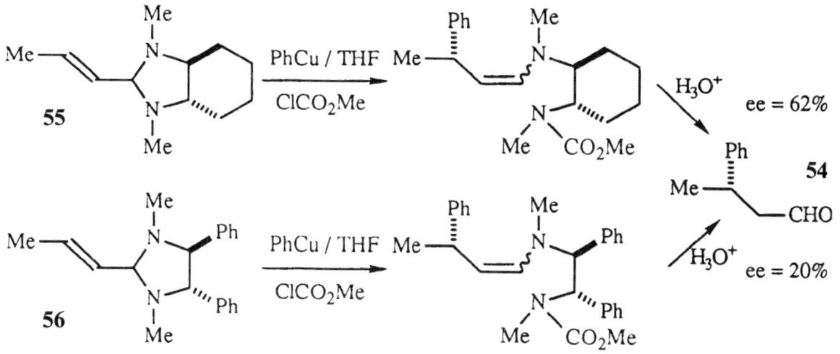

Scheme 5.29

5.9 Conclusions

This short overview shows that chiral aminals are promising new auxiliaries in asymmetric synthesis. Most often, very high diastereoselectivities are attained either by a steric control or by a chelation control. These factors may be modulated by fine tuning of the *N*-substituent. Another salient feature of this functional group is its ease of introduction and removal. Moreover, the basic nature of the diamines allows their very easy recovery, by a simple acid–base treatment. Finally it should be recalled that the aminal group also acts as an efficient protective group of the aldehyde functionality.

References

1. For a recent review see: Duhamel, L., in *The Chemistry of amino, nitroso and nitro compounds and their derivatives*, Suppl. F, S. Patai (ed.), J. Wiley, New York, 1982, pp. 849–907.
2. For a brief survey of this work see: Mukaiyama, T., *Tetrahedron*, 1981, **37**, 4111–4119.
3. Fiorini, M. and Giongo, G.M. *J. Mol. Cat.*, 1979, **5**, 303; see also ref. 6 for experimental details.
4. Mangeney, P., Alexakis, A., Grojean, F. and Normant, J.F. *Tetrahedron Lett.*, 1988, **29**, 2675.
5. Kanemasa, S., Hayashi, T., Tanaka, J., Yamamoto, H. and Sakuraï, T., *J. Org. Chem.*, 1991, **56**, 4473.
6. Alexakis, A., Mutti, S. and Mangeney, P., *J. Org. Chem.*, 1992, **57**, 1224.
7. Cuvinot, D., Mangeney, P., Alexakis, A., Normant, J.F. and Lellouche, J.-P., *J. Org. Chem.*, 1988, **54**, 2420.
8. (a) Alexakis, A., Mangeney, P., Marek, I., Rose-Munch, F., Rose, E., Semra, A. and Robert, F., *J. Am. Chem. Soc.*, 1992, **114**, 8288. (b) Alexakis, A., Kanger, T., Mangeney, P., Rose-Munch, F., Perroley, A. and Rose, E., *Tetrahedron: Asymmetry*, 1995, **6**, 47.

9. Mangeney, P., Alexakis, A. and Normant, J.F., *Tetrahedron Lett.*, 1988, **29**, 2677.
10. Pinsard, P., Lellouche, J.-P., Beaucourt, J.-P. and Grée, R., *Tetrahedron Lett.*, 1990, **31**, 1137.
11. Alexakis, A., Frutos, J.C. and Mangeney, P., *Tetrahedron: Asymmetry*, 1994, **4**, 2431.
12. Hiemstra, H. and Wynberg, H., *Tetrahedron Lett.*, 1977, 2183.
13. Alexakis, A., Sedrani, R., Mangeney, P. and Normant, J.F., *Tetrahedron Lett.*, 1988, **29**, 4411.
14. Alexakis, A., Sedrani, R. and Mangeney, P., *Tetrahedron Lett.*, 1990, **31**, 345.
15. Asami, M. and Mukaiyama, T., *Chem. Lett.*, 1979, 569.
16. Alexakis, A., Sedrani, R., Lensen, N. and Mangeney, P. in *Organic Synthesis via Organometallics*, D. Enders, H.J. Gais and W. Keim (eds), Vieweg, Wiesbaden, 1993.
17. Sedrani, R., Thèse de Doctorat, University P. and M. Curie, Paris, 1990.
18. Asami, M. and Mukaiyama, T., *Chem. Lett.*, 1980, 17.
19. Commerçon, M., Mangeney, P., Tejero, T. and Alexakis, A., *Tetrahedron: Asymmetry.*, 1990, **1**, 287.
20. Alexakis, A., Sedrani, R., Normant, J.F. and Mangeney, P., *Tetrahedron: Asymmetry.*, 1990, **1**, 283.
21. Mukaiyama, T., Sakito, Y. and Asami, M., *Chem. Lett.*, 1978, 1253.
22. Mukaiyama, T., Sakito, Y. and Asami, M., *Chem. Lett.*, 1979, 705.
23. Sakito, Y. and Mukaiyama, T., *Chem. Lett.*, 1979, 1027.
24. Sakito, Y., Tanaka, S., Asami, M. and Mukaiyama, T., *Chem. Lett.*, 1980, 1223.
25. Sakito, Y., Asami, M. and Mukaiyama, T., *Chem. Lett.*, 1980, 455.
26. Asami, M. and Mukaiyama, T., *Chem. Lett.*, 1983, 93.
27. Lensen, N., Thèse de Doctorat, University P. and M. Curie, Paris, 1992.
28. Alexakis, A., Tranchier, J.-P., Lensen, N. and Mangeney, P., *J. Am. Chem. Soc.*, 1995, **117**, in press.
29. (a) Alexakis, A., Lensen, N. and Mengeney, P., *Tetrahedron Lett.*, 1991, **32**, 1171. (b) Alexakis, A., Lensen, N., Tranchier, J.-P., Mangeney, P., Feneau-Dupont, T. and Declecq, J.P., *Synthesis*, 1995, 1038.
30. Alexakis, A., Lensen, N., Tranchier, J.-P. and Mangeney, P., *J. Org. Chem.*, 1992, **57**, 4563.
31. Alexakis, A., Lensen, N. and Mangeney, P., *Synlett*, 1991, 625.
32. (a) Gosmini, R., Mangeney, P., Alexakis, A., Commerçon, M. and Normant, J.F., *Synlett*, 1991, 111. (b) Raussou, S., Gosmini, R., Mangeney, P. and Gommerçon, M., *Tetrahedron Lett.*, 1994, **35**, 5433.
33. Mangeney, P., Gosmini, R., Raussou, S., Commerçon, M. and Alexakis, A., *J. Org. Chem.*, 1994, **59,** 1877.
34. Mangeney, P., Gosmini, R. and Alexakis, A., *Tetrahedron Lett.*, 1990, **31**, 3981.
35. Mangeney, P., Gosmini, R., Raussou, S., Commerçon, M. and Alexakis, A., unpublished work.
36. Kanemasa, S. and Yamamoto, H., *Tetrahedron Lett.*, 1990, **31**, 3633.
37. Kanemasa, S., Yamamoto, H., Wada, E., Sakuraï, T. and Urushido, K., *Bull. Chem. Soc. Jpn.*, 1990, **63**, 2857.
38. Kanemasa, S., Hayashi, T., Tanaka, J., Yamamoto, H., Wada, E. and Sakuraï, T., *Bull. Chem. Soc. Jpn.*, 1991, **64**, 3274.
39. Gruseck, U. and Heuschmann, M., *Tetrahedron Lett.*, 1987, **28**, 2681.
40. Heuschmann, M., *Chem. Ber.*, 1988, **121**, 39.
41. Mangeney, P., Beruben, D. and Alexakis, A., unpublished work.
42. Alexakis, A. and Mangeney, P., *Tetrahedron: Asymmetry*, 1990, **1**, 477.

6 Asymmetric deprotonation reactions using enantiopure lithium amide bases

N.S. SIMPKINS

6.1 Introduction

Lithium diisopropylamide (LDA), one of a family of dialkyllithium amides, is firmly established as the most important strong base used in organic synthesis. Bearing in mind the significance of such bases in deprotonation reactions of weak carbon acids, especially carbonyl compounds, sulphones and sulphoxides, and the tremendous effort that has been applied to the development of asymmetric reactions, it is surprising that asymmetric deprotonation chemistry has been so little developed. Indeed, prior to the report in 1986 of the asymmetric deprotonation of *cis*-2,6-dimethylcyclohexanone using enantiopure lithium amide bases*, only a handful of reports concerning the chemistry of these reagents had appeared. Since that time, enantiopure lithium amides have been developed extensively as reagents for organic synthesis, both as powerful bases and as chiral nucleophiles. This brief review will focus on reactions of enantiopure lithium amides as bases in which the deprotonation reaction is the key asymmetric step and will deal primarily with reactions of ketones.[1]†

6.2 Asymmetric deprotonation reactions

6.2.1 Prochiral ketones

By far the best developed area of enantiopure lithium amide base chemistry involves reactions in which a symmetrically substituted, cyclic prochiral ketone is converted directly into a non-racemic product via kinetically controlled discrimination between a pair of enantiotopic hydrogens. This process, which has been dubbed 'asymmetric deprotonation', was first applied to conformationally anchored cyclohexanones, since in these cases the chiral base would be expected to select between the two axial α-hydrogens for stereoelectronic reasons.[2,3] The bases that have been employed in this type

* These bases are often referred to by the term HCLA bases, standing for 'homochiral lithium amide' base, but since 'homochiral' is used in this book in its original meaning (of the same absolute configuration), the abbreviation HCLA is not employed here.

† Applications of enantiopure lithium amides as nucleophiles[1b] and involving non-covalent auxiliary-type chemistry.[1c]

of chemistry are usually prepared *in situ* by treatment of the appropriate enantiopure secondary amine with butyllithium; important examples include (**1**–**5**) (and their enantiomers).

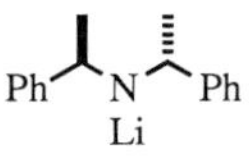

Of these bases, the bisphenylethylamide (**1**) offers the advantages of C_2 symmetry, resulting in very good levels of asymmetric induction in many cases, and is very easily prepared in either enantiomeric form. Even simpler phenylethylamine-derived bases, of general structure (**2**), (e.g. R = i-Pr, cyclohexyl), are also effective in some instances but do not usually give such good levels of asymmetric induction as base (**1**). Some very sterically hindered bases (**3**) are available via reductive amination of camphor and have proved uniquely useful in the sulphoxide chemistry described in Section 2.5. Koga's group have prepared a wide range of bases of general formula (**4**), starting from phenyl glycine (the circle on nitrogen indicating a ring—usually a piperazine). These bases give high levels of asymmetric induction in many reactions involving cyclohexanones but have been demonstrated to be inferior to base (**1**) in several instances. Proline derivatives (**5**) have, in common with the phenylglycine-derived lithium amides, the possibility for internal ligation of the lithium atom because of the presence of one or more additional heteroatoms. The amino acid derived compounds (**4**) and (**5**) are rather less accessible than the simple phenylethylamine derivatives (**1**) and (**2**), their synthesis involving longer and more difficult sequences with an increased risk of partial racemisation.

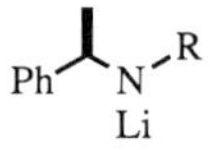

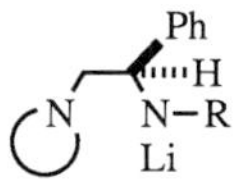

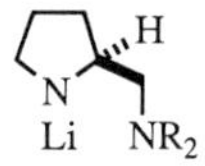

Some representative examples of the asymmetric deprotonation reactions of prochiral cyclohexanones are illustrated in Schemes 6.1–6.3.[2-4]

Early experiments with *cis*-2,6-dimethylcyclohexanone and 4-t-butylcyclohexanone revealed some important features of the reaction. The asymmetric induction is clearly the result of kinetically controlled deprotonation; effects owing to the coordination of the chiral amine with the intermediate lithium enolate are not important in the synthesis of enol derivatives,* and enolate

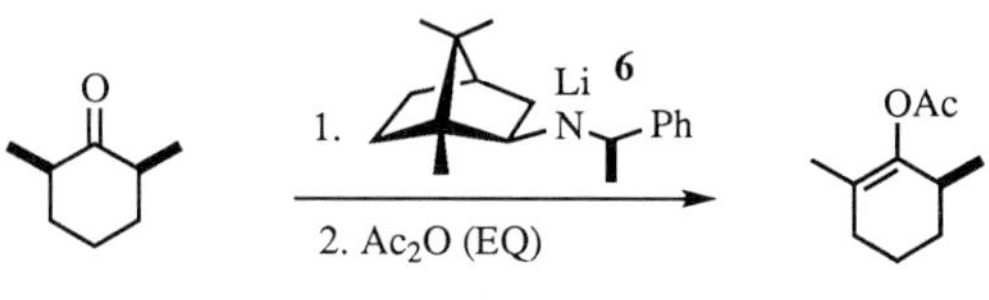

1. **6** ; 2. Ac_2O (EQ)

Scheme 6.1

* The diastereocontrol in subsequent *C*-alkylations can be influenced by the presence of a chiral amine.[5]

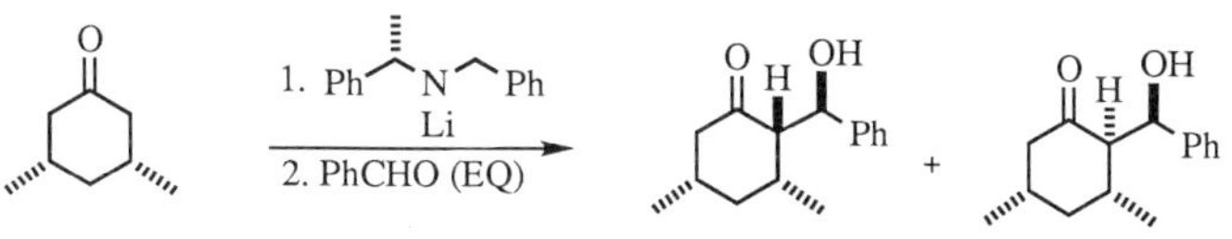

1. Ph N Ph Li
2. PhCHO (EQ)

7 74% ee

8 57% ee

Scheme 6.2

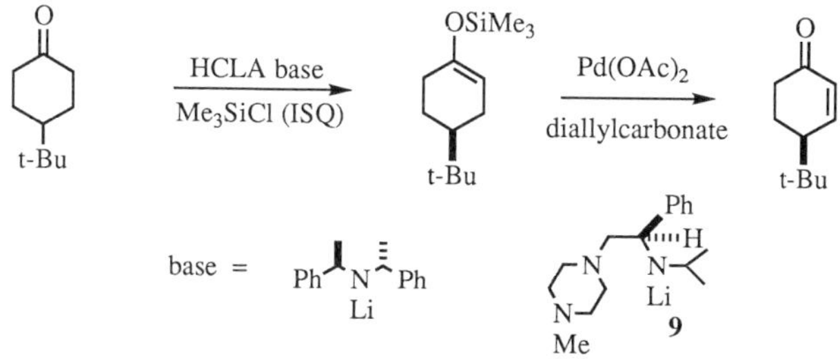

HCLA base
Me_3SiCl (ISQ)

$OSiMe_3$

$Pd(OAc)_2$
diallylcarbonate

t-Bu

base = Ph N Ph Li

up to 88% ee

see ref. 2,3 and 6

Scheme 6.3

equilibration can be ruled out. In deprotonation reactions of *cis*-2,6-dimethylcyclohexanone with the camphor derived base (**6**), enol silane, enol acetate and *C*-allylated derivatives could each be formed in 65% ee under external quench (EQ) conditions (electrophile added after enolisation), e.g. Scheme 6.1, but *in situ* quenching with Me_3SiCl (Me_3SiCl added to lithium amide prior to addition of ketone: ISQ) gave the enol silane derivative in significantly higher ee (83%), although in lower chemical yield in this case. These general trends have been verified by the subsequent work of Majewski and Gleave, using *cis*-3,5-dimethylcyclohexanone (Scheme 6.2).[4] Somewhat surprisingly, the enantiomeric excesses of the two aldol products (**7**) and (**8**), obtained by reaction with benzaldehyde, were significantly different in many cases. In this case, rapid addition of benzaldehyde to the reaction mixture resulted in the best levels of asymmetric induction, presumably as a result of some kind of ISQ effect.

The Me_3SiCl–ISQ has been used in much of the authors' work, for example the asymmetric enolisation of 4-t-butylcyclohexanone (Scheme 6.3).[2] Lithium amide (**1**) gave enol silane up to 88% ee, the optical rotation data for this compound being in conflict with the earlier results of Koga employing base (**9**).[3] A subsequent revision of the rotation data for the enol silanes derived from several 4-substituted cyclohexanones has appeared, which requires re-evaluation of the ee values in Koga's original report.[6] More recently, a modified family of phenylglycine-derived bases, such as (**10**) have been reported to give good levels of asymmetric induction in the same types of enolisation reactions.[7]

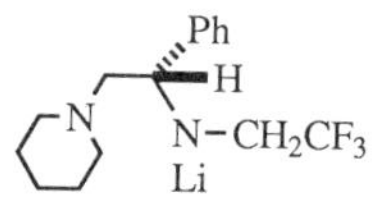

10

The likely aggregation of the lithium amide bases employed in this work is worthy of comment. In most of these studies lithium amides without additional coordination sites have been employed, whereas the phenylglycine-derived systems have either one or two additional nitrogens which could be involved in internal chelation. Simple lithium amides would be expected to be largely dimeric in solution, i.e. of general structure (**11**), and the structure of the important chiral base (**1**) in solution is likely to reflect the dimeric solid-state structure recently determined, Figure 6.1.[8]

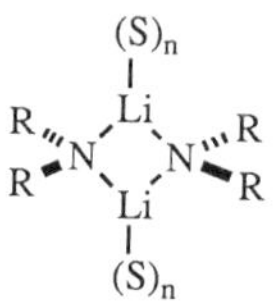

11 S = solvent e.g. THF
n = 1 or 2

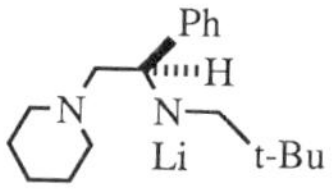

12

It is worth noting that the conformation adopted by each lithium amide unit in the dimer is very similar, and that the overall conformation is different to that predicted by some molecular modelling studies.[4]

Koga *et al.* have shown that the aggregation state of the piperidyl base (**12**) in solution is dependent on the solvent. In THF or DME the base is monomeric, because of the presence of the additional coordination site, whereas in toluene or ether the base is dimeric.[9] Somewhat surprisingly, considering the known behaviour of LDA, the addition of HMPA to the toluene solution of dimeric lithium amide was shown to result in conversion into the internally chelated monomer. The solvent conditions that generate

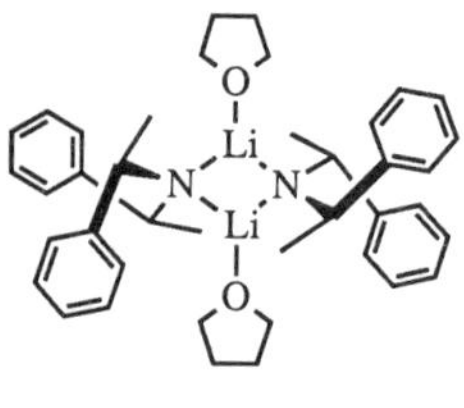

Figure 6.1 Representation of the solid-state structure of dimeric **1**.

monomeric chiral base (as observed by NMR) were also shown to give the best levels of asymmetric induction in deprotonation of 4-t-butylcyclohexanone. Since these studies demonstrate that monomeric lithium amides are generated in THF with bases having only one internal chelating nitrogen, the requirement for additional chelation sites (as in **9**) or for the addition of HMPA, for optimal asymmetric induction, is difficult to rationalise.

The asymmetric deprotonation protocol, using symmetrical cyclohexanone substrates has been applied in a number of synthetic studies (Schemes 6.4–6.6).[10–12]

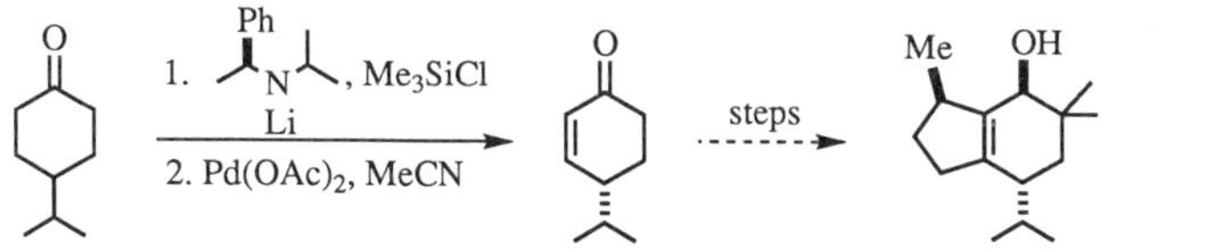

1. Ph, N, Li, Me_3SiCl
2. $Pd(OAc)_2$, MeCN
steps

13 (*R*)-(-)-cryptone
65% ee

14 (+)-brasilenol

Scheme 6.4

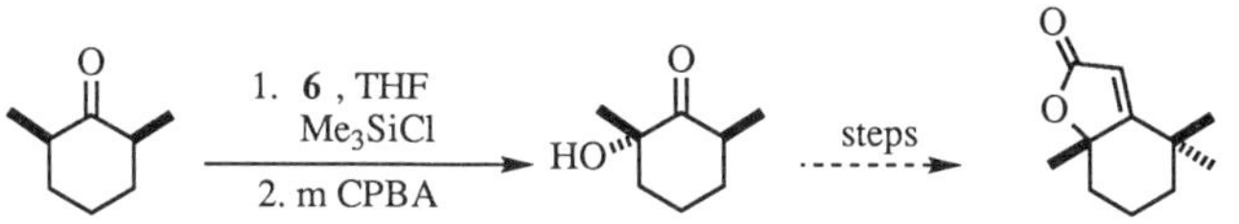

1. **6**, THF
Me_3SiCl
2. m CPBA
steps

15 65% ee

16

Scheme 6.5

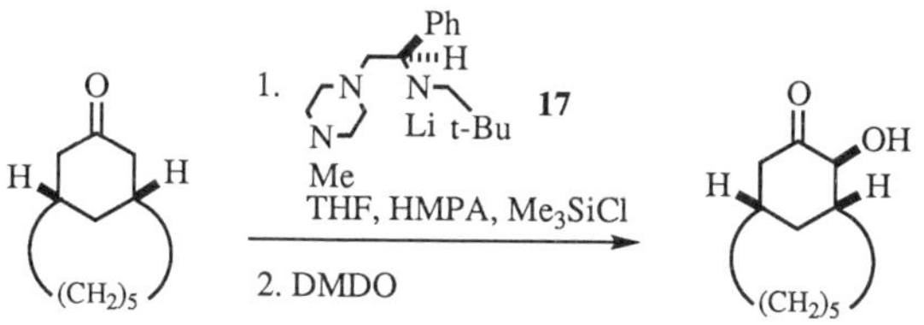

1. **17**
THF, HMPA, Me_3SiCl
2. DMDO

Scheme 6.6

Both (*R*)-(−)-cryptone (**13**) and the α-hydroxy ketone (**15**) were prepared in about 65% ee by the asymmetric deprotonation approach and converted into the natural products (+)-brasilenol (**14**), and (5*S*)-dihydroactinidiolide (**16**), respectively. In each case, recrystallisation of a synthetic intermediate allowed isolation of the final product in optically pure form.

Similar use of the symmetry-breaking approach has also been applied in the case of cyclobutanones and the ring-fused cyclopentanone (**20**) (Schemes 6.7 and 6.8).[13,14]

Several enantiopure lithium amide bases were examined for optimal

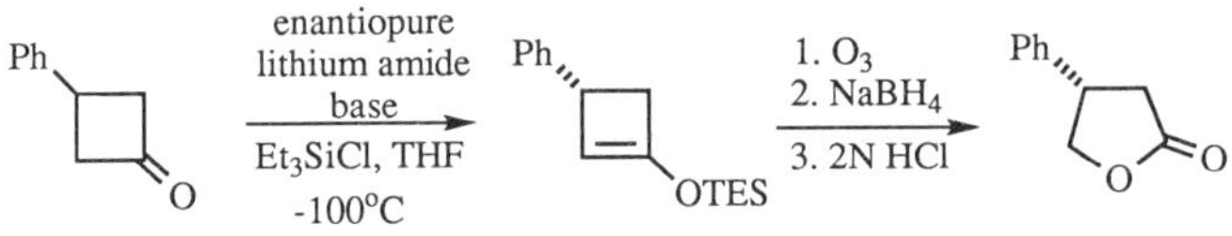

Scheme 6.7

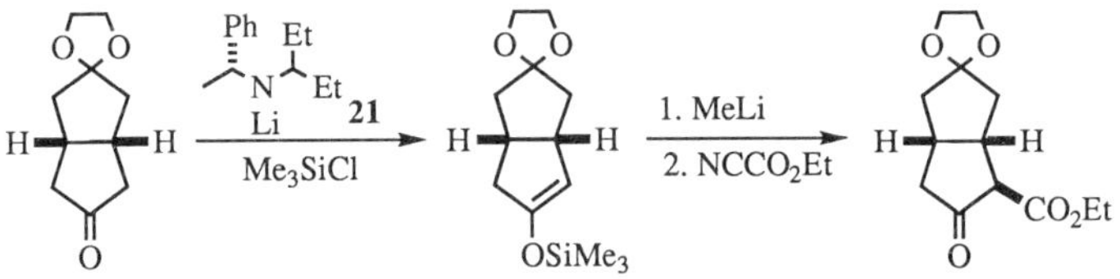

Scheme 6.8

asymmetric induction in the transformation of 3-phenylcyclobutanone into the chiral enol silane derivative (**18**), Et_3SiCl being employed as the ISQ because of the instability of the trimethylsilyl analogues. Best results with this ketone (92% ee) were achieved by using base (**1**), lithium amide (**17**) giving only 47% ee under comparable conditions, although with a different substrate, 3-methyl-3-phenylcyclobutanone, this situation was reversed, with the latter base proving more effective. The resulting enol silanes, such as (**18**) were straightforwardly converted into synthetically useful lactones, e.g. (**19**). In the conversion of ketone (**20**) into enol silane (**22**), a simple base (**21**) again proved highly efficient, allowing preparation of the carbacyclin intermediate (**23**) via enolate regeneration and carboxyethylation with Mander's reagent.

Schemes 6.9 and 6.10 show the chiral base chemistry of certain bridged bicyclic ketones, for example the oxabicyclo[3.2.1]octan-3-ones (**24**) and (**27**).[15]

Using the enantiopure base (**1**) with the usual Me_3SiCl-ISQ, the enol silane derivatives (**25**) and (**28**) were prepared in 88 and 85% ee, respectively (Schemes 6.9 and 6.10). These intermediates are useful for the asymmetric synthesis of certain cycloheptenones such as (**26**), and *cis*-substituted tetrahydrofurans via carbocyclic ring cleavage, including (**29**)—a known showdomycin precursor.

Analogous deprotonations of the corresponding azabicyclic ketones, e.g. tropinone (**30**), have also been carried out by several research groups (Schemes 6.11 and 6.12).[16–18]

The deprotonation reaction was clearly demonstrated to involve only the *exo*-orientated hydrogens, subsequent aldol reaction with benzaldehyde

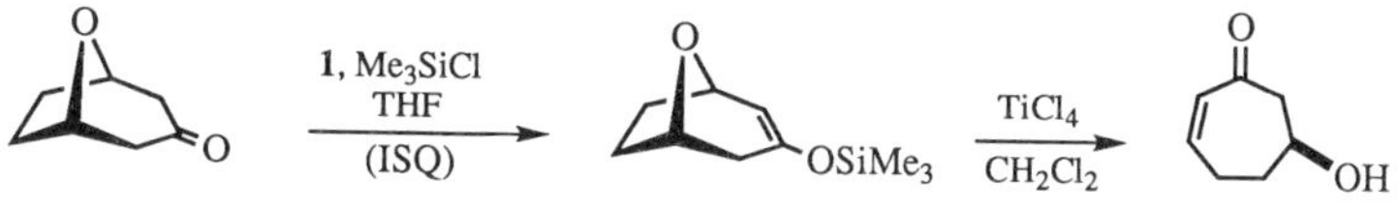

24 → 1, Me_3SiCl, THF (ISQ) → 25 79%, 88%ee → $TiCl_4$, CH_2Cl_2 → 26

Scheme 6.9

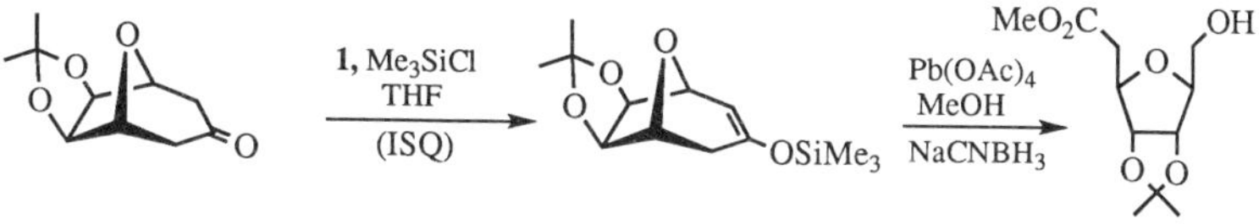

27 → 1, Me_3SiCl, THF (ISQ) → 28 88%, 85%ee → $Pb(OAc)_4$, MeOH, $NaCNBH_3$ → 29

Scheme 6.10

resulting in the formation of a single *exo-anti*-product (**31**) (Scheme 6.11).[17] Such asymmetric transformations of tropinone and its derivatives, using enantiopure lithium amide bases, have allowed the enantioselective synthesis of the alkaloid (+)-monomorine 1 (**32**) (Scheme 6.12).[18]

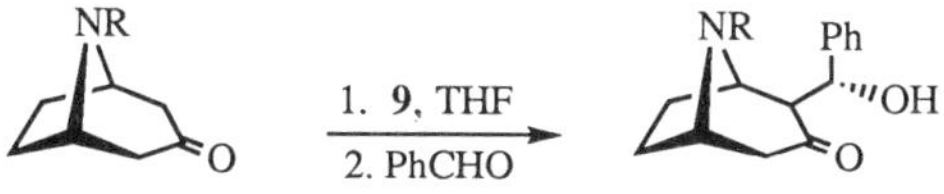

30 → 1. 9, THF; 2. PhCHO → 31 64%, 60% ee

Scheme 6.11

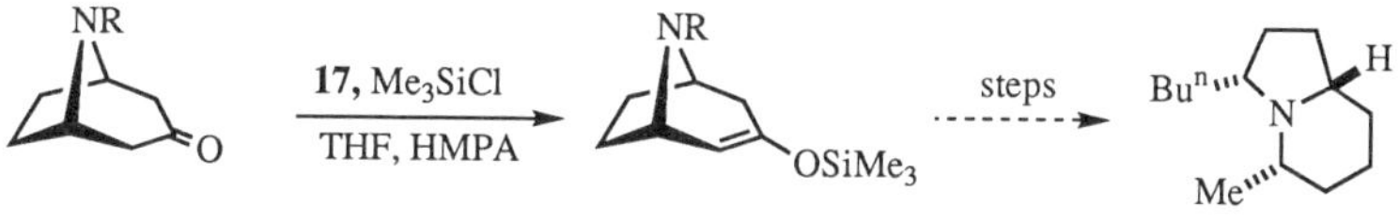

17, Me_3SiCl, THF, HMPA → 89%, 90% ee → steps → 32

Scheme 6.12

6.2.2 *Salt effects and the* in situ *quench protocol*

Studies with the oxabicyclic ketones (**24**) and (**27**) highlighted the importance of the Me_3SiCl-ISQ protocol for achieving the best levels of asymmetric induction, and lead us to question the origins of the ISQ effect. A range of ketones was included in a detailed study of the contrasting levels of ee obtained in deprotonations under EQ- and ISQ-type reaction conditions.[19] For example, in conversion of 4-t-butylcyclohexanone into the corresponding chiral enol silane (Scheme 6.3) using base (**1**), the ISQ procedure gave product of 69% ee (at −78°C) whereas the EQ method gave only 23% ee. Similar results were found for conversions of (**24**) (Scheme 6.9: ISQ 82% ee, EQ 33% ee) and (**27**) (Scheme 6.10: ISQ 70% ee, EQ 27% ee) into their enol silanes by base (**1**). An enhanced level of asymmetric induction in such reactions was also observed using other enantiopure bases such as (**6**), making it clear that this is a general effect.

Since enolate equilibration should not be a factor in such low temperature reactions, the different results must be caused by the different composition of the reaction mixture under the two sets of reaction conditions. Under EQ conditions, the enolisation is allowed to proceed to completion before the addition of Me_3SiCl, thus allowing the lithium enolate to accumulate as the reaction progresses. In the ISQ reactions, the enolate is presumably quenched immediately so that enolate does not accumulate, *but LiCl is liberated as the enolisation proceeds.* Since both lithium enolates and lithium halides are known to form mixed aggregates with lithium amides,* the formation of such species in the enolisation mixture could lead to modified enantioselectivity.

In a study on the effect of LiCl on the enantioselectivity of a range of ketone enolisations, a dramatic improvement in the ee of products was obtained from EQ reactions if LiCl was added. Addition of 0.5 equiv. LiCl to the EQ reactions involving ketones (**24**) and (**27**) (Schemes 6.9 and 6.10) resulted in substantial increases in the ee of products obtained (for (**24**): EQ 33% ee, EQ + LiCl 83% ee; for (**27**): EQ 27% ee, EQ + LiCl 58% ee). The same trend was also observed with a range of other ketones and enantiopure lithium amide bases. This finding is particularly significant in that the addition of LiCl allows very good levels of asymmetric induction to be achieved in chiral base reactions under EQ-type conditions, thus permitting direct use of a wide range of electrophiles (which would not be compatible with the ISQ protocol). This point is highlighted by the finding that the aldol reaction shown in Scheme 6.11 can be carried out using base (**1**) in the presence of LiCl to give the product (**31**) in 80% ee (compared with only 24% ee without added salt), a significant improvement on previous results.

It is probable that mixed aggregates, derived from lithium amide and LiCl, are implicated in these 'ISQ + LiCl' reactions, the modified reagent being more stereoselective than the dimeric lithium amide reagent presumably involved in EQ-type reactions. The improved level of asymmetric induction seen in the ISQ reactions can also be reasonably attributed to the LiCl generated in these reactions.

In subsequent studies it was shown that $ZnCl_2$ can also improve the enantioselectivity in EQ-type reactions, enabling conversion of tropinone (**30**) into the aldol derivative (**31**) in up to 85% ee (Scheme 6.11).

* The diastereocontrol in subsequent *C*-alkylations can be influenced by the presence of a chiral amine.[5]

6.2.3 Rationalisation of stereochemical results

The sense of asymmetric induction with some of the bases described above is quite predictable, for example with bisphenylethylamide (**1**), the selectivity seen with a range of prochiral and chiral (see below) ketones involves preferential removal of the highlighted hydrogen shown in Figure 6.2. The hydrogen preferred in each case is clearly in a similar orientation relative to the carbonyl group, this trend being illustrated in generalised form by (**33**) for a ketone in which the base selects between a pair of hydrogens on the β-face for steric or stereoelectronic reasons. That the same sense of deprotonation is observed (and at reasonable levels of induction in every case) with such a diverse range of substrates is quite remarkable.

The *sense* of selectivity can be rationalised by considering a least hindered approach of a conformationally constrained ketone to the monomeric lithium amide (**1**) in the conformation found in the dimeric crystal structure (but ignoring THF), as shown in Figure 6.3.

The ketone approaches the base through one of the empty quadrants not occupied by a phenyl substituent; coordination of lithium to the carbonyl oxygen then precedes deprotonation by the basic nitrogen. Twisting of the plane of the ketone, needed to facilitate the alternative sense of deprotonation, appears to involve a less favourable transition state. This simplistic picture of the deprotonation process ignores the possible involvement of lithium amide dimers or mixed aggregates with LiCl, although it might accommodate

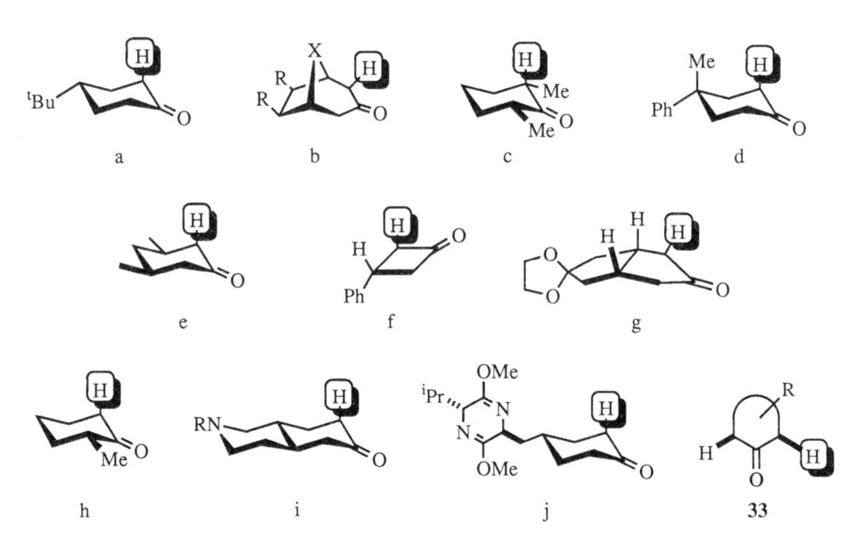

Figure 6.2 Pattern of asymmetric deprotonations using the enantiopure lithium amide base **1**. a, Scheme 6.3; b, Schemes 6.9–6.11; c, Scheme 6.1; d, ref. 20; e, Scheme 6.2; f, Scheme 6.7; g, ref. 21; h, ref. 22; i, Scheme 6.14; j, ref. 23.

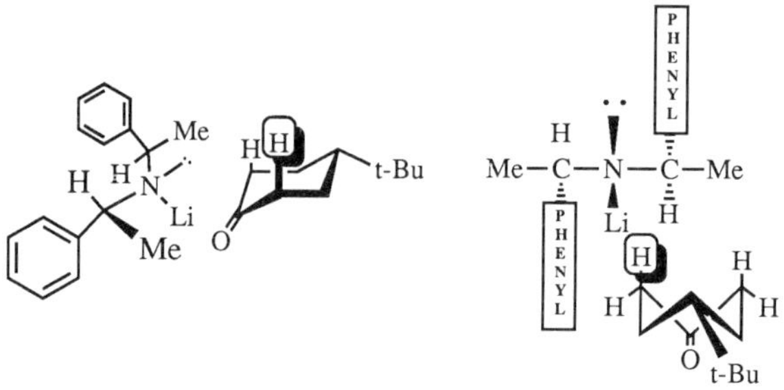

Figure 6.3 Rationalisation of base selectivity.

a group attached to lithium (not shown), which could be either LiCl or another lithium amide unit.*

The enantiopure lithium amide base-mediated deprotonations described above illustrate how useful this approach can be for the asymmetric synthesis of a varied range of products. The levels of induction seen are usually of the order of 90% ee, and, with methods for enantiomeric enrichment available, this approach can give enantiomerically pure products in good yield.

6.2.4 *Unsymmetrical ketone substrates*

In addition to the reactions outlined above, which involve the enantiopure base in reactions with *prochiral* substrates, these bases can also be effective in enantiomer recognition, i.e. kinetic resolution reactions, and in controlling the regiochemistry in enolisations of enantiopure substrates. The latter type of reaction was first described by Koga *et al.* using enantiopure 3-keto steroids such as (**34**) (Scheme 6.13).[25]

Under usual conditions of kinetic or thermodynamic control, ketone (**34**) provides the Δ^2 enol silane derivative (**35**) (deprotonation distal with respect to the ring junction) as the major product, although the regioselectivity is not particularly high. However, by using enantiopure bases of type (**4**), either

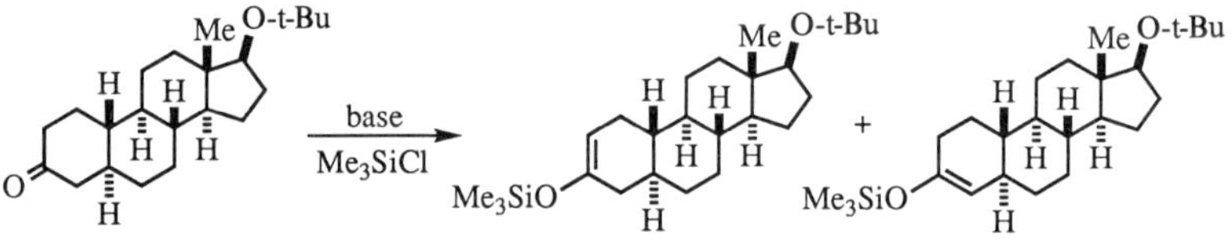

Scheme 6.13

* Ketone deprotonation by a lithium amide dimer has been proposed.[24]

Table 6.1 Control of the regiostereochemistry of the enolisation of enantiopure 3-keto steroids (**34**) (Scheme 6.13)

Conditions	Yield (%)	(**35**):(**36**)
Thermodynamic ($(HNSiMe_3)_2$, Me_3SiI)	63	80:20
Kinetic (LDA)	99	64:36
Kinetic ((*R*)-**17**)	97	95:5
Kinetic ((*S*)-**17**)	98	24:76

enhancement or reversal of the normal selectivity (matched and mismatched situations) can be achieved, thus allowing highly regioselective formation of either proximal (**36**) or distal (**35**) enol derivatives (Table 6.1). Even better control was seen in some other examples, but here the sense of deprotonation was difficult to reconcile with that seen in reactions of simple prochiral cyclohexanones—until labelling studies showed that ketones disubstituted at the 4-position (relative to C=O at C-1) undergo deprotonation via a boat or skew-boat conformation.[26]

Similar studies of the regiocontrol in deprotonations of each of the two enantiomers of perhydroisoquinolone (**37**) gave very similar matched and mismatched results to those shown in Scheme 6.13 by using the base (**1**) (Scheme 6.14).[27] Examination of the reaction of (*rac*-**37**) showed that each of the two enantiomers in the initial mixture is converted predominantly into one of the two regioisomeric enol silanes, i.e. ((*R*,*R*)-**37**) gives mainly distal enol silane (**38**), while ((*S*,*S*)-**37**) gives mainly the proximal isomer (**39**) (Scheme

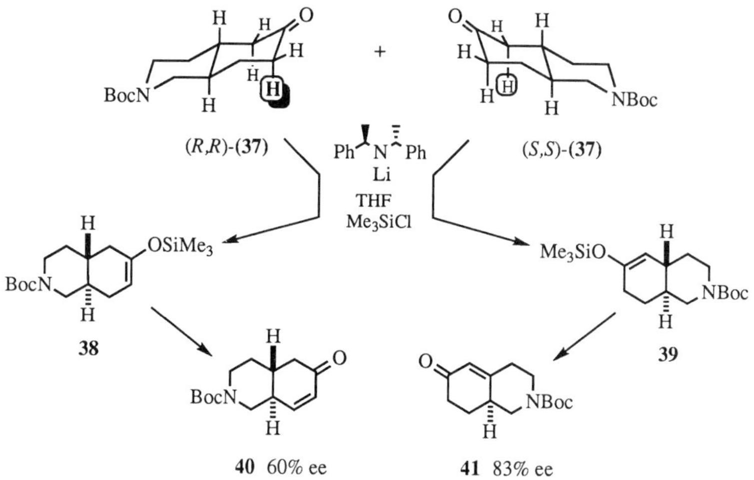

(*R*,*R*)-(**37**) + (*S*,*S*)-(**37**)

Ph N Ph, Li, THF, Me_3SiCl

38 **39**

40 60% ee **41** 83% ee

Scheme 6.14

6.14). The ee and absolute configurations of these compounds were determined following conversion into the corresponding enones (**40**) and (**41**). This unprecedented type of reaction, in which enantiomers are converted into regioisomers, is called regiodivergent resolution.* The reaction differs markedly from the classical kinetic resolution, which relies on *partial* conversion of a racemate, in that all of the starting material is converted into products.

Traditional kinetic resolution reactions, involving deprotonation by enantiopure lithium amide bases have also been applied to cyclic ketones. In one such study, racemic 2-substituted ketones were reacted with bases such as (**17**) and Me_3SiCl, allowing isolation of both enol silane product and unreacted ketone in good ee (Scheme 6.15).[29]

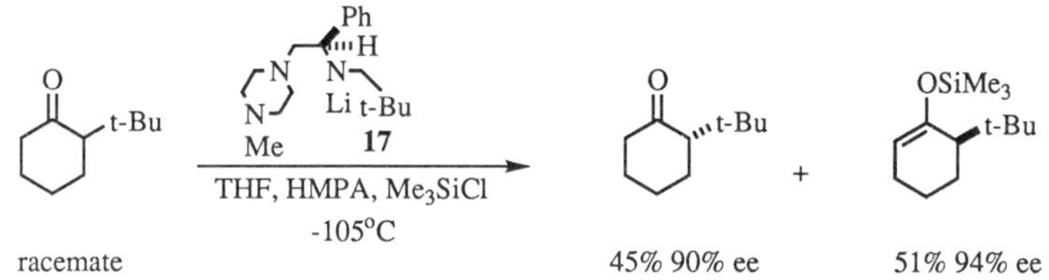

Scheme 6.15

6.2.5 *Non-ketone substrates*

One of our main areas of interest has involved expanding the scope of asymmetric deprotonation beyond applications involving cyclic ketones. A selection of results from our own investigations is shown in Schemes 6.16–6.19.[30–33]

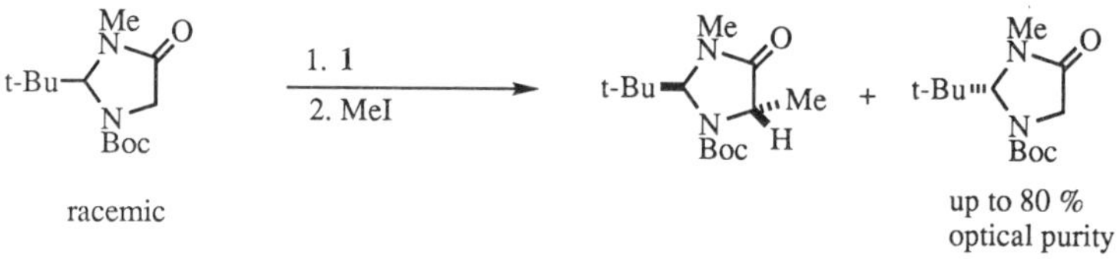

Scheme 6.16

The first two sequences illustrate the possibility of kinetic resolution of lactams and related systems, although the relative rates of deprotonation of the two enantiomers were fairly modest (k_{rel} *c.* 5–10).[30,31]

Schemes 6.18 and 6.19 show that the idea of breaking a plane of symmetry by enantiotopic hydrogen discrimination can be extended to cyclic sulphoxides[32] and monosubstituted tricarbonyl(η^6-arene)chromium complexes.[33] These

* Certain enzymatic reactions also effect this type of regiodivergent resolution.[28]

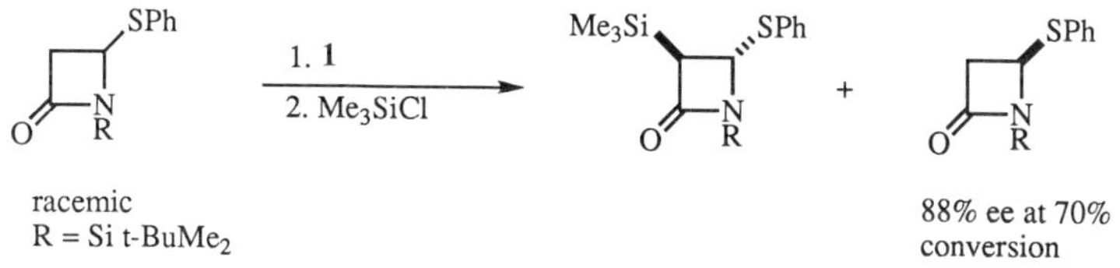
SPh
1. 1
2. Me3SiCl
Me3Si
SPh
+
SPh
racemic
R = Si t-BuMe2
88% ee at 70% conversion

Scheme 6.17

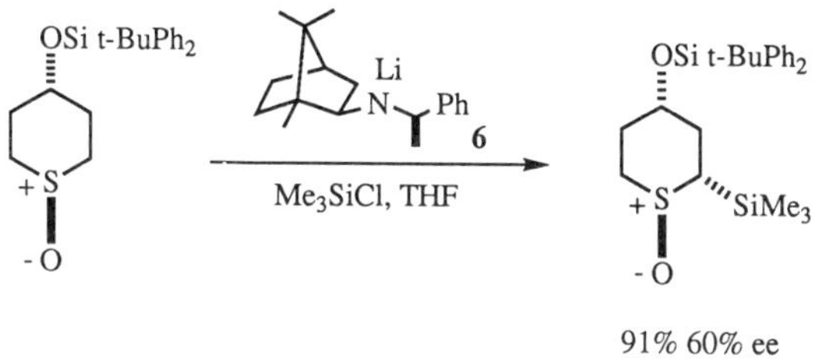
OSi t-BuPh2
Li
N
Ph
6
Me3SiCl, THF
OSi t-BuPh2
SiMe3
91% 60% ee

Scheme 6.18

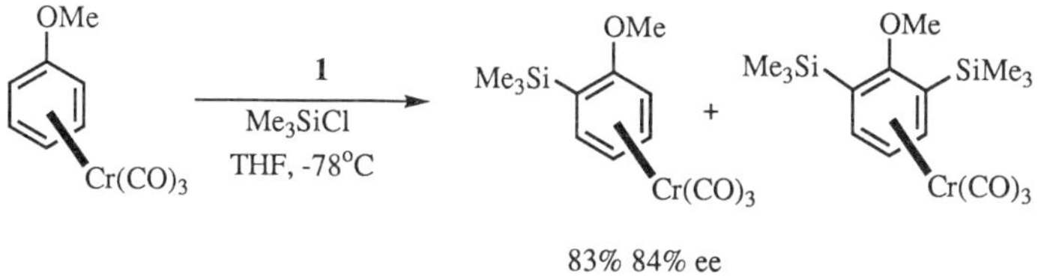
OMe
Cr(CO)3
1
Me3SiCl
THF, -78°C
Me3Si
OMe
+
Cr(CO)3
Me3Si
OMe
SiMe3
Cr(CO)3
83% 84% ee

Scheme 6.19

reactions have not been studied in detail, and in general the chemical yields and enantioselectivities obtained tend to be somewhat lower than with prochiral ketones. In the case of cyclic sulphoxides, the reactions are complicated by the possibility of multiple deprotonation and by the tendency of the initially formed products to undergo Pummerer-type reactions under the Me_3SiCl-ISQ reaction conditions. Multiple silylation is also observed in the reactions of some of the tricarbonyl(η^6-arene)chromium complexes, although here a more serious problem is the rapid racemisation of the deprotonated complex in the absence of an *in situ* quench.

A number of other types of asymmetric transformation using enantiopure lithium amide have been described, which would be regarded as involving 'enantioselective deprotonation'. These areas have not progressed significantly since the previous review[1a] and will not be described here. They include the rearrangement of epoxides into allylic alcohols,[1a,34] asymmetric [2,3]-Wittig rearrangement[35] and dehydrohalogenation of certain β-halogenated carboxylic acids.[36]

6.3 Conclusions

The use of enantiopure lithium amide bases for asymmetric deprotonation of cyclic ketones has now been firmly established as a useful synthetic method. Studies in other areas are less well advanced, but the examples illustrated above highlight the potential generality and power of this approach in organic synthesis. Further developments in understanding these complex reactions, and in chiral base design, should lead to further applications in this area.

References

1. (a) Review: Cox, P.J. and Simpkins, N.S., *Tetrahedron: Asymmetry*, 1991, **2**, 1; (b) Bunnage, M.E., Burke, A.J., Davies, S.G. and Goodwin, C.J., *Tetrahedron: Asymmetry*, 1994, **5**, 203; (c) Juaristi, E., Beck, A.K., Hansen, J., Matt, T., Mukhopadhyay, T., Simson, M. and Seebach, D., *Synthesis*, 1993, 1271.
2. Simpkins, N.S., *J. Chem. Soc., Chem. Commun.*, 1986, 88; Cousins, R.P.C. and Simpkins, N.S., *Tetrahedron Lett.*, 1989, **30**, 7241; Cain, C.M., Cousins, R.P.C., Coumbarides, G. and Simpkins, N.S., *Tetrahedron*, 1990, **46**, 523.
3. Shirai, R., Tanaka, M. and Koga, K., *J. Am. Chem. Soc.*, 1986, **108**, 543.
4. Majewski, M. and Gleave, D.M., *J. Org. Chem.*, 1992, **57**, 3599.
5. Hasegawa, Y., Kawasaki, H. and Koga, K., *Tetrahedron Lett.*, 1993, **34**, 1963.
6. Aoki, K., Nakajima, M., Tomioka, K. and Koga, K., *Chem. Pharm. Bull.*, 1993, **41**, 994.
7. Aoki, K., Noguchi, H., Tomioka, K. and Koga, K., *Tetrahedron Lett.*, 1993, **34**, 5105.
8. Edwards, A.S., Hockey, S., Mair, F.S., Raithby, P.R. and Snaith, R., *J. Org. Chem.*, 1993, **58**, 6942.
9. Sato, D., Kawasaki, H., Shimada, I., Arata, Y., Okamura, K., Date, T. and Koga, K., *J. Am. Chem. Soc.*, 1992, **114**, 761.
10. Greene, A.E., Serra, A.A., Barreiro, E.J. and Costa, P.R.R., *J. Org. Chem.*, 1987, **52**, 1170.
11. Cain, C.M. and Simpkins, N.S., *Tetrahedron Lett.*, 1987, **28**, 3723.
12. Underiner, T.L. and Paquette, L.A., *J. Org. Chem.*, 1992, **57**, 5438.
13. Honda, T., Kimura, N. and Tsubuki, M., *Tetrahedron: Asymmetry*, 1993, **4**, 1475; Honda, T. and Kimura, N., *J. Chem. Soc., Chem. Commun.*, 1994, 77.
14. Izawa, H., Shirai, R., Kawasaki, H., Kim, H. and Koga, K., *Tetrahedron Lett.*, 1989, **30**, 7221.
15. Bunn, B.J., Cox, P.J. and Simpkins, N.S., *Tetrahedron*, 1993, **49**, 207.
16. Simpkins, N.S., *Chem. Soc. Rev.*, 1990, **19**, 335.
17. Majewski, M. and Zheng, G-Z., *Can. J. Chem.*, 1992, **70**, 2618.
18. Momose, T., Toyooka, N., Seki, S. and Hirai, Y., *Chem. Pharm. Bull.*, 1990, **38**, 2072; Momose, T., Toyooka, N. and Hirai, Y., *Chem. Lett.*, 1990, 1319.
19. Bunn, B.J. and Simpkins, N.S., *J. Org. Chem.*, 1993, **58**, 533; Bunn, B.J., Simpkins, N.S., Spavold, Z. and Crimmin, M.J., *J. Chem. Soc., Perkin Trans.* 1, 1993, 3113.
20. Honda, T., Kimura, N. and Tsubuki, M., *Tetrahedron: Asymmetry*, 1993, **4**, 21.
21. Leonerd, J., Hewitt, J.D., Ouali, D., Rahman, S.K., Simpson, S.J. and Newton, R.F., *Tetrahedron: Asymmetry*, 1990, **1**, 699.
22. Cousins, R.P.C. and Simpkins, N.S., unpublished results.
23. Wild, H. and Born, L., *Angew. Chem., Int. Ed. Engl.*, 1991, **30**, 1685.
24. Willard, P.G. and Liu, Q-Y., *J. Am. Chem. Soc.*, 1993, **115**, 3380.
25. Sobukawa, M., Nakajima, M. and Koga, K., *Tetrahedron: Asymmetry*, 1990, **1**, 295.
26. Sobukawa, M. and Koga, K., *Tetrahedron Lett.*, 1993, **34**, 5101.
27. Bambridge, K., Simpkins, N.S. and Clark, B.P., *Tetrahedron Lett.*, 1992, **33**, 8141.
28. Grogan, G., Roberts, S.M. and Willetts, A.J., *J. Chem. Soc., Chem. Commun.*, 1993, 699.
29. Kim, H., Kawasaki, H., Nakayima, M. and Koga, K., *Tetrahedron Lett.*, 1989, **30**, 6537.
30. Coggins, P. and Simpkins, N.S., *Synlett*, 1991, 515.
31. Coggins, P. and Simpkins, N.S., *Synlett*, 1992, 313.
32. (a) Armer, R., Begley, M.J., Cox, P.J., Persad, A. and Simpkins, N.S., *J. Chem. Soc., Perkin*

Trans. 1, 1993, 3099; (b) Maercker, A., Schuhmacher, R., Buchmeier, W. and Lutz, H.D., *Chem. Ber.*, 1991, **124**, 2489.
33. Price, D.A., Simpkins, N.S., MacLeod, A.M. and Watt, A.P., *J. Org. Chem.*, 1994, in press.
34. (a) Milne, D. and Murphy, P.J., *J. Chem. Soc., Chem. Commun.*, 1993, 884; (b) Bhuniya, D. and Singh, V.K., *Synth. Commun.*, 1994, **24**, 375.
35. Marshall, J.A. and Lebreton, J., *J. Am. Chem. Soc.*, 1988, **110**, 2925.
36. Duhamel, L., Ravard, A., Plaquevent, J.C., Plé, G. and Davoust, D., *Bull. Soc. Chim. Fr.*, 1990, **127**, 787; Duhamel, L., Ravard, A. and Plaquevent, J-C., *Tetrahedron: Asymmetry*, 1990, **1**, 347.

7 Asymmetric Diels–Alder reactions

A. WHITING

7.1 Introduction

Since the first examples were carried out,[1] the Diels–Alder reaction has become one of the corner stones of organic synthesis, in large part because of the ability to construct up to four new asymmetric centres in one reaction. Early attempts[2] at accomplishing asymmetric Diels–Alder reactions were poor compared with contemporary examples, but the seeds of success had been sown, and in the intervening years the number of examples of Diels–Alder reactions occurring with good to high levels of asymmetric induction has increased almost exponentially. The advances made in asymmetric Diels–Alder reactions have been the subject of many general reviews,[3-8] with catalysis of asymmetric Diels–Alder reactions receiving particular attention.[9-13] The aim of this chapter is to survey some of the more important recent advances in the area of asymmetric Diels–Alder reactions, with particular emphasis on applications and future developments.

7.2 Diels–Alder reactions of C=C

Diels–Alder reactions involving C=C are readily transformed into asymmetric processes generally by one of three methods: (i) attaching a chiral auxiliary to the dienophile; (ii) attaching a chiral auxiliary to the diene; (iii) using a chiral catalyst, usually a Lewis acid. Over the last few years, attachment of a chiral auxiliary remains[4] the most common method for effecting asymmetric Diels–Alder reactions. However, the application of chiral, catalytic Lewis acids has enjoyed a surge in popularity with several excellent catalysts now readily available.

Most available chiral pool compounds have been utilised in one way or another for attaching either to a dienophile, diene or Lewis acid. This chapter will discuss recent methods for achieving good levels of asymmetric induction and gathers structurally related sources of chirality altogether.

7.2.1 Chiral dienophiles

The most common point of attachment of a chiral auxiliary to a dienophile is via an ester or amide linkage, i.e. R^4 on (**1**) (Scheme 7.1).

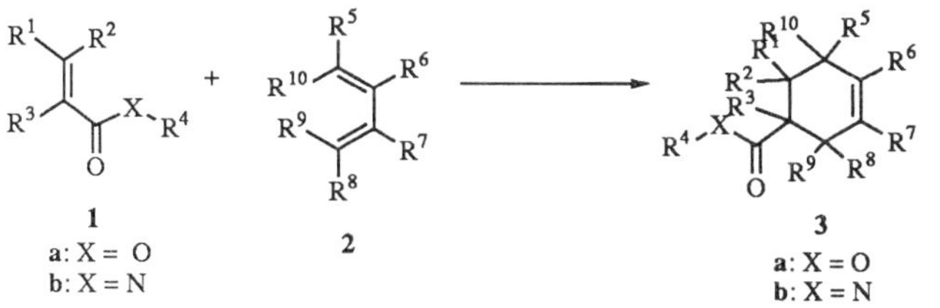

Scheme 7.1

Perhaps the most used type of chiral auxiliary is that derived from menthol or from related terpene-derived chiral pool materials. Direct attachment of menthol to an acrylate, i.e. (**4a**) affords modest levels of asymmetric induction,[4] which can be improved on γ-alumina.[14] However, better selectivities are found by increased functionalisation of the acrylate. For example, the *N*-acetoxy menthyl ester (**4b**) has been used in asymmetric Diels–Alder reactions with cyclopentadiene,[15] employing $TiCl_4$ as catalyst, to give high yields of a 69:31 mixture of *endo*- and *exo*-adducts (**5**) and (**6**) and (**7**) and (**8**), respectively (Scheme 7.2). Values of de were higher (95:5) for the *exo*-adducts (**7**) and (**8**) than for the *endo*-adducts (**5**) and (**6**) (77:23).

The asymmetric synthesis of nucleoside derivatives (**11**) has been achieved

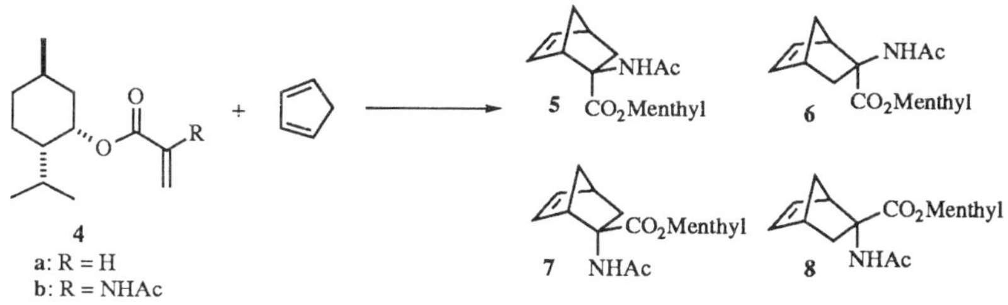

Scheme 7.2

via Diels–Alder adduct (**10**), which was prepared using the reaction of di-menthyl ester (**8**) with cyclopentadiene under Lewis acid conditions (Scheme 7.3).[16] The *endo*:*exo* ratio of adduct (**10**) depends very much upon the conditions used, as does the de, which varies from 54–61%.

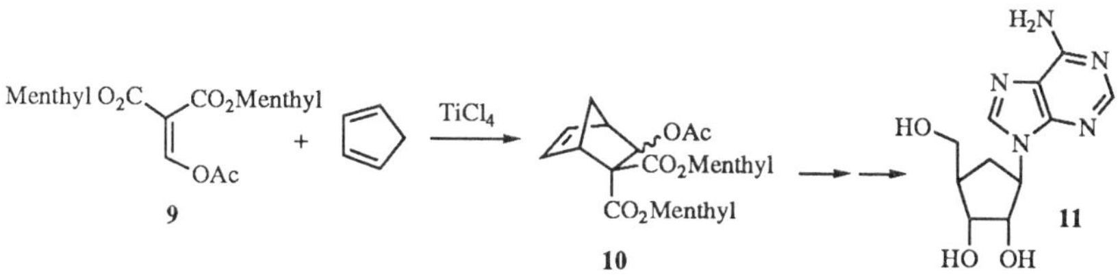

Scheme 7.3

Similarly, pseudo-sugars have been prepared conveniently from Diels–Alder adducts of types (**13a**) and (**13b**),[17] which are both available from the reactive vinyl sulfoxide (**12**) with high de values (>96%). Showdomycin (**14**) has also been prepared by similar methodology.[17]

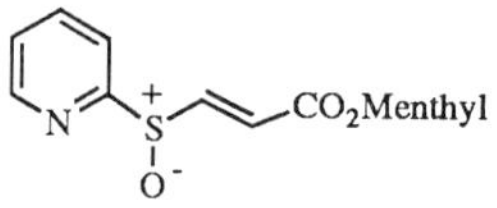

12

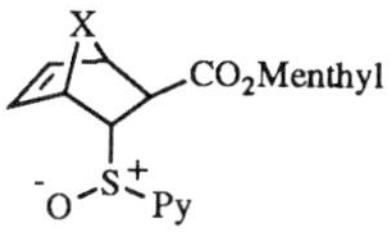

13
a: X = CH_2
b: X = O

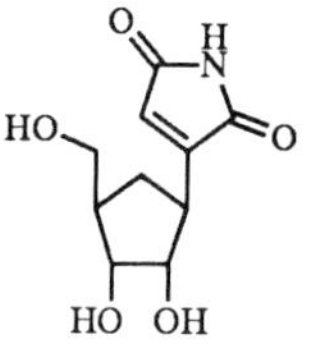

14

Isoborneol-10-sulfinyl auxiliaries have also been reported to give exceptionally high de values (approximately 100%) when used on doubly activated dienophiles (**15**) and (**16**).[18] Accordingly, reaction of (**15**) with cyclopentadiene affords adducts of types (**17**) and (**18**). Remarkably high values are also obtained from dienophiles (**19**),[19] (**20**)[20] and (**21**).[21]

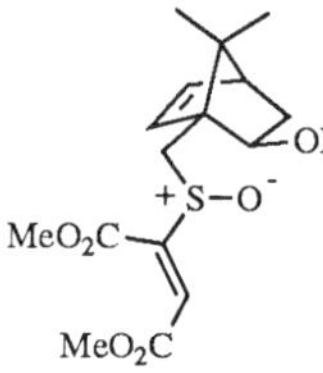

15 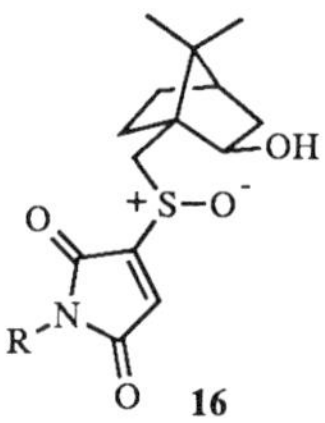 16 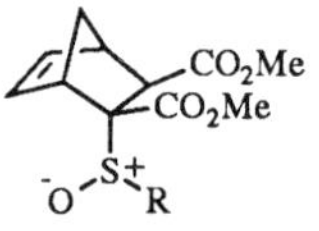 17 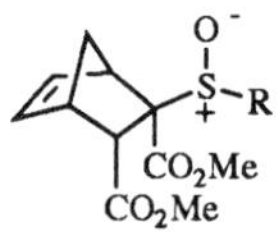 18

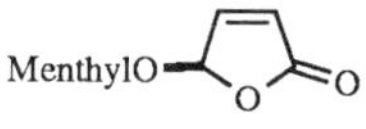

19 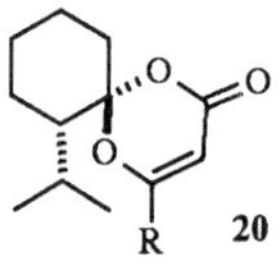 20 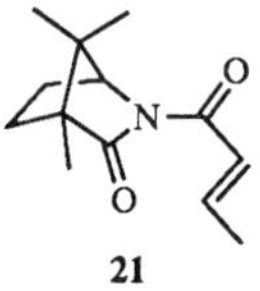 21

One of the most useful chiral auxiliaries has been the sultam derivative (**22**), which was introduced by Oppolzer[4] and has been recently reviewed elsewhere.[22] The X-ray structure of the $TiCl_4$ complex of (**23**) has been reported,[23] supporting previously postulated mechanisms for achieving excellent diastereocontrol in a wide range of examples (Figure 7.1).

Amino acid derivatives have been utilised more recently as sources of chirality in Diels–Alder reactions. The *N*-acryloyl amides (**24**) and (**25**) give variable de values under a range of Lewis acid-catalysed conditions.[24] The de can be significantly improved by reducing ester (**25a**) and using ether (**25b**) for the Lewis acid-catalysed cycloaddition with cyclopentadiene, to afford an adduct in up to 94% de (Figure 7.2).[25] It was also observed that the sense of asymmetric induction depended upon the Lewis acid used to catalyse the cycloaddition; models have been proposed to explain the results in terms of

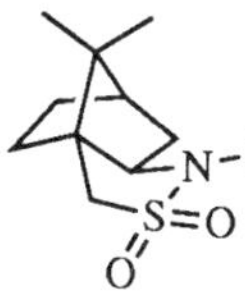

22

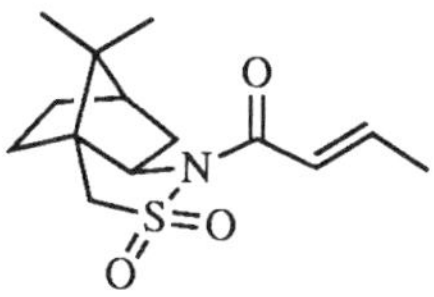

23

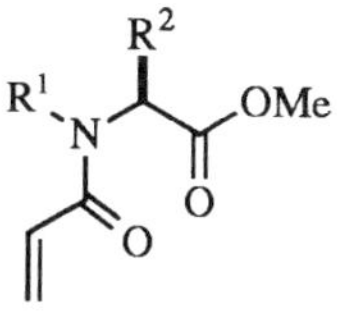

24

a: $R^1 = H, R^2 = Bn$
b: $R^1 = H, R^2 = Me$
c: $R^1 = Me, R^2 = Me$

Figure 7.1 Examples of dienophiles containing cyclic sultams.

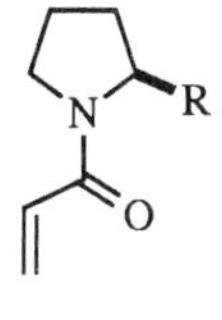

25

a: $R = CO_2Me$
b: $R = CH_2OBn$
c: $R = CHMe_2$

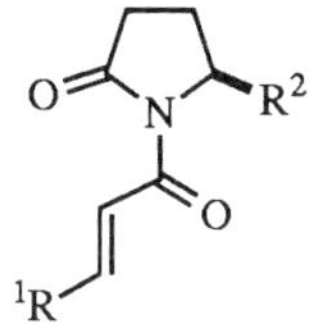

26

a: $R^2 = CH_2CH_2OMe$
b: $R^2 = CH_2OBn$

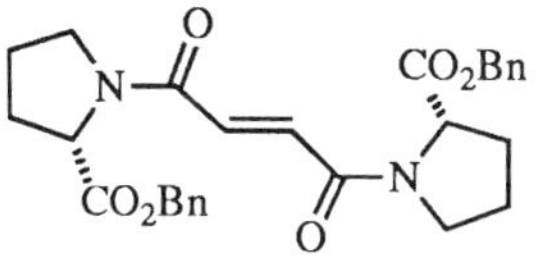

27

Figure 7.2 Examples of dienophiles containing chiral pyrrolidines.

single (for Zn, B and Al) and double (for Ti, and Sn) Lewis acid chelation. Similarly, related pyrrolidine derivatives (**25c**) and (**26**) afford high de values (>74%) with cyclopentadiene when activated by strong alkylating agents (for **25c**) or by Lewis acids (Figure 7.2).[26] The sense of asymmetric induction may be readily deduced by assuming approach of the diene from the opposite face to the R group of the planar imminium ion intermediate or activated amide.

In contrast, incorporation of two amino acid auxiliaries onto fumaric acid, i.e. diamide (**27**), results in a highly diastereoselective (>87%) dienophile under both thermal and Lewis acid-catalysed conditions.[27] An alternative to this approach has been pioneered by Meyers *et al.* by introducing oxazolidine dienophile derivatives of type (**28**).[28] Reaction of (**28**) with various dienes has been achieved, affording adducts (**29**) as single diastereoisomers which were readily transformed into carbocycles of type (**30**).

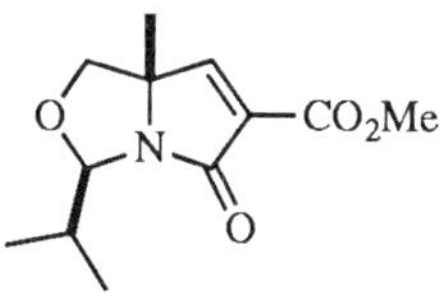

28

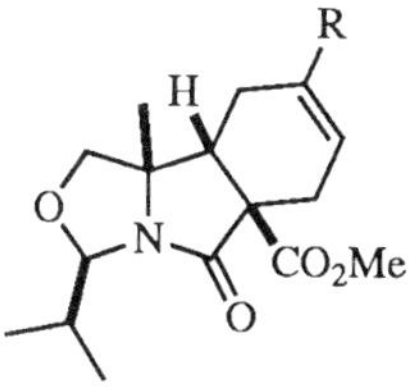

29

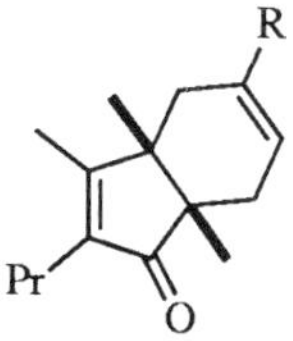

30

The α-hydroxyester derivatives related to (**31**) have been extensively investigated as highly efficient, recoverable dienophile auxiliaries, providing >94% de values in most cases.[29] Pantolactone derivatives (**31**) have become widely used in asymmetric synthesis.[30] The mode of action of (**31**) under Lewis acid conditions involves double coordination of, for example, titanium(VI) chloride to both carbonyl groups, as in complex (**32**). The titanium is therefore responsible for the shielding of one face of the dienophile, forcing the diene to approach from the *Si*-face.

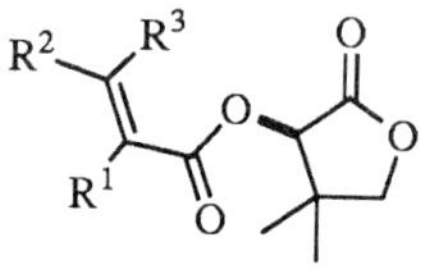

31 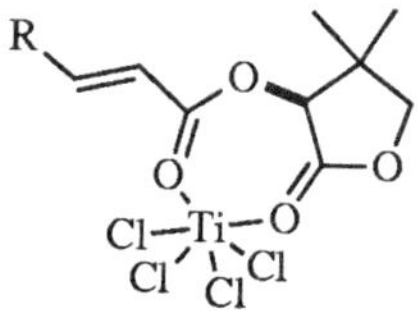 **32**

Derivatives of carbohydrates have been demonstrated to be useful chiral auxiliaries. Use has been made of arabinose acrylate (**33**),[31] which directs dienes to the *Re*-face in up to 94% de under Lewis acid-catalysed conditions. Also, open-chain carbohydrate analogues (**34**), (**35**) and (**36**)[32] produce good to high levels (>95% de) of diastereoselectivity with a range of dienes. The juglone-derived dienophile (**37**) also gives high diastereoselectivity when reacted with cyclopentadiene.[33]

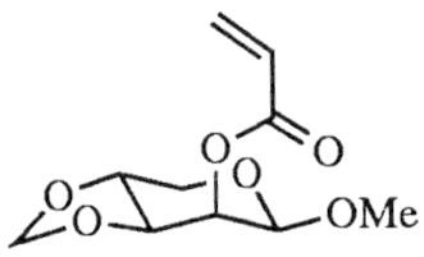

33 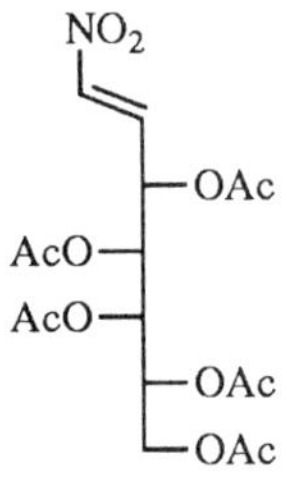 **34** 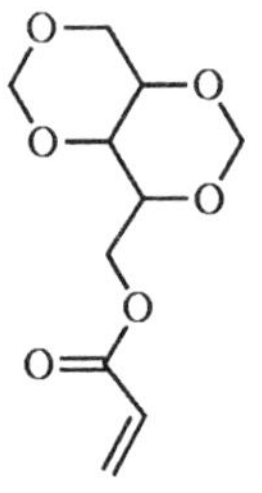 **35** 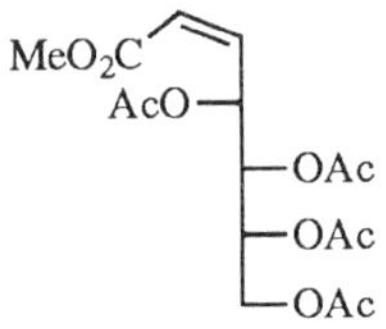 **36**

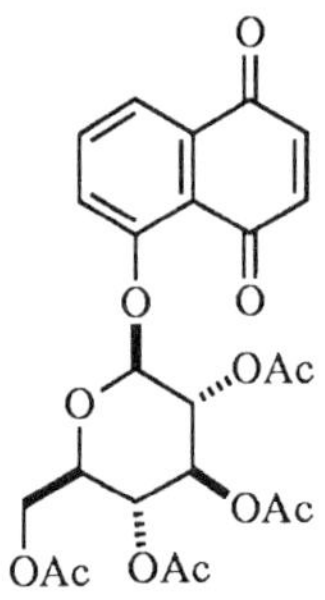

37

Chiral sulfoxides have been reported to be good controlling substituents when attached to electron-deficient alkenes. Among the examples reported,[34] the nitroalkenes (**38**) and (**39**) provide high ee values (>95%) when reacted with Danishefsky's diene, affording adducts of type (**40**) and (**41**).[35] High π-face selectivity is also observed with chiral quinone sulfoxide derivatives (**42**) and (**43**).[36]

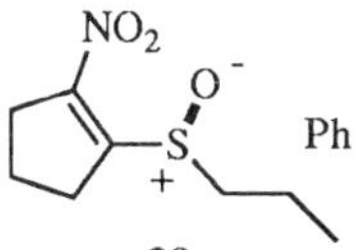

NO_2 O⁻ S + Ph

38

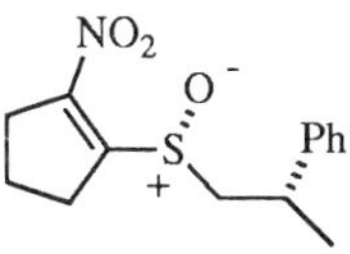

NO_2 O⁻ S + Ph

39

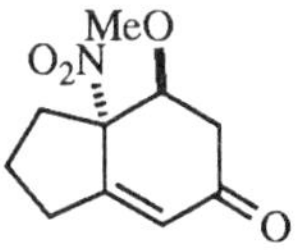

MeO O_2N O

40

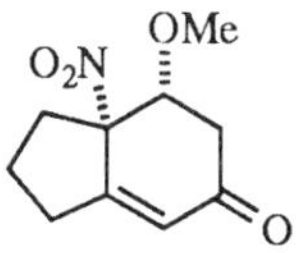

OMe O_2N O

41

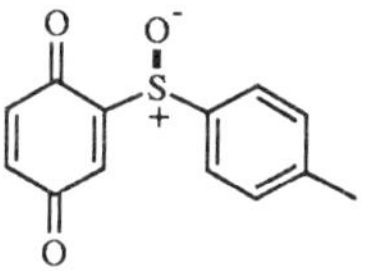

O O⁻ S + O

42

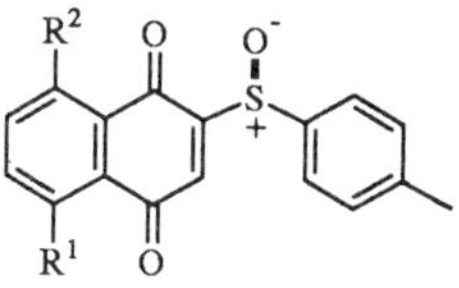

R^2 O O⁻ S + R^1 O

43

7.2.2 Chiral dienes

Perhaps the most accessible and, therefore, widely studied class of chiral diene are the sugar-substituted dienes (**44, 44a** and **45**) reported by Stoodley.[37] The dioxygenated glucose-derived dienes of type (**44**) react via conformation (**44a**), because of the *exo*-anomeric effect. Levels of de vary from moderate to good, depending on the solvent and dienophile. The adducts are, however, readily crystallised to give pure adducts such as (**45**). The scope of these types of reaction has been extensively studied, in terms of the butadiene structure,[38] anomeric configuration[39] and the sugar substituent.[40]

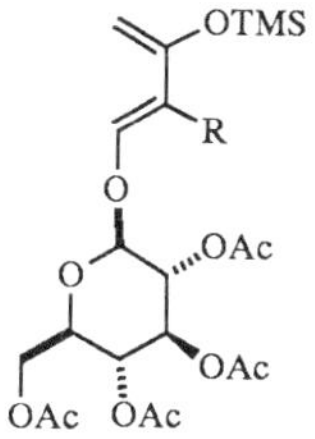

OTMS R O O OAc OAc OAc OAc

44

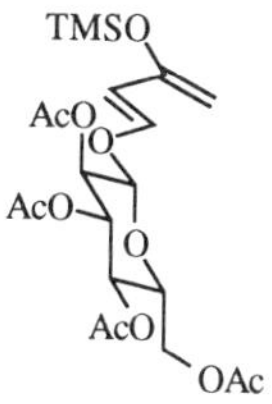

TMSO AcO O AcO O AcO OAc

44a

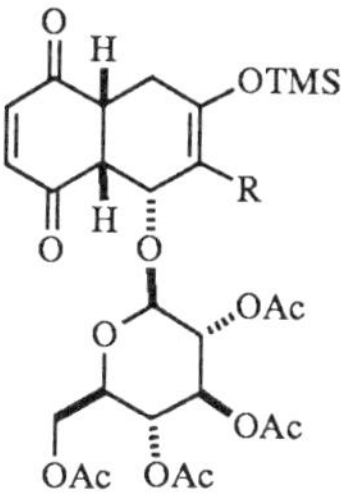

O H OTMS R H O O O OAc OAc OAc OAc

45

Amino acid-derived dienes recently reported include the butadiene derivative (**46**), which reacts with maleic anhydride to afford adduct (**47**) in 73% yield.[41] Similarly, the proline-derived diene (**48**) reacts with nitroalkenes to give adducts with high (75–95%) de.[42]

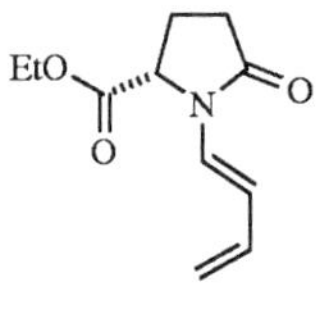

EtO O N O **46** EtO O N O H O O H O **47** N OMe **48**

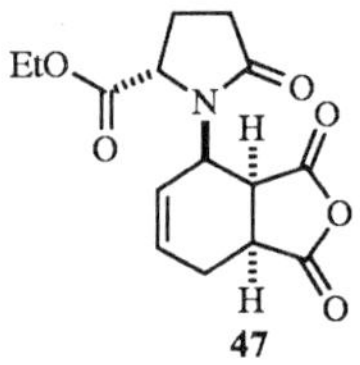

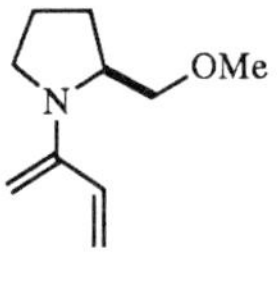

Thornton has reported that *O*-methylmandeloxy dienes (**49**) and inverse electron-demand diene (**50**) react with a range of dienophiles and vinyl ethers (under pressure), respectively. The electron-rich diene (**49**) provides high (>95%) de, whereas diene (**50**) gives up to 76% de. This work was followed by an examination of dienes (**51a**) and (**51b**),[44] which provide de values of 90% and 9%, respectively, with *N*-ethylmaleimide. These results were rationalised in terms of conformational locking in (**51a**) by hydrogen bonding between the hydroxyl and carbonyl groups.

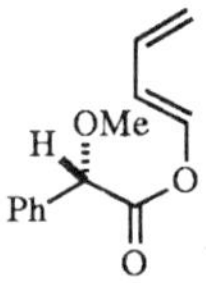

H OMe Ph O O **49** H OMe Ph O O O O **50** Me OR O O **51**

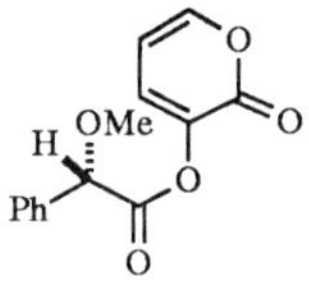

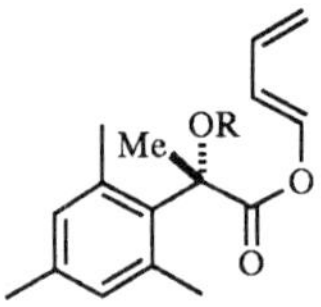

a: R = H
b: R = TMS

A precursor of hydroxyvitamin D_3 (**52**) has been prepared from the Diels–Alder adduct (**53**),[45] which was obtained from the highly diastereoselective inverse electron-demand Diels–Alder reaction of diene (**54**) with benzyl vinyl ether. The reaction gave >96% de (98% yield) using double stereodifferentiation to give optimum de, i.e. when using a catalyst [(−)-Pr(hfc)$_3$] which best matched the absolute stereochemistry of the diene. When the mismatched catalyst (+)-Pr(hfc)$_3$ was employed, the de dropped to 89% and the reaction was three to five times slower.

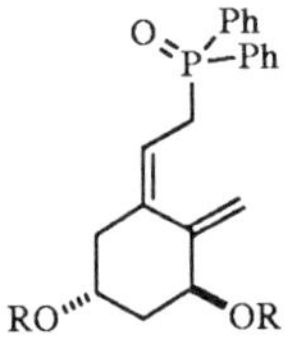

O P Ph Ph RO OR **52** O O CO_2Me OBn H **53** O O O O OMe H Me **54**

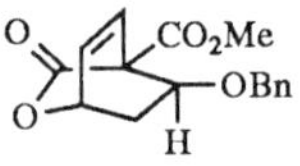

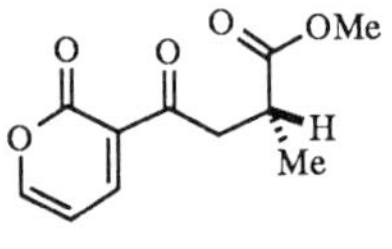

7.2.3 *Chiral catalysts*

Perhaps the most attractive method of inducing enantioselectivity into the Diels–Alder reaction is the use of a chiral catalyst, in the form of a Lewis acidic metal complex. In recent years, this area has seen the greatest advances made, with many excellent, truly catalytic processes being reported.

The most common sources of chirality employed for making a chiral Lewis acid complex are C_2 symmetric chiral diols. Indeed the early examples[46] of high asymmetric induction in the chiral Lewis acid-facilitated Diels–Alder reaction were of this type but suffered from not being catalytic. However, there are several general types of Lewis acid that have now been employed in a catalytic manner to give good to high asymmetric induction, for example (**55**) and (**56**).

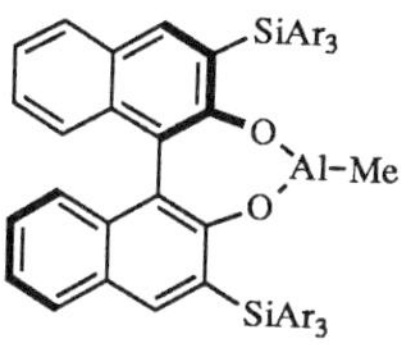

55

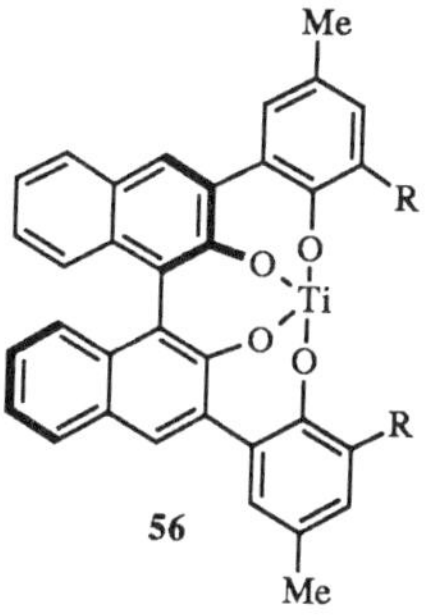

56

Chiral 1,1′-bi-2-naphthol derivatives have become widely used Lewis acid ligands for asymmetric Diels–Alder reactions,[9–13] because they can be readily functionalised in the 3,3′-positions to yield extremely hindered Lewis acid complexes when complexed to a suitable metal. Yamamoto has been particularly active in this regard and has recently reported that bis(tri-arylsilyl)binaphthol complex (**55**) catalyses the reaction of cyclopentadiene and methylacrylate between −78°C and 0°C in non-polar solvents, when used in 10 mol.%.[47] The resulting reaction is unusually *endo*-selective (>95%), with ee values ranging from 36–77%. The highest ee is obtained using the triphenylsilyl derivative of (**55**).

Following on from this work, the helical titanium complex (**56**) was reported as an efficient catalyst for the Diels–Alder reaction of unsaturated aldehydes with cyclopentadiene at the level of 10 mol.% of catalyst,[48] giving high *endo*-control, and ee values up to 98% were obtained for catalyst (**56**; R = tri-*o*-tolylsilyl).

Structurally similar to the binaphthols is biphenanthrol (**57a**),[49] which as the aluminium chloride complex (**57b**) is capable of 200 turnovers at −80°C in the reaction of methacrolein with cyclopentadiene. The adduct, (**58**), was obtained in 98% ee.

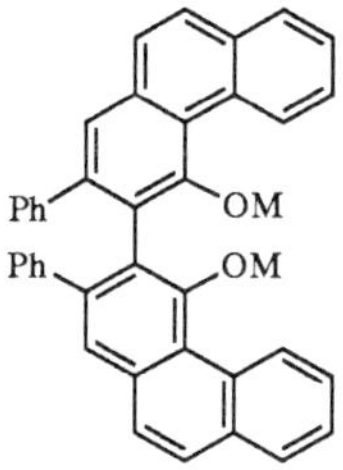

57

a: M = H
b: M = AlCl

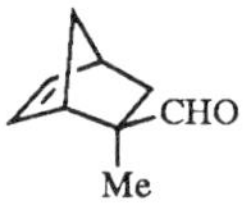

58

One class of chiral diols has received considerable attention as Lewis acid ligands, largely because of their ease of preparation from tartrate acetals. Carbinols of general structure (**59**) are excellent ligands for titanium and aluminium. Treatment of the carbinol (**59a**) with titanium(IV) isopropoxide generates complex (**60**) after treatment with $SiCl_4$. The adduct (**60**) catalyses (20 mol.%) the cycloaddition of unsaturated imides (**61**) with cyclopentadiene derivatives, affording high yields of *endo*-adducts (**62**) with ee values typically higher than 95%.[50] The transition state involved in these reactions has been discussed to explain the high asymmetric induction and may involve π–π interaction between the aryl groups on the ligand and the dienophile.

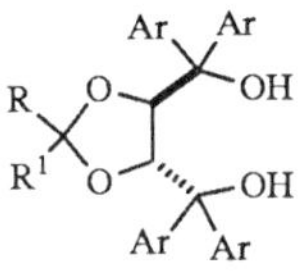

59

a: R = R^1 = Et, Ar = 3,5-dimethylphenyl
b: R = Me, R^1 = Ar = Ph
c: R = R^1 = Me, Ar = Ph

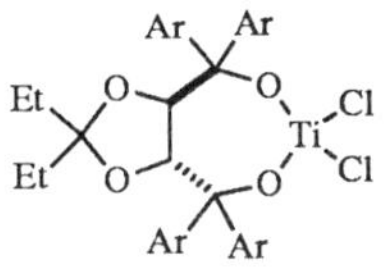

60

Ar = 3,5-dimethylphenyl

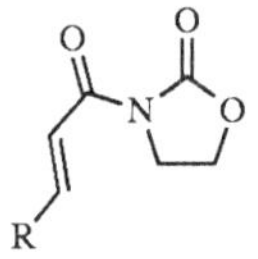

61

a: R = Me

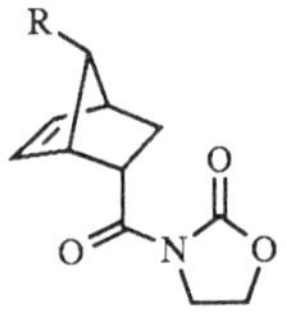

62

Other workers have also used ligands of type (**59**) to good effect.[51] The dichlorotitanium derivative of diol (**59b**) catalyses the reaction of acrylimide (**61a**) with acyclic dienes giving high yields of adducts with ee values in the range of 88–96%.[51a] Similarly, titanium and aluminium complexes of (**59c**) catalyse the reaction of dienophile (**63**) with cyclopentadiene to give a mixture of *endo*- and *exo*-adducts, in up to 70% ee.[51c] The Diels–Alder reaction of quinones (**64**) can also be catalysed by the dichlorotitanium complex of (**59b**), providing adducts of type (**65**) in 48–92% ee.[51d]

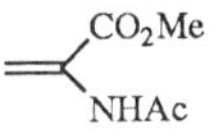

CO_2Me NHAc

63

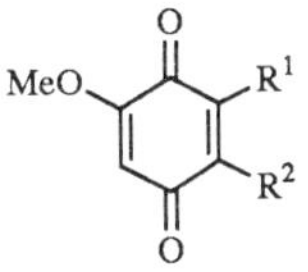

MeO R^1 R^2

64

MeO Me R^4 R^5 H

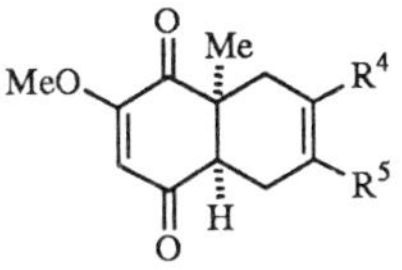

65

Structurally related to Lewis acid derivatives of diols (**59**), (*R*,*R*)-hydrobenzoin (**66**) has been used to generate a titanium dichloride complex which catalyses the reaction of 2,3-dimethylbuta-1,3-diene and dimethyl fumarate, providing the (*S*,*S*)-adduct in 92% ee.[52] Values of ee of up to 86% have also been obtained for *exo*-Diels–Alder adduct (**58**) using the ethyl chloroaluminium derivative of (**67**).[53]

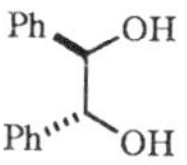

Ph OH Ph OH

66

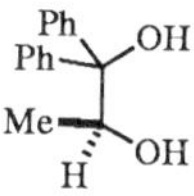

Ph Ph OH Me H OH

67

Sulfonamido derivatives of α-amino-acids can be used to prepare reactive Lewis acid complexes (**68**–**70**). Early examples of complexes typified by (**68**) gave moderate to good ee values (up to 74%) for a range of different dienes and dienophiles.[54] An example of an amino acid in this class of chiral Lewis acid followed from the group of Corey, which showed higher asymmetric induction, together with a rationale for the activity of complex (**69**).[55] Complex (**69**) is prepared from *N*-tosyl-(*S*)-tryptophan and butylboronic acid, followed by azeotropic removal of water.

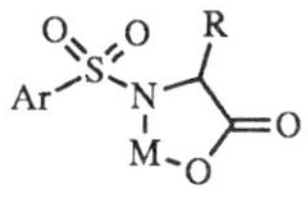

Ar S N R M O O

68

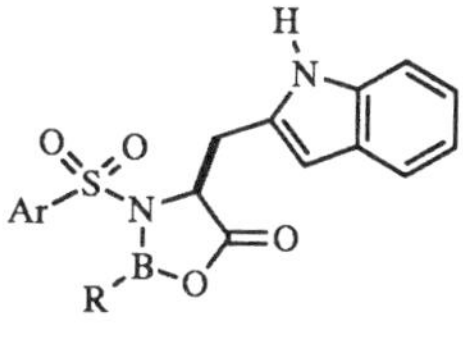

Ar S N B R O O H N

69

R S O O Ph N M–X Ph N S O R O

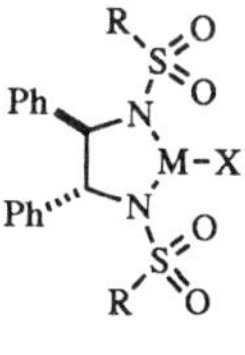

70

a: M = B
b: M = Al

Similar to the sulfonamido amino acids are the bissulfonamides (**70**) reported by Corey.[56] The bromoboron complex (**70a**) can be used at the level of 10 mol.% to catalyse the reaction of cyclopentadiene and the imide (**61a**), giving not only an *endo:exo* ratio of >50:1, but 91% ee at −78°C, which increases to 95% ee when the reaction is carried out at −90°C. The related aluminium complex (**70b**) is very similar in reactivity to the boron-based system at −78°C, but generally higher ee values are obtained, typically >95%, for the reaction of cyclopentadiene derivatives with the imide (**61a**). For example, reaction of the diene (**71**) with imide (**61a**) in the presence of 10 mol.% of (**70b**) affords the adduct (**73**) in 95% ee (Scheme 7.4). A model for the transition state (**72**) to explain the outcome of these processes has been proposed, with evidence coming from X-ray analysis of the dimer of (**70b**) and n.O.e. experiments on the complex of (**70b**) with (**61a**).

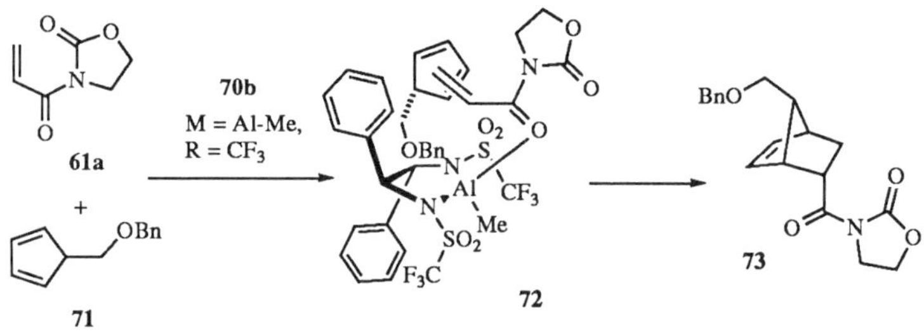

61a + 71 → (70b, M = Al-Me, R = CF_3) → 72 → 73

Scheme 7.4

Chiral hydroxy acid derivatives also provide excellent ligands for Lewis acids. Yamamoto has demonstrated the use of boron complexes of tartaric acid derivatives (**74**),[57] generated by treatment of (**75**) with BH_3·THF. The Lewis acids (**74**) catalyse the reaction of acrylic acid with cyclopentadiene to give (**76**) in 35–78% ee when used in 10 mol.%. However, higher ee values are obtained for the reaction of cyclopentadiene with methacrolein, providing the *exo*-adduct (**58**) in 96% ee. An intramolecular Diels–Alder reaction was also catalysed with high ensuing asymmetric induction.[57c]

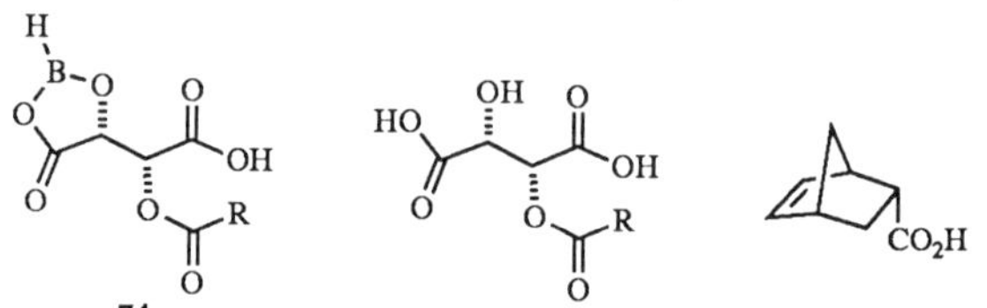

74 **75** **76**

The C_2 symmetric bis(oxazoline) (**77**) has recently been shown to be an effective ligand for both iron(III) and magnesium.[58] The resulting complexes catalyse the reaction of cyclopentadiene and imide (**61a**), providing high *endo*-selectivity and ee values up to 91%.

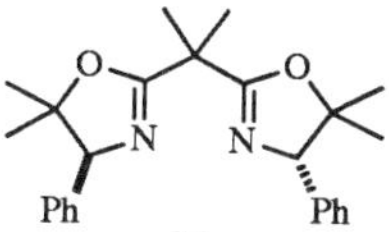

77 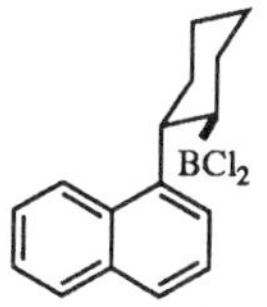 78 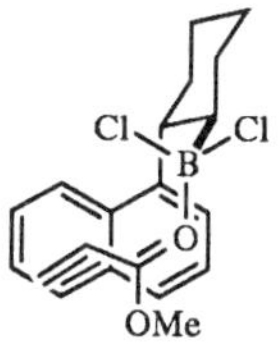 79

There are few examples of high asymmetric induction for Lewis acid-catalysed reactions of methyl acrylate; however, a new dichloroborane catalyst (**78**) catalyses the reaction of methyl acrylate with cyclopentadiene, providing the *endo*-adduct in 97% ee,[59] possibly through a complex of type (**79**).

7.3 Diels–Alder reactions of C=O

Carbon–oxygen double bonds readily undergo Diels–Alder-like reactions or cyclocondensations when aldehydes are employed, generally with an electron-withdrawing group or in the presence of a Lewis acid promoter. This process has been exploited by Danishefsky *et al.* for the synthesis of a wide range of saccharide derivatives.[60]

7.3.1 *Chiral dienophiles*

The most frequently employed method of achieving asymmetric induction in the hetero-Diels–Alder reaction of aldehydes is to attach a chiral auxiliary to the dienophile. Chiral glyoxylate ester derivatives have been reported,[61] although de values were poor (generally < 10%). Oppolzer's sultam auxiliary, however, is much more efficient as a chiral dienophile[62] when attached as a glyoxylate derivative (**80**). Reaction of (**80**), under high pressure, in uncatalysed conditions or in the presence of $Eu(fod)_3$ with diene (**81**) provides adducts such as (**82**) in high de (> 90%). Models to explain the sense of both catalysed and uncatalysed reactions were proposed and possibly involve a Lewis acid complex of type (**83**), with the diene attacking the *Si*-face of the dienophile.

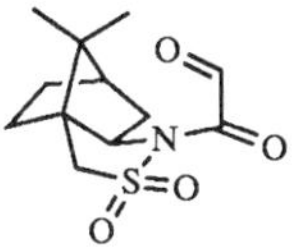

80  81 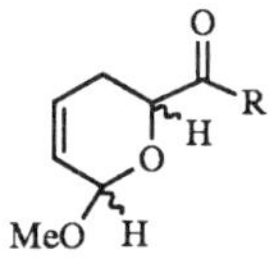 82 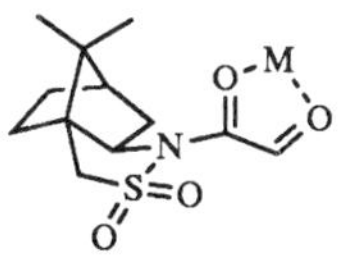 83

Chiral α-alkoxy aldehydes (**84**) react under Lewis acid-catalysed conditions (using $Eu(hfc)_3$, $MgBr_2$ and Et_2AlCl) to yield low to high de values of adducts (**86**) when reacted with Brassard's diene (**85**).[63] Similarly, α-amino aldehyde derivatives of type (**87**) react with diene (**81**) in the presence of $Eu(fod)_3$ under pressure, providing adduct such as (**88**) with low to moderate de.[64]

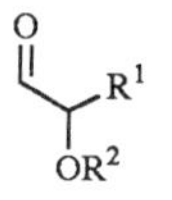

84 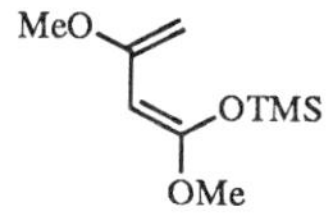 **85** 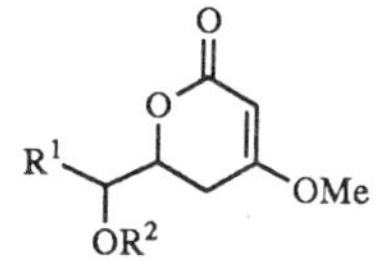 **86**

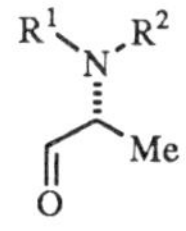

87

a: R^1 = Bn, R^2 = BOC
b: R^1 = R^2 = phthalimide

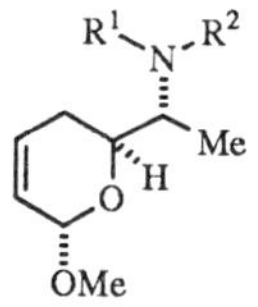

88

7.3.2 *Chiral catalysts*

Danishefsky demonstrated that the cycloaddition of an aldehyde with an oxygenated diene could be catalysed by chiral shift reagents.[60] Values of ee were generally moderate, unless other chiral centres were present in the dienophile component. Examination of a range of chiral non-lanthanide Lewis acids showed that other Lewis acids could also be used to catalyse the cycloaddition reaction with moderate asymmetric induction.[65] However, application of hindered 3,3′-triarylsilyl-substituted binaphthalenes of type (**89**), as their aluminium complexes, provided a method of catalysing the reaction of dienes (**90**) with benzaldehyde, affording high asymmetric induction in the major adducts (**91**) (Scheme 7.6).[66] What is particularly interesting about this work is the fact that catalyst (**89**) was employed at the level of 10 mol.%, which provided the adduct (**91**) in 77% yield and 95% ee.

Titanium complexes of binaphthols are also efficient chiral Lewis acid

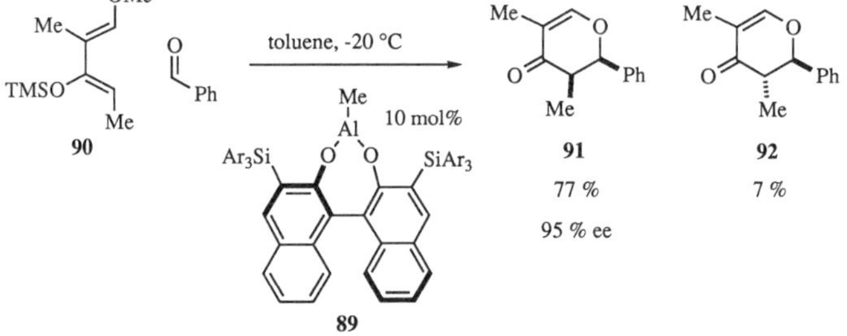

toluene, -20 °C; 10 mol% **89**

90 **91** 77 %, 95 % ee **92** 7 %

Scheme 7.5

catalysts for the reaction of glyoxylate esters with oxygenated dienes.[67] Thus, reaction of methyl glyoxylate with the diene (**81**) in the presence of catalytic (**93**) at −55°C provides the major adduct (**94**) in 96% ee.

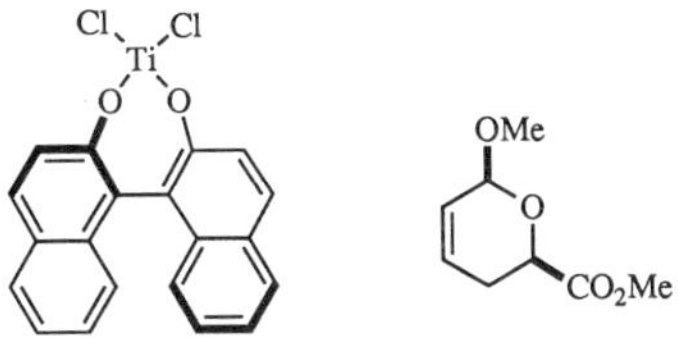

Stable acyl boronates (**95**), derived from tartaric acid, catalyse the reaction of dioxygenated dienes with benzaldehyde at room temperature.[68] When R = 2-MeOC$_6$H$_4$, catalyst (**95**) produces a 95% yield of the pyrone (**91**) from diene (**90**) and benzaldehyde with an ee of 97%. Similarly, the vanadium complex (**96**) efficiently catalyses the reaction of dioxygenated dienes with benzaldehyde when used at the level of 5 mol.% at −78°C, providing good levels of asymmetric induction.[69] Complex (**96**) also produces higher asymmetric induction when used to catalyse the reaction of the diene (**90**) with (*R*)-isopropylidene glyceraldehyde to provide the pyrone (**97**) with 93% de (versus 71% de for the BF$_3$-catalysed reaction). Double stereodifferentiation was, therefore, demonstrated by application of the corresponding mismatched pair, i.e. by using the enantiomer of catalyst (**96**) with (*R*)-isopropylidene glyceraldehyde, which resulted in 0% de.

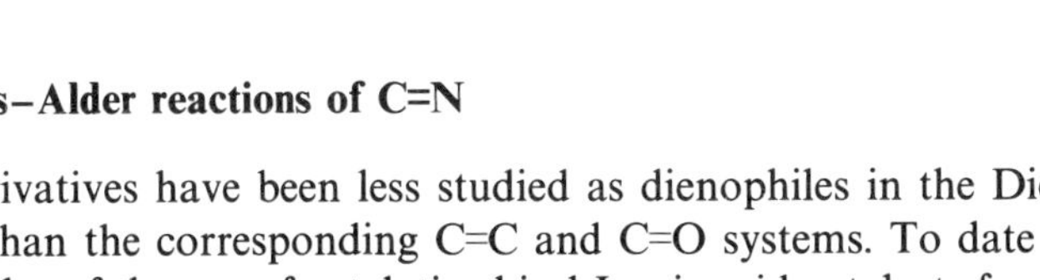

7.4 Diels–Alder reactions of C=N

Imine derivatives have been less studied as dienophiles in the Diels–Alder reaction than the corresponding C=C and C=O systems. To date there are no examples of the use of catalytic chiral Lewis acid catalysts for achieving asymmetric induction in the Diels–Alder reaction of imines; the most common method of achieving induction is by use of a chiral dienophile derivative.

7.4.1 *Chiral dienophiles*

Ester-activated iminium ions (**97**) and (**98**) react readily with cyclopentadiene, affording de values of up to 80% for the *exo*-adducts.[70a] Values of de were improved up to 94% by use of imine (**99**) under Lewis acid-catalysed conditions, with Brassard's diene (**85**),[70b] affording adducts such as (**100**).

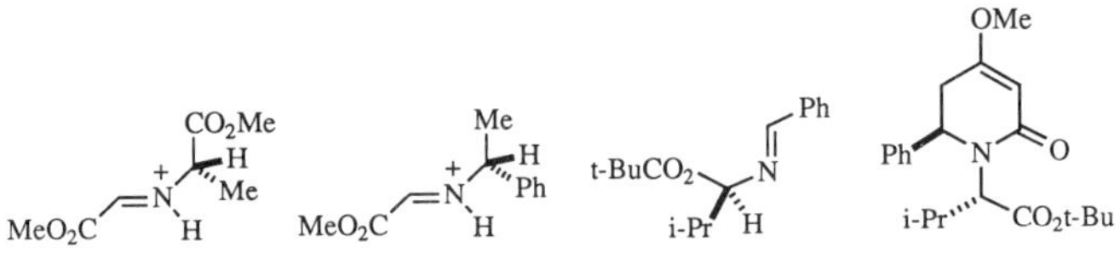

97 **98** **99** **100**

Similarly, the dienophile (**101**) has been reported to react with simple dienes in polar solvents, giving de values of up to 70%.[71] *N*-Camphorsulfonyl imines such as (**102**) show only moderate de values (up to 40%) under Lewis acid-catalysed conditions, affording only *exo*-adducts.[72]

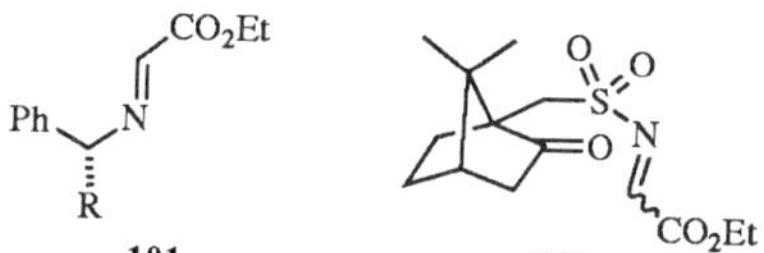

101 **102**

The chiral ester derivatives (**103**) and (**104**) show improved de values of 76 and 70%, respectively, under Lewis acid-catalysed conditions with cyclopentadiene, providing access to adducts of type (**105**).[73]

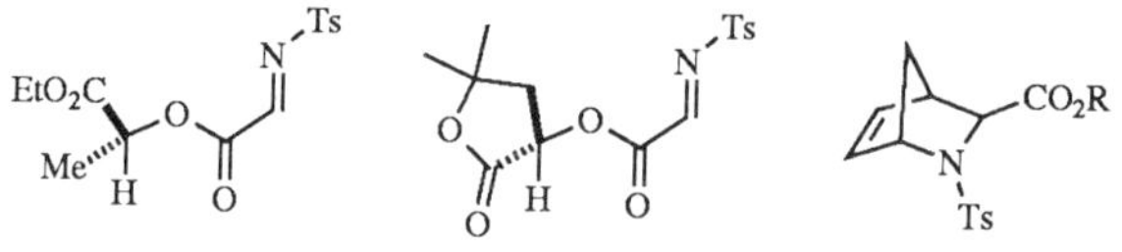

103 **104** **105**

7.4.2 *Chiral dienes*

The chiral diene (**106**) reacts with dienophile (**107**) to provide the adduct (**108**) as a single diastereoisomer in 81% yield. The adduct (**108**) was then transformed into the alkaloid cannabisativine (**109**) (Scheme 7.6).[74]

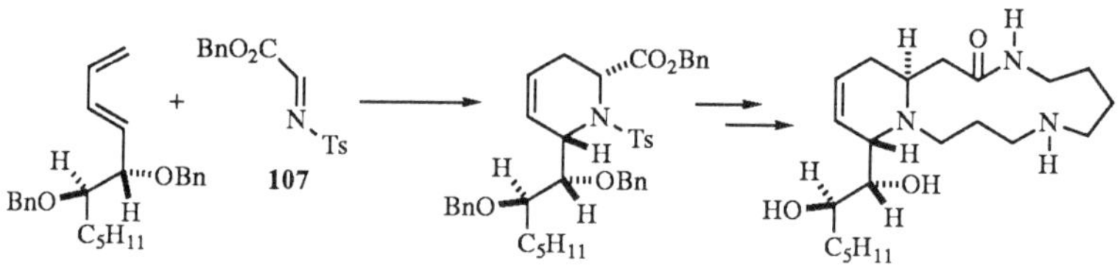

106 **108** **109**

Scheme 7.6

7.4.3 Chiral catalysts

Stoichiometric amounts of binaphthol-derived chiral Lewis acids of general structure (**110**) activate aromatic imines (**111**), affording adducts such as (**112**) in up to 86% ee from the reaction of (**111**) with Danishefsky's diene (**113**). Double asymmetric induction using chiral imines can improve induction; thus, using the matched pair of the (*R*)-binaphthol catalyst (**110**) and an imine derived from (*S*)-α-methylbenzylamine, induction improves up to 98% de. However, the mismatched pair of (*R*)-catalyst (**110**) and an (*R*)-α-methylbenzylamine-derived imine gave only 86% de and in a lower yield.[75]

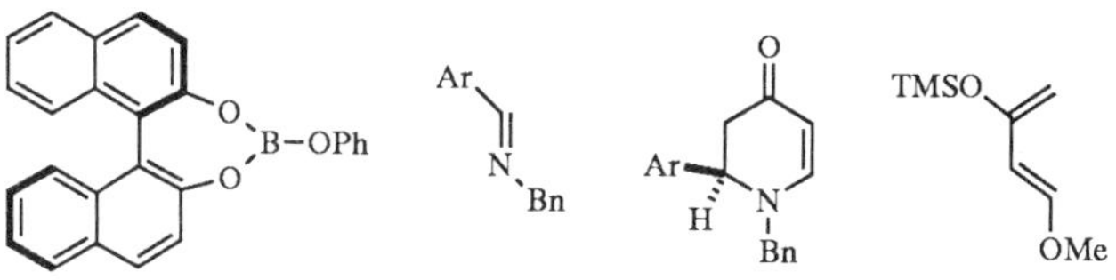

B–OPh Ar N Bn O Ar H N Bn TMSO OMe

110 111 112 113

7.5 Diels–Alder reactions of N=O

Both N=O and C=O readily undergo Diels–Alder-like reactions when activated by an electron-withdrawing group, generally in the form of a conjugated carbonyl group.

7.5.1 Chiral dienophiles

Oppolzer's sultam is an excellent chiral auxiliary for the Diels–Alder reaction of acylnitroso compounds, without the addition of a Lewis acid catalyst. Thus, (**114**) reacts with both cyclopentadiene and cyclohexadiene affording adducts such as **115** in >98% de (Scheme 7.7).[76]

Amino acid derivatives and related chiral amines have been successfully used to induce asymmetry in the Diels–Alder reactions of acyl nitroso compounds. Proline-derived nitroso compounds (**116**) and (**117**) provide poor to moderate de values with cyclohexadiene.[77] However, de values are

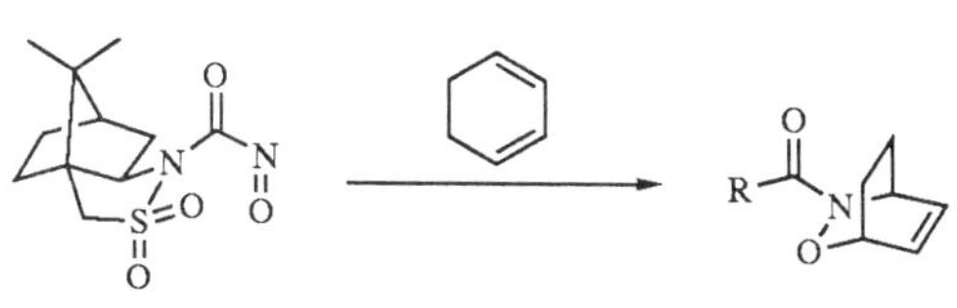

O N N O S=O O R O N O

114 115

Scheme 7.7

considerably improved by use of the related C_2 symmetric auxiliaries (**118**),[78] which react with a variety of dienes providing adducts such as (**119**) in >98% de.

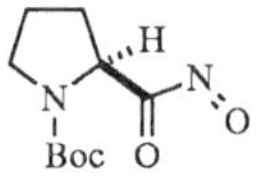

116 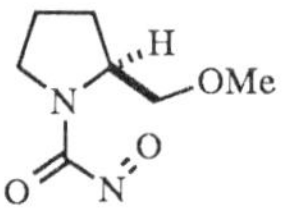 **117** 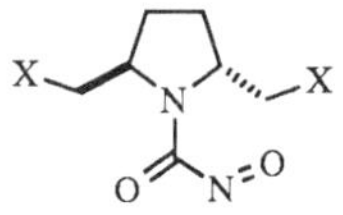 **118**

a: X = OMe
b: X = H

The utility of amino sugar compounds as precursors of nitroso dienophiles has been demonstrated.[79] The dienophile (**120**) reacts with dienes of type (**121**), affording *endo*-adducts (**122**) with >97% ee. Cleavage of the N–O bond of (**122**) provides a rapid and highly stereoselective route to chiral amino poly-hydroxylated cyclohexane compounds.

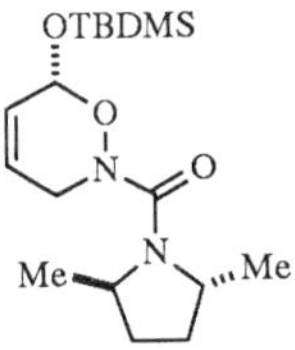

119 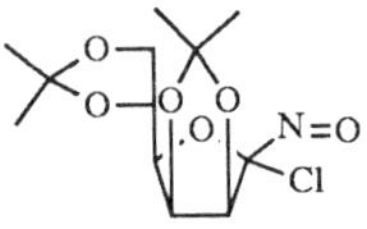 **120** 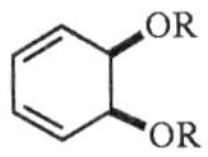 **121** 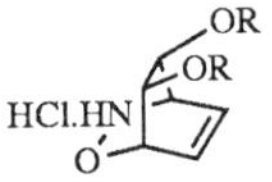 **122**

7.6 Conclusions

Since the early 1980s, there has been a rapid development of this field of asymmetric Diels–Alder reactions, especially in the area of chiral Lewis acid-mediated processes. There are several chiral auxiliaries which are easily attached to substrates, are readily removed and provide high asymmetric induction. These can be used on a wide variety of systems. In terms of chiral catalysts, there are still many reactions where catalytic methods for achieving high levels of asymmetric induction have yet to be devised and where such improvements would be highly desirable.

References

1. Diels, O. and Alder, K., *L. Ann. Chem.*, 1928, **460**, 98.
2. (a) Walborsky, H.M., Barash, L. and Davis, T.C., *J. Org. Chem.*, 1961, **26**, 4778; (b) *Tetrahedron*, 1963, **19**, 2333; (c) Guseinov, M.M., Akhmedov, I.M. and Mamedov, E.G., *Azerb. Khim. Zh.*, 1976, 46 (*Chem. Abstr.*, 1976, **85**, 176925z).

3. Paquette, L.A. in *Asymmetric Synthesis*, Vol. 3B, Morrison, J.D. (ed.), Academic Press, New York, 1984, p. 455.
4. Oppolzer, W., *Angew. Chem., Int. Ed. Engl.*, 1984, **23**, 876.
5. Helmchen, G. in *Modern Synthetic Methods*, Vol. 4, Scheffold, R. (ed.), Springer-Verlag, Berlin, 1986, p. 262.
6. Nogradi, M., *Stereoselective Synthesis*, Verlag-Chemie, Weinheim, 1987, p. 261.
7. Taschner, M.J., *Org. Synth. Theory Appl.*, 1989, **1**, 1.
8. Krohn, K., *Organic Synthesis Highlights*, Verlag-Chemie, Weinheim, 1991, p. 54.
9. Noyori, R. and Kitamura, M., *Mod. Synth. Methods*, 1989, **5**, 115.
10. Tomioka, K., *Synthesis*, 1990, 541.
11. Narasaka, K., *Synthesis*, 1991, 1.
12. Kagan, H.B. and Riant, O., *Chem. Rev.*, 1992, **92**, 1007.
13. Deloux, L. and Srebnik, M., *Chem. Rev.*, 1993, **93**, 763.
14. Hondrogiannis, G., Pagni, R.M., Kabalka, G.W., Kurt, R. and Cox, D., *Tetrahedron Lett.*, 1991, **32**, 2303.
15. (a) Catriviela, C., Lopez, P. and Mayoral, J.A., *Tetrahedron: Asymmetry*, 1990, **1**, 379; (b) *Tetrahedron: Asymmetry*, 1990, **1**, 61; (c) *Tetrahedron: Asymmetry*, 1991, **2**, 449.
16. (a) Katagiri, N., Akatsuka, H., Kaneko, C. and Sera, A., *Tetrahedron Lett.*, 1988, **29**, 5397; (b) Katagiri, N., Haneda, T., Watanabe, N., Hayasaka, E. and Kaneko, C., *Chem. Pharm. Bull.*, 1988, **36**, 3867; (c) Katagiri, N., Watanabe, N. and Kaneko, C., *Chem. Pharm. Bull.*, 1990, **38**, 69.
17. (a) Takahashi, T., Namiki, T., Takeuchi, Y. and Koizumi, T., *Chem. Pharm. Bull.*, 1988, **36**, 3213; (b) Arai, Y., Hayashi, Y., Yamamoto, M., Takayema, H. and Koizumi, T., *J. Chem. Soc., Perkin Trans. 1*, 1988, 3133; (c) Arai, Y., Takadoi, M. and Koizumi, T., *Chem. Pharm. Bull.*, 1988, **36**, 4162; (d) Takahashi, T., Jabe, A., Arai, Y. and Koizumi, T., *Synthesis*, 1988, 189; (e) Takahashi, T., Katsubo, H., Iyobe, A., Namiki, T. and Koizumi, T., *J. Chem. Soc., Perkin Trans. 1*, 1990, 3065; (f) Yang, T.K., Teng, T.F. and Lee, D.S., *J. Chin. Chem. Soc.*, 1991, **38**, 375; (g) Takahashi, T., Katsubo, H. and Koizumi, T., *Tetrahedron: Asymmetry*, 1991, **2**, 1035.
18. (a) Arai, Y., Hayashi, K., Matsui, M., Koizumi, T., Shiro, M. and Kuriyama, K., *J. Chem. Soc., Perkin Trans. 1*, 1991, 1709; (b) Arai, Y., Matsui, M., Koizumi, T. and Shiro, M., *J. Org. Chem.*, 1991, **56**, 1983.
19. De Jong, J.C., Janoor, J.F.C.A. and Feringa, B.L., *Tetrahedron Lett.*, 1990, **31**, 3047.
20. Soto, M., Orii, C., Sakaki, J. and Kaneko, C., *J. Chem. Soc., Chem. Commun.*, 1989, 1435.
21. Boeckman, R.K., Nelson, S.G. and Gaul, M.D., *J. Am. Chem. Soc.*, 1992, **114**, 2258.
22. Kim, B.H. and Curran, D.P., *Tetrahedron*, 1992, **49**, 293.
23. Oppolzer, W., Podriguez, I., Bragg, J. and Bernardinelli, G., *Helv. Chim. Acta.*, 1988, **72**, 123.
24. (a) Bueno, M.P., Cativiela, C., Mayoral, J.A., Avenaza, A., Chawo, P., Roy, M.A. and Andres, J.M., *Can. J. Chem.*, 1988, **66**, 2826; (b) Bueno, M.P., Catioviela, C.A., Mayoral, J.A. and Avenaza, A., *J. Org. Chem.*, 1991, **56**, 6551.
25. Waldmann, H., *J. Org. Chem.*, 1988, **53**, 6133.
26. (a) Jung, M.E., Vaccaro, W.D. and Buszek, K.R., *Tetrahedron Lett.*, 1989, **30**, 1893; (b) Ikota, N., *Chem. Pharm. Bull.*, 1989, **37**, 2219; (c) Tanioka, K., Hamada, W., Suenaga, T. and Koga, K., *J. Chem. Soc., Perkin Trans. 1*, 1990, 426.
27. Waldmann, H. and Draeger, M., *Tetrahedron Lett.*, 1989, **30**, 4227.
28. (a) Meyers, A.I. and Busacca, C.A., *Tetrahedron Lett.*, 1989, **30**, 6973; (b) *Tetrahedron Lett.*, 1989, **30**, 6977; (c) *J. Chem. Soc., Perkin Trans. 1*, 1991, 2299.
29. (a) Helmchen, B., Abdel, H.A.F., Hartmann, H., Karge, R., Katz, A., Sartor, K. and Urmann, M., *Pure Appl. Chem.*, 1989, **61**, 409; (b) Poll, T., Abdel Hady, A.F., Karge, R., Linz, G., Weetman, J. and Helmchen, G., *Tetrahedron Lett.*, 1989, **30**, 5595; (c) Linz, G., Weetman, J., Abdel Hady, A.F. and Helmchen, G., *Tetrahedron Lett.*, 1989, **30**, 5599.
30. (a) Avenaza, A., Catrioviela, C., Mayoral, J.A. and Peregrina, J.M., *Tetrahedron: Asymmetry*, 1992, **3**, 913; (b) Knol, J., Jansen, J.F.G.A., van Bolhmis, F. and Feringa, B., *Tetrahedron Lett.*, 1991, **32**, 7465; (c) Catioviela, C., Mayoral, J.A., Avenaza, A., Peregrina, J.M., Lahaz, F.J. and Gimena, S., *J. Org. Chem.*, 1992, **57**, 4664; (d) Cativiela, C., Figueras, F., Fraile, J.M., Garcia, J.I. and Mayoral, J.A., *Tetrahedron: Asymmetry*, 1993, **4**, 223; (e) Trost, B.M. and Kondo, Y., *Tetrahedron Lett.*, 1991, **32**, 1613; (f) Miyaji, K., Ohara, Y., Takahashi, Y., Tsuruda, T. and Arai, K., *Tetrahedron Lett.*, 1991, **32**, 4557.
31. Nouguier, R., Gras, J.L., Giruad, B. and Virgili, A., *Tetrahedron Lett.*, 1991, **32**, 5529.

32. (a) Serrano, J.A., Caceres, L.E. and Roman, E., *J. Chem. Soc., Perkin Trans. 1*, 1992, 941; (b) Gras, J.L., Poncet, A. and Nouguier, R., *Tetrahedron Lett.*, 1992, **33**, 3323; (c) Horton, D. and Koh, D., *Tetrahedron Lett.*, 1993, **34**, 2283.
33. Beagley, B., Curtis, A.D.M., Pritchard, R.G. and Stoodley, R.J., *J. Chem. Soc., Perkin Trans. 1*, 1992, 1981.
34. (a) Alonso, I., Carretero, J.C., Garcia, R. and Jose, L., *J. Org. Chem.*, 1993, **58**, 3231; (b) Carretero, J.C., Garcia, R., Lorente, A. and Yuste, F., *Tetrahedron: Asymmetry*, 1993, **4**, 177; (c) Arai, Y., Yamamoto, M. and Koizumi, T., *Bull. Chem. Soc. Jpn.*, 1988, **61**, 467.
35. (a) Fuji, K., Tanaka, K., Abe, H., Itoh, A., Node, M., Taga, T., Miwa, Y. and Shiro, M., *Tetrahedron: Asymmetry*, 1991, **2**, 179; (b) *Tetrahedron: Asymmetry*, 1991, **2**, 1319.
36. (a) Carmen, C.M., Garcia, R., Jose, L. and Urbano, A., *Tetrahedron Lett.*, 1989, **30**, 4003; (b) *J. Org. Chem.*, 1992, **57**, 6870.
37. (a) Gupta, R.C., Larsen, D.S., Stoodley, R.J., Slawin, A.M.Z. and Williams, D.J., *J. Chem. Soc., Perkin Trans. 1*, 1989, 739; (b) Gupta, R.C., Raynor, C.M., Stoodley, R.J., Slawin, A.M.Z. and Williams, D.J., *J. Chem. Soc., Perkin Trans. 1*, 1988, 1773.
38. Larsen, D.S. and Stoodley, R.J., *J. Chem. Soc., Perkin Trans. 1*, 1989, 1841.
39. Larsen, D.S. and Stoodley, R.J., *J. Chem. Soc., Perkin Trans. 1*, 1990, 1339.
40. Beagley, B., Larsen, D.S., Pritchard, R.G. and Stoodley, R.J., *J. Chem. Soc., Perkin Trans. 1*, 1990, 3113.
41. Menezes, R.F., Zezza, C.A., Shea, J. and Smith, M.B., *Tetrahedron Lett.*, 1989, **30**, 3295.
42. Enders, D., Meyer, O. and Raabe, G., *Synthesis*, 1992, 1242.
43. (a) Siegel, C. and Thornton, E.R., *Tetrahedron Lett.*, 1988, **29**, 5225; (b) Prapansiri, V. and Thornton, E.R., *Tetrahedron Lett.*, 1991, **32**, 3147.
44. Tripathy, R., Carroll, P.J. and Thornton, E.R., *J. Am. Chem. Soc.*, 1990, **112**, 6743.
45. Posner, G.H., Carry, J.C., Anjeh, T.E.W. and French, A.N., *J. Org. Chem.*, 1992, **57**, 7012.
46. (a) Kelly, T.R., Whiting, A. and Chandrakumar, N.S., *J. Am. Chem. Soc.*, 1986, **108**, 3510; (b) Maruoka, K., Sakurai, M., Fujiwara, J. and Yamamoto, H., *Chem. Lett.*, 1986, 4895.
47. Maruoka, K., Concepcion, A.B. and Yamamoto, H., *Bull. Chem. Soc. Jpn*, 1992, **65**, 3501.
48. Maruoka, K., Murase, N. and Yamamoto, H., *J. Org. Chem.*, 1993, **58**, 2938.
49. Bao, J., Wulff, W.D. and Rheingold, A.L., *J. Am. Chem. Soc.*, 1993, **115**, 3814.
50. Corey, E.J. and Matsumura, Y., *Tetrahedron Lett.*, 1991, **32**, 6289.
51. (a) Narasaka, K., Tanaka, H. and Kanai, F., *Bull. Chem. Soc. Jpn*, 1991, **64**, 387; (b) Narasaka, K., Iwasawa, N., Inoue, M., Yamada, T., Nakashima, M. and Sugimori, J., *J. Am. Chem. Soc.*, 1989, **111**, 5340; (c) Cativiela, C., Lopez, P. and Mayoral, J.A., *Tetrahedron: Asymmetry*, 1991, **2**, 1295; (d) Engler, T.A., Letavic, M.A. and Takusagawa, F., *Tetrahedron Lett.*, 1992, **33**, 6731.
52. Devine, P.N. and Oh, T., *J. Org. Chem.*, 1992, **57**, 396.
53. Rebiere, F., Riant, O. and Kagan, H.B., *Tetrahedron: Asymmetry*, 1990, **1**, 199.
54. (a) Takasu, M. and Yamamoto, H., *Synlett*, 1990, 194; (b) Santor, D., Saffrich, J. and Helmchen, G., *Synlett*, 1990, 197; (c) Santor, D., Saffrich, J., Helmchen, G., Richards, C.J. and Lambert, H., *Tetrahedron: Asymmetry*, 1991, **2**, 639.
55. (a) Corey, E.J., Loh, T.P., Roper, T.D., Azimioara, M.D. and Noe, M.C., *J. Am. Chem. Soc.*, 1992, **114**, 8290; (b) Corey, E.J. and Loh, T.P., *J. Am. Chem. Soc.*, 1991, **113**, 8966.
56. (a) Corey, E.J., Imwinkelried, R., Pikul, S. and Xiang, Y.B., *J. Am. Chem. Soc.*, 1989, **111**, 5493; (b) Corey, E.J., *Tetrahedron Lett.*, 1991, **32**, 7517; (c) Corey, E.J., Sarshar, S. and Bordner, J., *J. Am. Chem. Soc.*, 1992, **114**, 7938.
57. (a) Furuta, K., Miwa, Y., Iwanaga, K. and Yamamoto, H., *J. Am. Chem. Soc.*, 1988, **110**, 6254; (b) Furuta, K., Shimiza, S., Miwa, Y. and Yamamoto, H., *J. Org. Chem.*, 1989, **54**, 1481; (c) Furuta, K., Kanematsu, A., Yamamoto, H. and Takaoka, S., *Tetrahedron Lett.*, 1989, **30**, 7231.
58. Corey, E.J. and Ishihara, K., *Tetrahedron Lett.*, 1992, **33**, 6807.
59. Hawkins, J.M. and Loren, S., *J. Am. Chem. Soc.*, 1991, **113**, 7794.
60. Danishefsky, S.J. and De Ninno, M.P., *Angew. Chem. Int. Edn. Engl.*, 1987, **26**, 15.
61. Cervinka, O., Svatos, A., Triska, P. and Pech, P., *Collect. Czech. Chem. Commun.*, 1990, **55**, 230.
62. Baur, T., Chapuis, C. and Kozak, J., *Helv. Chim. Acta*, 1989, **72**, 482.
63. Midland, M.M. and Koops, R.W., *J. Org. Chem.*, 1990, **55**, 5058.
64. Jurczak, J., Golebiowski, A. and Raczko, J., *Tetrahedron Lett.*, 1988, **29**, 5975.
65. Jankowski, K., *J. Chem. Soc., Chem. Commun.*, 1987, 676.

66. Maruoka, K., Itoh, T., Shirasaka, T. and Yamamoto, H., *J. Am. Chem. Soc.*, 1988, **110**, 310.
67. Terada, M., Mikami, K. and Nikai, T., *Tetrahedron Lett.*, 1991, **32**, 935.
68. Gao, Q., Maruyama, T., Mouri, M. and Yamamoto, H., *J. Org. Chem.*, 1992, **57**, 1951.
69. (a) Togni, A., *Organometallics*, 1990, **9**, 3106; (b) Togni, A., Rist, G., Rihs, G. and Schweiger, A., *J. Am. Chem. Soc.*, 1993, **115**, 1908.
70. (a) Waldmann, H. and Braun, M., *L. Ann. Chem.*, 1991, 1045; (b) Waldmann, H., Braun, M. and Draeger, M., *Tetrahedron: Asymmetry*, 1991, **2**, 1231.
71. Bailey, P.D., Brown, G.R., Korber, F., Reed, A. and Wilson, R.D., *Tetrahedron: Asymmetry*, 1991, **2**, 1263.
72. McFarlane, A.K., Thomas, G. and Whitting, A., *Tetrahedron Lett.*, 1993, **34**, 2379.
73. Hamley, P., Helmchen, G., Holmes, A.B., Marshall, D.R., MacKinnon, J.W.M., Smith, D.F. and Ziller, J.W., *J. Chem. Soc., Chem. Commun.*, 1992, 786.
74. Hamada, T., Zenkoh, T., Sato, H. and Yonemitsu, O., *Tetrahedron Lett.*, 1991, **32**, 1649.
75. (a) Hattori, K. and Yamamoto, H., *Synlett*, 1993, 129; (b) *J. Org. Chem.*, 1992, **57**, 3264; (c) *Tetrahedron*, 1993, **49**, 1749.
76. Gouverneur, V., Dive, G. and Ghosez, L., *Tetrahedron: Asymmetry*, 1991, **2**, 1173.
77. (a) Brouillard-Poichet, A., Defoin, A. and Streith, J., *Tetrahedron Lett.*, 1989, **30**, 7061; (b) Defoin, A., Brouillard-Poichet, A. and Streith, J., *Helv. Chim. Acta.*, 1992, **75**, 109.
78. (a) Gouverneur, V. and Ghosez, L., *Tetrahedron Lett.*, 1991, **32**, 5349; (b) Defoin, A., Pires, J., Tissot, I., Tschamber, T., Bur, D., Zehnder, M. and Streith, J., *Tetrahedron: Asymmetry*, 1991, **2**, 1209; (c) Defoin, A., Brouillard-Poichet, A. and Streith, J., *Helv. Chim. Acta.*, 1991, **74**, 103.
79. (a) Werbitzky, O., Klier, K. and Felber, H., *L. Ann. Chem.*, 1990, 267; (b) Schuerrle, K., Beier, B. and Piepersberg, W., *J. Chem. Soc., Perkin Trans. 1*, 1991, 2407.

8 Transition metal catalysts for asymmetric reduction

J.-P. GENET

A great number of optically active compounds contain a hydrogen atom at the stereogenic centre. As this hydrogen atom can be introduced into appropriate unsaturated precursors by hydrogenation reactions, asymmetric hydrogenation is of particular importance to access highly enantiomerically pure compounds.

Enantioselective catalysis using chiral transition metal complexes (among several possibilities) appears as one of the most efficient methods since a small amount of material can, in principle, produce a large amount of optically active product. In this chapter, only enantioselective hydrogenation reactions catalysed by optically active transition metal catalysts are included. Several reviews have been devoted to enantioselective catalysis.[1]

8.1 Transition metal complexes for asymmetric hydrogenation reactions

In transition metal-catalysed enantioselective processes, the chirality is commonly introduced by the presence of chiral ligands bound to the transition metal. In the first studies of homogeneous asymmetric hydrogenations, a modification of the achiral Wilkinson's catalyst was used in the hydrogenation of prochiral alkenes. Horner[2] and Knowles[3] replaced triphenyl phosphine by a chiral phosphine having the phosphorus atom as the stereogenic centre, and modest optical yields were observed. An important improvement was introduced when Kagan demonstrated that a chiral phosphorus atom is not necessary if a chiral bidentate ligand, such as DIOP, with the chirality at the carbon skeleton is used.[4] The basic idea was that a suitable functionality and skeletal rigidity or flexibility of the phosphine ligand would contribute the differentiation of transition states needed to accomplish enantioselective catalysis. In order to increase optical yields, a range of bidentate phosphines were synthesized and became the most important type of ligand.

8.2 Chiral ligands

There are reviews for the preparation[5] and classification of optically active ligands used in enantioselective catalysis.[6] Most of the diphosphines used

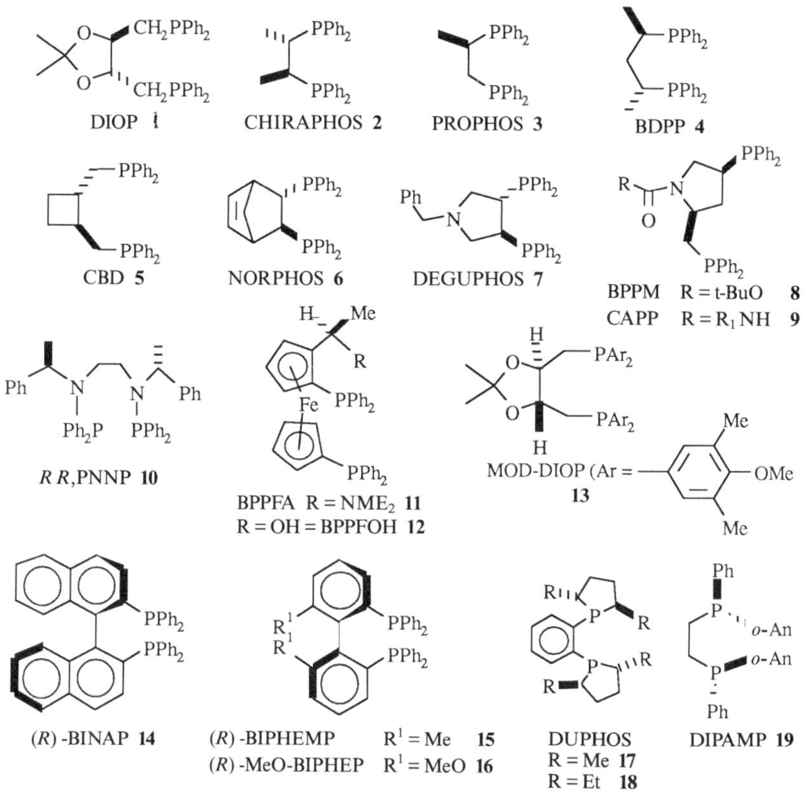

DIOP 1 CHIRAPHOS 2 PROPHOS 3 BDPP 4

CBD 5 NORPHOS 6 DEGUPHOS 7

BPPM R = t-BuO 8
CAPP R = R_1NH 9

R R,PNNP 10

BPPFA R = NME_2 11
R = OH = BPPFOH 12

MOD-DIOP (Ar = 3,5-dimethyl-4-methoxyphenyl) 13

(*R*)-BINAP 14

(*R*)-BIPHEMP R^1 = Me 15
(*R*)-MeO-BIPHEP R^1 = MeO 16

DUPHOS
R = Me 17
R = Et 18

DIPAMP 19

Figure 8.1 Ligands for asymmetric hydrogenation catalysts.

are 1,2-diphosphines or 1,4-diphosphines. Some of the familiar chiral diphosphines that are mentioned in this chapter are shown in Figure 8.1.

The diphosphines can be classified according to their structure, the asymmetry being in a side-chain or at the phosphorus. Phosphines having a chiral side-chain are amongst the most easily accessible because their syntheses start with compounds from the chiral pool.[5] In contrast, the family of disphosphines having two stereogenous phosphorus atoms is much more difficult to obtain. A general method to create an asymmetric phosphorus centre was introduced by Mislow.[7] A general method is also now available. Oxazaphospholidines derived from ephedrine are efficient starting materials for the large-scale preparation of DIPAMP (**19**)[8] and analogues.[9] Atropisomeric ligands such as BINAP (**14**),[10] BIPHEMP (**15**)[11] and MeO-BIPHEP (**16**)[12] have been obtained by resolution. A new class of chiral C_2 symmetric bis(phospholane) ligand DUPHOS (**17** and **18**) has been prepared recently by Burk *et al.*[13]

8.3 Preparation of chiral catalysts

8.3.1 Chiral rhodium catalysts

For the preparation of the so-called *in situ* rhodium catalysts, the commercially available $[RhCl_2(alkene)_2]_2$ [alkene = 1,5-cyclooctadiene (cod), norbornadiene (nbd)] is used as starting material, adding the correct amount of chiral diphosphines. However, cationic chiral complexes (**20**) and (**21**)[14] $[(alkene)_2Rh\ P{*}P]^+$, X^- ($X^- = PF_6^-$, ClO_4^-, $CF_3SO_3^-$) are more active and give reliable results.[15,16] Scheme 8.1 illustrates typical ways to prepare chiral rhodium catalysts. These methods are general and very convenient for screening of chiral mono- and diphosphines. Numerous rhodium complexes have been prepared according to this technology.

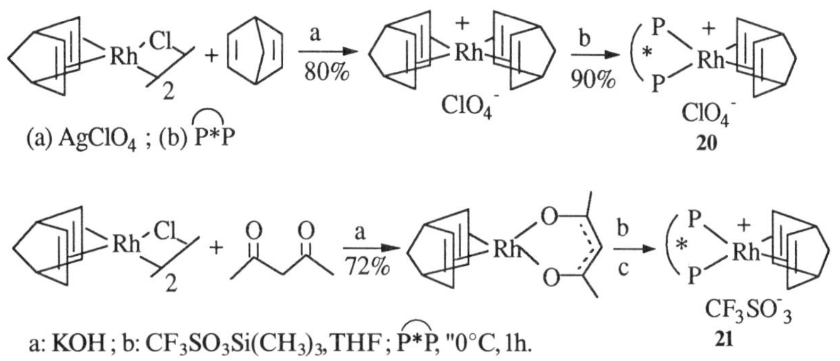

(a) $AgClO_4$; (b) P*P

a: KOH; b: $CF_3SO_3Si(CH_3)_3$, THF; P*P, "0°C, 1h.

Scheme 8.1

8.3.2 Chiral ruthenium catalysts

In contrast to rhodium chemistry, there were few reports on chiral ruthenium complexes. To our knowledge, the first chiral Ru(II) catalyst was discovered by James in 1975.[17] With the advent of $Ru(BINAP)(OAc)_2$ (**22**) introduced by Noyori[18] and $[Ru_2Cl_4(BINAP)_2]{\cdot}NEt_3$ (**23**), discovered by Ikariya and Saburi,[19] the situation changed. These catalysts were prepared from the polymeric complex $[RuCl_2(cod)]_n$ by adding BINAP in toluene reflux in the presence of triethylamine (Scheme 8.2).

The development of ruthenium chemistry for homogeneous asymmetric hydrogenation requires mild and reliable synthesis of chiral ruthenium catalysts. A major improvement was introduced in 1991 when Mallart and Genet demonstrated that $P{*}PRu(2\text{-methylallyl})_2$ complexes (**25**) are available (Scheme 8.3).[20] This synthesis uses easily accessible $Ru(cod)(2\text{-methylallyl})_2$ (**24**) complex as starting material. The methodology is general and several chiral diphosphines have been used, including DIPAMP.[21]

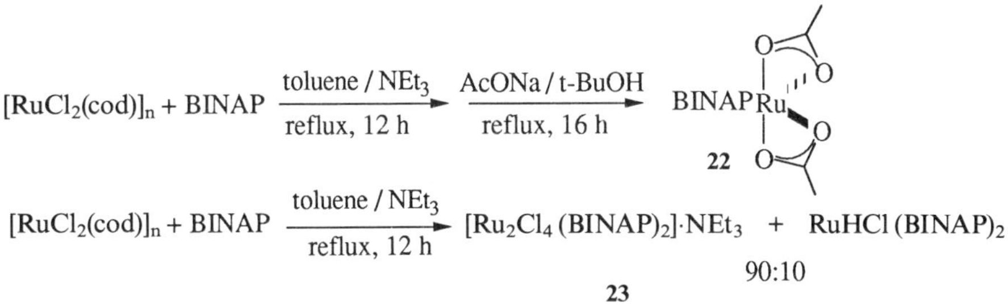

$[RuCl_2(cod)]_n$ + BINAP $\xrightarrow[\text{reflux, 12 h}]{\text{toluene / NEt}_3}$ $\xrightarrow[\text{reflux, 16 h}]{\text{AcONa / t-BuOH}}$ BINAPRu$(OAc)_2$ **22**

$[RuCl_2(cod)]_n$ + BINAP $\xrightarrow[\text{reflux, 12 h}]{\text{toluene / NEt}_3}$ $[Ru_2Cl_4(BINAP)_2]\cdot NEt_3$ + RuHCl(BINAP)$_2$
23 90:10

Scheme 8.2

24 $\xrightarrow[\text{hexane, 50°C, 5h or toluene}]{\text{P*P}}$ **25** $\xrightarrow{\text{HX}}$ **26**

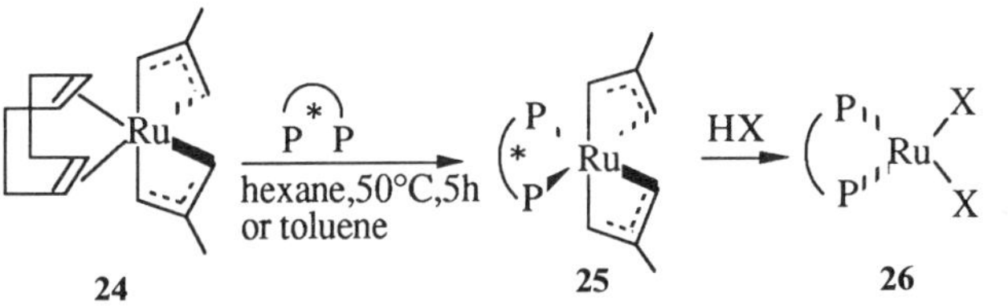

P*P = DIOP, CHIRAPHOS, PROPHOS, BPPM, DIPAMP, CBD, NORPHOS, DEGUPHOS, BINAP, BIPHEMP

Scheme 8.3

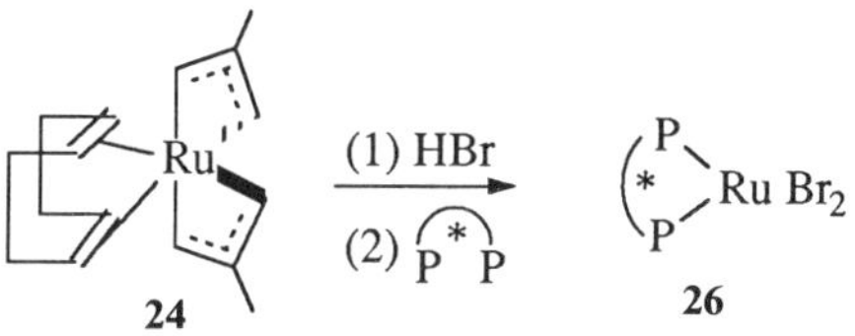

24 $\xrightarrow[\text{(2) P*P}]{\text{(1) HBr}}$ (P*P)Ru Br_2 **26**

P*P = DUPHOS, BINAP, DIOP, BIPHEMP, etc.

Scheme 8.4

These preformed (P*P)Ru(2-methylallyl)$_2$ complexes are useful precursors, after protonation with HX, of the corresponding chiral dihalides (P*P)RuX_2.[22] Interestingly, the same dibromide catalysts were prepared directly *in situ* from Ru(cod)(2-methylallyl)$_2$ by adding HBr in the presence of the appropriate chiral ligand (Scheme 8.4).[23] This method allows the screening of a large set of new chiral phosphines with Ru(II) complexes in asymmetric hydrogenation.

Heiser has also found a synthetically useful method for the preparation of atropisomeric diphosphine Ru(II) biacetate complexes (**27**), including BINAP, BIPHEMP and MeO-BIPHEP (Scheme 8.5).[24] Brown has also described a facile preparation of new chiral complexes P*P Ru(allyl)

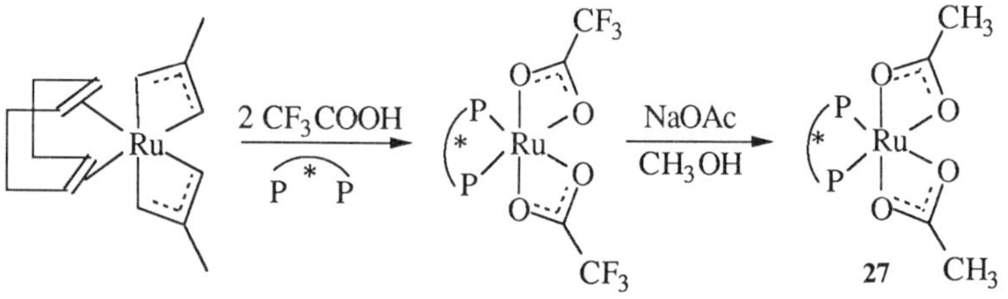

Ru; 2 CF_3COOH, P * P; P, P, Ru, O, O, O, O, CF_3, CF_3; NaOAc, CH_3OH; P, P, Ru, O, O, O, O, CH_3, CH_3; **27**

Scheme 8.5

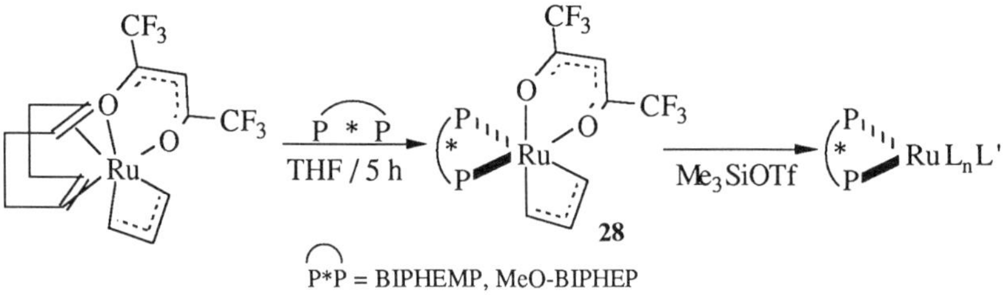

CF_3, O, O, Ru, CF_3; P * P, THF / 5 h; P, P, Ru, O, O, CF_3, CF_3, **28**; Me_3SiOTf; P, P, RuL_nL'

P*P = BIPHEMP, MeO-BIPHEP

Scheme 8.6

hexafluoroacetylacetonate (**28**) (Scheme 8.6).[25] Noyori has also developed a novel *in situ* preparation of $RuCl_2$(arene)(BINAP)[26a] and cationic [Ru-X(arene)BINAP]$^+$Y$^-$ (**29**) (Scheme 8.7).[26]

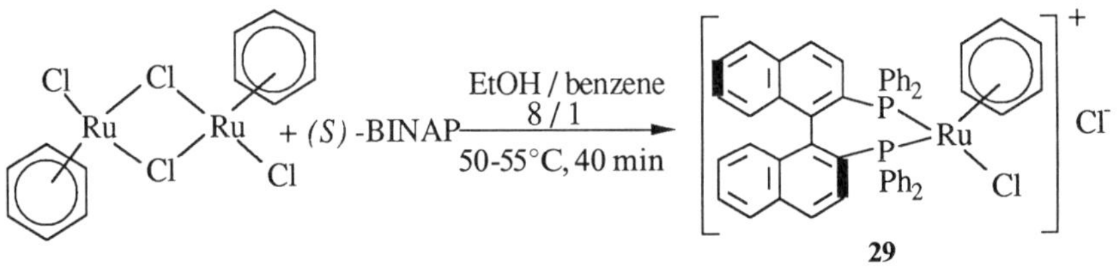

Cl, Cl, Ru, Cl, Ru, Cl + (*S*)-BINAP; EtOH / benzene 8 / 1, 50-55°C, 40 min; Ph_2, P, P, Ph_2, Ru, Cl, Cl^-, +; **29**

Scheme 8.7

Very recently some groups have described some improvement for the synthesis of (BINAP)Ru(II) catalysts.[27]

8.3.3 *Chiral iridium and cobalt catalysts*

Homogeneous hydrogenation is not limited to rhodium and ruthenium catalysts. Iridium complexes have also been used in the reduction of alkenes.[28] A recent novel *in situ* preparation of chiral iridium complexes of type (**30**) has been described by Spingler,[29] and Osborn has recently reported a new family of air-stable chiral Ir(III) anionic complexes (**31**) (Scheme 8.8).[30]

Pfaltz has devised a (semicorrinato)cobalt(I) complex (**32**). This is prepared

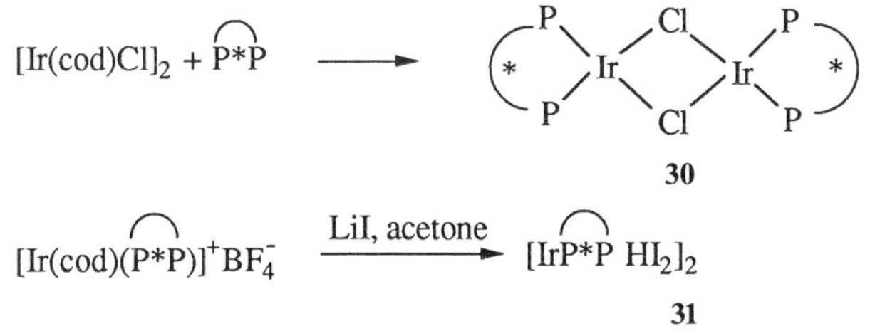

$[Ir(cod)Cl]_2$ + P*P ⟶ **30**

$[Ir(cod)(P*P)]^+BF_4^-$ $\xrightarrow{\text{LiI, acetone}}$ $[IrP*P\ HI_2]_2$ **31**

Scheme 8.8

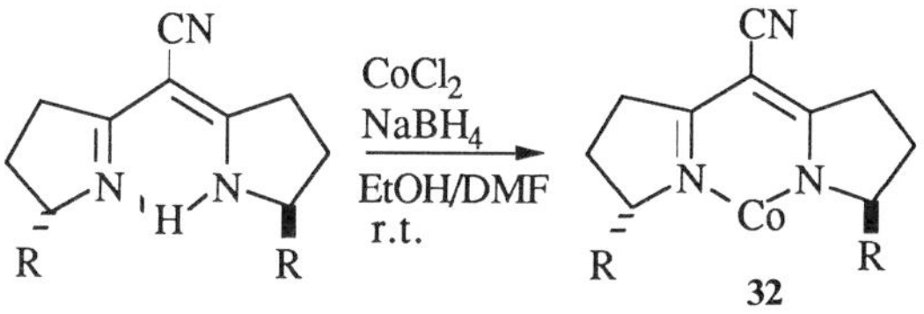

$CoCl_2$, $NaBH_4$, EtOH/DMF, r.t. ⟶ **32**

Scheme 8.9

in situ from cobalt(II) chloride and a C_2-symmetric semicorrin ligand as shown in Scheme 8.9.[31]

8.4 Asymmetric hydrogenation of alkenes

It is impossible to consider here all the alkenes and chiral phosphines which have been used in asymmetric reduction using transition metal catalysts. The results have been summarized in many reviews covering the whole field.[32,16a] In this chapter, enantioselective hydrogenations catalysed by rhodium, ruthenium, cobalt and iridium are taken into account. Many alkenes have been reduced in the presence of transition metal complexes. Optical yields in the asymmetric hydrogenation of isolated prochiral alkenes are low. In order to have high optical induction, it is important that the substrate has a functional group next to the carbon–carbon double bond, because this group is able to coordinate the metal providing a rigid system with special chiral phosphines.

8.4.1 Hydrogenation of dehydroamino acids and related substrates

(a) *Dehydroamino acids.* The α-acylamino acrylic acids were the first successful substrates used in homogeneous asymmetric hydrogenation. Kagan reported in 1971 the asymmetric hydrogenation of (*Z*)-α-acetylamino cinnamic acid with Rh[(−)-DIOP].[4] Knowles has also demonstrated that chiral phosphines such as (*R*,*R*)-DIPAMP are very good ligands for the asymmetric hydrogenation of various dehydroamino acids.[33]

These substrates are easily available by many methods and they are excellent precursors for the preparation of optically active α-amino acids. Presently, a great number of metal–chiral phosphine complexes are known to produce *N*-acetylalanine and *N*-acetylphenylalanine from dehydroalanine (**33**) and dehydrophenylglycine (**34**), some of which are presented in Table 8.1. The bidendate ligands such as DIOP,[34] CHIRAPHOS,[15] DIPAMP,[33] BINAP,[36] PROPHOS,[35] BDPP[38] and BPPFA[40] provide very efficient catalysts. The hydrogenation reactions can be conducted under mild conditions at room temperature and at normal or low pressure of hydrogen. The acetamido acrylic acid or ester coordinates to the metal[44] thus forming a chelate ring, which assures exceptionally high ee values of up to 98% (see Table 8.1). However, the geometry of dehydroamino acid has been shown to dramatically influence the rate and selectivity of hydrogenation with rhodium catalysts. In cases where (*Z*)-isomers are hydrogenated with relatively high enantioselectivity, the corresponding (*E*)-dehydroamino acids are often reduced with much lower selectivity (see Table 8.1).[33] Burk has reported very recently a new class of C_2-symmetric bis(phospholane) ligand (DUPHOS) and its use in asymmetric hydrogenation of dehydroamino acids, with enantioselectivities approaching 100%. Interestingly, in these hydrogenation mixtures (*Z*)- and (*E*)-acetamido acrylic esters (**35**) were reduced with (*R*,*R*)-Pr-DUPHOS in exceptionally high ee, up to 99.5%.[13b] These results are particularly important for substrates which are difficult to prepare in isomerically pure form.

Ruthenium(II) catalysts also serve as effective catalysts for the hydrogenation of dehydroamino acids with ee varying between 76 and 92%.[19,37,41] Interestingly, the chirality induced in the product by the use of ruthenium catalysts with BINAP ligand is opposite to that obtained with rhodium. The carboxylic functionality in dehydroamino acids can be replaced by several other electron-withdrawing groups[32c] including cyano, alkoxycarbonyl and keto groups.

Dipeptides and oligopeptides have been obtained with high selectivity by asymmetric hydrogenation of the corresponding prochiral dehydropeptides.[45]

Rhodium catalysis is of practical significance;[1g] (*S*)-dopa, a drug for the treatment of Parkinson's disease, has been prepared from (**36**) (Scheme 8.10)[46] at Monsanto Co. USA and by UEB Isis Chemie (Zwickan, Germany).[47] The (*S*)-product thus prepared is sold in combination with carbidopa under the trade name Isicon. Another efficient process for this asymmetric hydrogenation was designed by Beck and Nagel at Degussa Co., using DEGUPHOS as the ligand. A catalytic ratio of $1{:}10^4$ is enough for performing the hydrogenation with 100% ee under 10–15 bars at 50°C within 4 h (Scheme 8.10).[48]

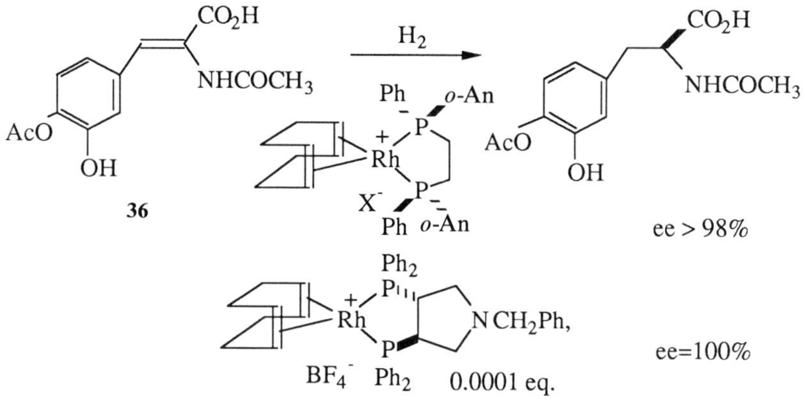

Scheme 8.10

Another industrial application using the Rh-PNNP catalyst (**37**) was recently developed in Italy by ANIC S.P.A. for preparing (*S*)-phenylalanine, a component of the sweetener aspartam (Scheme 8.11).[49]

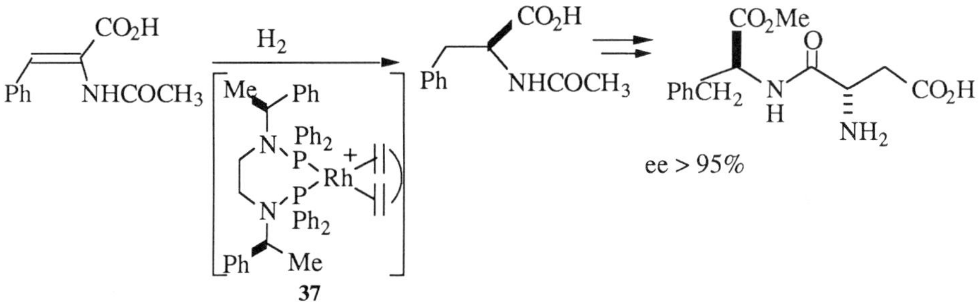

Scheme 8.11

Chiral rhodium catalysts are impressive for giving remarkable enantioselectivities. The behaviour of these man-made catalysts rivals that of enzymatic catalysts. The mechanism of the reaction and the origin of the enantioselectivity were examined in detail by Brown[44,54b] and Halpern[50] and are set out in Scheme 8.12. It was thought that the preferential binding of the substrate determined the sense of the asymmetric induction. On the basis of combined kinetic and NMR studies, Halpern showed that the pre-equilibrium between the two diastereomeric complexes (**38**) and (**39**) is not the important step for controlling the enantioselectivity. The system P*P = CHIRAPHOS that gives the predominant (*R*)-enantiomer (**40**) is actually formed from the less abundant diastereomer (**38**), which reacts about 10^3 times faster than the major diastereomer. Since (**38**) and (**39**) interconvert rapidly, the major

Table 8.1 Asymmetric hydrogenation of dehydroamino acids with various chiral phosphines as ligands of rhodium and ruthenium catalysts

Substrate	R^1	R^2	Isomer	Catalyst	ee (%)	Configuration	Ref.
33 ($CH_2=C(CO_2R^1)NHCOR^2$)	H	Me		Rh[(*R*,*R*)-DIOP]	73	(*R*)	34
	H	CH_2Cl		Rh[(*R*,*R*)-DIPAMP]	95	(*S*)	33
	H	Me		Rh[(*S*,*S*)-CHIRAPHOS]	88	(*R*)	15
	H	Me		Rh[(*R*)-PROPHOS]	90	(*S*)	35
	H	Ph		Rh[(*S*)-BINAP]	98	(*R*)	36
	Me	Me		Rh[(*R*,*R*)-Pr-DUPHOS]	99.8	(*R*)	13b
	H	Me		Ru_2Cl_4[(*R*)-$BINAP_2$]NEt_3	76	(*R*)	19
	H	Me		$RuBr_2$[(*R*,*R*)-BIPHEMP]	77	(*S*)	37
	H	Me		$RuBr_2$[(*R*,*R*)-DIPAMP]			
34 ($PhCH=C(CO_2R^1)NHCOR^2$)	H	Me	*Z*	Rh[(*S*,*S*)-BDPP]	92	(*R*)	38
	Me	Ph	*Z*	Rh[(*R*)-BINAP]	99	(*S*)	39
			Z	Rh[(*S*,*R*)-BPPFA]	93	(*S*)	40
	H	Ph	*Z*	Ru_2Cl_4[(*R*)-$BINAP_2$]NEt_3	86	(*R*)	41
	Me	Me	*Z*	Ru[(*R*)-BINAP]	85	(*R*)	42
	Me	Me	*Z*	Rh[(*S*,*S*)-Et-DUPHOS]	99	(*R*)	13b
	H	Ph	*Z*	Rh[(*R*,*R*)-DIPAMP]	93	(*S*)	33
	H	Ph	*E*	Rh[(*R*,*R*)-DIPAMP]	39	(*S*)	33
35 ($EtCH=C(CO_2R^1)NHCOR^2$)	Me	Me	*Z*	Rh[(*R*,*R*)-NORPHOS]	79	(*R*)	43
	Me	Me	*Z*:*E* 3:1	Rh[(*R*,*R*)-Pr-DUPHOS]	99.5	(*R*)	13b

diastereomer is continually converting into the minor as the hydrogenation reaction depletes the concentration of the minor diastereomer (Scheme 8.12).

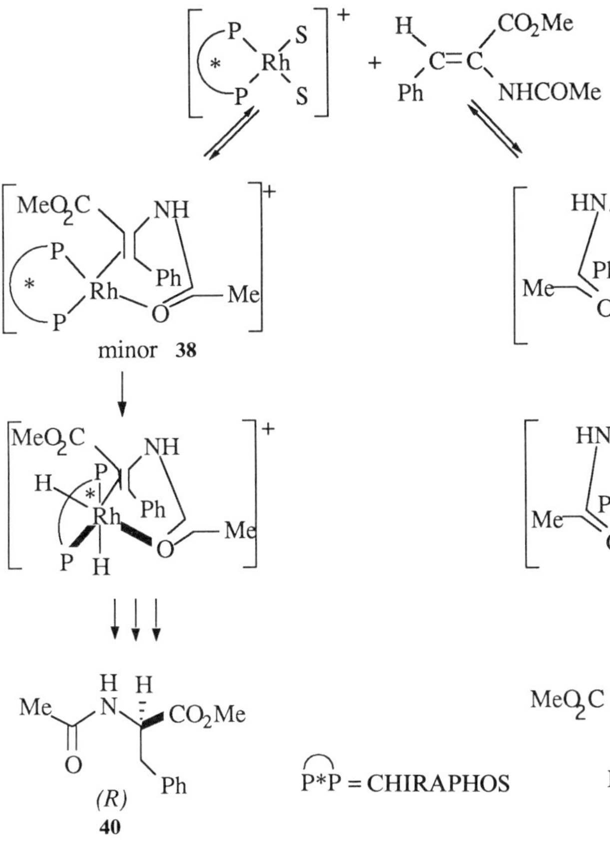

minor **38**

major **39**

(R) **40**

(S)

P*P = CHIRAPHOS

Scheme 8.12

(b) *Enamides.* Enamides without any other functional groups are also interesting prochiral substrates (see Table 8.2). Several chiral rhodium catalysts have been employed giving fair to moderate enantioselectivities.[1a,32] Thus, the maximum optical yield, during the hydrogenation of (**41**), in the presence of Rh(I)DIOP, was only 60%.[51] Replacement of the rhodium by ruthenium proved to be more successful in the hydrogenation of (*E*)-β-acylaminoacrylic acids (**42**) and (**43**), for example the use of Ru(BINAP)$(OAc)_2$ as shown in Table 8.2.[52] Here, as in the dehydroamino acids, the transition from Rh(I) to Ru(II) in the catalyst with the same BINAP leads to the change in the direction of asymmetric induction. A variety of *N*-acyl-(*Z*)-1-benzylidene-1,2,3,4-tetrahydroisoquinolines (**44**) were hydrogenated under

Table 8.2 Asymmetric hydrogenation of enamides catalysed by chiral rhodium and ruthenium catalysts

Substrate	Catalyst	ee (%)	Product	Ref.
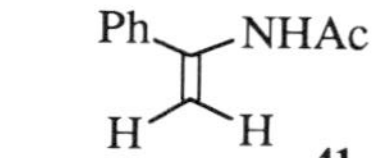 Ph, NHAc, H, H **41**	Rh[(*S*,*S*)-DIOP]	68	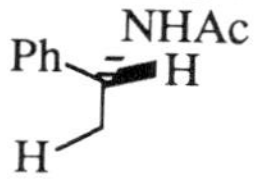 Ph, NHAc, H, H	51
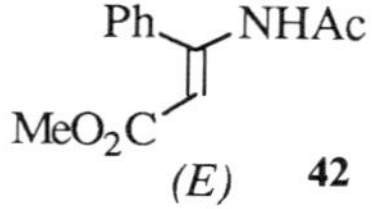 Ph, NHAc, MeO_2C (*E*) **42**	$[Rh(R)\text{-BINAPMeOH}_2]^+ClO_4^-$	60	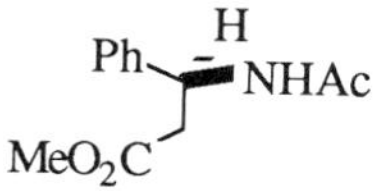 Ph, H, NHAc, MeO_2C	52a
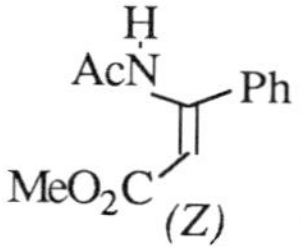 H, AcN, Ph, MeO_2C (*Z*) **43**	$Ru[(R)\text{-BINAP}](OAc)_2$	90	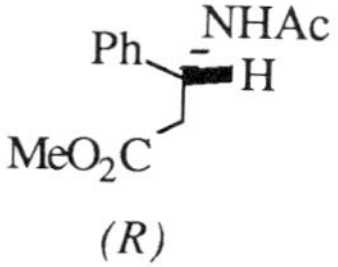 Ph, NHAc, H, MeO_2C	52a
	$Ru[(R)\text{-BINAP}](OAc)_2$	9	(*R*)	52a
MeO, MeO, N-COR, OMe, OMe **44**	$Ru[(R)\text{-BINAP}](OAc)_2$	95	MeO, MeO, N-COR, *, OMe, OMe **45** (*R*)	
	$Ru[(S)\text{-BINAP}](OAc)_2$	95	(*S*)	52
	$RuCl_2[(R)\text{-BINAP}]$, DMF	99	(*R*)	52b
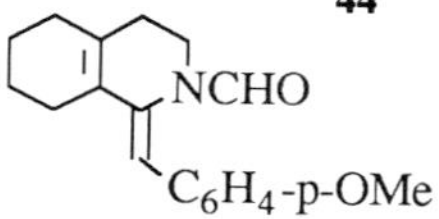 NCHO, C_6H_4-p-OMe **46**	$Ru[(S)\text{-BIPHEMP}](OAc)_2$	98	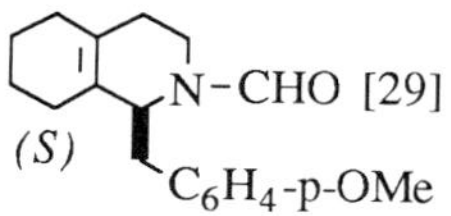 N-CHO [29], C_6H_4-p-OMe (*S*)	24

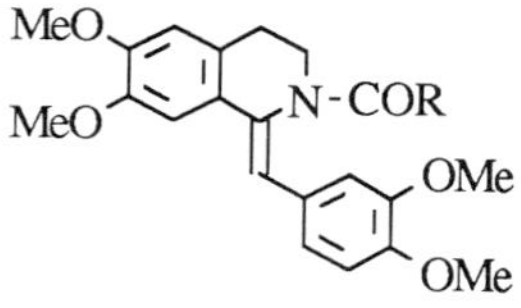

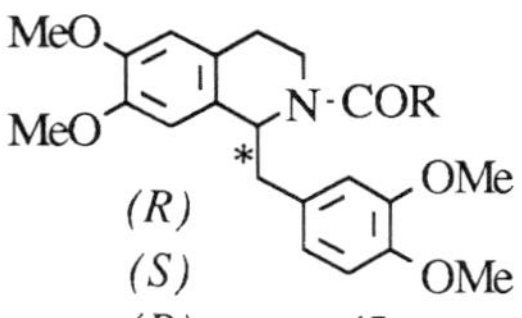

mild conditions (4 atm., 30°C, 48 h) with various chiral ruthenium catalysts in almost quantitative yield to (**45**) and with 95–99% ee.[53] Remarkably, the hydrogenation of (*Z*)-enamide (**46**) with the catalyst either prepared *in situ* or preformed proceeded with high enantioselectivity.[24]

(c) *α,β-Unsaturated acids and esters.* The efficiency in the asymmetric hydrogenation of dehydroamino acids is the result of the presence of the acylamino moiety, which is able to coordinate the metal.[54] The α,β-unsaturated acids or esters without the amide group could not be catalytically hydrogenated with high ee. For example, the enantioselective hydrogenation of α-arylpropenoic acids (**47**) proceeded with moderate enantioselectivity with rhodium catalysts bearing the CBD diphosphine (**11**) (see Table 8.3).[55a] Comparatively good results (up to 94% ee) were obtained during the hydrogenation of itaconic acid (**48**) or the corresponding ester in the presence of various Rh(I) catalysts with the diphosphines BPPM (**8**),[55b] CAPP (**9**)[56] and DIPAMP.[57] Achiwa has designed a modified DIOP (**13**) with neutral rhodium or cationic complexes, and β-aryl-substituted itaconic acids (**50**) were reduced under mild conditions (1 atm. H_2, 30°C, methanol) with 95% ee.[58] These optically active acids (**51**) are valuable structural units in the synthesis of lignans[59] and renin inhibition.[60] The high ee in the asymmetric hydrogenation of itaconic acid is the result of the presence of the carboxylic function on the bond, which is able to coordinate the catalysts. Most asymmetric hydrogenations had been accomplished using chiral rhodium complexes, but with the discovery by Noyori and Takaya,[18] and Ikariya and Saburi[19] of BINAP ruthenium catalysts (**22**) and (**23**) the situation changed.

Table 8.3 Asymmetric hydrogenation of α,β-unsaturated acids with rhodium complexes

Substrate	Catalyst	ee (%)	Product	Ref.
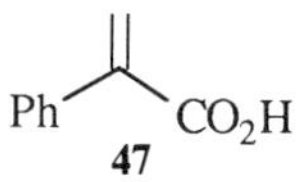 Ph–C(=CH2)–CO2H **47**	Rh[(*R*,*R*)-CBD]	70	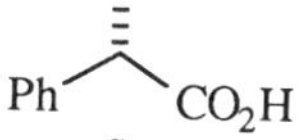 Ph–CH(CH3)–CO2H (*S*)	55a
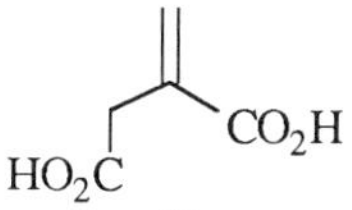 HO2C–CH2–C(=CH2)–CO2H **48**	Rh[(*S*,*S*)-BPPM]	94	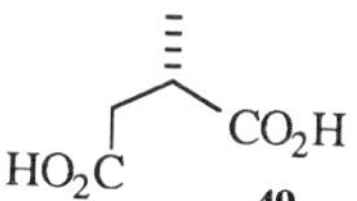 HO2C–CH2–CH(CH3)–CO2H (*S*) **49**	55b
	Rh[(*S*,*S*)-CAPP]	95	(*R*)	56
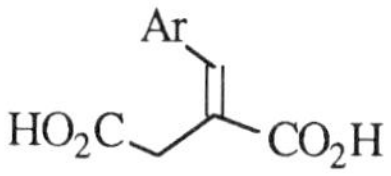 HO2C–CH2–C(=CH–Ar)–CO2H **50**	Rh[(*R*,*R*)-MOD-DIOP]	94	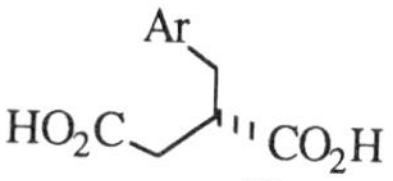 HO2C–CH2–CH(CH2Ar)–CO2H **51**	58

These hexacoordinated ruthenium catalysts containing axially symmetric ligands are efficient for a wide range of substrates.[61] However, various types of Ru(II) complex, (**25**–**29**), are used in practice.

The hydrogenation of itaconic acid (**48**) to methyl succinic acid (**49**) was possible with good enantioselectivity (up to 91% ee) with the binuclear ruthenium complex $Cl_4Ru_2[(R)\text{-BINAP}]_2 \cdot NEt_3$[41] and the monohydride complex RuHCl[(*R*)-BINAP]. However, a better selectivity was obtained (up to 98% ee)[37] using $RuBr_2[(R)\text{-BINAP}]$ or $RuBr_2[(R)\text{-MeO-BIPHEP}]$ prepared *in situ*.[23] Ruthenium-catalysed hydrogenation finds spectacular generality. Several examples for a wide range of α,β-unsaturated acids are shown in Table 8.4. It should be noted that catalysts bearing atropisomeric ligands such as BINAP,[37,41,61] BIPHEMP[37] and MeO-BIPHEP[37] are highly

Table 8.4 Asymmetric hydrogenation of α,β-unsaturated acids with ruthenium catalysts

Substrate	Catalyst	ee (%)	Product	Ref.
52	$Ru_2Cl_4[(R)\text{-BINAP}]_2NEt_3$	88	49 (S)	41
	$Ru[(R)\text{-BINAP}]Br_2$	98		37
	$Ru[(R)\text{-MeO-BIPHEMP}]Br_2$	93		37
	$Ru[(R)\text{-BINAP}](OAc)_2$	91		63
53	$Ru[(R)\text{-BINAP}](allyl)_2$	90	54	20
	$RuBr_2[(R)\text{-MeO-BIPHEP}]$	92		37
	$Ru(R)\text{-DIOP}(allyl)_2$	50		20
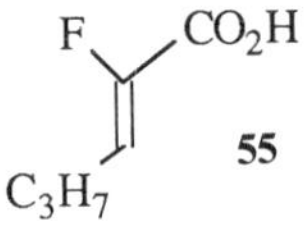 55	$Ru_2Cl_4(BINAP)_2NEt_3$	90	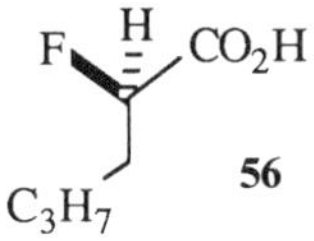 56	64
57	$Ru[(S)\text{-BINAP}](OAc)_2$	97	58	63
	$Ru[(S)\text{-BINAP}](allyl)_2$	85		37
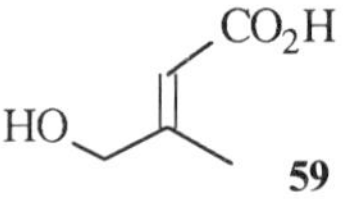 59	$Ru[(S)\text{-BINAP}](OAc)_2$		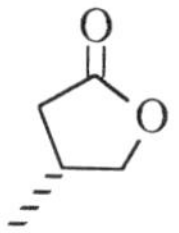 60	63
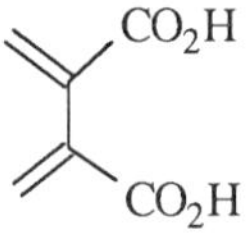 61	$Ru_2Cl_4[(R)\text{-BINAP}] \cdot NEt_3$	96	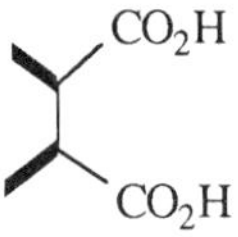 62	64b
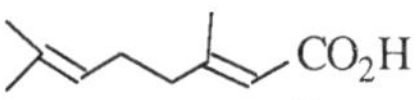 63	$Ru[(R)\text{-BINAP}](OAc)_2$	87	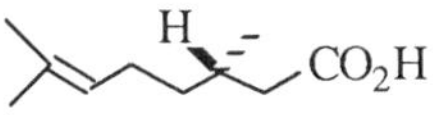 64	63

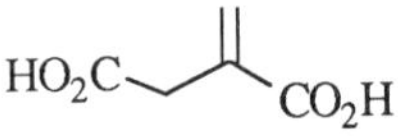

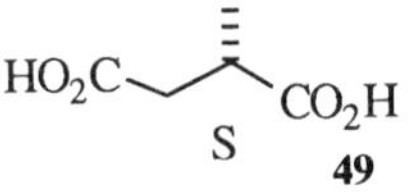

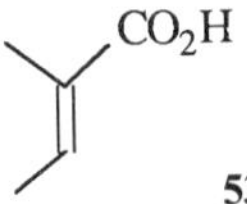

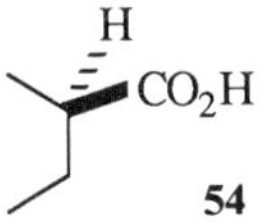

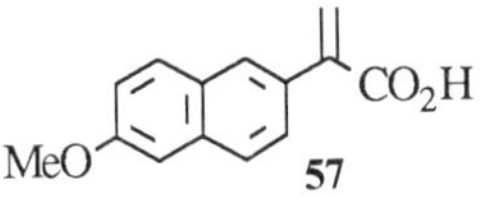

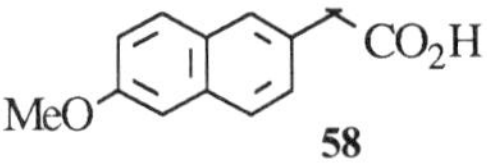

effective for the asymmetric hydrogenation of tiglic acid (**53**) and have obvious advantages over other phosphines (e.g. DIOP, CHIRAPHOS, etc.)[20,37] Under high pressure of H_2 with Ru[(*S*)-BINAP]$(OAc)_2$ [62,63] and Ru(BINAP)$(allyl)_2$, 2-(6-methoxy-2 naphthyl)propenoic acid (**57**) was hydrogenated to (*S*)-naproxen (**58**), an important anti-inflammatory drug marketed by Syntex. This approach has also been used by Montsanto[65] and Merck[27b] for another non-stereoidal analgesic, ibuprophen. Acrylic acids of type (**55**) having fluorine[64] and hydroxyalkyl[61] substituents can also be hydrogenated highly enantioselectively with the same Ru–BINAP complexes. An interesting double asymmetric hydrogenation of the conjugated diene 1,3-butadien-2,3-dicarboxylic acid (**61**) has been performed in the presence of a Ru catalyst to give (*S,S*)-2,3-dimethylsuccinic acid (**62**) with 96% ee.[64b]

Methyl esters of α,β-unsaturated acids are inert to the hydrogenation with ruthenium catalysts. In contrast, using (semicorrinato)Co(I) complexes (**32**), the enantioselective reduction of acrylic esters is a very efficient process,[31,66] as shown by the highly selective conversion of geranic ester (**65**) and the (*Z*)-isomer (**67**) to (*R*)- and (*S*)-ethyl citronellate (**66** and **68**) with high enantioselectivity (up to 94%) (Table 8.5).[66]

The stereochemical course and mechanism of the asymmetric hydrogenation of unsaturated carboxylic acids has been investigated by Takaya[67] and Halpern.[68] The reaction is characterized by operation of the monohydride mechanism, in contrast to the Rh(I)-catalysed reaction.

(d) *Allylic and homoallylic alcohols.* Some chiral rhodium and iridium complexes have been used for the diastereoselective hydrogenation of chiral allylic and homoallylic alcohols, which leads to useful hydroxylated-directed hydrogenations and kinetic resolution procedures.[69] In the hydrogenation of nerol (**70**) and geraniol (**69**), Ru-BINAP dicarboxylate complexes exhibit a very high efficiency (Table 8.6).[70a] (*R*)- or (*S*)-citronellol (**69** and **71**) was obtained in nearly quantitative yield using pressure greater than 30 atm. The values of the enantioselectivity range from 96 to 99% with high chemoselectivity since only the C-2–C-3 double bond is hydrogenated. The C-6–C-7 double

Table 8.5 Asymmetric hydrogenation of α,β-unsaturated esters with cobalt(I) catalysts

Substrate	Catalyst	ee (%)	Product	Ref.
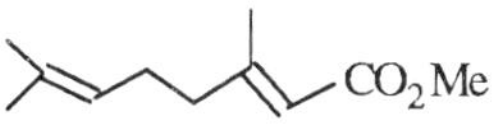 CO_2Me **65**	Semicorrinato Co(I) NaBH$_4$ (**32**)	94	 CO_2Me **66**	66
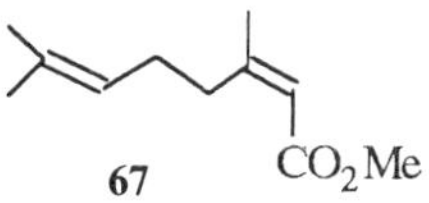 **67** CO_2Me	Semicorrinato Co(I) (**32**)	94	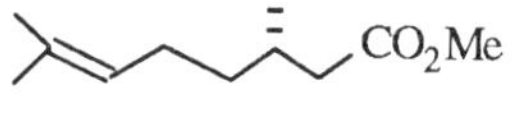 CO_2Me **68**	66

Table 8.6 Asymmetric hydrogenation of allylic alcohols with ruthenium catalysts

Substrate	Catalyst	Product	Ref.
OH **69**	Ru[(*S*)-BINAP]; Ru[(*R*)-BINAP]	OH (*R*) **69** 96-99% ee	70a
OH **70**	Ru[(*S*)-BINAP]; Ru[(*R*)-BINAP]	OH (*S*) **71**	
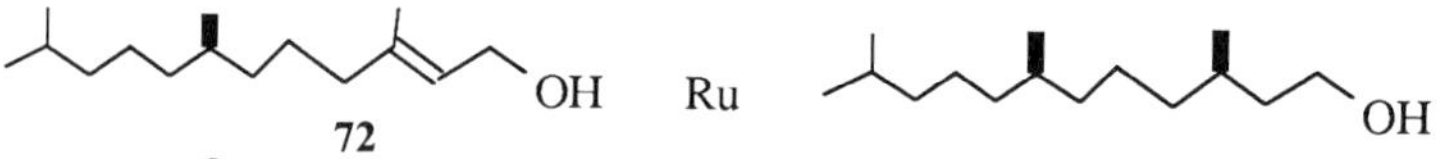 OH **72**	Ru	OH	24
O, HO (±) **73**	Ru[(*S*)-BINAP] $(OAc)_2$	O, HO **74** 98% ee; O, HO	70d

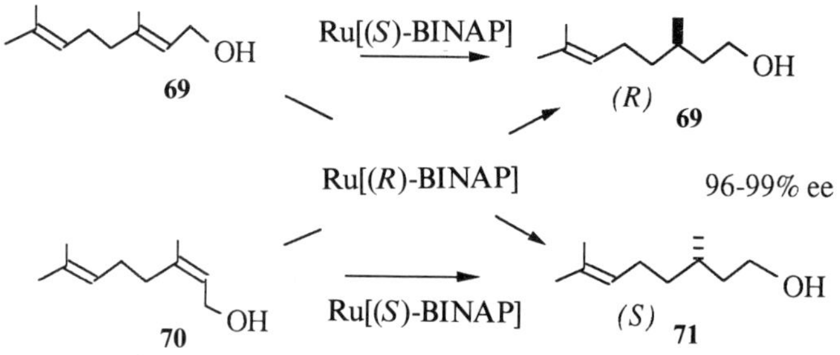

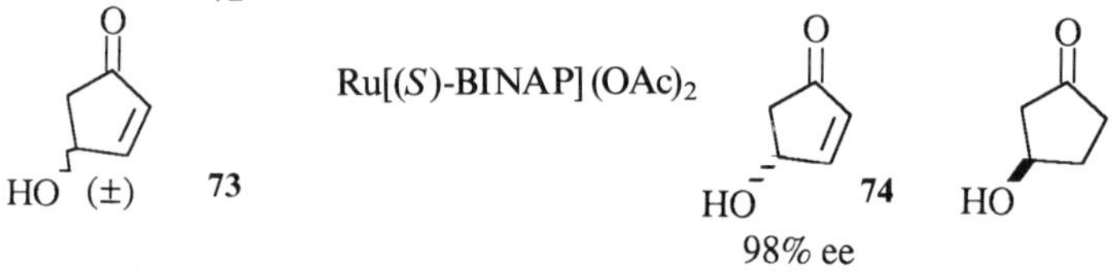

bond remains intact and the product was contaminated by less than 0.5% dihydrocitronellol. In the near future, it is proposed to conduct the process on an industrial scale.[61b] The atropisomeric ligands such as tol-BINAP, BIPHEMP and MeO-BIPHEP have been used successfully in the ruthenium-catalysed hydrogenation of (2*E*,7*R*)-tetrahydrofarnesol (**72**) under mild conditions with a selectivity of up to 96% ee.[24] The catalytic hydrogenation of racemic allylic alcohols with chiral ruthenium catalysts provide an efficient kinetic resolution of the two enantiomers. It was possible from racemic 4-hydroxycyclopentenone (**73**) to generate the hydroxycyclopentenone ((*R*)-**74**) with an optical purity of 98% ee;[70d] this derivative is an important unit for the synthesis of prostaglandins.

In the multistage synthesis of complex molecules of biological interest containing several chiral centres, a new stereogenic centre often appears in the presence of one or several previously formed centres. In such cases, it is possible to improve the diastereoselectivity of the reaction by the use of chiral catalysts (double stereodifferentiation). A practical example of double asymmetric induction is shown in Table 8.7. Hydrogenation of the chiral azetidinone (**75**) appears as an interesting solution for the preparation of 1β-methylcarbapenem. In the presence of the (*S*)-atropisomeric ligand BINAP[70c] or BIPHEMP,[37] the allylic alcohol was hydrogenated to give as major product the α-methyl derivative (**76**), as shown in Table 8.7. Interestingly the

Table 8.7 Asymmetric hydrogenation of a chiral azetidinone with chiral ruthenium complexes

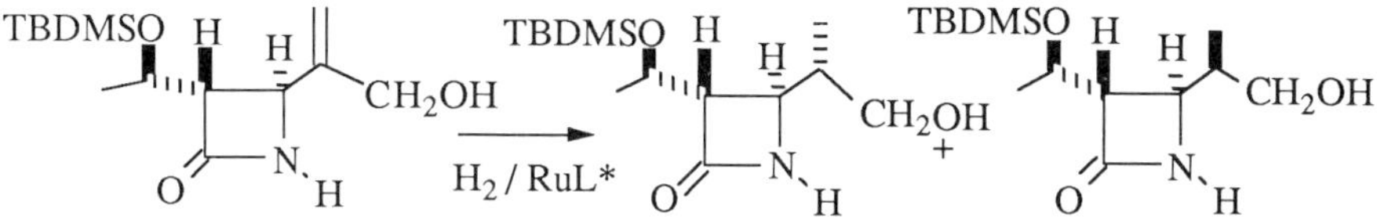

TBDMSO ... CH_2OH **75** —H_2 / RuL*→ **76** + **77**

Catalyst	Conditions: Pressure (atm.)	Conditions: Temperature (°C)	Yield (%)	de (%)	Product	Ref.
Ru[(*S*)-tol-BINAP](OAc)$_2$	1	25	100	78	**76**	70c
Ru[(*R*)-tol-BINAP](OAc)$_2$	2		100	99.9	**77**	70c
Ru[(*S*)-BIPHEMP](allyl)$_2$	15	25	80	56	**76**	37
Ru[(*R*)-BINAP](allyl)$_2$	15	25	100	93	**77**	37

(*R*)-configuration of the atropisomeric ligand produced β-methyl product (**77**) with high diastereoselectivity. The chiral azetidone backbone and the chiral catalyst are acting in concert.

One of the serious problems in homogeneous catalytic hydrogenation is the separation of the reaction products from the active catalyst, which is expensive. Two methods have been devised to overcome these difficulties. The first one involves the proper design of a polymer-bound catalyst.[71,72] Usually immobilization of the catalyst involves the loss of some degree of enantioselectivity.[71] However, optical yields of up to 100% are obtained in the hydrogenation of α-(acetylamino) cinnamic acid and its methyl ester in the presence of rhodium complexes of (3*R*,4*R*)-bis(diphenylphosphino)pyrrolidine (**78**) bound covalently to silica (Scheme 8.13).[72]

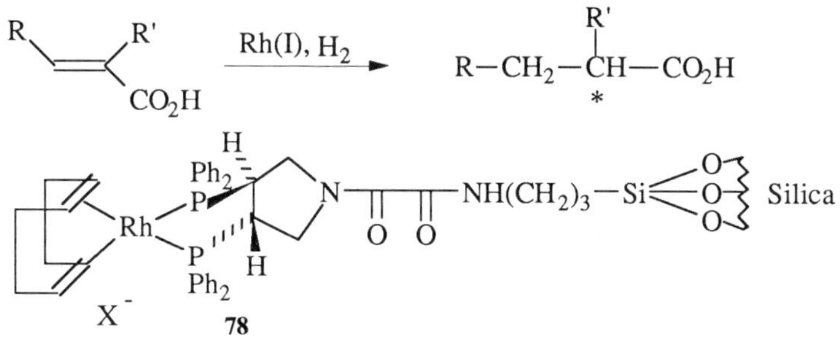

R, R', CO_2H —Rh(I), H_2→ R—CH_2—CH(R')—CO_2H (*)

Rh, X$^-$, Ph_2P, PPh_2, H, N, O, O, NH$(CH_2)_3$—Si(O)$_3$ Silica **78**

Scheme 8.13

A second very elegant solution to this problem consists of using a water-soluble ligand which is poorly soluble in organic media. The catalysis could be carried out in a two-phase system and decantation would allow an easy separation and recovery of the catalyst. Water-soluble rhodium complexes with *meta*-sulfonated diphosphines (*R*,*R*)-CBD (**79**) have been used with some success for the hydrogenation of the dehydroamino acids (ee up to 88%) (Scheme 8.16).[73] Very recently, a Ru(II) complex of sulfonated BINAP (**80**) has been synthesized and this novel water-soluble complex led to excellent selectivities for 2-acylamino acid and methylenesuccinic acid in both aqueous and methanolic media (Scheme 8.14).[74] Schemes 8.13 and 8.14 show methods of asymmetric hydrogenation using Rh(I) and Ru(II) complexes anchored to polymers, and as water-solubilized complexes in aqueous media.

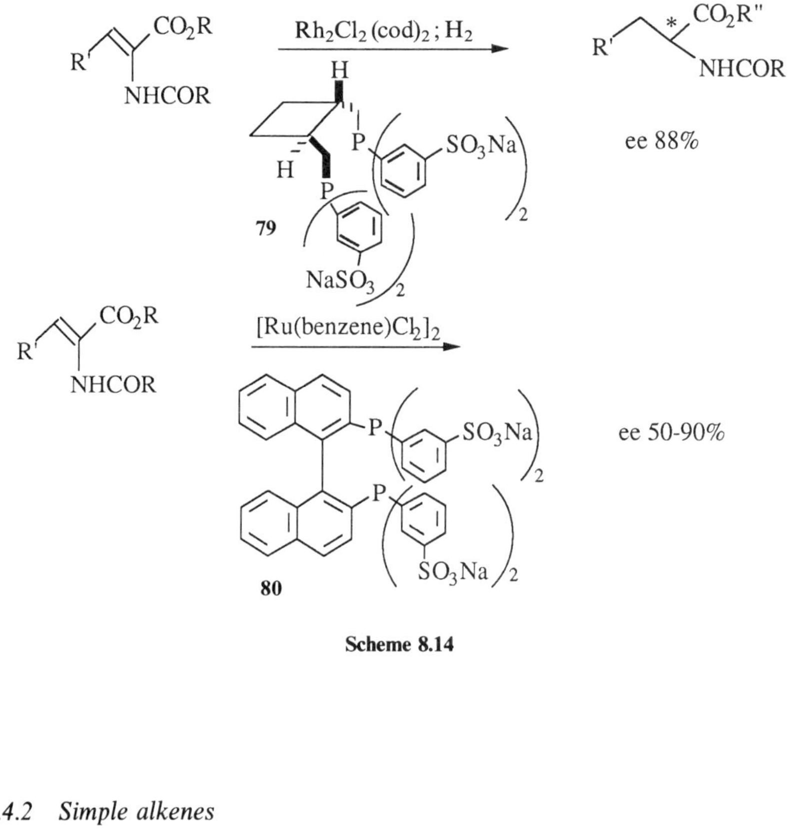

CO_2R R' NHCOR $Rh_2Cl_2(cod)_2$; H_2 → R' * CO_2R'' NHCOR

ee 88%

H H P P SO_3Na 2 $NaSO_3$ 2 **79**

CO_2R R' NHCOR $[Ru(benzene)Cl_2]_2$ →

P SO_3Na 2 P SO_3Na 2 **80**

ee 50-90%

Scheme 8.14

8.4.2 Simple alkenes

Hydrogenation of non-functionalized alkenes in high ee has not been achieved with chiral mononuclear rhodium or ruthenium complexes.[75] Some more encouraging results were attained with polynuclear complexes of ruthenium.[76] However, the hydrogenation of 2-phenyl-1-butene (**81**) to 2-phenylbutane (**82**) in the presence of a cyclopentadienyl C_2-symmetric-titanocene complex was achieved with 96% ee (Scheme 8.15).[77]

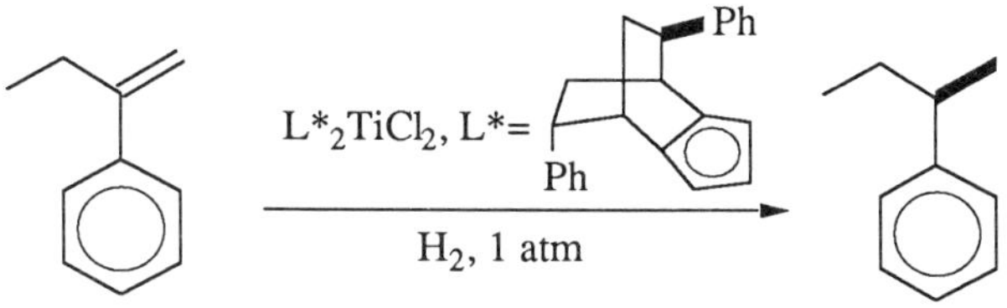

Scheme 8.15

8.5 Asymmetric hydrogenation of keto groups

8.5.1 Ketones

The homogeneous asymmetric hydrogenation of ketones presents some difficulties. Ketones are not reduced with the classical Wilkinson catalyst. However, the use of cationic rhodium complexes, or addition of base to a neutral rhodium complex, affords efficient catalysts for ketone reduction.[78] Marko *et al.* were able to reduce acetophenone (**83**) to 1-phenylethanol (**84**) with 80% enantioselectivity at 69 atm. of H_2 using Rh[(*S*)-BPPP] (Table 8.8). All the ketones that can be reduced with high ee have an additional functional group α or β to the carbonyl groups such as amino,[79] or carboalkoxy, as shown in Table 8.8. Several alkylamino aryl ketones, important intermediates for pharmaceutical products, were reduced with chiral rhodium catalysts to the corresponding secondary alcohols in high optical yields. Thus, the (*S,R*)-BPPFOH ligand (**12**) in the rhodium(I) complex hydrogenates the aminoacetophenone derivative (**85**) to (**86**), and (**87**) to (**88**) ($\alpha\beta$-adrenoceptor agonist), with quantitative yield and 98% enantioselectivity.[80] Another interesting application of asymmetric hydrogenation is shown in Table 8.8; here (**89**) is reduced with Rh(I) catalysts to (*S*)-propanolol (**90**) a β-adrenergic-blocking agent.[81]

The Ru-BINAP-catalysed hydrogenation exhibits a wider scope than the reaction with rhodium complexes[61] and is capable of producing a variety of functionalized alcohols (**91**–**93**) in high ee (Table 8.9). The dicarboxylate complexes, RuBINAP(OCOR)$_2$ (**22**)[62] or the RuX$_2$(BINAP) (**26**) may be used as catalysts. As shown in Table 8.9, excellent levels of enantioselection (up to 96% ee) have been obtained. Some nitrogen- and oxygen-containing directing groups include hydroxyl, dialkylamino and carboxyl.[82] Interestingly, a halogen atom facilitates the carbonyl hydrogenation. Thus *o*-bromoacetophenone (**94**) afforded the (*R*)-alcohol (**95**) with 92% enantioselectivity, although unsubstituted acetophenone failed to be hydrogenated in a satisfactory manner under comparable conditions.[26,82] The hydrogenation of an unsymmetrical α,α'-dialkoxy derivative of acetone (**96**) takes place with

Table 8.8 Catalytic asymmetric hydrogenation of ketones using rhodium catalysts

Substrate	Catalyst	ee (%)	Product	Ref.
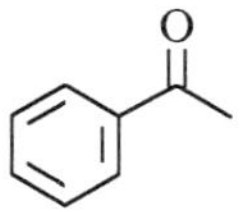 O **83**	Rh[(*S*,*S*)-DIOP]NEt_3	80	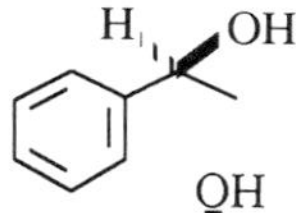 H OH **84**	78
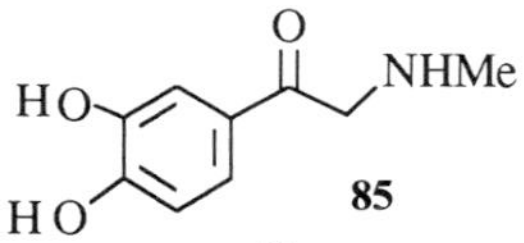 HO HO O NHMe **85**	[Rh[(*R*,*S*)-BPPFOH](nbd)]$^+ClO4^-$	95	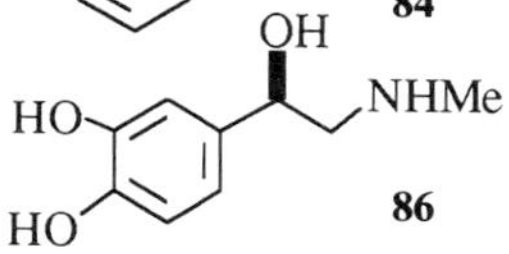 HO HO OH NHMe **86**	79
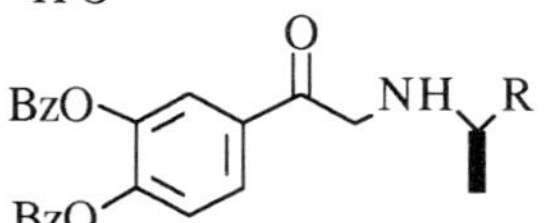 BzO BzO O NH R **87**	Rh[(*S*,*R*)-BPPFOH] 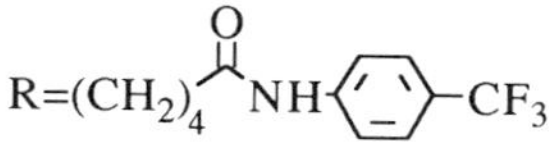 R=$(CH_2)_4$C(O)NH-C_6H_4-CF_3	98	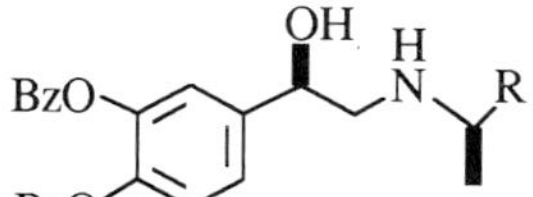 BzO BzO OH H N R **88**	80
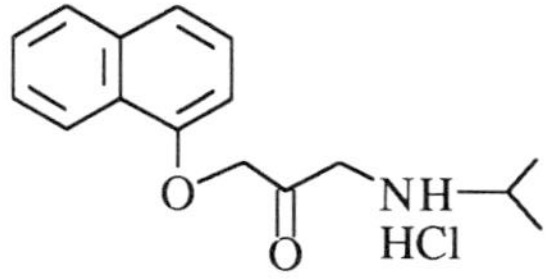 O O NH HCl **89**	Rh[(2*S*,4*S*)-MCCPMB]	91	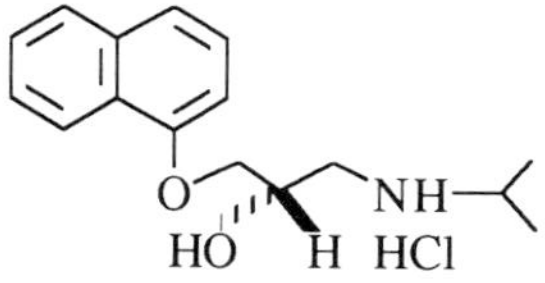 O HO H NH HCl **90**	81

Table 8.9 Catalytic asymmetric hydrogenation of ketones using ruthenium catalysts

Substrate	Catalyst	ee (%)	Product	Ref.
$MeCOCH_2NMe_2$	Ru[(*S*)-BINAP]$(OAc)_2$	96	**91**	82
$MeCOCH_2OH$	$RuCl_2$[(*R*)-BINAP]	92	**92**	82
2-acetylbenzoic acid	$RuBr_2$[(*R*)-BINAP]	92	**93**	82
94	$RuBr_2$[(*R*)-BINAP]	92	**95**	82
96	Ru_2Cl_4[(*S*)-BINAP]$_2NEt_3$	96	$ROCH_2CH(OH)CH_2O{-}CPh_3$	83

high enantioselectivity using the binuclear complex Ru_2Cl_4[(*S*)-BINAP]$_2NEt_3$.[83]

8.5.2 *Alpha-keto esters, acids, amides, lactones*

It was found that the rhodium catalysts are also effective for the reduction of a ketone α to an ester or an acid group. This reaction provides access to optically active α-hydroxycarboxylic acids or esters, which are important precursors for several biologically active products. Thus α-keto esters (**97**) and (**98**) were hydrogenated with rhodium complexes with moderate to good enantioselectivity (Table 8.10).[84,85] Again an excellent level of enantioselectivity has been observed in the reduction of $MeCOCO_2Me$ (**98**) with $RuCl_2$[(*R*)-BINAP].[82] The effectiveness of *in situ* Ru catalysts in the asymmetric hydrogenation of phenylglyoxylic methyl ester (**99**) to (**100**) has been demonstrated recently.[37] Atropisomeric ligands like BINAP and MeO-BIPHEP gave optical yields up to 86%. Interestingly with an *in situ* ruthenium catalyst containing (*R*,*R*)-Me-DUPHOS (**17**), the asymmetric hydrogenation was also carried out with good enantioselectivity (80% ee).[37]

Takaya *et al.* showed that asymmetric hydrogenation is possible with the cationic complex $[RuCl(BINAP)benzene]^{+}Cl^{-}$. Thus hydrogenation of methyl *N*-phthaloyl-3-aminooxopropanoate (**102**) afforded the corresponding (*S*)-alcohol (**103**) in 81% ee, a key step for the preparation of isoserine (Table 8.10).[86]

Many research groups have investigated the asymmetric hydrogenation of dihydro-4,4-dimethyl-2,3-furandione (ketopantolactone) (**104**) to pantolactone (**105**), an important intermediate in the synthesis of the calcium salt of pantothenic acid (vitamin B_5) or D-(+)-pantothenic acid (vitamin B_3)[87] (Table 8.11). Achiwa *et al.* demonstrated that with Rh complexes bearing modified BPPM ligands (e.g. BCPM (**106**), BCCP) the optical yield reached 90%.[88] The catalyst:substrate ratios reached 10000:1. A further increase in enantioselectivity has been observed at Hoffmann la Roche.[89] Interestingly, fluorinated acetic acid derivatives accelerate the reaction with an increase of the selectivity up to 95%.[89] $RuBr_2$[(*R*,*R*)-DIPAMP] catalyst used in the reaction under quite drastic conditions (100 atm. at 50°C) gave 78% optical yield.[21] The importance of the synthesis of D-(−)-pantolactone makes it necessary to renew attempts to improve the catalyst of this process. Mortreux *et al.* have recently found that using (*S*)-Cp-5-oxo-PRONOP (**107**) as its rhodium complex is the most effective catalyst, with the reduction

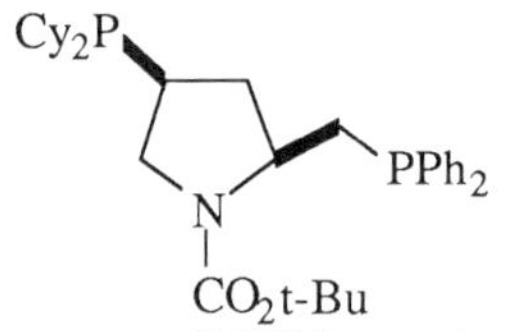

Cy_2P PPh$_2$ N CO_2t-Bu
BCPM

106

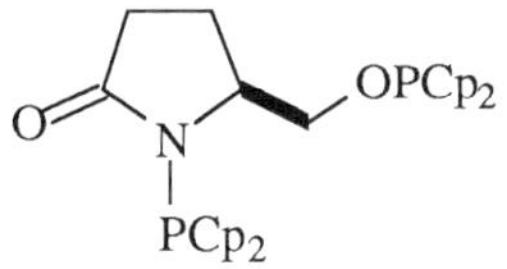

O N OPCp$_2$ PCp$_2$
(*S*) -Cp-5-oxo-PRONOP

107

proceeding at 1 atm. of H_2, 25°C, 1.5 h with 96% ee.[90] The effect of different catalysts and reaction conditions on the efficiency of asymmetric induction is illustrated in Table 8.11.

8.5.3 *Hydrogenation of β-keto derivatives*

The asymmetric hydrogenation of *β*-keto carboxylic esters is particularly useful for enantioselective access to *β*-hydroxycarboxylic esters, which serve as key intermediates for natural product synthesis. Catalytic systems based on Ru(II) complexes containing atropisomeric ligands (e.g. BINAP, BIPHEMP, MeO-BIPHEP) are very efficient and universal in the hydrogenation of a wide range of *β*-keto esters[61] with extremely high enantioselectivities (up to 100%), as shown in Table 8.12.

Table 8.10 Hydrenation of α-keto esters and amides with rhodium and ruthenium catalysts

Substrate	Catalyst	ee (%)	Product	Ref.
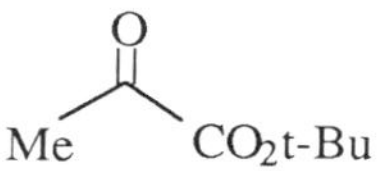 Me, O, CO_2t-Bu **97**	RhCl(BPPM)	70	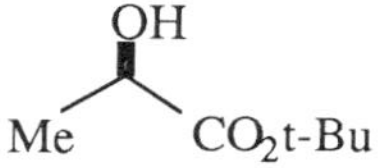 OH, Me, CO_2t-Bu	84
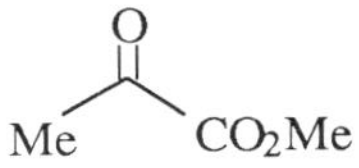 Me, O, CO_2Me **98**	$RuBr_2$[(*R*)-BINAP]	83	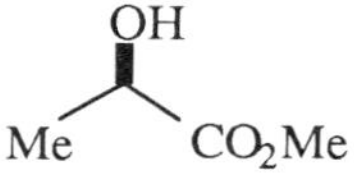 OH, Me, CO_2Me	82
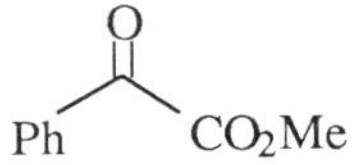 Ph, O, CO_2Me **99**	$RuBr_2$[(*S*)-MeO-BIPHEP]	86	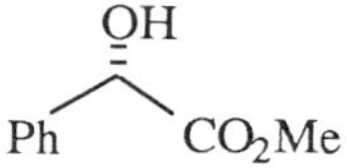 OH, Ph, CO_2Me	37
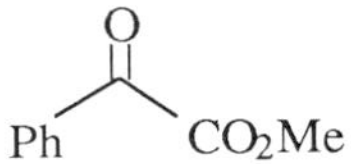 Ph, O, CO_2Me	$RuBr_2$[(*R*,*R*)-Me-DUPHOS]	80	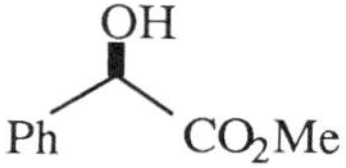 OH, Ph, CO_2Me **100**	37
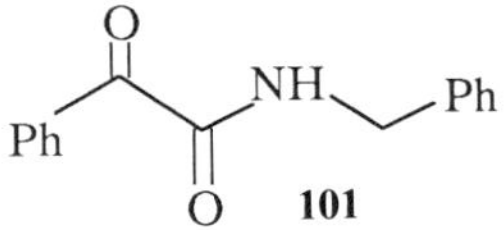 Ph, O, NH, Ph, O **101**	Rh[(*S*)-ProNOP]	70	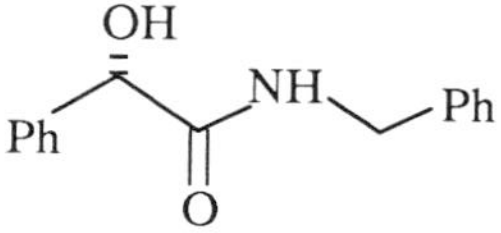 OH, Ph, NH, Ph, O	85
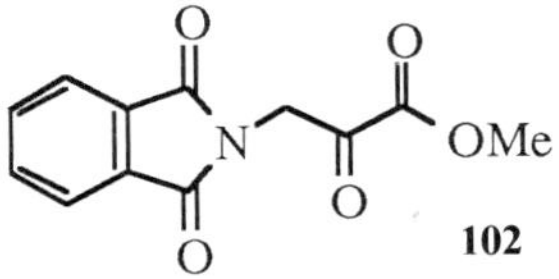 O, N, O, O, OMe, O **102**	$[Ru(Cl)[(R)\text{-BINAP}]benzene]^+Cl^-$	81	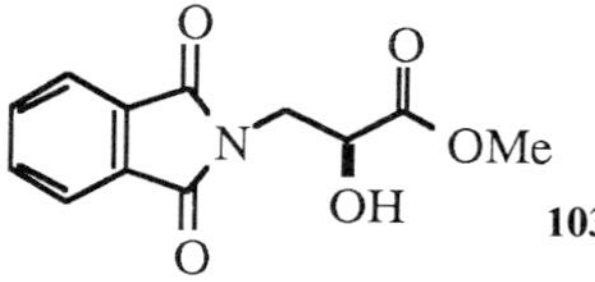 O, N, O, OMe, OH, O **103**	86

Table 8.11 Hydrogenation of ketopantolactone (**104**) with chiral rhodium and ruthenium catalysts

104 $\xrightarrow[H_2]{M_T P * P}$ **105**

Catalyst	Conditions		ee (%)	Ref.
	Pressure (atm.)	Temperature (°C)		
Rh-BPPM	50	50	86	88
Rh-BCPM	50	50	92	88
$RuBr_2$[(*R,R*)-DIPAMP]	100	50	78	21
Rh(Cp-5-oxo-ProNOP)	1	25	96	90

Hydrogenation of β-keto esters catalysed by Ru(II)BINAP dicarboxylate proceeds slowly. However, the related chiral-containing halogen (X = Cl, Br, I) ruthenium complex is highly active for the reduction of β-keto esters. Thus, during the hydrogenation of methyl-3-oxobutanoate[91] at high pressure (100 atm. 30°C, r.t.) in the presence of $RuCl_2$[(*R*)-BINAP], the (*R*)-methyl-3-hydroxybutyrate (**109**) was obtained with an optical yield >99% and almost a quantitative chemical yield (Table 8.12). This catalytic asymmetric hydrogenation using Ru(II)BINAP complexes can be further extended to a variety of β-keto esters (R = Et, i-Pr, Ph, etc.).[91b] The possibilities for the practical use of Ru(II)BINAP in the hydrogenation of β-keto esters were somewhat restricted both by the complexity of the synthesis of the catalyst and by the use of a high pressure of hydrogen.

Recently these difficulties were overcome using a simplified procedure for the *in situ* preparation of the catalysts.[23,24,26] These *in situ* catalysts proved to be extremely active. Thus, it was possible at higher temperature to reduce the pressure of hydrogen to 10–35 atm. Excellent results were also obtained in the hydrogenation of methyl isopropyl acetoacetate (**110**) (ee>99%) and β-keto esters having a long alkyl chain (**111** and **112**: R = $C_{11}H_{23}$, ee = 97%;[24] R = $C_{15}H_{31}$, ee = 96%[37]). An air-stable Ru(II) complex $[RuCpBINAP]^+Cl^-$ has also been found to be effective at high pressure in this reaction.[27e] The low pressure hydrogenation is obviously desirable for industrial production. An interesting result of optimization using the Ikariya complex as crude catalyst in the reduction β-keto esters has been recently described.[92] Noyori demonstrated that the Ru(II)BINAP complex (**29**) at 100°C, 4 atm. H_2 is a highly efficient catalyst for asymmetric hydrogenation of β-keto esters.[26a] The $RuBr_2$[(*S*)-BIPHEMP] prepared from Ru(2-methylallyl)$_2$[(*S*)-BIPHEMP] (Scheme 8.3) showed high efficiency (ee up to 99%) under mild conditions (5 atm. 50°C).[37] Recently, hydrogenation of β-keto esters has been conducted

Table 8.12 Ruthenium asymmetric hydrogenation of β-keto esters using chiral Ru(II) complexes

Substrate	Catalyst	Conditions: Pressure (atm.)	Conditions: Temperature (°)	Product	ee (%)	Ref.
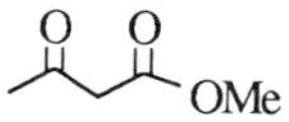 **108** (O O, OMe)	$RuCl_2[(R)\text{-BINAP}]$	100	30	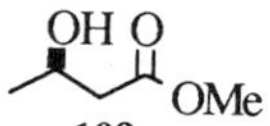 **109** (OH O, OMe)	>99	91
	$RuBr_2[(R)\text{-BINAP}]$ *in situ*	10	80		98	37
	$RuBr_2[(R)\text{-Me-DUPHOS}]$ *in situ*	20	50		87	37
	$RuBr_2[(R)\text{-MeO-BIPHEP}]$	20	50	**109**	>99	37
	$RuCl_2[(S)\text{-BINAP}](benzene)$	4	100		98	26
	$RuBr_2[(S)\text{-BIPHEMP}]$	5	50	**109** 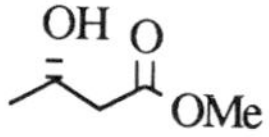 (OH O, OMe)	99	37
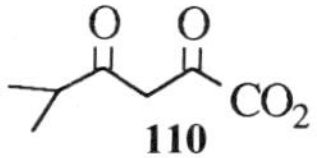 **110** (O O, CO_2)				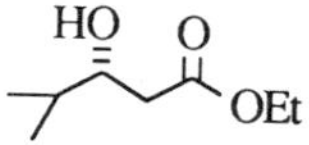 (HO O, OEt)		
	$RuCl_2[(R)\text{-BINAP}]$	100	30		>99	91
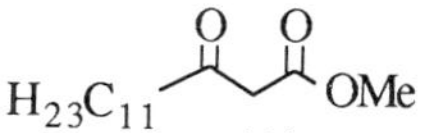 **111** ($H_{23}C_{11}$, O O, OMe)	$RuCl_4(cod)_2, CH_3CN$ (R)-BIPHEMP	30	80	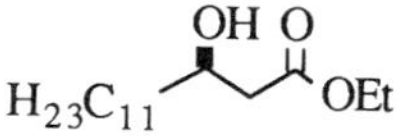 $H_{23}C_{11}$ (OH O, OEt)	97.3	24
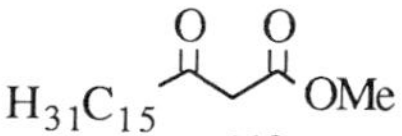 **112** ($H_{31}C_{15}$, O O, OMe)	*in situ* $RuBr_2[(R)\text{-BINAP}]$	50	50	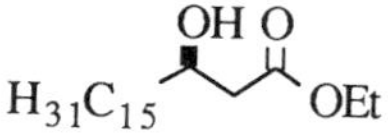 $H_{31}C_{15}$ (OH O, OEt)	96	37

at 3 atm. H_2, 40°C using $[RuCl_2(BINAP)_2]\cdot NEt_3$ in the presence of 0.1% HCl.[27b] The same procedure was used for a practical synthesis of carnitine from the γ-chloro-β-hydroxybutyrate ((*R*)-**114**). Several kinds of Ru(II) catalysts Ru(II)L*$(OAc)_2$, Ru(II)L*$(allyl)_2$ and RuL*Br_2(L*: BINAP and BIPHEMP) are highly efficient at high temperature during the hydrogenation of (**113**), as shown in Table 8.13.

This procedure was successfully employed in the synthesis of (*S*)-carnitine (**116**) (ee = 95%) by reduction of the ammonium salt (**115**) with $[Ru_2Cl_4(BINAP)_2]\cdot NEt_3$ at 100 atm. (Scheme 8.16)[37]

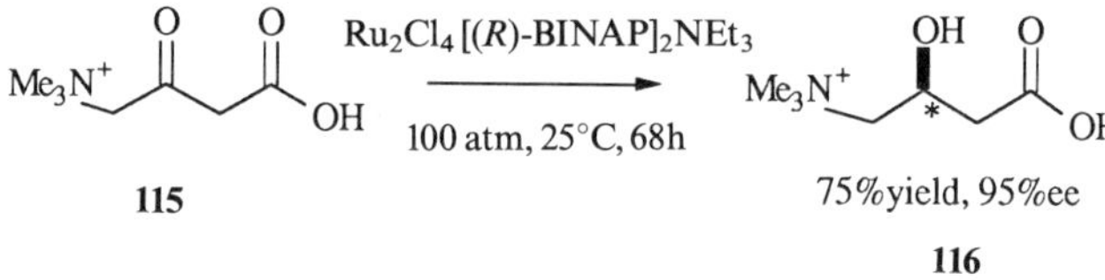

Ru_2Cl_4[(*R*)-BINAP]$_2$NEt$_3$, 100 atm, 25°C, 68h

115 → **116** (75%yield, 95%ee)

Scheme 8.16

Interestingly, the β-keto esters (**117**) and (**119**), having a disubstituted double bond, are chemoselectively hydrogenated under controlled conditions using $RuBr_2$BIPHEMP and $RuBr_2$BINAP to form unsaturated chiral alcohols (**118**) and (**120**) with excellent optical purities under quite mild and controlled conditions (6 atm. H_2, 80°C, 5–30 min, ee up to 99%) (Table 8.14). The Ru(II) complex BINAP discovered by Ikariya and Saburi was also used as crude catalyst in the chemoselective hydrogenation of (**117**) and (**118**).[92]

The advantages of enantioselective ruthenium-mediated hydrogenation are obvious and have been used in several multistage syntheses. An example is

Table 8.13 Hydrogenation of γ-chloro-β-ketobutyrate (**113**) with ruthenium catalyst

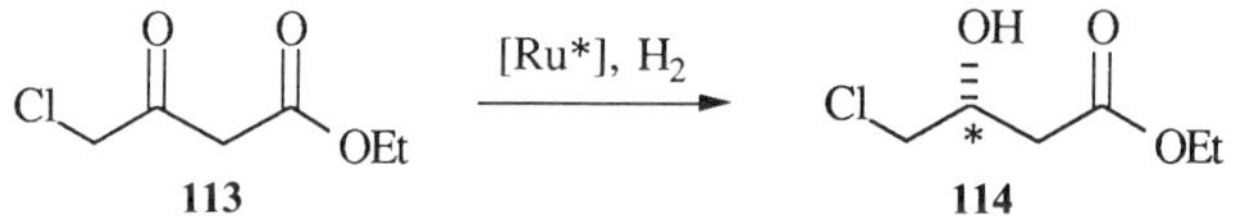

113 —[Ru*], H_2→ **114**

Ru*	Conditions					
	Pressure (atm.)	Temperature (°C)	Time (min)	Yield (%)	ee (%)	Ref.
$RuBr_2$[(*S*)-BINAP]	100	90	4	97	97	93
Ru[(*S*)-BINAP]$(OAc)_2$	100	90	4	100	97	93
Ru[(*S*)-BINAP]$(allyl)_2$	100	80	45	95	90	37
Ru[(*S*)-BIPHEMP]$(allyl)_2$	70	90	60	100	80	37

Table 8.14 Hydrogenation with Ru(II) complexes of β-keto esters having an unsaturated chain

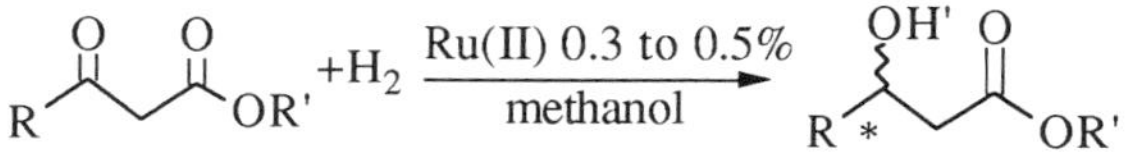

R–CO–CH$_2$–COOR' + H_2 $\xrightarrow[\text{methanol}]{\text{Ru(II) 0.3 to 0.5\%}}$ R–*CH(OH')–CH$_2$–COOR'

R	Conditions	Product	ee (%)	Ref.
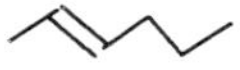 **117**	$RuBr_2$[(*S*)-BIPHEMP] 0.3%, 40°C, 65 h, 20 atm.	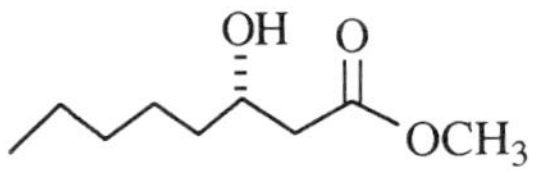 OH O, OCH_3	>99	37
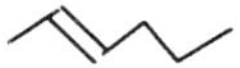 **117**	$RuBr_2$[(*S*)-BIPHEMP] 0.5%, 80°C, 5 min, 6 atm.	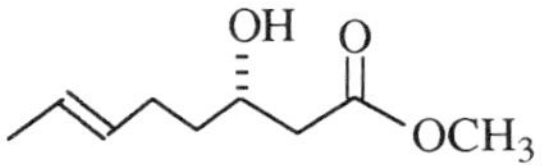 OH O, OCH_3	99	37
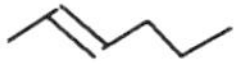 **117**	Ru_2Cl_4[(*S*)-BINAP$_2$]·NEt_3	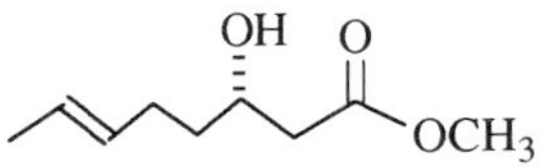 OH O, OCH_3 **118**	>95	92
 **119**	$RuBr_2$[(*S*)-BIPHEMP] 0.3%, 80°C, 35 min, 6 atm.	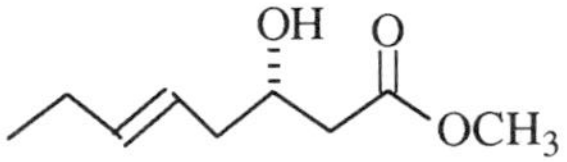 OH O, OCH_3 **120**	99	37

the production of one of the fragments of the natural macrolide antibiotic FK-506.[94] The binuclear complex (**23**) was used in the synthesis of the macrolide brefeldine A.[95] A similar approach was used for the key stage in the total synthesis of the HMG-CoA synthase inhibitor 1233A[96] and statine.[97] (−)-Roxaticin has been synthesized recently by a convergent route. Each one of the optically pure building blocks was prepared using RuBINAP asymmetric hydrogenation.[98]

In the asymmetric hydrogenation of α-substituted-β-keto esters having a labile C-2 stereogenic centre in a 1,3-dicarbonyl environment, there is the possibility of stereoselective hydrogenation to afford a single enantiomerically pure hydroxyester among the four that are possible. The conditions for the realization of this tempting possibility require an increase in the isomerization rate of the C-2 stereogenic centre above the hydrogenation rate of the keto group and high chiral recognition of the catalyst.

An example of excellent *syn*-selectivity with an ideal dynamic kinetic resolution has been realized during the hydrogenation of 2-acylamino-substituted acetoacetate (**121**) using Ru(II)Br_2BINAP[99] and Ru(II)Br_2-CHIRAPHOS[22,100] catalysts (Table 8.15). The *syn*:*anti* ratio in the derivatives of threonine (**122** or **124**) and allothreonine (**123** or **125**) was up to 99:1 using the appropriate solvent (CH_2Cl_2) and ligand, as shown in Table 8.15. It should be noted that in the change to Rh(I) complexes with a large range of chiral phosphines (DIOP, NORPHOS, DIPAMP, etc.), the *syn*-selectivity was high, up to 94%, but there was a significant decrease of enantioselectivity; 18 and 39% ee, respectively.[22] This approach represents a practical synthesis of *syn*-β-hydroxy-α-amino acids such as (L)- and (D)-threonine (Table 8.15).

The present procedure of dynamic kinetic resolution was successfully used in the synthesis of biphenomicine A.[101] With alkyl-substituted acetoacetic esters, the Ru–BINAP complexes gave high C-3 chiral recognition (ee up to 93%) but no appreciable kinetic resolution of enantiomers was observed. An equimolecular mixture of *syn*- and *anti*-derivatives was obtained. Use of cyclic keto esters, however, such as (2-methoxycarbonyl)cyclopentanone, results in satisfactory diastereoselectivity. Thus, in the presence of Ru_2[(*R*)-BINAP], an excellent *anti*-diastereoselectivity is seen, 99:1, with optical purity greater than 90%.[102a] The hydrogenation of (2-methoxycarbonyl)cyclopentanone and 2-(methoxycarbonyl)cycloheptanone with the catalytic system RuBINAP proceeds with high enantioselectivity; however, the ultimate diastereoselectivity is not satisfactory.[102b] Interestingly, the reduction of racemic β-keto esters having the tetralone structure (2-ethoxycarbonyltetralone) by $RuBr_2$MeO-BIPHEP and $RuBr_2$BINAP catalysts is realized with an ideal kinetic dynamic resolution, affording remarkably high *anti*-selectivity, approaching 100%, and enantioselectivity up to 97%.[103]

Two new C_2 symmetric chiral phosphines *m*-BITAP and *o*-BITAP have been prepared in optically pure forms based on asymmetric hydrogenation of diaryl-2-oxocyclopentane carboxylates catalysed by Ru(II)BINAP complexes.[104]

Table 8.15 Hydrogenation of 2-acylamino-substituted acetoacetates (**121**) using chiral ruthenium and rhodium catalysts

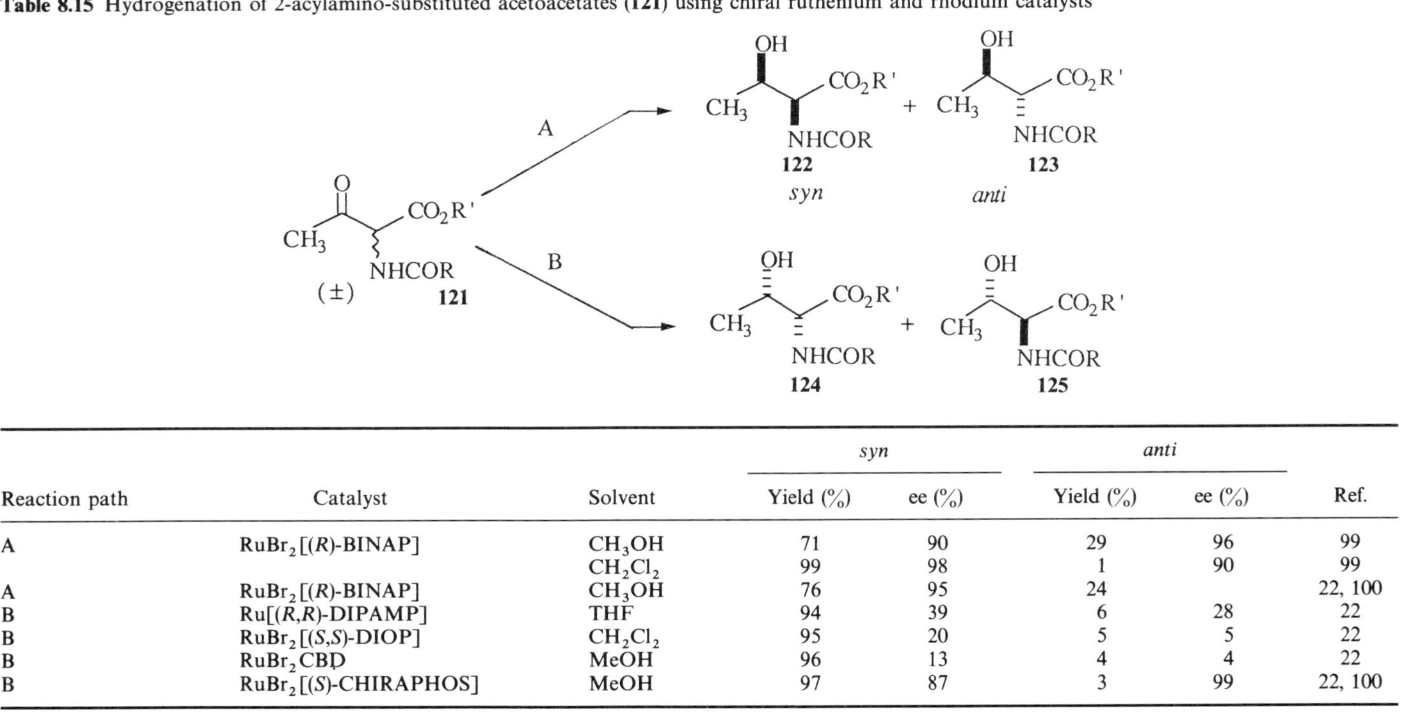

Reaction path	Catalyst	Solvent	*syn* Yield (%)	*syn* ee (%)	*anti* Yield (%)	*anti* ee (%)	Ref.
A	$RuBr_2$[(*R*)-BINAP]	CH_3OH	71	90	29	96	99
		CH_2Cl_2	99	98	1	90	99
A	$RuBr_2$[(*R*)-BINAP]	CH_3OH	76	95	24		22, 100
B	Ru[(*R*,*R*)-DIPAMP]	THF	94	39	6	28	22
B	$RuBr_2$[(*S*,*S*)-DIOP]	CH_2Cl_2	95	20	5	5	22
B	$RuBr_2$CBD	MeOH	96	13	4	4	22
B	$RuBr_2$[(*S*)-CHIRAPHOS]	MeOH	97	87	3	99	22, 100

This interesting dynamic kinetic resolution acquires enormous practical significance.[61] For example, asymmetric hydrogenation under controlled conditions of racemic 2-benzamidomethyl-3-oxobutanoate (**126**) catalysed by cationic Ru(II)[(*S*)-BINAP] (50°C, 100 atm. H_2, 40 h) led to the *syn*-(2*S*,3*R*)-isomer of the hydroxyamidocarboxylic ester (**127**) with almost complete diastereoselectivity (98%) and enantioselectivity (99% ee).[105] This derivative (**127**) is a versatile intermediate in the synthesis of the key acetoxy azetidinone (**128**), an important carbapenem intermediate (Scheme 8.17).

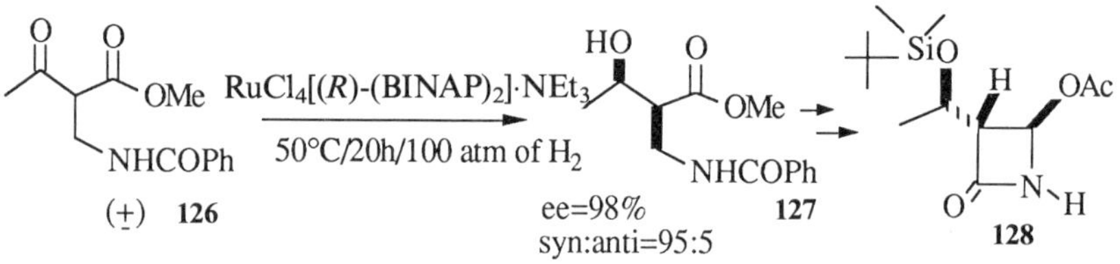

$RuCl_4[(R)\text{-}(BINAP)_2]\cdot NEt_3$; 50°C/20h/100 atm of H_2

(±) **126** → **127** (ee=98%, syn:anti=95:5) → **128**

Scheme 8.17

8.6 Asymmetric hydrogenation of C=N

(a) *Imines, hydrazones, sulfonimines.* Although much is known about the asymmetric hydrogenation of alkenes and ketones catalysed by chiral metal complexes, the asymmetric hydrogenation of prochiral C=N has received much less attention, in spite of the obvious preparative value of the process, which provides a simple route to chiral secondary amino derivatives. Some typical examples are presented in Table 8.16.

The first attempts to hydrogenate prochiral imines were made with Rh(I) chiral phosphine catalysts. These catalysts exhibit low catalytic activity and the ee values of the corresponding amines are moderate.[106] During an investigation of the behaviour of Rh(I) complexes with chiral diphosphines in the hydrogenation of a wide range of substituted imines (e.g. **129**), James has demonstrated that it is possible to increase optical yields from moderate to high with a Rh(I)[(*R*)-CYPHOS] system in the presence of iodide as the catalyst.[107] An extremely efficient enantioselective hydrogenation of the C=N group developed by Burk involved the hydrogenation of *N*-aroylhydrazones of structure (**130**) using cationic complexes $[Rh(cod)(DUPHOS)]^+CF_3SO_3^-$ as catalyst precursors.[108]

The carbonyl function of the hydrazones appears to be the crucial structural feature required for these hydrogenations to proceed. As shown in Table 8.16, Et-DUPHOS proved to be superior in terms of enantioselectivity (up to 96% ee) under mild conditions (4 atm. H_2, 0°C, 12 h). This asymmetric hydrogenation of *N*-benzoylhydrazines was completed by a facile N–N bond

Table 8.16 Hydrogenation of C=N bonds with chiral transition metal catalysts

Substrate	Catalyst	Product	ee (%)	Ref.
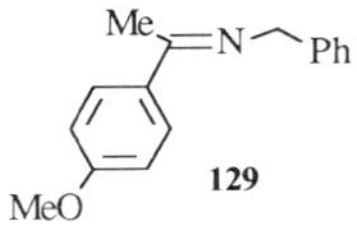 Me, =N, Ph, MeO, **129**	Rh[(*R*)-CYCPHOS] Bu_4NI	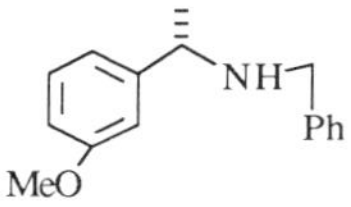 NH, Ph, MeO	91	107
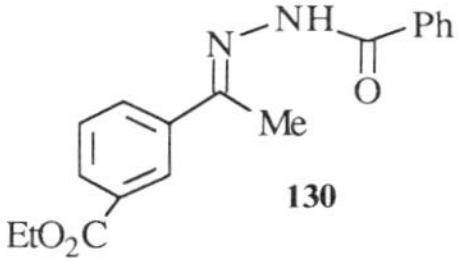 N, NH, Ph, O, Me, EtO_2C, **130**	$[Rh(cod)[(R,R)\text{-EtDUPHOS}]]^+$ HO^- (0°C, 12 h)	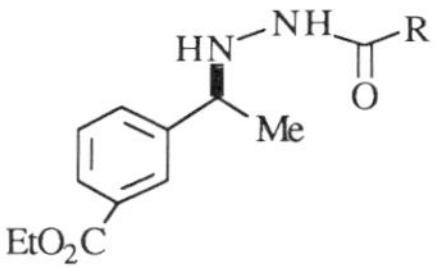 HN, NH, R, O, Me, EtO_2C	96	108
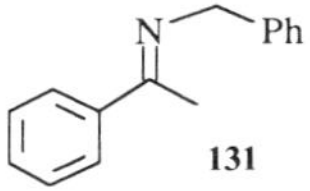 N, Ph, **131**	$[RhCl_2(cod)]_2$, sulfonated BDPP	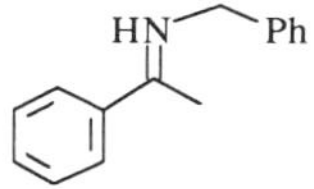 HN, Ph	91	110
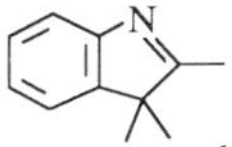 N, **132**	$[IrI_2(BDPP)]_2$	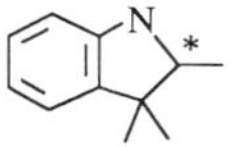 N, *	80	30
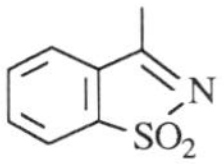 N, SO_2	$Ru_2Cl_4[(R)\text{-BINAP}]NEt_3$	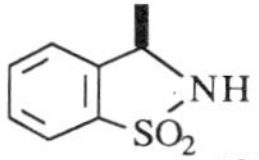 NH, SO_2, **134**	99	111
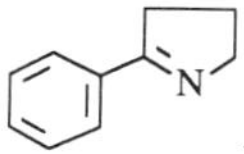 N, **133**	L_2TiX_2, 140 atm. X_2 = 1,1′-binaphthyl-2,2′-diolate	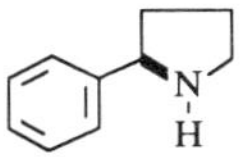 N, H	91	112

cleavage with no loss of optical purity, using SmI_2.[108] An *in situ* Rh(I) catalyst formed from $RhCl_2$(nbd) and CYCPHOS effects asymmetric hydrogenation of some commercially important imines with optical yields up to 91%.[109]

Imines ArC(Me)=NCH_2Ph (**131**) are hydrogenated to the corresponding amines with extremely high enantioselectivities of up to 96% under fairly mild conditions (70 atm. H_2, 20°C, 6 h) using Rh(I) sulfonated BDPP in a two-phase system.[110] This success is extremely important at the practical level since the amines were recovered in almost quantitative yield from the reaction mixture in the two-phase system.

Chiral auxiliaries toluene-2α-sultams introduced by Oppolzer were selectively prepared via Ru[(*R*)-BINAP]- or Ru[(*S*)-BINAP]-catalysed asymmetric hydrogenation of cyclic sulfonated imines (4 atm. H_2, 22°C, 12 h) in 72% yield (ee > 99%).[111] Good results in the hydrogenation of prochiral imines were obtained by the use of Ir(I) catalysts of type (**31**). It was possible to obtain high enantioselectivity (up to 84% ee) in the hydrogenation of *N*-arylketimine (**131**) using BDPP as the phosphine ligand.[29] Substantially greater catalytic activity is exhibited by the binuclear Ir(III) complexes (**31**).[30] Using this complex bearing BDPP as chiral ligand, an enantioselectivity of 80% ee was obtained during the hydrogenation of the cyclic imine (**132**) (Table 8.16).[30] Another solution for the synthesis of enantiopure amines from ketimines has been reported recently by Buchwald and uses asymmetric chiral titanocene catalysts. The 1,1′-binaphth-2,2-diolate derivative serves as a useful pre-catalyst. Reduction of cyclic *N*-benzyl imines (**133**) proceeded with moderate to good enantioselectivity, and cyclic imines were transformed to the corresponding amines (**134**) with excellent enantioselectivity (Table 8.16).[112]

References

1. (a) Kagan, H.B., in *Comprehensive Organometallic Chemistry*, G. Wilkinson (ed.), 1982, Pergamon. Press., **8**, 463; *Chem. Rev.*, **89**, 257; (b) Pino, P. and Consiglio, G., *Pure Appl. Chem.*, 1983, **55**, 11, 1781; (c) Morrison, J.D. (ed.), *Asymmetric Synthesis*, 1985, Vol. 5, Academic Press, New York; (d) Bosnich, B., *Asymmetric Catalysis NATO AST Series E*, 1986, Vol. 103, Martinus Nijhoff, Dordercht; (e) Apsimon, J.W. and Collier, T., *Tetrahedron*, 1986, **42**, 5157; (f) Consiglio, G. and Waymouth, R.M., *Chem. Rev.*, 1989, **89**, 257; (g) Kagan, H.B., *Bull. Soc. Chim. Fr.*, 1988, **846**, 9; (h) Blystone, S.I., *Chem. Rev.*, 1989, **89**(8), 1663; (i) Ojima, J. and Clos, N., Bastos, C., *Tetrahedron*, 1989, **45**, 6901.
2. (a) Horner, I., Siegel, H. and Buthe, H., *Angew. Chem., Int. Ed. Engl.*, 1968, **7**(12), 942; (b) *Tetrahedron Lett.*, 1968, **37**, 4023.
3. Knowles, W.S. and Sabacky, M.J., *J. Chem. Soc., Chem. Commun.*, 1968, 1445.
4. Dang, T.P. and Kagan, H.B., *J. Chem. Soc., Chem. Commun.*, 1971, 481.
5. Kagan, H.B., in *Asymmetric Synthesis*, 1985, Vol. 5, J.D. Morrison (ed.), Academic Press, New York, p. 1.
6. Brunner, H., in *Topics in Stereochemistry*, 1988, Vol. 18, E.L. Eliel and S.H. Wilen, (eds), p. 541.
7. Mislow, K., *et al.*, *J. Am. Chem. Soc.*, 1968, **90**, 4842.
8. (a) Jugé, S., Genet, J.-P., Stéphan, M., Laffitte, J.A., *Tetrahedron Lett.*, 1990, **31**, 44, 6357; (b) For a recent synthesis see Corey, E.J., Chen, Z. and Tanoury, G.J., *J. Am. Chem. Soc.*, 1993, **115**, 11000.

9. Merdes, R., unpublished work, 1993, Thèse Université P.M. Curie, Paris.
10. (a) Miyashita, A., Yasuda, A., Takaya, H., Torium, K., Ito, K., Souchi, T. and Noyori, R., *J. Am. Chem. Soc.*, 1980, **102**, 7932; (b) Miyashita, A., Takaya, H., Souchi, T. and Noyori, R., *Tetrahedron*, 1984, **40**, 8, 1245.
11. Knierzinger, A. and Schönholzer, P., *Helv. Chem. Acta*, 1992, **75**, 1211.
12. Schmidt, R., Foricher, J., Cereghetti, M. and Schönholzer, P., *Helv. Chim. Acta*, 1991, **74**, 370.
13. (a) Burk, M.J., Feaster, J.E. and Harlow, R.L., *Organometallics*, 1990, **9**, 2653; (b) Burk, M.J., Feaster, J.E., Nugent, W.A. and Harlow, R.L., *J. Am. Chem. Soc.*, 1993, **115**, 10125; (c) Burk, M.J., *J. Am. Chem. Soc.*, 1991, **113**, 8518; (d) Burk, M.J., Feaster, J.E. and Harlow, R.L., *Tetrahedron: Asymmetry*, 1991, **2**, 7, 569.
14. (a) Schrock, R.R. and Osborn, J.A., *J. Am. Chem. Soc.*, 1971, **93**, 2397; (b) Sharpley, J.R., Schrock, R.R. and Osborn, J.A., *J. Am. Chem. Soc.*, 1969, **91**, 2816.
15. Frysuk, M.D. and Bosnich, B., *J. Am. Chem. Soc.*, 1977, **99**, 6262.
16. (a) Brown, J.M., *Angew. Chem., Int. Ed. Engl.*, 1987, **26**, 190; (b) Hayashi, T., Konishi, M., Fukashima, M., Mise, T., Kagotani, M., Tojika, M. and Kumeda, M., *J. Am. Chem. Soc.*, 1982, **104**, 180.
17. James, B.R., Wang, D.K.W. and Voight, R.F., *J. Chem. Soc., Chem. Cummun*, 1975, 574.
18. (a) Noyori, R., Ohta, M., Hsiao, Y., Kitamura, M., Ohta, T. and Takaya, H., *J. Am. Chem. Soc.*, 1986, **108**, 7117; (b) Ohta, T., Takaya, H. and Noyori, R., *Inorg. Chem.*, 1988, **27**, 566; (c) Kitamura, M., Tokunaga, M. and Noyori, R., *J. Org. Chem.*, 1992, **57**, 4053.
19. Ikariya, T., Ischii, Y., Kawano, H., *et al.*, *J. Chem. Soc., Chem. Commun.*, 1985, 922.
20. Genet, J.P., Mallart, S., Pinel, C. and Laffitte, J.A., *Tetrahedron: Asymmetry*, 1991, **2**, 43.
21. Genet, J.P., Mallart, S., Pinel, C., *et al.*, *Tetrahedron Lett.*, 1992, **33**, 5343.
22. Genet, J.P., Mallart, S., Pinel, C., Juge, S. and Laffitte, J.A., *Tetrahedron: Asymmetry*, 1991, **2**, 555.
23. Genet, J.P., Pinel, C., Ratovelomanana-Vidal, V., *et al.*,, *Tetrahedron: Asymmetry*, 1994, **5**, 665.
24. Heiser, B., Broger, E.A. and Crameri, Y., *Tetrahedron: Asymmetry*, 1991, **2**, 47.
25. Alcock, N.W., Brown, J.M., Rose, M. and Wienand, A., *Tetrahedron: Asymmetry*, 1991, **2**, 47.
26. (a) Kitamura, M., Tokunaga, M., Ohkuma, T. and Noyori, R., *Tetrahedron. Lett.*, 1991, **32**, 4163; (b) Mashima, K., Kusamo, K.H., Noyori, R. and Takaya, H., *J. Chem. Commun.*, 1989, 1208.
27. (a) Mashima, K., Hino, T. and Takaya, H., *Tetrahedron. Lett.*, 1991, **32**, 3101; (b) King, S.A., Thompson, A.S., King, A.O. and Verhoeven, T.R., *J. Org. Chem.*, 1992, **57**, 6689; (c) Fronczek, F.R., Watkins, S.E., Stahly, G., *et al.*, *Organometallics*, 1993, **12**, 1467; (d) James, B.R., Pacheco, A., Rettig, S.J., *et al.*, *J. Mol. Cat.*, 1987, **41**, 147; (e) Hoke, J.B., Hollis, L.S. and Stern, E.W., *J. Organomet. Chem.*, 1993, **455**, 193.
28. Deutsch, P.P. and Eisenberg, R, *Chem. Rev.*, 1988, **88**, 1147.
29. Spindler, F., Pugin, B. and Blaser, H.U., *Angew Chem., Int. Ed. Engl.*, 1990, **29**, 558.
30. Cheong Chan, Ng and Osborn, J.A., *J. Am. Chem. Soc.*, 1990, **112**, 9400.
31. Leutenegger, U., Madin, A. and Pfaltz, A., *Angew Chem. Int. Ed. Engl.*, 1989, **101**, 61.
32. (a) Knowles, W.S., *Acc. Chem. Res.*, 1983, **16**, 106; (b) Klabunovskii, E.I., *Russ. Chem. Rev.*, 1982, **51**, 1103; (c) Koenig, K.E., in *Asymmetric Catalysis*, 1985, Vol. 5, J.E. Morrison (ed.), Academic Press, New York, p. 71; (d) Caplar, V., Comisso, G. and Sunjic, V., *Synthesis*, 1981, 85; (e) Dunina, V.V. and Beletskaya, I.P., *Zhurnal Organicheskoi Khimii*, 1992, **28**(9), 1547; (f) Inoguchi, K., Sakuraba, S. and Achiwa, K., *Synlett*, 1992, 169.
33. (a) Knowles, W.S., Sabacky, M.J. and Vineyard, B.D., *J. Chem. Soc., Chem. Commun.*, 1972, 10; (b) Vineyard, B.D., Knowles, W.S., Sabacky, M.J., *et al.*, *J. Am. Chem. Soc.*, 1977, **99**, 5946.
34. (a) Kagan, H.B. and Dang, T.P., *J. Am. Chem. Soc.*, 1972, **94**, 6429; (b) Dang, T.P., Poulin, J.C. and Kagan, H.B., *J. Organomet. Chem.*, 1975, **91**, 105.
35. Frysuk, M.D. and Bosnich, B., *J. Am. Chem. Soc.*, 1978, **100**, 5491.
36. Miyachita, A., Yasuda, A., Takaya, H., Noyori, R., *et al.*, *J. Am. Chem. Soc.*, 1980, **102**, 7932.
37. Genet, J.P., Ratovelomanana-Vidal, V., Pfister, X., *et al.*, *Tetrahedron: Asymmetry*, 1994, in press.
38. MacNeil, P.A., Roberts, N.K. and Bosnich, B., *J. Am. Chem. Soc.*, 1981, **103**, 2273.
39. Miyachita, A., Takaya, H., Souchi, T. and Noyori, R., *Tetrahedron*, 1984, **40**, 1245.
40. Hayashi, T. and Kumada, M., in *Fundamental Research in Homogeneous Catalysis*, 1978, Vol. 2, Y. Tshii and M. Tsutsui, (eds.), p. 159.

41. Kawano, H., Ikariya, T., Ishii, Y., *et al.*, *J. Chem. Soc., Perkin Trans 1*, 1989, 1571.
42. Noyori, R., Ikeda, T., Ohkuma, T., *et al.*, *J. Am. Chem. Soc.*, 1989, **111**, 9134.
43. Scott, J.W., Keith, D.D., Nix, G., *et al.*, *J. Org. Chem.*, 1981, **46**, 5086.
44. (a) Brown, J.M. and Chaloner, P.A., *J. Am. Chem. Soc.*, 1980, **102**, 3040; (b) Brown, J.M. and Evans, P.L., *Tetrahedron*, 1988, **44**(15), 4905; (c) Brown, J.M., *Chemical Soc. Rev.*, 1993, 25.
45. (a) Meyer, D., Poulin, J.C., Kagan, H.B., *et al.*, *J. Org. Chem.*, 1980, **45**, 4680; (b) El Baba, Y., Nuzillard, J.M., Poulin, J.C. and Kagan, H.B., *Tetrahedron*, 1986, **42**, 3851, and references cited therein; (c) Yamagishi, T., Ikeda, S., Yatagai, M., Yamagishi, M. and Hida, M., *J. Chem. Soc., Perkin Trans. 1*, 1988, 1787.
46. Knowles, W.S., *J. Chem. Educ.*, 1986, **63**, 222.
47. Vocke, W., Hunel, R. and Flöther, F.U., *Chemische Technik (Berlin)*, 1987, **39**, 123.
48. (a) Nagel, V., Kinzel, E., Antrade, J. and Prescher, G., *Chem. Ber.*, 1986, **119**, 3326; (b) Nagel, U., Kinzel, E., Antrade, J. and Prescher, G., *Angew. Chem. Int. Ed. Engl.*, 1984, **23**, 435; (c) Nagel, V. and Beck, W., E.P. 0151282.
49. Fiorini, M. and Giongo, G.M., *J. Mol. Catal.*, 1979, **5**, 303.
50. Halpern, J., *Pure Appl. Chem.*, 1983, **55**, 99; (b) Landis, C.R. and Halpern, J., *J. Am. Chem. Soc.*, 1987, **109**, 1746; (c) Halpern, J., *Asymmetric Catalysis*, 1985, Vol. 5, J.D. Morrison, (ed.), Academic Press, New York, p. 41.
51. Sinou, D. and Kagan, H.B., *J. Organomet. Chem.*, 1976, **114**, 325.
52. (a) Noyori, R., Ohta, M., Hsiao, Y., *et al.*, *J. Am. Chem. Soc.*, 1986, **108**, 7117; (b) Kitamura, M., Hsiao, Y., Noyori, R. and Takaya, H., *Tetrahedron Lett.*, 1987, **28**, 41, 4829; (c) Lubell, W.O., Kitamura, M. and Noyori, R., *Tetrahedron: Asymmetry*, 1991, **2**, 543.
53. Kitamura, M., Hsiao, Y., Ohta, M., Noyori, R., *et al.*, *J. Org. Chem.*, 1994, **59**, 297.
54. (a) Brown, J.M. and Chaloner, P.A., *J. Chem. Soc., Chem. Commun.*, 1979, 613; (b) Brown, J.M., Chaloner, P.A. and Morris, G.A., *J. Chem. Soc., Chem. Commun.*, 1983, 664.
55. (a) Aviron-Violet, P., Coleuille, Y. and Varagnat, J., *J. Mol. Cat.*, 1979, **5**, 41; (b) Achiwa, K., *Chem. Lett.*, 1978, 561.
56. Ojima, I. and Yoda, N., *Tetrahedron Lett.*, 1980, **21**, 1051.
57. Christofiel, W.C. and Vineyard, B.D., *J. Am. Chem. Soc.*, 1979, **101**, 4406.
58. Morimoto, T., Chiba, M. and Achiwa, K., *Tetrahedron Lett.*, 1989, **30**, 735.
59. Morimoto, T., Chiba, M. and Achiwa, K., *Heterocycles*, 1992, **33**, 435.
60. Jendralla, H., *Tetrahedron Lett.*, 1991, **31**, 3671.
61. (a) Noyori, R. and Kitamura, M., *Modern Synthetic Methods*, 1989, R. Scheffold, (ed.), Springer Verlag, 128; (b) Noyori, R., *Chem. Soc. Rev.*, 1989, **18**, 187; (c) Noyori, R. and Takaya, H., *Acc. Chem. Res.*, 1990, **23**, 345; (d) Noyori, R., *Science*, 1990, **248**, 1194; (e) Noyori, R., *Organic Synthesis in Japan Past, Present and Future*, 1992, R. Noyori, (ed.), Kagaku Dozin, Tokyo, p. 301.
62. Fluka Prize 'Reagent of year 1989', *J. Am. Chem. Soc.*, **111**, 8.
63. Ohta, T., Takaya, H., Kitamura, M., Nagai, K. and Noyori, R., *J. Org. Chem.*, 1987, **52**, 3174.
64. (a) Saburi, M., Shao, L., Sakurai, T. and Uchida, Y., *Tetrahedron Lett.*, 1992, **33**, 51, 7877; (b) Muramatsu, H., Saburi, Y., *et al.*, *J. Chem. Soc. Chem. Commun.*, 1969, 768.
65. Chan, A.S.C., *Chemtech*, 1993, 46.
66. Pfaltz, A., *Modern Synthetic Methods*, 1989, Scheffold (ed.), Springer Verlag, 231.
67. Ohta, T., Takaya, H. and Noyori, R., *Tetrahedron Lett.*, 1990, **31**, 49, 7189.
68. (a) Halpern, J., *Organometallics*, 1991, **10**, 2011; (b) Ashby, M.T. and Halpern, J., *J. Am. Chem. Soc.*, 1991, **113**, 589.
69. (a) Evans, D.A. and Morrissey, M.A., *J. Am. Chem. Soc.*, 1984, **106**, 3866; (b) Brown, J.M. and Cutting, I., *J. Chem. Soc., Chem. Commun.*, 1985, 578; (c) Brown, J.M., *Angew. Chem., Int. Ed. Engl.*, 1987, **26**, 190; (d) Brown, J.M., Cutting, I. and James, A.P., *Bull. Soc. Chim. Fr.*, 1988, 211.
70. (a) Kasahara, I. and Noyori, R., *J. Am. Chem. Soc.*, 1987, **109**, 5, 1596; (b) Takaya, H., Ohta, T., Inone, S.I., Tokunaga, M., Kitamura, M. and Noyori, R., *Org. Synth.*, 1993, **72**, 74; (c) Kitamura, M., Nagai, K., Hsiao, Y. and Noyori, R., *Tetrahedron Lett.*, 1990, **31**, 549; (d) Kitamura, M., Kasahara, I., Manabe, K., Noyori, R. and Takaya, H., *J. Org. Chem.*, 1988, **53**, 708.
71. (a) Baker, G.L., Fritschel, S.J. and Stille, J.K., *J. Org. Chem.*, 1981, **46**, 2960; (b) Deschenaux, R. and Stille, J.K., *J. Org. Chem.*, 1985, **50**, 2299.
72. Nagel, U. and Kinzel, E., *J. Chem. Soc., Chem. Commun.*, 1986, 1098.

73. (a) Alario, F., Amrani, Y., Coleuille, Y., Dang, T.P., Jenck, J., Morel, D. and Sinou, D., *J. Chem. Soc., Chem. Commun.*, 1986, 202; (b) Amrani, Y., Leconte, I., Bakos, J., Toth, I., Heil, B. and Sinou, D., *Organometallics*, 1989, **8**, 542; (c) Wan, K.T. and Davis, M.C., *J. Chem. Soc., Chem. Commun.*, 1993, 1262.
74. Wan, K. and Davis, M.E., *Tetrahedron: Asymmetry*, 1993, **4**, 2461.
75. (a) Dumont, W., Poulin, J.C., Dang, T.P. and Kagan, H.B., *J. Am. Chem. Soc.*, 1973, **95**, 8295; (b) Samuel, O., Couffignal, R., Lauer, M., Zang, S.Y. and Kagan, H.B., *Nouv. J. Chim.*, 1981, **5**, 15; (c) Hayashi, T., Tanaka, M. and Ogata, I., *Tetrahedron Lett.*, 1977, 295.
76. Jenke, T. and Suss-Fink, G., *J. Organometal. Chem.*, 1991, **405**, 383.
77. (a) Halterman, R.L. and Vollhardt, K.P.C., *Organometallics*, 1988, **7**, 883; (b) Halterman, R.L., Vollhardt, K.P.C., Welker, M.E., *et al.*, *J. Am. Chem. Soc.*, 1987, **109**, 8105.
78. (a) Toros, S., Kollar, L., Heil, B. and Marko, L., *J. Organomet. Chem.*, 1982, **17**, 232; (b) Heil, B., Toros, S., Bakos, J. and Marko, L., *J. Organomet. Chem.*, 1979, **175**, 229; (c) Bakos, J., Ioth, I., Heil, B. and Marko, L., *J. Organomet. Chem.*, **279**, 1985, 23; (d) Takeda, H., Takeshi, H., Tchinami, T., Achiwa, K., *et al.*, *Tetrahedron Lett.*, 1989, **30**, 363.
79. Hayashi, T., Katsumara, A., Konishi, M. and Kumada, M., *Tetrahedron Lett.*, 1979, **20**, 425.
80. Marki, H.P., Crameri, Y., Eigenmann, R., Krasso, A., Ramuz, H., Bernauer, K., Goodman, M. and Melmon, K.L., *Helv. Chim. Acta.*, 1988, **71**, 320.
81. Takahashi, H., Sakuraba, S., Takeda, H. and Achiwa, K., *J. Am. Chem. Soc.*, 1990, **112**, 5876.
82. Kitamura, M., Ohkuma, T., Inoue, S., Sayo, N., Kumobayashi, H., Akutagawa, S., Ohta, T., Takya, H. and Noyori, R., *J. Am. Chem. Soc.*, 1988, **110**, 629.
83. (a) Cesarotti, E., Prati, L., Pallavicini, M. and Villa, L., *Tetrahedron Lett.*, 1991, **32**, 4381; (b) Cesarotti, E., Antognazza, P., Mauri, A., Pallavicini, M. and Villa, L., *Helv. Chim. Acta.*, 1992, **75**, 2563.
84. Hatat, C., Karim, A., Kokel, N., Mortreux, A. and Petit, F., *Nouv J. Chem.*, 1990, **14**, 141.
85. (a) Ojima, H., Kogure T. and Achiwa, K., *J. Chem. Soc., Chem. Commun.*, 1977, 428; (b) Takahashi, T., Morimoto, T. and Achiwa, K., *Chem. Lett.*, 1987, 855.
86. Nozaki, K., Sato, N. and Takaya, H., *Tetrahedron: Asymmetry*, 1993, **4**, 2179.
87. (a) Purko, M., Nelson, W.O. and Wood, W.A., *J. Biol. Chem.*, 1954, **207**, 51; (b) Brown, G.M. and Reynolds, J.J., *Annu. Rev. Biochem.*, 1963, **32**, 419.
88. (a) Takahashi, H., Hattori, M., Chiba, M., Morimoto, T. and Achiwa, K., *Tetrahedron Lett.*, 1986, **27**, 4477; (b) Morimoto, T., Takahashi, H. and Achiwa, K., *Chem. Lett.*, 1986, **12**, 2061; (c) Achiwa, K., *Heterocycles*, 1978, **9**, 1539.
89. Broger, E. and Crameri, Y., 1985, E.P. 0158875 and E.P. 0218970.
90. Roucoux, A., Agbossou, F., Mortreux, A. and Petit, F., *Tetrahedron: Asymmetry*, 1993, **4**, 2279.
91. (a) Noyori, R., Ohkuma, T., Kitamura, M., *et al.*, *J. Am. Chem. Soc.*, 1987, **109**, 5856; (b) Kimatura, M., Tokunaga, M., Ohkuma, T. and Noyori, R., *Org. Synth.*, 1992, **71**, 1.
92. Taber, D.F. and Silverberg, L.J., *Tetrahedron Lett.*, 1991, **32**, 34, 4227.
93. Kitamura, M., Ohkuma, T., Takaya, H. and Noyori, R., *Tetrahedron Lett.*, 1989, **29**, 13, 1555.
94. Jones, A.B., Yamaguchi, M., Patten, A., Danishefsky, S.J., Ragan, J.A., Smith, D.B. and Schreiber, S.L., *J. Org. Chem.*, 1989, **54**, 17.
95. Taber, D.F., Silverberg, L.J. and Robinson, E.D., *J. Am. Chem. Soc.*, 1991, **113**, 6339.
96. Wovkulich, P.M., Shankaran, K., Kiegiel, J. and Uskokovic, M.R., *J. Org. Chem.*, 1993, **58**, 832.
97. Nishi, T., Kitamuma, M., Ohkuma, T. and Noyori, R., *Tetrahedron Lett.*, 1988, **29**, 48, 6327.
98. Rychnovsky, S.D. and Hoye, R.C., *J. Am. Chem. Soc.*, 1994, **116**, 1753.
99. Noyori, R., Ikeda, T., Ohkuma, T., *et al.*, *J. Am. Chem. Soc.*, 1989, **111**, 9134.
100. Genet, J.P., Mallart, S. and Jugé, S., Fr. Pat. 8911159, 1989.
101. Schmidt, U., Leitenberger, V., Griesser, H., Schmidt, J. and Meyer, R., *Synthesis*, 1992, 1248.
102. (a) Kitamura, M., Ohkuma, T., Tokunaga, M. and Noyori, R., *Tetrahedron: Asymmetry*, 1990, **1**, 1; (b) M. Kitamura, M. Tokunaga and R. Noyori, (1993), *J. Am. Chem. Soc.*, 1993, **115**, 144.
103. Genet, J.P., Pfister, X., Ratovelomanana-Vidal, V., Pinel, C. and Lafftte, J.A., *Tetrahedron Lett.*, 1994, **35**, 26, 4559.
104. Fukada, N., Mashima, Y., Matsumura, Y. and Takaya, H., *Tetrahedron Lett.*, 1990, **31**, 6327.
105. (a) Mashima, K., Matsumura, Y.I., Kusano, K.H., *et al.*, *J. Chem. Soc., Chem. Commun.*, 1991, **9**, 609; (b) Nishi, T., Kitamura, M., Ohkuma, T. and Noyori, R., *Tetrahedron Lett.*, 1988, **29**, 6327.
106. (a) Kagan, H.B., Langlois, N. and Dang, T.P., *J. Organomet. Chem.*, 1975, **279**, 283;

(b) Levi, A., Modena, G. and Scorrano, G., *J. Chem. Soc., Chem. Commun.*, 1975, 6; (c) Vastag, S., Heil, B., Toros, S. and Marko, L., *Transition Met. Chem.*, 1977, **2**, 58; (d) Vatsag, S., Bakos, J., Toros, S., Takach. N.E., King, R.B., Heil, B. and Marko, L., *J. Mol. Catal.*, 1984, **22**, 283; (e) Bakos, J., Toth, I., Heil, B. and Marko, K., *J. Organomet. Chem.*, 1985, **279**, 23.

107. Kang, G.J., Cullen, W.R., Fryzuk, M.D., James, B.R. and Kutney, J.P., *J. Chem. Soc., Chem. Commun.*, 1988, 1466.
108. Burk, M.J. and Feaster, J.E., *J. Am. Chem. Soc.*, 1992, **114**, 6267.
109. Becalski, A.G., Cullen, W.R., Fryzuk, M.D., James, B.R., Kang, G.J. and Rettig, S.J., *Inorg. Chem.*, 1991, **30**, 5002.
110. Bakos, J., Oros, Z.A., Heil, B., Laghmari, M., Lhoste, P. and Sinou, D., *J. Chem. Soc., Chem. Commun.*, 1991, 1684.
111. Oppolzer, W., Wills, M., Starkeman, C. and Bernardinelli, G., *Tetrahedron Lett.*, 1990, **31**, 4117.
112. Willoughby, C.A. and Buchwald, S.L., *J. Am. Chem. Soc.*, 1992, **114**, 7562.
113. Bolm, C., *Angew. Chem., Int. Ed. Engl.*, 1993, **32**, 232.

9 Asymmetric hydroboration

K. BURGESS and W.A. VAN DER DONK

9.1 Introduction

Hydroboration has become indispensable in synthetic organic chemistry because of several factors. First, organoboranes from hydroborations of alkenes are produced by *cis*-stereospecific addition of the boron hydride across the carbon–carbon double bond. This characteristic makes the transformation useful for creation of stereocentres with high fidelity. Second, a large arsenal of hydroborating reagents has been developed, so that a wide spectrum of reactivities is available.[1,2] Finally, hydroboration products can be converted into a variety of functionalities including alcohols, amines,[3] halides[4] and homologated materials (Figure 9.1).[5]

This chapter focuses on the scope and limitations of *stereoselective* hydroborations, i.e. additions of boranes to produce new asymmetric centres. These chiral centres can be formed at the position where the boron atom adds, at the adjacent carbon atom or both (Scheme 9.1). An inherent problem encountered in hydroborations of unsymmetrical, 1,2-disubstituted alkenes

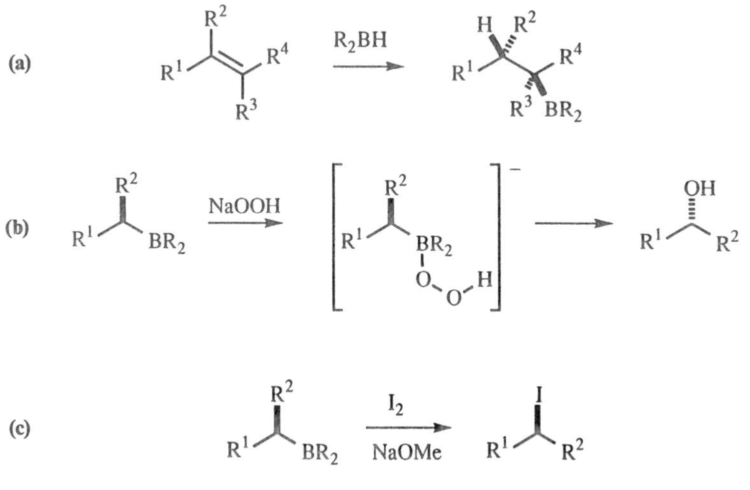

(a) R^1, R^2, R^3, R^4 alkene $\xrightarrow{R_2BH}$ H, R^2, R^1, R^4, R^3, BR_2

(b) R^2, R^1, BR_2 $\xrightarrow{NaOOH}$ $[R^2, R^1, BR_2, O, O, H]^-$ $\longrightarrow$ OH, R^1, R^2

(c) R^2, R^1, BR_2 $\xrightarrow[NaOMe]{I_2}$ I, R^1, R^2

Figure 9.1 Organoboranes. (a) Formation of organoboranes from hydroboration of alkenes. (b) Conversion into alcohols generally occurs with retention of configuration at carbon. (c) Halogenation involves inversion of stereochemistry.

is regioselectivity (Scheme 9.1, reaction (b)). Two products may be formed, hence purification of the crude reaction mixtures may be required, reducing the yield of the desired regioisomers. This is not an issue, however, for symmetrically disubstituted 1,2-alkenes.

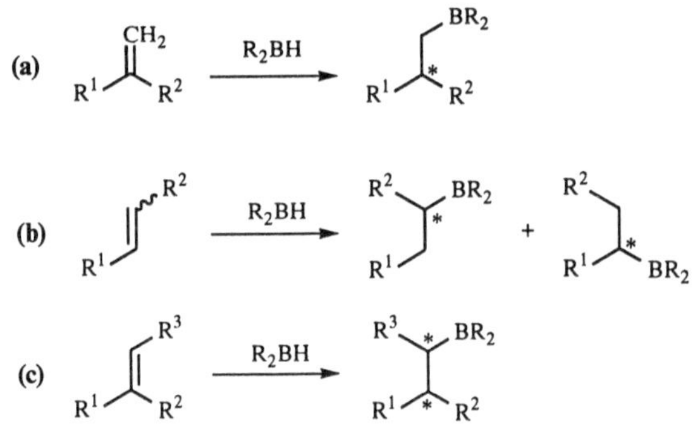

(a) $R^1C(=CH_2)R^2$ —R_2BH→ $R^1C^*H(R^2)CH_2BR_2$

(b) $R^1CH_2CH=CHR^2$ —R_2BH→ $R^2CH(BR_2)^*CH_2CH_2R^1$ + $R^2CH_2CH_2C^*H(BR_2)R^1$

(c) $R^1C(R^2)=CHR^3$ —R_2BH→ $R^3C^*H(BR_2)C^*H(R^1)R^2$

Scheme 9.1

Face selectivities in hydroboration reactions can be controlled by use of chiral hydroborating agents (reagent-controlled diastereoselectivity), chiral ligands in metal-catalysed hydroborations (enantioselectivity) or by asymmetric centres already present in the substrate (substrate-controlled diastereoselectivity). These three aspects of asymmetric hydroborations will be discussed in Sections 9.2–9.4, respectively.

9.2 Reagent-controlled diastereoselectivity in hydroborations

9.2.1 Scope and limitations

Reagent-controlled diastereoselective reactions of prochiral substrates rely on a chiral reagent to induce asymmetry. In general, reaction of a prochiral alkene with a chiral borane can lead to two diastereomeric organoboranes, and effective asymmetric induction will occur if the transition state for production of the desired isomer is favoured over that for the undesired isomer (Scheme 9.2).

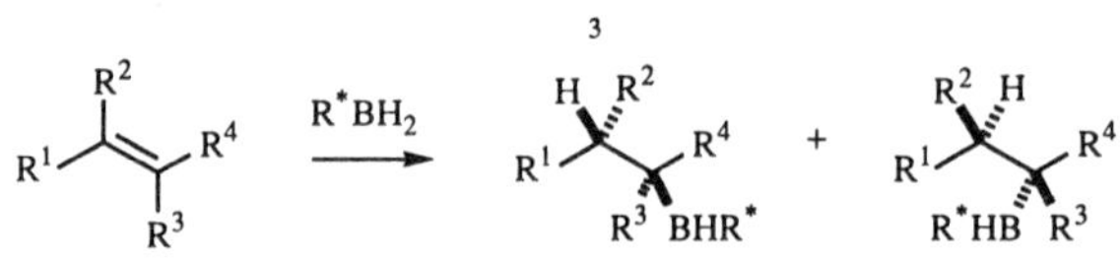

3

$R^1R^2C=CR^3R^4$ —R^*BH_2→ $R^1C(H)(R^2)$–$C(R^4)(R^3)BHR^*$ + $R^1C(R^2)(H)$–$C(R^4)(R^3)BHR^*$

Scheme 9.2

Reagent-controlled diastereoselective hydroborations have one inherent disadvantage: they require a stoichiometric amount of a chiral auxiliary. Besides the obvious cost factors, this means that the products obtained after oxidation are contaminated with materials derived from the chiral auxiliary. Consequently, large-scale reactions are difficult, and tedious purifications of the crude reaction products may be required.

9.2.2 *Stereodiscriminating hydroborating reagents from optically active alkenes*

(*a*) *Syntheses and properties of the reagents.* Most of the chiral hydroborating reagents used to date are alkylboranes produced via stereoselective hydroborations of naturally occurring optically active alkenes. Terpenes and terpene derivatives have been used extensively because of their ready availability and the high face selectivities observed in the hydroboration of these cyclic alkenes. Figure 9.2 shows several chiral dialkylboranes prepared in this way.

In 1961, Brown described the first synthesis of a chiral hydroborating agent, diisopinocampheylborane (Ipc_2BH).[6] In the subsequent three decades, Ipc_2BH has become the most widely used chiral hydroborating agent in organic synthesis. One compelling advantage of Ipc_2BH over some similar reagents that have been developed is the relative simplicity of its synthesis. Both enantiomers of the reagent can be conveniently prepared from borane and the readily available terpene α-pinene. Commercial α-pinene is not optically pure, but it was found that reaction mixtures that are left for extended periods of time produced Ipc_2BH of higher enantiomeric purity than the α-pinene from which it was derived.[7] This behaviour was attributed to preferential incorporation of the major pinene enantiomer into the crystalline complex in an equilibrium process (Scheme 9.3), while the minor optical isomer accumulates in solution. During the synthesis, BH_3 and α-pinene reacted rapidly to form triisocampheyldiborane ($Ipc_2B(\mu\text{-H})_2BHIpc$). Unfortunately, conversion of ($Ipc_2B(\mu\text{-H})_2BHIpc$) to *sym*-tetraisopinocampheyldiborane ($Ipc_2B(\mu\text{-H})_2BIpc_2$) was slow, especially with dilute BH_3 solutions. Concentrated BH_3 in THF provided faster reaction times, but these solutions are not commercially available and are relatively hard to prepare. This problem was overcome by use of the stable borane–dimethylsulfide complex (BH_3SMe_2), when the dimethylsulfide liberated during the reaction

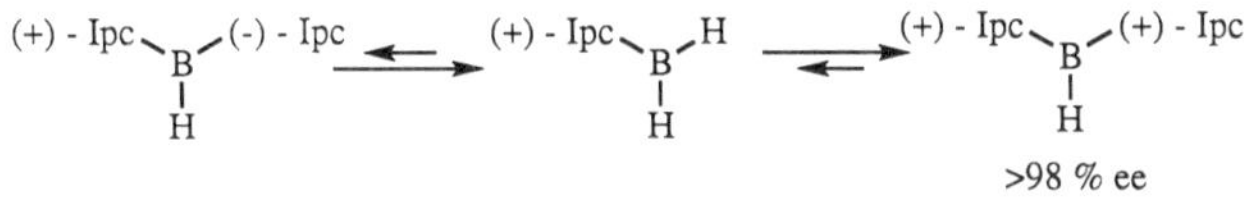

(+) - Ipc–B(H)–(-) - Ipc ⇌ (+) - Ipc–B(H)–H ⇌ (+) - Ipc–B(H)–(+) - Ipc

>98 % ee

Scheme 9.3

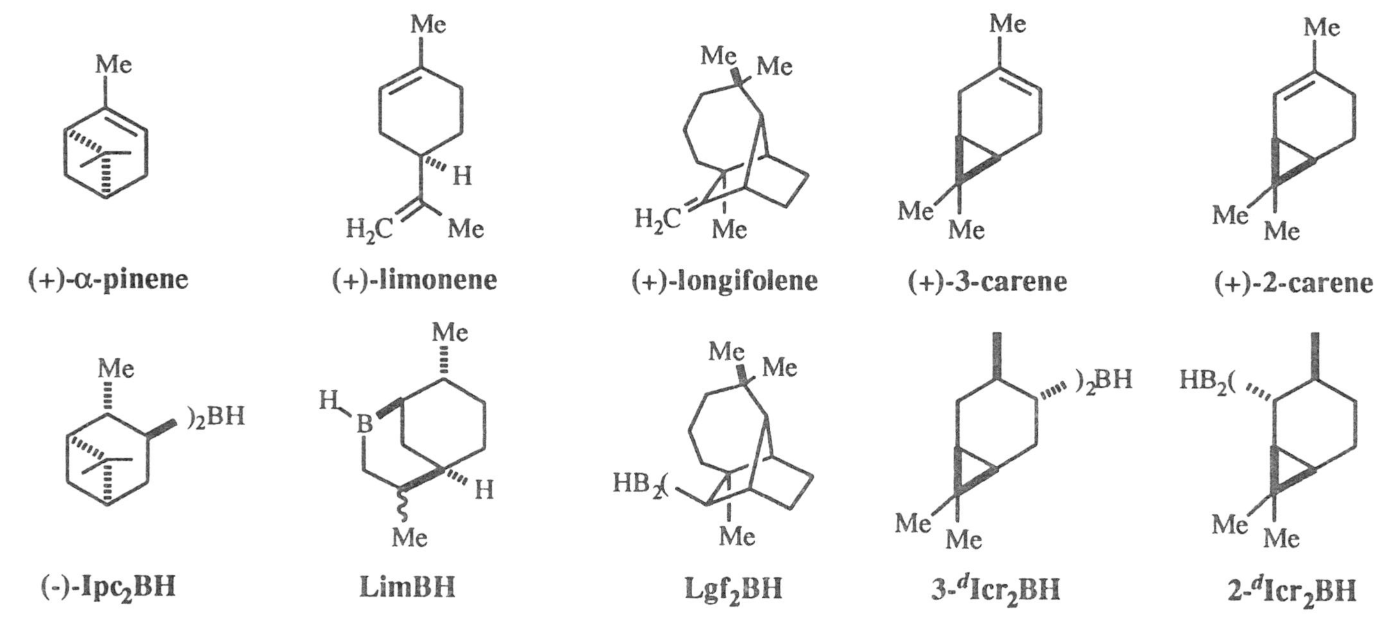

Figure 9.2 Asymmetric hydroborating agents and the terpenes from which they were derived.

is removed by applying a vacuum. Using this procedure, the solid product can be obtained in 98% ee or higher from α-pinene of only 92% ee.[8]

The size of Ipc_2BH causes very sluggish additions to trisubstituted alkenes and otherwise sterically hindered substrates.[9] Over the prolonged periods of these reactions, Ipc_2BH partially disproportionates into $IpcBH_2$ and α-pinene. Unfortunately, the $IpcBH_2$ produced reacts more rapidly with the substrates than Ipc_2BH, and gives products with the *opposite* configuration, so reducing the optical yields of the products (Scheme 9.4).[9] This difficulty is partially alleviated using diglyme rather than THF as solvent, because disproportionation of Ipc_2BH is slower in the former medium.

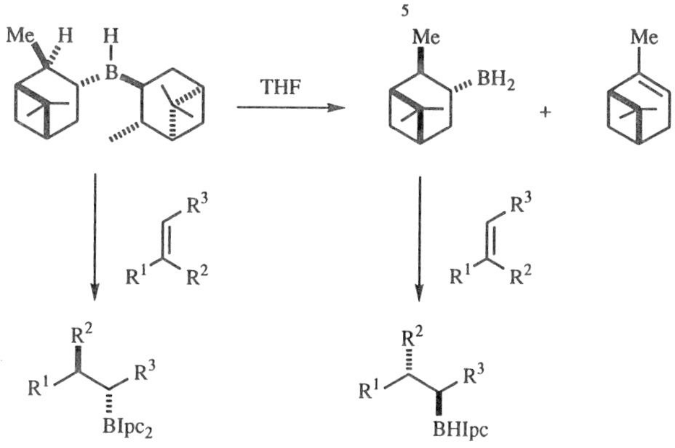

Scheme 9.4

There have been numerous attempts to improve on the performance of Ipc_2BH. For instance, the chiral boraheterocycle limonylborane (LimBH)[10] was prepared from the cyclic hydroboration of limonene with monochloroborane etherate ($ClBH_2{\cdot}Et_2O$) followed by reaction with $LiAlH_4$ (Scheme 9.5). The reagent exists as a relatively stable dimer. It is usually formed *in situ* from LimBCl, and, since limonene is readily available in both optical antipodes, both enantiomers of limonylborane can be prepared.

The other chiral dialkylboranes shown in Figure 9.2 were prepared via procedures similar to that described for Ipc_2BH. Only one optical isomer of

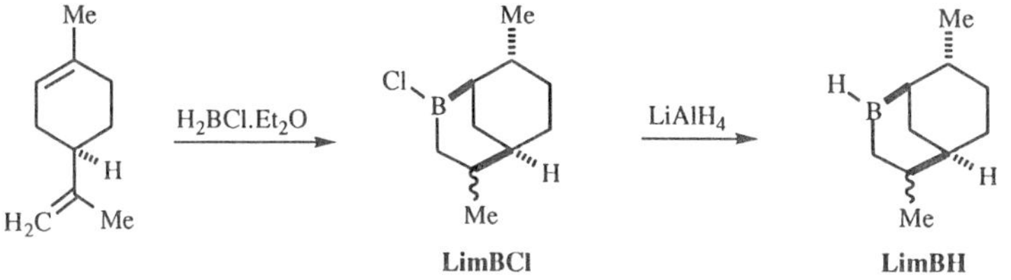

Scheme 9.5

longifolene, 2-carene, and of 3-carene is commercially available; therefore, both enantiomeric forms of Lgf_2BH, 2-${}^{d}Icr_2BH$ and 4-${}^{d}Icr_2BH$ cannot be obtained, limiting the potential of these reagents in asymmetric syntheses.

Relatively reactive monoalkylboranes, R^*BH_2, have been used to circumvent the problems encountered in additions of chiral dialkylboranes to sterically hindered alkenes. One such reagent is monoisopinocampheylborane, $IpcBH_2$. The reduced steric requirements of $IpcBH_2$ compared with Ipc_2BH indeed facilitated the hydroboration of trisubstituted and other hindered alkenes, and elimination of α-pinene was not a problem in these reactions.[11]

Unfortunately, $IpcBH_2$ produced by controlled addition of 1 equiv. BH_3·THF to α-pinene contains significant amounts of Ipc_2BH. For instance, equilibration of BH_3·THF solutions with α-pinene for prolonged periods (*c.* 96 h at 25°C) or at elevated temperatures (*c.* 50°C) gives about 80% $IpcBH_2$ and 5% Ipc_2BH and BH_3·THF.[12] Better, but more difficult, procedures for preparation of pure $IpcBH_2$ have been described. Thus, treatment of Ipc_2BH with TMEDA generated the adduct (**1**) as a crystalline solid with 100% optical purity (Scheme 9.6). This material was converted into $IpcBH_2$ by reaction with BF_3·Et_2O, and the insoluble BF_3-amine adducts were easily removed by filtration (Scheme 9.7).[13]

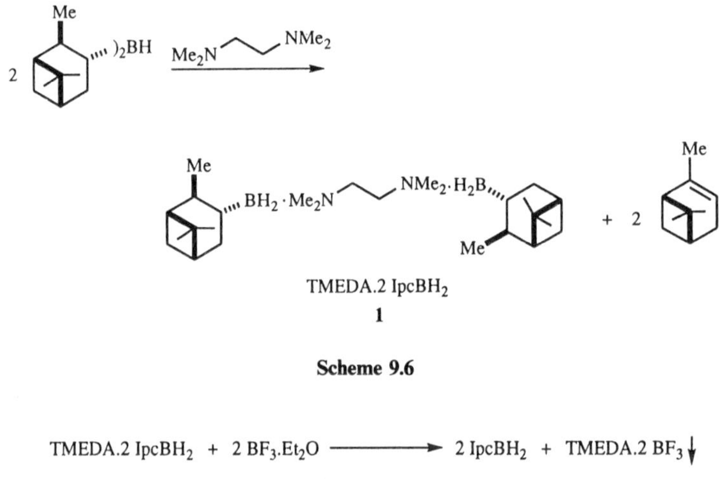

TMEDA.2 $IpcBH_2$

1

Scheme 9.6

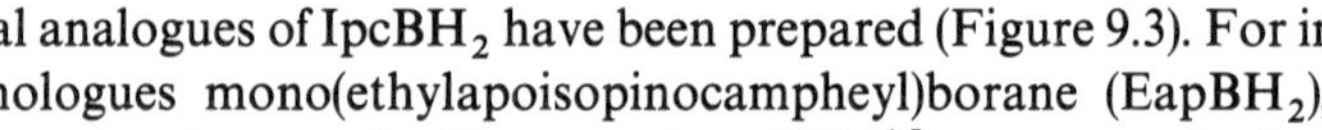

TMEDA.2 $IpcBH_2$ + 2 $BF_3.Et_2O$ ⟶ 2 $IpcBH_2$ + TMEDA.2 BF_3↓

Scheme 9.7

Several analogues of $IpcBH_2$ have been prepared (Figure 9.3). For instance, the homologues mono(ethylapoisopinocampheyl)borane ($EapBH_2$),[14] and mono(phenylapoisocampheyl)borane ($PapBH_2$)[15] were synthesized from (−)-2-ethylapopinene and (−)-2-phenylapopinene; these alkenes are not readily available and were prepared from (−)-nopol and (−)-β-pinene, respectively (Scheme 9.8).

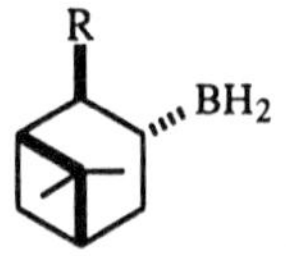

IpcBH$_2$, R = Me
EapBH$_2$, R = Et
PapBH$_2$, R = Ph
MDBH$_2$, R =

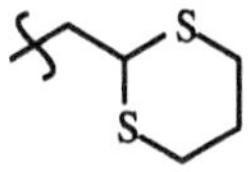

Figure 9.3 Monoalkylboranes derived from hydroboration of α-pinene derivatives.

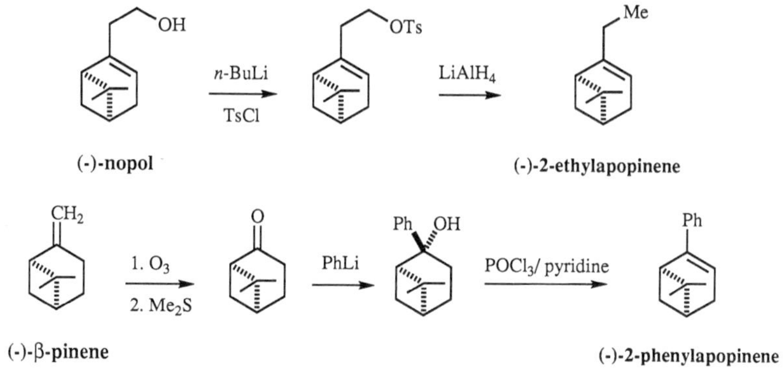

(-)-nopol → (n-BuLi, TsCl) → ($LiAlH_4$) → (-)-2-ethylapopinene

(-)-β-pinene → (1. O_3, 2. Me_2S) → (PhLi) → ($POCl_3$/ pyridine) → (-)-2-phenylapopinene

Scheme 9.8

Another hydroborating reagent, 2-(1,3-dithianyl)myrtanylborane ($MDBH_2$), was obtained via hydroboration of the alkene (**2**) which was itself prepared in two steps from myrtenol, in 62% yield (Scheme 9.9).[16] Unlike the α-pinene derivatives discussed so far, the BH_3 hydroboration of (**2**) stopped at the monohydroboration stage. Optically pure $MDBH_2$ was obtained by

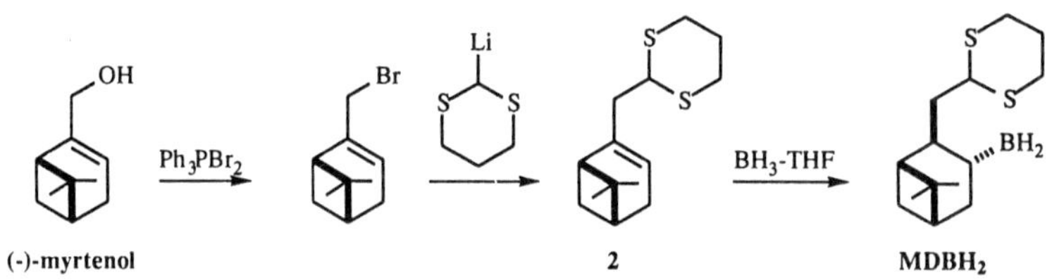

(-)-myrtenol → (Ph_3PBr_2) → (2-lithio-1,3-dithiane) → **2** → (BH_3-THF) → **MDBH$_2$**

Scheme 9.9

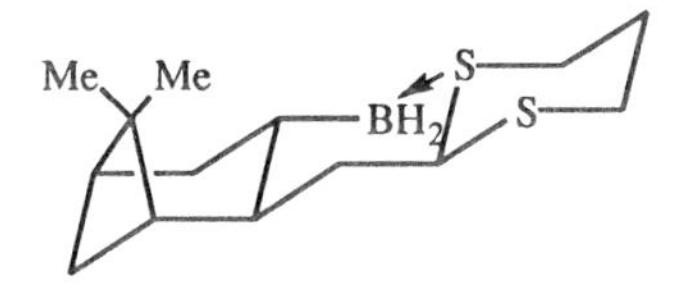

Figure 9.4 Possible coordination of sulfur to boron in $MDBH_2$.

recrystallization of the crude reaction product as a stable, monomeric solid, which is not very sensitive to moisture even in the uncomplexed form. The ^{11}B NMR signal for this compound indicates significantly more shielding than expected for an uncomplexed monoalkylborane. These observations strongly implicate coordination of one of the sulfur lone pairs to the boron (Figure 9.4). Like other borane–sulfide complexes, $MDBH_2$ is a hydroborating reagent with good stability; it can be stored under an inert atmosphere at 0°C for more than a year. A disadvantage is that while (−)-myrtenol is a commercially available natural product, (+)-myrtenol is not, although it can be prepared from α-pinene.[16]

(*b*) *Effectiveness of the reagents in diastereoselective reactions.* Table 9.1 presents data for the hydroboration of five alkenes with chiral mono- and dialkylboranes. Excellent results are obtained for the reaction of *cis*-2-butene

Table 9.1 Asymmetric hydroboration with chiral mono- and dialkylboranes

Reagent	% ee (configuration) a	b	c	d	e
(+)-Ipc_2BH	98 (*R*)	13 (*S*)	22 (1*S*,2*S*)	14 (*S*)	21 (*R*)
LimBH	55 (*R*)	59 (*R*)	45 (1*R*,2*R*)	67 (*R*)	5 (*R*)
(+)-Lgf_2BH	78 (*R*)	25 (*S*)	63 (1*R*,2*R*)	70 (*R*)	1.5 (*S*)
2-^{d}Icr_2BH	93 (*S*)	30 (*R*)	3 (1*R*,2*R*)	37 (*S*)	15 (*S*)
4-^{d}Icr_2BH	50 (*R*)	40 (*S*)	3 (1*R*,2*R*)	0	5 (*R*)
(+)-$IpcBH_2$	24 (*S*)	73 (*S*)	66 (1*R*,2*R*)	53 (*R*)	1 (*S*)
$EapBH_2$	30 (*R*)	76 (*R*)	68 (1*R*,2*R*)	68 (*R*)	2 (*R*)
$PapBH_2$	12 (*R*)	37 (*R*)	20 (1*R*,2*R*)	31 (*R*)	1 (*R*)
(−)-$MDBH_2$	30 (*R*)	–	36 (1*S*,2*S*)	40 (*R*)	–

with Ipc_2BH and 2-$^dIcr_2BH$[17,18] and both reagents perform well with a variety of other *cis*-alkenes.

Comparison of the enantioselectivities in hydroborations of *trans*-2-butene (Table 9.1) reveals that the best results are obtained with $IpcBH_2$ and $EapBH_2$,[11,13,14,19,20] but the optimal performance for *trans*-alkenes does not reach the same levels of asymmetric induction that are obtained in reactions of *cis*-alkenes with Ipc_2BH.[13,19] Overall, easier access to α-pinene compared with 2-ethylapopinene makes $IpcBH_2$ the reagent of choice for *trans*-alkenes.

The trisubstituted alkenes 1-methylcyclopentene and 2-methyl-2-butene are hydroborated with fair levels of asymmetric induction using Lgf_2BH, $IpcBH_2$ or $EapBH_2$. Optimal enantiomeric purities in reactions of trisubstituted alkenes with $IpcBH_2$ are obtained for substrates bearing a phenyl substituent (Scheme 9.10).[21] The extent of chiral induction is comparable for both the (*E*)- and (*Z*)-isomers. (*E*)-Alkenes provide *anti*-products, while (*Z*)-alkenes lead to the formation of *syn*-products. With both isomers of the starting alkenes in hand, it is, therefore, possible to synthesize all four product diastereomers in high optical yields using both enantiomers of $IpcBH_2$ (Scheme 9.10). Moreover, dialkylboranes produced in reactions of 1:1 mixtures of $IpcBH_2$ and hindered alkenes commonly exist as crystalline solids, so their optical purities can be upgraded, often to near 100% by crystallization.[22] Chemical yields may be compromised in this procedure.[22]

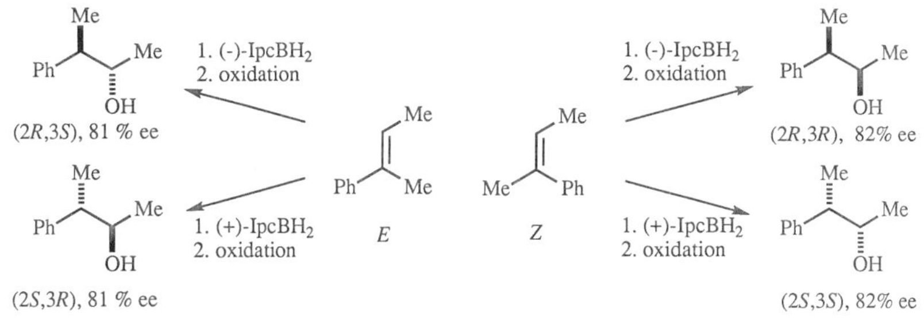

Me Ph Me OH (2*R*,3*S*), 81 % ee ← 1. (-)-$IpcBH_2$ 2. oxidation — Ph Me Me *E* — 1. (+)-$IpcBH_2$ 2. oxidation → Me Ph Me OH (2*S*,3*R*), 81 % ee

Me Me Ph *Z* — 1. (-)-$IpcBH_2$ 2. oxidation → Me Ph Me OH (2*R*,3*R*), 82% ee; 1. (+)-$IpcBH_2$ 2. oxidation → Me Ph Me OH (2*S*,3*S*), 82% ee

Scheme 9.10

All the asymmetric hydroborations of 2-methyl-1-butene shown in Table 9.1 resulted in poor asymmetric induction. This is a general observation for 1,1-disubstituted alkenes,[12,14–16,18] so the high selectivity obtained in the reaction of the alkene (**3**) with Ipc_2BH is surprising (Scheme 9.11). Hydroboration of this substrate with (−)-Ipc_2BH afforded the *syn*-product. The *anti*-isomer was produced predominantly when the reaction was performed with (+)-Ipc_2BH,[23] hence the stereoselectivity is genuinely reagent controlled.

Asymmetric hydroborations of *cis*- and *trans*-alkenes by Ipc_2BH and $IpcBH_2$, respectively, have been rationalized using the diastereomeric transition

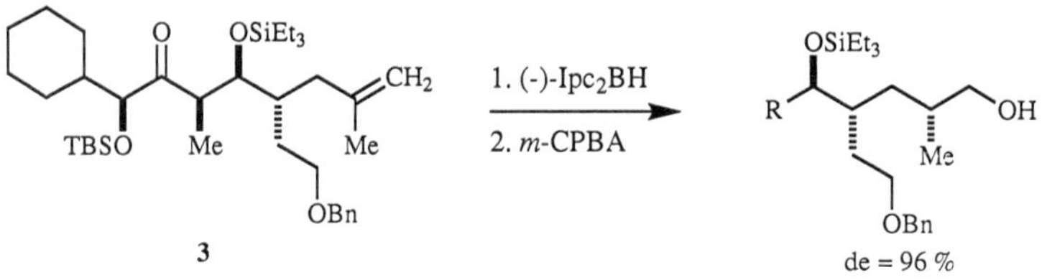

Scheme 9.11

state models depicted in Figure 9.5. The geometries for the reactive conformations in these reactions were calculated using semiempirical methods.[24] Stereochemical preferences for the hydroboration of *trans*-alkenes with $IpcBH_2$ were postulated to arise from steric repulsions between the substrate and the methyl group on the cyclohexyl ring of the reagent in the transition state, leading to the minor isomers. In the preferred conformation (calculated $\Delta G = 1.0\,\text{kcal mol}^{-1}$) this methyl group points away from the alkene, i.e. the observed selectivities result from the size difference between the CH_2- and CHMe- moieties (Figure 9.5). This model cannot always be applied to other related systems; for instance, it is inconsistent with the observation that the phenyl-substituted $IpcBH_2$-analogue ($PapBH_2$) gave lower degrees of asymmetric induction than the parent compound (Table 9.1).[15]

Steric interactions also were postulated to determine the lowest energy transition state conformations for reaction of *cis*-alkenes with Ipc_2BH (Figure 9.5b). The major product isomer is generated from the conformation that places the largest substituent on the isopinocamphenyl group closest to the alkene in the *anti*-orientation.

(*c*) *Carbon–carbon bond formation.* Stereocontrol of carbon–carbon bond formation is one of the major challenges in organic synthesis, and optically active organoboranes from asymmetric hydroborations of alkenes have become valuable compounds for such transformations.[1,2] The fundamental chemical step in these processes is the migration of alkyl groups in four-coordinate 'ate'-complexes, which occurs with retention of configuration at carbon (e.g. Scheme 9.12). The major obstacle for application of this methodology to the organoboron products from asymmetric hydroboration was competitive migration of the chiral auxiliary.[12] This problem has been overcome by selective displacement of α-pinene from the product organoboranes Ipc_2BR^* and $IpcBHR^*$ by treatment with excess acetaldehyde (Scheme 9.13). The resulting borinic esters ($IpcBR^*OEt$) and boronic esters ($R^*B(OEt)_2$) can be used for chain extension reactions.[25,26] Thus, ketones of high optical purity have been prepared from the products of asymmetric hydroborations with Ipc_2BH and $IpcBH_2$ (e.g. Scheme 9.14).[12] Incidentally, elimination of α-pinene from the products of asymmetric hydroborations using acetaldehyde is also effective for recovery of the chiral auxiliary in high optical purity.

Scheme 9.12

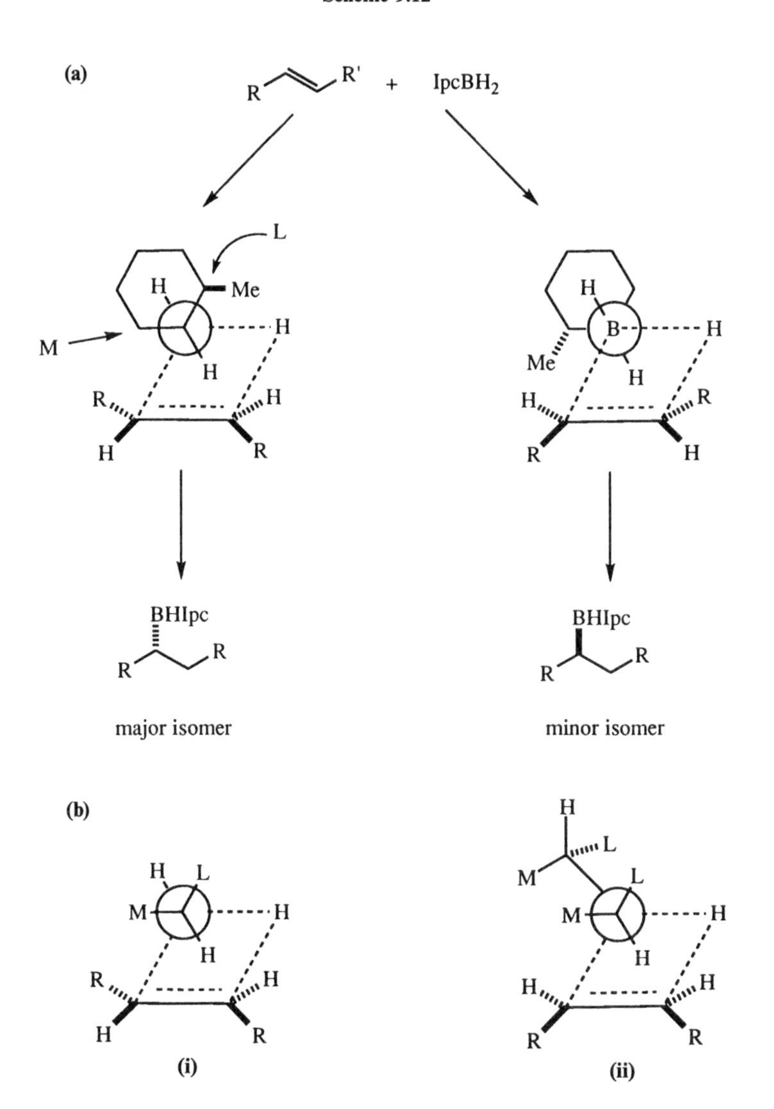

Figure 9.5 (a) Two lowest energy conformations for hydroborations of *trans*-alkenes with $IpcBH_2$. (b) Schematic representation of the lowest energy transition states for reaction of (i) $IpcBH_2$ with *trans*-alkenes, and, (ii) Ipc_2BH with *cis*-alkenes. In these models L represents the sterically encumbered CHMe moiety of the cyclohexyl ring, and M depicts the CH_2 group.

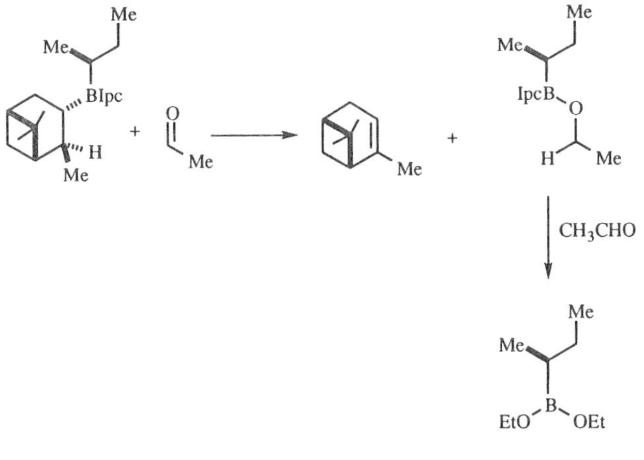

Scheme 9.13

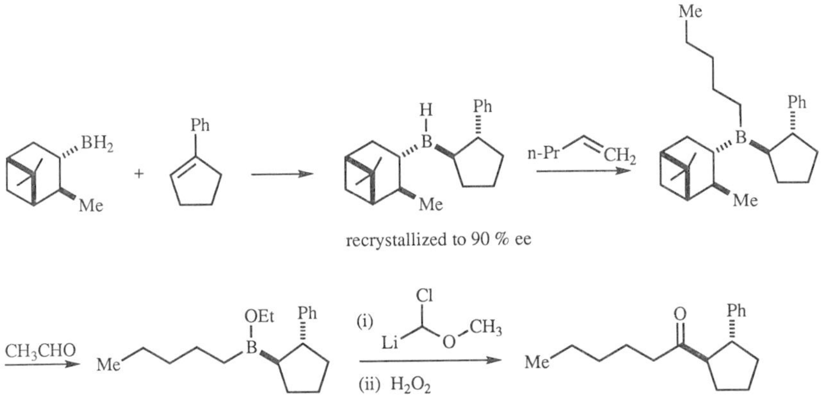

recrystallized to 90 % ee

Scheme 9.14

9.2.3 *Resolution of a chiral borane*

All the chiral hydroborating reagents discussed so far were obtained via face selective hydroborations of chiral alkenes. An alternative approach is depicted in Scheme 9.15, which shows the first resolution of a chiral borane.[27] Reaction of the Grignard reagent derived from 2,5-dibromohexane with (diethylamino)dichloroborane (Cl_2BNEt_2) and subsequent methanolysis gave a mixture of (2,5-dimethyl)methoxyborolanes (**4**). Precipitation of compound (**5**) formed from selective complexation of *N*,*N*-dimethylethanolamine with (*cis*-**4**) was followed by distillation of the volatile, uncomplexed *trans*-isomer. The two enantiomers of (*trans*-**4**) were resolved by treatment of the racemate with substoichiometric (*S*)-prolinol. This led to preferential formation of the adduct (**6**) from the (*R*,*R*)-borolane. The volatile fraction was collected by distillation, providing essentially pure ((*S*,*S*)-**4**) and the (*R*,*R*)-methoxyborolane (**4**) was obtained by methanolysis of complex (**6**). Both enantiomers were transformed into the desired products by treatment with $LiAlH_4$, followed by reaction of the crystalline borates (**7**) with excess methyl iodide. Scheme 9.15 illustrates the synthesis of both enantiomers of borolane (**8**).[27]

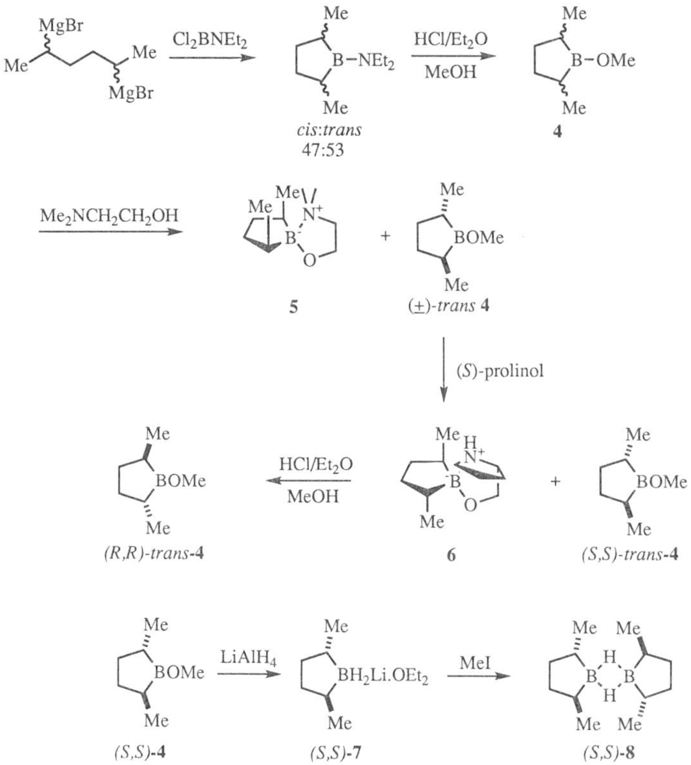

Scheme 9.15

The C_2-symmetric borolane (**8**) gave uniformly good diastereoselection with all classes of prochiral alkenes except 1,1-disubstituted ones (Table 9.2). In terms of chiral induction, this reagent was superior to any of the reagents derived from terpenes.[27] However, a relatively elaborate synthesis is required, and (**8**) is thermally unstable in solution, rearranging to the dimer (**9**) (half-life of several days) (Scheme 9.16). For this reason, the borane is usually produced *in situ* from precursor (**7**) by treatment with 2 equiv. MeI (Scheme 9.15). Overall, reagent (**8**) is of limited practical value since it is so hard to prepare.

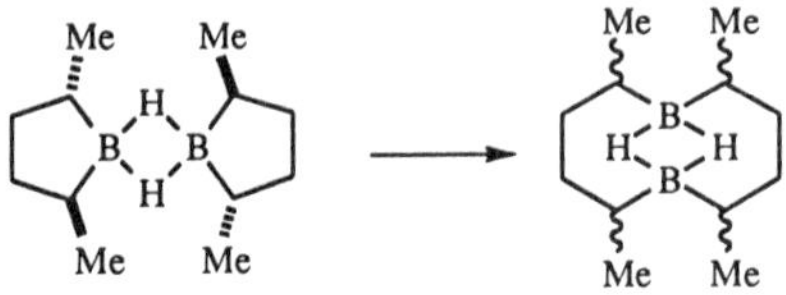

Me Me Me Me
H B B H
Me Me
8
→
Me Me
B H H B
Me Me
9

Scheme 9.16

Table 9.2 Asymmetric hydroboration with borolane ((*R*,*R*)-**8**) of >95% ee

Entry	Substrate	Product	ee (%)	Configuration
1	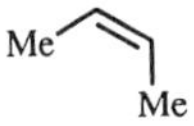 Me, Me	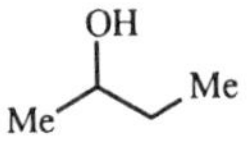 OH, Me, Me	98	*S*
2	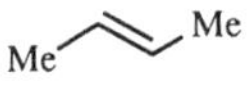 Me, Me	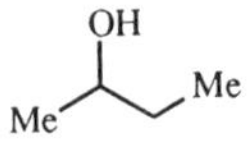 OH, Me, Me	100	*S*
3	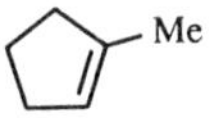 Me	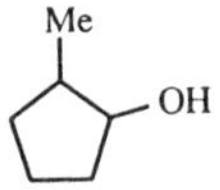 Me, OH	100	1*S*,2*S*
4	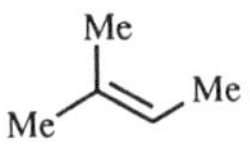 Me, Me, Me	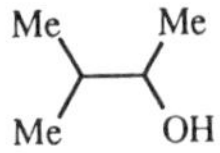 Me, Me, Me, OH	98	*S*
5	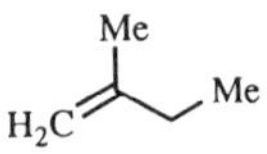 Me, H_2C, Me	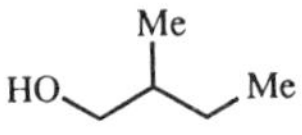 Me, HO, Me	1	*S*

9.3 Enantioselective hydroborations

Asymmetric hydroborations of prochiral alkenes can be effected using normally unreactive catecholborane in the presence of a catalyst. This approach relegates the requisite diastereoselective steps to a catalytic cycle, and the overall process is *enantioselective* (e.g. Figure 9.6). Asymmetric hydroborations of this kind are conceptually superior to reagent-controlled diastereoselective variants because: (i) an achiral boron-hydride is used; (ii) the only optically active material required is a relatively small quantity of chiral ligand for the catalyst; and (iii) the product is not contaminated with large quantities of substances formed from chiral auxiliaries, since the by-product catechol can be removed by simple extraction with aqueous base.

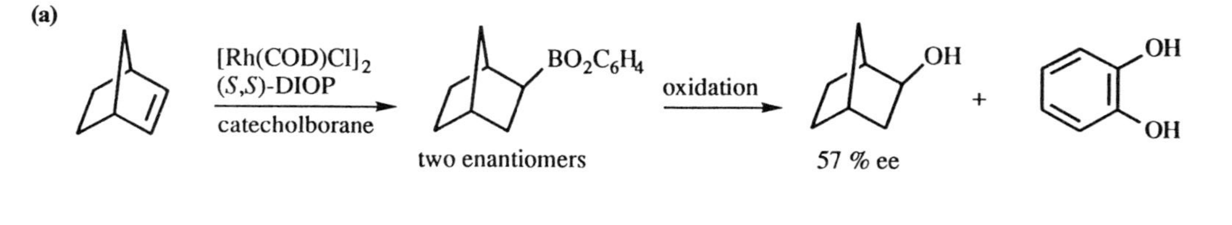

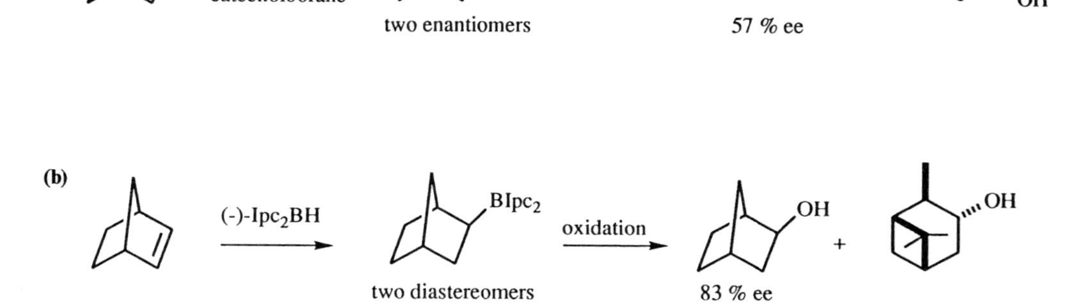

Figure 9.6 Hydroboration of norbornene. (a) Enantioselectivity in catalysed reactions. (b) Diastereoselectivity in reagent-controlled reactions.

9.3.1 Mechanism of rhodium-catalysed hydroboration

It is necessary to understand the reaction pathways for catalysed hydroborations to appreciate the origins of enantioselectivities in this process. Unfortunately, the mechanism of rhodium-catalysed hydroboration is still under investigation.[28,29] Scheme 9.17 shows a possible catalytic cycle for $RhCl(PPh_3)_3$, involving oxidative addition of the boron–hydrogen bond of catecholborane to the dissociation complex (**10**)[30] coordination of the alkene and migratory insertion of the alkene into the rhodium–hydride bond. There are alternative possibilities, however. For instance, insertion into the rhodium–boron bond leading to complex (**11**) is also feasible.[29,31,32] If intermediate (**11**) is formed, this nicely accounts for the formation of dehydrogenative borylation products (e.g. **12**) that are sometimes observed.[29,31,32]

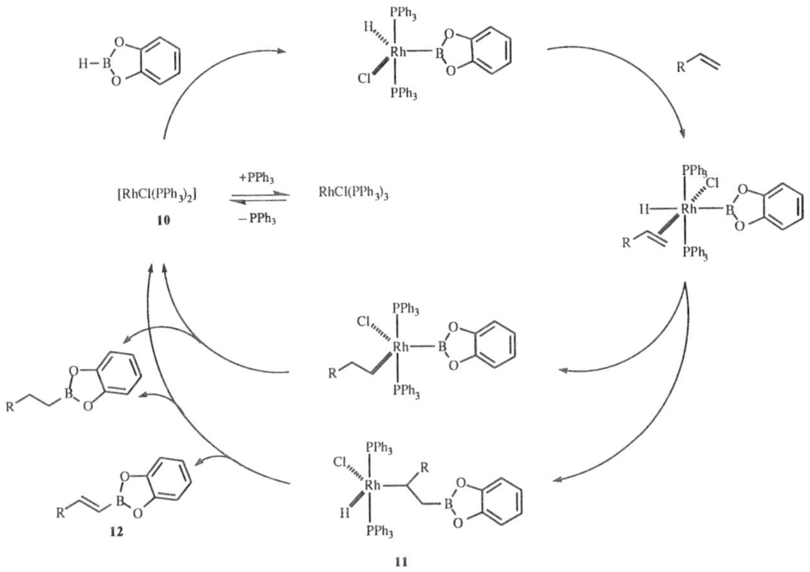

Scheme 9.17

9.3.2 Scope and limitations

Table 9.3 presents the best results obtained for asymmetric rhodium-catalysed hydroborations in the presence of chiral phosphine ligands (*R*,*R*)-DIOP, (*S*)-BINAP and ((*S*)-**13**).[33] Enantioselectivities of >95% were achieved in hydroborations with a cationic rhodium catalyst containing BINAP or the related ligand (**13**).[36] However, these optical yields could only be obtained with arylethene derivatives at −78°C (Table 9.3, entries 6 and 7).[35,38]

Remarkably, these substrates afford Markovnikov hydroboration products under most catalysed conditions (Scheme 9.18).

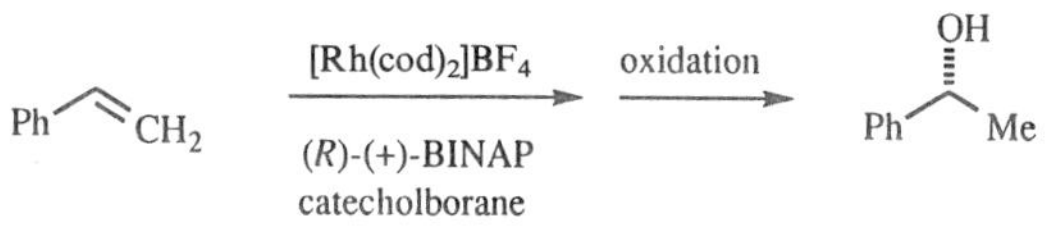

Ph⁀CH₂ —[$[Rh(cod)_2]BF_4$, (R)-(+)-BINAP, catecholborane]→ —oxidation→ Ph–CH(OH)–Me

Scheme 9.18

Table 9.3 Highest enantioselectivities for catalysed asymmetric hydroboration

Entry	Substrate	Ligand	Temperature (°C)	Solvent	% ee (configuration)	Ref.
1	norbornene	(*R*,*R*)-DIOP	−25	THF	57 (1*R*,2*R*)	34
2	norbornene	(*S*)-BINAP	−25	THF	64 (1*R*,2*R*)	34
3	norbornadiene	(*R*,*R*)-DIOP	−25	THF	76 (1*S*,2*R*)	34
4	2,3,3-trimethyl-1-propene	(*R*,*R*)-DIOP	−5	THF	69 (*R*)	34
5	phenylethene	(*R*)-BINAP	−30	THF	76 (*R*)	35
6	phenylethene	(*R*)-BINAP	−78	DME	96 (*R*)	35
7	*p*-methoxy-phenylethene	(*R*)-BINAP	−78	DME	89 (*R*)	35
8	indene	(*S*)-**13**	20	THF	91 (*S*)	36
9	indene	(*R*,*R*)-DIOP	−30	MePh	74 (*R*)	37

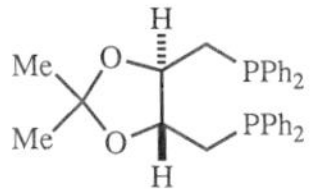

(*R*,*R*)-DIOP 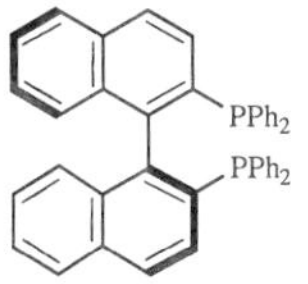 (*S*)-BINAP 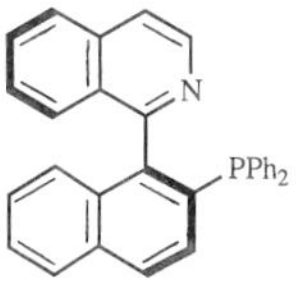 (*S*)-13

The possibility of Rh(I)-catalysed hydroboration with a chiral borane was also established, but poor asymmetric induction was obtained in the presence of an achiral catalyst.[39] Combinations of a chiral catalyst and a chiral borane also have been investigated for hydroborations of prochiral alkenes. Double asymmetric induction was observed in these transformations,[39–41] but the optical yields were not improved compared with the corresponding catecholborane reactions (cf. Table 9.3, entry 7 and Scheme 9.19).

Rhodium-catalysed asymmetric hydroborations have several disadvantages. High enantioselectivities have been limited to reactive substrates such as phenylethene and norbornene, while less reactive alkenes have only been hydroborated with modest optical purities so far. This might be a consequence of alternative pathways by which hydroboration products can be formed.[29] For instance, significant amounts of BH_3-derived products have been

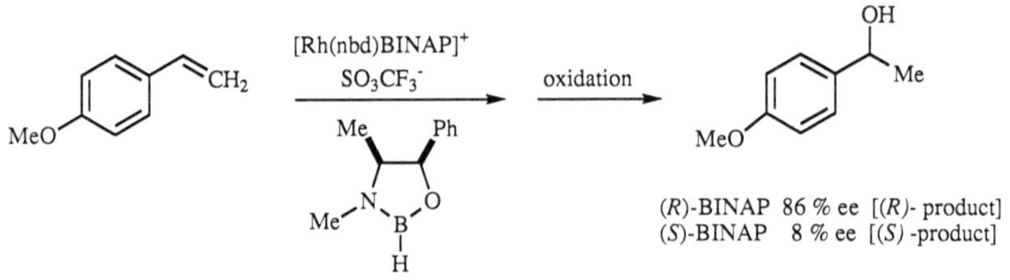

(*R*)-BINAP 86 % ee [(*R*)- product]
(*S*)-BINAP 8 % ee [(*S*) -product]

Scheme 9.19

observed in reactions with relatively hindered substrates. In these cases the BH_3 is produced by metal-catalysed and phosphine-promoted disproportionation of catecholborane into $B_2(C_6H_4O_2)_3$ (**14**) and $BH_3{\cdot}PR_3$ (Scheme 9.20).[42] Metal-catalysed hydrogenation of dehydrogenative borylation by-products formed in some rhodium-mediated processes (see Scheme 9.17) also could diminish optical yields (Scheme 9.21).[29,32]

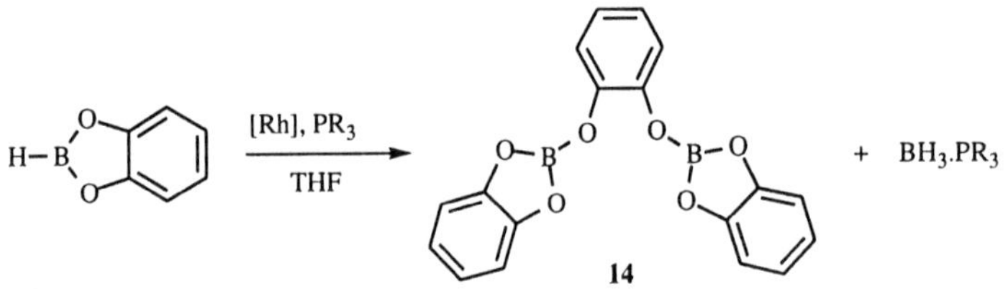

Scheme 9.20

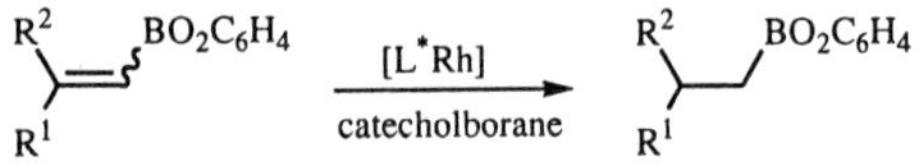

Scheme 9.21

Mechanistically simple systems for catalysed hydroboration are required to achieve optimal enantioselectivities. One very recent development is addition of catecholborane to alkenes accelerated by organolanthonide complexes (**15**) and (**16**).[43] The proposed reaction mechanism for this transformation is shown in Scheme 9.22. No asymmetric version has been developed yet, but chiral cyclopentadienyl analogues are available and may be suitable ligands for enantioselective variants.[44]

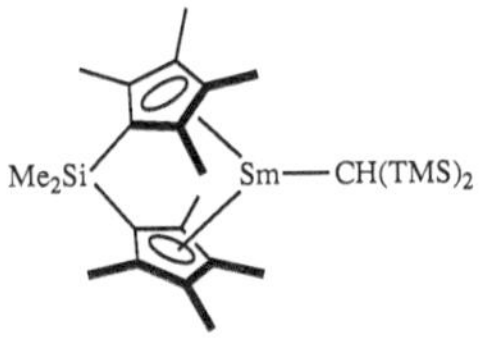

Me$_2$Si Sm—CH(TMS)$_2$

15

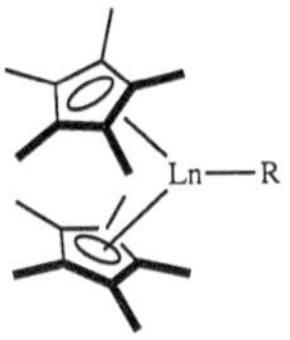

Ln—R

16
Ln = Sm, La
R = H, CH(TMS)$_2$

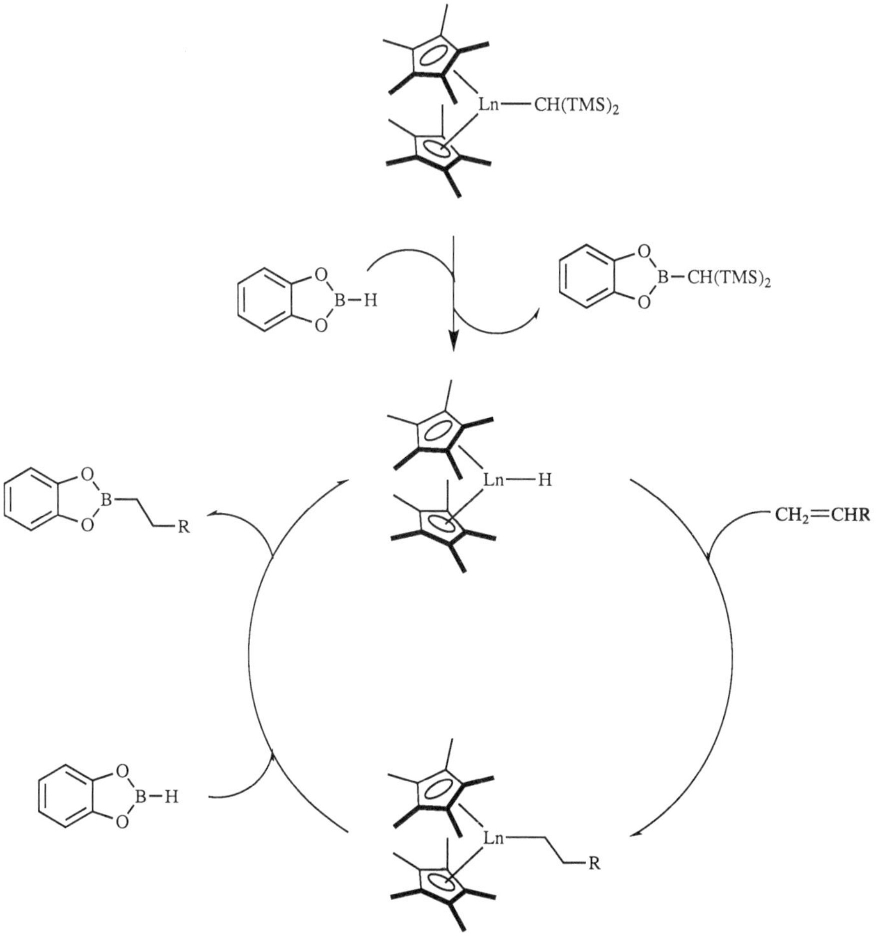

Ln—CH(TMS)$_2$

B—H

B—CH(TMS)$_2$

Ln—H

B—R

CH$_2$=CHR

B—H

Ln—R

Scheme 9.22

9.4 Substrate-controlled diastereoselectivity in hydroborations

In substrate-controlled diastereoselective hydroborations a chiral centre in the alkene determines the face selectivity of the addition of the borane (Scheme 9.23). Therefore, this type of reaction can often be performed with achiral hydroborating reagents.

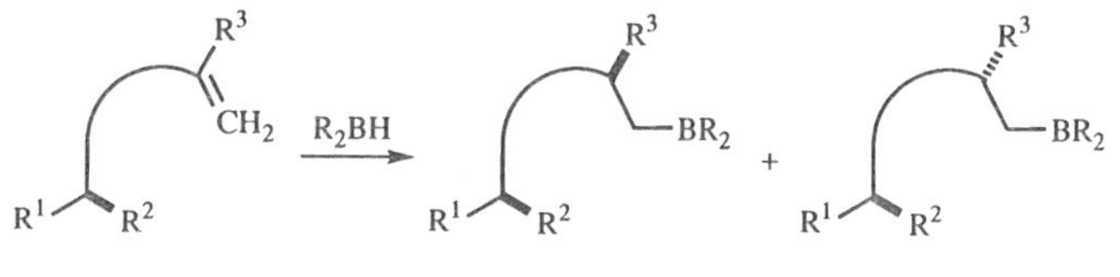

Scheme 9.23

9.4.1 *Intermolecular reactions*

(*a*) *Cyclic alkenes.* Diastereoselective hydroborations of cyclic chiral alkenes are more easily accomplished than for acyclic ones, because the former tend to have better defined reactive conformations. In the absence of exceptional electronic or coordinative effects, boranes tend to add to the least hindered face of a cyclic alkene. Thus, high face selectivities have been observed for reactions of terpenes with BH_3 (cf. syntheses of the chiral hydroborating agents Ipc_2BH, Lgf_2BH, $^dIcr_2BH$, and 3-$^dIcr_2BH$). Similarly, hydroborations of steroid derivatives often display good levels of substrate-controlled diastereoselectivity (e.g. Scheme 9.24).[45-47]

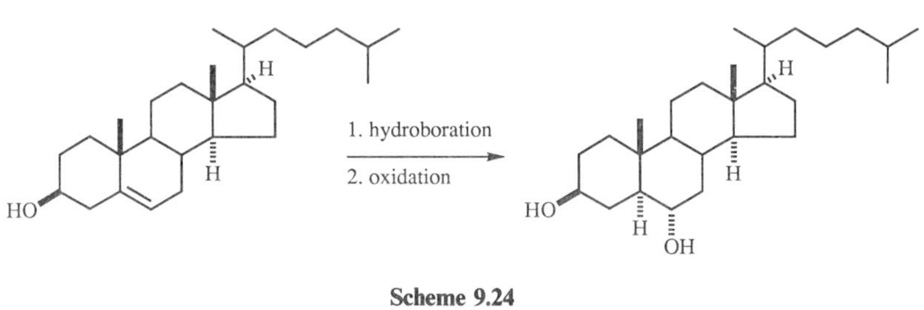

Scheme 9.24

Directing effects have been observed in metal-catalysed transformations. For instance, Crabtree's iridium catalyst ([Ir(cod)(PCy_3)]PF_6) in hydroborations of unsaturated *N*-alkyl, or *N*,*N*-dialkylamides lead predominantly to product isomers derived from transient coordination of the metal complex (e.g. Scheme 9.25).[48]

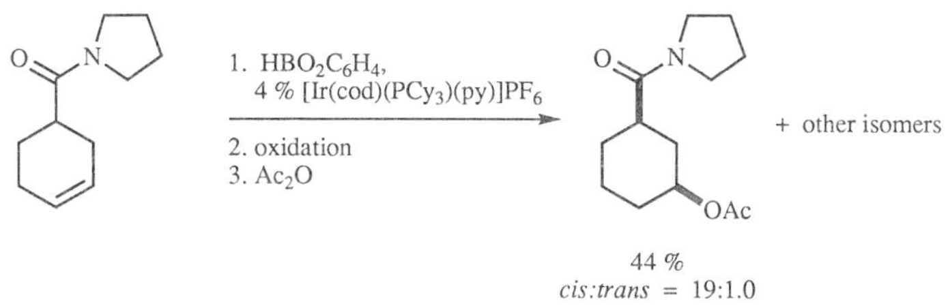

1. $HBO_2C_6H_4$, 4 % $[Ir(cod)(PCy_3)(py)]PF_6$
2. oxidation
3. Ac_2O

+ other isomers

OAc

44 %
cis:*trans* = 19:1.0

Scheme 9.25

(*b*) *1,2-Induction for acyclic substrates.* Early examples of 1,2-asymmetric induction were presented by the Kishi group.[49,50] For instance, the major diastereomer (~8:1) from hydroboration of the alkene (**17**) had the configuration shown in Scheme 9.26.[49] Several other alkenes with a methyl group and a sterically more encumbered moiety at the allylic chiral centre were hydroborated, giving predominantly products with a *syn* relationship between this methyl group and the newly introduced alcohol functionality.

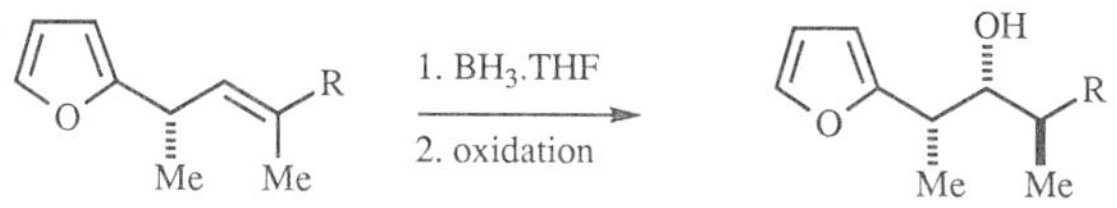

1. BH_3.THF
2. oxidation

17, R = CO_2Et

85 %

Scheme 9.26

Sterically hindered boranes can be used to hydroborate allylic alcohol derivatives with high *anti*-selectivity (Scheme 9.27).[51] The degree of induction is enhanced by more electronegative *O*-protecting groups. For instance, acetate gives a 7.5:1 ratio of *anti* to *syn* products, while this ratio is increased to 14:1 for the corresponding trifluoroacetate.

An elegant application of diastereoselective hydroboration of allylic alcohol derivatives is shown in Scheme 9.28.[52] The alkene (**18**) was hydroborated with dicyclohexylborane, providing an intermediate azido borane, which then underwent a cycloalkylation. The migrating alkyl moiety and the departing diazonium group are positioned antiperiplanar in the proposed intermediate **A** facilitating the rearrangement.

Good *anti*-selectivity is not limited to alkyl-substituted allylic alcohol derivatives. Several vinyl ethers produced predominantly *anti*-products when reacted with thexylborane (e.g. Scheme 9.29).[53] However, *syn*-elimination of the intermediate β-alkoxy boranes can complicate these reactions (e.g. Scheme 9.30).

Houk *et al.* used *ab initio* calculations to evaluate transition states for

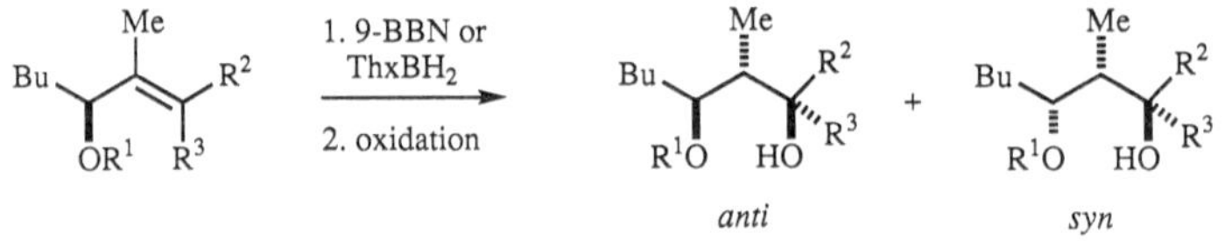

R^1	R^2	R^3	*anti:syn*
H	H	Bu	>15:1
$COCH_3$	H	H	7.5:1
$COCF_3$	H	H	14:1

Scheme 9.27

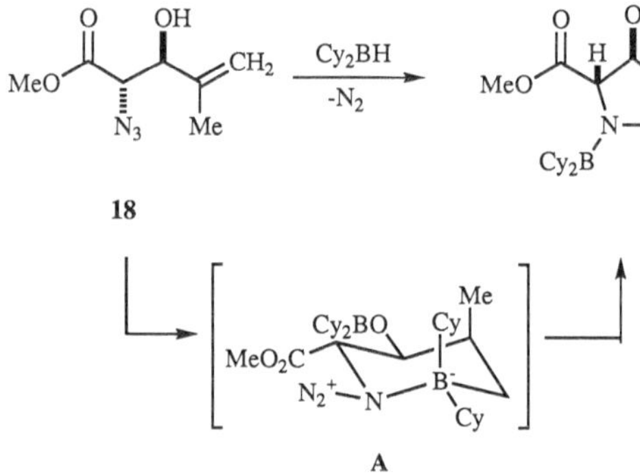

Scheme 9.28

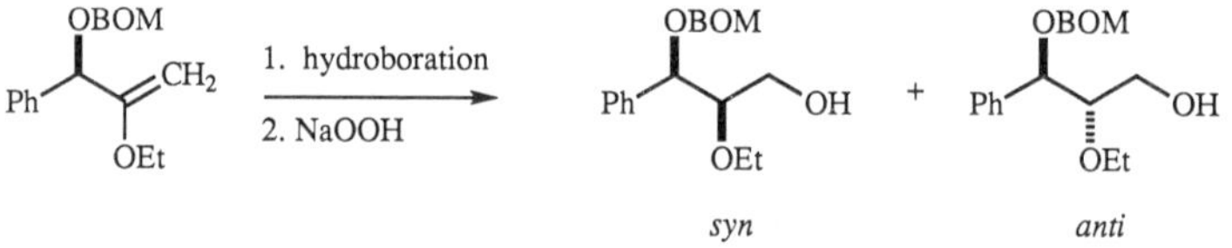

reagents	*syn:anti*
BH_3-DMS, DME	75:25
$ThxBH_2$, THF	3:97

Scheme 9.29

uncatalysed hydroborations of α-substituted alkenes.[54] The lowest energy conformation calculated had the allylic substituents staggered with respect to the forming bond (Figure 9.7). When the substrates are allylic alcohol derivatives, the oxygen preferentially occupies either the inside or the outside

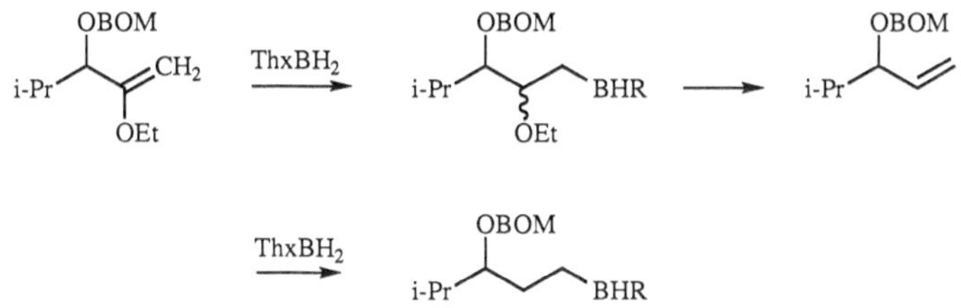

OBOM i-Pr CH_2 OEt $\xrightarrow{ThxBH_2}$ OBOM i-Pr BHR OEt $\longrightarrow$ OBOM i-Pr

$\xrightarrow{ThxBH_2}$ OBOM i-Pr BHR

Scheme 9.30

position, and the *anti*-position is occupied by the most electron-donating group (Figure 9.7). For sterically demanding boranes, the most reactive conformation **B** has the smallest substituent on the chiral centre at the inside position, resulting in predominant formation of the *anti*-product. Reactions with less encumbered reagents like BH_3 can occur via either conformer **A** or **B** depending on more subtle steric and electronic effects. The results of these theoretical studies agree with the experimentally observed selectivities (Schemes 9.26–9.29).

An interesting development in substrate-controlled diastereoselective hydroborations of allylic alcohol derivatives is the predominant production of *syn*-products in rhodium-catalysed addition of catecholborane to these substrates.[55-59] Thus, while the hydroboration of substrate (**19**) with 9-BBN gives good *anti*-selectivity, the catalysed reaction provides predominantly the *syn*-alcohol (Scheme 9.31). This selectivity is enhanced by electronegative and/or sterically hindered *O*-protecting groups.

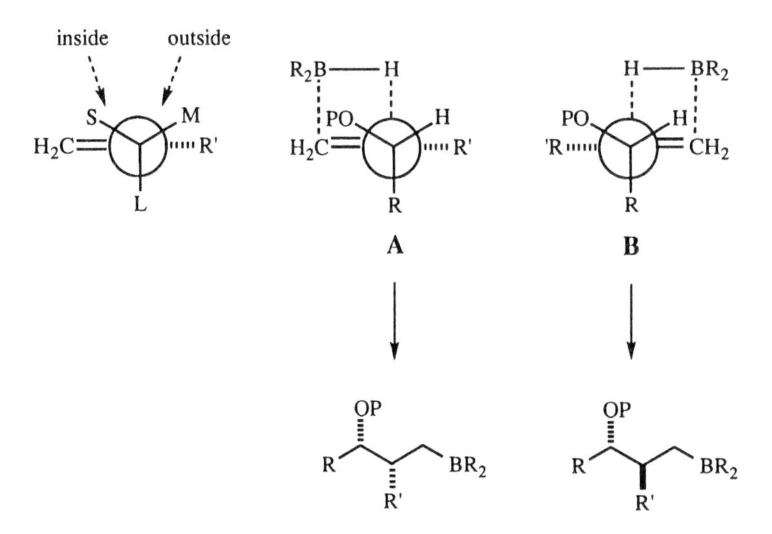

Figure 9.7 Calculated geometries for the transition states of hydroboration of chiral allylic alcohol derivatives.

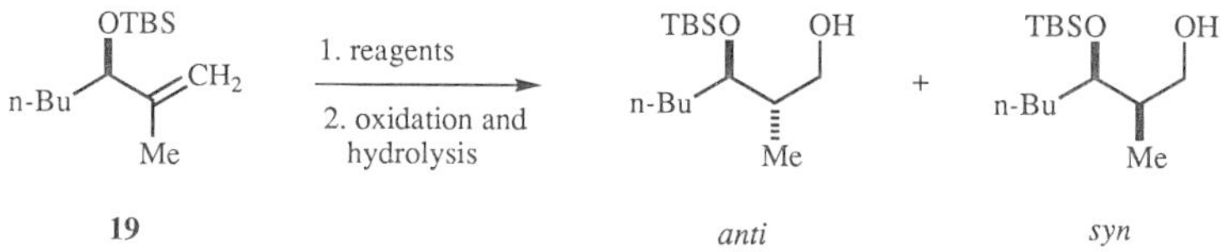

reagents	*anti*:*syn*
$RhCl(PPh_3)_3$, catecholborane	4:96
9-BBN	90:10

Scheme 9.31

Optically active allylamine derivatives prepared from α-amino acids also exhibited *syn*-selectivity in catalysed hydroborations (e.g. Scheme 9.32).[57,60] The bulky *N*-(benzyltosyl)amines gave better selectivities than the less encumbered mono-protected allylamines. Presumably the sterically demanding groups in these reactions are positioned away from the approaching metal complex in the reactive conformer; increased size of the substituents, therefore, improves the *syn*-selectivity. Reactions of these substrates with BH_3·THF yield *anti*-products with high selectivities, consistent with the model described above for uncatalysed hydroborations of allylic alcohol derivatives, whereas hydroborations with 9-BBN are *anti*-selective.[57,60]

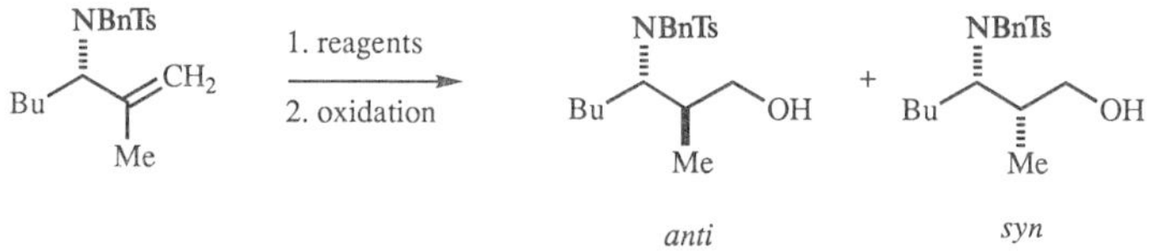

reagents	*anti*:*syn*
$[Rh(cod)Cl]_2$/4 PPh_3, catecholborane	91:9
BH_3	5:95
9-BBN	87:13

Scheme 9.32

A few examples of stereocontrolled hydroborations of acyclic alkenes lacking heteroatom functionalities have been documented. One such case is the highly selective reaction of the steroid derivative (**20**) with sterically encumbered boranes like disiamylborane and dicyclohexylborane (Scheme 9.33).[61] The observed selectivity can be rationalized using the reactive conformation shown in Figure 9.8. The less reactive *endo*-cyclic double bond was not hydroborated in these reactions.

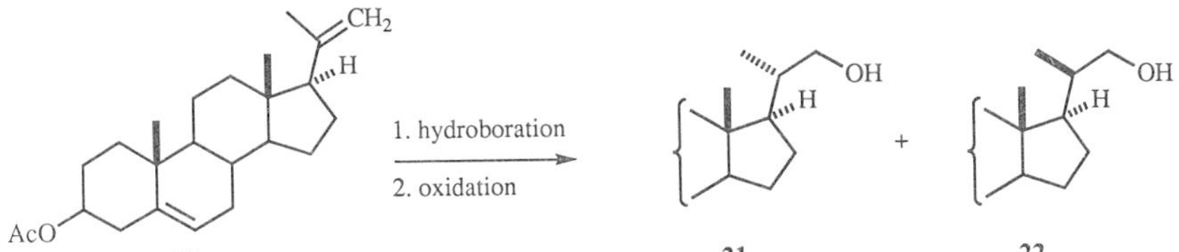

reagents	21:22
9-BBN	14:1
$(Siam)_2BH$	22:1
$(Cy)_2BH$	26:1

Scheme 9.33

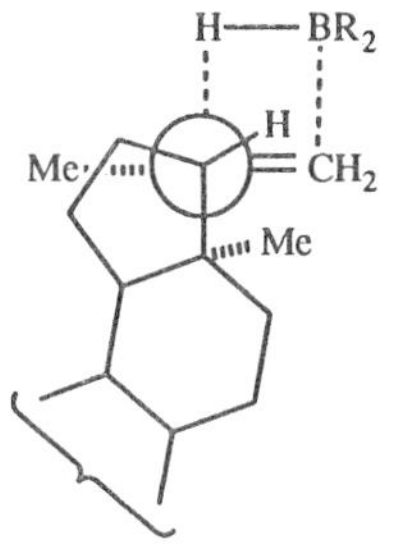

Figure 9.8 Preferred conformation for the hydroboration of (**20**) with hindered dialkylboranes.

(*c*) *1,3-Induction for acyclic substrates.* In contrast to 1,2-asymmetric induction, only a few useful cases of substrate-controlled 1,3-induction have been reported for hydroboration reactions. For alkenes with a β-chiral centre bearing a methyl group and a sterically more demanding substituent, only the stereochemistry of the proximal chiral centre determines the outcome of the reaction: in Schemes 9.34 and 9.35, the stereochemistry at C-4 governs the face selectivity.[62,63] Double asymmetric induction was not observed in the corresponding reactions with the chiral hydroborating reagents (+)- and (−)-$IpcBH_2$ because both enantiomers gave similar levels of asymmetric induction and provided the same major product.

9.4.2 *Intramolecular reactions*

Still *et al.* showed the potential of stereocontrolled intramolecular delivery of boron hydrides to chiral alkenes. These reactions involve cyclic intermediates with relatively rigid transition states, hence good stereoselection could be achieved from remote chiral centres (Scheme 9.36).[64–66] In reactions of several

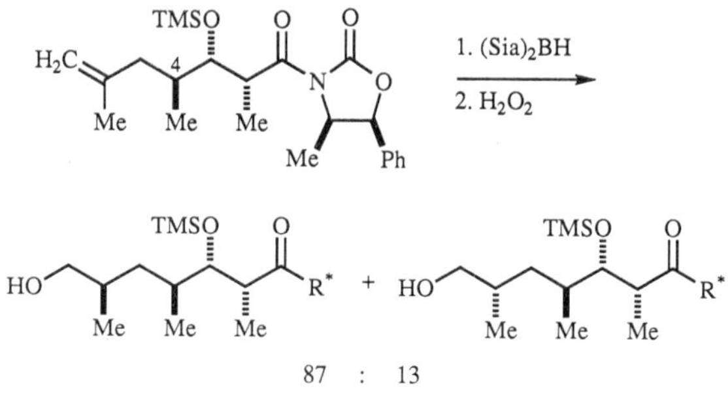

87 : 13

Scheme 9.34

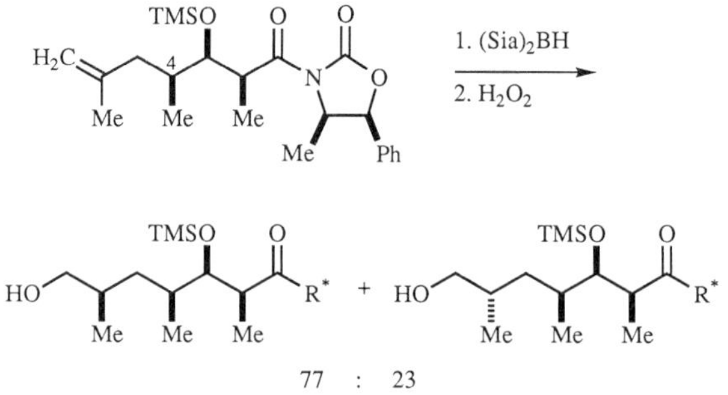

77 : 23

Scheme 9.35

non-conjugated dienes, a chiral centre was created in the first hydroboration of the most reactive terminal alkene. Subsequently, the thexyl group was lost prior to cyclization, presumably via β-elimination, producing a monoalkylborane, which then reacted with the remaining alkene to form the second chiral centre with excellent 1,3- and 1,4-asymmetric induction (Schemes 9.36 and 9.37, respectively). The observed preference was rationalized in terms of a boat-like transition state that positions ring substituents in equatorial positions and eclipses the boron–hydrogen bond with respect to the alkene π-system.

Cyclic hydroboration of the allylic vinylether (**23**) produced the alcohol (**25**) with >200:1 selectivity via initial reaction of the more reactive vinyl ether and subsequent intramolecular addition of the second boron–hydrogen

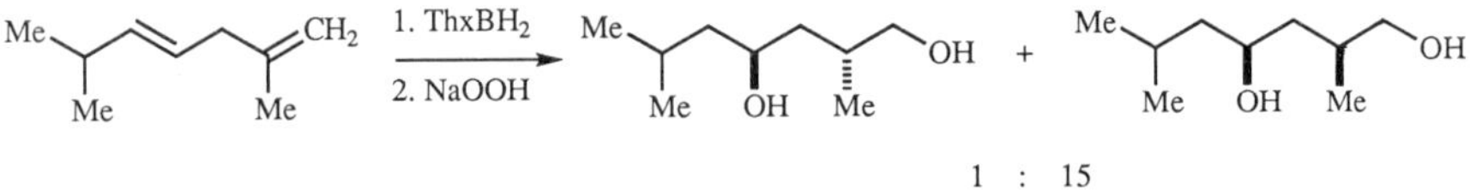

1 : 15

Scheme 9.36

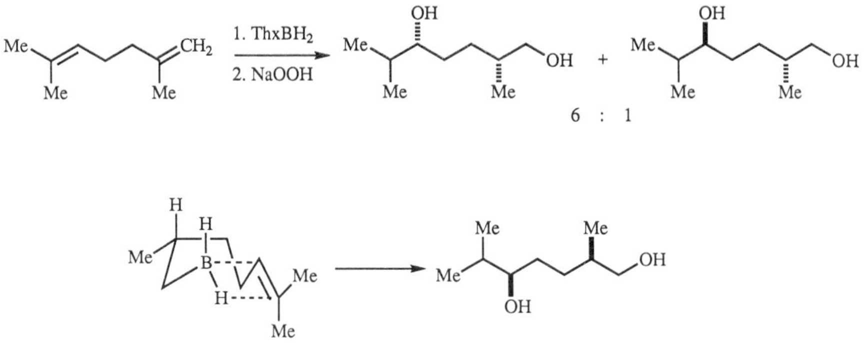

Scheme 9.37

bond to the other alkene (Scheme 9.38). The 7-membered boraheterocycle (**24**) produced undergoes *syn*-elimination of ethene, hence oxidation of the resulting cyclic borinic ester yields the major product. The overall transformation provides products with opposite relative stereochemistry compared with intermolecular hydroborations of allylic alcohol derivatives (see Section 9.4.1b).

Another example of intramolecular asymmetric hydroboration of allylic alcohol derivatives is shown in Scheme 9.39.[67,68] Addition of a hindered borane like thexylborane to diene (**26**) occurred with good *anti*-selectivity in the first hydroboration step, providing dialkylborane (**27**). Reaction of this intermediate with the second alkene furnished the 7-membered boracycle (**28**) with high selectivity, and diol (**30**) as the predominant product after oxidation. In contrast, hydroboration of alkene (**26**) with 9-BBN took place via two intermolecular processes that produced almost exclusively the diol (**29**) after oxidation (Section 9.4.1b).[68]

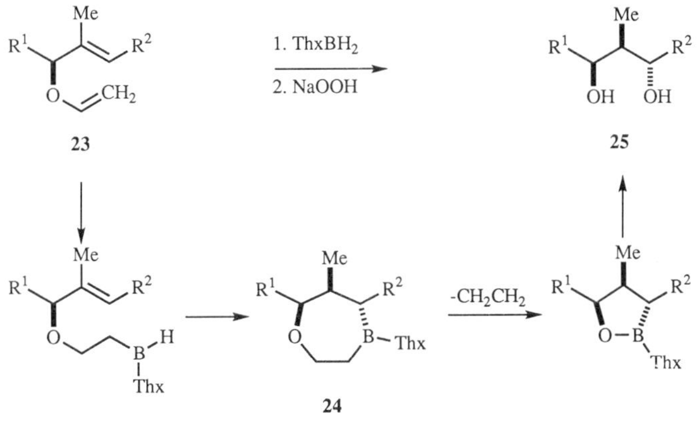

Scheme 9.38

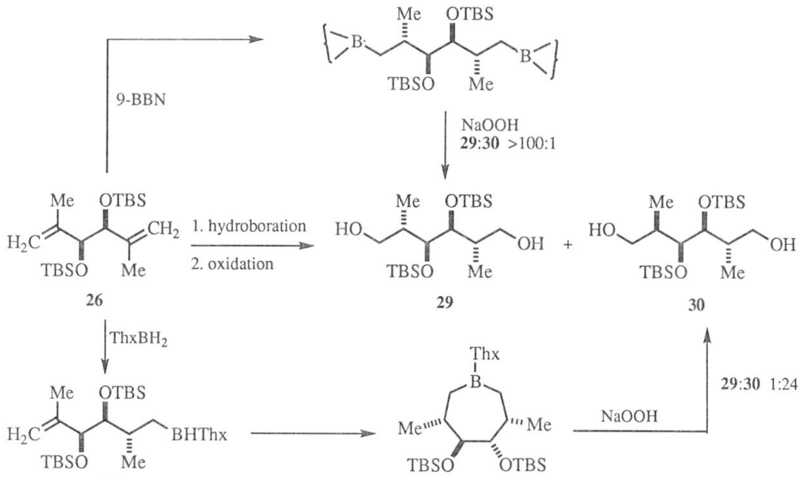

Scheme 9.39

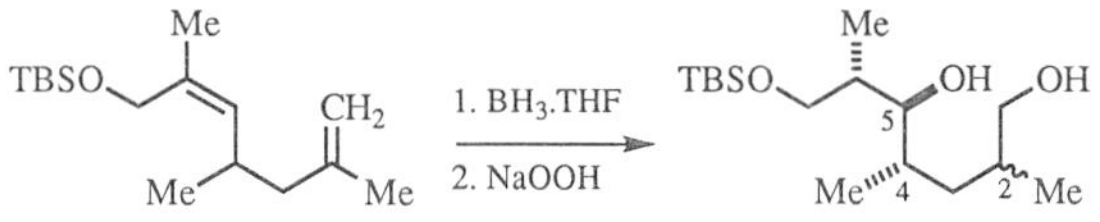

Scheme 9.40

Another highly stereoselective intramolecular hyroboration is the reaction of diene (**31**) with BH_3, providing a 1:1 mixture of the diols (**32**) (Scheme 9.40). The random stereochemistry at C-2 is the result of the non-stereoselective initial hydroboration, but the stereocentres at C-4 and C-5 arise from preferential reaction via conformation **B**, avoiding the sterically unfavourable conformation **A** (Figure 9.9).[65,66]

Intramolecular hydroboration of α-alkoxy-β,γ-unsaturated esters apparently

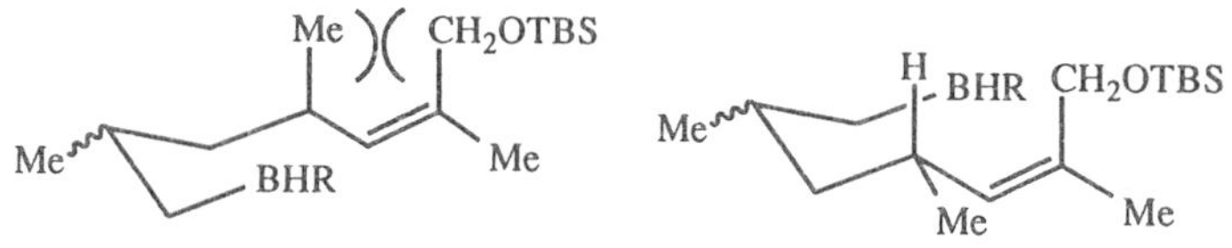

Figure 9.9 Minimization of 1,3-allylic strain governs the face selectivity in cyclic hydroboration of diene (**31**).

occurs via initial reduction of the ester functionality.[69] Alkoxyboranes such as (**34**) are believed to be insufficiently reactive to hydroborate alkenes, but several observations support the postulated intermediacy of (**34**) in these reactions, such as the isolation of cyclic boronate ester (**36**) from the reaction with substrate (**33**) (Scheme 9.41). The stereochemistry at C-2 was shown to control the face selectivity of the hydroboration of the alkene, as illustrated by the comparison of the two reactions shown in Scheme 9.42. Once more, the observed diastereoselectivity for these hydroborations can be explained on the basis of minimized 1,3-allylic strain.[70]

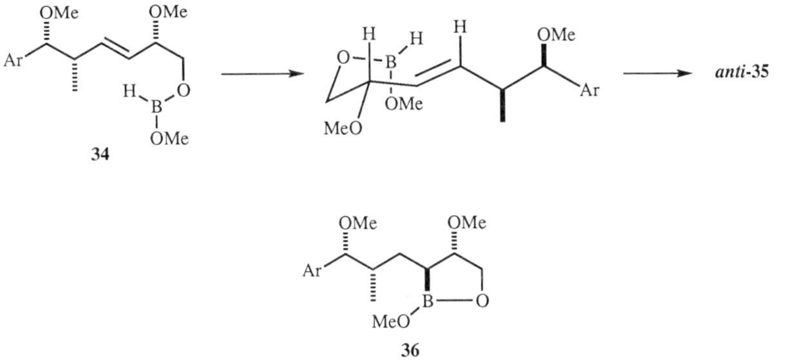

Scheme 9.41

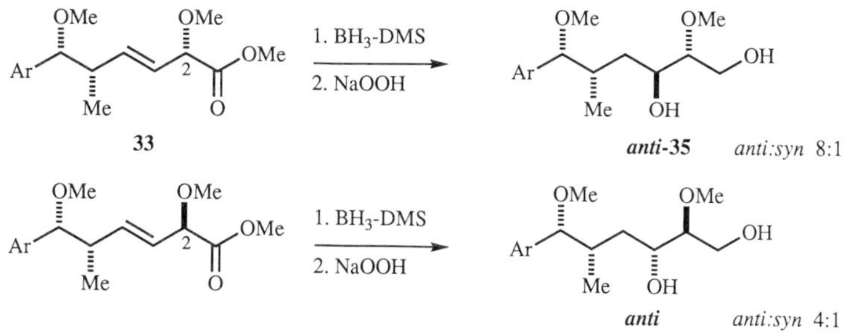

Scheme 9.42

9.5 Conclusions

Reagent-controlled diastereoselectivity has been the most widely investigated approach to control of absolute stereochemistry in hydroboration reactions. Ipc_2BH and $IpcBH_2$ are still the most useful chiral hydroborating reagents derived from asymmetric alkenes. In isolated cases, some of the other chiral mono- and dialkylboranes provide higher degrees of asymmetric induction, but the ready availability of both isomers of α-pinene and the well-established synthetic procedures for Ipc_2BH and $IpcBH_2$ compensate overall. The rate of progress in this field has slowed since Ipc_2BH and $IpcBH_2$ were developed, and enantioselective hydroborations may offer the greatest potential now for further development. With regard to substrate-controlled diastereoselectivities, methods for 1,2-induction are now common, but there is much room for improvement in methods for long-range stereocontrol and directed hydroborations.

References

1. Brown, H.C., Kramer, G.W., Levy, A.B. and Midland, M.M., *Organic Syntheses via Boranes*, New York, Wiley-Interscience, 1975.
2. Pelter, A., Smith, K. and Brown, H.C., *Borane Reagents*. Academic Press, New York, 1988.
3. Rathke, M.W., Inoue, N., Varma, K.R. and Brown, H.C., *J. Am. Chem. Soc.*, 1966, **88**, 2870.
4. Kabalka, G.W. and Hedgecock, H.C., *J. Am. Chem. Soc.*, 1976, **98**, 1290.
5. Brown, H.C. and Singaram, B., *Pure Appl. Chem.*, 1987, **59**, 879.
6. Brown, H.C. and Zweifel, G., *J. Am. Chem. Soc.*, 1961, **83**, 486.
7. Brown, H.C. and Yoon, N.M., *Israel J. Chem.*, 1977, **15**, 12.
8. Brown, H.C. and Singaram, B., *J. Org. Chem.*, 1984, **49**, 945.
9. Brown, H.C., Ayyanger, N.R. and Zweifel, G., *J. Am. Chem. Soc.*, 1964, **86**, 1071.
10. Jadhav, P.K. and Kulkarni, S.U., *Heterocycles*, 1982, **18**, 169.
11. Brown, H.C. and Yoon, N.M., *J. Am. Chem. Soc.*, 1977, **99**, 5514.
12. Brown, H.C. and Jadhav, P.K., in *Asymmetric Synthesis*, Vol. 2, Morrison, J.D. (ed.), Academic Press, New York, 1983, 1.
13. Brown, H.C., Jadhav, P.K. and Mandal, A.K., *J. Org. Chem.*, 1982, **47**, 5074.
14. Brown, H.C., Randad, R.S., Bhat, K.S., Zaidlewicz, M., Weissman, S.A., Jadhav, P.K. and Perumal, P.T., *J. Org. Chem.*, 1988, **53**, 5513.
15. Brown, H.C., Weissman, S.A., Perumal, P.T. and Dhokte, U.P., *J. Org. Chem.*, 1990, **55**, 1217.
16. Richter, R.K., Bonato, M., Follet, M. and Kamenka, J.-M., *J. Org. Chem.*, 1990, **55**, 2855.
17. Brown, H.C., Desai, M.C. and Jadhav, P.K., *J. Org. Chem.*, 1982, **47**, 5065.
18. Brown, H.C., Prasad, J.V.N.V. and Zaidlewicz, M., *J. Org. Chem.*, 1988, **53**, 2911.
19. Brown, H.C. and Jadhav, P.K., *J. Org. Chem.*, 1981, **46**, 5047.
20. Jadhav, P.K. and Brown, H.C., *J. Org. Chem.*, 1981, **46**, 2988.
21. Mandal, A.K., Jadhav, P.K. and Brown, H.C., *J. Org. Chem.*, 1980, **45**, 3543.
22. Brown, H.C. and Singaram, B., *J. Am. Chem. Soc.*, 1984, **106**, 1797.
23. Masamune, S., Lu, L.D.-L., Jackson, W.P., Kaiho, T. and Toyoda, T., *J. Am. Chem. Soc.*, 1982, **104**, 5523.
24. Houk, K.N., Rondan, N.G., Wu, Y., Metz, J.T. and Paddon-Row, M.N., *Tetrahedron*, 1984, **40**, 2257.
25. Sadhu, K.M. and Matteson, D.S., *Organometallics*, 1985, **4**, 1687.
26. Brown, H.C., Naik, R.G., Singaram, B. and Pyun, C., *Organometallics*, 1985, **4**, 1925.
27. Masamune, S., Kim, B., Petersen, J.S., Sato, T., Veenstra, S.J. and Imai, Z., *J. Am. Chem. Soc.*, 1985, **107**, 4549.
28. Evans, D.A., Fu, G.C. and Anderson, B.A., *J. Am. Chem. Soc.*, 1992, 114, 6679.
29. Burgess, K., van der Donk, W.A., Westcott, S.A., Marder, T.B., Baker, R.T. and Calabrese, J.C., *J. Am. Chem. Soc.*, 1992, **114**, 9350.
30. Männig, D. and Nöth, H., *Angew. Chem., Int. Ed. Engl.*, 1985, **24**, 878.
31. Brown, J.M. and Lloyd-Jones, G.C., *J. Chem. Soc., Chem. Commun.*, 1992, 710.
32. Westcott, S.A., Marder, T.B. and Baker, R.T., *Organometallics*, 1993, **12**, 975.
33. Burgess, K. and Ohlmeyer, M.J., *Chem. Rev.*, 1991, **91**, 1179.
34. Burgess, K. and Ohlmeyer, M.J., *J. Org. Chem.*, 1988, **53**, 5178.
35. Hayashi, T., Matsumoto, Y. and Ito, Y., *Tetrahedron: Asymmetry*, 1991, **2**, 601.

36. Brown, J.M., Hulmes, D.I. and Layzell, T.P., *J. Chem. Soc., Chem. Commun.*, 1993, 1673.
37. Sato, M., Miyaura, N. and Suzuki, A., *Tetrahedron Lett.*, 1990, **31**, 231.
38. Hayashi, T., Matsumoto, Y. and Ito, Y., *J. Am. Chem. Soc.*, 1989, **111**, 3426.
39. Brown, J.M. and Lloyd-Jones, G.C., *Tetrahedron: Asymmetry*, 1990, **1**, 869.
40. Masamune, S., Choy, W., Peterson, J.S. and Sita, L.R., *Angew. Chem., Int. Ed. Engl.*, 1985, **24**, 1.
41. Burgess, K., van der Donk, W.A. and Ohlmeyer, M.J., *Tetrahedron: Asymmetry*, 1991, **2**, 613.
42. Westcott, S.A., Blom, H.P., Marder, T.B., Baker, R.T. and Calabrese, J.C., *Inorg. Chem.*, 1993, **32**, 2175.
43. Harrison, K.N. and Marks, T.J., *J. Am. Chem. Soc.*, 1992, **114**, 9220.
44. Halterman, R., *Chem. Rev.*, 1992, **92**, 965.
45. Wechter, W.J., *Chem. Ind.*, 1959, 294.
46. Nussim, M. and Sondheimer, F., *Chem. Ind.*, 1960, 400.
47. Sieff, D., Sondheimer, F. and Nussim, M., *J. Org. Chem.*, 1961, **26**, 630.
48. Evans, D.A. and Fu, G.C., *J. Am. Chem. Soc.*, 1991, **113**, 4042.
49. Schmid, G., Fukuyama, T., Akasaka, K. and Kishi, Y., *J. Am. Chem. Soc.*, 1979, **101**, 259.
50. Johnson, M.R., Nakata, T. and Kishi, Y., *Tetrahedron Lett.*, 1979, **45**, 4343.
51. Still, W.C. and Barrish, J.C. *J. Am. Chem. Soc.*, 1983, **105**, 2487.
52. Evans, D.A. and Weber, A.E., *J. Am. Chem. Soc.*, 1987, **109**, 7151.
53. McGarvey, G.J. and Bajwa, J.S., *Tetrahedron Lett.*, 1985, **26**, 6297.
54. Paddon-Row, M.N., Rondan, N.G. and Houk, K.N., *J. Am. Chem. Soc.*, 1982, **104**, 7162.
55. Evans, D.A., Fu, G.C. and Hoveyda, A.H., *J. Am. Chem. Soc.*, 1988, **110**, 6917.
56. Burgess, K. and Ohlmeyer, M.J., *Tetrahedron Lett.*, 1989, **30**, 395.
57. Burgess, K., Cassidy, J. and Ohlmeyer, M.J., *J. Org. Chem.*, 1991, **56**, 1020.
58. Burgess, K. and Ohlmeyer, M.J., *Am. Chem. Soc., Symp. Ser.*, 1992, 163.
59. Evans, D.A., Fu, G.C. and Hoveyda, A.H., *J. Am. Chem. Soc.*, 1992, **114**, 661.
60. Burgess, K. and Ohlmeyer, M.J., *Tetrahedron Lett.*, 1989, **30**, 5857.
61. Midland, M.M. and Kwon, Y.C., *J. Am. Chem. Soc.*, 1983, **105**, 3725.
62. Evans, D.A. and Bartroli, J., *Tetrahedron Lett.*, 1982, **23**, 807.
63. Evans, D.A., Bartroli, J. and Godel, T., *Tetrahedron Lett.*, 1982, **23**, 4577.
64. Still, W.C. and Darst, K.P., *J. Am. Chem. Soc.*, 1980, **102**, 7385.
65. Still, W.C. and Shaw, K.R. *Tetrahedron Lett.*, 1981, **22**, 3725.
66. Morgans, D.J., *Tetrahedron Lett.*, 1981, **22**, 3721.
67. Harada, T., Matsuda, Y., Uchimura, J. and Oku, A., *J. Chem. Soc., Chem. Commun.*, 1989, 1429.
68. Harada, T., Matsuda, Y., Wada, I., Uchimura, J. and Oku, A., *J. Chem. Soc., Chem. Commun.*, 1990, 21.
69. Panek, J.S. and Xu, F., *J. Org. Chem.*, 1992, **57**, 5288.
70. Hoffmann, R.W., *Chem. Rev.*, 1989, **89**, 1841.

10 Asymmetric hydrosilylation and hydroformylation reactions

E. CAMPI and P. PERLMUTTER

10.1 Introduction

In this chapter, recent advances in selected areas of asymmetric hydrosilylation and hydroformylation reactions will be described. The mechanism will only be briefly discussed here as the focus of this chapter is on the significant progress being made in ligand design. Both processes most likely proceed via hydrometallation of the alkene as outlined in Scheme 10.1. The intermediate metalalkyl (**I**) can then undergo overall formylation or silylation depending on the reaction conditions. At least one new chiral centre is generated in these reactions. As will be described below, to a very significant degree the absolute stereochemistry of these new centres may now be controlled by the incorporation of any of a number of new chiral, non-racemic ligands in the transition metal catalyst. Scheme 10.1 also highlights the problem of control of regioselectivity often associated with these reactions.

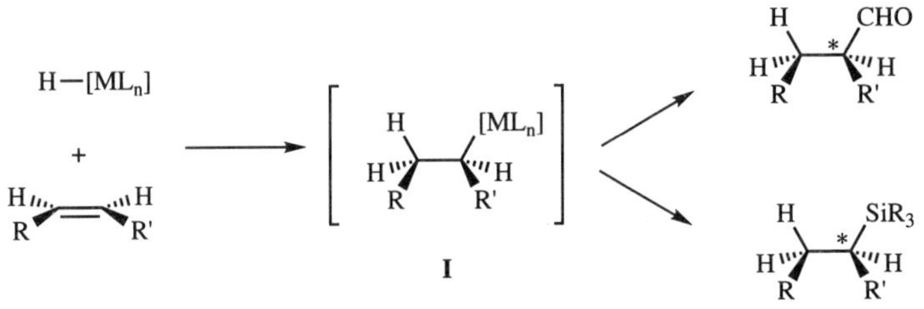

H—[ML_n] + alkene → [I] → CHO product / SiR_3 product

Scheme 10.1

10.2 Asymmetric hydrosilylation

10.2.1 *Asymmetric hydrosilylation of ketones*

The hydrosilylation of ketones is generally carried out using a rhodium, palladium or platinum catalyst and a silane (e.g. diphenylsilane) at relatively low temperatures (room temperature or below) in benzene or toluene solution, sometimes neat (Scheme 10.2). The use of chiral, non-racemic ligands has the potential to render these reactions asymmetric. Since the early 1980s, through

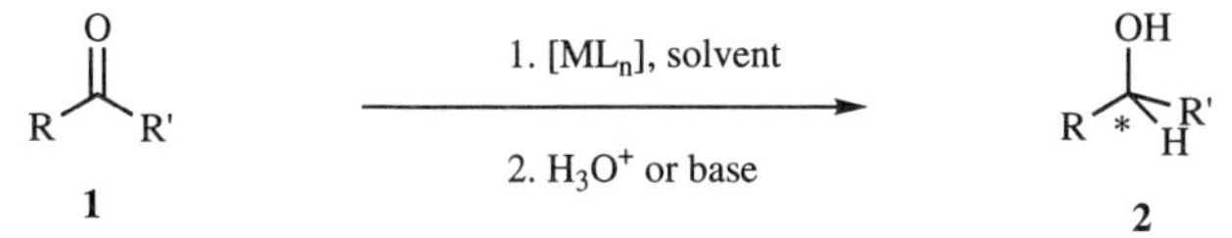

1. $[ML_n]$, solvent

2. H_3O^+ or base

Scheme 10.2

the design and synthesis of more effective ligands, the enantioselectivity for this process has improved dramatically and ee values >90% can now be achieved.[1] In some cases the enantioselectivity approaches that of the best available reductive methods (see Chapter 8). However, this is still not a completely reliable procedure, and much variation in enantioselection is still encountered in reactions with different ketones. Figure 10.1 contains a collection of ligands which have given good to excellent enantioselection in the rhodium-catalysed hydrosilylation of acetophenone (Scheme 10.2, R = Ph, R′ = CH_3). The absolute stereochemistry of the product alcohols is included as well.

The asymmetric hydrosilylation of other arylketones is also efficient. An outstanding example is the hydrosilylation of α-tetralone with (**5**) which gave a remarkable 99% ee in excellent chemical yield (Scheme 10.3).[4]

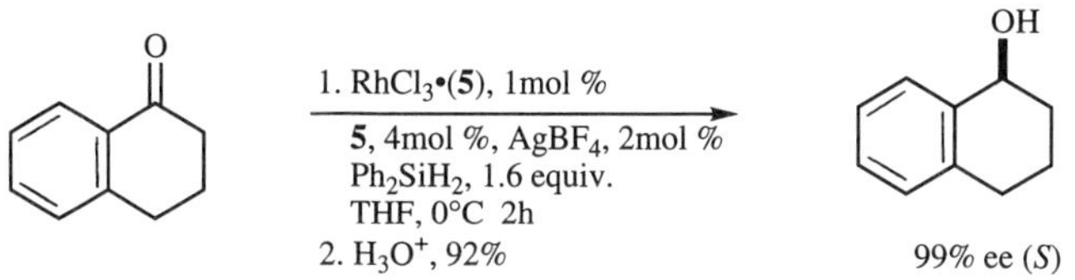

1. $RhCl_3$•(**5**), 1mol %
5, 4mol %, $AgBF_4$, 2mol %
Ph_2SiH_2, 1.6 equiv.
THF, 0°C 2h
2. H_3O^+, 92%

99% ee (*S*)

Scheme 10.3

Fewer studies have been carried out on the asymmetric hydrosilylation of *dialkyl* ketones. Nevertheless, the results are very encouraging and some examples are given in Schemes 10.4,[9] 10.5 and 10.6,[7] 10.7[10] and 10.8.[4a]

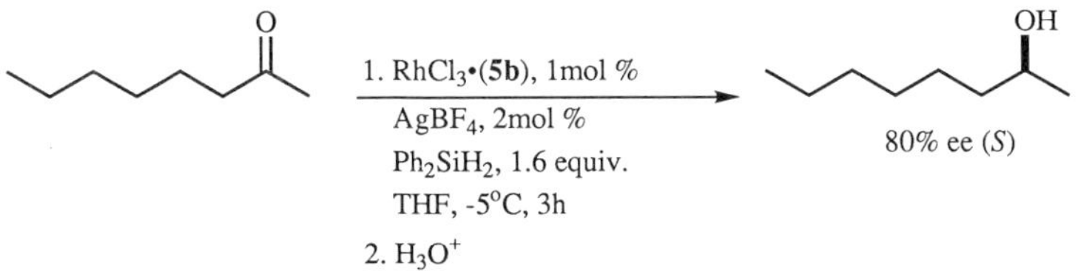

1. $RhCl_3$•(**5b**), 1mol %
$AgBF_4$, 2mol %
Ph_2SiH_2, 1.6 equiv.
THF, -5°C, 3h
2. H_3O^+

80% ee (*S*)

Scheme 10.4

10.2.2 *Asymmetric hydrosilylation of alkenes*

The hydrosilylation of alkenes is generally carried out under similar conditions to those used with ketones.[11] The demonstration that alkylsilanes may be

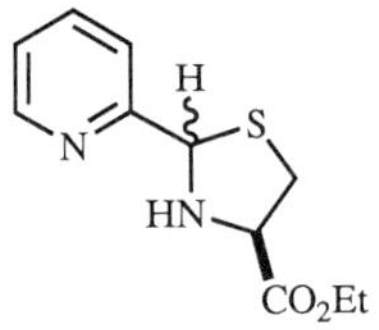

H S N HN CO_2Et

3 (*R*)-Pythia
99% yld, 97.6% ee (*R*)

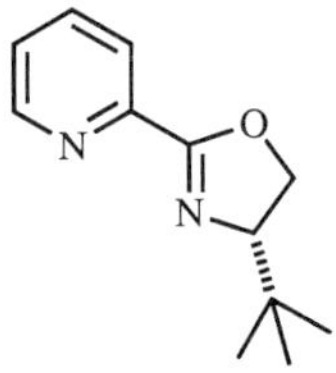

N O N

4 (*S*)-Pymox-tb
91% ee (*R*)

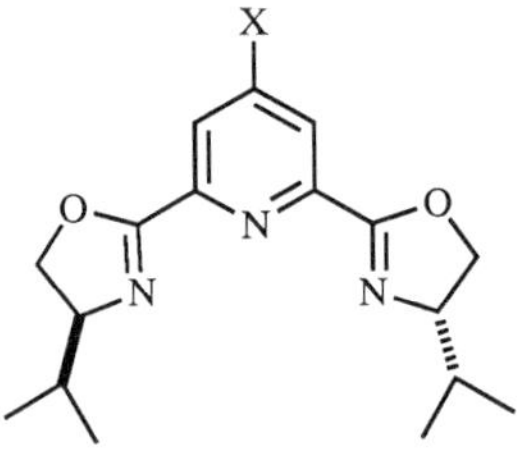

X O N N N O

5a X=H (*S,S*)-Pybox-ip
94% yld, 95% ee (*S*)
5b X=Cl, 4-Cl-(*S,S*)-Pybox-ip
90% yld, 94% ee (*S*)

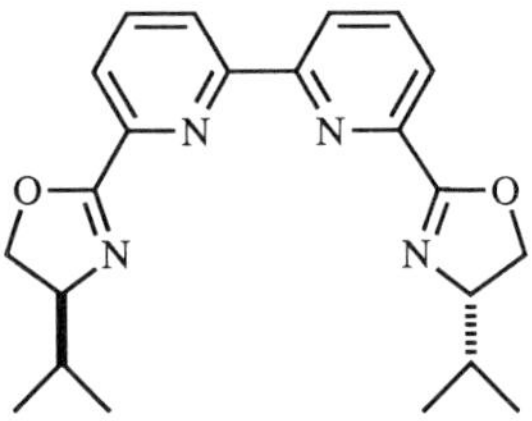

N N O N N O

6 (*S,S*)-Bipymox-ip
98% yld, 90% ee (*S*)

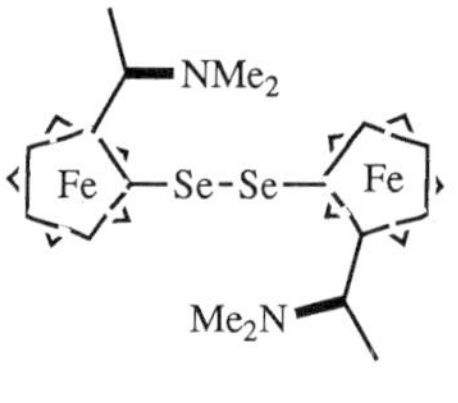

NMe_2 Fe Se-Se Fe Me_2N

7
31% yld, 85% ee (*R*)

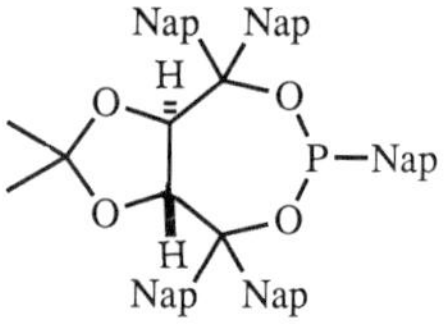

Nap Nap H O O P—Nap O O H Nap Nap

8 TADDOL
99% yld, 84% ee (*R*)

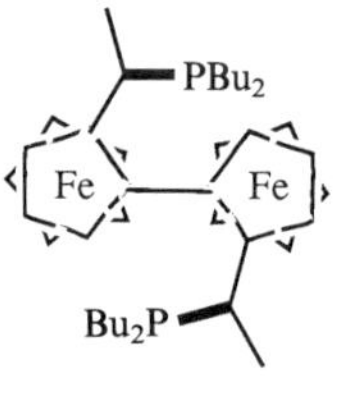

PBu_2 Fe Fe Bu_2P

9 n-BuTRAP
89% yld, 92% ee (*S*)

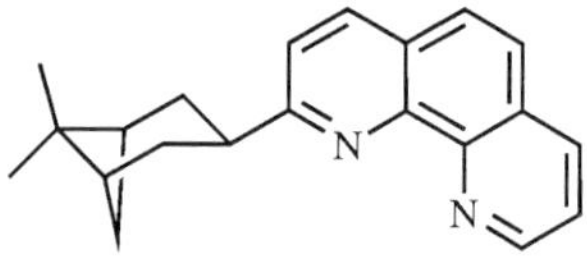

N N

10
76% ee (*R*)

Figure 10.1 Selected chiral, non-racemic ligands used in rhodium-catalysed hydrosilylation of acetophenone: (**3**).[2] (**4**)[3], (**5**),[4b] (**6**),[3] (**7**)[5], (**8**),[6] (**9**)[7] and (**10**).[8]

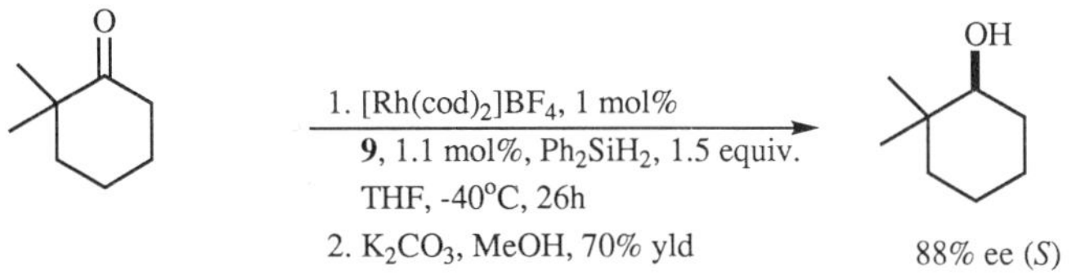

1. $[Rh(cod)_2]BF_4$, 1 mol%
9, 1.1 mol%, Ph_2SiH_2, 1.5 equiv.
THF, -40°C, 26h
2. K_2CO_3, MeOH, 70% yld

88% ee (*S*)

Scheme 10.5

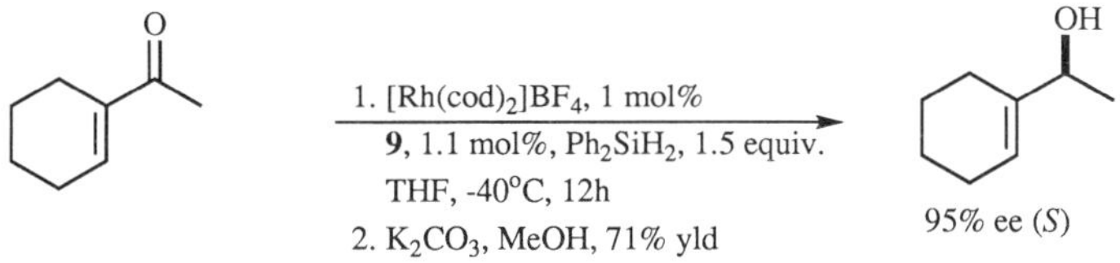

1. $[Rh(cod)_2]BF_4$, 1 mol%
9, 1.1 mol%, Ph_2SiH_2, 1.5 equiv.
THF, -40°C, 12h
2. K_2CO_3, MeOH, 71% yld

95% ee (*S*)

Scheme 10.6

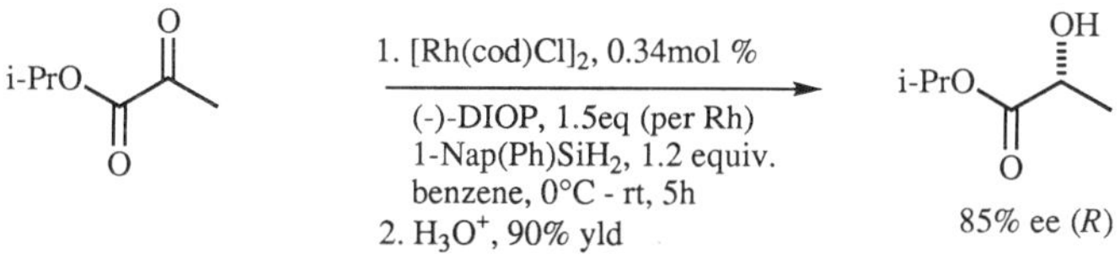

i-PrO

1. $[Rh(cod)Cl]_2$, 0.34mol %
(-)-DIOP, 1.5eq (per Rh)
1-Nap(Ph)SiH_2, 1.2 equiv.
benzene, 0°C - rt, 5h
2. H_3O^+, 90% yld

i-PrO

85% ee (*R*)

Scheme 10.7

EtO

1. $RhCl_3$•(4), 1mol %
4, 4mol %, $AgBF_4$, 2mol %
Ph_2SiH_2, 1.6 equiv.,THF
0°C, 7h
2. MeOH, 91% yld

EtO

95% ee (*S*)

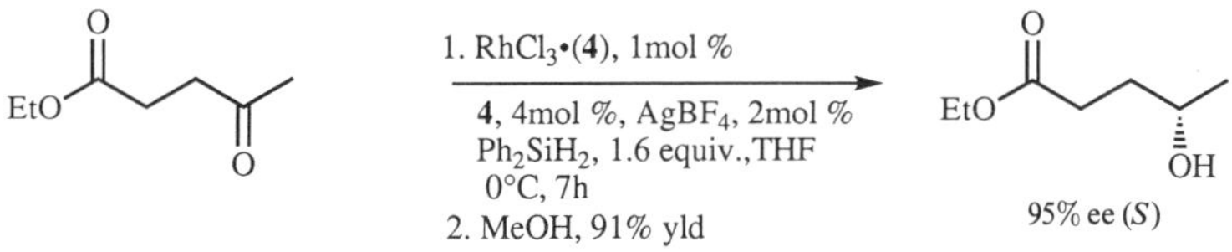

Scheme 10.8

simply and efficiently converted into the corresponding alcohols[12] has stimulated considerable activity in the development of regio- and enantioselective hydrosilylations of alkenes. This conversion of silanes into alcohols is based on Kumada's observation that, although the carbon–silicon bond in organosilanes is resistant to oxidation, the corresponding bond in organopentafluorosilicates is easily cleaved oxidatively.[12,29] Thus treatment of an organosilanes with potassium fluoride (to generate the organopentafluorosilicate), followed by oxidation, yields the required alcohol (as shown in Schemes 10.9–10.12).

Over the past few years the enantioselectivity of such hydrosilylations has improved dramatically.[1a] In particular, Hayashi's group has shown that the use of the optically pure binaphthol-derived ligands (*R*)- or (*S*)-MOP (2-methyl-2′-diphenylphosphino-1,1′-binaphthyl) (**11**) and (**12**)† in conjunction

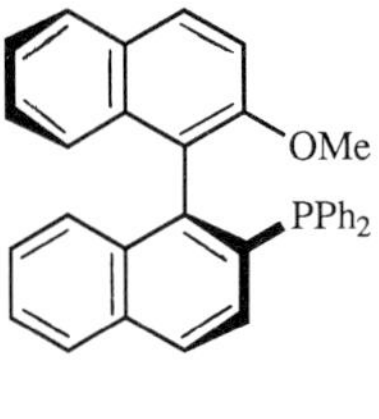

OMe
PPh_2

(*S*)-MOP **11**

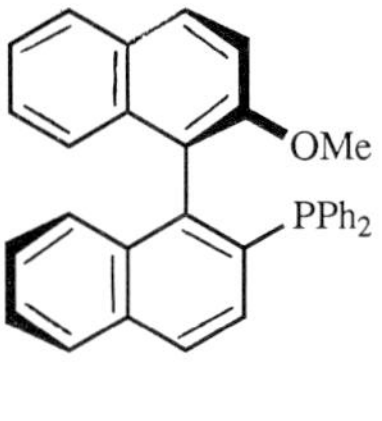

OMe
PPh_2

(*R*)-MOP **12**

with a palladium catalyst in the hydrosilylation of a range of alkenes leads to products of high enantiomeric purity. Terminal alkenes (Scheme 10.9),[14] cycloalkenes (Schemes 10.10[15] and 10.11[16]) and styrenes (Scheme 10.12)[17] all give high ee values, although the problem of regioselectivity has not been completely overcome.

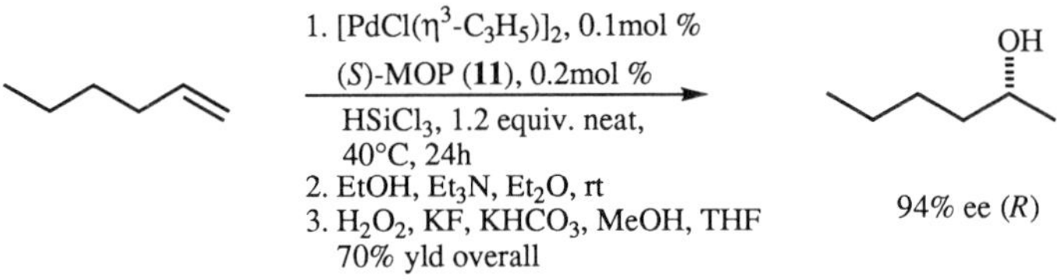

1. $[PdCl(\eta^3\text{-}C_3H_5)]_2$, 0.1mol %
(*S*)-MOP (**11**), 0.2mol %
$HSiCl_3$, 1.2 equiv. neat,
40°C, 24h
2. EtOH, Et_3N, Et_2O, rt
3. H_2O_2, KF, $KHCO_3$, MeOH, THF
70% yld overall

OH

94% ee (*R*)

Scheme 10.9

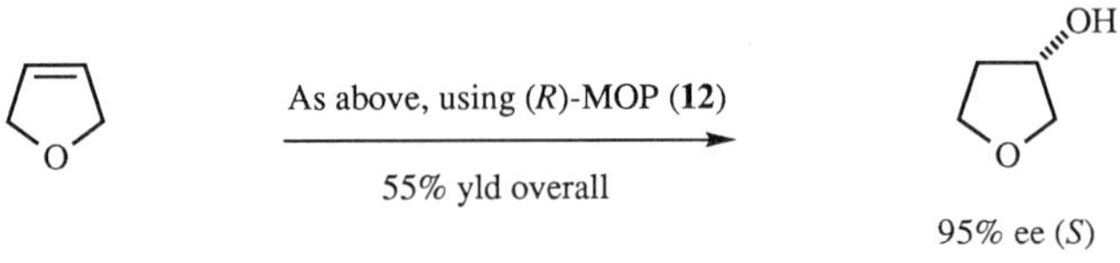

O

As above, using (*R*)-MOP (**12**)

55% yld overall

OH
O

95% ee (*S*)

Scheme 10.10

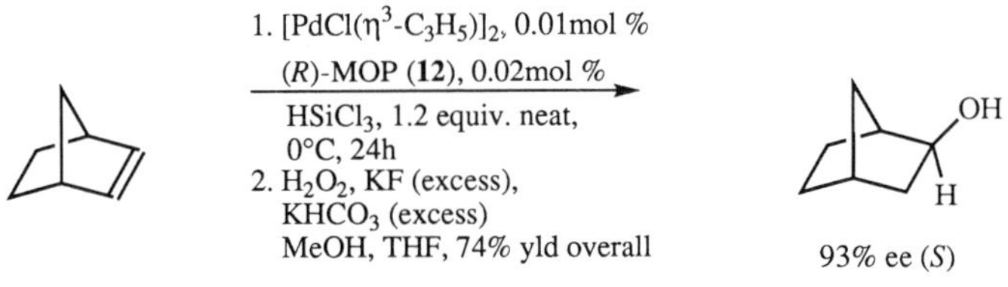

1. $[PdCl(\eta^3\text{-}C_3H_5)]_2$, 0.01mol %
(*R*)-MOP (**12**), 0.02mol %
$HSiCl_3$, 1.2 equiv. neat,
0°C, 24h
2. H_2O_2, KF (excess),
$KHCO_3$ (excess)
MeOH, THF, 74% yld overall

OH
H

93% ee (*S*)

Scheme 10.11

† Full experimental details on the preparation of these ligands in ref. 13.

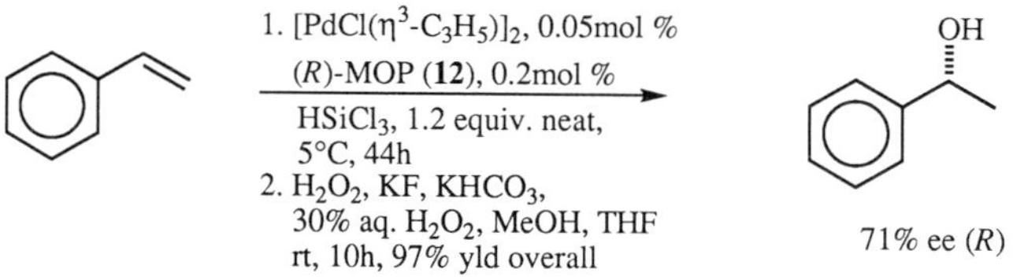

1. $[PdCl(\eta^3\text{-}C_3H_5)]_2$, 0.05mol %
(*R*)-MOP (**12**), 0.2mol %
$HSiCl_3$, 1.2 equiv. neat,
5°C, 44h
2. H_2O_2, KF, $KHCO_3$,
30% aq. H_2O_2, MeOH, THF
rt, 10h, 97% yld overall

OH

71% ee (*R*)

Scheme 10.12

10.3 Asymmetric hydroformylation of alkenes

The hydroformylation of alkenes (Schemes 10.13 and 10.14) forms the basis of the industrial 'oxo' process.[18] The process involves reacting an alkene with synthesis gas (CO/H_2) in an inert solvent with a transition metal complex at moderate to high pressures. The catalysts used in this process were

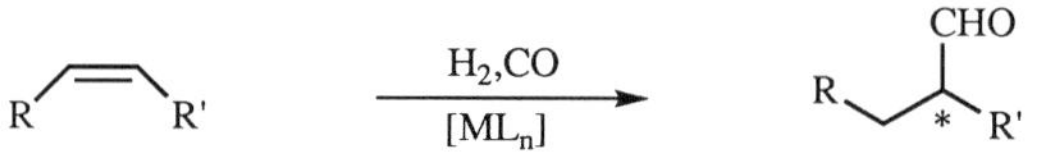

R R' H_2,CO [ML_n] R CHO * R'

Scheme 10.13

R H_2,CO [ML_n] CHO R * + R CHO

R R' H_2,CO [ML_n] CHO * R R' + R * CHO R'

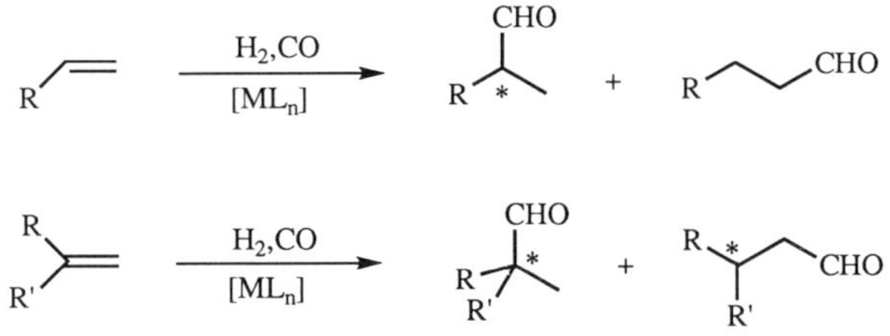

Scheme 10.14

originally based on cobalt; however these have largely been replaced by rhodium catalysts. The latter, in the presence of triphenylphosphine, operate under much milder conditions than the corresponding cobalt catalysts.

Certain classes of alkene, especially styrenes, often undergo hydroformylation with high regioselectivity (e.g. Scheme 10.13 above, R = H, R′ = alkyl or aryl).[19] As with hydrosilylation, the use of non-racemic, chiral ligands promises to provide an asymmetric version of hydroformylation.

As opposed to the impressive results being obtained in asymmetric hydrosilylations, the development of reliable conditions for obtaining high ee values in asymmetric hydroformylations has proved more difficult. One of the most intensively studied areas of asymmetric hydroformylation has been the reactions of substituted styrenes and related aryl alkenes. This is because the products are advanced intermediates in the synthesis of clinically useful

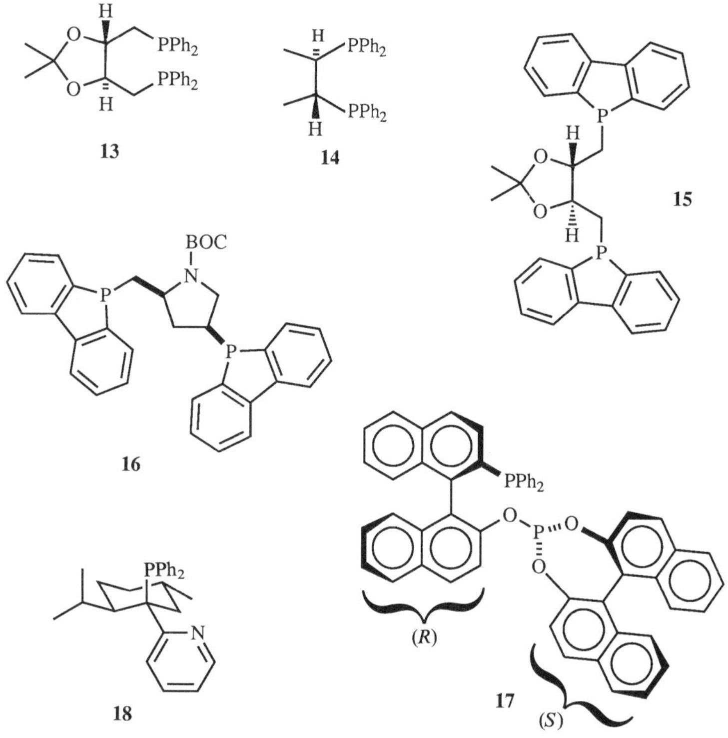

Figure 10.2 Chiral, non-racemic ligands for asymmetric hydroformylation.

drugs.[20] A wide range of chiral, non-racemic bidentate phosphines (Figure 10.2) has been examined as ligands in rhodium-catalysed hydroformylations and ee values in the hydroformylation of styrene itself, for example, usually fall into the range of 2 to ~35%: for example, (**13**) DIOP (25% ee),[21] (**14**) CHIRAPHOS (~25% ee)[22] and (**15**) (−)-DBP-DIOP (33% ee).[23] To date, Takaya's group has achieved the most impressive results with rhodium-based systems.[24,25] Using the new ligand (*R*,*S*)-BINAPHOS (**17**)[26] and $Rh(acac)(CO)_2$ as the catalyst, hydroformylation of styrene produced (*S*)-2-phenylpropanal in 94% ee (Scheme 10.15). The ratio of branched to linear products was only 88:12. The relative stereochemistry of the ligand was found to be crucial as only the (*R*,*S*)- and (*S*,*R*)-ligands, and not the (*R*,*R*)- or (*S*,*S*)-ligands, gave high enantioselectivities.

Using these conditions, a variety of terminal and internal alkenes was shown to hydroformylate with impressive enantioselectivities (73 to 95%). Only (*E*)-2-butene gave a disappointing result (48% ee). Asymmetric hydroformylation of the corresponding (*Z*)-isomer was much more efficient (82% cc).

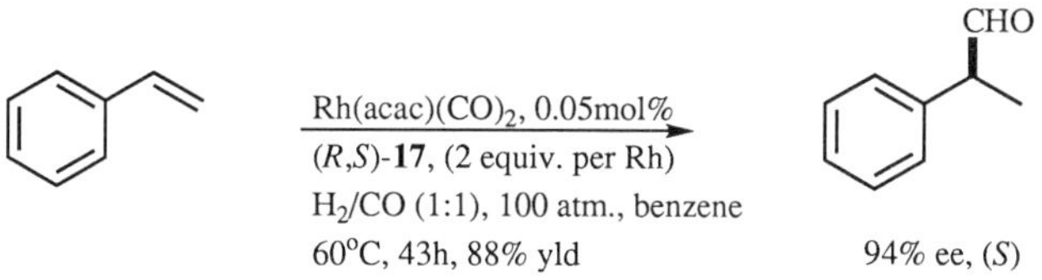

CHO

Rh(acac)$(CO)_2$, 0.05mol%
(*R*,*S*)-**17**, (2 equiv. per Rh)
H_2/CO (1:1), 100 atm., benzene
60°C, 43h, 88% yld

94% ee, (*S*)

Scheme 10.15

Another new ligand which shows much promise is the bidentate P,N ligand (**18**).[27] Although results with styrenes were very disappointing (6% ee), hydroformylation of methyl acrylate provided the (*R*)-aldehyde in 92% ee and 97% regioselectivity (Scheme 10.16). This only serves to illustrate how sensitive this process can be to any variation in the structures of the catalyst *and* the substrate.

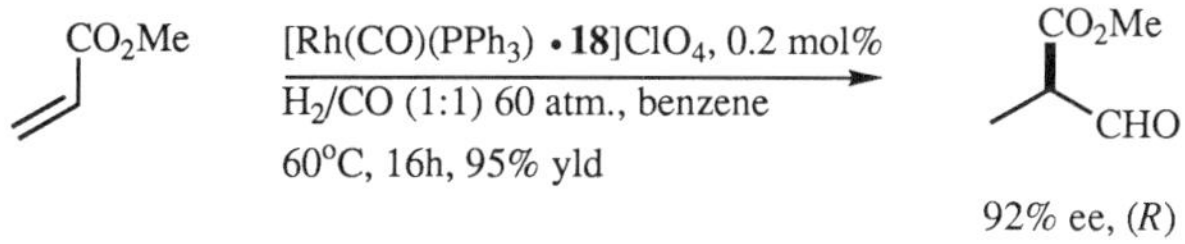

CO_2Me

[Rh(CO)(PPh_3) • **18**]ClO_4, 0.2 mol%
H_2/CO (1:1) 60 atm., benzene
60°C, 16h, 95% yld

CO_2Me
CHO

92% ee, (*R*)

Scheme 10.16

Finally, Stille and Hegedus have shown that platinum-based complexes (which were regarded as poorer hydroformylation catalysts because of low reaction rates, poor regioselectivity and problems with competing hydrogenation) can, in fact, function as excellent hydroformylation catalysts if the reactions are carried out in the presence of stannous chloride.[28] In particular, they demonstrated the efficacy of the platinum complex [(**16**)·$PtCl_2/SnCl_2$] in reactions with a variety of arylethenes. They also demonstrated that the low ee values in these reactions were the result of *in situ* racemization. By the simple expedient of including triethyl orthoformate in the reaction mixture to trap the initially formed aldehyde, acetals of virtually complete enantiomeric purity were obtained (Schemes 10.17 and 10.18).

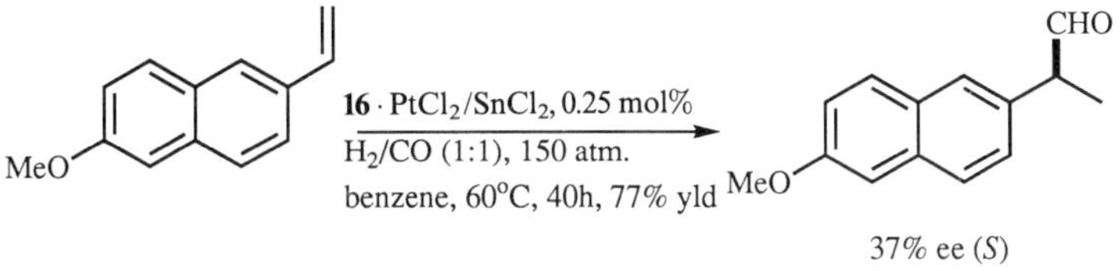

MeO

16 · $PtCl_2/SnCl_2$, 0.25 mol%
H_2/CO (1:1), 150 atm.
benzene, 60°C, 40h, 77% yld

CHO
MeO

37% ee (*S*)

Scheme 10.17

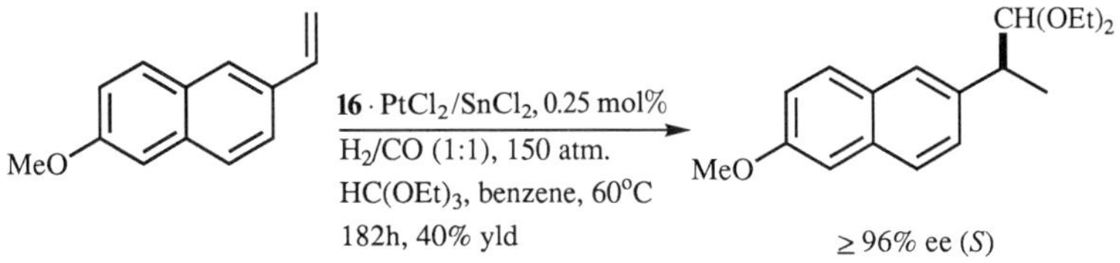

MeO — $\xrightarrow[\text{H}_2\text{/CO (1:1), 150 atm. HC(OEt)}_3\text{, benzene, 60°C 182h, 40\% yld}]{\mathbf{16}\cdot \text{PtCl}_2\text{/SnCl}_2\text{, 0.25 mol\%}}$ — MeO, $CH(OEt)_2$

≥ 96% ee (*S*)

Scheme 10.18

10.4 Conclusions

The ability to add the elements of HX across a double bond, where X = SiR_3 or CHO, with high levels of regio- and enantioselectivity is now a reality for some specific substrates. Much experimentation still needs to be done before these processes can be carried out with complete control and predictability. However, it is clear from the examples cited in this chapter that very significant progress has been made towards this goal.

References

1. (a) For an excellent recent review, see: Brunner, H., Nishiyama, H. and Itoh, K., in *Catalytic Asymmetric Synthesis*, Ojima, I. (ed.), VCH, New York, 1993, pp. 303–322; For earlier reviews see (b) Ojima, I. in *The Chemistry of Silicon Compounds*, Part 2, Patai, S. and Rappoport, Z. (eds); Wiley, New York, 1989, pp. 1479–1526; (c) Ojima, I., Clos, N. and Bastos, C., *Tetrahedron*, 1989, **45**, 6901; (d) Brunner, H., *Synthesis*, 1988, 645; (e) Ojima, I. and Hirai, K., in *Asymmetric Synthesis*, Vol. 5, Morrison, J.D. (ed.), Academic Press, Orlando, 1985, pp. 103–146.
2. Brunner, H., Becker, R. and Riepl, G., *Organometallics*, 1984, **3**, 1354.
3. Nishiyama, H., Yamaguchi, S., Park, S.-B. and Itoh, K., *Tetrahedron: Asymmetry*, 1993, **4**, 143.
4. (a) Nishiyama, H., Kondo, M., Nakamura, T. and Itoh, K., *Organometallics*, 1991, **10**, 500; (b) Nishiyama, H., Yamaguchi, S., Kondo, M., Nakamura, T. and Itoh, K., *J. Org. Chem.*, 1992, **57**, 4306.
5. Nishibayashi, Y., Singh, J.D., Segawa, K., Fukuzawa, S. and Uemura, S., *J. Chem. Soc., Chem. Commun.*, 1994, 1375.
6. Sakaki, J., Schweizer, W.B. and Seebach, D., *Helv. Chim. Acta*, 1993, **76**, 2654.
7. Sawamura, M., Kuwano, R. and Ito, Y., *Angew. Chem., Int. Ed. Engl.*, 1994, **33**, 111.
8. Gladiali, S., Pinna, L., Delogu, G., Graf, E. and Brunner, H., *Tetrahedron: Asymmetry*, 1990, **1**, 937.
9. Nishiyama, H., Yamaguchi, S., Wakamatsu, S., Kondo, M., Nakamura, T. and Itoh, K., in *6th IUPAC Symp. Organomet. Chem. Directed Towards Org. Synth.*, Utrecht, Aug. 25–29, 1991, No. S-14.
10. (a) Ojima, I., Kogure, T. and Kumagai, M., *J. Org. Chem.*, 1977, **42**, 1671; (b) Ojima, I., Tanaka, T. and Kogure, T., *Chem. Lett.*, 1981, 823.
11. Hiyama, T. and Kusumoto, T., in *Comprehensive Organic Synthesis*, Vol. 8, Trost, B.M. and Fleming, I. (eds), Pergamon Press, Oxford, 1991, pp. 763–792.
12. Tamao, K., Kakui, T. and Kumada, M., *J. Am. Chem. Soc.*, 1978, **100**, 2268.
13. Uozomi, Y., Tanahashi, A., Lee, S.-Y. and Hayashi, T., *J. Org. Chem.*, 1993, **58**, 1945.
14. Hayashi, T. and Uozomi, Y., *J. Am. Chem. Soc.*, 1991, **113**, 9887.
15. Uozomi, Y. and Hayashi, T., *Tetrahedron Lett.*, 1993, **34**, 2335.

16. Uozomi, Y., Lee, S.-Y. and Hayashi, T., *Tetrahedron Lett.*, 1992, **33**, 7185.
17. Uozomi, Y., Kitayama, K. and Hayashi, T., *Tetrahedron Lett.*, 1993, **34**, 2419.
18. Weissermel, K. and Arpe, H.-J., *Industrial Organic Chemistry*, VCH, Weinheim, 1993, pp. 123–134.
19. Stille, J.K., in *Comprehensive Organic Synthesis*, Vol. 4, Trost, B.M. and Fleming, I. (eds), Pergamon Press, Oxford, 1991, pp. 913–950.
20. Botteghi, C., Paganelli, S., Schionato, A. and Marchetti, M., *Chirality*, 1991, **3**, 355.
21. (a) Consiglio, G. and Pino, P., *Top. Curr. Chem.*, 1982, **105**, 77; (b) Ojima, I. and Hirai, K., in *Asymmetric Synthesis*, Vol. 5, Morrison, J.D. (ed.), Academic Press, New York, 1985, p. 103.
22. Consiglio, G., Morandini, F., Scalone, M. and Pino, P., *J. Organomet. Chem.*, 1985, **279**, 193.
23. Haelg, P., Consiglio, G. and Pino, P., *J. Organomet. Chem.*, 1985, **296**, 281.
24. Sakai, N., Mano, S., Nozaki, K. and Takaya, H., *J. Am. Chem. Soc.*, 1993, **115**, 7033.
25. Sakai, N., Nozaki, K. and Takaya, H., *J. Chem. Soc., Chem. Commun.*, 1994, 395.
26. Higashizima, T., Sakai, N., Nozaki, K. and Takaya, H., *Tetrahedron Lett.*, 1994, **35**, 2023. "Note that '(*R*,*S*)' in the context of compound **17** refers to the absolute configuration of *each* binaphthyl moiety, i.e. **17** is enantiomerically pure."
27. Arena, C.G., Nicolo, F., Drommi, D., Bruno, G. and Faraone, F., *J. Chem. Soc., Chem. Commun.*, 1994, 2251.
28. Stille, J.K., Su, H., Brechot, P., Parrinello, G. and Hegedus, L.S., *Organometallics*, 1991, **10**, 1183; see also Consiglio, G. and Pino, P., *Helv Chim. Acta*, 1976, **59**, 642.
29. See also Fleming, I. and Sanderson, P.E.J., *Tetrahedron Lett.*, 1987, **28**, 4229; and Fleming, I., Henning, R. and Plant, H., *J. Chem. Soc. Chem. Commun.*, 1984, **29**.

11 Asymmetric conjugate addition reactions

P. PERLMUTTER

11.1 Introduction

The field of conjugate addition reactions is a vast one. A book[1] and several excellent reviews[2] have appeared over the last few years, which cover most aspects of conjugate addition reactions. The focus in this chapter will be mainly on the use of non-racemic[3] additives, e.g. ligands and/or catalysts, in asymmetric conjugate additions. The chapter begins with a brief introduction to the chemistry of conjugate additions. After this it is organized according to the class of nucleophile or incipient nucleophile.

11.2 Stereoselective conjugate addition reactions

Conjugate addition reactions involve the addition of a nucleophile to a centre remote from, and in conjugation with, a functional group (FG) capable of stabilising an adjacent negative charge. Subsequent 'quenching' of the anion (usually by protonation but also by addition of other electrophiles, E^+)[4] leads to the conjugate adduct (Scheme 11.1). A wide variety of nucleophiles participate in this reaction and many functional groups can serve as the activating (and stabilising) remote functional group.

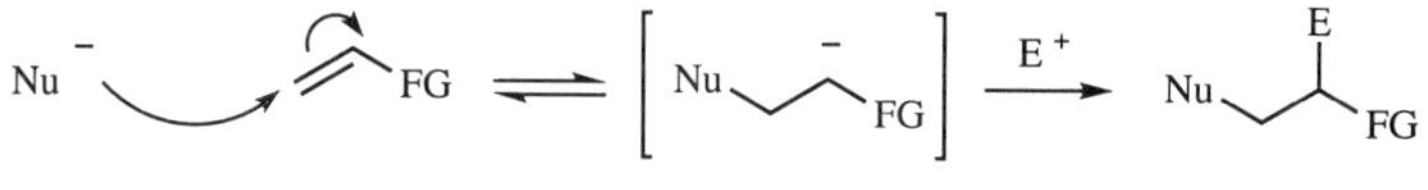

Scheme 11.1

By utilising non-racemic nucleophiles and/or prochiral (or racemic) conjugate acceptors, a number of potential stereoselective processes become possible. When one of the reaction partners is non-racemic by virtue of a chiral centre resident in its structure (i.e. covalently attached) and the other is prochiral, the reactions are diastereoselective in nature. Such reactions have been extensively reviewed and will not be discussed further here.[2] When a non-racemic reactant is generated *in situ*, transiently, the reactions are enantioselective because the product only possesses one stereogenic centre (Figure 11.1a–d).

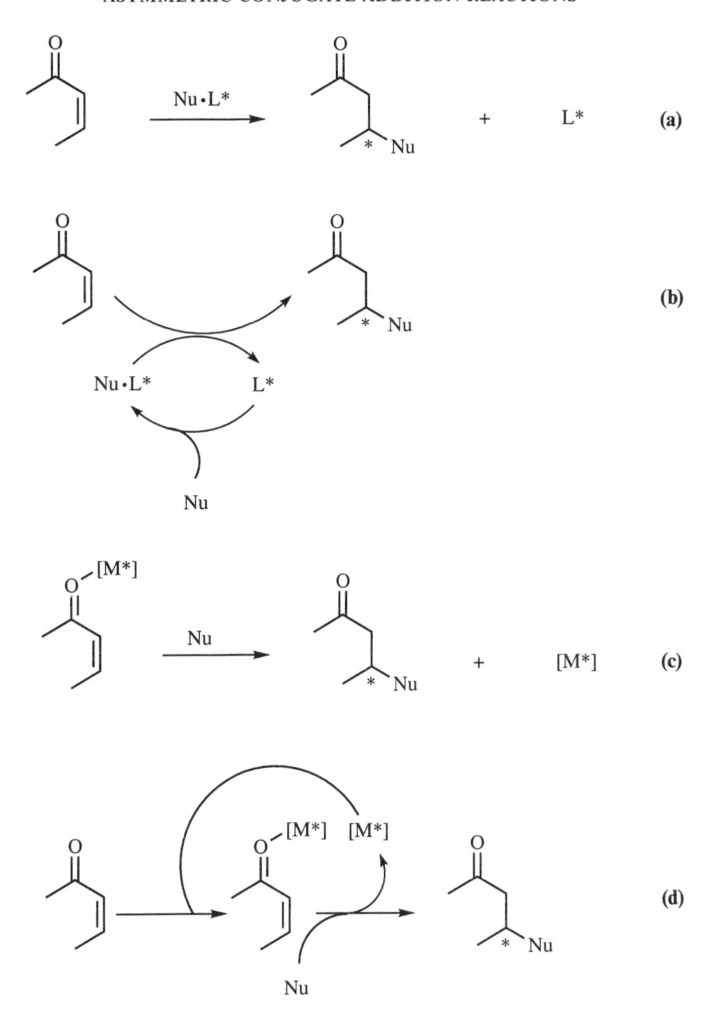

Figure 11.1 (a) Stoichiometric asymmetric modification. (b) Catalytic asymmetric modification of the nucleophile. (c) Stoichiometric asymmetric modification of the electrophile. (d) Catalytic asymmetric modification of the electrophile.

This chapter will describe progress being made in developing the processes summarised by reactions (a) and (b) in Figure 11.1. (To the best of the author's knowledge, no attempts have yet been made to develop the processes outlined in reactions (c) and (d)). Only organocuprate, organomagnesium (including Grignard) and organozinc reagents have been studied so far. The conjugate acceptors used in these studies have been almost exclusively cycloalkenones of various ring sizes.

11.3 Organocuprate reagents

Enantioselective conjugate additions of organocuprate reagents have been the most intensively studied so far.

11.3.1 Lithium heterocuprates

Non-racemic lithium heterocuprates are formed when an organocopper reagent (RCu) is mixed with the lithium salt of a non-racemic alcohol, amine or mercaptan (X*: Scheme 11.2). Naturally, it is essential that only the organic

$$X^*Li + RCu \rightarrow [R(X^*)CuLi]$$

Scheme 11.2

ligand ('R') be transferable. Up to the present time, the use of monoanionic, monodentate non-racemic non-transferable ligands has met with only modest success. The best of these have proved to be non-racemic amine derivatives.[5] Values of ee as high as 50% have been achieved in additions to 2-cyclohexenone carried out in dimethyl sulfide (Scheme 11.3).

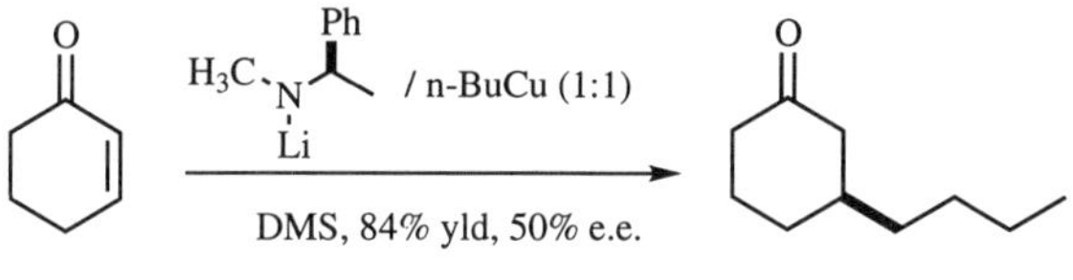

Ph
H_3C N Li / n-BuCu (1:1)
DMS, 84% yld, 50% e.e.

1

Scheme 11.3

Remarkably, the use of diethyl ether or mixtures of ether and dimethyl sulfide *reverses* the enantioselectivity. Much better results are obtained with *bidentate*, non-racemic monoanionic ligands. Both alkoxo- and amido-ligands have been studied. Some of the most successful ligands (i.e. those whose use has led to high ee values) are shown in Figure 11.2. Their respective ee values for conjugate additions to 2-cyclohexenone are also given along with the associated transfer ligand (R in Scheme 11.2).

Although the use of diamine (**5**) provided an excellent ee value of 97%, the chemical yield was disappointing (30%). Attempts to improve the yields in additions to cyclohexenone have so far not been successful. However, in additions to cycloheptenone and cyclooctenone, the chemical yields are considerably improved, although the very high ee values are maintained only with cycloheptenone (e.g. Scheme 11.4).

Little is known about the actual structures of these heterocuprates either in solution or the solid state. Rossiter's group has drawn an analogy with

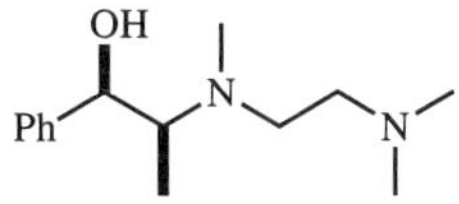

2 (Et, 90% yld, 92% ee, *R*
n-Bu, 90% yld, 89% ee, *R*)

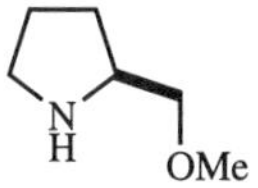

3 (Me, 68% yld, 83% ee, *R*)

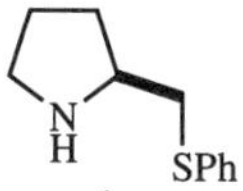

4 (Me, 71% yld, 80% ee, *R*)

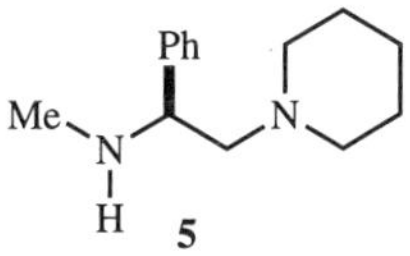

5 (n-Bu, 92% yld, 83% ee, *S*
Ph, 30% yld, 97% ee, *S*)

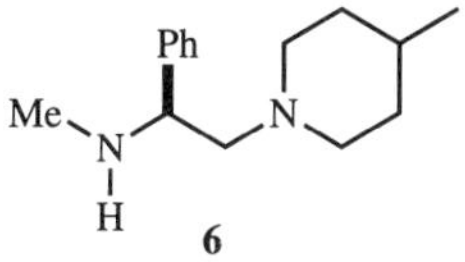

6 (n-Bu, 69% yld, 84% ee, *S*)

Figure 11.2 Bidentate monoanionic ligands used in conjugate additions of lithium heterocuprates to 2-cyclohexenone.

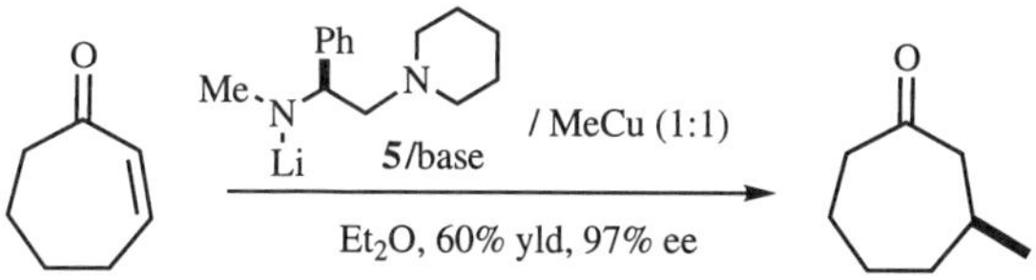

5/base / MeCu (1:1)

Et_2O, 60% yld, 97% ee

7

Scheme 11.4

the known rectangular structure of $[Me_2CuLi]_2$ to explain the selectivity observed with the use of (**5**) and a related series of tetraamines.[6] In this model, the preferred mode of complexation places one 'R' group close to the *Re*-face of cyclohexenone (which is situated 'below' the dimer to avoid steric interactions with the phenyl groups of the ligand, as shown in Figure 11.3). At least three other modes of complexation are possible; however, they all suffer from unfavourable steric interactions.

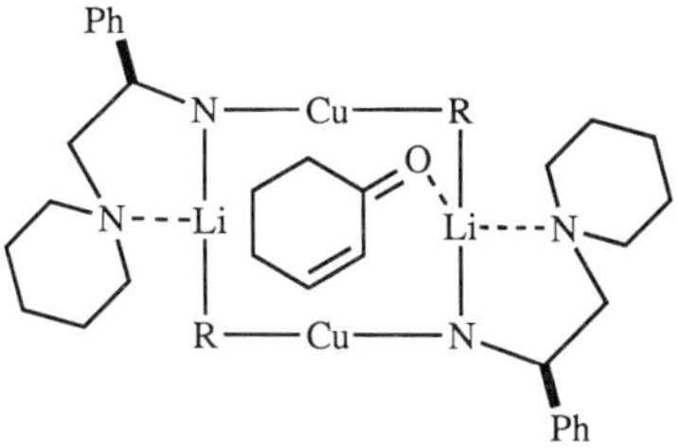

Figure 11.3 Possible mechanism for enantioface recognition in cyclohexenone using chiral auxiliary (**5**). R is the associated transferable ligand in Scheme 11.2.

Undoubtedly the most spectacular results so far have been obtained with the use of norbornyl-derived amino alcohols (**8**) and (**9**) in conjugate additions to (*E*)-2-cyclopentadecenone (Scheme 11.5).[7] The best results were obtained when running the reactions in toluene using several (up to ten) equivalents of THF. It remains to be established whether these ligands are as efficient in additions to other systems, including smaller cycloalkenones.

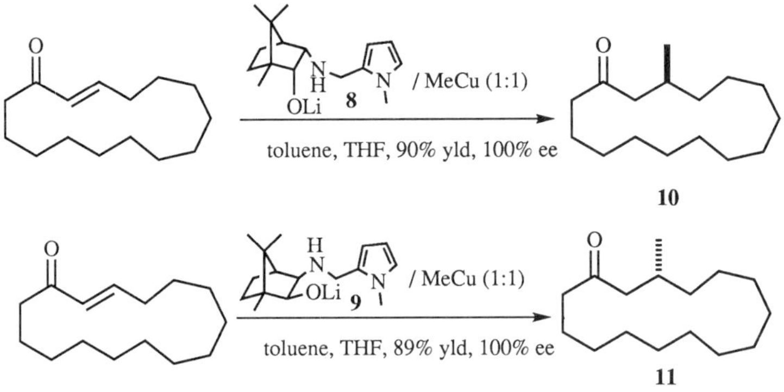
O
N
H
OLi
8
N
/ MeCu (1:1)
toluene, THF, 90% yld, 100% ee
O
10
H
N
OLi
9
N
/ MeCu (1:1)
toluene, THF, 89% yld, 100% ee
O
11

Scheme 11.5

11.3.2 Lithium homocuprates

By definition, homocuprates contain only organic ligands bound to copper. Very often these are formed in the presence of non-covalently bound, neutral ligands based on sulfur or phosphorus. Therefore, the opportunity exists to design non-racemic versions of such ligands and employ them in asymmetric conjugate addition reactions. In the early reports, the use of a hydroxyproline-derived ligand appeared the most promising.[8] Systematic variation of the nitrogen substituent culminated in ligand (**12**), which gave impressive results in conjugate additions to chalcone (Scheme 11.6).

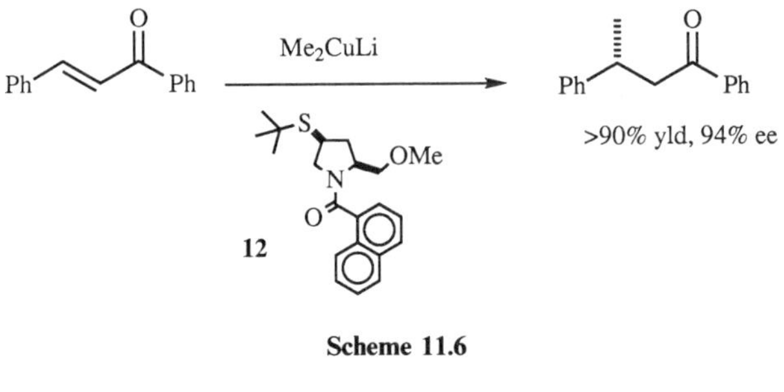
O
Ph
Ph
Me2CuLi
S
OMe
N
O
12
O
Ph
Ph
>90% yld, 94% ee

Scheme 11.6

More recently, a series of non-racemic phosphaoxazines, e.g. (**13**), has been prepared.[9] Each compound formed a soluble complex with CuX (X = Br or

I) in ether or THF. Their effectiveness as asymmetric ligands was then tested with organolithium-derived copper reagents of varying stoichiometry (Scheme 11.7). Some of the ee values obtained with 'medium-order' cuprates ($R_5Cu_3Li_2$)[10]

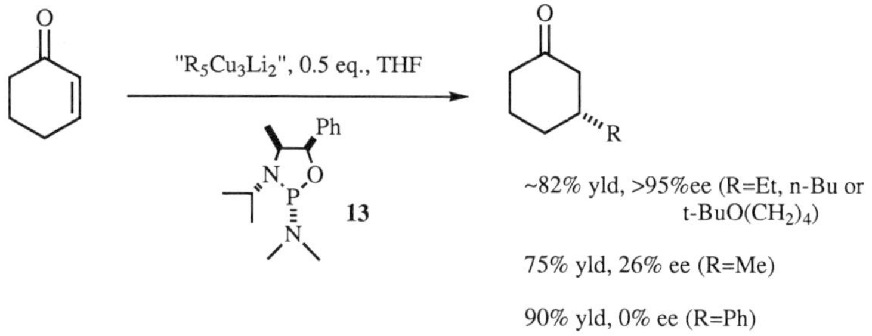

"$R_5Cu_3Li_2$", 0.5 eq., THF

13

~82% yld, >95%ee (R=Et, n-Bu or t-BuO(CH_2)$_4$)

75% yld, 26% ee (R=Me)

90% yld, 0% ee (R=Ph)

Scheme 11.7

are excellent (>95%). However, results with other transferable groups (R = Me, Ph) were disappointing, as were additions to cyclopentenone and cycloheptenone (ee values ~70–76%). The choice of 'medium-order' cuprates for these additions arose from the observation that chemical yields and ee values were relatively modest with additions of organocoppers (RCu). With lower-order cuprates (R_2CuLi) chemical yields were consistently high, but the asymmetric induction was very unreliable. The use of higher-order cuprates or cyanocuprates gave only racemic adducts. The authors speculate that the use of 'medium-order' cuprates provides sufficient excess CuI to prevent the n-BuOLi, found in commercial solutions of n-BuLi, forming higher-order heterocuprates.

11.4 Organomagnesium reagents

Only a small number of studies has been carried out using organomagnesium-derived reagents in asymmetric conjugate additions. Perhaps the most significant observation has been that good enantioselectivity can be achieved using a non-racemic *catalyst* (as opposed to ligand) in such additions. Lippard's group has designed and tested a variety of copper complexes, the more successful of which have been those based on non-racemic aminotropane imines such as (**14**).[11] Although the results reported are variable and additives such as chlorosilane and HMPA are necessary, this approach is a very promising one (Scheme 11.8). More recently Pfaltz's group has shown that Cu(I) thiolate complexes derived from mercaptoaryloxazolines also catalyse enantioselective conjugate additions of organomagnesium reagents.[12] Of these, (**15**) is the preferred ligand. The use of THF as solvent and HMPA as an additive gave optimal results. Conjugate additions to cycloheptenone gave

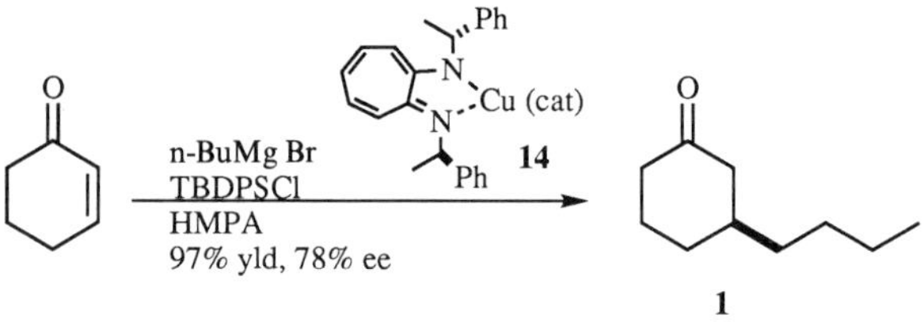
Ph
N
Cu (cat)
N
Ph
14
n-BuMg Br
TBDPSCl
HMPA
97% yld, 78% ee
O
O
1

Scheme 11.8

the highest ee values (Scheme 11.9). Additions to cyclopentenone and cyclohexenone gave poorer results.

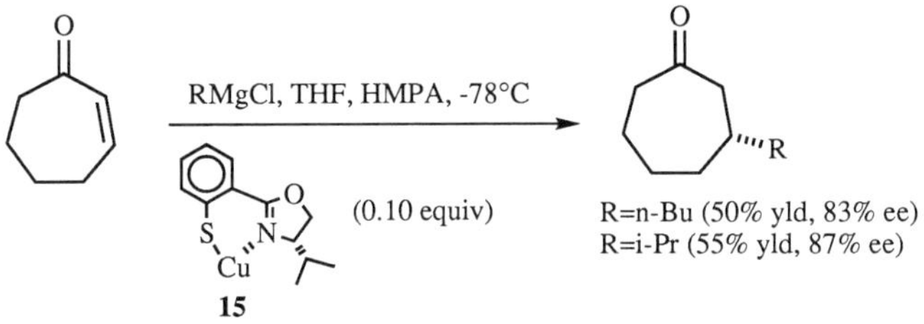
O
RMgCl, THF, HMPA, -78°C
O
R
O
S
N
Cu
(0.10 equiv)
15
R=n-Bu (50% yld, 83% ee)
R=i-Pr (55% yld, 87% ee)

Scheme 11.9

Very encouraging results have also been obtained with the related 2-(1-aminoethyl)phenylthiolatecopper(I) complex introduced by van Koten's group.[13] It was found that the method of addition had a profound effect on the outcome of the reaction. Rather than adding one of the reactants to a mixture of the other and the catalyst, simultaneous addition of the enone and organomagnesium reagent to a solution of the catalyst gave the best results (Scheme 11.10).

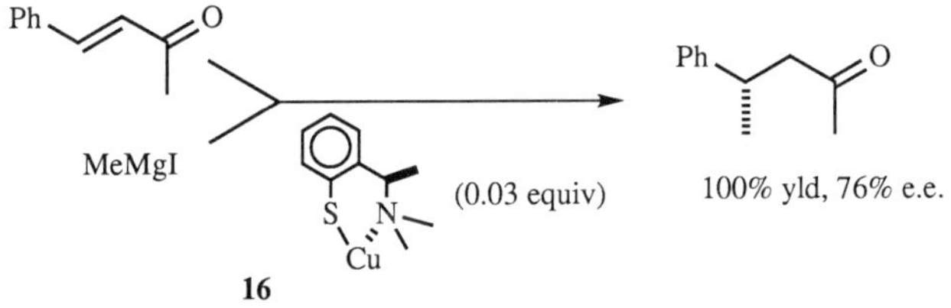
Ph
O
MeMgI
S
N
Cu
(0.03 equiv)
16
Ph
O
100% yld, 76% e.e.

Scheme 11.10

11.5 Organozinc reagents

In 1941, Gilman and Kirby demonstrated that diphenylzinc delivers a phenyl group to benzalacetophenone in conjugate fashion.[14] However, it was not until 1985, when Luche's group[15] described the ability of Ni(II) salts to promote such conjugate additions, that the possibility that organozinc reagents could participate in asymmetric conjugate additions was recognised. Since then, several reports have appeared detailing the use of a variety of Ni(II) complexes of non-racemic ligands. The complex is formed by heating (usually), a solution of $Ni(acac)_2$ with the non-racemic ligand (usually a β-amino alcohol) prior to use.

Most studies have focused on conjugate additions to acyclic alkenones. Chemical yields are usually excellent and the ee values are often ~90% (e.g. Scheme 11.11).

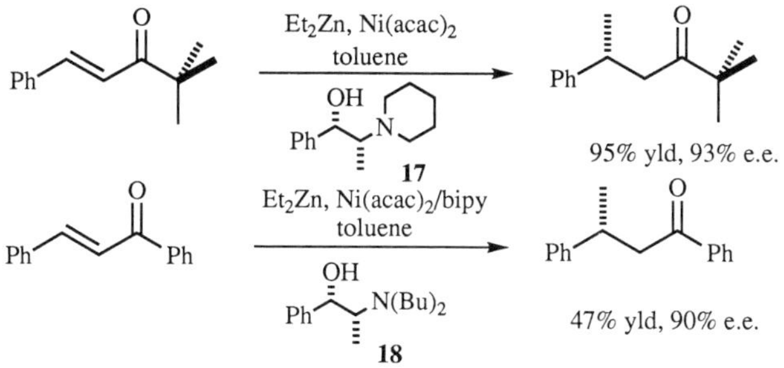

Et_2Zn, $Ni(acac)_2$, toluene, **17** — 95% yld, 93% e.e.

Et_2Zn, $Ni(acac)_2$/bipy, toluene, **18** — 47% yld, 90% e.e.

Scheme 11.11

Very recently a survey of ten ligands was carried out in reactions of this type and it was concluded that isoborneol- and borneol-derived ligands were the most promising, with ee values approaching those of (**17**) and (**18**).[16]

11.6 Conclusions

Clearly this field is rapidly developing and has reached an exciting stage. Both stoichiometric and catalytic asymmetric conjugate additions which proceed with good to excellent enantioselectivity have been demonstrated. However, the range of nucleophiles and conjugate acceptors is quite small. The next few years should see the development of more powerful predictive models as well as the extension of these reactions to other chemical classes.

References

1. Perlmutter, P., *Conjugate Addition Reactions in Organic Synthesis*, Pergamon Press, Oxford, 1992.
2. (a) Rossiter, B.E. and Swingle, N.M., *Chem. Rev.*, 1992, **92**, 771; (b) Lee, V.J., in *Comprehensive Organic Synthesis*, Trost, B.M. and Fleming, I. (eds), Pergamon Press, Oxford, 1991.
3. Heathcock, C.H., Finkelstein, B.L., Jarvi, E.T., Radel, P.A. and Hadley, C.R., *J. Org. Chem.*, 1988, **53**, 1922.
4. Chapdelaine, M.J. and Hulce, M., *Org. React.*, 1990, **38**, 225.

5. Rossiter, B.E. and Eguchi, M., *Tetrahedron Lett*, 1990, **31**, 965.
6. Swingle, N.M., Reddy, K.V. and Rossiter, B.E., *Tetrahedron*, 1994, **50**, 4455.
7. (a) Tanaka, K. and Suzuki, H., *J. Chem. Soc., Chem. Commun.*, 1991, 101; (b) Tanaka, K., Ushio, H. and Suzuki, H., *J. Chem. Soc., Chem. Commun.*, 1990, 795.
8. (a) Leyendecker, F. and Laucher, D., *Nouv. J. Chim.*, 1985, **9**, 13; (b) Leyendecker, F. and Laucher, D., *Tetrahedron Lett.*, 1983, **24**, 3517.
9. (a) Alexakis, A., Frutos, J. and Mangeney, P., *Tetrahedron: Asymmetry*, 1993, **4**, 2427; (b) Alexakis, A., Mutti, S. and Normant, J.F., *J. Am. Chem. Soc.*, 1991, **113**, 6332.
10. Ashby, E.C., Lin, J.J. and Watkins, J.J., *J. Org. Chem.*, 1977, **42**, 1099.
11. (a) Ahn, K.-H., Klassen, R.B. and Lippard, S.J., *Organometallics*, 1990, **9**, 3178; (b) Villacorta, G.M., Pulla Rao, Ch. and Lippard, S.J., *J. Am. Chem. Soc.*, 1988, **110**, 3175.
12. Zhou, Q.-L. and Pfaltz, A., *Tetrahedron*, 1994, **50**, 4467.
13. van Koten, G., *Pure Appl. Chem.*, 1994, **66**, 1455.
14. Gilman, H. and Kirby, R.H., *J. Am. Chem. Soc.*, 1941, **63**, 2046.
15. Petrier, C., Barbosa, J.C. de S., Dupuy, C. and Luche, J.-L., *J. Org. Chem.*, 1985, **50**, 5761.
16. de Vries, A.H.M., Jansen, J.F.G.A. and Feringa, B.L., *Tetrahedron*, 1994, **50**, 4479.

12 Stereoselective transformations mediated by chiral titanium complexes

A. HAFNER and R.O. DUTHALER

12.1 Introduction

After many years of academic endeavour, stereocontrol in organic synthesis has also become a major issue for the chemical industry.[1] One of the basic processes for inducing optical activity, the stereoselective addition of nucleophiles to prochiral carbonyl groups can, in the case of organometallics, be controlled efficiently with chiral ligands. The necessary coordinative robustness of such chiral complexes can be obtained either with metalloids, such as boron or tin, or with early transition metals (mainly titanium and zicronium).[2] The major advantages of titanium and also of zicronium chemistry are the high abundance of these elements, the possibility of adjusting reactivity and selectivity by the correct choice of ligands and the relative inertness towards redox processes. With the exception of a structurally narrow group of cytotoxic titanocenes and bis-β-diketonato-complexes, titanium and zirconium compounds show no intrinsic biological activity; so far, toxic effects can only be related to the ligand.[2a]

The stability, tendency for aggregation and the reactivity of Ti complexes is related to the steric and electronic nature of the ligands. In the case of η^1-bound groups, the Lewis acidity is electronically governed by the bonding atom and decreases in the sequence: halogen > carbon > nitrogen $\approx$ oxygen. A much greater impact on reactivity and selectivity is, however, exerted by η^5-bound cyclopentadienyls (Cp), as the formal electronic configuration of titanium is augmented from 8 to 12 upon introduction of one Cp ligand, and to 16 for titanocenes. While chiral titanium complexes with 8 formal valence electrons have been successfully applied as chiral Lewis acids[3] for the addition of alkyl groups and other d^1-nucleophiles to carbonyls[2b,c,4] and for asymmetric epoxidation,[3c] the main application of chiral biscyclopentadienyl compounds is stereoregular polymerization.[3d] The chiral monocyclopentadienyl complexes, however, turned out to be ideal templates for the stereoselective addition of allyl groups and ester enolates to aldehydes.[2a,5,6] This chapter is restricted to chiral titanium complexes which transfer titanium-bound carbon nucleophiles to carbonyl groups (Scheme 12.1).

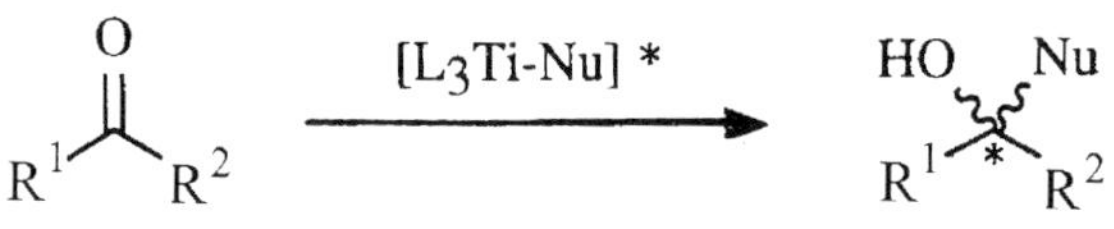

$$R^1C(=O)R^2 \xrightarrow{[L_3Ti\text{-}Nu]^*} R^1C^*(OH)(Nu)R^2$$

Scheme 12.1

12.2 Synthesis and structures of chiral complexes

Because of the stability of the Ti–O bond, chiral alcohols are well suited as Ti ligands. The general routes by which the chiral Ti(IV) complexes are prepared are summarized in Figure 12.1 (see also Refs. 2 and 4). Protic ligands LH such as alcohols can displace a titanium-bound chloride after deprotonation, silylation (Figure 12.1b) or stannylation. With more acidic ligands HCl gas is evolved spontaneously. The equilibrium, which is usually on the side of the titanium chloride, can be shifted by evaporation of HCl or by neutralization with a weak base (Figure 12.1c). A very efficient method for the preparation of titanium alkoxides is the 'transesterification' with a free alcohol. The

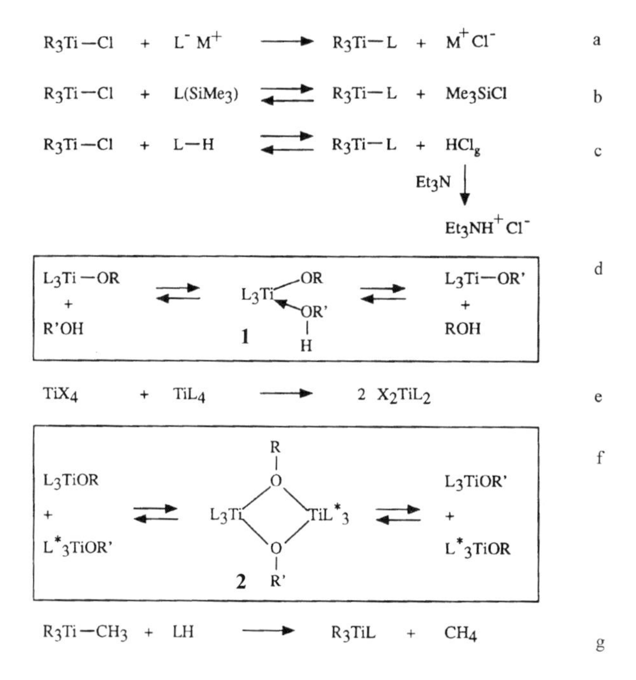

Figure 12.1 Synthesis and structures of chiral titanium complexes. (a) Basic ligands displace Cl^-. (b) Displacement of Cl^- by silylation. (c) Displacement of equilibrium by removing HCl. (d) Displacement of equilibrium by removing a volatile by-product. (e) Ligand redistribution. (f) Interchange of alkoxide ligands via bridged dimers. (g) Dealkylation in protic solvents under mild conditions.

equilibrium is thereby controlled by distilling off the volatile ligand (Figure 12.1d). A *caveat* has, however, to be entered, as these transesterifications never go to completion,[5] and variable amounts of alcohol are retained via adduct (**1**). This has to be kept in mind if such mixtures are used for further transformations without purification. Interestingly, molecular sieves can affect such equilibria quite substantially.[5a,b] In certain cases (X = halide; L = OR or cyclopentadienyl), ligand redistribution reactions result in comproportionation (Figure 12.1e).[6] Alkoxide ligands can be interchanged without the need of free ROH via bridged dimers of type (**2**) (Figure 12.1f). Such polynuclear aggregates are often the favoured form of such complexes, even in solution.[7] Furthermore, the equilibria of the reactions (b) to (f) in Figure 12.1 are also very much dependent on electronic steric factors. Finally, a very mild method to introduce protic ligands is displacement of an alkyl ligand from titanium (Figure 12.1g).

Many compounds, and among them some of the most efficient reagents, have not been characterized beyond their stoichiometric composition or the symmetry exhibited by their NMR spectra. For clarity, structures are given in brackets if no X-ray data are published for the specific compound, a precursor or otherwise closely related complex.

12.2.1 Titanium complexes with chiral monodentate ligands

Optically active alcohols are the favoured monodentate ligands for the preparation of chiral titanium reagents. Most convenient is the use of natural products or their derivatives. Alkoxy–Ti complexes having 8 valence electrons have been used for further transformations without purification and characterization. Equilibria involving different polynuclear species and fast ligand exchange can be assumed for their solutions. This tendency can be suppressed by the electronic and steric influence of cyclopentadienyl ligands. Consequently, complex (**3**)[8], obtained from $CpTiCl_3$ and 1:2,5:6-di-*O*-isopropylidene-α-D-glucofuranose according to Figure 12.1c, is monomeric and its structure was confirmed by ^1H and ^{13}C NMR spectroscopy, as well as by X-ray diffraction.[8b] Structures with comparable properties could be obtained using different acetal protection for the D-glucose-derived ligands (e.g. **4** and **5**).[9] The complexes (**3**–**5**) show two sets of ^1H and ^{13}C NMR signals for the diastereotopic alkoxy groups; an indication that ligand exchange, if occurring at all, is slow. Unfortunately, this is not a general feature of dialkoxycyclopentadienyltitanium chlorides, and even seemingly minor changes of the sugar skeleton lead to extremely labile compounds. Complex (**6**)[9] obtained from 1:2,5:6-di-*O*-isopropylidene-α-L-idofuranose (the C-5 epimer of glucose) could not be analysed by NMR spectroscopy, as only the signals of the free ligand were observed. Similar observations were made with ligands derived from D-xylose or D-allose.[1a,10d]

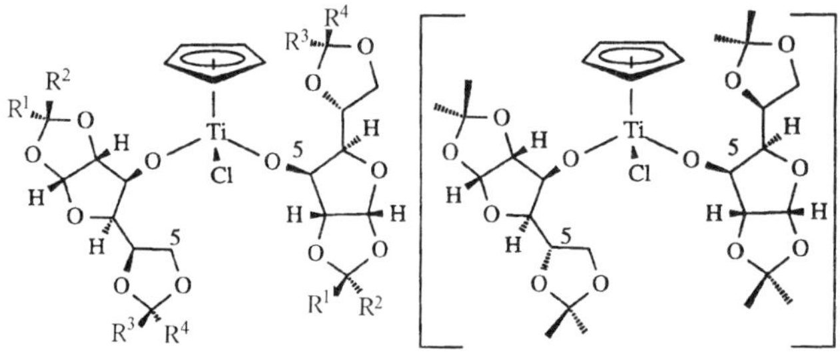

3 R^1–R^4=Me
4 R^1/R^2, R^3/R^4=H/CCl_3
5 R^1, R^2=Me; R^3, R^4=H/alkyl

6

While the chiral ligands of the complexes (**3**–**6**) remain bound to titanium, chirality of the reacting nucleophilic ligand is another effective means to control stereoselectivity in carbonyl additions. If titanium is attached to a chiral carbon atom, stereoselective transformations are needed for the preparation of such reagents. This route has been pursued successfully by Hoppe *et al.* (Scheme 12.2).[11] The chiral carbamate (**7**) can be lithiated with retention of configuration, and, depending on the method of transmetallation, titanium can be introduced either with retention (**8**) or inversion (**9**) of the configuration.[11a] Alternatively, the organolithium reagent (**10**) can be obtained from racemic (**7**) by enantiomer-differentiating deprotonation with s-BuLi·sparteine.[11c] If the achiral carbamate (**11**) is lithiated with s-BuLi·sparteine, the chiral organolithium compound (**12**) is obtained in high optical purity by crystallization of the equilibrating racemate. Titanation without loss of optical activity is possible as illustrated by the formation of (**13**).[11b] The structure of an organolithium compound related to (**12**) has been elucidated by X-ray analysis.[12]

12.2.2 Titanium complexes with chiral bidentate ligands

The idea of using bidentate chiral ligands originates from the expectation that the chelate rings stabilize such complexes and that the rigidity gained should improve the stereoselectivity of the corresponding reagents. Since the methods for the preparation of such structures rely on thermodynamically controlled equilibria (cf. Figure 12.1), the stabilities of the target complexes should be carefully evaluated. In addition to entropic factors, the crucial parameters for ring formation are the Ti–X–C (X = O, N) and X–Ti–X bond angles, as well as the Ti–X bond lengths. These values depend not only on the atoms X but also, to a very large extent, on the number of η^5-bound Cp ligands (for a detailed discussion see Ref. 2a).

Following the general trend, the Lewis acidity of alkoxy and aryloxy titanates lacking cyclopentadienyl ligands is enhanced to such an extent that,

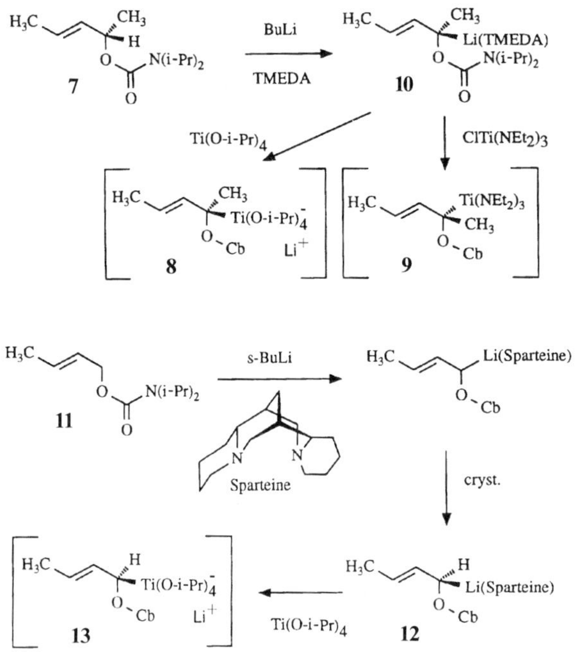

Scheme 12.2

in the absence of other ligands, the coordinative unsaturation is overcome by the formation of aggregates (Scheme 12.3).[7b,d] The strain of cyclic monomers (**14**) is thereby reduced, and even 5-membered rings can be incorporated in dimers of type (**15**) or similar higher oligomers. Further complications can arise because of an equilibrium between the tricyclic species (**15**) and the macrocycle (**16**), which is influenced by subtle structural differences.

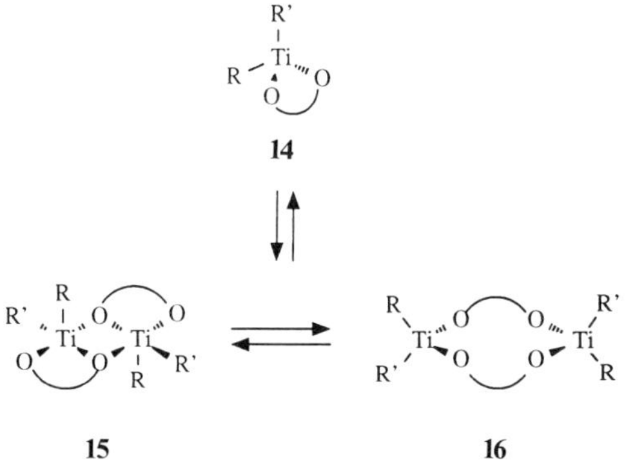

Scheme 12.3

The complexes (**17**–**20**)[13–17] obtained from precursors lacking Cp ligands, are again stabilized by aggregation. Nevertheless, these rather ill-defined structures have been successfully applied as chiral Lewis acids for ene reactions,[3a] cycloadditions[3b] and allylstannylations.[18] Aggregate formation is also held responsible for the non-linear correlation between optical purity and enantioselectivity of reagent (**17**)[15b] (see also Chapter 2).

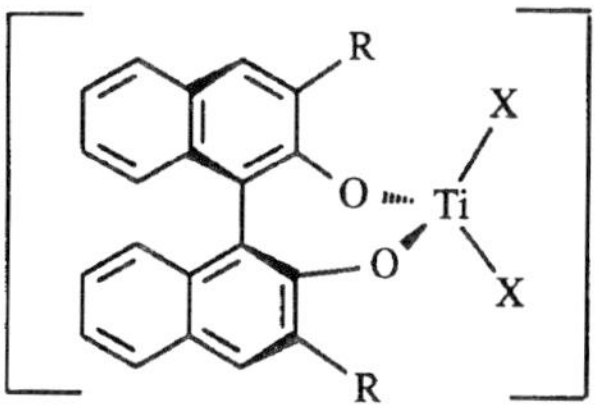

17 R=H X=Cl
18 R=H X=O-i-Pr
19 R=H X=Cl, O-i-Pr
20 R=Ph X=Cl

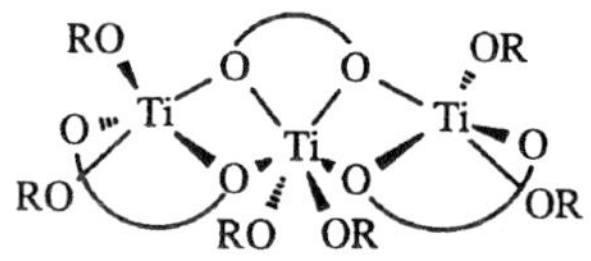

18 (Crystal structure;[14] schematic representation)

Spontaneous chelation occurs with the diols (**21**) ($R^2 \neq H$) (Scheme 12.4)[10] and titanium alkoxides (Scheme 12.5).[5] The Ingold–Thorpe effect is obviously responsible for an ideal 'chelation conformation'. Using the tetraphenyl-substituted 1,3-dioxolane-4,5-dimethanol ligand (**26**) together with $Ti(OPr^i)_4$ or $TiCl_2(O\text{-i-Pr})_2$, equilibrium reactions are occurring.[7] The equilibrium can be shifted towards the complex by addition of molecular sieves (Scheme 12.5). Further complexes are obtained either from $TiCl(O\text{-i-Pr})_3$ (**29**–**31**)[19] or from $Ti(O\text{-i-Pr})_4$ (**32**–**34**).[20,21] The structure of the spirocyclic titanate (**34**) was determined by X-ray diffraction.[21] If (**34**) is mixed with 1 equiv. $Ti(O\text{-i-Pr})_4$, a comproportionation according to Figure 12.1e affords two monocyclic complexes (**32**) (Scheme 12.6).[21] At present, it is not clear whether the

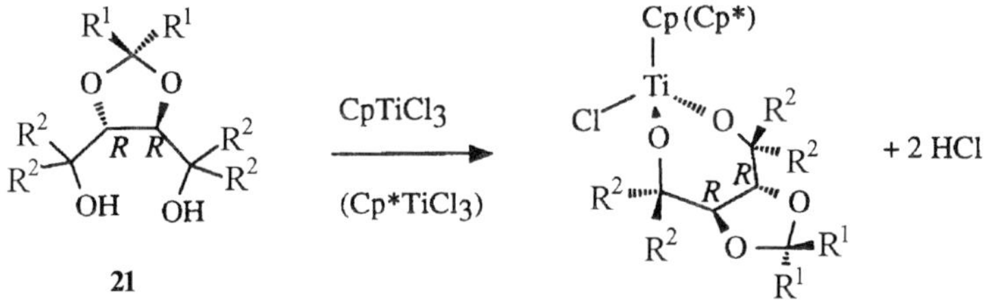

21

$CpTiCl_3$ (Cp^*TiCl_3)

+ 2 HCl

22 $R^1=CH_3, R^2=Ph$ Cp
23 $R^1, R^2=CH_3$ Cp
24 $R^1=H, R^2=Ph$ Cp
25 $R^1, R^2=CH_3$ Cp*

Scheme 12.4

26 + $(i\text{-}PrO)_2TiX_2$ ⇌ (Toluene; − 2 i-PrOH)

13	:	87	**27** X=Cl
6	:	94	(mol sieves)
25	:	75	**28** X=i-PrO

Scheme 12.5

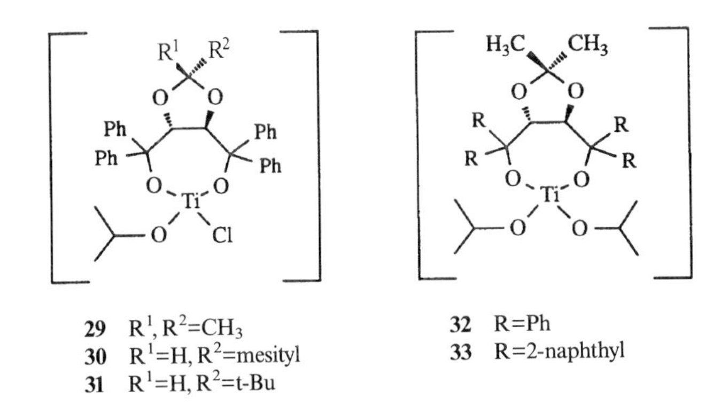

complexes (**27**–**33**) are monomeric structures or aggregates. Single resonances in the NMR spectra point to monomers or highly symmetrical dimers.

34 —$Ti(O\text{-}i\text{-}Pr)_4$→ 2 **32**

Scheme 12.6

If cyclic cyclopentadienyldialkoxytitanium chlorides should be obtained from $CpTiCl_3$ and chiral diols under thermodynamic control, then the resulting titanacycles should be free of strain. According to a force field calculation based on Ti–O bond length and O–Ti–O and Ti–O–C(α) bond angles, obtained from the crystal structure of (**22**),[8b] a 7-membered ring is optimal.[10a] The 1,4-relation of hydroxy functions should, therefore, be chosen for such ligands, rather than 1,2- or 1,3-diols, which generally lead to complex polymeric or highly clustered titanium complexes.[2a] Several C_2-symmetric chiral diols were treated with $CpTiCl_3$ and 2 equiv. Et_3N in Et_2O at room temperature or in toluene at 100°C (Scheme 12.4). After filtration from the precipitated ammonium chloride, NMR analysis of the crude products was again a useful method for assessing the conversions. In addition to distinct complexation shifts, doubling of most signals was indicative of the formation of cyclopentadienylchlorotitanium derivatives, lacking C_2 symmetry. It turned out that the geometrical constraints for the formation of monomeric and stable complexes were more severe than anticipated. Many promising ligands, such as binaphthol (**35**), D-mannitol bisacetonide (**36**), anhydromannitol (**37**)[22] and the 1,3-dioxolane-4,5-dimethanol (**38**),[23] gave either mixtures of polymeric complexes or extremely sensitive derivatives displaying only the ligand signals in the NMR spectra (Figure 12.2).[24] It has, however, to be noted, that binaphthol (**35**) is a successful ligand for titanocenes.[24] Remarkably stable monomeric cyclopentadienyl–chlorotitanium complexes, amenable to NMR and X-ray analysis, were obtained from the tetramethyl- and tetraphenyl-substituted 1,3-dioxolane-4,5-dimethanols (**39**) and (**26**),[10] as well as from (**40**) and (**41**).

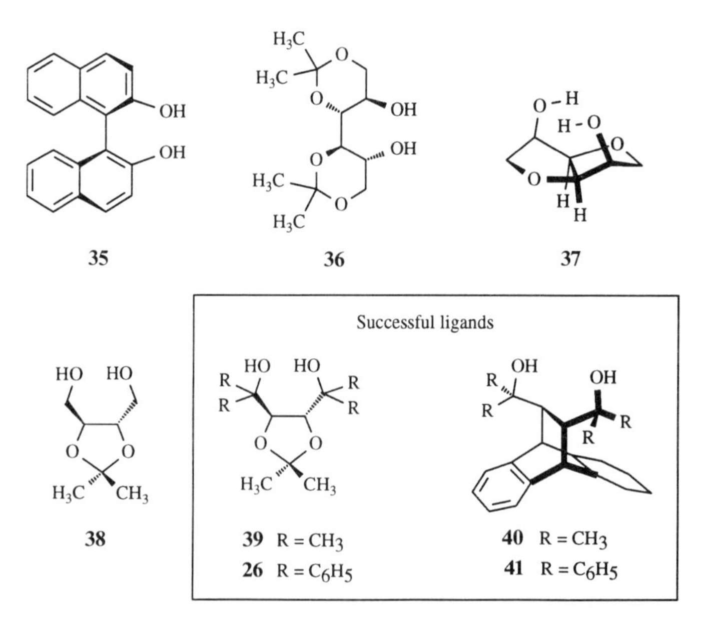

Figure 12.2 Structures of successful and unsuccessful 1,4-diol ligands in reactions with $CpTiCl_3$.

12.3 Addition of d^3-nucleophiles to carbonyl compounds

Discussion of the chemical transformations using chiral titanium complexes are divided into two parts. The first is dedicated to the addition of d^1-nucleophiles (e.g. methyltitanium reagents), the second to reactions of d^3-nucleophiles (e.g. allyltitanium reagents or titanium enolates) with carbonyl compounds.[25a,b] Such a classification accounts for the different mechanisms by which d^1- and d^3-nucleophiles are added to carbonyl groups (Scheme 12.7).[26b] For d^1-nucleophiles the situation is rather complex, as the four-centre transition state (**42**), implied by a bimolecular mechanism, is considered unfavourable in many cases. Instead, a lesser strained 6-centre transition state (**43**), including a third molecule M′X as mediator, has been invoked to explain such processes (Scheme 12.7).[26] In ideal cases such a 'mediator', which may be a second molecule of the d^1-reagent, can be used catalytically. For d^3-nucleophiles, however, a bimolecular 6-centre cyclic transition state (**44**) is well established as a mechanistic notion of the 1,2-addition to carbonyls (Scheme 12.8).[27] As the titanium ligands, which are transferred as nucleophiles, are covalently bound to the metal centre, open transition states, proposed for highly ionic reagents, can be neglected.

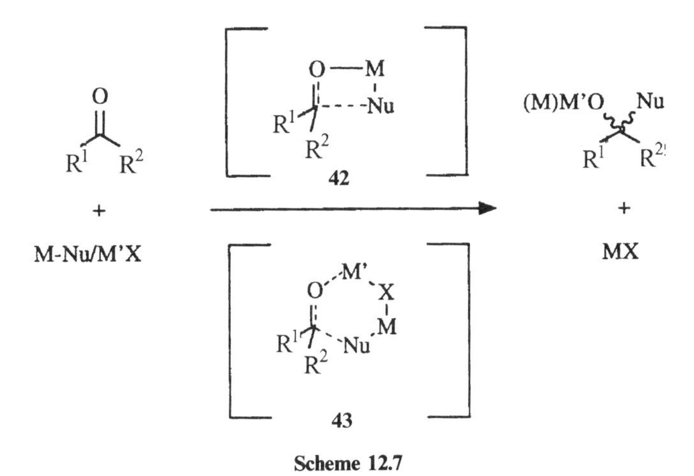

Scheme 12.7

12.3.1 Enantioselective aldol reaction with titanium enolates

The aldol reaction has been developed into one of the most useful synthetic methods, primarily because the stereoselective formation of two new asymmetric centres can usually be controlled with high predictability and selectivity[28c,29] (see also Chapter 3).

A pivotal role in controlling the stereoselectivity of aldol additions is played by the cation. It influences the geometry of the transition state sterically, via metal ligands, and electronically, via its Lewis acidity and number of vacant

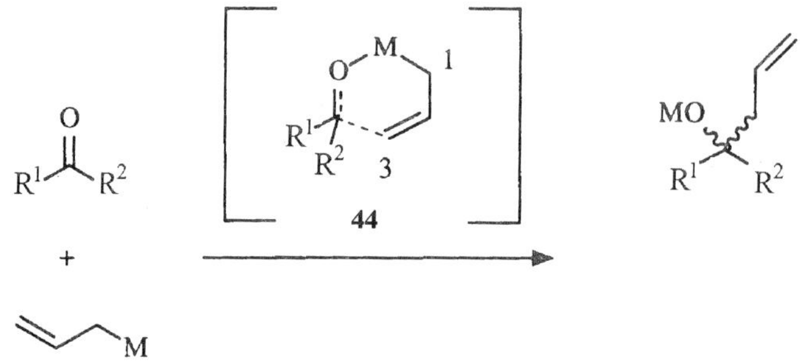

Scheme 12.8

coordination sites. Titanium enolates have become very popular in this respect, not only for improving selectivity[30] but also for inverting the face selectivity obtained by the use of other enolates.

The enantiofacial differentiation by enolates can be conveniently directed by internal chiral auxiliaries, which are covalently bound as ester or amide derivatives. The synthetic potential of this 'first-generation' approach has been demonstrated by Evans *et al.* using titanium enolates.[31] Isomerically pure compounds can be obtained even from reactions of moderate stereodifferentiation, as the diastereomeric products can often be separated. The sometimes cumbersome extra steps needed to introduce and remove these auxiliaries are, however, an inherent handicap of this principle, and the 'second-generation' approach, which makes use of chiral enolate counterions, is preferable, if the enantioselectivity is high.

Titanium complexes with chiral ligands have emerged as excellent templates for the chiral modification of enolates. As opposed to the chiral auxiliary approach, the chiral ligands are recovered easily upon hydrolytic work-up and no additional chemical transformations of the products are required. Only mediocre enantiofacial selectivity was observed when cyclopentadienyl-free titanium enolates bearing chiral ligands were used.

Much better results were obtained when the cyclopentadienyltitanium chloride (**3**), with diacetone–glucose ligands[8a,c,d] was used for aldol reactions. Lithium enolates can be transmetallated at low temperature (−30 to −78°C) with cyclopentadienyldialkoxytitanium chlorides. Thus, ester enolates (**45**) afford the chiral titanium endates (**46**), when treated with the diacetone–glucose complex (**3**) (Scheme 12.9). As one of the main driving forces of the allyltitanation (the transfer of titanium from carbon to oxygen) is missing, titanium enolates are less reactive than allyltitanium compounds. Therefore, ketone enolates and enamides prepared from chloride (**3**) are unreactive, and only the more nucleophilic ester enolates (**46**) react with aldehydes at −78°C. The acetate and propionate aldols, (**47**)[8c,d] and (**48**),[18f] are obtained with excellent enantioselectivity, the *Re*-side of the aldehyde carbonyl being attacked preferentially. As opposed to the allyltitanation (cf. Scheme 12.11, below), a boat transition state (**49**) has to be assumed to explain the high *syn*-selectivity of pure (*E*)-enolates (**46**). More astonishing is the fact that (*Z*)-enolates, such as the propionate enolate (**50**), obtained by equilibrium of the corresponding (*E*)-enolate (**46**),[8f] also react via a boat transition state (**51**), affording *anti*-aldols (**52**) of high optical purity. In the case of benzaldehyde and other substrates with a branched sp^2 α-carbon, a chair transition state (**53**) giving the *syn*-aldol (**54**) with low enantioselectivity (47% ee), accounts for 77% of the product mixture together with only 23% of *anti*-aldol (**52**) of 94% ee (Scheme 12.9). The enolate (**50**) was the first example of an *anti*-selective (*Z*)-enolate.[8f]

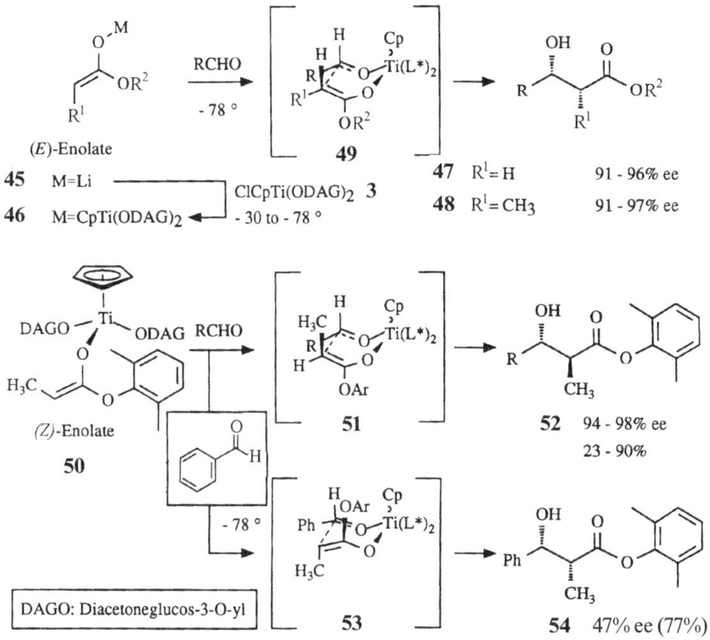

Scheme 12.9

However, as L-glucose is not readily available and the titanium enolates derived from the chloride (**22**) react less selectively (ee ⩽ 78%) than those obtained from (**3**), an efficient method for the corresponding enantiomeric aldol products based on similar methodology is still lacking.[9]

Aldol reactions of glycine enolates give β-hydroxy-α-amino acids. These are important synthetic targets, as β-hydroxylated derivatives of many proteinogenic and non-proteinogenic amino acids are found in biologically interesting peptides of microbial origin. Transmetallation of lithium enolate (**56**) derived from the 'stabase'-protected glycine ethylester with titanium complex (**3**), gives the chiral titanium enolate (**57**), which is again a highly stereoselective reagent (Scheme 12.10). Addition to various aldehydes and subsequent mild acidic hydrolysis gives (*R*)-configurated *threo*-β-hydroxy-α-amino acid ethyl esters, e.g. (D-**58**) of very high diastereomeric and enantiomeric purity.[8d] In contrast to other methods, where demasking is often a major problem, these esters are versatile intermediates.

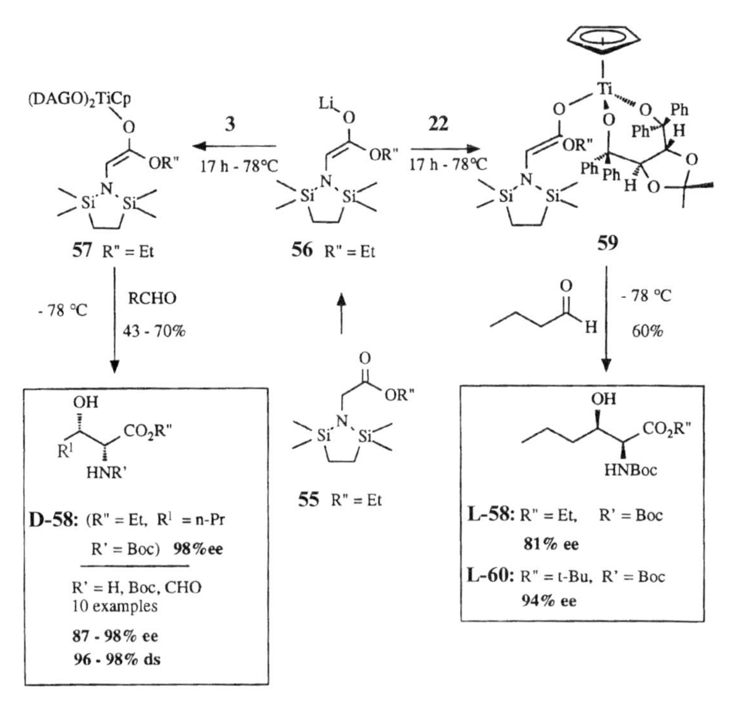

Scheme 12.10

(*S*)-Configurated *threo*-β-hydroxy-α-amino acids can be obtained by transmetallation of the lithium enolate (**56**) with the (*R*,*R*)-configurated chiral titanium complex (**22**) (Scheme 12.10). The optical purity of the product (L-**58**) is again considerably lower (81% ee) than that obtained with the diacetone–glucose reagent (**57**) (98% ee) for the same substrate. In this case, however, excellent stereoselectivity can be obtained by using the lithium enolate of the analogous glycine t-butyl ester derivative (L-**60**, 94% ee).[10b]

In conclusion, reagents prepared from (**3**) belong at present to the most useful chiral enolates known. Diacetone–glucose is one of the most readily available chiral auxiliaries, and enantioselectivities surpassing the results of most other methods[32,33a,b,34] are easily achieved.

The first successful steps toward a 'third generation' of stereoselective aldolizations have been achieved with the gold-catalysed addition of isocyanoacetate to aldehydes[35] with the use of chiral Lewis acids for the Mukaiyama reaction with silyl-enolates[32,33].

12.3.2 Enantioselective allyltitanation of aldehydes

The regio- and stereoselective allylmetallation of aldehydes to give acyclic structures with one or two new stereogenic centres is synthetically very useful.[28] Many possibilities for further transformations are provided by the double bond, and, depending on the substitution, these reagents can be considered as aldol-[28a,c,d] or homoaldol-synthons.[28e] A 6-centre cyclic transition state (**69**) (Scheme 12.11) with chair conformation explains the diastereoselectivity and also accounts for the higher reactivity of allyltitanium compounds, when compared with the alkyl counterparts. Therefore, Cp- and even Cp_2-substituted complexes can be reacted even with ketones.[36] With achiral allyltitanium reagents, impressive regio- and diastereocontrol was achieved[2,4,37] but, until recently, the enantiofacial selectivity was difficult to control with chiral titanium complexes.

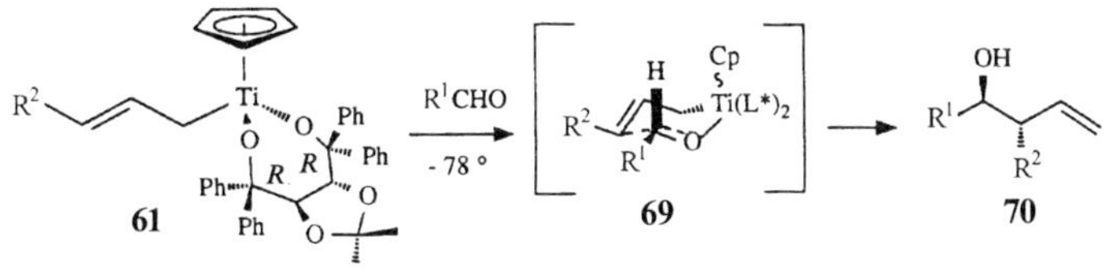

61 $\xrightarrow[-78\,°]{R^1CHO}$ [**69**] → **70**

62 R^2=H; **63** R^2=CH_3;
64 R^2=C_6H_5; **65** R^2=OC_6H_5;
66 R^2=Me_3Si; **67** R^2=OEt;
68 R^2=$OC_6H_4OCH_3$.

95–98% isomeric purity
95–98% isomeric purity

Scheme 12.11

Excellent enantiofacial control of nucleophilic additions to carbonyl groups is possible if the chirality is incorporated in the nucleophilic ligand. In the case of enolates of carboxylic acid derivatives, chiral auxiliaries can be attached via ester or amide linkages. This approach is generally not possible for allylic nucleophiles. However, the α-carbon of substituted allyl ligands is a stereogenic centre, and if its configuration can be controlled, highly stereoselective reagents result.[38] Hoppe *et al.* have prepared such reagents without covalently bound auxiliaries.[11] The addition of the enantiomeric reagents (**8**) and (**9**) (cf. Scheme 12.2)[11a] to isobutyraldehyde gives (**71**) and its enantiomer (ent-**71**) with excellent diastereoselectivity and moderate to good enantioselectivity (Scheme 12.12).[1b] The stereoselectivity of the chiral titanium complex (**13**) (cf. Scheme 12.2) lacking the methyl substituent at C-1 is much better, and the lactaldehyde-adduct (**72**) is obtained with high enantiomeric purity (>95% ee).[11b]

Monocyclopentadienyltitanium complexes with chiral ligands have emerged as extremely efficient templates for the enantioselctive allyltitanation of aldehydes.[8a,9,10] The Lewis acidity and, therefore, also the reactivity and the tendency to aggregation are considerably reduced for cyclopentadienyl-

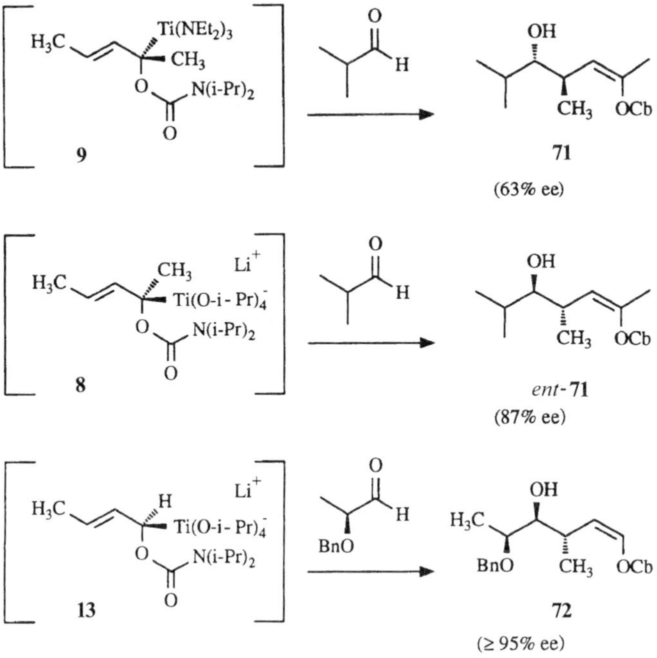

Scheme 12.12

substituted compounds. This appears to be an important prerequisite for a clean bimolecular reaction path. Transmetallation of allylmagnesium chloride, terminally monosubstituted allyl Grignard compounds or α-lithiated alkenes with the chlorides (**22**–**24**) (Scheme 12.4)[10] affords the allyltitanium compounds (**61**) (Scheme 12.11). Very high *Si*-face selectivity ($\geqslant 95\%$ ee) is induced by the chelating ligands derived from (*R*,*R*)-tartrate[10] (Table 12.1). *Re*-face preference, however, is exhibited by the analogous reagents derived from the diacetone–glucose complexes (**3**–**5**) (Scheme 12.13),[8a,9] but in these cases the enantioselectivities are somewhat lower.

A fast 1,3-titanium shift and bond rotation, evidenced by the degenerate ^1H and ^{13}C NMR signals in case of the unsubstituted allyl complex (**61**) ($R_2 = H$, Scheme 12.11), is responsible for rapid equilibration to the thermodynamically-favoured *trans*-isomers with titanium bound to the unsubstituted terminal allylic position, independent of the isomeric status of the organometallic precursor.

The almost exclusive formation of *anti*-adducts (**70**) (Table 12.1) is best rationalized by a cyclic 6-membered transition state (**69**) with chair conformation. The mechanistic picture is further corroborated by reacton of the titanated 2,4-dihydro-1,3-dioxin (**75**), which affords exclusively the *syn*-isomer (**76**) upon addition to benzaldehyde (Scheme 12.14). The relative configuration of (**76**)

Table 12.1 Diastereoselective and enantioselective allyltitanation of aldehydes (**73**) with the reagents (**62**–**68**) (Scheme 12.11)[1a,10d]

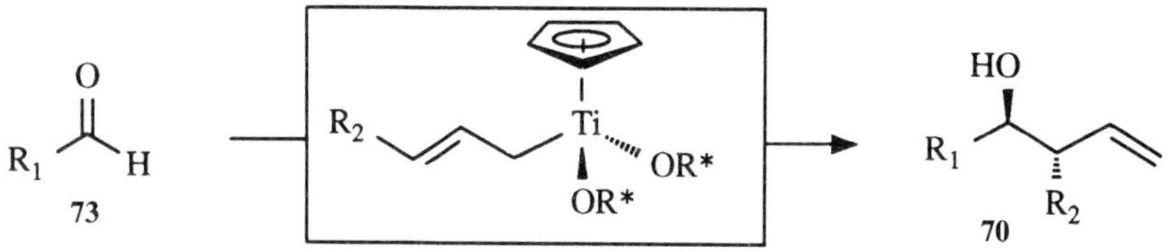

Aldehyde R^1	Reagent	R^2	ee (%)	de (%)	Yield (%)
C_6H_5	**62**	H	95	–	93
$CH_3(CH_2)_8$	**62**	H	94	–	92
$(CH_3)_2CH$	**62**	H	97	–	90
$(CH_3)_3C$	**62**	H	97	–	88
CH_2=CH	**62**	H	95	–	86
C_6H_5	**63**	CH_3	98	97	89
C_6H_5	**64**	C_6H_5	97	≥98[a]	54
C_6H_5	**66**	Me_3Si	≥98[a]	≥98[a]	68
C_6H_5	**67**	EtO	95	75	77
C_6H_5	**68**	PMPO[b]	≥98[a]	≥98[a]	93
$CH_3(CH_2)_8$	**63**	CH_3	≥98[a]	≥98[a]	86
$CH_3(CH_2)_8$	**66**	Me_3Si	≥98[a]	≥98[a]	68
$CH_3(CH_2)_8$	**67**	EtO	92	94	73
$CH_3(CH_2)_8$	**68**	PMPO[b]	≥95[c]	≥95[c]	69

[a] Only one isomer was detected by capillary GC (Chirasil-Val[R]).
[b] PMP: *p*-Methoxy-phenyl.
[c] Determined by ^1H NMR spectroscopy with 2,2-trifluoro-1-(9′-anthracenyl)-ethanol (TFAE).

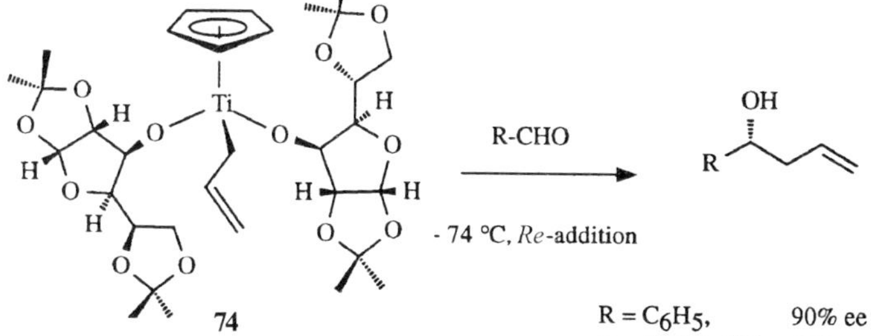

R-CHO

- 74 °C, *Re*-addition

74

R = C_6H_5, 90% ee
R = $(CH_2)_8CH_3$, 92% ee
R = $CH(CH_3)_2$, 90% ee
R = $C(CH_3)_3$, 88% ee
R = $CH=CH_2$, 86% ee

Scheme 12.13

can again be rationalized by a chair transition state (**77**). The absolute configuration of (**76**) is directly related to the C-4 configuration of reagent (**75**). The stereochemistry of (**75**) and the enantiofacial preference of the addition are both steered by the chiral ligand, and the slightly inferior enantioselectivity (86% ee), when compared with the results with monosubstituted allylreagents, might reflect an antagonism of the two influences.

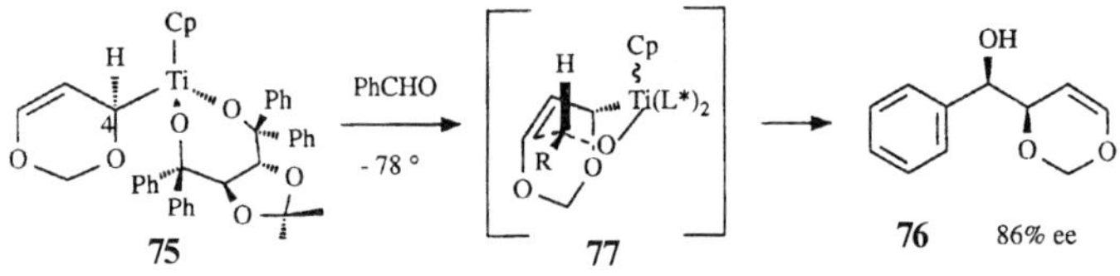

Scheme 12.14

Excellent results are also obtained with more complex substrates, as exemplified by conversions of chiral aldehydes with (**62**) and (**63**) (Scheme 12.15).[10d,39] Nucleophilic additions to α-phenylbutyraldehyde (**78**) follow with high preference the Cram rule, affording the (*S*,*S*)- or (*R*,*R*)-diastereomers.[10d,39] It is, therefore, difficult to obtain the epimers by reagent

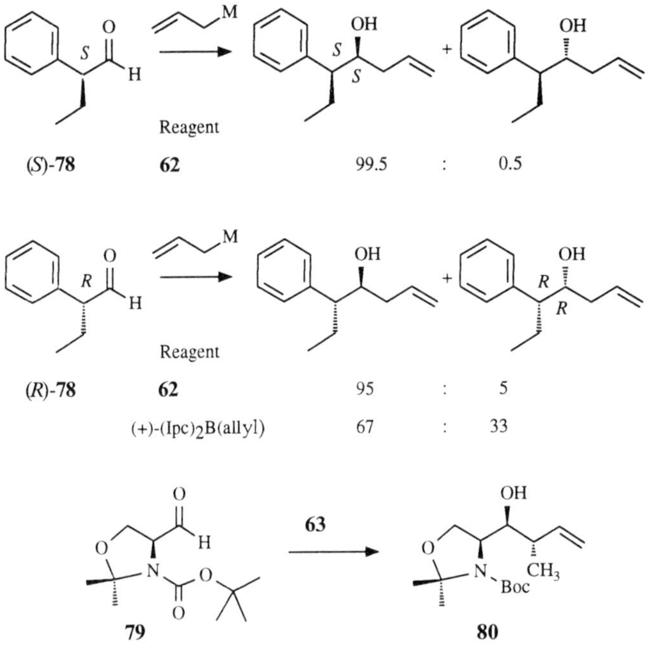

	Reagent		
(*S*)-**78**	**62**	99.5 :	0.5
(*R*)-**78**	**62**	95 :	5
	(+)-(Ipc)$_2$B(allyl)	67 :	33

Scheme 12.15

control. As expected, the stereoselectivity is excellent for matched combinations, and the de obtained from the transformation of ((*S*)-**78**) with the chiral allyltitanium reagent (**62**) is 99%. Astonishingly, however, a 95:5 product ratio (90% de) is observed from the mismatched reaction with ((*R*)-**78**), even more surprising if compared with the 34% de obtained from diisopinocampheylborane,[39] one of the most selective allylboron reagents. Reaction of the protected serine aldehyde (**79**) with the crotyltitanium complex (**63**) affords the *anti*-diastereomer (**80**) in isomerically pure form (Scheme 12.15).[10d]

From these results, it is clear that chirally modified cyclopentadienyltitanium complexes are excellent allyl-transfer reagents and are amongst the most efficient known. Products are formed with exceptionally high stereoselectivities (typically ee and/or de is in the range 95–98%).[10d] Therefore, the complexes compete favourably with allylboron reagents,[40] with respect to stereoselectivity and ease of preparation. In all cases *anti*-diastereomers are formed exclusively using substituted allyltitanium reagents. For the preparation of *syn*-isomers the allylboron reagents are still the best choice.

Allyl reagents prepared from the chlorides (**4**) or (**5**), with different acetal protection of their glucose ligands, show similar stereoselectivity as (**74**), but compounds with undefined structure, like the extremely labile L-idose complex (**6**), afford, according to a quite general rule, unselective reagents.[9,10d] In the case of the tartrate-derived ligands, the influence of the acetal groups on the stereoselectivity is low, and replacement of the acetonide of reagent (**62**) by other substituents gave inductions ranging from 80% ee for the 2,2-unsubstituted dioxolane to 91% ee with the bulky fluorenone acetal.[10d] Although the substituents of the acetal carbon have only a marginal influence on the stereoselectivity, the size of the geminal substituents of the α-carbons is critical for good induction (Schemes 12.13 and 12.16). The enantioselectivity of (**62**) (95% ee) is almost completely lost if the four phenyl groups of the ligand are replaced by methyls (**81**) (12% ee). A very interesting synergism between the chiral chelating ligand and the cyclopentadienyl substituent is apparent. Very good enantioselectivity can also be obtained when the small tetramethylthreitol ligand is combined with the bulky pentamethylcyclopentadienyl group (**82**) (88% ee) (Scheme 12.16).[10d]

A straightforward interpretation of this effect would relate the induction to an asymmetrical arrangement of these substituents around the reacting centre, i.e. formation of an enzyme-like chiral cavity defined by the bulk of these residues. While such an asymmetric environment was exhibited by the crystal structure of the diacetone–glucose complex (**3**),[8b] the four phenyl groups of (**22**) are oriented quite symmetrically around titanium.[9,10d]

New X-ray analysis data show that dioxolane-fixed 7-membered 1,3-dioxa-2-titanacycles are conformationally quite flexible. Explanations for the mechanism of asymmetric induction of various titanium complexes with such 'TADDOL' ligands (e.g. **26**), based on conformational analysis of the ligands alone,[21] have, therefore, to be rated as unreliable. Another plausible explanation

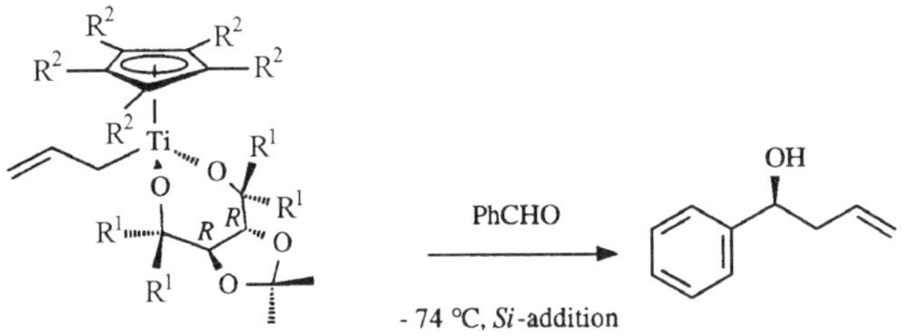

			PhCHO, -74 °C, *Si*-addition	
62	R^1= Ph	R^2= H		**95% ee** (93%)
81	R^1= Me	R^2= H		12% ee
82	R^1= Me	R^2= Me		**88% ee**

Scheme 12.16

involves a chiral distortion of the coordination geometry, resulting in a stereoelectronically governed induction by an asymmetrically hybridized metal centre. Such a distortion effect, which could be named 'enantiocontrol by *convex* asymmetry', as opposed to the notion *concave* used for enzyme-like chiral cavities, could also be operative for these titanium reagents. The distortion could be caused by interactions of the cyclopentadienyl substituent with parts of the asymmetric ligands, i.e. one of the 5,6-dioxolane rings in the cases of the diacetone–glucose system (**3**) and the geminal (*R*)-substituents in case of the complexes (**62**) and (**82**).[2a,9,10d] This would give a plausible explanation for the unexpected enantioselectivity enhancement (12% ee to 88% ee) observed upon replacing the Cp-ligand of (**82**) by the larger pentamethyl-Cp in (**83**) (cf. Scheme 12.16). This hypothesis was sustained by ^{49}Ti NMR spectroscopy of these complexes. The symmetry of the electron distribution at the Ti atom can be assessed qualitatively, but with high sensitivity, by the ^{49}Ti line width, which is broadened considerably by electronic dissymmetry.[10b,d] Much broader lines were indeed observed for the distorted structures (**3**) ($\nu_{1/2} \geqslant 4500$ Hz) and (**22**) ($\nu_{1/2} = 3460$ Hz) than for (**23**) ($\nu_{1/2} = 1080$ Hz).[10d]

The electronic disymmetry was initially also related to different Ti–O–C(α) and Ti–O′–C(α′) bond angles, but it was shown later that crystal packing effects may be the reason for these observations.

The qualitative correlation of enantioselectivity with the electronic distortion of (**3**), (**22**) and (**23**) may reflect a casual relation. Similar arguments have been used recently by Faller *et al.* for explaining the stereoselectivity of a chiral crotylmolybdenum reagent.[41]

Recently, a very efficient catalytic asymmetric preparation of homoallylic alcohols was introduced by two different groups[18] using simple aldehydes together with allyltri-*n*-butylstannane and a catalytic amount of a known chiral Lewis acid prepared from binaphthol and $TiCl_2(O\text{-}i\text{-}Pr)_2$.[3a] High enantiofacial selectivities are observed with aromatic and aliphatic aldehydes. However no data are available using the same protocol with more complex substrates (e.g. chiral or functionalized aldehydes).

(a) *Applications to the stereoselective synthesis of C-2-modified pentoses.*[10c] A lot of research effort is presently invested in the possibility of controlling gene-expression with 'antisense' oligonucleotides. 2′-*O*-Alkylation of ribosides is a most useful method of backbone modification, as both the duplex energy and the stability against degradation of such oligonucleotides are enhanced. While simple *O*-alkylated derivatives (e.g. 2′-*O*-methyl or 2′-*O*-allyl) are best prepared by direct alkylation of ribonucleosides or other ribose derivatives, different C-2′ modifications, which might also be of interest, can only be introduced by *de novo* synthesis. D-Pentoses are easily prepared by partial synthesis from D-glyceraldehyde (**83**), a chiral building block, which is also conveniently prepared from D-mannitol on a large scale. However, the stereoselectivity of additions to (**83**) is difficult to control, as *Si*-attack is the favoured mode of addition for most nucleophiles.[42] Substituted allylreagents (**61**) and (ent-**61**) (Scheme 12.17) (obtained by transmetallation with the

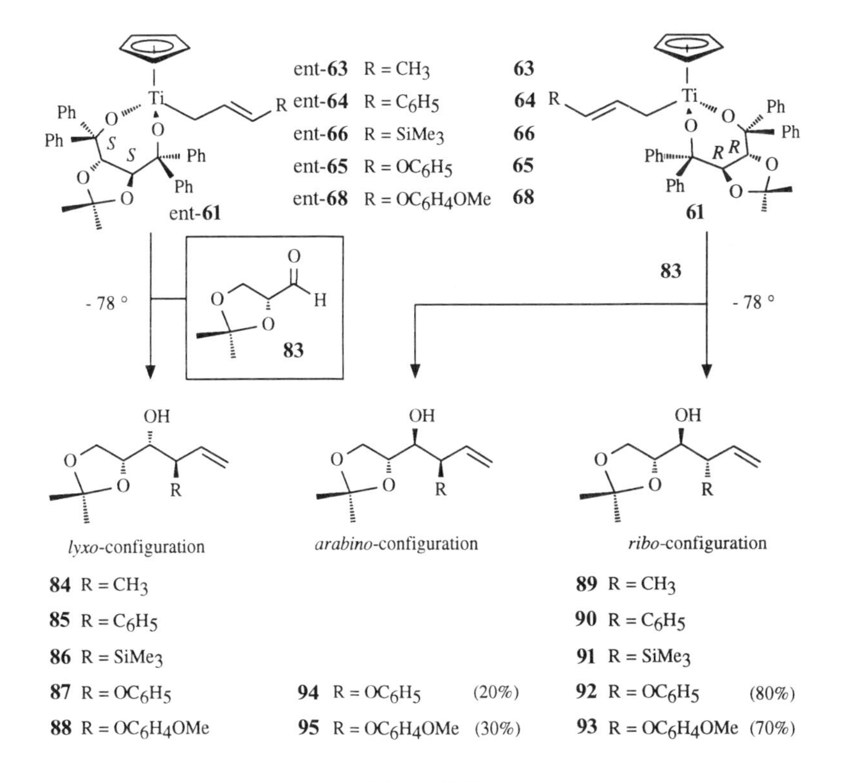

Scheme 12.17

(*R*,*R*)-configurated complex (**22**), or with the enantiomeric complex (ent-**22**)) add to D-glyceraldehyde acetonide (**83**) with excellent diastereofacial selectivity, affording *anti*-products with *ribo*-configuration using (**61**) reagents, or the corresponding *lyxo*-isomers in the case of (ent-**61**) reagents. All reagents with the (*S*,*S*)-configuration and most of the (*R*,*R*)-configurated complexes yielded essentially one product with an *anti*-configuration between the newly formed stereocentres (compounds **84**–**91**). However *syn*-diastereomers (**94**) and (**95**) were formed as by-products of the major *anti*-isomers (**92**) and (**93**) in reactions of the phenoxy-substituted allyltitanium compounds (**65**) and (**68**).

The yields of the methyl- or phenyl-substituted adducts (**84**), (**85**), (**89**) and (**90**) are high, and distillation of the crude material at high vacuum suffices for purification. In the other cases, the yields are lower, but improvements should be possible either by using more than 1.3 equiv. of the titanium reagents or by optimization of the lithiation and/or transmetallation step.

As shown for the crotyl adduct (**89**), these glyceraldehyde derivatives could be transformed straightforwardly into 2-deoxy-2-substituted furanoses with

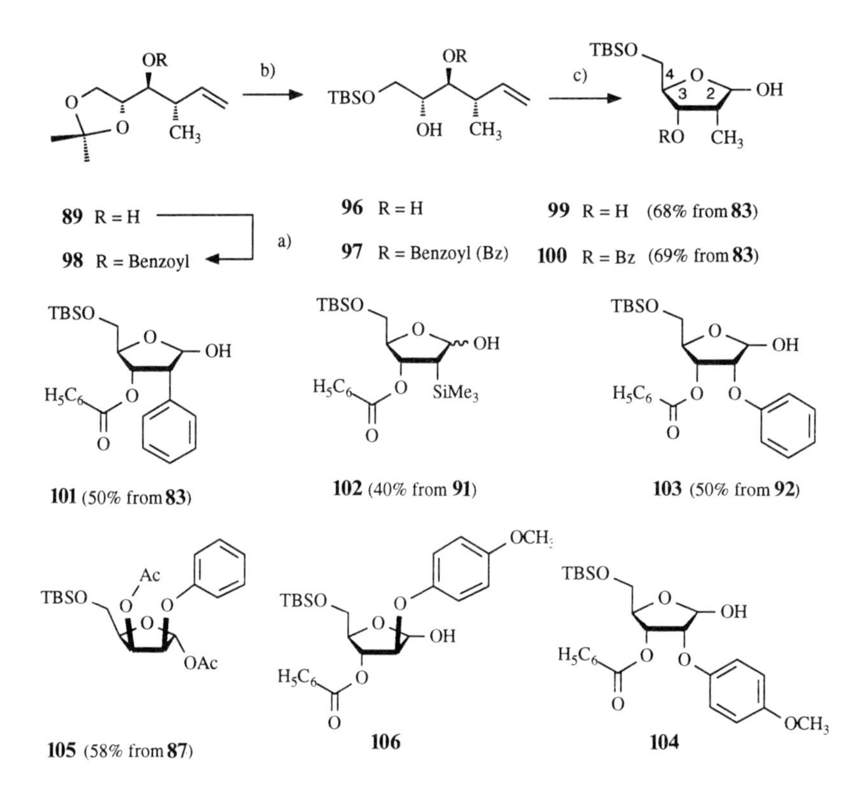

Scheme 12.18 (a) Benzoyl chloride/pyridine; (b) 1. CH_3OH/CF_3OH (or Amberlyst-15), 2. t-Butyldimethylsilyltriflate/Et_3N, CH_2Cl_2; (c) 1. O_3/CH_3OH, −78°C, 2. Me_2S, −78°C to room temperature.

D-*ribo*-configuration. Especially rewarding is the possibility of selective *O*-3- and *O*-5-protection (Scheme 12.18). The target ribofuranoses (**99**) and (**100**) were isolated in excellent overall yield after ozonolysis and reductive work-up. The 2-deoxy-2-phenyl-, 2-trimethylsilyl-, 2-phenoxy- and 2-(4′-methoxy-phenoxy)-ribose, in addition to the 2-deoxy-2-phenoxy-lyxose and 2-deoxy-2-(4′-methoxyphenoxy)-arabinose derivatives (**101**–**106**)[10] were all obtained analogously (Scheme 12.18). Because of the high selectivities and the potential for further modifications, this scheme for the preparation of unusual pentoses from glyceraldehyde compares favourably with other closely related methods.[43]

12.4 Addition of d^1-nucleophiles to carbonyl compounds

12.4.1 Stoichiometric use of chiral alkyltitanium complexes

Chiral methyltitanium reagents can be prepared by two routes as shown in Scheme 12.19. Displacement of methyl groups by chiral alcohols (R*OH) gives the salt-free reagents (**108**) (cf. Figure 12.1g). Because of the instability of other di-, tri- and tetraalkyltitanium compounds, this access is limited to methyl reagents. Alternatively, a chiral chlorotitanium complex (**107**) is transmetallated with MeLi or other alkyllithium, Grignard or organometallic reagents. Depending on the conditions, solvent and precursors, either ‘ate’-complexes, bimetallic aggregates or, if LiCl is precipitated, a pure organotitanium compound is formed. After the reaction with an aldehyde

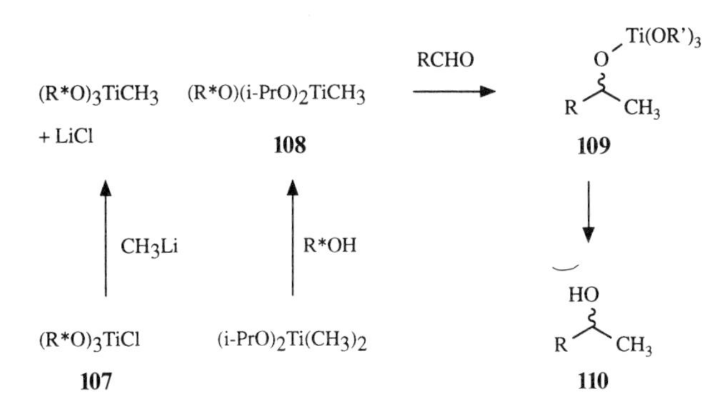

Scheme 12.19

(ketones react at best sluggishly with alkyltitanium compounds[2,4]) the product (**110**) is obtained by hydrolysis of the titanate (**109**).

The enantiofacial discrimination achieved with reagents bearing monodentate chiral ligands is in all cases low to mediocre. Much better results were observed by using bidentate chiral ligands. Values of ee exceeding 90% are, however, rare and are limited to very specific reagent–substrate combinations.[13,19,44]

The most reliable process with chiral titanium nucleophiles is the addition of phenyl groups to aromatic aldehydes, using 1,1′-binaphthol as ligand (Scheme 12.20).[13,16] The resulting diarylmethanols (**111**) are thereby obtained with optical purities of 90% or higher. A most impressive example of practical value is the addition of a highly substituted phenyl group to phytenal, a chiral α,β-unsaturated aldehyde, affording the vitamin E precursor (**112**) with 82% de (Scheme 12.20).[45] A special case, with the chiral ligand derived from prolinol attached via a sulfonamide linkage to the aryl nucleophile, is the reagent obtained from (**113**). Addition to aldehydes gives secondary alcohols (e.g. **114**) with 62–82% de[46] (Scheme 12.20).

If the tetraalkoxytitanium compounds (**32**) and (**34**) are treated with alkyllithium or alkylmagnesium reagents, the formation of 'ate'-complexes is assumed. (See also the results with (**18**) (Scheme 12.10).[16b] Whatever the nature of the resulting species might be, the methylation of simple aromatic

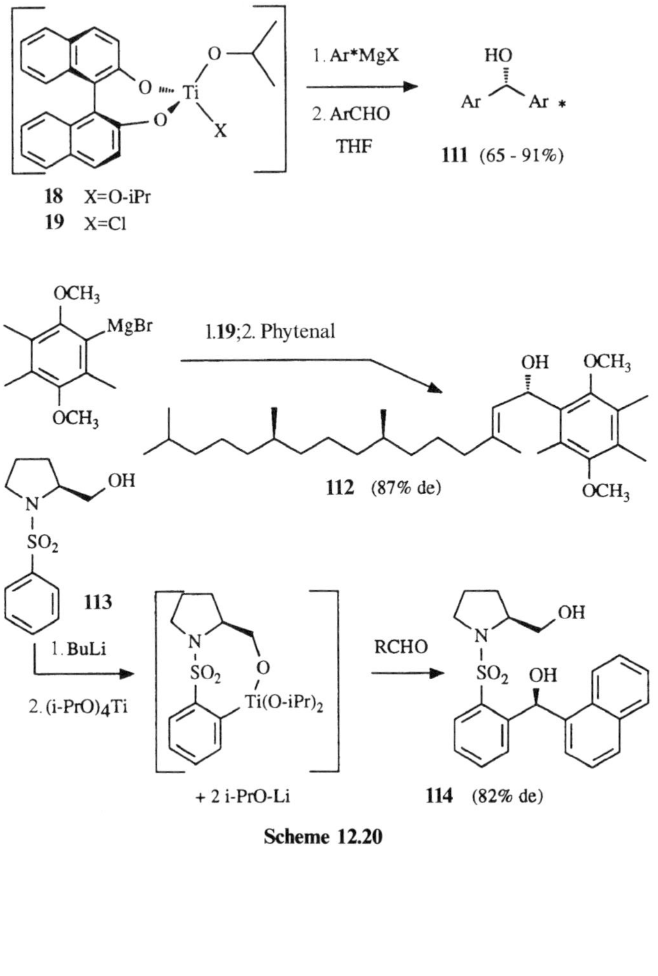

Scheme 12.20

aldehydes is achieved with good enantioselectivity (Scheme 12.21).[20a,b] In the case of the spirocyclic titanate (**34**), MeLi gives much better results than Grignard reagents.[20b]

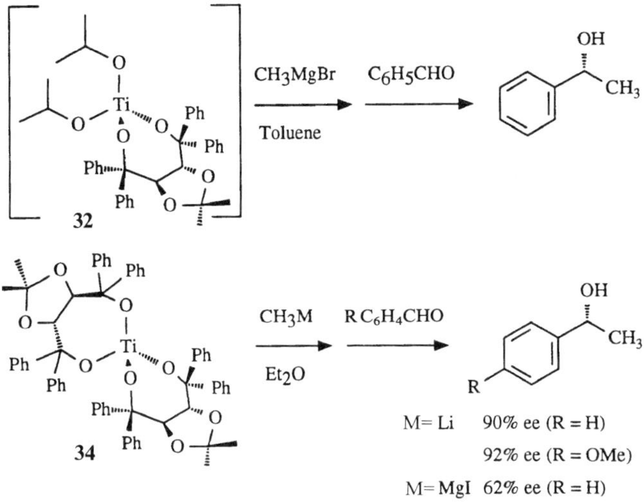

32

CH_3MgBr Toluene C_6H_5CHO

OH CH_3

34

CH_3M Et_2O RC_6H_4CHO

OH CH_3 R

M= Li 90% ee (R = H)
92% ee (R = OMe)
M= MgI 62% ee (R = H)

Scheme 12.21

Alkyltitanium complexes with cyclopentadienyl ligands have much higher stability than the analogous complexes with η^1-bound ligands. Because of the reduced Lewis acidity, their reactivity is also much lower, and while $Cl_2CpTiCH_3$ still reacts with aldehydes at room temperature, the dialkoxy–$CpTiCH_3$ complexes derived from (**22**) as well as analogous compounds obtained from (**3–5**) are completely unreactive. This can be overcome by replacing Ti(IV) with Zr(IV) or Hf(IV), as their reactivities are much higher (Scheme 12.22). Alkylzirconium or alkylhafnium compounds, obtained from the chlorides (**115**) and (**116**), therefore react with aldehydes even at −78°C. The enantioselectivity of these reagents is also much higher than that of any of the alkyltitanium complexes described above, reaching 97–98% ee for the addition of a methyl group to benzaldehyde. In contrast to the titanium reagents lacking a Cp ligand, Grignard and organolithium compounds give equally good results. However, the enantioselectivity drops to mediocre values, if alkyl groups other than methyl are transferred and if non-aromatic aldehydes are used instead of benzaldehyde.[10b,c]

12.4.2 *Chiral titanium complexes as catalysts for enantioselective additions using dialkylzinc compounds*

The enantioselective addition of dialkylzinc compounds to aldehydes, mediated by a catalytic amount of a chiral amino alcohol, was considered

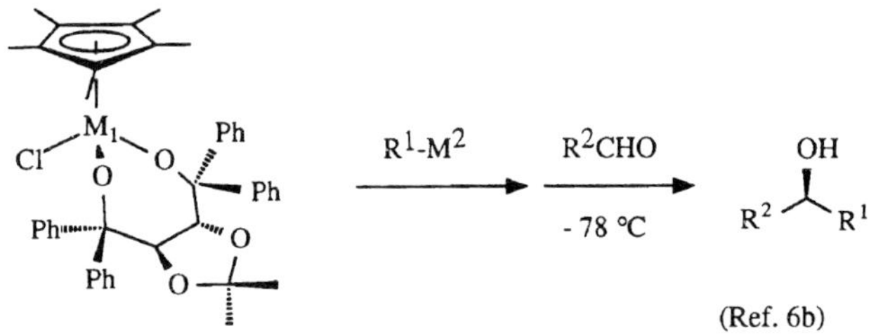

R^1-M^2 → R^2CHO, - 78 °C → R^2CH(OH)R^1

(Ref. 6b)

115 M_1=Zr	CH_3MgBr	R^2: C_6H_5	**98% ee**
	CH_3Li	R^2: C_6H_5	**97% ee**
	EtMgBr	R^2: C_6H_5	52% ee
	n-BuLi	R^2: C_6H_5	68% ee
	CH_3Li	R^2: C_9H_{19}	**80% ee**
116 M_1=Hf	CH_3MgBr	R^2: C_6H_5	**97% ee**
	EtMgBr	R^2: C_6H_5	47% ee
	n-BuLi	R^2: C_6H_5	65% ee

Scheme 12.22

a breakthrough in asymmetric synthesis.[26c,d,47] It was soon found that the actual catalyst of this process is not the amino alcohol itself, but a zinc chelate formed *in situ* from the zinc reagent and the amino alcohol. As a consequence, the variety of new catalyst systems increased. Lithium salts[48] and boron compounds[49] derived from such amino alcohols were introduced as mild Lewis acids in place of the original zinc complexes. Still further away from the original amino alcohols is the combination of 0.3–1.2 equiv. $Ti(O\text{-}i\text{-}Pr)_4$ with 0.5–4 mol.% of a chiral bissulfonamide ligand, introduced by Yoshioka *et al.* as a very efficient catalyst for alkylations using Et_2Zn (Scheme 12.23).[50] The aggregate (**118**) (Scheme 12.23) was postulated as a reactive species in the catalytic cycle.[50] In addition, Seebach *et al.* showed that tetraalkoxytitanium complexes (**32**) and (**33**) with bulky tartrate-derived ligands can activate Et_2Zn better than $(\text{-}i\text{-PrO})_4Ti$ and are, therefore, excellent chiral catalysts for aldehyde ethylations.[20b,21]

The results of the Et_2Zn additions catalysed by titanium complexes are compiled in Table 12.2. As far as comparisons are possible, complexes prepared from (**117**)[50] and (**33**)[20c] appear to be somewhat superior to (**32**).[20b] Excellent inductions (91–99% ee) are, therefore, obtained even in reactions with non-aromatic substrates. However, good results with aliphatic aldehydes and functionalized aliphatic aldehydes[51] have been reported recently for amino alcohol catalysis.[47h,n,o,p,48a]

Until recently, a limitation of the reaction was the lack of an easy general

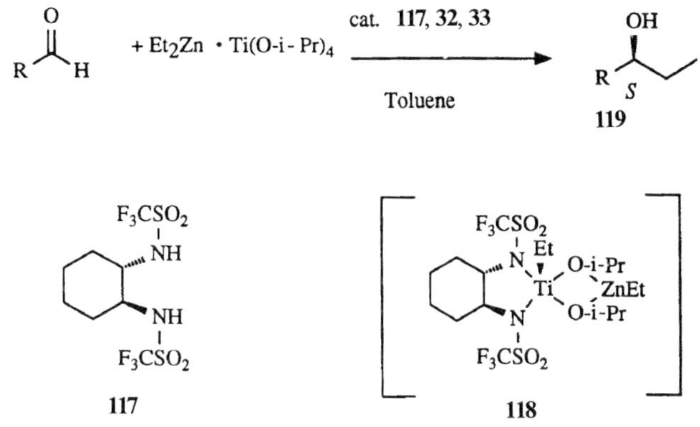

Scheme 12.23

Table 12.2 Ethylation of aldehydes with Et_2Zn/(-i-PrO)$_4$Ti, catalysed by the titanates (**117**), (**32**) and (**33**) (Scheme 12.23)[20b,c,50]

Aldehyde R	[L*]Ti(OR)$_2$ No.	[L*]Ti(OR)$_2$ (mol.%)	Et_2Zn (mol.%)	Ti(OR)$_4$ (mol.%)	Temperature °C	Product (**119**) Yield %	Product (**119**) ee (%)	Ref.
C_6H_5	**117**	0.5	120	120	0	96	99 (*S*)	50a
C_6H_5	**32**	20	120	120	−75 to 0	75[a]	99[a] (*S*)	20b
C_6H_5	**33**	5	180	120	−25	95	92 (*S*)	20c
C_6H_5-CH=CH	**117**	2	220	30	−50	85	99 (*S*)	50b
C_6H_5-CH=CH	**33**	5	180	120	−25	87	91 (*S*)	20c
C_6H_5-C≡C	**33**	5	180	120	−25	83	99 (*S*)	20c
C_6H_5-CH_2CH_2	**117**	1	220	60	0	95	92 (*S*)	50b
C_6H_5-CH_2CH_2	**32**	20	120	120	−75 to 0	82	85 (*S*)	20b
C_6H_5-CH_2CH_2	**33**	5	180	120	−25	87	98 (*S*)	20c
$CH_3(CH_2)_4$	**117**	4	220	60	−20	78	99 (*S*)	50b
$CH_3(CH_2)_5$	**32**	20	120	120	−75 to 0	75	92 (*S*)	20b
$CH_3(CH_2)_5$	**33**	5	180	120	−25	70	97 (*S*)	20c
Cyclohexyl	**32**	20	120	120	−75 to 0	67	82 (*S*)	20b
Cyclohexyl	**96**	5	180	120	−25	77	99 (*S*)	20c

[a] In Et_2O the yield is 62% and the ee is 98%.[19b]

access to dialkylzinc compounds. Seebach *et al.* showed that, in the case of the titanium catalyst (**32**) (cf. Table 12.2, 2nd entry), various dialkylzinc reagents can be used in form of dioxane complexes (prepared from Grignard solutions and $ZnCl_2$, followed by precipitation of $MgCl_2$) (Scheme 12.24).[52]

Knochel reported that a wide variety of different functionalized diorganozinc compounds can be used in this type of reaction (prepared either by transmetallation methods[53] or by reaction of primary iodides and diethyl zinc[54]). With this modification, the scope of this method has been considerably

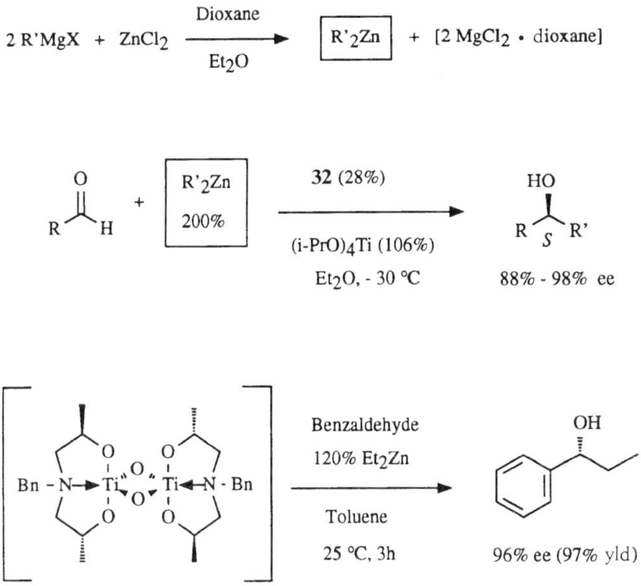

$$2\,R'MgX + ZnCl_2 \xrightarrow[Et_2O]{Dioxane} \boxed{R'_2Zn} + [2\,MgCl_2 \cdot dioxane]$$

R-CHO + R'$_2$Zn (200%) → [32 (28%), (i-PrO)$_4$Ti (106%), Et$_2$O, - 30 °C] → R-CH(OH)-R' (S), 88% - 98% ee

120 5 mol% (10% Ti) + Benzaldehyde, 120% Et$_2$Zn, Toluene, 25 °C, 3h → 1-phenylpropanol, 96% ee (97% yld)

Scheme 12.24

extended. Limitations are mainly the result of steric influences; slow reactions and low yields are observed for alkyl nucleophiles with branched β-carbons. In the absence of $Ti(O\text{-}i\text{-}Pr)_4$, chiral titanium complexes can also mediate the enantioselective alkylation of aldehydes with dialkylzinc compounds, but these reactions are usually sluggish. An exception in this respect is the chiral titanium oxide (**120**), which was recently prepared by Nugent and was subsequently shown to be a very efficient catalyst for the ethylation of benzaldehyde (Scheme 12.24).[55]

References

1. (a) Scott, J.W., in *Topics in Stereochemistry*, Vol. 19, Eliel, E.L. and Wilen, S.H. (eds), Wiley, New York, 1989, pp. 209–226; (b) Crosby, J., *Tetrahedron*, 1991, **47**, 4789; (c) Ramos Tombo, G.M. and Belluš, D. *Angew. Chem., Int. Ed. Engl.*, 1991, **30**, 1193.
2. (a) Duthaler, R.O. and Hafner, A., *Chem. Rev.*, 1992, **92**, 807; (b) Reetz, M.T., in *Topics in Current Chemistry*, Vol. 106, Boschke, F.L. (ed.), Springer Verlag, Berlin, 1982, pp. 3–54; (c) Seebach, D., Weidmann, B. and Widler, L., in *Modern Synthetic Methods*, Vol. 3, Scheffold, R. (ed.), Salle, Frankfurt, 1983, pp. 217–353; (d) Reetz, M.T., *Organotitanium Reagents in Organic Synthesis*, Springer Verlag, Berlin, 1986.
3. (a) Mikami, K., Terada, M. and Nakai, T., *J. Am. Chem. Soc.*, 1990, **112**, 3949; (b) Narasaka, K., Iwasawa, N., Inoue, M., Yamada, Y., Nakashima, M. and Sugimori, J.J., *J. Am. Chem. Soc.*, 1989, **111**, 5340; (c) Finn, M.G. and Sharpless, K.B., in *Asymmetric Synthesis*, Vol. 5,

Morrison, J.D. (ed.), Academic Press, New York, 1985, p. 247; (d) Brintzinger, H.H., in *Transition Metals and Organometallics as Catalysts for Olefin Polymerization*, Kaminsky, W. and Sinn, H. (eds), Springer Verlag, Berlin, 1988, p. 249.

4. (a) Weidmann, B. and Seebach, D., *Angew. Chem., Int. Ed. Engl.*, 1983, **22**, 31; *Angew. Chem.*, 1983, **95**, 12; (b) Seebach, D., Beck, A.K., Schiess, M., Widler, L. and Wonnacott, A., *Pure Appl. Chem.*, 1983, **55**, 1807; (c) Reetz, M.T., *Pure Appl. Chem.*, 1985, **57**, 1781; (d) Reetz, M.T., *S. Afr. J. Chem.*, 1989, **42**, 49.
5. (a) Iwasawa, N., Hayashi, Y., Sakurai, H. and Narasaka, K., *Chem. Lett.*, 1989, 1581; (b) Narasaka, K., Kanai, F., Okuda, M. and Miyoshi, N., *Chem. Lett.*, 1989, 187.
6. (a) Marsella, J.A., Moloy, K.G. and Coulton, K.G., *J. Organomet. Chem.*, 1980, **201**, 389; (b) Gorsich, R.D., *J. Am. Chem. Soc.*, 1960, **82**, 4211.
7. (a) Bradley, D.C. and Holloway, C.E., *J. Chem. Soc. (A)*, 1968, 1316; (b) Watenpaugh, K. and Caughlan, Ch.N., *Inorg. Chem.*, 1966, **5**, 1782; (c) Schmidt, F., Feltz, A., Colditz, R. and Gustav, K.Z., *Anorg. Allg. Chem.*, 1989, **574**, 218; (d) Bachaud, B. and Wuest, J.D., *Organometallics*, 1991, **10**, 2015.
8. (a) Riediker, M. and Duthaler, R.O., *Angew. Chem. Int. Ed. Engl.*, 1989, **28**, 494; (b) Riediker, M., Hafner, A., Piantini, U., Rihs, G. and Togni, A., *Angew. Chem. Int. Ed. Engl.*, 1989, **28**, 499; (c) Duthaler, R.O., Herold, P., Lottenbach, W., Oertle, K. and Riediker, M., *Angew. Chem. Int. Ed. Engl.*, 1989, **28**, 495; (d) Bold, G., Duthaler, R.O. and Riediker, M., *Angew. Chem., Int. Ed. Engl.*, 1989, **28**, 497; (e) Oertle, K., Beyeler, H., Duthaler, R.O., Lottenbach, W., Riediker, M. and Steiner, E., *Helv. Chim. Acta.*, 1990, **73**, 353; (f) Duthaler, R.O., Herold, P., Wyler-Helfer, S. and Riediker, M., *Helv. Chim. Acta*, 1990, **73**, 659.
9. Duthaler, R.O., Hafner, A., Alsters, P.L., Bold, G., Rihs, G., Rothe-Streit, P. and Wyss, B., *Inorg. Chim. Acta*, in press.
10. (a) Duthaler, R.O., Hafner, A. and Riediker, M., *Pure Appl. Chem.*, 1990, **62**, 631; (b) Duthaler, R.O., Hafner, A. and Riediker, M., in *Organic Synthesis via Organometallics*, Dötz, K.H. and Hoffmann, R.W. (eds), Vieweg, Braunschweig, 1991, pp. 285–309; (c) Duthaler, R.O., Hafner, A., Alsters, P.L., Rother-Streit, P. and Rihs, G., *Pure Appl. Chem.*, 1992, **64**, 1897; (d) Hafner, A., Duthaler, R.O., Marti, R., Rihs, G., Rothe-Streit, P. and Schwarzenbach, F., *J. Am. Chem. Soc.*, 1992, **114**, 2321; (e) Complex (**63**) is commercially available: *Fluka AG*, CH-9470 Buchs, Switzerland.
11. (a) Krämer, Th. and Hoppe, D., *Tetrahedron Lett.*, 1987, **28**, 5149; (b) Hoppe, D. and Zschage, O., *Angew. Chem., Int. Ed. Engl.*, 1989, **28**, 69; (c) Zschage, O., Schwark, J.-R. and Hoppe, D., *Angew. Chem. Int. Ed. Engl.*, 1990, **29**, 296; (d) Hoppe, D., Krämer, Th., Schwark, J.-R. and Zschage, O., *Pure Appl. Chem.*, 1990, **62**, 1999; (e) Hoppe, D. and Zschage, O., in *Organic Synthesis via Organometallics*, Dötz, K.H. and Hoffmann, R.W. (eds), Vieweg, Braunschweig, 1991, pp. 267–283; (f) Dreller, S., Dyrbusch, M. and Hoppe, D., *Synlett*, 1991, 397.
12. Marsch, M., Harms, K., Zschage, O., Hoppe, D. and Boche, G., *Angew. Chem., Int. Ed. Engl.*, 1991, **30**, 321.
13. Olivero, A.G., Weidmann, B. and Seebach, D., *Helv. Chim. Acta*, 1981, **64**, 2485.
14. Martin, Cheryl Ann, PhD Thesis, Massachusetts Institute of Technology, 1988.
15. (a) Reetz, M.T., Kyung, S.-H., Bolm, C. and Zierke, T., *Chem. Ind. (Lond.)*, 1988, 824; (b) Terada, M., Mikami, K. and Nakai, T., *J. Chem. Soc., Chem Commun.*, 1990, 1623.
16. (a) Seebach, D., Beck, A.K., Roggo, S. and Wonnacott, A., *Chem. Ber.*, 1985, **118**, 3673; (b) Wang, J.-T., Fan, X. and Qian, Y.-M., *Synthesis*, 1989, 291.
17. Chapius, Ch. and Jurczak, J., *Helv. Chim. Acta*, 1987, **70**, 436.
18. (a) Costa, A.L., Piazza, M.G., Tagliavini, E., Trombini, C. and Umani-Ronchi, A., *J. Am. Chem. Soc.*, 1993, **115**, 7001; (b) Keck, G.E., Tarbet, K.H. and Geraci, L.S., *J. Am. Chem. Soc.*, 1993, **115**, 8467; (c) Keck, G.E., Krishnamurthy, D. and Grier, M.C., *J. Org. Chem.*, 1993, **58**, 6543.
19. Seebach, D., Beck, A.K., Imwinkelried, R., Roggo, S. and Wonnacott, A., *Helv. Chim. Acta*, 1987, **70**, 954.
20. (a) Takahashi, H., Kawabata, A., Niwa, H. and Higashiyama, K., *Chem. Pharm. Bull.*, 1988, **36**, 803; (b) Schmidt, B. and Seebach, D., *Angew. Chem., Int. Ed. Engl.*, 1991, **30**, 99; (c) Schmidt, B. and Seebach, D., *Angew. Chem., Int. Ed. Engl.*, 1991, **30**, 1321.
21. Seebach, D., Plattner, D.A., Beck, A.K., Wang, Y.M., Hunziker, D. and Petter, W., *Helv. Chim. Acta*, 1992, **75**, 2171.
22. Goodwin, J.C., Hodge, J.E. and Weisleder, D., *Carbohydr. Res.*, 1980, **79**, 133.

23. Carmack, M. and Kelly, Ch.J., *J. Org. Chem.*, 1968, **33**, 2171.
24. Wild, F.R.W.P., Zsolnai, L., Huttner, G. and Brintzinger, H.H., *J. Organomet. Chem.*, 1982, **232**, 233.
25. (a) Seebach, D., *Angew. Chem., Int. Ed. Engl.*, 1979, **18**, 239; (b) Fuhrhop, J. and Penzlin, G., *Organic Synthesis, Concepts, Methods, Starting Materials*, Verlag Chemie, Weinheim, 1983, p. 1.
26. (a) Solladié, G., in *Asymmetric Synthesis*, Vol. 2, Morrison, J.D. (ed.), Academic Press, New York, 1983, pp. 157–199; (b) Evans, D.A., *Science*, 1988, **240**, 420; (c) Noyori, R., Suga, S., Kawai, K., Okada, S. and Kitamura, M., *Pure Appl. Chem.*, 1988, **60**, 1597; (d) Noyori, R. and Kitamura, M., *Angew. Chem., Int. Ed. Engl.*, 1991, **30**, 49.
27. (a) Li, Y., Paddon-Row, M.N. and Houk, K.N., *J. Org. Chem.*, 1990, **55**, 481; (b) Yamago, S., Machii, D. and Nakamura, E., *J. Org. Chem.*, 1991, **56**, 2098; (c) Denmark, S.E. and Henke, B.R., *J. Am. Chem. Soc.*, 1991, **113**, 2177.
28. (a) Hoffmann, R.W., *Angew. Chem., Int. Ed. Engl.*, 1982, **21**, 555; (b) Yamamoto, Y. and Maruyama, K., *Heterocycles*, 1982, **18**, 357; (c) Hoffmann, R.W., *Angew. Chem., Int. Ed. Engl.*, 1987, **26**, 489; (d) Yamamoto, Y., *Acc. Chem. Res.*, 1987, **20**, 243; (e) Hoppe, D., *Angew. Chem., Int. Ed. Engl.*, 1984, **23**, 932; (f) Mulzer, J., Kattner, L., Strecker, A.R., Schröder, Ch., Buschmann, J., Lehmann, Ch. and Luger, P., *J. Am. Chem. Soc.*, 1991, **113**, 4218.
29. (a) Evans, D.A., Nelson, J.V. and Taber, T.R., *Topics Stereochem.*, 1982, **13**, 1; (b) Heathcock, C.H., in *Asymmetric Synthesis*, Vol. 3, Morrison, J.D. (ed.), Academic Press, New York, 1984, pp. 111–212; (c) Masamune, S., Choy, W., Petersen, J.S. and Sita, L.R., *Angew. Chem., Int. Ed. Engl.*, 1985, **24**, 1; (d) Heathcock, C.H., *Aldrich Chim. Acta*, 1990, **23**, 99.
30. (a) Shibasaki, M., Ishida, Y. and Okabe, N., *Tetrahedron Lett.*, 1985, **26**, 2217; (b) Devant, R. and Braun, M., *Chem. Ber.*, 1986, **119**, 2191.
31. Evans, D.A., Clark, J.St., Metternich, R., Novack, V.J. and Shepard, G.S., *J. Am. Chem. Soc.*, 1990, **112**, 866.
32. Mukaiyama, T., Inobushi, A., Suda, Sh., Hara, R. and Kobayashi, Sh., *Chem. Lett.*, 1990, 1015.
33. (a) Mukaiyama, T., Takashima, T., Kusaka, H. and Shimpuka, T., *Chem. Lett.*, 1990, 1777; (b) Kobayashi, Sh., Furuya, M., Ohtsubo, A. and Mukaiyama, T., *Tetrahedron: Asymmetry*, 1991, **2**, 635.
34. (a) Braun, M., *Angew. Chem., Int. Ed. Engl.*, 1987, **26**, 24; (b) Liebeskind, L.S. and Welker, M.E., *Tetrahedron Lett.*, 1984, **25**, 4341; (c) Davies, S.G., Dordor, I.M. and Warner, P., *J. Chem. Soc., Chem. Commun.*, 1984, 956; (d) Helmchen, G., Leikauf, U. and Taufer-Knöpfel, I., *Angew. Chem., Int. Ed. Engl.*, 1985, **24**, 874; (e) Oppolzer, W. and Marco-Contelles, J., *Helv. Chim. Acta*, 1986, **69**, 1699; (f) Paterson, I., Goodman, J.M., Lister, M.A., Schumann, R.C., McClure, C.K. and Norcross, R.D., *Tetrahedron*, 1990, **46**, 4663; (g) Corey, E.J. and Choi, S., *Tetrahedron Lett.*, 1991, **32**, 2857.
35. (a) Ito, Y., Sawamura, M. and Hayashi, T., *J. Am. Chem. Soc.*, 1986, **108**, 6405; (b) Hayashi, T., Uozumi, Y., Yamazaki, A., Sawamura, M., Hamashino, H. and Ito, Y., *Tetrahedron Lett.*, 1991, **32**, 2799; (c) Pastor, S.D. and Togni, A., *J. Am. Chem. Soc.*, 1989, **111**, 2333.
36. (a) Seebach, D. and Widler, L., *Helv. Chim. Acta*, 1982, **65**, 1972; (b) Reetz, M.T. and Wenderoth, B., *Tetrahedron Lett.*, 1982, **23**, 5259; (c) Reetz, M.T., Steinbach, R., Westerman, J., Peter, R. and Wenderoth, B., *Chem. Ber.*, 1985, **118**, 1441.
37. (a) Sato, F., Iijima, S. and Sato, M., *Tetrahedron Lett.*, 1981, **22**, 243; (b) Kobayashi, Y., Umeyama, K. and Sato, F., *J. Chem. Soc., Chem. Commun.*, 1984, 621; (c) Widler, L. and Seebach, D., *Helv. Chim. Acta*, 1982, **65**, 1085; (d) Reetz, M.T. and Sauerwald, M., *J. Org. Chem.*, 1984, **49**, 2292; (e) Hanko, R. and Hoppe, D., *Angew. Chem., Int. Ed. Engl.*, 1982, **21**, 372; (f) Hoppe, D., Gonschorrek, Ch., Schmidt, D. and Egert, E., *Tetrahedron*, 1987, **43**, 2457; (g) Ikeda, Y., Furuta, K., Meguriya, N., Ikeda, N. and Yamamoto, H., *J. Am. Chem. Soc.*, 1982, **104**, 7663; (h) Collins, S., Dean, W.P. and Ward, D.G., *Organometallics*, 1988, **7**, 2289; (i) Martin, St.F. and Li, W., *J. Org. Chem.*, 1989, **54**, 6129; (j) Hoffmann, R.W. and Sander, Th., *Chem. Ber.*, 1990, **123**, 145.
38. Roder, H., Helmchen, G., Peters, E.-M., Peters, K. and von Schnering, H.-G., *Angew. Chem., Int. Ed. Engl.*, 1984, **23**, 898.
39. (a) Brown, H.C., Bhat, K.S. and Randad, R.S., *J. Org. Chem.*, 1987, **52**, 319; (b) Brown, H.C., Bhat, K.S. and Randad, R.S., *J. Org. Chem.*, 1989, **54**, 1570.
40. (a) Herold, Th. and Hoffmann, R.W., *Angew. Chem., Int. Ed. Engl.*, 1978, **17**, 768; (b) Andersen, M.W., Hildebrandt, B. and Hoffmann, R.W., *Angew. Chem., Int. Ed. Engl.*, 1991, **30**, 97;

(c) Brown, H.C. and Jadhav, P.K., *J. Am. Chem. Soc.*, 1983, **105**, 2092; (d) Racherla, U.S. and Brown, H.C., *J. Org. Chem.*, 1991, **56**, 401; (e) Roush, W.R. and Kageyama, M., *Tetrahedron Lett.*, 1985, **26**, 4327; (f) Roush, W.R., Ando, K., Powers, D.B., Palkowitz, A.D. and Halterman, R.L., *J. Am. Chem. Soc.*, 1990, **112**, 6339; (g) Reetz, M.T. and Zierke, T., *Chem. Ind. (Lond.)*, 1988, 663; (h) Short, R.P. and Masamune, S., *J. Am. Chem. Soc.*, 1989, **111**, 1892; (i) Corey, E.J., Yu, Ch.-M. and Kim, S.S., *J. Am. Chem. Soc.*, 1989, **111**, 5495; (j) Stürmer, R., *Angew. Chem., Int. Ed. Engl.*, 1990, **29**, 59.

41. Faller, J.W., John, J.A. and Mazzieri, M.R., *Tetrahedron Lett.*, 1989, **30**, 1769.
42. (a) Reetz, M.T., *Angew. Chem., Int. Ed. Engl.*, 1984, **23**, 556; (b) McGarvey, G.J., Kimura, M., Oh, T. and Williams, J.M., *J. Carbohydr. Chem.*, 1984, **3**, 125.
43. (a) Yamaguchi, M. and Mukayama, T., *Chem. Lett.*, 1981, 1005; (b) McGarvey, G.J., Kimura, M., Oh, T., Williams, J.M., *J. Carbohydr. Chem.*, 1984, **3**, 125; (c) Mulzer, J., Kattner, L., Strecker, A.R., Schröder, C.H., Buschmann, J., Lehmann, C.H. and Luger, P., *J. Am. Chem. Soc.*, 1991, **113**, 4218; (d) Roush, W.R. and Grover, P.T., *Tetrahedron*, 1992, **48**, 1981; (e) Mikami, K., Terada, M. and Nakai, T., *Tetrahedron: Asymmetry*, 1991, **2**, 993; (f) Hoffmann, W., Endesfelder, A. and Zeiss, H.-J., *Carbohydr. Res.*, 1983, **123**, 320; (g) Tamao, K., Nakajo, E. and Ito, J., *J. Org. Chem.*, 1987, **52**, 957.
44. Reetz, M.T., Kükenhöhner, Th. and Weining, P., *Tetrahedron Lett.*, 1986, **27**, 5711.
45. Hübscher, J. and Barner, R., *Helv. Chim. Acta*, 1990, **73**, 1068.
46. Takahashi, H., Tsukubi, T. and Higashiyama, K., *Chem. Pharm. Bull.*, 1991, **39**, 260.
47. (a) Oguni, N. and Omi, T., *Tetrahedron Lett.*, 1984, **25**, 2823; (b) Kitamura, M., Suga, S., Kawai, K. and Noyori, R., *J. Am. Chem. Soc.*, 1986, **108**, 6071; (c) Kitamura, M., Okada, S., Suga, S. and Noyori, R., *J. Am. Chem. Soc.*, 1989, **111**, 4028; (d) Noyori, R., *Science*, 1990, **248**, 1194; (e) Noyori, R., Suga, S., Kawai, K., Okada, S., Kitamura, M., Oguni, N., Hayashi, M., Kaneko, T. and Matsuda, Y., *J. Organomet. Chem.*, 1990, **382**, 19; (f) Smaardijk, A.A. and Wynberg, H., *J. Org. Chem.*, 1987, **52**, 135; (g) Soai, K., Ookawa, A., Ogawa, K. and Kaba, T., *J. Chem. Soc., Chem. Commun.*, 1987, 467; (h) Soai, K., Yokoyama, S. and Hayasaka, T., *J. Org. Chem.*, 1991, **56**, 4264; (i) Soai, K. and Watanabe, M., *Tetrahedron: Asymmetry*, 1991, **2**, 97; (j) Itsuno, Sh. and Fréchet, J.M.J., *J. Org. Chem.*, 1987, **52**, 4140; (k) Corey, E.J. and Hannon, F.J., *Tetrahedron Lett.*, 1987, **28**, 5237; (l) Corey, E.J., Yuen, P.-W., Hannon, F.J. and Wierda, D.A., *J. Org. Chem.*, 1990, **55**, 784; (m) Oppolzer, W. and Radinov, R.N., *Tetrahedron Lett.*, 1988, **29**, 5645; (n) Watanabe, M., Araki, S., Butsugan, Y. and Uemura, M., *J. Org. Chem.*, 1991, **56**, 2218; (o) Shono, T., Kise, N., Shirakawa, E., Matsumoto, H. and Okazaki, E., *J. Org. Chem.*, 1991, **56**, 3063; (p) Chaloner, P.A. and Langadinaou, E., *Tetrahedron Lett.*, 1990, **31**, 5185; (q) Asami, M. and Inoue, S., *Chem. Lett.*, 1991, 685; (r) Oppolzer, W. and Radinov, R.N., *Helv. Chim. Acta.*, 1992, **75**, 170.
48. (a) Soai, K., Ookawa, A., Kaba, T. and Ogawa, K., *J. Am. Chem. Soc.*, 1987, **109**, 7111; (b) Soai, K., Niwa, S., Yamada, Y. and Inoue, H., *Tetrahedron Lett.*, 1987, **28**, 4841; (c) Corey, E.J. and Hannon, F.J., *Tetrahedron Lett.*, 1987, **28**, 5233.
49. Joshi, N.N., Srebnik, M. and Brown, H.C., *Tetrahedron Lett.*, 1989, **30**, 5551.
50. (a) Yoshioka, M., Kawakita, T. and Ohno, M., *Tetrahedron Lett.*, 1989, **30**, 1657; (b) Takahashi, H., Kawakita, T., Yoshioka, M., Kobayashi, S. and Ohno, M., *Tetrahedron Lett.*, 1989, **30**, 7095; (c) Takahashi, H., Kawakita, T., Ohno, M., Yoshioka, M. and Kobayashi, S., *Tetrahedron*, 1992, **48**, 5691; (d) Soai, K., Hivose, Y. and Ohno, Y., *Tetrahedron: Asymmetry*, 1993, **4**, 1473; (e) Ito, K., Kimura, Y., Okamura, H. and Katsuki, T., *Synlett.*, 1992, 573.
51. (a) Knochel, P., Brieden, W., Roezema, M.J. and Eisenberg, C.H., *Tetrahedron Lett.*, 1993, **34**, 5881; (b) Brieden, W., Ostwald, R. and Knochel, P., *Angew. Chem.*, 1993, **105**, 629.
52. Seebach, D., Behrendt, L. and Felix, D., *Angew. Chem., Int. Ed. Engl.*, 1991, **30**, 1008.
53. (a) Knochel, P. and Singer, R.D., *Chem. Rev.*, 1993, **93**, 2117; (b) Langer, F., Waas, J. and Knochel, P., *Tetrahedron Lett.*, 1993, **34**, 5261.
54. (a) Rozema, M.J., Eisenberg, C.H., Lütjens, H., Ostwald, R., Belyk, K. and Knochel, P., *Tetrahedron Lett.*, 1993, **34**, 3115, (b) Knochel, P., Rozema, M.J., Tucker, Ch.E., Retherford, C., Furlong, M. and Achyutha Rao, S., *Pure Appl. Chem.*, 1992, **64**, 361; (c) Rozema, M.J., Achyutha Rao, S. and Knochel, P., *J. Org. Chem.*, 1992, **57**, 1956; (d) Knochel, P. and Achyutha Rao, S., *J. Am. Chem. Soc.*, 1990, **112**, 6164; (e) Knöss, H.P., Furlong, M.T., Rozema, M.J. and Knochel, P., *J. Org. Chem.*, 1991, **56**, 5974.
55. Nugent, W.A., personal communication.

13 Asymmetric synthesis using enzymes and whole cells

N.J. TURNER

13.1 Introduction

In view of their inherent chirality and well-known ability to catalyse reactions with high selectivity, enzymes are increasingly exploited for asymmetric synthesis. In this context, enzyme-catalysed reactions possess a number of attractive features. For example, the reactions usually occur under mild conditions and with attendant selectivity (e.g. chemo-, regio- and stereoselectivity) leading to the synthesis of single isomers. The rate of development of this field is such that, whereas in 1980 much of this methodology was confined to specialised laboratories, nowadays the kinetic resolution of racemic esters using lipases and esterases has become routine. This rapid development has been considerably aided by the appearance of a number of excellent books[1] and reviews[2] and, for the practical work, an organic synthesis-style laboratory manual containing validated experiments.[3]

The aim of this chapter is to give an overview of the various types of biotransformation that are available to the synthetic organic chemist. Despite the diverse range of enzyme-catalysed reactions that have been documented, in practice only certain types of transformation have emerged as synthetically useful. Attention will therefore be focused on the following reactions:

- hydrolysis
- reduction
- oxidation
- formation of C–C bonds

13.2 Hydrolysis

The most widely used biocatalytic reactions are those that involve the hydrolysis of an organic substrate, usually an ester or an amide, although attention is now being directed towards other functional groups (e.g. nitriles, epoxides and glycosides). The two main reasons for the successful emergence of this methodology are (i) the ease of carrying out the reactions, no cofactors are required; and (ii) the commercial availability of many of the required catalysts. For the hydrolysis of esters, the general rule governing the selection of either a lipase or an esterase is given in Figure 13.1, although there are

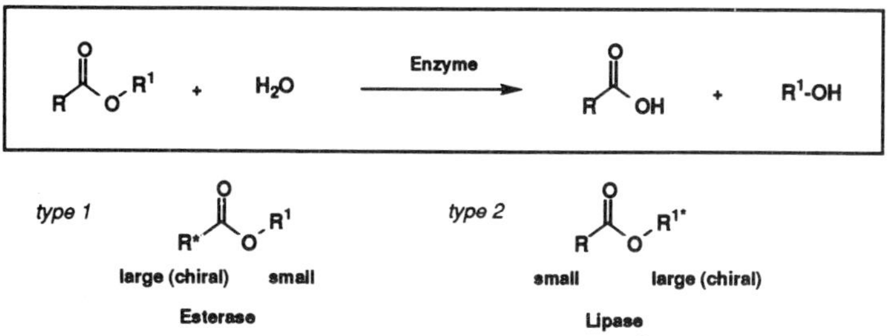

Figure 13.1 Lipase/esterase-catalysed hydrolysis of esters.

many examples now in the literature of exceptions in which esterases act on type 1 esters normally associated with lipases and *vice versa.* The majority of the cited examples involve the kinetic resolution of racemic substrates, although there is increasing emphasis on the development of asymmetric processes using either prochiral or *meso* substrates (Figure 13.2).

With respect to the particular choice of lipase or esterase, it is now well established that certain hydrolase enzymes have a wide substrate specificity, i.e. pig-liver esterase, porcine pancreatic lipase, *Pseudomonas* spp. lipase, *Candida cylindracea* lipase and *Mucor miehei* lipase. These five enzymes are, therefore, usually screened initially, although if there is no success it is straightforward to widen the screen to other commercially available hydrolases (Table 13.1). Increasingly, it is becoming possible to predict the absolute stereochemistry of the derived chiral carboxylic acid or alcohol either by the

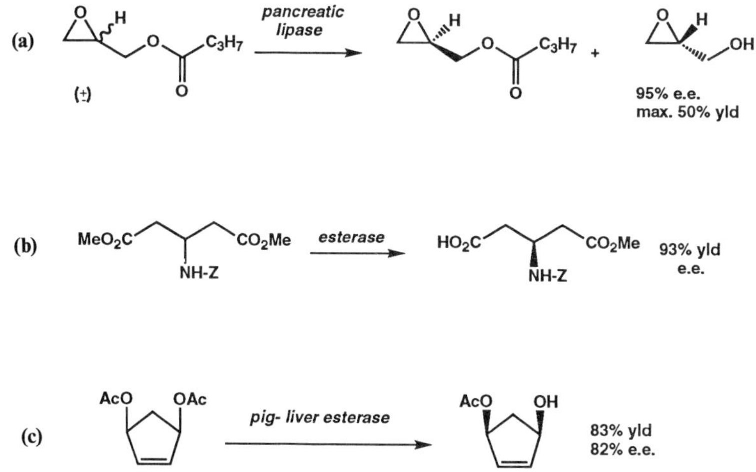

Figure 13.2 Strategies for using hydrolases to prepare chiral intermediates. (a) Kinetic resolution.[4] (b) Prochiral subtrates.[5] (c) *Meso* substrates.[6]

Table 13.1 Esterases, lipases, and proteases that are useful in synthetic reactions

Source of enzyme	Supplier (trade name®)
Esterases	
Pig-liver and horse-liver esterase (ple and hle)	Sigma, Fluka (immobilised on Eupergit C)
Electric eel cholinesterase	Sigma, Fluka
Lipases	
Candida cylindracea	Sigma, Amano (OF-360), Fluka
Chromobacterium viscosum	Sigma, Toyo Jozo
Geotrichum candidum	Sigma, Amano (Amano GC-4)
Mucor javanicus	Amano (Amano-M-10), Fluka
Mucor miehei	Novo (Lipozyme-immobilised)
Porcine pancreas (ppl)	Sigma, Amano, Boehringer
Pseudomonas spp.	Sigma, Boehringer
Pseudomonas fluorescens	Amano (Amano-P), Fluka (SAM-2)
Rhizopus arrhizus	Sigma, Boehringer, Fluka
Rhizopus delemar	Sigma, Amano, Tanabe seiyaku
Rhizopus japonicus	Amano, Osaka Salken Lab. (Lusepase)
Rhizopus javanicus	Amano (Amano F-AP-15)
Rhizopus niveus	Osaka Salken Lab. (Newlase), Fluka
Proteases	
Hog-kidney acylase 1	Sigma, Fluka
Aspergillus acylase 1	Sigma, Amano, Fluka (Plexazym AC—immobilised)
α-Chymotrypsin	Sigma, Fluka, Boehringer
Subtilisin	Sigma, Fluka, Boehringer
Papain	Sigma, Fluka, Boehringer
Thermolysin	Sigma, Fluka, Boehringer
Carboxypeptidase A	Sigma

use of empirical models[7] or, more recently, by a knowledge of the 3-dimensional structure of the enzyme.[8]

In certain cases it is possible to modify the basic kinetic resolution reaction so that the unreactive enantiomer of the substrate undergoes epimerisation *in situ* to replenish the reactive enantiomer, thereby leading to conversions greater than 50% and hence avoiding the wastage of large quantities of the unwanted isomer. Such dynamic resolutions are relatively rare but have found important application in the synthesis of α-amino acids, as illustrated by the synthesis of L-(*S*)-*t*-leucine (**1**) (Scheme 13.1).[9]

Other hydrolytic enzymes have been widely used, especially those that catalyse the hydrolysis of amides (acylases, proteases, peptidases) or lactams (lactamases). An important example of the second class of biocatalysts is the

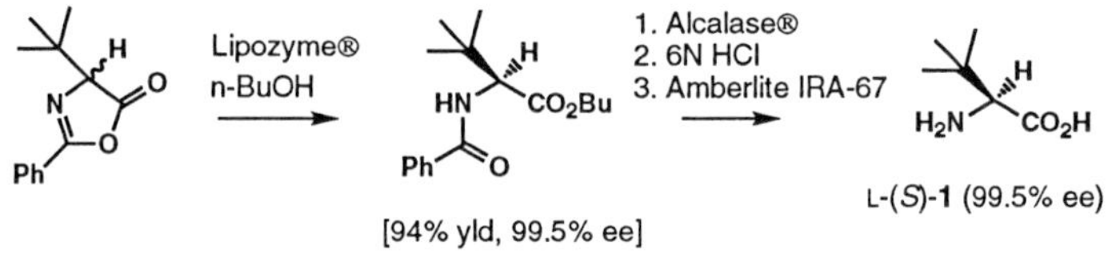

Lipozyme® n-BuOH

[94% yld, 99.5% ee]

1. Alcalase®
2. 6N HCl
3. Amberlite IRA-67

L-(*S*)-**1** (99.5% ee)

Scheme 13.1

hydrolysis of the racemic bicyclic lactam (**2**), shown in Scheme 13.2, using a whole-cell system (*Rhodococcus equi*) which contains the requisite γ-lactamase. The optically pure γ-amino acid (**3**) that is derived is a valuable synthon for the synthesis of carbocyclic nucleosides.[10]

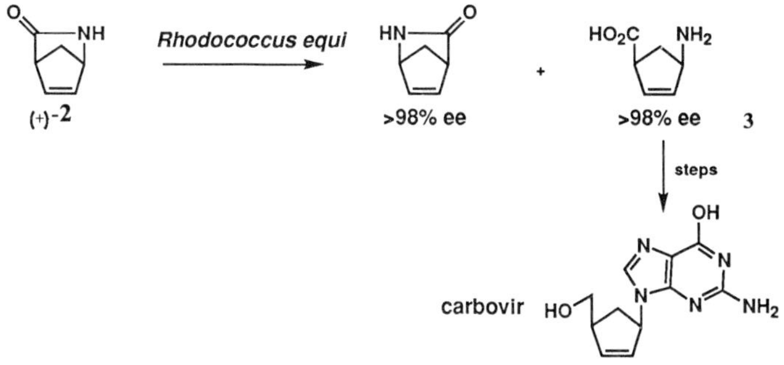

Scheme 13.2

In 1985, Klibanov demonstrated that it was possible to use hydrolytic enzymes under low water conditions to catalyse the *formation* of esters.[11] This approach has become very popular, to the extent that it is often the method of choice for the resolution of alcohols.[12] Moreover, it has been shown that other hydrolytic enzymes can also function well under low water conditions, e.g. the DSM-Toyo Soda synthesis of aspartame (**4**) which utilises the protease thermolysin for construction of the peptide bond (Scheme 13.3).

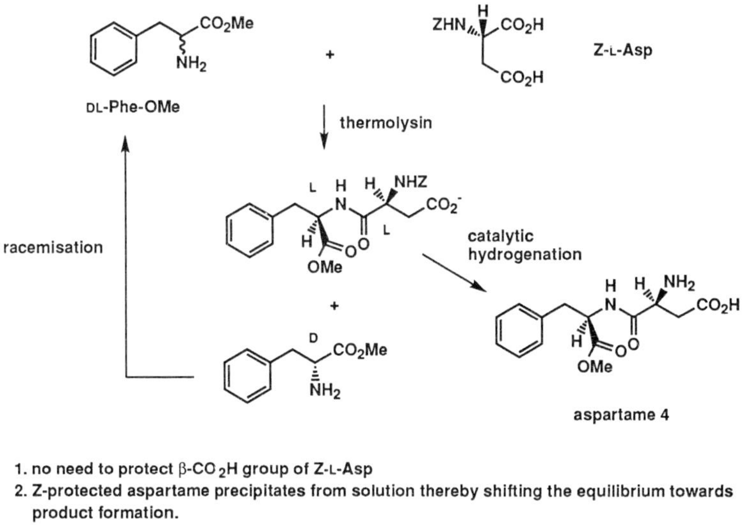

1. no need to protect β-CO_2H group of Z-L-Asp
2. Z-protected aspartame precipitates from solution thereby shifting the equilibrium towards product formation.

Scheme 13.3

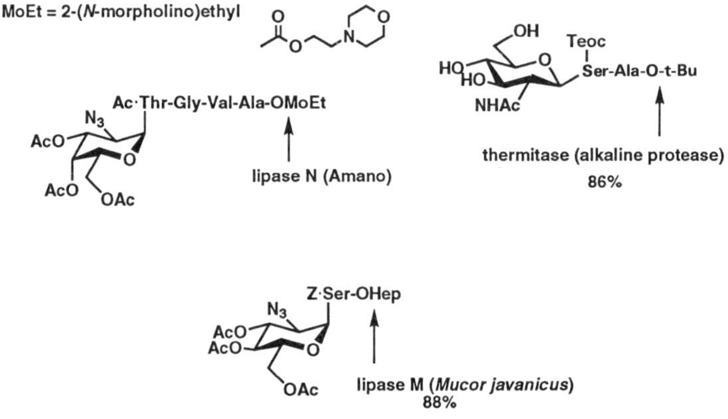

MoEt = 2-(N-morpholino)ethyl

Ac·Thr-Gly-Val-Ala-OMoEt

lipase N (Amano)

Teoc Ser-Ala-O-t-Bu

thermitase (alkaline protease) 86%

Z·Ser-OHep

lipase M (*Mucor javanicus*) 88%

Figure 13.3 Lipase-catalysed deprotection of glycopeptides.

Since the application of hydrolytic enzymes is now firmly established, attention is naturally turning to more difficult systems, e.g. organometallic substrates,[13] sterically hindered substrates[14] and compounds containing acid- or base-sensitive functionality, e.g. glycopeptides (Figure 13.3).[15]

Aside from lipases, esterases and proteases, other hydrolytic systems have been less well explored. Recently a number of groups have demonstrated that the hydrolysis of nitriles to amides and/or carboxylic acids can be carried out under mild conditions; in certain circumstances the transformations are stereo- and regioselective.[16] The reactions are best carried out using whole-cell systems (e.g. *Rhodococcus* spp.) that contain the necessary nitrile-hydrolysing enzymes.

Another potentially useful transformation is the asymmetric hydrolysis of epoxides using microbial systems (Scheme 13.4).[17] This reaction has been known for some time, although the only available source of enzymes was

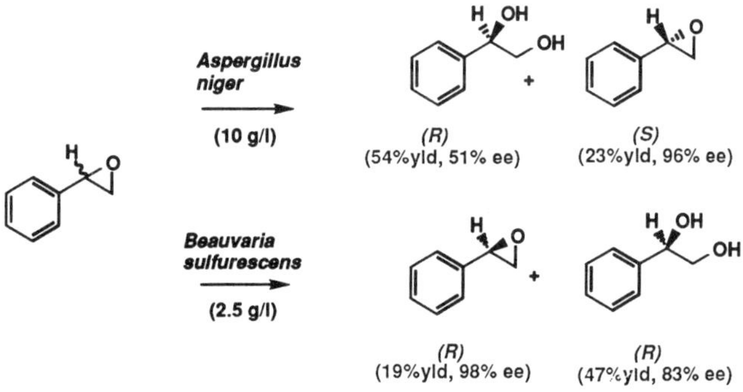

Aspergillus niger (10 g/l)

(R) (54%yld, 51% ee) + (S) (23%yld, 96% ee)

Beauvaria sulfurescens (2.5 g/l)

(R) (19%yld, 98% ee) + (R) (47%yld, 83% ee)

Scheme 13.4

mammalian tissues (e.g. rat liver microsomes). The initial results indicate that the hydrolysis can either proceed via a 'retention pathway' by the use of *Aspergillus niger* or by an inversion pathway using *Beauvaria sulfurescens*, in which hydrolysis of the reactive epoxide occurs at the benzylic position thereby leading to products of the same configuration (Scheme 13.4).

13.3 Reduction

The asymmetric reduction of prochiral functional groups is an extremely useful transformation and has been extensively investigated using biocatalytic methods. Figure 13.4 summarises the variety of groups that can be subjected to reduction.

In contrast to enzyme-catalysed hydrolytic reactions, in biocatalytic reductions there is an important difference between using isolated enzymes and whole cells for the transformation, in that recycling of the essential nicotinamide adenine dinucleotide (phosphate) [NAD(P)H] cofactor must be considered. For the use of isolated enzymes, a cofactor-recycling system must be introduced, thereby allowing the addition of only catalytic (5 mol.%) amounts of NAD(P)H. With whole cells, cofactor recycling is achieved automatically, thereby overcoming one problem, although additional problems

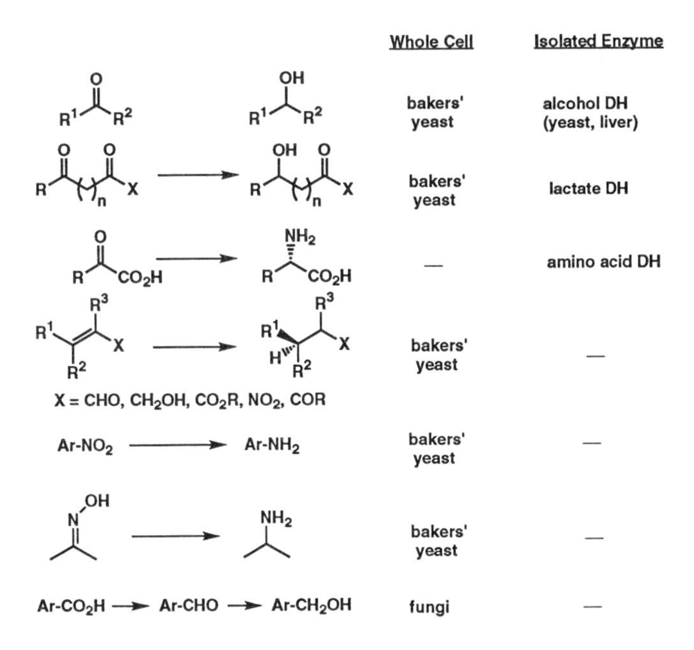

Figure 13.4 Overview of biocatalytic reductions.

Table 13.2 Comparison of whole cells versus isolated enzymes for bioreductions

	Advantages	Disadvantages
Whole cells		
	* no cofactor recycling necessary	* low substrate concentration * high biomass:substrate * side reactions
fermenting	* high activities	* large biomass, more bi-products
non-fermenting	* easier work-up	* lower activities
immobilised	* purification of products easier	* lower activities
Isolated enzymes		
	* easy to use * simple work-up * higher subtrate concentration	* cofactor recycling necessary
immobilised	* enzyme recovery easy	* loss of activity during immobilisation

are introduced, namely the need to use lower substrate concentrations and difficulties encountered with recovery of the products (Table 13.2). Both isolated enzymes and whole cell systems can be easily immobilised thereby facilitating the separation of the substrates and products from the biocatalyst.

Reduction of simple ketones to optically active secondary alcohols is best carried out with semi-purified alcohol dehydrogenases (ADHs). Several ADHs are commercially available (e.g. yeast, horse-liver, *Thermoanaerobium brockii*) and the choice of the particular ADH depends upon the steric requirements of the substrate (Figure 13.5). Thus sterically hindered ketones are best reduced using hydroxysteroid dehydrogenase whereas methyl ketones are efficiently reduced with high e.e. using *Thermoanaerobium brockii* ADH, as illustrated by Scheme 13.5, which shows the method applied to the synthesis of (+)-(*S*)-sulcatol.[18] It should be noted that the (*R*)-isomer of sulcatol can be obtained by using a complementary whole cell reduction with *Aspergillus niger*.

The reduction of β-keto esters to the corresponding β-hydroxy esters has been widely investigated and successfully achieved using yeasts (especially *Saccharomyces cerevisiae* — baker's yeast). The reaction is of wide scope and can be used with a range of β-keto esters substituted in either the α- or

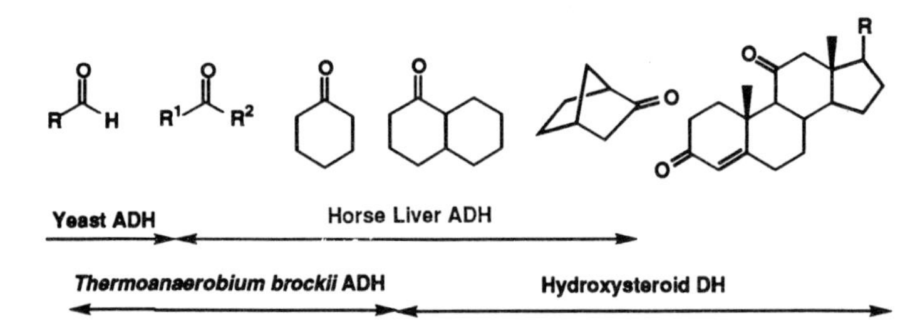

Figure 13.5 Preferred substrate size for dehydrogenases.

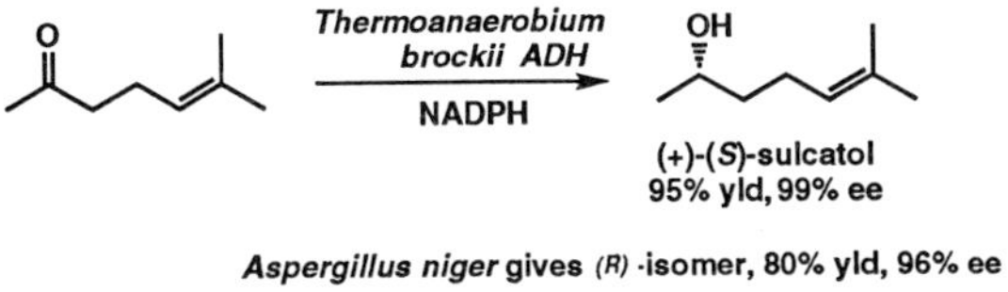

Thermoanaerobium brockii ADH

NADPH

OH

(+)-(*S*)-sulcatol
95% yld, 99% ee

Aspergillus niger gives (*R*)-isomer, 80% yld, 96% ee

Scheme 13.5

γ-positions, e.g. for the synthesis of L-carnitine (**5**) (Scheme 13.6)[19] In comparison with the use of isolated enzymes, the yeast-mediated reductions are simpler to perform, requiring only the addition of tap water and freeze-dried yeast to the substrate. However, they suffer from the disadvantage that lower substrate concentrations must be used (e.g. $< 5\,\mathrm{g\,l^{-1}}$) and that the work up is generally more complicated because of the presence of the whole cells.

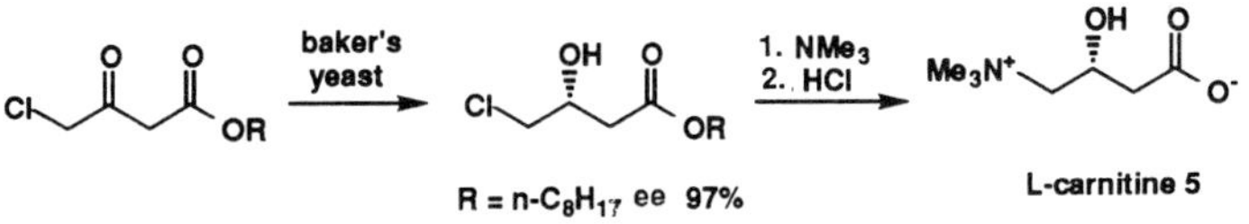

baker's yeast

1. NMe_3
2. HCl

$R = n\text{-}C_8H_{17}$ ee 97%

L-carnitine **5**

Scheme 13.6

This reaction is not limited to acyclic β-keto esters but also has been shown to work well with a variety of cyclic compounds.[20] The facial selectivity of the reduction of both isolated ketones and β-keto esters can be reliably determined by the application of Prelog's rule, first proposed in the early 1960s (Table 13.3.).[21]

Other carbonyl-based reductions include the conversion of α-keto acids to either α-hydroxy acids using lactate dehydrogenase[22] or α-amino acids using amino acid dehydrogenases.[23] An example of the latter reaction is the synthesis of L-t-leucine (**1**) by Degussa, which incorporates large-scale recycling of the NADH cofactor (Scheme 13.7).

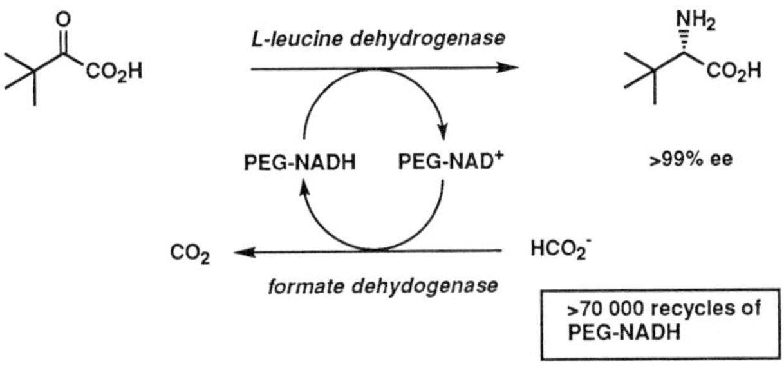

L-leucine dehydrogenase

PEG-NADH PEG-NAD$^+$

CO_2 HCO_2^-

formate dehydogenase

>99% ee

>70 000 recycles of PEG-NADH

Scheme 13.7

Table 13.3 Prelog's rule

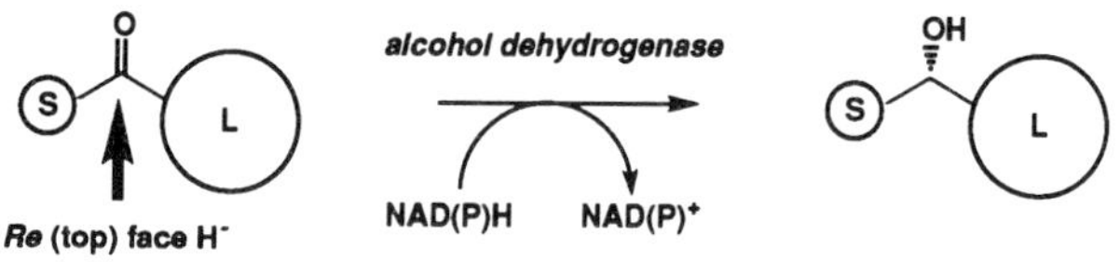

Dehydrogenase	Specificity	Cofactor	Commericial availability
Yeast ADH	Prelog	NADH	+
Horse-liver ADH	Prelog	NADH	+
Thermoanaerobium brockii ADH	Prelog	NADPH	+
Hydroxysteroid dehydrogenase	Prelog	NADH	+
Curvularia falcata ADH	Prelog	NADPH	+
Mucor javanicus ADH	Anti-Prelog	NADPH	–
Lactobacillus kefir ADH	Anti-Prelog	NADPH	–
Pseudomonas sp. ADH	Anti-Prelog	NADH	–

The reduction of carbon–carbon double bonds can only be achieved if one or more electron-withdrawing groups are present to increase the reactivity of the double bond. Treatment of the cyclohexenedione (**6**) with baker's yeast results in regio- and stereoselective reduction of the double bond providing a useful intermediate (**7**) for the synthesis of (3*R*,3′*R*)-zeaxanthin (Scheme 13.8).[24]

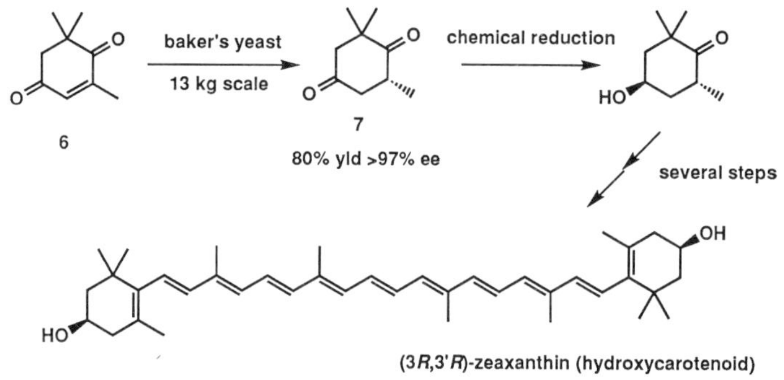

Scheme 13.8

13.4 Oxidation

The use of biocatalysts for the oxidation of organic substrates represents a challenging opportunity. Included in this section are the oxidation of aromatic substrates, Baeyer–Villiger oxidation of cyclic ketones and hydroxylation of alkanes at ostensibly unactivated carbon–hydrogen bonds.

The reaction involving the oxidation of benzene and its derivatives to the corresponding cyclohexadienediols using *Pseudomonas putida* has been known since the 1950s but has only recently been exploited by synthetic organic chemists (Scheme 13.9). The *P. putida* organisms were initially selected on their ability to use benzene as their sole carbon source and were thereafter mutated to prevent further metabolism of the cyclohexadienediols that are produced.

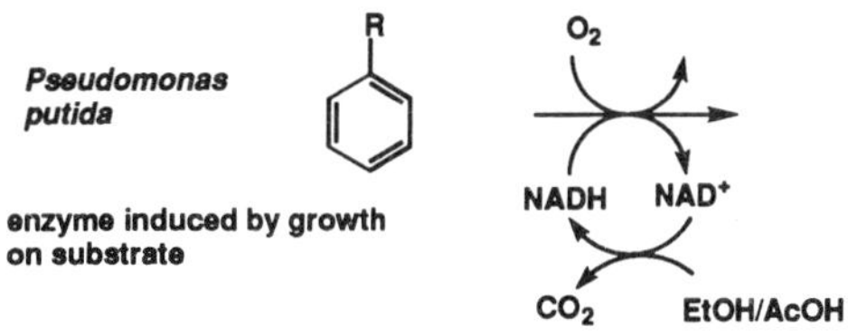

Pseudomonas putida

enzyme induced by growth on substrate

O_2 NADH NAD^+ CO_2 EtOH/AcOH

R OH OH

Scheme 13.9

With the exception of benzene, the corresponding diols are chiral and usually enantiopure. Moreover, they have provided a rich source of chiral starting materials for natural and unnatural product synthesis.[25] The *meso*-diol (**8**) derived from benzene has been elegantly converted into an optically active chiral building block (**9**) according to the route shown in Scheme 13.10 devised by Carl Johnson.[26] This intermediate (**9**) plays a key role in a synthesis of (+)-conduritol C.

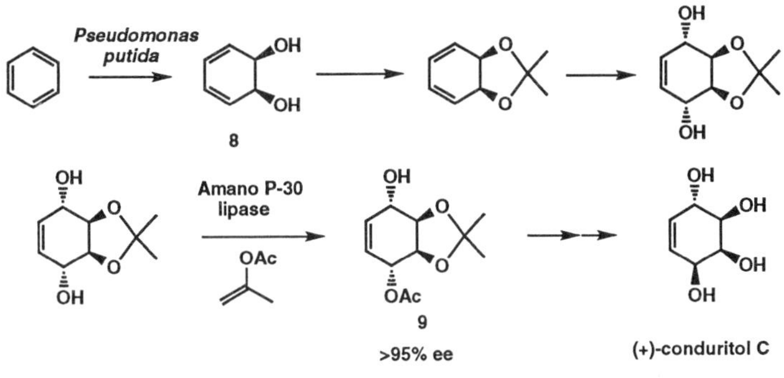

Pseudomonas putida

8

Amano P-30 lipase

OAc

9

>95% ee

(+)-conduritol C

Scheme 13.10

An extensive amount of work has been carried out on the Baeyer–Villiger type oxidation of cyclic ketones to lactones using monooxygenase enzymes. In the early work, it was only possible to use whole cell systems as shown in Scheme 13.11 for the Baeyer–Villiger oxidation of bicyclic racemic ketones. Recently, isolated enzymes have been demonstrated to work with the appropriate cofactor recycling systems.[27,28]

Finally in this section, the hydroxylation of unactivated C–H bonds will

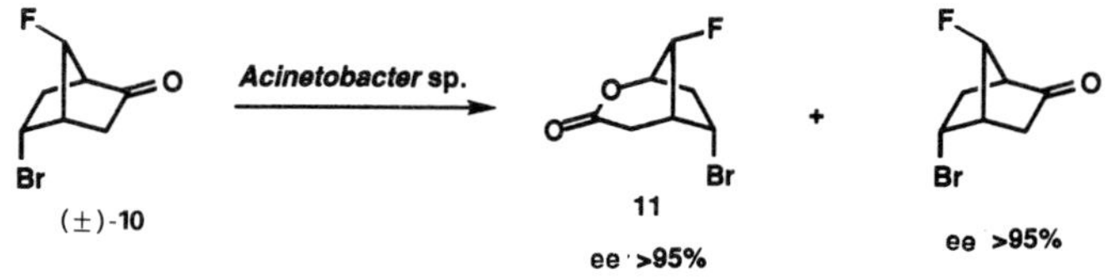

Scheme 13.11

be discussed. This biocatalytic reaction has many obvious attractions, not least of which is the fact that the equivalent reaction is difficult to achieve selectively in the absence of an enzyme. The example shown in Scheme 13.12 illustrates the combination of regio- and enantioselectivity that can be achieved, in this case by the use of *Beauvaria sulfurescens*.[29]

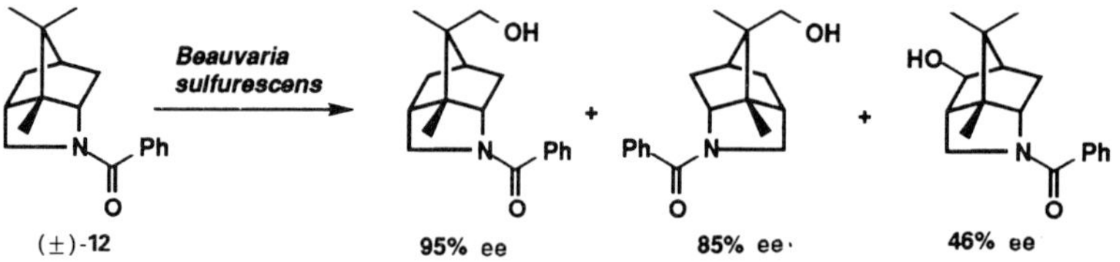

Scheme 13.12

13.5 Formation of C–C bonds

A number of enzymatic reactions have been described for the asymmetric construction of carbon–carbon bonds. The enzyme oxynitrilase, isolated either from almonds (*R*-specific)[30] or a microorganism (*S*-specific),[31] catalyses the enantioselective addition of cyanide ion to a range of aromatic aliphatic aldehydes (Scheme 13.13), generating cyanohydrins with ee values up to 99%. Two features of this reaction are noteworthy: (i) it is necessary to add ethyl acetate as a co-solvent to the reaction in order to suppress the non-enzyme-catalysed reaction which leads to a reduction in the ee value, and (ii) in order to avoid using potassium or sodium cyanide, acetone cyanohydrin can be added as a cyanide donor.[32]

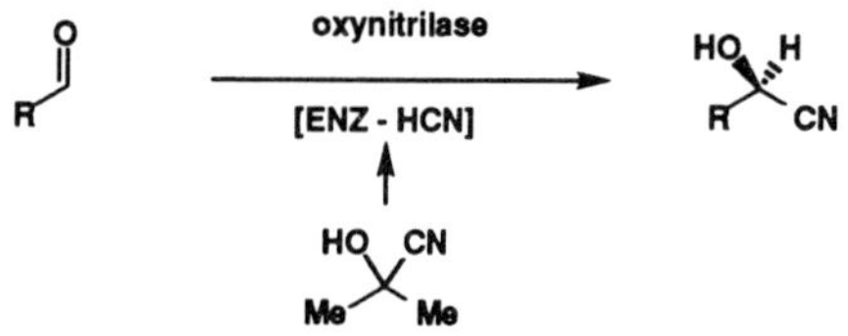

Scheme 13.13

The asymmetric addition of two carbon atoms has been achieved by the use of yeasts which are able to carry out acyloin-type condensations (Scheme 13.14). Typically the addition of an α,β-unsaturated aldehyde to fermenting yeast containing ethanol leads to a two carbon extension by incorporation of ethanol.[33]

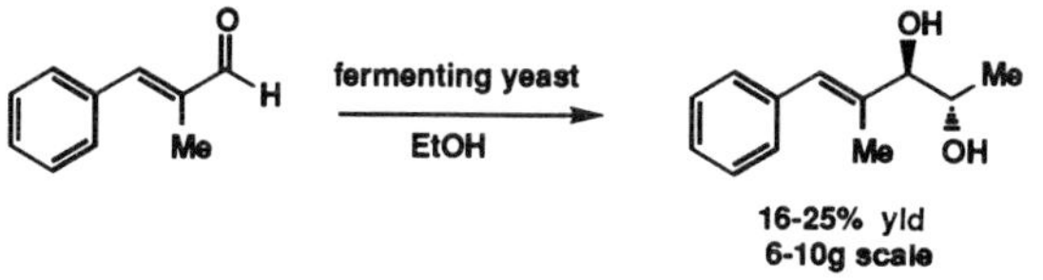

fermenting yeast
EtOH
16-25% yld
6-10g scale

Scheme 13.14

The majority of work in the area of asymmetric C–C bond synthesis has involved the use of aldolases and related enzymes. Fructose-1,6-bisphosphate aldolase from rabbit muscle has been shown to catalyse the condensation of dihydroxyacetone phosphate with a wide range of structurally different aldehydes leading, after dephosphorylation with phosphatase, to enantiopure triols with *threo*-stereochemistry (Scheme 13.15).[34] Such triols have proved to be efficient building blocks for the synthesis of polyhydroxylated natural products, e.g. deoxynojirimycin (**13**) (Scheme 13.15).[35]

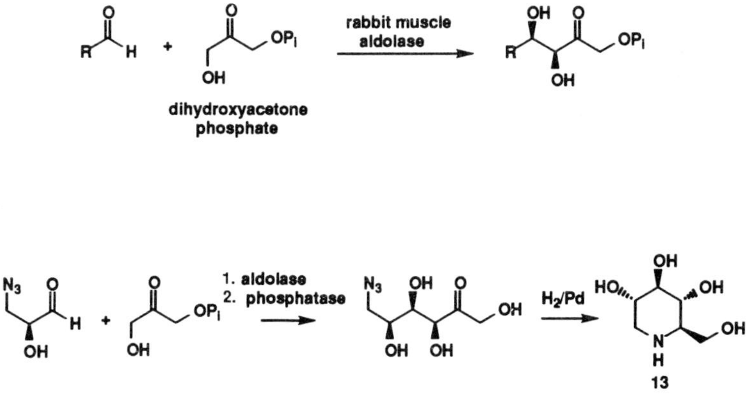

rabbit muscle aldolase
dihydroxyacetone phosphate

1. aldolase
2. phosphatase
H_2/Pd
13

Scheme 13.15

Aldolases that generate all three other possible combinations of stereochemistry have also been described, although less is known about their substrate specificities (Scheme 13.16).[36]

A related enzyme, transketolase, also utilises aldehydes but has the advantage that the cosubstrate hydroxypyruvate does not need to be

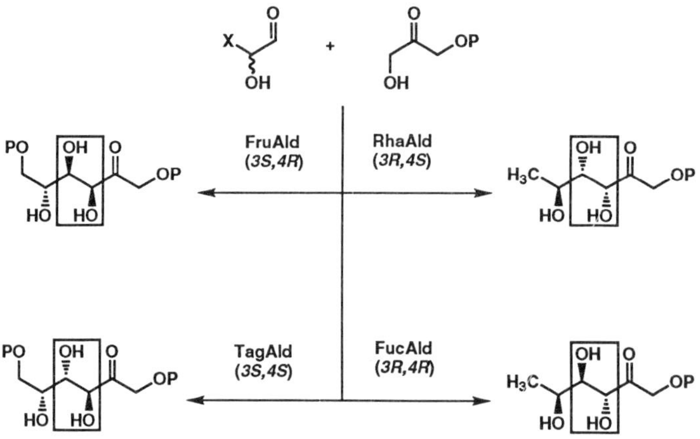

FruAld (3S,4R)
RhaAld (3R,4S)
TagAld (3S,4S)
FucAld (3R,4R)

Scheme 13.16

phosphorylated (Scheme 13.17). The enzyme is available from yeast[37] or spinach[38] and has recently been over-expressed at a high level in *Escherichia coli*.[39] The release of carbon dioxide ensures that the reaction proceeds to equilibrium.

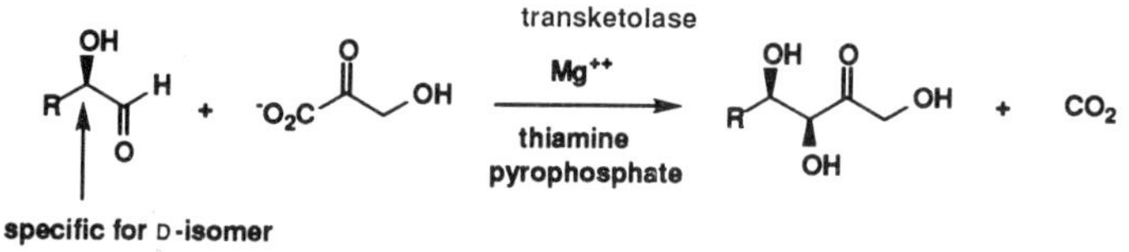

transketolase
Mg^{++}
thiamine pyrophosphate
specific for D-isomer

Scheme 13.17

The enzyme *N*-acetylneuraminic acid (sialic acid) aldolase has been successfully used for the synthesis of a wide range of analogues of sialic acid (**14**) that are difficult to prepare via alternative procedures (Scheme 13.18).[40] Neuraminic acid (**14**) is an important constituent of the oligosaccharide component of glycoproteins.

13.6 Conclusions

It can be seen from the work presented in this chapter that the use of certain enzymes (lipases, esterases, proteases, yeasts for reduction) for asymmetric synthesis has become well established and in many cases the reactions are so straightforward to perform that their application is commonplace. Other biotransformations, namely oxidations, carbon–carbon bond synthesis and certain reductive processes, are the subject of intense investigation and are

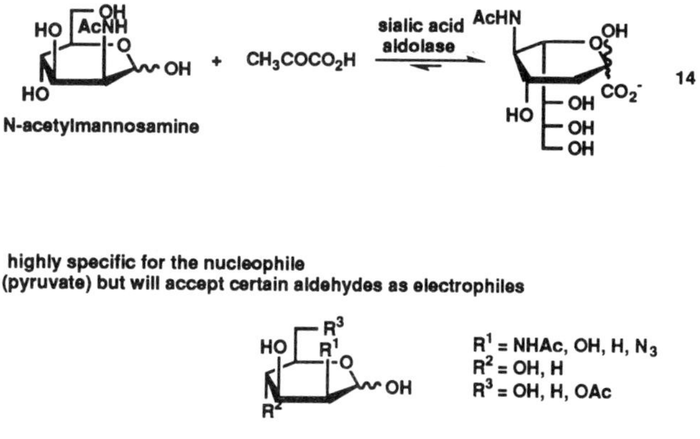

N-acetylmannosamine + CH_3COCO_2H ⇌ (sialic acid aldolase) 14

highly specific for the nucleophile
(pyruvate) but will accept certain aldehydes as electrophiles

R^1 = NHAc, OH, H, N_3
R^2 = OH, H
R^3 = OH, H, OAc

Scheme 13.18

undergoing rapid advances such that they will undoubtedly be in a 'user-friendly' format in the near future. The techniques of molecular biology that allow the almost routine over-expression of a protein in a host organism are contributing greatly to the advances being made, as are other important disciplines, such as biochemical engineering and X-ray crystallography. In combination with organic synthesis and microbiology, the ultimate aim is both to increase the range of readily available biocatalysts and to have a greater understanding of the selectivity that they exhibit.

References

1. Faber, K., *Biotransformations in Organic Chemistry*, Springer Verlag, 1992; Wong, C.-H. and Whitesides, G.M., *Enzymes in Synthetic Organic Chemistry*, Pergamon, 1994.
2. Turner, N.J., *Nat. Prod. Rep.*, 1994, **11**, 1; Santaniello, E., Ferraboschi, F., Grisenti, P. and Manzocchi, A., *Chem. Rev.*, 1992, **92**, 1071.
3. Roberts, S.M. (ed.), Wiggins, K., Casey, G. and Phythian, S.J. (assoc. eds), *Preparative Biotransformations; Whole Cell and Isolated Enzymes in Organic Synthesis*, Wiley, Chichester, 1993.
4. Ladner, W.E. and Whitesides, G.M., *J. Am. Chem. Soc.*, 1984, **106**, 7250.
5. Ohno, M., Kobayashi, S., Limori, T., Wang, Y.-F. and Izawa, T., *J. Am. Chem. Soc.*, 1981, **103**, 2405.
6. Wang, Y.-F., Chen, C.-S., Girdaukas, G. and Sih, C.J., *J. Am. Chem. Soc.*, 1984, **106**, 3695.
7. Hultin, P.G. and Jones, J.B., *Tetrahedron Lett.*, 1992, **33**, 1399; Provencher, I. and Jones, J.B., *J. Org. Chem.*, 1994, **59**, 2729.
8. Cygler, M., Grochulski, P., Kazlauskas, R.J., Schrag, J.D., Bouthillier, F., Rubin, B., Serreqi, A.N. and Gupta, A.K., *J. Am. Chem. Soc.*, 1994, **116**, 3180.
9. Turner, N.J., Winterman, J.R., McCague, R., Parratt, J.S. and Taylor, S.J.C., *Tetrahedron Lett.*, 1995, **36**, 1113.
10. Taylor, S.J.C., Sutherland, A.G., Lee, C., Wisdom, R., Thomas, S., Roberts, S.M. and Evans, C., *J. Chem. Soc., Chem. Commun.*, 1990, 1120.
11. Kirchner, G., Scollar, M.P. and Klibanov, A.M., *J. Am. Chem. Soc.*, 1985, **107**, 7072.
12. Weidner, J., Theil, F. and Schick, H., *Tetrahedron: Asymmetry*, 1994, **5**, 751.
13. Bergbreiter, D.E. and Momongan, M., *Appl. Biochem. Biotechnol.*, 1992, **32**, 55.
14. Holdgrun, X.K. and Sih, C.J., *Tetrahedron Lett.*, 1991, **32**, 3465.

15. Braum, G., Braun, P., Kowalczyk, D. and Kunz, H., *Tetrahedron Lett.*, 1993, **34**, 3111; Braun, P., Waldmann, H. and Kunz, H., *Synlett*, 1992, 39.
16. Crosby, J., Moilliet, J., Parratt, J.S. and Turner, N.J., *J. Chem. Soc., Perkin Trans. 1*, 1994, 1679 and references cited therein.
17. Chen, X.J., Archelas, A. and Furstoss, R., *J. Org. Chem.*, 1993, **58**, 5528, 5533; Hechtberger, P., Wirnsberger, G., Mischitz, M., Klempier, N. and Faber, K., *Tetrahedron: Asymmetry*, 1993, **4**, 1161.
18. Belan, A., Bolte, J., Fauve, A., Gourcy, J.G. and Veschambre, H., *J. Org. Chem.*, 1987, **52**, 256.
19. Gopalan, A.S. and Sih, C.J., *Tetrahedron Lett.*, 1984, **25**, 5235.
20. Cooper, J., Gallagher, P.T. and Knight, D.W., *J. Chem. Soc., Chem. Commun.*, 1988, 509.
21. Prelog, V., *Pure Appl. Chem.*, 1964, **9**, 119.
22. Kim, M.-J. and Whitesides, G.M., *J. Am. Chem. Soc.*, 1988, **110**, 2959.
23. Schutte, H., Hummel, W., Tsai, H. and Kula, M.R., *Appl. Microbiol. Biotechnol.*, 1985, **22**, 306.
24. Leuenberger, H.G.W., Boguth, W., Barner, R., Schmid, M. and Zell, R., *Helv. Chim. Acta*, 1976, **59**, 1832.
25. Hudlicky, T., Olivo, H.F. and McKibben, B., *J. Am. Chem. Soc.*, 1994, **116**, 5108.
26. Johnson, C.R., Ple, P.A. and Adams, J.P., *J. Chem. Soc., Chem Commun.*, 1991, 1006.
27. Roberts, S.M. and Willetts, A.J., *Chirality*, 1993, **5**, 334.
28. Levitt, M.S., Newton, R.F., Roberts, S.M. and Willetts, A.J., *J. Chem. Soc., Chem. Commun.*, 1990, 619.
29. Archelas, A., Fourneron, J.D. and Furstoss, R., *J. Org. Chem.*, 1988, **53**, 1797.
30. Effenberger, F., *Angew. Chem., Int. Ed. Engl.*, 1987, **26**, 458.
31. Klempier, N., Griengl, H. and Hayn, M., *Tetrahedron Lett.*, 1993, **34**, 4769.
32. Ognyanov, V.I., Datcheva, V.K. and Kyler, K.S., *J. Am. Chem. Soc.*, 1991, **113**, 6992.
33. Fuganti, C., Grasselli, P., Servi, S., Spreafico, F., Zirotti, C. and Casati, P., *J. Org. Chem.*, 1984, **49**, 4087.
34. Bednarski, M.D., Simon, E.S., Bishofberger, N., Fessner, W.D., Kim, M.-J., Lees, W., Saito, T., Waldmann, H. and Whitesides, G.M., *J. Am. Chem. Soc.*, 1989, **111**, 62.
35. van der Osten, C.H., Sinskey, A.J., Barbas, C.F., Pederson, R.L., Wang, Y.F. and Wong, C.-H., *J. Am. Chem. Soc.*, 1989, **111**, 3924; Turner, N.J. and Whitesides, G.M., *J. Am. Chem. Soc.*, 1989, **111**, 624.
36. Fessner, W.D., Sinerius, G., Schneider, A., Dreyer, M., Sculz, G.E., Badia, J. and Aguilar, J., *Angew. Chem., Int. Ed. Engl.*, 1991, **30**, 555; Henderson, I., Wong, C.-H., and Sharpless, K.B., *J. Am. Chem. Soc.*, 1994, **116**, 558.
37. Ziegler, T., Straub, A. and Effenberger, F., *Angew. Chem., Int. Ed. Engl.*, 1988, **27**, 716; Myles, D.C., Andrulis, P.J. and Whitesides, G.M., *Tetrahedron Lett.*, 1991, **32**, 4835.
38. Valentin, M.L. and Bolte, J., *Tetrahedron Lett.*, 1993, **34**, 8103.
39. Hobbs, G.R., Lilly, M.D., Turner, N.J., Ward, J.M., Willetts, A.J. and Woodley, J.M., *J. Chem. Soc., Perkin Trans. 1*, 1993, 165.
40. Kragl, U., Godde, A., Wandrey, C., Lubin, N. and Augé, C., *J. Chem. Soc., Perkin Trans. 1*, 1994, 909.

14 Asymmetric palladium-catalysed coupling reactions

G.R. STEPHENSON

14.1 Introduction

Organometallic chemistry offers special effects in bond-forming reactions. Of all the new metal-mediated processes to be applied in synthesis, it is palladium-catalysed coupling reactions that have found the most immediate and enthusiastic reception in industrial laboratories. The key to the attraction of this type of chemistry is that two components that would normally be combined can be linked in a single step, often in a highly selective and efficient way. What is more, in many cases and in contrast to the use of organocuprate, lithium and magnesium methodologies, protection of hydroxylic functionality in the substrates is often unnecessary with palladium-catalysed coupling. There are two main classes of coupling reaction: the Heck coupling reaction and cross-coupling reactions of the types pioneered by Stille (organotin precursors) and Suzuki (organoboron precursors). In 1989, the first examples of enantioselective Heck coupling were described independently by the groups of Shibasaki[1] and Overman[2]. Since then, there has been a rapid development of the asymmetric modification of all classes of palladium-catalysed coupling reaction, providing a synthetically direct and versatile new approach to the enantioselective preparation of intermediates for use in organic synthesis. It is these developments that are the subject of this chapter.

Heck coupling combines substituted vinyl or aryl derivatives (typically bromides, iodides or triflates) with an alkene (typically an α,β-unsaturated carbonyl compound or styrene derivative, though many other types of alkene can also be employed). The two reactants are coupled to form (**1**) without consuming the unsaturation in the alkene portion (Scheme 14.1). Cross-coupling reactions also combine unsaturated reactants in a way that retains the unsaturation in the product. For example, the vinyl arene target molecule (**2**) can be prepared by a cross coupling between an aryl bromide (ArX) and a vinyltin reagent (Scheme 14.2). In terms of disconnections retrosynthesis design, consideration of the preparation of both (**1**) ($\varepsilon = CO_2Me$) and (**2**), for example, indicates that in both cases an aryl bromide and a substituted alkene would be appropriate starting materials.

Schemes 14.1 and 14.2 show the similarity between the disconnections afforded by the two bond-forming processes. Although giving rise to products that are similar in structural type, these two coupling procedures follow different mechanisms. These are illustrated in Schemes 14.3 and 14.4. In both

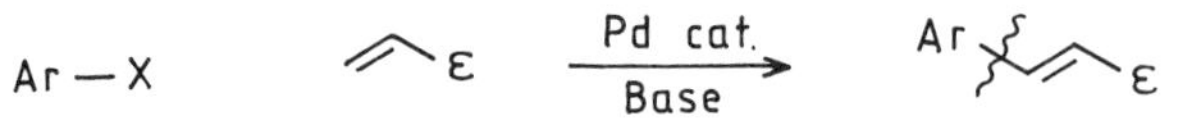

Ar—X

E

Pd cat.

Base

Ar

E

1

Scheme 14.1

Ar—X

Y

Pd cat.

Ar

2

Scheme 14.2

Heck and cross-coupling reactions, vinyl coreactants CH_2=CHX can be used to good effect in place of the aryl-X derivative.

14.1.1 *Mechanism of Heck coupling*

Heck coupling with aryl halides begins with an oxidative addition into the Ar–X bond. In this way, the aryl group becomes attached to the transition metal catalyst. The alkene approaches initially as a side-on η^2 ligand. Carbon–carbon bond formation is achieved by σ-bond migration, displacing the palladium to the far end of the alkene. The metal is lost by β-elimination, so reforming the alkene in the organic product, which is liberated in this step. Heck coupling is typically performed in the presence of the base, so that loss of HX (which is picked up by the base to form, for example, $Et_3NH^+X^-$) completes the catalytic cycle.

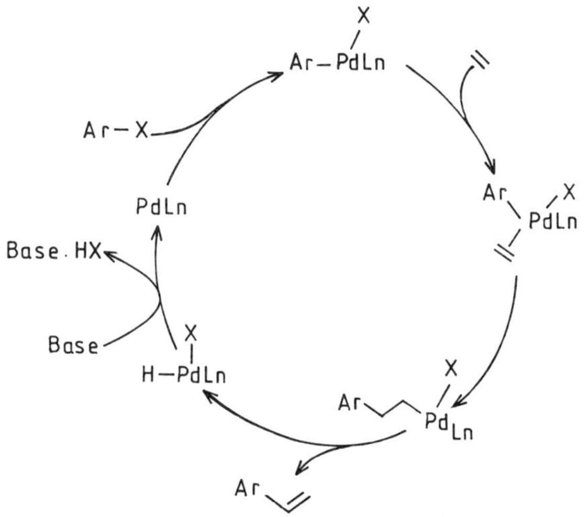

X

Ar–PdLn

Ar–X

PdLn

Ar

X

PdLn

Base. HX

Base

X

H–PdLn

X

Ar

Pd Ln

Ar

Scheme 14.3

14.1.2 Mechanism of cross coupling

Cross coupling begins with an oxidative addition step and so initially resembles the mechanism for the Heck coupling reaction. In the cross-coupling version, however, the coupling partner, although typically unsaturated, contains a further reactive σ-bond, and a second oxidative addition can occur. Because of this sequence of two oxidative additions, catalysts for cross-coupling reactions must be capable of adopting a wide variety of oxidation states at the metal. The catalytic cycle for cross coupling differs in the way the organic product is liberated from its attachment to the transition metal centre. Since both coreactants are attached as σ-bonded ligands, the organic product is formed by reductive elimination. In principle, a further reductive elimination step can close the catalytic cycle, though the products from this step are not normally isolated. A simplified version of this process is shown in Scheme 14.4. In practice, other ligand-exchange steps (e.g. displacement of X^- by HO^- in the case of Suzuki coupling) are often also thought to be involved.

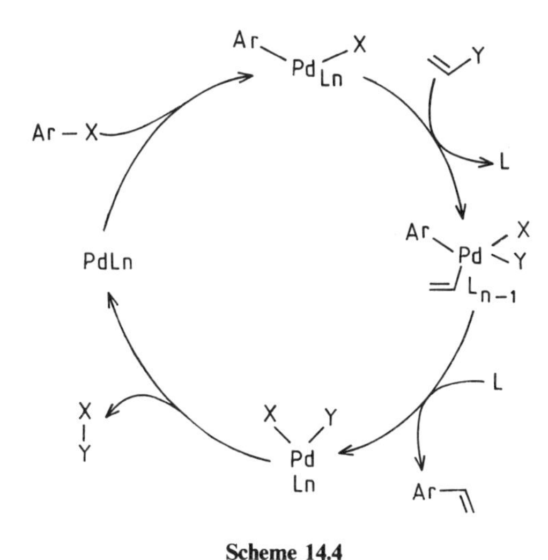

Scheme 14.4

14.2 Enantioselective modification of palladium-catalysed coupling reactions

At first sight, the reaction sequences shown in Schemes 14.3 and 14.4 look unpromising for asymmetric modification, since no new chiral centre is produced. To achieve an enantioselective version of this chemistry, two changes are necessary. An optically pure chiral auxiliary must be introduced, and the reaction pathway must be diverted so that a chiral product can be obtained. Most developments in enantioselective palladium-catalysed coupling reactions have concentrated on the Heck reaction, so this will provide the best illustration of the role of the chiral auxiliary. The chirality properties of side-on alkene ligands (see Chapter 16) make the influence of the chiral auxiliary less easy to understand than in many other asymmetric reactions. This is illustrated in Figures 14.1 and 14.2.

Several strategies have been developed to manipulate the course of the coupling reactions so that chiral products are obtained. Reactions based on either enantioface or enantiotopos differentiation have been developed. These modifications to the normal coupling processes are each considered in turn later in the chapter.

14.2.1 Role of chiral auxiliaries in Heck coupling reactions

Figure 14.1 illustrates the stereochemical properties of an η^2-organopalladium intermediate. When an unsymmetrical alkene binds to a metal centre, the enantiofaces of the alkene are differentiated. The face attached to the metal becomes different to the other face, so that a planar stereogenic unit is formed upon complexation. In the catalytic cycle, σ-bond migration moves the Ar group to the alkene. The reaction takes place within the coordination sphere of the metal, and the aromatic group becomes attached to the face of the alkene which is bound to the metal. In this way, the carbon atom which has received the alkene group has become chiral. The stereochemistry at this centre is thus determined by the face-recognition of the alkene in the original complexation step. When no chiral influence is present, a racemic product mixture is formed via the two enantiomerically related intermediates shown in Figure 14.1. A simple monosubstituted alkene is used in this illustration,

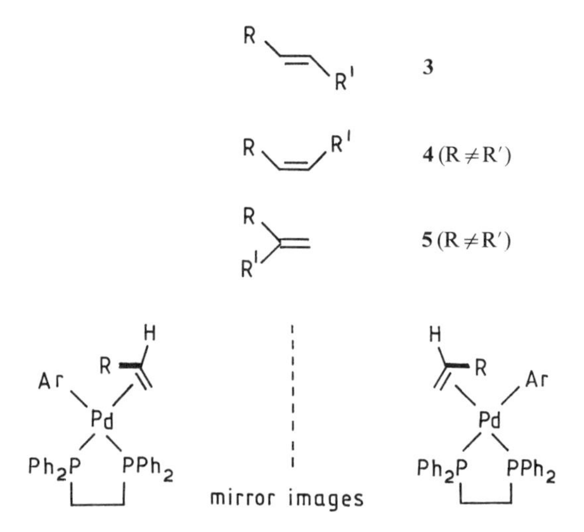

Figure 14.1 Enantiomeric relationship between intermediates in Heck coupling to unsymmetrical alkenes in the absence of a chiral ligand at the metal.

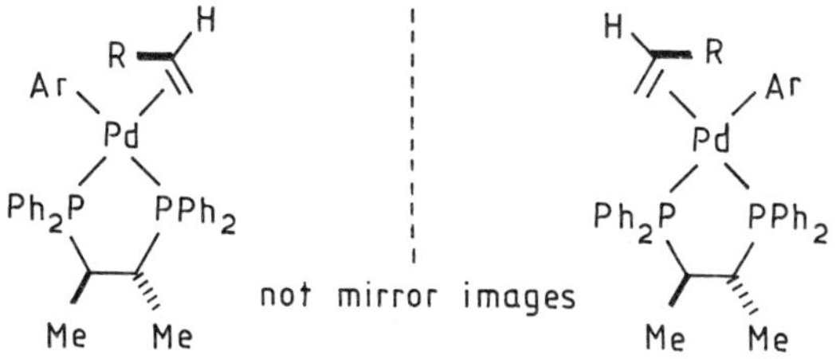

Figure 14.2 Diastereomerically related transition states for Heck coupling in the presence of a C_2 symmetric diphosphine ligand.

but the same arguments hold true for other unsymmetrically substituted prochiral alkenes (see Chapter 16) such as (**3**), (**4**) or (**5**) (R and/or $R^1 \neq H$). In a similar way, the stereochemical properties of more highly substituted alkenes determine the outcome attempts at asymmetric coupling reactions.

When an enantiopure chiral auxiliary is introduced into this structure, the stereochemical relationship between the two substituents is changed. This is illustrated in Figure 14.2 in the case of a C_2 symmetric phosphine analogue of the dppe ligand shown in Figure 14.1. The C_2 symmetric ligand is present in the same form in both structures, so the mirror image relationship between these intermediates has been destroyed. The structures are now two diastereoisomers. In practice, the most widely utilised ligand for the asymmetric modification of Heck coupling is BINAP. This ligand is also C_2 symmetric but the stereogenic unit has the axial chirality of a binaphthyl moiety. Nonetheless, the principles are the same as those illustrated in Figure 14.2.

Aryl transfer to the π-bound alkenes takes place stereoselectively. When the two transition states become diastereomerically related, as in Figure 14.2, the ease of the two σ-bond migration steps can be different, since the preferred conformations of the ligand in the two diastereomeric metal complexes can be different. If one diastereoisomer is better aligned for σ-bond migration than the other, the rate of one reaction will be faster than the other, and so an asymmetric induction will be achieved in the formation of the new chiral centre by the carbon–carbon bond-making step.

Four distinct strategies for the utilisation of this stereodifferentiation will be presented. The objective is to illustrate how these different strategies open up quite different opportunities for application in asymmetric synthesis, and to show how to select a strategy that is well suited for a particular application.

14.3 Strategy 1: enantioface differentiation

A reaction that makes a chiral centre from a flat (prochiral) portion of a molecule will form one or other of the two possible enantiomers, depending on which face (enantioface) of the structure participates in the reaction.

Selective reaction at only one of the enantiofaces affords an enantiomerically pure product. Since Heck coupling re-forms an alkene, this does not usually build a chiral centre. However, β-elimination need not replace the alkene at the location of the original unsaturation in the starting material. This opens the door for asymmetric modification.

The stereochemistry of β-elimination requires that the carbon–metal bond and carbon–hydrogen bond are aligned in a *syn*-planar fashion. With an acyclic ligand, conformational mobility after σ-bond migration allows the hydrogen at the carbon that received the new substituent to become correctly aligned and so transfer to the metal. When the alkene is within a ring, this conformational change cannot occur, but if an adjacent carbon–hydrogen bond is present in the correct orientation, the direction of β-elimination is altered and the alkene is placed at a position in the carbon skeleton different to that in the starting material. As seen in Scheme 14.5, this can form a new chiral centre at the carbon atom which receives the aryl group. In synthesis design, the disconnection shown in Scheme 14.6 reveals the possibility of enantioface differentiation in the functionalisation of the prochiral alkene (**6**).

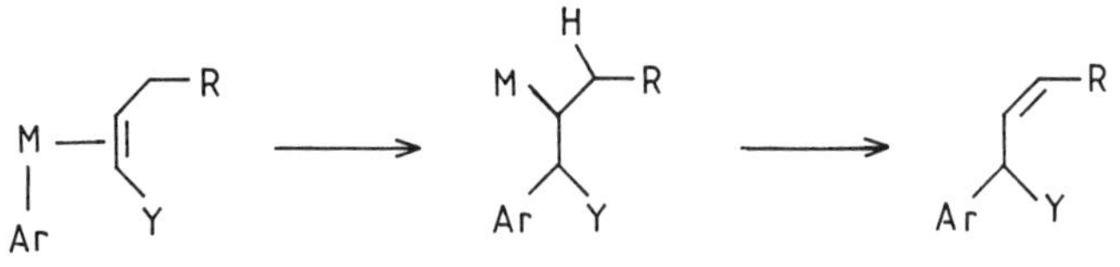

Scheme 14.5

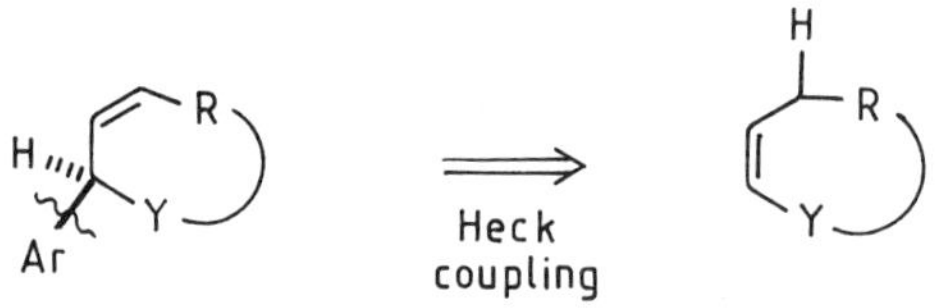

6

Scheme 14.6

14.3.1 The intermolecular case

A simple intermolecular example (Scheme 14.7) is best to illustrate how this principle can be put into practice. The palladium-catalysed arylation of 4*H*-1,3-dioxin is suitably straightforward because the carbon–oxygen bond within the heterocyclic ring prevents further metal-mediated steps, and β-elimination affords only the chiral aryl-substituted structure (**7**).[3] Apart from the inclusion of (*R*)-BINAP, reaction conditions are typical for the Heck coupling process. Asymmetric modification has been examined with both aryl triflate and aryl iodide reagents. Both the nature of the X group and the choice of base and solvent proved important in optimising asymmetric induction, and the best chemical and optical yields were achieved with silver carbonate in DMF. The optical purity of the product was shown to be 43% ee by hydrogenation of the alkene, hydrolysis of the acetal, protection of the primary alcohol as a TBDMS derivative and esterification with (*S*)-(+)-α-methoxyphenyl acetate. In this way, the absolute configuration of the product was also proved.

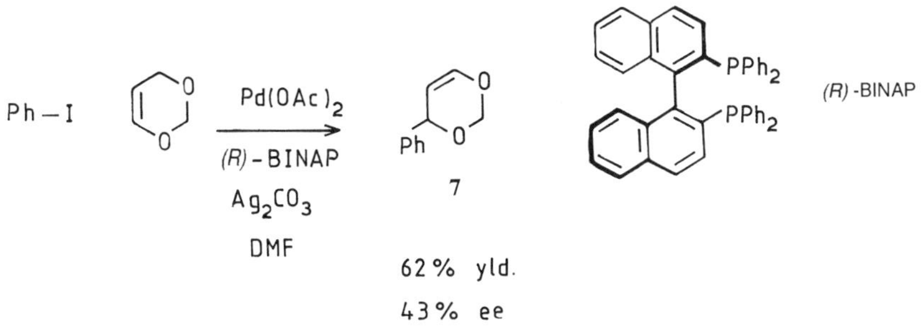

Ph–I
$Pd(OAc)_2$
(*R*)-BINAP
Ag_2CO_3
DMF
Ph
7
62% yld.
43% ee
PPh_2
PPh_2
(*R*)-BINAP

Scheme 14.7

Both the Overman and Shibasaki groups have been examining intermolecular versions of the arylation process. Much of the work in the Overman laboratory has concerned the enantiomeric synthesis of spirooxindoles and related spirocycles. An aryl halide and an α,β-unsaturated amide are the two key functional groups in the coupling reaction. These are linked through an amide bond to a 2-iodoaniline. The result of the palladium-catalysed cyclisation is to form a 5-membered ring.

14.3.2 The intramolecular case

In this study, BINAP has again been used as the chiral auxiliary and reaction conditions are also similar, but a less basic silver salt has been employed in place of the silver carbonate. The choice of base and the strategy for halide removal are crucial variables for the optimisation of reactions of this type. In this case, switching to an amine base (1,2,2,6,6-pentamethylpiperidine, PMP) reverses the enantioselectivity (Scheme 14.8). Thus either (*R*) or (*S*)-enantiomers are available using the same chiral ligand. Values of ee in the range 60–75% are typical, but the best examples were 89–95%.[4]

This methodology has been applied in a catalytic asymmetric synthesis of either enantiomer of physostigmine (**8**). This is based on the disconnection shown in Figure 14.3.

The two key features of this synthesis are the formation of the central ring by palladium-catalysed Heck coupling to a trisubstituted alkene and the

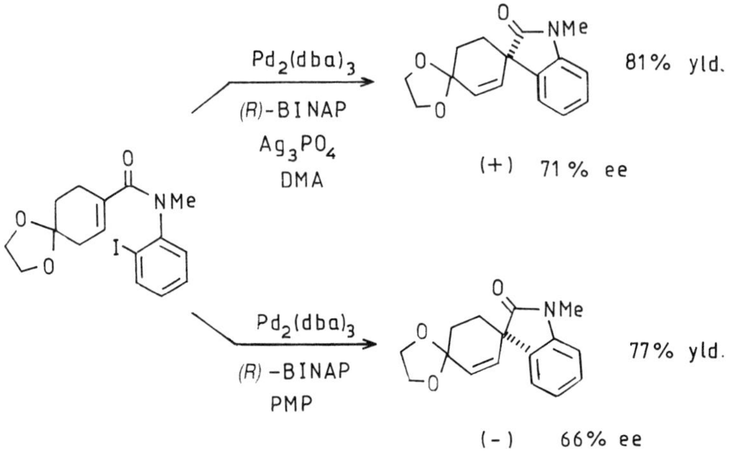

Scheme 14.8

placement of an allylic substituent so that migration of the double bond forms an enol ether, so giving access to aldehyde functionality at the far right-hand portion of the target molecule. In this way, closure of the heterocyclic ring can be achieved. Because of the intramolecular character of the reaction, the σ-bond migration step takes place α not β to the amide carbonyl group, despite this also being the more substituted position on the alkene (Scheme 14.9). The coupling step affording (**9**) was performed with (*S*)-BINAP (and with PMP as the base) to afford a product >95% ee. One recrystallisation achieved an enantiopure product.[5]

In Shibasaki's group, intramolecular Heck coupling with an aryltriflate and a longer spacer has given access to a key intermediate in an efficient synthesis of (−)-eptazocine (**11**).[6] A model study (Scheme 14.10) showed that the stereochemistry of the alkene exerted a considerable influence on the efficiency of asymmetric induction. An (*E*)-starting material gave 51% ee, which was improved to 87% when the (*Z*)-stereoisomer was used. The key steps of the synthesis of eptazocine are shown in Scheme 14.11. Cyclisation

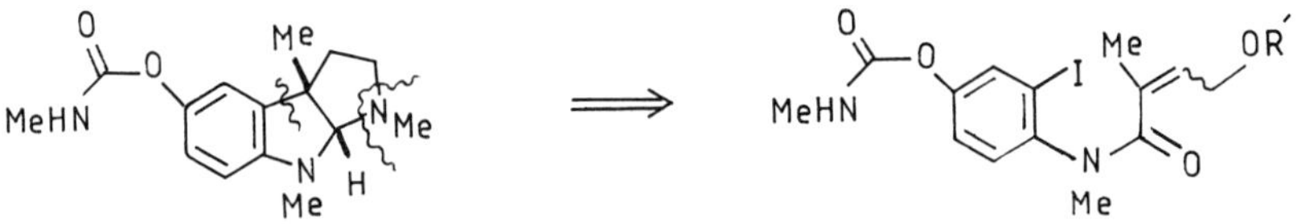

8

Figure 14.3 Retrosynthetic analysis reveals enantioface differentiation in the asymmetric Heck coupling to form the central ring in physostigmine.

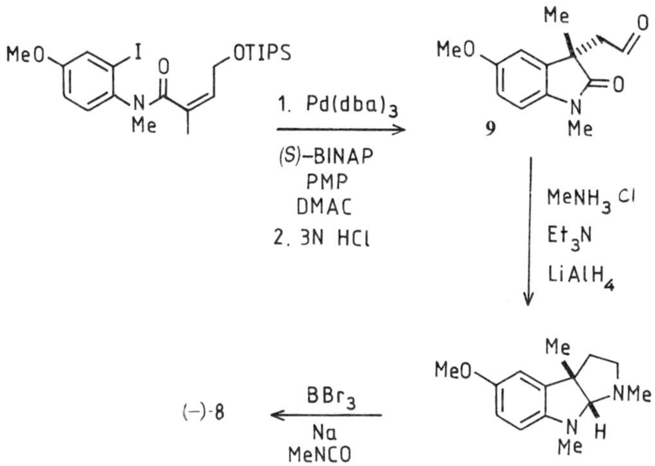

Scheme 14.9

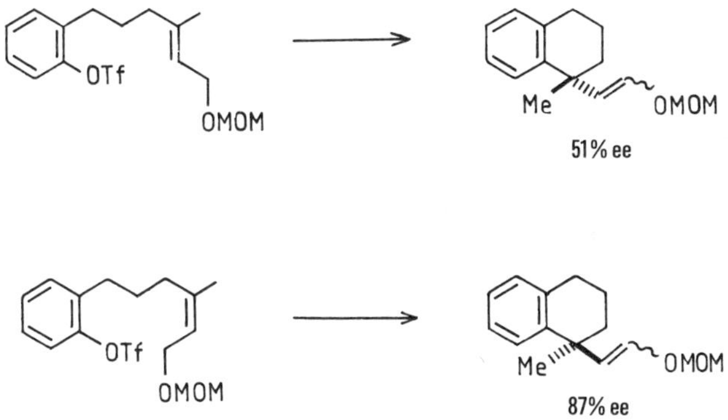

Scheme 14.10

with (*R*)-BINAP in the (*Z*)-alkene series afforded the product (**10**) in 90% yield and 90% ee. In this case, potassium carbonate was employed as the base. Again, the use of an allylic oxygen substituent in the alkene allows simple access to an aldehyde after cyclisation. In this case, hydrolysis of the enol ether and reductive amination completes the lower portion of the target structure. The remaining steps are straightforward, requiring benzylic oxidation with chromium trioxide followed by addition of paraformaldehyde and cyclisation to nitrogen. The key disconnections that underpin this strategy are shown in Figure 14.4.

Asymmetric induction in the cyclisation reactions of vinyl halides has also been examined.[7] In this case, a survey of different chiral auxiliaries has been made. The best results (74% ee) were obtained with the chiral ferrocene-based

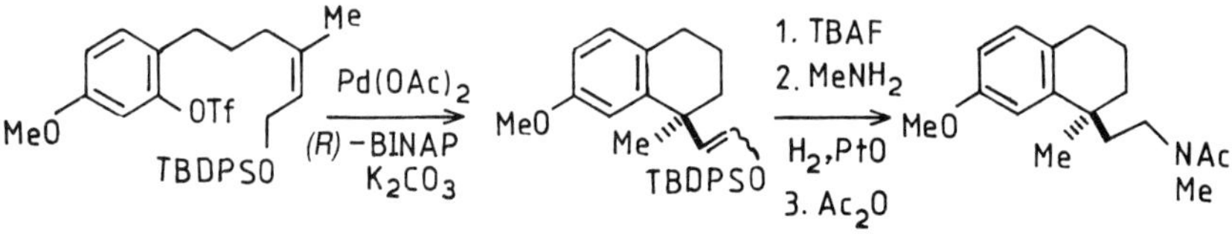

Me
MeO OTf TBDPSO
$Pd(OAc)_2$
(R)-BINAP
K_2CO_3
MeO Me TBDPSO
1. TBAF
2. $MeNH_2$
H_2, PtO
3. Ac_2O
MeO Me NAc Me

10

Scheme 14.11

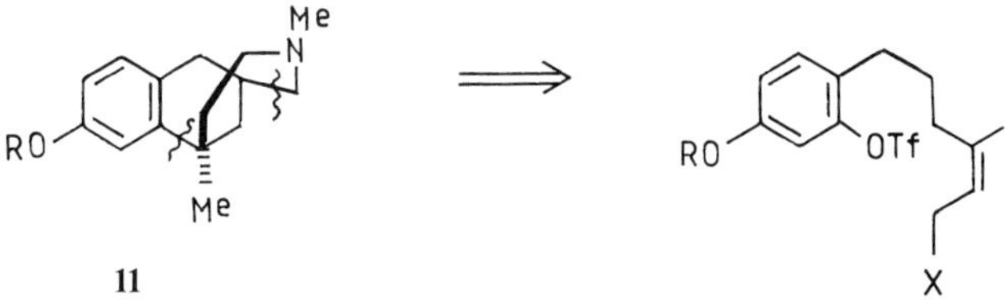

RO Me N Me

11

RO OTf X

Figure 14.4 Retrosynthetic analysis indicates enantioface differentiation in a palladium-catalysed Heck coupling to gain access to eptazocine.

auxiliary (*R*,*S*)-BPPFOH. Variation of the base and the solvent used for the reaction again has a considerable effect on the efficiency of asymmetric induction. In this cyclisation reaction, the position of the alkene following the coupling step is not locked (there is no second heteroatom within the 6-membered ring) and a number of regioisomers can result from the action of metal-mediated hydrogen shifts. The product mixture can be controlled by a judicious choice of reaction conditions (Scheme 14.12) Complete selectivity for the α,β-unsaturated ketone is possible, but only at the expense of yield and optical purity. The best asymmetric induction was achieved at low-temperature conditions, affording a 2:3 mixture of the two regioisomers in 94% yield. In this case, the enone was formed in 86% ee. The product structure shows promise for elaboration into the glycosidase inhibitor castanospermine.

Scheme 14.13 illustrates a version of the iodoarene cyclisation in which a

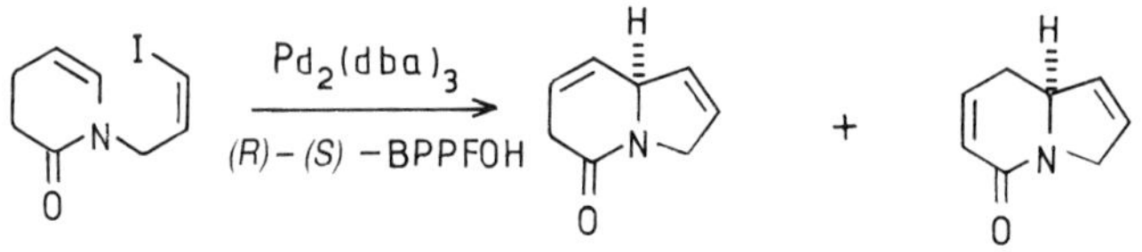

I N O
$Pd_2(dba)_3$
(R)-(S)-BPPFOH
H N O + H N O

Ag_3PO_4	0 : 1	45% yld.	74% ee
Silver exchanged zeolite	2 : 3	94% yld.	86% ee

Scheme 14.12

heteroatom tether connects the arene and alkene portions of the molecule. In this case, an allylsilane has been used as the alkene coreactant. With triphenylphosphine, and silver oxide as the base, only β-hydrogen elimination was observed. With BINAP present to induce asymmetry, under the silver phosphate conditions, the reaction pathway changed, and desilylation predominated to form (**12**) as the major product in 90% ee.[8]

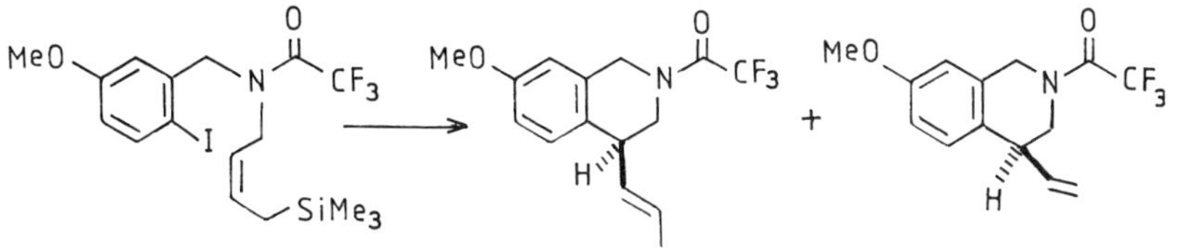

17 : 83 (90% ee)

Scheme 14.13

14.3.3 Combining asymmetric induction and kinetic resolution

Work from the groups of Ozawa and Hayashi in the Catalysis Research Centre of Hokkaido University has demonstrated how asymmetric induction and kinetic resolution can be combined (Scheme 14.14) in the asymmetric coupling of aryl triflates with 2,3-dihydrofuran.[9] The first product from β-elimination is the 3,4-dihydro adduct (**13**). However, in this reaction (as

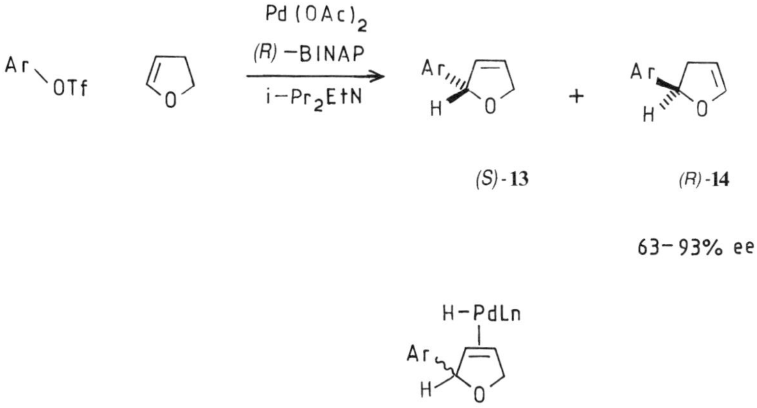

$Pd(OAc)_2$, (R)-BINAP, i-Pr_2EtN

(S)-**13** (R)-**14**

63–93% ee

H–PdLn

15

Scheme 14.14

was the case in Scheme 14.12), the metal hydride intermediate (**15**) can further rearrange by hydrogen transfer to the η^2-alkene complex and the second β-elimination takes the hydrogen from the CH_2 group next to oxygen. If this reaction is allowed to proceed, a new regioisomer (**14**) appears as the product. An important observation in this work is that (**13**) and (**14**) are formed in opposite absolute configurations. Consequently, if asymmetric induction in the first step of Heck coupling is imperfect, the optical purity can be improved because the predominant enantiomer from the Heck coupling step reacts more slowly in the alkene migration process. In this circumstance, the optical purity of both regioisomers will be enhanced by kinetic resolution during the alkene migration reaction. Controlling these reactions is complicated. Not only does the degree of conversion and the ratio of regioisomers influence the optical yield, but also the reaction is susceptible to the nature of the substituents on the aromatic unit. In some cases, very high selectivities can be obtained. For example, as shown in Scheme 14.15, with 1,8-bis(dimethylamino)naphthalene as the base, considerable selectivity for (**16**) and very good optical purity can be achieved. The rearranged product (**16**) predominated under these reaction conditions and was formed in 96% ee.[10] Because a mixture of regioisomers is formed, however, the chemical yield of (**16**) is reduced.[11] A study of the effect of pressure on this enantioselective Heck coupling reaction has also been performed. The original ethyldiisopropylamine base was employed, but with phenyl nonaflate in place of the aryl triflate. At normal pressure (**16**) was formed in 47% ee. This was improved to 89% at 10 kbar.[12]

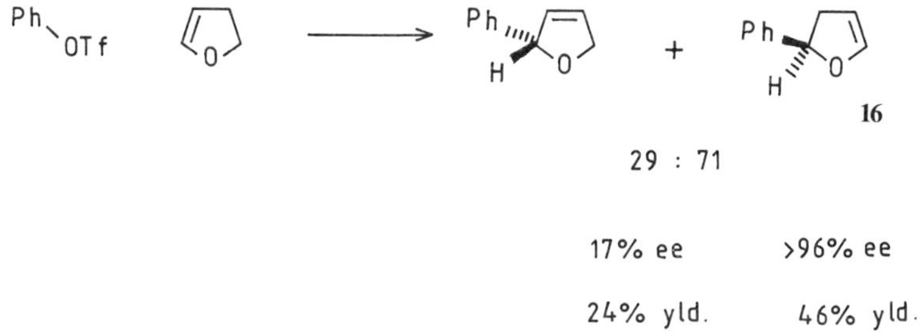

Ph–OTf + (2,3-dihydrofuran) ⟶ (product) + **16**

29 : 71

17% ee | >96% ee

24% yld. | 46% yld.

Scheme 14.15

Asymmetric coupling of the type shown in Scheme 14.14 can also be achieved with *N*-substituted 2-pyrrolines.[13] With ethyldiisopropylamine as the base (Scheme 14.16), the rearranged product (**17**) can be obtained in reasonable yields and ee. As before, the 1,8-bis(dimethylamino)naphthalene base gave the best ee value (83%), but in this case at the expense of yield, which was reduced to 9%, with 58% of the aryl triflate being recovered.

Since their discovery, considerable optimisation of these coupling processes

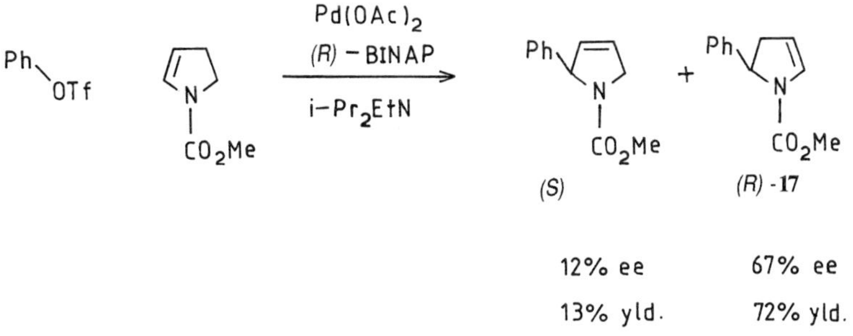

Scheme 14.16

has been made in the Ozawa/Hayashi groups. In particular, it has been found that alkenyl triflates gave better enantioface selectivity, and this allowed the reaction to be taken through to optimise the yield of the rearranged product. Unless there is a mechanism for epimerisation of the chiral centre built in the carbon–carbon bond formation step, the role of kinetic resolution during the alkene migration is presumably negligible in these more fully optimised coupling reactions (Scheme 14.17).[14] In the case of the aryl group transfer to 2,3-dihydrofuran, a detailed discussion of the mechanism of asymmetric

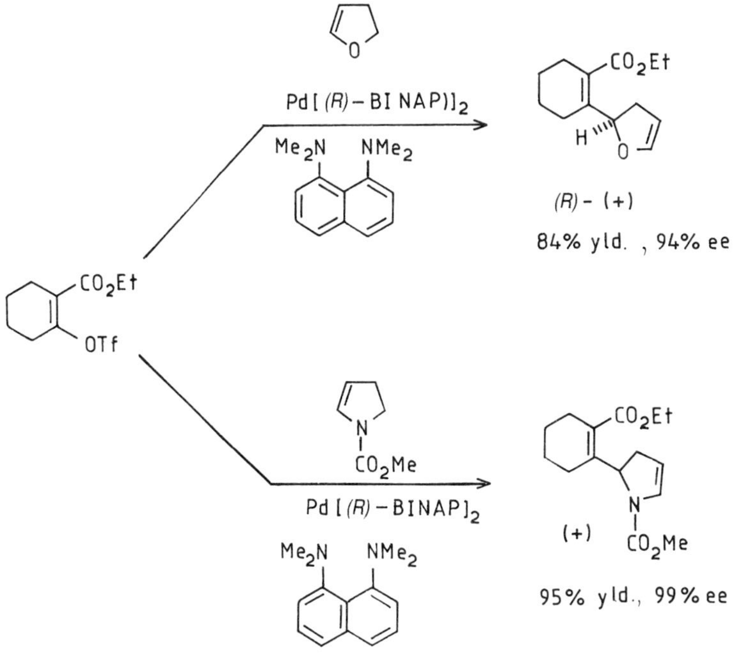

Scheme 14.17

induction has been presented.[11] In these reactions, the palladium hydride species are assumed to remain attached to the organic ligand until β-elimination affords the final products.

14.4 Strategy 2: enantiotopos differentiation

When the plane of symmetry in a prochiral starting material lies between two chiral centres with identical substituents, a *meso*-structure is present. The two groups related to one another by the mirror plane are termed enantiotopic. A reaction that changes one or other (but not both) of these groups will give a chiral product as a racemic mixture of enantiomers. If reaction occurs at only one of the enantiotopic groups, an optically pure product will result.

14.4.1 Enantiotopic ends of a symmetrical alkene

An opportunity for enantiotopic group differentiation arises when a symmetrical alkene is employed in a Heck coupling reaction. In the coupling step, the two ends of the alkene are differentiated. One receives a new carbon–carbon bond in the coupling step, while the other carries the palladium atom and so participates in subsequent metal-mediated reactions. As seen in the preceding section, normally this next step is β-elimination. A common variant, however, is to include formic acid in the reaction mixture so that the σ-bound metal complex can be intercepted, following the formation of a metal hydride. This allows the metal to become detached by reductive elimination, placing a hydrogen at the carbom atom that carried the metal. Overall, in this variant (Figure 14.5), a carbon–carbon bond is formed at one end of the alkene while a carbon–hydrogen bond is formed at the other. The reaction, however, is limited in scope since two of the substituents at the chiral centre in the target must be homologues of one another (e.g. Ar′ and CH_2Ar').

Asymmetric hydroarylation of norbornene with aryl iodides has been examined by Brunner and Cramler.[15] Stereoselectivity has recently been improved to around 70% ee by the use of an aryl triflate with the aminophosphene derivative VALPHOS.[16] Good results have also been obtained with alkenyl halides as the coupling partner. A typical example is shown in Scheme 14.18. Alkenyl triflates are also good coupling partners, and, with the bis(dimethylamino)naphthalene base, an excellent 93% ee was

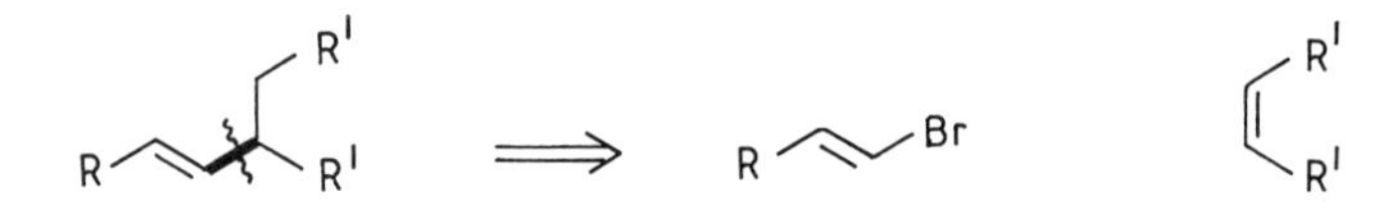

Figure 14.5 A chiral target structure can be formed using enantiotopos differentiation in the coupling process.

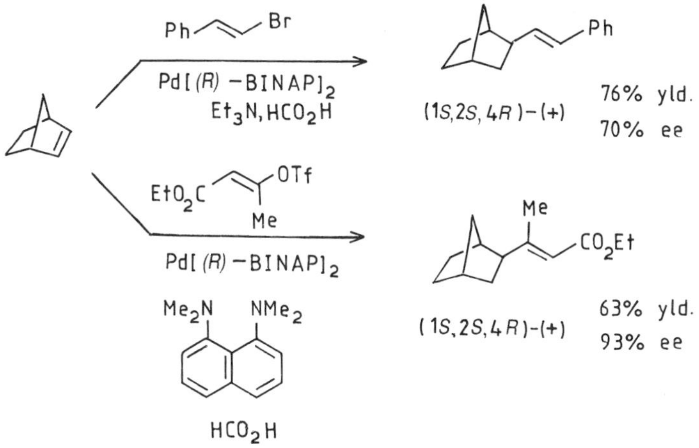

Scheme 14.18

achieved.[17] Similar asymmetric coupling has also been performed with 7-oxa- and 7-azanorbornene derivatives.

14.4.2 *Enantiotopos differentiation with two alkenes*

Heck coupling can offer opportunities for enantiotopos differentiation when *two* enantiotopic alkenes are provided in the substrate. (Two enantiotopic halide groups are another possibility, but there is not an example of this at present.) With identical alkenes as substituents at a carbon atom carrying two different R groups (R^1 and R^2 in Figure 14.6), a symmetry plane lies through the centre of structure (**18**). The carbon atom between the two alkenes is not a chiral centre. Heck coupling at one of the two alkenes alters the structure of one substituent and so destroys the symmetry plane. This induces asymmetry in the product, and if the reaction can be performed selectively at only one of the two enantiotopic alkenes, an enantiopure product will result. Only intramolecular examples have so far been examined, with vinyl halides being coupled by cyclisation to one of a pair of prochiral alkenes (Figure 14.7). The reaction has been applied to the synthesis of chiral hydrindans.[18] In the example shown in Scheme 14.19, an alkenyl iodide is employed for the oxidative addition step.[19] As in earlier cases, the silver phosphate base (used in the presence of calcium carbonate) gave better reuslts than silver oxide.

This approach is easily adjusted to gain access to chiral decalin systems.[18] If the carbon–heteroatom bond for oxidative addition is placed at the far end of the alkene, a larger ring will be formed. The example shown in Scheme 14.20 employs an alkenyl triflate for this purpose.[20]

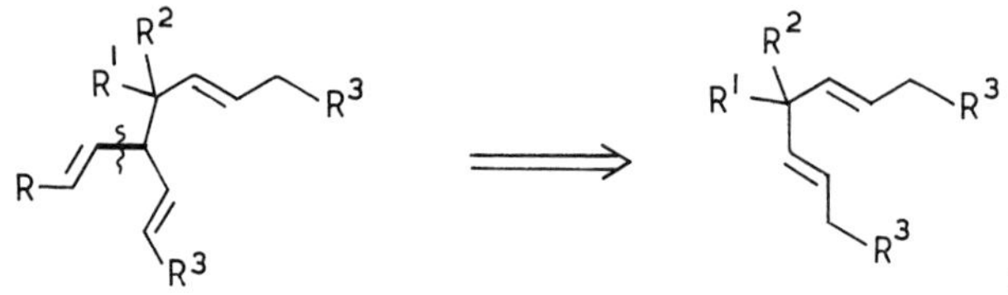

Figure 14.6 Formation of a chiral centre by differentiating enantiotopic alkene groups.

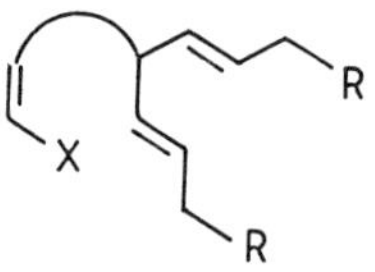

Figure 14.7 A substrate suitable for an intramolecular Heck coupling to differentiate prochiral alkenes.

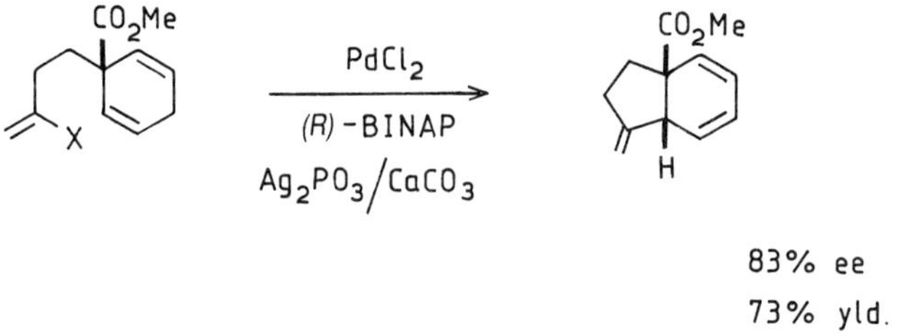

Scheme 14.19

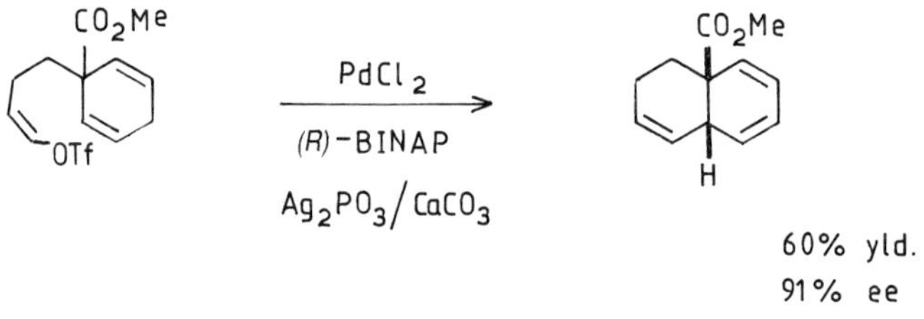

Scheme 14.20

The 5-membered ring-closure shown in Scheme 14.19 has been modified to gain access to key intermediates for synthetic routes to the capnellenol class of natural products.[21] In the cases shown in Schemes 14.21 and 14.22, the alkene participating in the metal-mediated carbon–carbon bond formation is located in a 5-membered ring. For enantiotopos differentiation, the prochiral alkenes must take the form of a conjugated 1,3-diene in a disubstituted cyclopentadiene. Beta-elimination following the palladium-catalysed carbon–carbon bond formation cannot now take place, so a combined Heck reaction/anion capture process can be examined. Asymmetric Heck coupling by palladium acetate and (*S*)-BINAP in the presence of tetrabutylammonium acetate afforded the anion capture product (**20**) in 77% yield and 80% ee. This intermediate has been taken on to the fused tricyclic system (**21**) through a series of steps. During this reaction sequence, the exocyclic alkene was moved into the 5-membered ring with DBU (Scheme 14.21). Cyclisation of the alkenyl halide (Scheme 14.22) was examined in the hope that this would replace the double bond directly in the 5-membered ring, as required for the target structures. Under the same conditions employed for (**19**), the expected cyclisation product (**22**) was formed, but in much lower yield and optical purity.

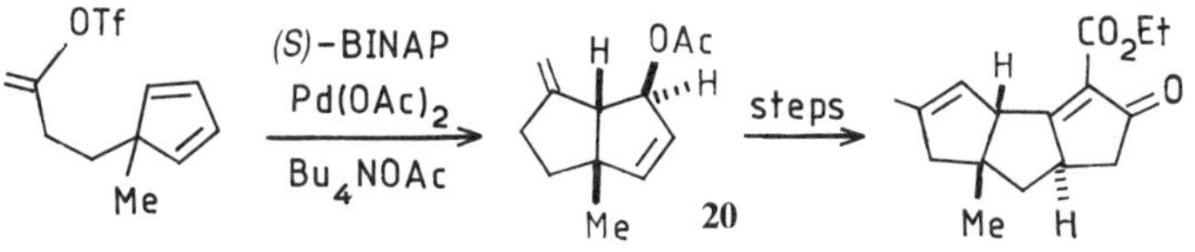

Scheme 14.21

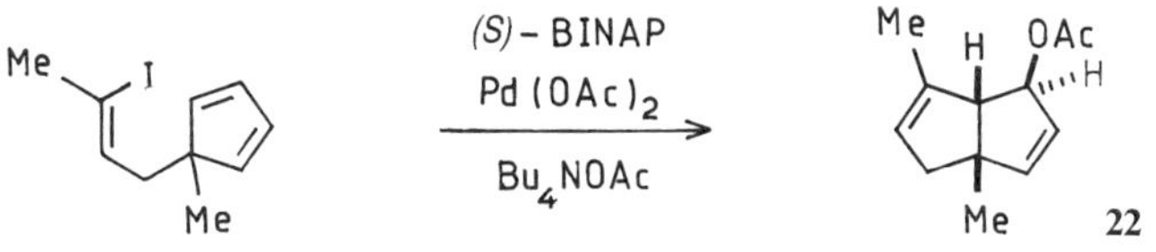

Scheme 14.22

Coupling reactions of the type shown in Scheme 14.20 have also been employed in natural-product synthesis. In this case, the target molecule was vernolepin,[22] and, consequently, an enone was required in one of the 6-membered rings. The approach adopted was to place an OH group as a substituent between the two alkenes, as shown in Scheme 14.23. The intention was that β-elimination would form an enol, so leading ultimately to the required ketone. It was, therefore, important for the OH group to be *trans* to the carbon chain carrying the alkenyl triflate. The required relative stereochemistry in this prochiral starting material could be obtained by allylic oxidation and reduction of the carbonyl of the dienone unit. Asymmetric

Heck coupling proved more efficient with $Pd_2(dba)_3$ than with palladium acetate. Again, BINAP was employed as the chiral auxiliary.

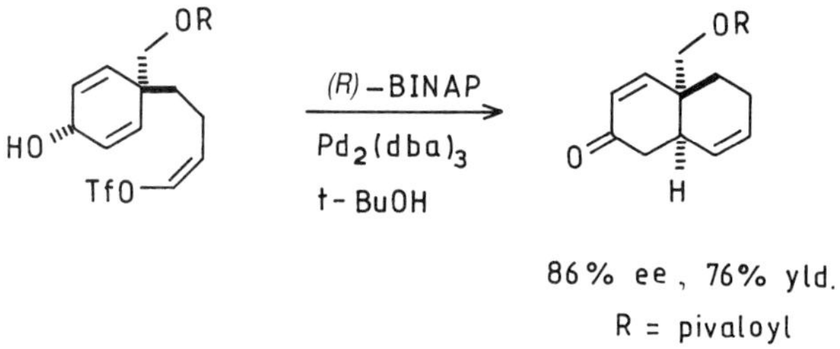

(R)-BINAP
$Pd_2(dba)_3$
t-BuOH

86% ee, 76% yld.
R = pivaloyl

Scheme 14.23

The mechanism for this reaction has been set out in detail in a full paper describing this work.[22] Palladium-catalysed cross coupling to the dienone[23] (Scheme 14.24) has also been examined. Since β-elimination requires a *syn*-transition state, exchange of palladium between the two faces of the ring must occur. In view of the adjacent carbonyl group, involvement of an η^3-oxallyl palladium intermediate has been proposed.[22] A second full paper has recently offered a mechanism to account for asymmetric induction in related routes to simple decalin systems.[24]

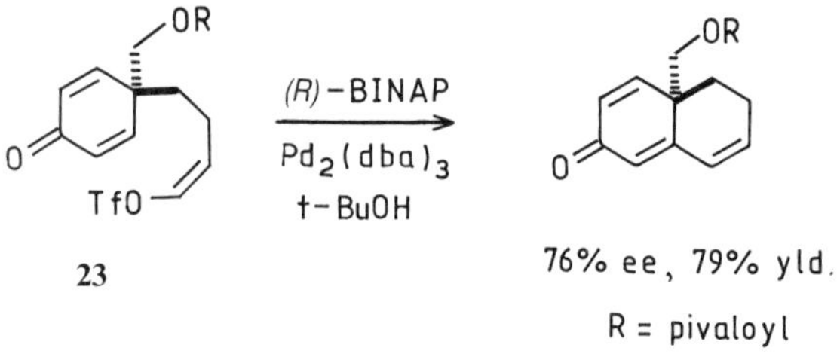

(R)-BINAP
$Pd_2(dba)_3$
t-BuOH

23

76% ee, 79% yld.
R = pivaloyl

Scheme 14.24

All the reactions discussed so far have been based on the asymmetric modification of Heck coupling. Cross-coupling reactions that differentiate prochiral dihalides offer opportunities for asymmetric cross coupling, but the cases examined to date have involved rather specialised techniques. The dichlorobenzene complex (**24**) is prochiral with the chlorine substituents lying to each side of the symmetry plane. Coupling with vinylzinc chloride in the presence of the ferrocene-based chiral phosphine (*S*)-(*R*)-PPFA afforded the (1*S*,2*R*)-isomer (**25**) in 37% yield, together with a smaller amount of the divinyl-substituted double coupling product (Scheme 14.25). The optical purity of (**25**) was 42% ee. Suzuki coupling with vinylboronic acid has also been examined. The process proved similar in terms of the efficiency of asymmetric induction but proceeded in a higher yield. A further improvement in yield was obtained with Stille coupling using tributylvinyltin, but a racemic product was obtained (Scheme 14.25).[25]

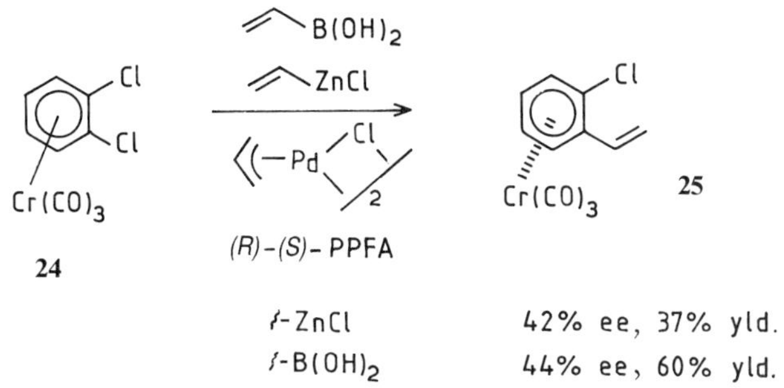

24 → 25; reagents: $\diagup\!\!\diagdown B(OH)_2$, $\diagup\!\!\diagdown ZnCl$; (R)-(S)-PPFA palladium complex

$Cr(CO)_3$

ZnCl	42% ee, 37% yld.
$B(OH)_2$	44% ee, 60% yld.

Scheme 14.25

When the alkene used in the coupling reaction contains two planes of symmetry, both enantiofaces and enantiotopic groups can be present together. This situation arises in the 4,7-dihydro-1,3-dioxepin (**26**), shown in Scheme 14.26. Coupling in the presence of (*S*)-BINAP with phenyl triflate afforded the adduct (**27**) in 84% ee. In this case, palladium acetate was the catalyst and potassium carbonate was employed as the base. Examples with substituents at the methylenedioxy group were also examined but reaction proceeded less efficiently, both in terms of yield and the degree of asymmetric induction.[26]

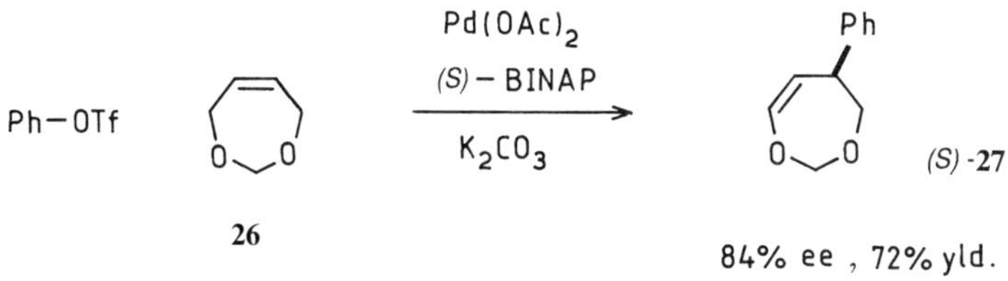

Ph–OTf + 26 → (*S*)-27 [$Pd(OAc)_2$, (*S*)-BINAP, K_2CO_3]

84% ee, 72% yld.

Scheme 14.26

14.5 Strategy 3: combined enantioface and enantiotopos recognition

An example of an approach that makes two skeletal bonds is shown in Figure 14.8. This shows how nucleophile addition at a metal alkene affords a σ-bound intermediate that can participate in a second coupling step. Two alkenes are involved but are differentiated in the reaction. A chiral centre has been built at one of the alkene positions in the starting material, so enantioface or diastereoface recognition is possible. However, since two

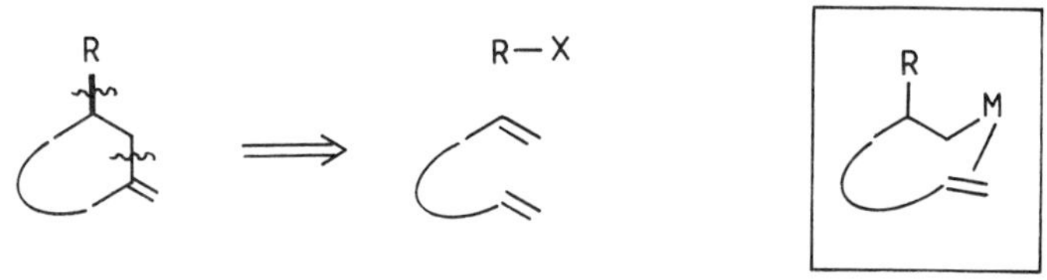

Figure 14.8 Two skeletal bonds can be made during combined enantioface/enantiotopos differentiation.

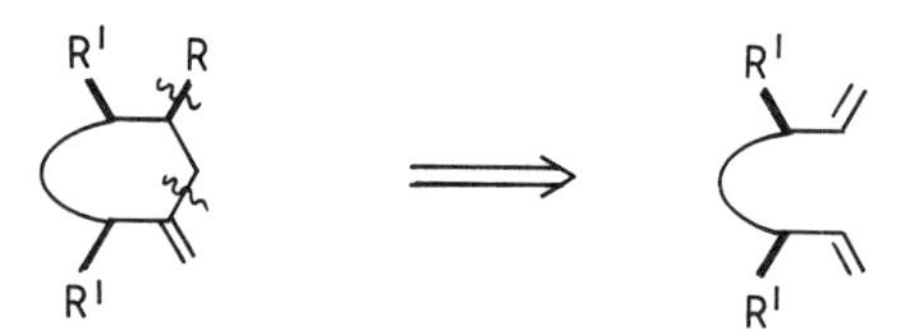

Figure 14.9 Diastereoface and enantiotopos differentiation induces asymmetry at three chiral centres.

alkenes are used, in principle, there is also an opportunity for enantiotopos differentiation. The two strategies can be combined (Figure 14.9).

14.5.1 Use of a chiral nucleophile in the differentiation of prochiral alkenes

In Scheme 14.27, a chiral nucleophile is used to induce asymmetry, but the key step from a stereochemical point of view is the selection between the enantiofaces of the alkenes and between the two enantiotopic reaction sites.[27]

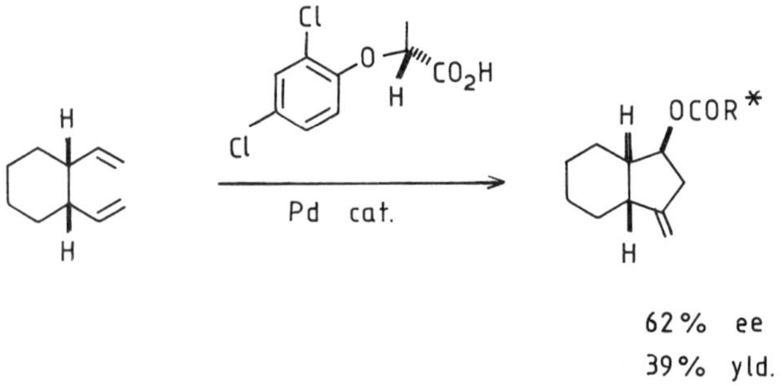

Pd cat.

62% ee
39% yld.

Scheme 14.27

14.6 Strategy 4: asymmetric transformation

Interconverting enantiomers with labile chirality attached to a metal centre offers a further strategy for asymmetric induction. The participation of the enantiomeric units must differ in a key reaction, and the product must not be enantiomerically labile. Palladium-catalysed cross coupling provides an opportunity of this type (Figure 14.10), by bringing together a stereochemically labile chiral nucleophile with an electrophile. Two situations should be considered.

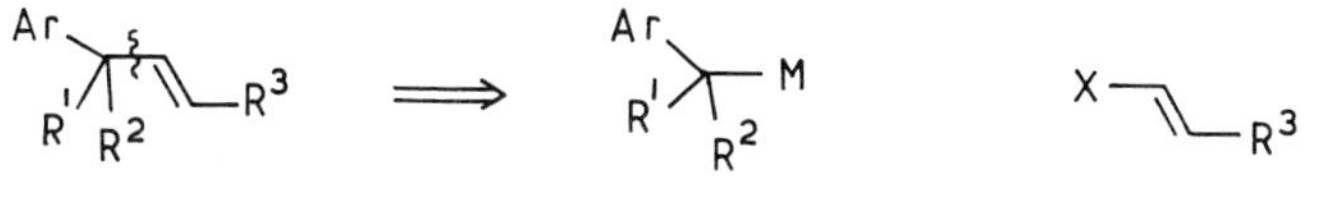

Ar R1 R2 R3 ⟹ Ar R1 R2 M X R3

Figure 14.10 Asymmetric transformation in coupling a stereochemically labile organometallic reagent.

14.6.1 Kinetic resolution at the coupling step

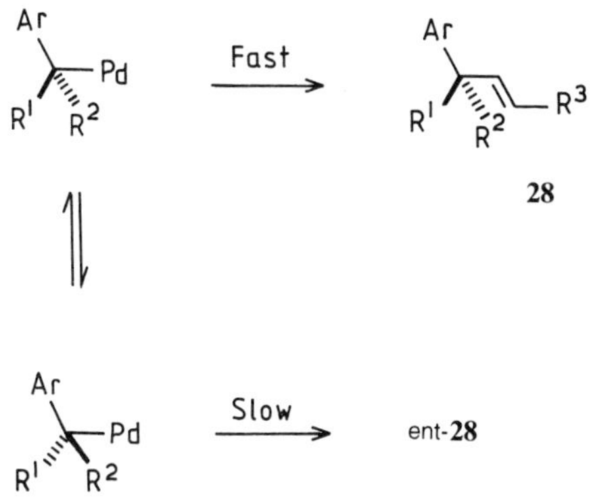

Ar R1 R2 Pd —Fast→ Ar R1 R2 R3 **28**

Ar R1 R2 Pd —Slow→ ent-**28**

Scheme 14.28

14.6.2 Kinetic resolution selecting between two interconverting precursors

An example of an asymmetric transformation based on this approach is provided by the work of Hayashi[28] who used a palladium catalyst and t-LEUPHOS or the ferrocene-based chiral auxiliary PPFA (**30**) to couple 1-phenylethylmagnesium chloride with vinyl bromide. Recently, Weissensteiner's group[29] has developed a new ferrocene-based auxiliary (PTFA) and improved the efficiency of this reaction to afford the product (3-phenyl-1-butene) in 95% yield and 79% ee. While ferrocene-based chiral auxiliaries are relatively common, organometallic species are not generally used within chiral auxiliaries.

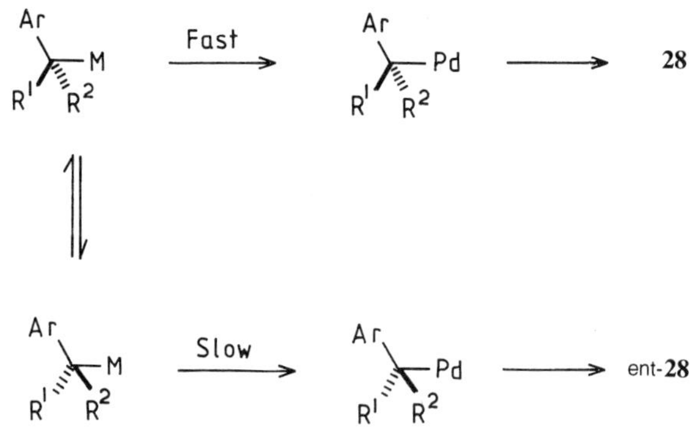

Scheme 14.29

A nice example of a chiral auxiliary based on a metal-bound aromatic ring is provided by a collaboration between the research groups of Uemura and Hayashi. The tricarbonylchromium complex (**29**) gave similar results to PPFA and produced better asymmetric induction than that obtained when the chiral benzylamine derivative (**31**) was used on its own. In this work, Grignard and organozinc reagents were compared, with slightly superior results being obtained in the zinc series (Scheme 14.30).[30]

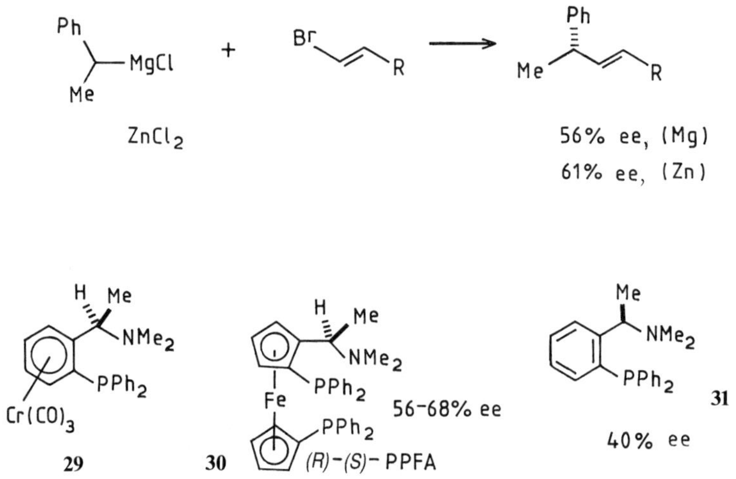

Scheme 14.30

14.6.3 Differentiation at the reductive elimination step

Two achiral precursors can be combined to make a chiral metal complex, with differentiation between enantiomers occurring at the reductive elimination step. An example of this type of coupling process is provided by the Italian research team of Musco and Santi.[31] Oxidative addition into the carbon–bromine bond of bromobenzene, followed by complexation of the silylketene acetal is proposed to afford a palladium complex of a chiral ester. A number of chiral auxiliaries have been examined in attempts to influence this reaction. The best results (Scheme 14.31) were again obtained with BINAP, which afforded the coupling product in 68% yield and 54% ee.[32]

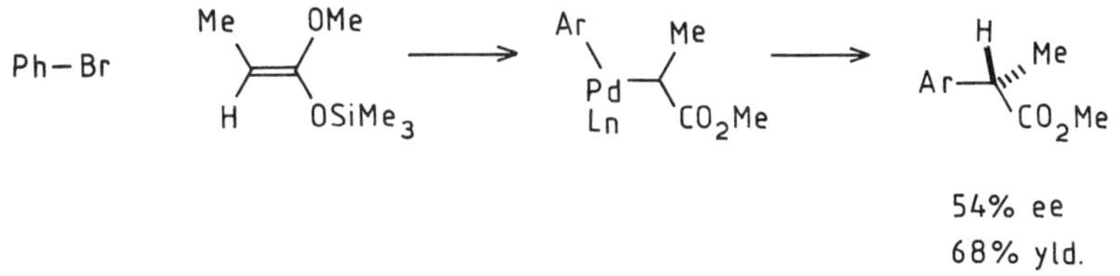

Ph–Br Me OMe H $OSiMe_3$ ⟶ Ar Pd Ln Me CO_2Me ⟶ H Ar Me CO_2Me

54% ee
68% yld.

Scheme 14.31

14.7 Conclusions

In this review, 4th generation-type asymmetric synthesis has been examined. Here, the advantage of employing only a catalytic amount of the chiral auxiliary comes to the fore. Recent developments have brought these reactions through to a point where they can be employed with some confidence in enantioselective synthesis. This research constitutes a significant development from the many examples of stereocontrolled palladium-catalysed bond-forming reactions that employ covalently bound chiral functionality with the substrate.

References

1. Sato, Y., Sodeoka, M. and Shibasaki, M., *J. Org. Chem.*, 1989, **54**, 4738.
2. Carpenter, N.E., Kucera, D.J. and Overman, L.E., *J. Org. Chem.*, 1989, **54**, 5846.
3. Sakamoto, T., Kondo, Y. and Yamanaka, H., *Tetrahedron Lett.*, 1992, **33**, 6845.
4. Ashimori, A. and Overman, L.E., *J. Org. Chem.*, 1992, **57**, 4571.
5. Ashimori, A., Matsuura, T., Overman, L.E. and Poon, D.J., *J. Org. Chem.*, 1993, **58**, 6949.
6. Takemoto, T., Sodeoka, M., Sasai, H. and Shibasaki, M., *J. Am. Chem. Soc.*, 1993, **115**, 8477.
7. Nukui, S., Sodeoka, M. and Shibasaki, M., *Tetrahedron Lett.*, 1993, **34**, 4965.
8. Tietze, L.F. and Schimpf, R., *Angew. Chem., Int. Ed. Engl.*, 1994, **33**, 1089.
9. Ozawa, F., Kubo, A. and Hayashi, T., *J. Am. Chem. Soc.*, 1991, **133**, 1417.
10. Ozawa, F., Kubo, A. and Hayashi, T., *Tetrahedron Lett.*, 1992, **33**, 1485.
11. Ozawa, F., Kubo, A., Matsumoto, Y. and Hayashi, T., *Organometallics*, 1993, **12**, 4188.

12. Hillers, S. and Reiser, O., *Tetrahedron Lett.*, 1993, **34**, 5265.
13. Ozawa, F. and Hayashi, T., *J. Organomet. Chem.*, 1992, **428**, 267.
14. Ozawa, F., Kobatake, Y. and Hayashi, T., *Tetrahedron Lett.*, 1993, **34**, 2505.
15. Brunner, H. and Kramler, K., *Synthesis*, 1994, 1324.
16. Sakuraba, S., Awano, K. and Achiwa, K., *Synlett*, 1994, 291.
17. Ozawa, F., Kobatake, Y., Kubo, A. and Hayashi, T., *J. Chem. Soc., Chem. Commun.*, 1994, 1323.
18. Sato, Y., Honda, T. and Shibasaki, M., *Tetrahedron Lett.*, 1992, **33**, 2593.
19. Sato, Y., Honda, T. and Shibasaki, M., *Tetrahedron Lett.*, 1992, **33**, 2595.
20. Sato, Y., Watanabe, S. and Shibasaki, M., *Tetrahedron Lett.*, 1992, **33**, 2589.
21. Kagechika, K., Ohshima, T. and Shibasaki, M., *Tetrahedron*, 1993, **49**, 1773.
22. Kondo, K., Sodeoka, M., Mori, M. and Shibasaki, M., *Synthesis*, 1993, 920.
23. Kondo, K., Sodeoka, M., Mori, M. and Shibasaki, M., *Tetrahedron Lett.*, 1993, **34**, 4219.
24. Sato, Y., Nukui, S., Sodeoka, M. and Shibasaki, M., *Tetrahedron*, 1994, **50**, 371.
25. Uemura, M., Nishimura, H. and Hayashi, T., *Tetrahedron Lett.*, 1993, **34**, 107.
26. Koga, Y., Sodeoka, M. and Shibasaki, M., *Tetrahedron Lett.*, 1994, **35**, 1227.
27. Tottie, L., Baeckström, P., Moberg, C., Tegenfeldt, J. and Haumann, A., *J. Org. Chem.*, 1992, **57**, 6579.
28. Hayashi, T., Fukushima, M., Konishi, M. and Kumada, M., *Tetrahedron Lett.*, 1980, **21**, 79; Hayashi, T., Hagihara, T., Katsuro, Y. and Kumada, M., *Bull. Chem. Soc. Jpn*, 1983, **56**, 363; Hayashi, T., Yamamoto, A., Hojo, M. and Ito, Y., *J. Chem. Soc., Chem. Commun.*, 1989, 495.
29. Jedicka, B., Kratky, C., Weissensteiner, W. and Widhalm, M., *J. Chem. Soc., Chem. Commun.*, 1993, 1329.
30. Uemura, M., Miyake, R., Nishimura, H., Matsumoto, Y. and Hayashi, T., *Tetrahedron: Asymmetry*, 1993, **3**, 213.
31. Galarini, R., Musco, A., Pontellini, R. and Santi, R., *J. Mol. Catal.*, 1992, **72**, L11.
32. Uemura, M., Miyake, R., Nishimura, H., Matsumoto, Y. and Hayashi, T., *Tetrahedron: Asymmetry*, 1992, **3**, 213.

15 Palladium allyl π-complexes in asymmetric synthesis

J.M.J. WILLIAMS

15.1 Introduction

This chapter discusses the use of palladium-catalysed allylic substitution in enantioselective and diastereoselective processes. The early sections will outline the mechanism and scope of the reaction, as well as the behaviour and properties of the π-allylpalladium intermediates. Based on recent results, it seems likely that future advances in the asymmetric version of the palladium-catalysed allylic substitution reaction will be made through an understanding of the intermediates involved in the catalytic cycle. There are several reviews covering the scope of this reaction in a general sense.[1]

15.2 Palladium allyl π-complexes

There are several methods for the preparation of palladium allyl π-complexes. All of the methods rely upon the fact that palladium(0) and palladium(II) complexes are able to coordinate to many simple alkenes. First, the reaction of palladium metal or a palladium(0) complex with an alkene possessing a suitably disposed allylic leaving group results in an oxidative insertion to provide a palladium π-allyl complex.[2] Second, coordination of palladium(II) to a diene, followed by attack of a nucleophile provides an alternative access to π-allyl complexes.[3] Finally, direct interaction between alkenes and some palladium(II) salts can lead directly to palladium π-allyl complexes (Scheme 15.1).[4] The structures of allylpalladium complexes are well defined by X-ray crystallography[5] and by NMR studies.[6] Additionally, a set of MM2 parameters has recently been developed for the π-allylpalladium moiety.[7]

15.2.1 Simple palladium-catalysed allylic substitution

In 1965, Tsuji demonstrated that nucleophiles could be added to palladium allyl complexes to liberate allylated organic products.[8] During the 1970s, a number of research groups[9] including those of Trost[10] and Tsuji[11] developed a catalytic process in which a substrate with an allylic leaving group could be substituted by an incoming nucleophile (Scheme 15.2). In Table 15.1, a representative selection of leaving groups, nucleophiles and palladium catalysts have been identified.

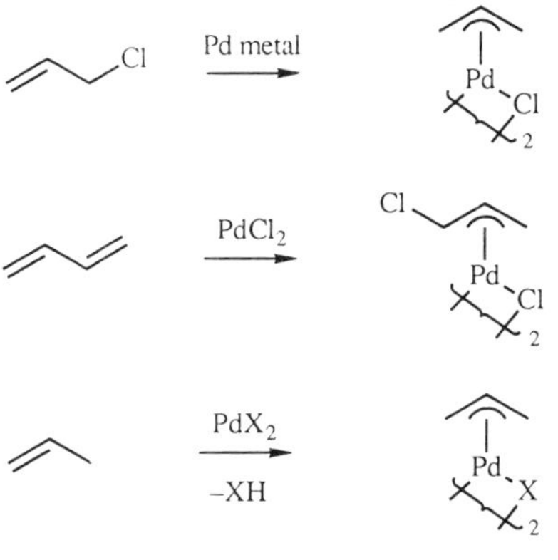

Scheme 15.1

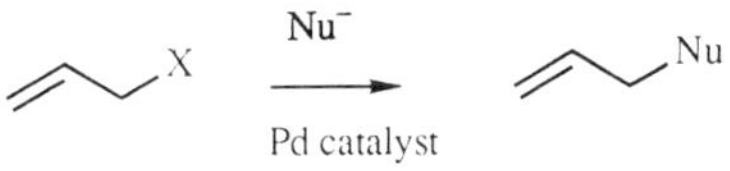

Scheme 15.2

The mechanism of palladium-catalysed allylic substitution proceeds initially as for the formation of π-allylpalladium complexes.[35] An association of the allyl substrate with the palladium catalyst forms an alkene complex with subsequent oxidative addition forming a π-allylpalladium complex. In the presence of phosphine ligand, a cationic allyl complex can form, which will be more reactive to nucleophilic addition.[36] The nucleophilic addition takes place at either terminus of the allyl group to afford the substitution product complexed to palladium. Dissociation of palladium affords the product and regenerates the palladium catalyst (Scheme 15.3). The intermediacy of the π-allylpalladium species creates many opportunities for stereocontrol as well as a multitude of synthetic possibilities.[37]

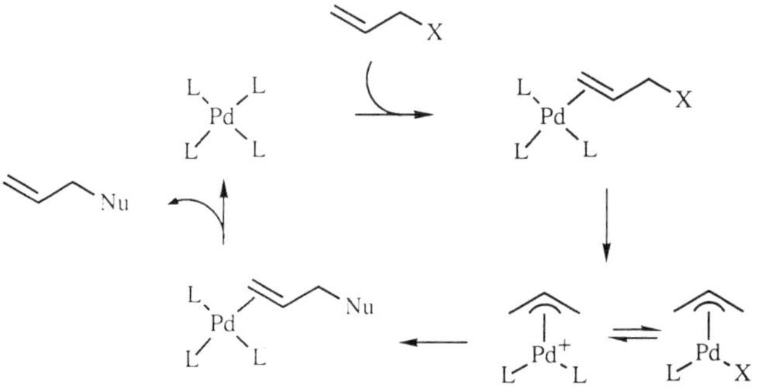

Scheme 15.3

Table 15.1 The range of leaving groups, nucleophiles and catalysts available for allylic substitution

Leaving groups

X = OAc. The acetate group is the most widely used of the possible leaving groups.
X = Hal. Many allyl halides react with nucleophiles in the absence of palladium, causing other stereochemical outcomes.
X = OCO_2R. The choice of carbonate has the advantage that alkoxide is generated, which can be employed to deprotonate pronucleophiles such as dimethylmalonate.[12]
X = OH. Although simple allyl alcohols are readily available, their use has been limited by the need for an activating additive.[13]
Allylic epoxides have been widely used. The oxygen atom remains attached to the substrate.[14]
Other leaving groups include phosphates[15] and sulfones.[16]

Nucleophiles

C-nucleophiles have been extensively investigated. Stabilised C-nucleophiles, such as those derived from malonate are popular. Many organometallics have also been employed. Recently, trimethylsilyl cyanide has been used.[17]
N-nucleophiles include azide,[18] secondary amines,[19] sulfonamides,[20] amides[21] and carbamates.[22]
Hydride sources including formate[23] and borohydrides[24] have been used.
Other heteroatom-based nucleophiles have been used successfully. Oxygen,[25] sulfur,[26] phosphorus[27] and silicon[28] nucleophiles have all been reported.

Palladium catalysts

$Pd(PPh_3)_4/2PPh_3$ is a particularly popular catalyst. The need for fresh catalyst is important, and although commercially available, freshly prepared material may be superior on some occasions.[29]
$Pd(dba)_2/2PPh_3$ makes use of stable dibenzylidene acetone (dba) complexes of palladium(0).[30] Otherwise the reactivity is similar to the above. $Pd_2(dba)_3 \cdot CHCl_3$ is also available.[31]
$[Pd(\pi\text{-}C_3H_5)Cl]$[32]$/2PPh_3$ is a further possibility. On the addition of a nucleophile, a palladium(0) catalyst is formed *in situ*. Palladium allyl chloride dimer is stable.
The use of other phosphines may be of benefit in terms of reactivity.[33] Bidentate ligand such as dppe can enhance the reactivity of the catalyst.[34] Enantiomerically pure ligands also have applications (*vide infra*).

15.2.2 *Palladium-catalysed reactions involving diastereomerically pure substrates*

There have been a variety of studies conducted to examine the mechanism of palladium-catalysed allylic substitution.[38] The use of cyclohexenyl acetates which incorporate stereochemical marker have revealed that the mechanism depends upon the choice of nucleophile employed. Therefore, treatment of the cyclohexenyl acetate (**1**) with the sodium salt of dimethylmalonate proceeds with overall retention of stereochemistry to afford the substitution product (**2**) (Scheme 15.4).[39] The reaction proceeds by two sequential inversion processes.

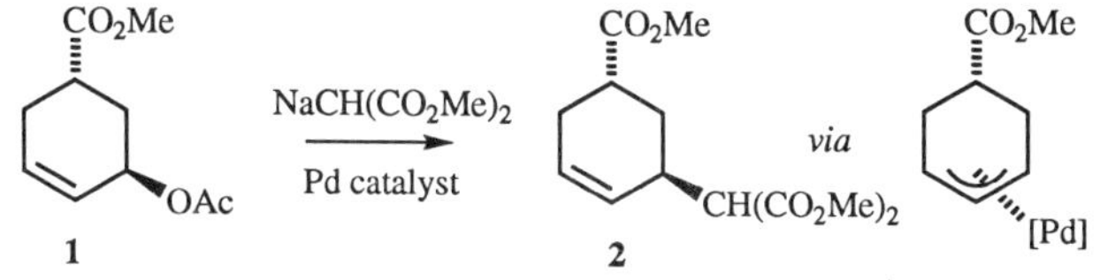

Scheme 15.4

For harder nucleophiles, such as phenylzinc chloride, the palladium-catalysed allylic substitution process occurs with overall inversion of stereochemistry.[40] The mechanism still occurs through a π-allylpalladium species, but the nucleophile is delivered via prior coordination to the palladium, resulting in the alternative outcome of overall retention of stereochemistry to afford substitution product (**3**) (Scheme 15.5).

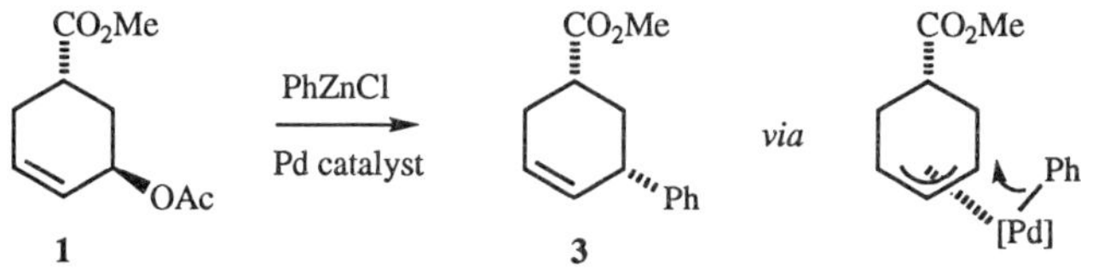

CO_2Me ... OAc **1** —PhZnCl / Pd catalyst→ CO_2Me ... Ph **3** *via* CO_2Me ... Ph [Pd]

Scheme 15.5

15.2.3 Palladium-catalysed reactions involving enantiomerically pure substrates

Enantiomerically pure allylic acetates may be employed as substrates for palladium-catalysed allylic substitution.[41] This is illustrated by the cyclisation of the enantiomerically pure substrate (**4**) to afford the enantiomerically pure product (**5**) (Scheme 15.6).[42] This reaction also illustrates the conversion of a single diastereomer of starting material into a single diastereomer of product. However, in many instances, the stereochemical purity is lost during the substitution process. For example, the reaction of an enantiomerically pure allyl acetate may proceed via an achiral allylpalladium intermediate. The nucleophile has no preference for attack at either end of the allyl moiety, and hence a racemic mixture is formed (Scheme 15.7). There is no memory of the original stereochemistry provided by the starting material. Even for palladium-catalysed allylic substitution processes which do not proceed via

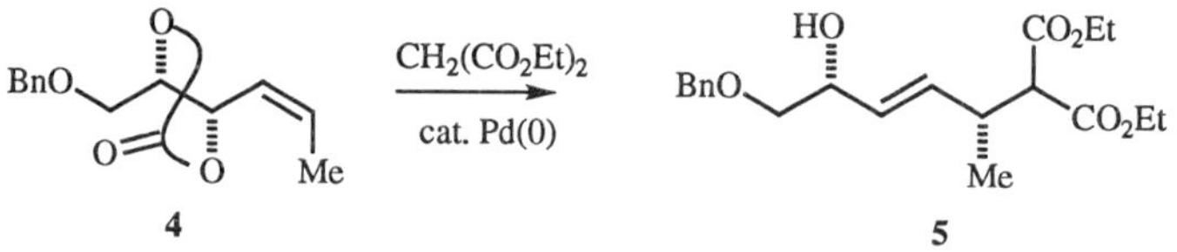

BnO ... O Me **4** —$CH_2(CO_2Et)_2$ / cat. Pd(0)→ BnO HO ... CO_2Et CO_2Et Me **5**

Scheme 15.6

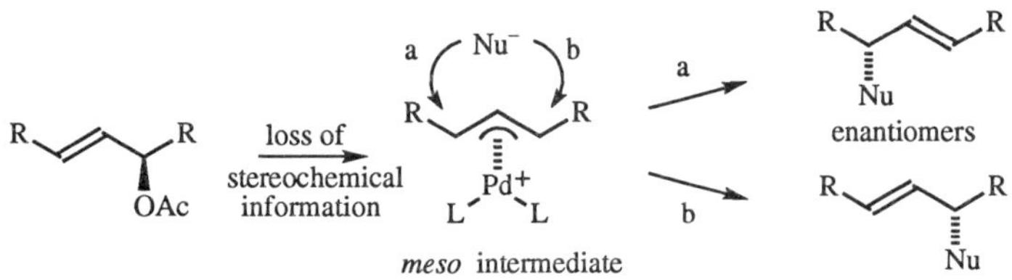

R R OAc —loss of stereochemical information→ a Nu⁻ b R R Pd^+ L L *meso* intermediate; a → R R Nu enantiomers; b → R R Nu

Scheme 15.7

a *meso*-intermediate, there can be a substantial erosion of stereochemical purity caused by other factors, including epimerisation of the allylpalladium intermediate by a $\pi-\sigma-\pi$ process[43] and epimerisation by the presence of acetate[44] or palladium salts.[45] Tsuji has demonstrated that the timing involved in the nucleophilic addition to the imtermediate allylpalladium complex is important for the preservation of stereochemistry. When the substrate (**6**) is enolised prior to the palladium reaction, then the enantiomeric excess of the cyclised product (**7**) reflects that of the starting material. However, when the enolate is formed during the reaction conditions, the intermediate allylpalladium complex has time to racemise prior to nucleophilic addition, and racemic product is observed (Scheme 15.8).[46]

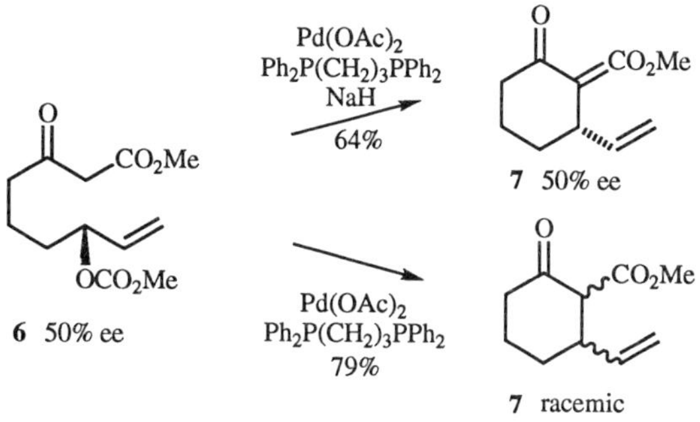

O
CO_2Me
OCO_2Me
6 50% ee
$Pd(OAc)_2$
$Ph_2P(CH_2)_3PPh_2$
NaH
64%
O
CO_2Me
7 50% ee
$Pd(OAc)_2$
$Ph_2P(CH_2)_3PPh_2$
79%
O
CO_2Me
7 racemic

Scheme 15.8

15.3 Diastereoselectivity with allylpalladium complexes

The formation of diastereomers during palladium-catalysed allylic substitution can occur in several ways. The use of enantiomerically pure nucleophiles[47] as well as enantiomerically pure allylic acetates have been employed to control the formation of a newly formed chirality centre.

A diastereoselective relay of stereochemistry has been demonstrated by Hiroi *et al.*[48] The enantiomerically pure enamine (**8**) undergoes palladium-catalysed allylic substitution reaction with a tethered allylic ester to afford the product (**9**) with 90% ee after hydrolysis (Scheme 15.9). The reaction is

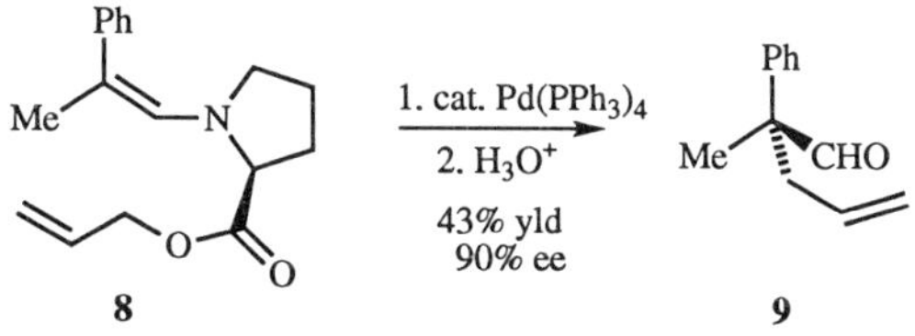

Ph
Me
N
O
O
8
1. cat. $Pd(PPh_3)_4$
2. H_3O^+
43% yld
90% ee
Ph
Me
CHO
9

Scheme 15.9

an example of diastereocontrol, although the first formed product is hydrolysed to form a product that is not attached to the auxiliary.

There have also been examples of the use of substrate control for diastereoselective reactions.[49] One example involves the cyclisation of the allylic acetate (**10**), in which a chirality centre is adjacent to the allylpalladium intermediate. Diastereocontrol is reasonably good, but highly dependent upon the reaction conditions, as reflected in the variation in the ratios of the products (**11**) and (**12**) observed (Scheme 15.10).[50]

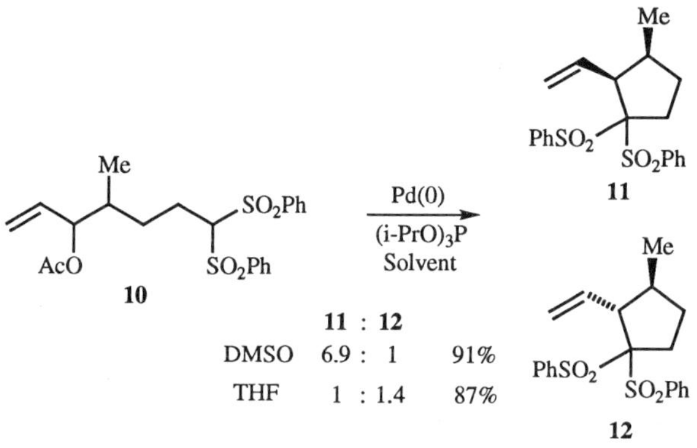

Pd(0), (i-PrO)$_3$P, Solvent

Solvent	**11** : **12**	Yield
DMSO	6.9 : 1	91%
THF	1 : 1.4	87%

Scheme 15.10

There are very few well-documented examples of reactions involving two achiral precursors which combine to form diastereomers, and control of such processes is still in its infancy. One example involves the reaction between the allylic acetate (**13**) (which reacts via a *meso*-intermediate) and the imino phosphonate (**14**). The product is a mixture of diastereomers (**15**) and (**16**) (Scheme 15.11).[51]

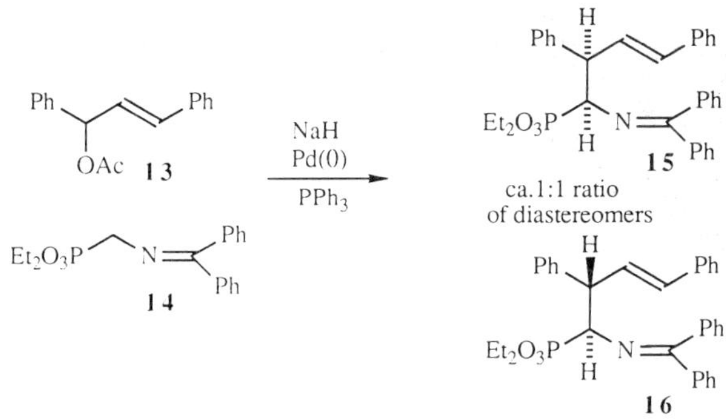

NaH, Pd(0), PPh$_3$

ca. 1:1 ratio of diastereomers

Scheme 15.11

15.4 Enantiocontrol of palladium-catalysed allylic substitution by the use of enantiomerically pure ligands

There have been many examples of enantiocontrol in palladium-catalysed allylic substitution. One of the main considerations is an appropriate choice of starting substrate. Substrates which proceed via symmetrical allyl-palladium complexes represent the most commonly used system for examining the efficacy of ligands in this process.

15.4.1 Enantiocontrol via meso-*complexes*

Treatment of the racemic allylic acetate (**13**) with the palladium(0) catalyst results in the *meso*-complex (**17**) as an intermediate. For nucleophiles which approach from the *exo*-face of this complex, attack may occur to either of the allylic termini to afford the enantiomers (**18**) and (*ent*-**18**) (Scheme 15.12). The product enantioselectivity is controlled by the regiochemistry of this addition. These reactions need to be under kinetic control. However, it has recently been reported that the addition of malonate nucleophiles may be a reversible process,[52] which would result in an erosion of product enantioselectivity.

Ph Ph OAc 13 → a Nu⁻ b Ph Ph Pd^+ L L 17 → a: Ph Ph Nu 18; b: Ph Ph Nu ent-18

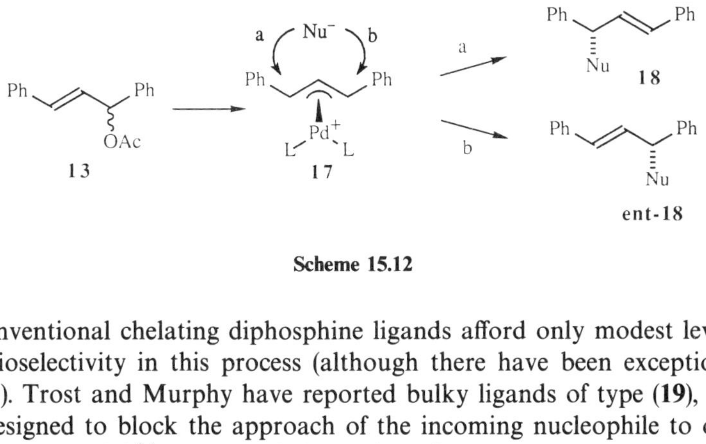

Scheme 15.12

Conventional chelating diphosphine ligands afford only modest levels of enantioselectivity in this process (although there have been exceptions to this[53]). Trost and Murphy have reported bulky ligands of type (**19**), which are designed to block the approach of the incoming nucleophile to one of the allylic termini.[54] Asymmetric induction of up to 69% ee was reported in the allylic substitution process.

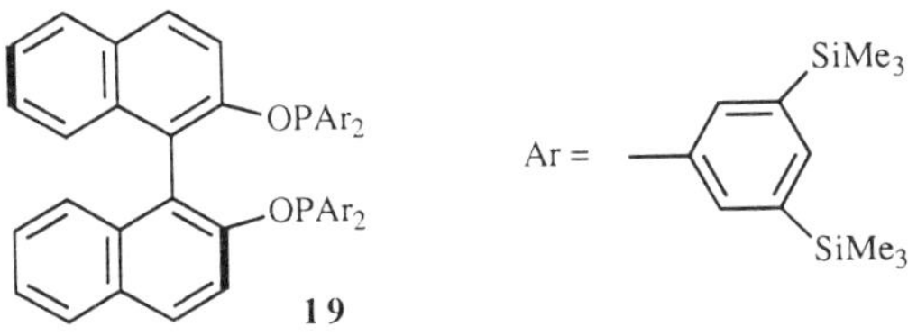

$OPAr_2$ $OPAr_2$ 19 Ar = $SiMe_3$ $SiMe_3$

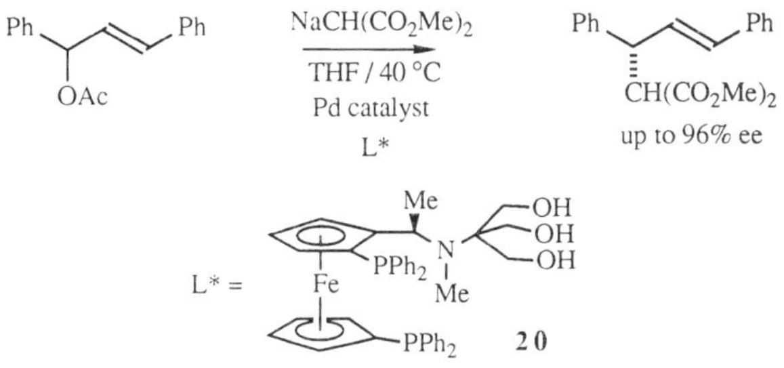

Ph OAc Ph → $NaCH(CO_2Me)_2$, THF / 40 °C, Pd catalyst, L* → Ph $CH(CO_2Me)_2$ Ph, up to 96% ee

L* = **20**

Scheme 15.13

Hayashi reported highly effective ligands (**20**), which were designed to chelate the palladium and to associate with the incoming nucleophile, thereby directing nucleophilic addition to one of the allylic termini. This approach was successful and enantioselectivities as high as 96% ee were recorded (Scheme 15.13).[55] Other ligands (**21**), (**22**),[56] (**23**) and (**24**)[57] have been designed with the same principle in mind. However, enantioselectivities obtained in the palladium-catalysed allylic substitution reaction were not as high as for ligand (**20**).

Alternative ligand designs are possible: Pfaltz *et al.* have shown that the ligands (**25**) and (**26**) are able to provide highly enantioselective palladium-catalysed allylic substitution reactions.[58] Figure 15.1 shows how the steric bulk of the ligand forces one of the allylic termini away from the palladium and affords more positive charge character at this position. Crystallographic studies lend support to this idea.[59]

There are many ligands which do not appear to be able to either associate with the incoming nucleophile or to sterically hinder the approach to one

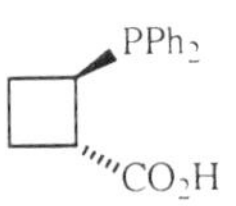

21 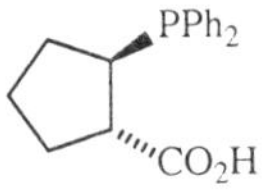 **22** 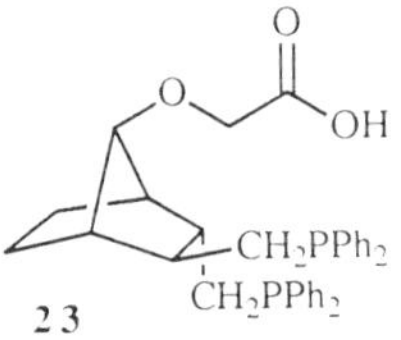 **23** 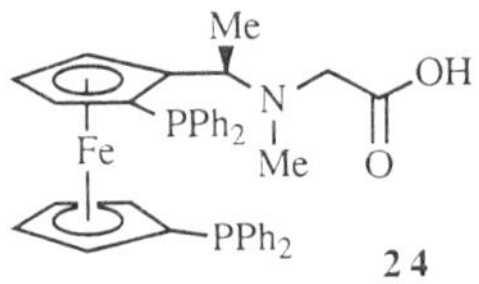 **24** 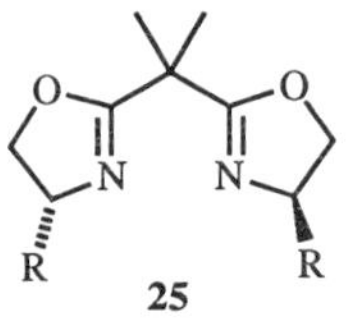 **25** 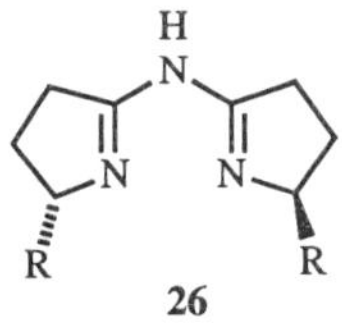 **26**

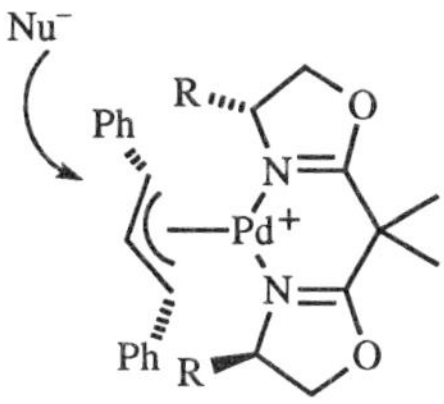

Nu⁻ Ph R O N Pd+ N Ph R O

Figure 15.1 Geometry of approach of nucleophile to the palladium complex of (**25**).

end of the allyl moiety. For example ligands (**27**)[60] and (**28**)[61]; these are, nevertheless, able to exert high levels of stereocontrol on the palladium-catalysed allylic substitution reaction (up to 85 and 91% ee, respectively). Presumably, these ligands also function by steric forces distorting the symmetry of the allyl unit.

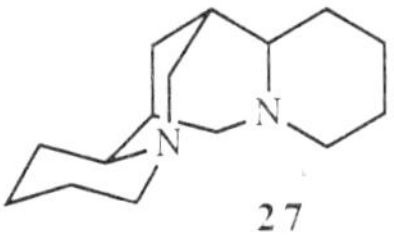

27

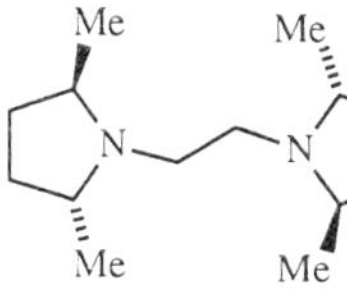

28

Several groups have reported the use of chelating ligands such as (**29**–**31**) which contain two distinct donor atoms for asymmetric palladium-catalysed allylic substitution reactions.[62] A likely explanation for the action of these ligands is both steric and electronic. Åkermark and Vitagliano *et al.* have demonstrated that phosphine ligands are able to generate positive charge character *trans* to themselves in allylpalladium complexes,[63] and presumably, this is where an incoming nucleophile is most likely to attack. Therefore, a ligand containing a phosphine in conjunction with another donor ligand can provide an electronically distorted allylic unit. In fact, the basic requirement is that one donor atom is a π-acceptor and that the other donor atom is not.

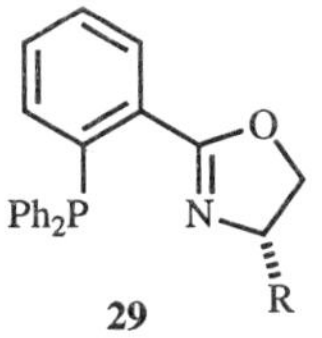

29 provides up to 99% ee (with **13**)

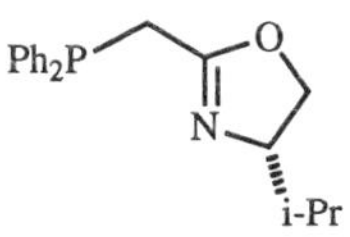

30 provides up to 73% ee (with **13**)

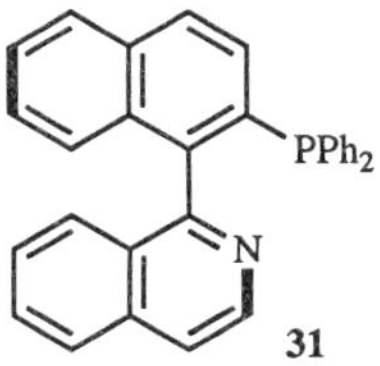

31 provides up to 98% ee (with **13**)

Helmchen has obtained a crystal structure of an allylpalladium complex coordinated to a phosphorus-containing oxazoline ligand.[64] The enantiomeric outcome is consistent with nucleophilic addition *trans* to the phosphine in this complex. It is somewhat surprising that complex (**32**) is the major species, since on steric grounds, the alternative diastereomer would appear to be more favoured.[65] Nevertheless, it is the oxazoline group which provides the source of asymmetry in the ligands, and it is the electronic disparity in the donor atoms which controls the course of the nucleophilic addition, as illustrated in Figure 15.2.

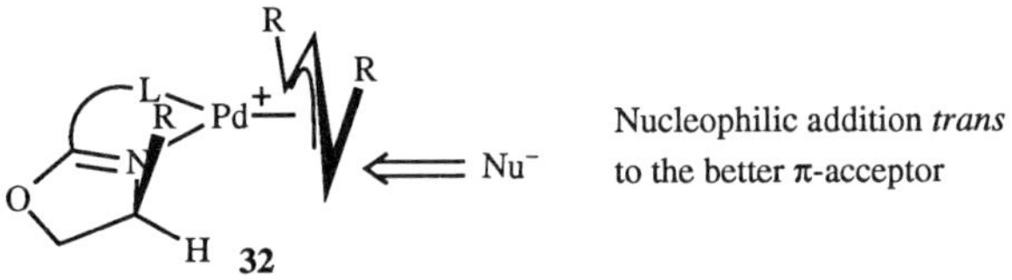

Nucleophilic addition *trans* to the better π-acceptor

Figure 15.2 Geometry of approach of nucleophile to the palladium complex of (**32**).

Variations on this theme also produce good levels of asymmetric induction in the allylic substitution process. Ligands containing sulfur[66] and selenium[64] (**33**–**37**) are also able to provide an asymmetric environment in conjunction with two electronically distinct donor atoms.

The importance of the electronic effect in controlling the enantioselectivity of the reaction is evident from comparison of several sterically similar, but

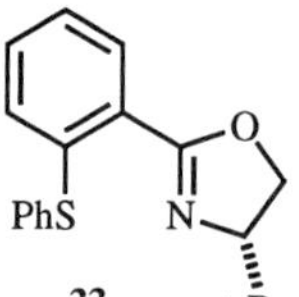

33 provides >96% ee (with **13**)

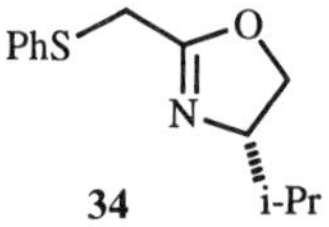

34 provides up to 76% ee (with **13**)

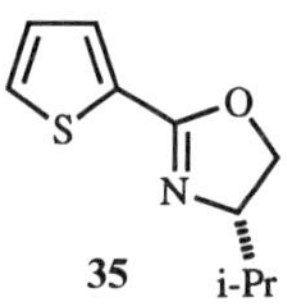

35 provides up to 81% ee (with **13**)

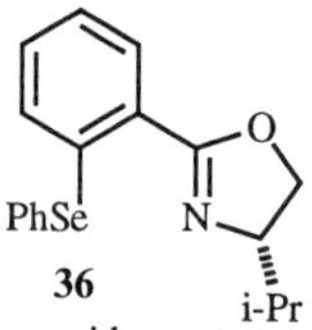

36 provides up to 95% ee (with **13**)

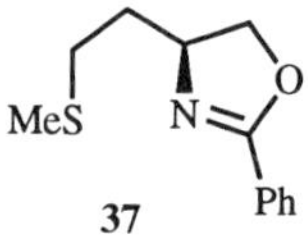

37 provides up to 88% ee (with **13**)

electronically altered ligands. Experimentally (Figure 15.3), the enantioselectivity observed in the palladium-catalysed allylic substitution reaction is found to be highest when the ligand contains a *p*-methoxy substituent. The origin of the variation in enantioselectivity is more likely to be electronic than steric.[67]

The oxazoline is not unique in providing a suitable steric environment for allylic substitution, and phosphine and sulfur donor groups attached to enantiomerically pure acetals, ligands (**38**) and (**39**), have also proved to be effective for the asymmetric palladium-catalysed allylic substitution process.[68] Hayashi has also reported the ligand (**40**) for palladium-catalysed allylic substitution, although a crystal structure suggests that this ligand is monodentate.[69] Presumably there is scope for further developments which involve a ligand possessing a suitably orientated asymmetric environment combined with two electronically distinct donor atoms.

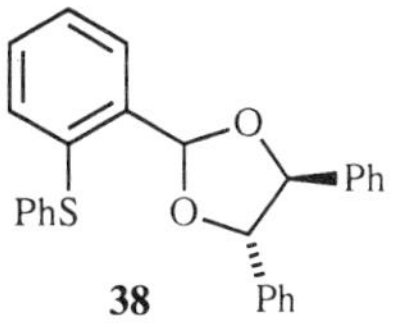

PhS, O, O, Ph, Ph

38

provides up to 82% ee (with **13**)

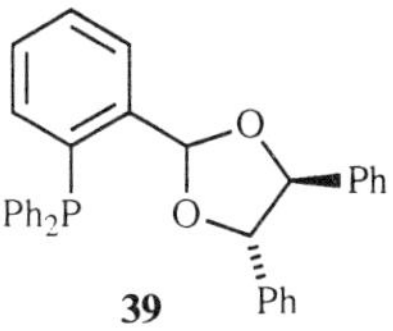

Ph_2P, O, O, Ph, Ph

39

provides up to 88% ee (with **13**)

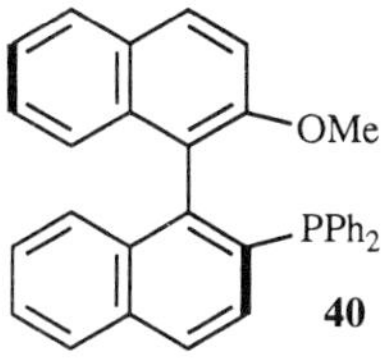

OMe, PPh_2

40

provides up to 87% ee in Pd catalysed allylic substitution

15.4.2 *Enantioselective palladium-catalysed allylic substitution reactions which do not proceed via* meso-*complexes*

It is not a requirement that the allylic precursor is designed to proceed through a symmetrically disposed allylpalladium complex. For example, the use of the allylic acetate (**41**) has been shown by Bosnich and others to undergo an enantioselective palladium-catalysed allylic substitution reaction to afford the product (**42**) (Scheme 15.14).[70] In the presence of enantiomerically pure ligands, there are two diastereomeric allylpalladium complexes. These complexes are able to interconvert rapidly, and one complex predominates at equilibrium. Nucleophilic addition must occur selectively to one of the diastereomeric complexes in order for asymmetric induction to be achieved.

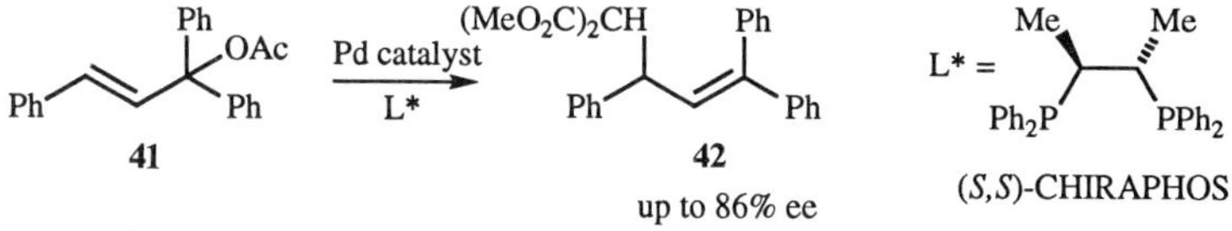

Ph, OAc, Ph, Ph — **41** —(Pd catalyst, L*)→ $(MeO_2C)_2CH$, Ph, Ph, Ph — **42** up to 86% ee

L* = Me, Me, Ph_2P, PPh_2 (*S,S*)-CHIRAPHOS

Scheme 15.14

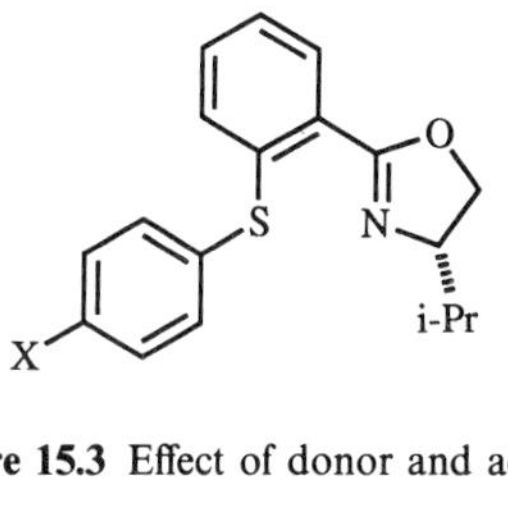

S, O, N, i-Pr, X

X	ee obtained from (**13**)
NO_2	81
H	90
Me	83
MeO	94
Me_2N	88

Figure 15.3 Effect of donor and acceptor substituents (X) on asymmetric induction.

Asymmetric induction in palladium-catalysed allylic substitution may also be achieved by the use of a substrate in which there is a choice of enantiotopic leaving group. This is illustrated by the reaction of (**43**), in which palladium-catalysed allylic substitution breaks the plane of symmetry within the substrate, affording a chiral product (**44**) with one enantiomer predominating when using the enantiomerically pure ligand (**45**) (Scheme 15.15).[71]

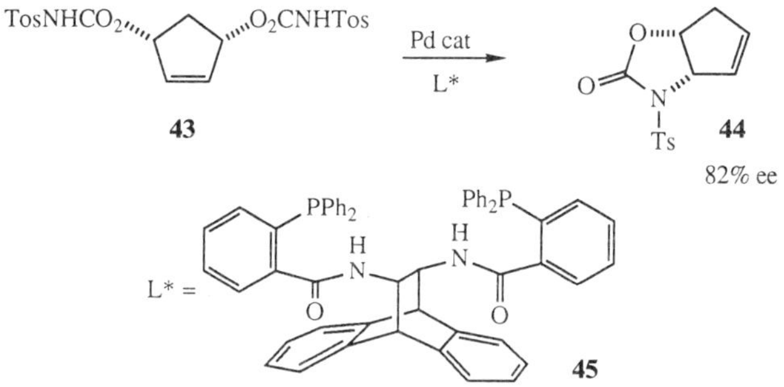

Scheme 15.15

Prochiral nucleophiles have been employed in palladium-catalysed allylic substitution reactions, and, in the presence of a suitable ligand, asymmetric induction has been achieved.[72] For example, treatment of diketone (**46**) with allyl acetate in the presence of a palladium catalyst and enantiomerically pure ligand (**47**) affords the substitution product (**48**) in 52% ee (Scheme 15.16).

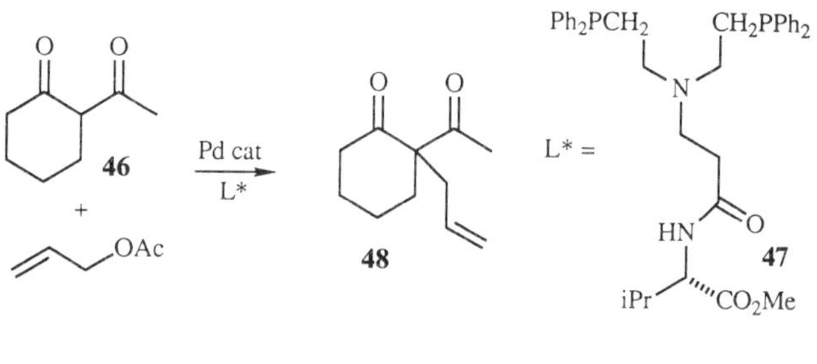

Scheme 15.16

15.5 Conclusions

Palladium-catalysed allylic substitution reactions are able to tolerate a wide range of functional groups in both the nucleophile and the substrate. Additionally, regiocontrol and diastereocontrol is possible in suitable systems. Since the mid-1980s, there have been many advances in enantioselective palladium-catalysed allylic substitution, and there have recently been many ligands devleoped which do not require functional groups to direct the incoming nucleophile. Whilst still in its infancy, it is now possible to design ligands which are capable of perturbing the sterics and electronics of allylpalladium intermediates and of introducing high levels of stereocontrol.

References

1. (a) Godleski, S.A., in *Comprehensive Organic Synthesis*, Vol. 4, Trost, B.M. (ed.), Pergamon Press, Oxford, 1991, p. 585; (b) Consiglio, G. and Waymouth, R.M., *Chem. Rev.*, 1989, **89**, 257; (c) Tsuji, J., *Pure Appl. Chem.*, 1986, **58**, 869; (c) Frost, C.G., Howarth, J. and Williams, J.M.J., *Tetrahedron: Asymmetry*, 1992, **3**, 1089.
2. Rieke, R.D., Kavaliunas, A.V., Rhyne, L.D. and Fraser, D.J.J., *J. Am. Chem. Soc.*, 1979, **101**, 246.
3. Robinson, S.D. and Shaw, B.L., *J. Chem. Soc.*, 1963, 4806.
4. Trost, B.M. and Metzher, P.J., *J. Am. Chem. Soc.*, 1980, **102**, 3572.
5. (a) Mason, R. and Whimp, P.O., *J. Chem. Soc. (A)*, 1969, 2709; (b) Smith, A.E., *Acta Cryst.*, 1965, **18**, 331; (c) Hegedus, L.S., Åkermark, B., Olsen, D.J., Anderson, O.P. and Zetterberg, K., *J. Am. Chem. Soc.*, 1982, **104**, 697.
6. Pregosin, P.S., Ruegger, H., Salzmann, R., Albinati, A., Lianza, F. and Kunz, R.W., *Organometallics*, 1994, **13**, 83.
7. Norrby, P.-O., Åkermark, B., Haeffner, F., Hansson, S. and Blomberg, M., *J. Am. Chem. Soc.*, 1993, **115**, 4859.
8. Tsuji, J., Takashashi, H. and Morikawa, M., *Tetrahedron Lett.*, 1965, 4387.
9. Atkins, K.E., Walker, W.E. and Manyik, R.M., *Tetrahedron Lett.*, 1970, 3821.
10. Trost, B.M. and Verhoeven, T.R., *J. Am. Chem. Soc.*, 1978, **100**, 3435.
11. Tsuji, J. and Yamakawa, T., *Tetrahedron Lett.*, 1976, 613.
12. Tsuji, J., Shimizu, I., Minami, I., Ohashi, Y., Takahashi, K. and Sugiura, T., *J. Org. Chem.*, 1985, **50**, 1523.
13. Stary, I., Stara, I.G. and Kocovsky, P., *Tetrahedron Lett.*, 1993, **34**, 179.
14. Trost, B.M. and Warner, R.W., *J. Am. Chem. Soc.*, 1982, **104**, 6112.
15. Tanigawa, Y., Nishimura, K., Kawasaki, A. and Murahashi, S., *Tetrahedron Lett.*, 1982, **23**, 5549.
16. Trost, B.M., Schmuff, N.R. and Miller, M.J., *J. Am. Chem. Soc.*, 1980, **102**, 5979.
17. Tsuji, Y., Yamada, N. and Tanaka, S., *J. Org. Chem.*, 1993, **58**, 16.
18. Murahashi, S., Tanigawa, Y., Imada, Y. and Taniguchi, Y., *Tetrahedron Lett.*, 1986, **27**, 227.
19. Baer, H.H. and Hanna, Z.S., *Can. J. Chem.*, 1981, **59**, 889.
20. Byström, S.E., Aslanian, R. and Bäckvall, J.E., *Tetrahedron Lett.*, 1985, **26**, 1749.
21. (a) Takagi, M. and Yamamoto, K., *Chem. Lett.*, 1989, 2123; (b) Inoue, Y., Taguchi, M., Toyofuku, M. and Hashimoto, H., *Bull. Chem. Soc. Jpn*, 1984, **57**, 3021.
22. (a) Inoue, Y., Taguchi, M., Toyofuku, M. and Hashimoto, H., *Bull. Chem. Soc. Jpn*, 1984, **57**, 3021; (b) Connell, R.D., Rein, T., Åkermark, B. and Helquist, P., *J. Org. Chem.*, 1988, **53**, 3845.
23. Mandai, T., Matsumoto, T. and Tsuji, J., *Synlett*, 1993, 113.
24. Greenspoon, N. and Keinan, E., *J. Org. Chem.*, 1988, **53**, 3723.
25. (a) Tsuji, J., Sakai, K., Nagashima, H. and Shimizu, I., *Tetrahedron Lett.*, 1981, **22**, 131; (b) Larock, R.C., Harrison, L.W. and Hsu, M.H., *J. Org. Chem.*, 1984, **49**, 3662.
26. Julia, M., Nei, M. and Saussime, L., *J. Organomet. Chem.*, 1979, **181**, C17.
27. Åkermark, B., Nystrom, J.E., Rein, T., Bläckvall, J.E., Helquist, P. and Aslanian, R., *Tetrahedron Lett.*, 1984, **25**, 5719.
28. (a) Trost, B.M., Yashida, J. and Lautens, M., *J. Am. Chem. Soc.*, 1983, **105**, 4494; (b) Tsuji, Y., Kajita, S., Isobe, S. and Funato, M., *J. Org. Chem.*, 1993, **58**, 3607.
29. (a) Coulson, D.R., *Inorg. Synth.*, 1972, **13**, 21; (b) Coulson, D.R., *Inorg. Synth.*, 1990, **28**, 107.
30. Bläckvall, J.-E., Nyström, J.-E. and Nordberg, R.E., *J. Am. Chem. Soc.*, 1985, **107**, 3676.
31. Ukai, T., Kawazura, H., Ishii, Y., Bonnet, J.J. and Ibers, J.A., *J. Organomet. Chem.*, 1974, **65**, 253.
32. Tatsuno, Y., Yoshida, T. and Otsuka, S., *Inorg. Synth.*, 1990, **28**, 342.
33. Mandai, T., Matsumoto, T., Tsuji, J. and Saito, S., *Tetrahedron Lett.*, 1993, **34**, 2513.

34. Takemoto, T., Nishikimi, Y., Sodeoka, M. and Shibasaki, M., *Tetrahedron Lett.*, 1992, **33**, 3527.
35. (a) Yamamoto, T., Akimoto, M., Saito, O. and Yamamoto, A., *Organometallics*, 1986, **5**, 1559; (b) Kurosawa, H., *J. Organomet. Chem.*, 1987, **334**, 243.
36. Åkermark, B., Hansson, S., Krakenberger, B., Vitagliano, A. and Zetterberg, K., *Organometallics*, 1984, **3**, 679.
37. Trost, B.M., *Angew. Chem., Int. Ed. Engl.*, 1989, **28**, 1173.
38. Fiaud, J.-C. and Legros, J.-Y., *J. Org. Chem.*, 1987, **52**, 1907.
39. Trost, B.M. and Verhoeven, T.R., *J. Org. Chem.*, 1976, **41**, 3215.
40. Matsushita, H. and Negishi, E., *J. Chem. Soc., Chem. Commun.*, 1982, 160.
41. Deardorff, D.R., Linde, R.G., Martin, A.M. and Shulman, M.J., *J. Org. Chem.*, 1989, **54**, 2759.
42. Kang, S.-K., Kim, S.-G. and Lee, J.-S., *Tetrahedron: Asymmetry*, 1992, **3**, 1139.
43. Corradini, P., Maglio, G., Musco, A. and Paiaro, G., *J. Chem. Soc., Chem. Commun.*, 1966, 618.
44. Trost, B.M., Verhoeven, T.R. and Fortunak, J.M., *Tetrahedron Lett.*, 1979, 2301.
45. Hayashi, T., Yamamoto, A. and Hagihara, T., *J. Org. Chem.*, 1986, **51**, 723.
46. Yamamoto, K., Deguchi, R., Ogimura, Y. and Tsuji, J., *Chem. Lett.*, 1984, 1657.
47. Genet, J.P., Kopola, N., Juge, S., Ruiz-Montes, J., Antunes, O.A.C. and Tanier, S., *Tetrahedron Lett.*, 1990, **31**, 3133.
48. Hiroi, K., Abe, J., Suga, K., Sato, S. and Koyama, T., *J. Org. Chem.*, 1994, **59**, 203.
49. Marshall, J.A., Andrews, R.C. and Lebioda, L., *J. Org. Chem.*, 1987, **52**, 2378.
50. Trost, B.M. and Lee, P.H., *J. Am. Chem. Soc.*, 1991, **113**, 5076.
51. Baldwin, I.C. and Williams, J.M.J., unpublished results.
52. Nilsson, Y.I.M., Andersson, P.G. and Bäckvall, J.-E., *J. Am. Chem. Soc.*, 1993, **115**, 6609.
53. Yamaguchi, M., Shima, T., Yamagishi, T. and Hida, M., *Tetrahedron Lett.*, 1990, **31**, 5049.
54. Trost, B.M. and Murphy, D.J., *Organometallics*, 1985, **4**, 1143.
55. (a) Hayashi, T., Yamamoto, A., Hagihara, T. and Ito, Y., *Tetrahedron Lett.*, 1986, **27**, 191; (b) Hayashi, T., *Pure and Appl. Chem.*, 1988, **60**, 7.
56. (a) Okada, Y., Minami, T., Sasaki, Y., Umezu, Y. and Yamaguchi, M., *Tetrahedron Lett.*, 1990, **31**, 3905; (b) Okada, Y., Minami, T., Umezu, Y., Nishikawa, S., Mori, R. and Nakayama, Y., *Tetrahedron: Asymmetry*, 1991, **2**, 667.
57. Yamazaki, A., Morimoto, T. and Achiwa, K., *Tetrahedron: Asymmetry*, 1993, **4**, 2287.
58. Leutenegger, V., Umbricht, G., Fahrni, C., von Matt, P. and Pfaltz, A., *Tetrahedron*, 1992, **48**, 2143.
59. Pfaltz, A., *Acc. Chem. Res.*, 1993, **26**, 339.
60. (a) Togni, A., *Tetrahedron: Asymmetry*, 1991, **2**, 683; (b) Togni, A., Rihs, G., Pregosin, P.S. and Ammann, C., *Helv. Chim. Acta*, 1990, **73**, 723.
61. Kubota, H., Nakajima, M. and Koga, K., *Tetrahedron Lett.*, 1993, **34**, 8135.
62. (a) Sprinz, J. and Helmchen, G., *Tetrahedron Lett.*, 1993, **34** , 1769; (b) von Matt, P. and Pfaltz, A., *Angew. Chem., Int. Ed. Engl.*, 1993, **32**, 566; (c) Dawson, G.J., Frost, C.G., Williams, J.M.J. and Coote, S.J., *Tetrahedron Lett.*, 1993, **34**, 3149; (d) Brown, J.M., Hulmes, D.I. and Guiry, P.J., *Tetrahedron*, in press.
63. Åkermark, B., Krakenberger, B., Hansson, S. and Vitagliano, A., *Organometallics*, 1987, **6**, 620.
64. Sprinz, J., Kiefer, M., Helmchen, G., Reggelin, M., Huttner, G., Walter, O. and Zsolnai, L., *Tetrahedron Lett.*, 1994, **35**, 1523.
65. Reiser, O., *Angew. Chem., Int. Ed. Engl.*, 1993, **32**, 547.
66. (a) Frost, C.G. and Williams, J.M.J., *Tetrahedron Lett.*, 1993, **34**, 2015; (b) Frost, C.G. and Williams, J.M.J., *Tetrahedron: Asymmetry*, 1993, **4**, 1785; (c) Dawson, G.J., Frost, C.G., Martin, C.J., Williams, J.M.J. and Coote, S.J., *Tetrahedron Lett.*, 1993, **34**, 7793.
67. Allen, J.V. and Williams, J.M.J., manuscript in preparation.
68. Frost, C.G. and Williams, J.M.J., manuscript in preparation.
69. Hayashi, T., Iwamura, H., Naito, M., Matsumoto, Y., Uozumi, Y., Miki, M. and Yanagi, K., *J. Am. Chem. Soc.*, 1994, **116**, 775.
70. (a) Mackenzie, P.B., Whelan, J. and Bosnich, B., *J. Am. Chem. Soc.*, 1985, **107**, 2046; (b) Farrar, D.H. and Payne, N.C., *J. Am. Chem. Soc.*, 1985, **107**, 2054.
71. Trost, B.M. and van Vranken, D.L., *Angew. Chem., Int. Ed. Engl.*, 1992, **31**, 228.
72. (a) Fiaud, J.C., DeGournay, A.H., Lachevéque, M. and Kagan, H.B., *J. Organomet. Chem.*, 1978, **154**, 175; (b) Hayashi, T., Kanechira, K., Tsuchiya, H. and Kumada, M., *J. Chem. Soc., Chem. Commun.*, 1982, 1162; (c) Hayashi, T., Kanehira, K., Hagihara, T. and Kumada, M., *J. Org. Chem.*, 1988, **53**, 113; (d) Sawamura, M., Nagata, H., Sakamoto, H. and Ito, Y., *J. Am. Chem. Soc.*, 1992, **114**, 2586.

16 Stoichiometric π-complexes in asymmetric synthesis

G.R. STEPHENSON

16.1 Introduction

When stoichiometric π-complexes are used in asymmetric synthesis, the metal centre can be brought into play in a series of reactions. In this way, the most efficient use is made of the metal/ligand system. In contrast, if asymmetric control is required at just a single synthetic step, the best way to proceed is to use a chiral catalyst system, so that only a small amount of the often expensive chiral auxiliary is then needed. However when catalytic control methods must operate in a series of separate reactions, the approach relies, at each step, on a new and highly selective recognition of an increasingly elaborate chiral substrate. After the first induction of asymmetry, subsequent asymmetric induction is diastereoselective, not enantioselective.

If a stoichiometric chiral auxiliary is introduced into the molecule at an early stage in the synthesis and imparts so powerful a control influence that it can overcome the stereodirecting effects of other chirality present in intermediates, a potentially attractive alternative to a sequence of catalytic steps becomes apparent. The auxiliary group must also be versatile in its mode of action, so that it can control a wide selection of bond-forming reactions. Stoichiometric auxiliaries capable of meeting these criteria can provide a stereochemical anchor point as a synthesis progresses towards its target structure. The same control group is present at each asymmetric step in the reaction sequence, and the same methodology can be employed to form each of a sequence of new bonds. In contrast, a series of catalytic steps frequently requires a different approach in each reaction.

16.1.1 Stoichiometric metal complexes in synthesis design

Stoichiometric transition-metal π-complexes are particularly suitable to be used in a synthetic strategy of this type. First, the attachment of the metal to an unsymmetrical ligand itself constitutes a stereogenic unit, so no expensive additional chiral auxiliary is needed. Second, attachment of an unsaturated ligand to a transition metal renders the ligand electrophilic. With cationic metal complexes, exceptionally powerful electrophiles are available. Therefore, in a reaction sequence, it is possible to switch to highly activated electrophilic forms of the metal complex (and so to switch on the possibility of powerfully controlled bond-forming reactions) while, in other steps of the synthesis, the metal in its neutral form can be carried through unchanged while conventional organic steps are performed at peripheral functional groups. Third, many transition metal complexes can impart stereocontrol at positions remote from the site of attachment to the ligand and so can influence the stereochemistry of conventional organic bond-forming processes at nearby functional groups, or can, in particular circumstances, offer the opportunity to perform specialised metal-mediated transformations, often forming several new carbon–carbon bonds at a single step.

Of these valuable attributes, the most important is the exceptional degree of stereocontrol that is typical of the reactions of nucleophiles to metal-bound electrophiles. With a wide variety of ligand sizes and types of nucleophile, only a single diastereoisomer is formed. It is rare to find 100% stereocontrol in bond-forming methodologies, and the general availability of complete diastereoselectivity is quite exceptional.

If full benefit is to be gained from these features, the utilisation of the metal must lie at the heart of the synthesis plan. Chemists need to learn new design skills to identify opportunities to bring this synthetic methodology into use. Although many of the individual reaction steps are well established (some of the key reactions were discovered in the early 1970s), the design principles which allow their efficient utilisation in target molecule synthesis are only now emerging. The objective of this review is to draw attention to the new methods of synthesis design[1] that are possible through the use of stoichiometric π-complexes in asymmetric synthesis.

16.1.2 Chirality properties of transition metal π-complexes

At the start, key structural and stereochemical features of transition-metal π-complexes need to be introduced. Simple alkene complexes are most familiar. The two carbon atoms in the alkene are both bound to the metal. The extent of this binding is referred to as the hapticity of the ligand, and is indicated by a superscript numeral to the Greek letter *eta* (η). An alkene complex is η^2. Larger ligands also form π-complexes. Allyl structures can have three carbon atoms bound to the metal, in which case they are η^3-complexes, for dienes it is η^4, and so on (Figure 16.1). These complexes are referred to as π-complexes because there is at least one antisymmetric component in the molecular orbitals that make up the metal ligand-bonding. However, the single line between the metal and the ligand represents a multiple bond with both σ- and π-components. In larger ligands such as η^6-arene complexes, several π character bonding interactions are present. Ligands in the range η^{2-6} are typical of the metal π-complexes used in asymmetric synthesis, but η^7- and η^8-coordination modes are also possible. The structural principles illustrated in Figure 16.1 apply across the whole range of ligand sizes. These are all 'side-on' metal complexes. If the ligands are represented within the plane of the paper, the thickened line to each metal atom in

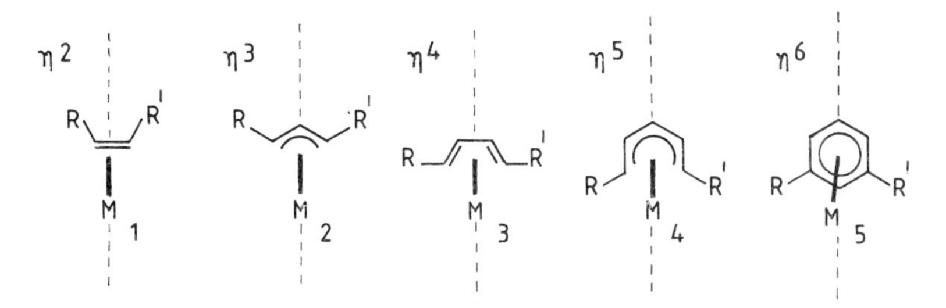

Figure 16.1 Potential symmetry planes in η^{2-6} π-complexes are indicated by dotted lines through the centres of the structures. In **(5)**, several potential symmetry planes exist, depending on the nature of R and R′. Metal π-complexes are chiral when the unsymmetrical substitution patterns differentiate the sides of the potential symmetry plane (R and/or R′ ≠ H and R ≠ R′ for **(1–4)**; R ≠ R′ ≠ H for **(5)**.

Figure 16.1 represents a structure in which the metal lies directly above the ligand. In an η^6-benzene complex, for example (**5**:R = R′ = H), the metal lies directly above the centre of the 6-membered ring. Depending on the nature of substituents on the ligands, metal complexes of this type can be chiral. For example, in the case of (**1**:R ≠ H, R′ = H), the structure with the metal up is the mirror image of the structure with the metal down. The metal has distinguished the two faces of the prochiral ligand, and the metal complex is chiral. The two structures, **(1)** and its mirror image, are enantiomers. The same feature is true for the η^{3-5}-complexes **(2–4)**, but in the η^6-case **(5)** the two substituents R and R′ are *needed* for the structure to be chiral, and R and R′ must be different.

16.1.3 Potential symmetry planes in transition metal π-complexes

Each of the ligands in Figure 16.1 has a potential symmetry plane. Suitable substituents can differentiate the sides of this plane so rendering the complex chiral. Symmetrical placement of substituents relative to this plane constitutes an achiral structure since there is an actual symmetry plane through the centre of the molecule. Potential symmetry planes for η^{2-6} structures are shown in Figure 16.1, but in the case of cyclic π-systems, as in **(5)**, several other potential symmetry planes exist. It is for this reason that two different unsymmetrically placed substituents are needed to make the complex of an arene chiral.

(a) *Odd and even π-systems.* Odd and even π-systems[2] have different chirality properties. The even complexes (η^2 and η^4) can never have a substituent located on the potential symmetry plane. In the disubstituted cases shown in Figure 16.2, it is enough that either R or R′ is not a hydrogen atom for the complexes to be chiral. In all cases except **(14)**, this chirality is retained when R and R′ are the same. With the odd complexes (η^3, η^5), a substituent

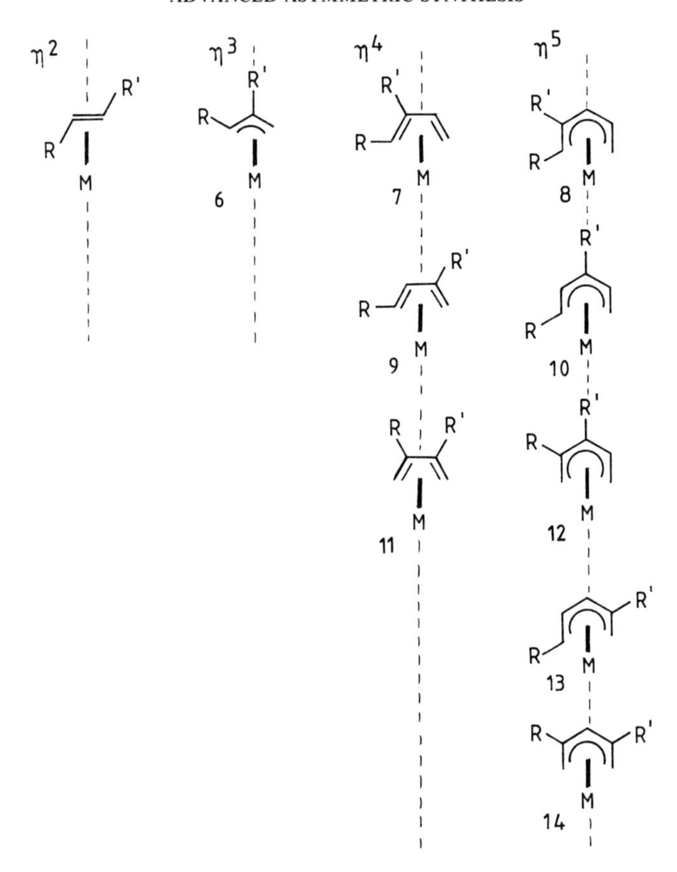

Figure 16.2 Chirality properties for examples of open π-systems. Chiral: η^2 R and/or R' $\neq$ H; η^3 (**6**) R $\neq$ H; η^4 (**7** and **9**) R and/or R' $\neq$ H; (**11**) R and/or R' $\neq$ H, R $\neq$ R'; η^5 (**8** and **13**) R and/or R' $\neq$ H; (**10** and **12**) R $\neq$ H; (**14**) R and/or R' $\neq$ H, R $\neq$ R'.

lying on the potential symmetry plane has no influence on the chirality properties of the molecule. Thus in (**6**) it is R not R' that differentiates the sides of the central plane. In the η^5-complexes, both situations arise.

The examples shown in Figure 16.2 do not constitute an exhaustive list of structural possibilities. In η^{3-5}-structures, antisubstitution patterns are also possible when substituents are at the terminus of the π-systems. This is a less common substitution pattern, and chirality properties can be determined by employing the same approach, by identifying potential symmetry planes and the placement of substituents relative to these features of the ligand. Similarly, the chirality of more highly substituted complexes depends on the distribution of substituents relative to the potential symmetry plane of the ligand.

(b) *Special chirality properties of closed π-systems.* As illustrated by our consideration of structure (**5**), closed π-systems require more careful analysis because several potential symmetry planes are possible. Once again, odd and even ligands are distinct. Chiral examples are defined in Figure 16.3, but for even π-systems, since it is possible for both substituents to lie on a potential symmetry plane of cyclobutadiene and arene ligands, some disubstituted complexes are achiral regardless of the nature of the substituent groups (Figure 16.4). With the odd closed π-systems, disubstituted η^3-complexes must always have one substituent on the potential symmetry plane and so will be chiral even when R and R′ are the same. In the η^5 case, in this 1,3 arrangement, neither of the two substituents can lie on the potential symmetry plane, so these complexes will be chiral in all cases *except* when the substituents are the same. It is important to keep in mind that substituted closed π-systems can be achiral even when substituted. Some examples are given in Figure 16.4.

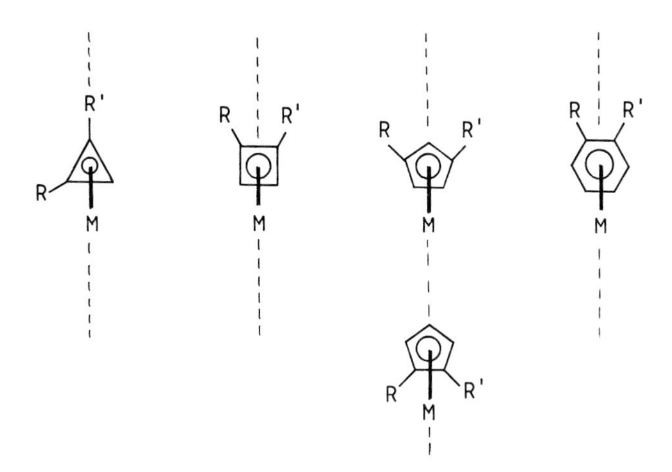

Figure 16.3 Disubstituted closed π-systems (chiral: R and R′ ≠ H, R ≠ R′).

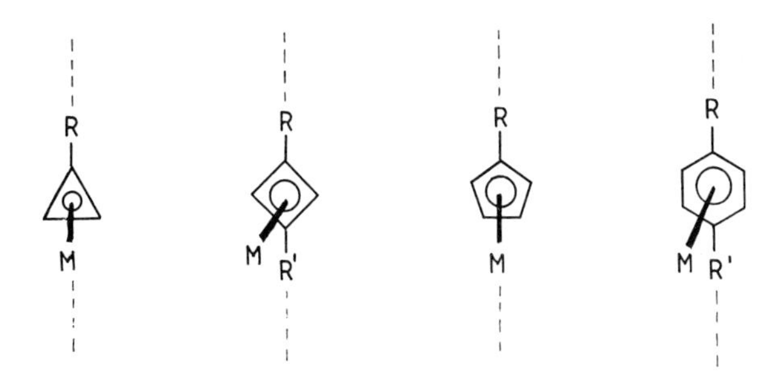

Figure 16.4 Examples of achiral substituted closed π-ligand (achiral: R and/or R′ = or ≠ H).

16.1.4 Stereochemistry of reactions of transition-metal π-complexes

Because the faces of the ligand are distinguished by the attachment of the metal, the chiral metal complexes shown in Figures 16.1–16.3 are examples of structures with planar chirality. This distinction between the faces is not only important because of its structural implications. There are also different reactivity properties on the face of the ligand carrying the metal and the face opposite to the metal. When the metal complex is electrophilic, nucleophiles add to a metal-bound carbon atom. This is illustrated in Figure 16.5, for a general case (η^n: $x = n - 2$) and for the typical reaction pathway for cationic π-systems involving nucleophile addition at the terminus of the η^n-unit. Nucleophile addition opposite to the face of the ligand carrying the metal is defined as *exo* while addition to the same face of the ligand (a much less usual result) is *endo*. Nucleophiles can also add to conventional electrophilic centres adjacent to metal-bound η^n structures. In this circumstance, the terms *ψ-exo* and *ψ-endo* are used, but the definition of these diastereoisomers is not straightforward since it relies on the identification of a designated substituent at the new chiral centre formed in the reaction. In the reaction shown in Figure 16.5, planar chirality in the η^n starting material induces carbon-centred chirality in the product. When nucleophile addition takes place at a metal-bound carbon atom, the extent of π-bonding in the product is reduced (an η^n-M^+ starting material affords a neutral η^{n-1}-M product). Of course, when nucleophiles add at a position adjacent to the metal complex, the hapticity of the metal/ligand unit remains unchanged. Many nucleophile

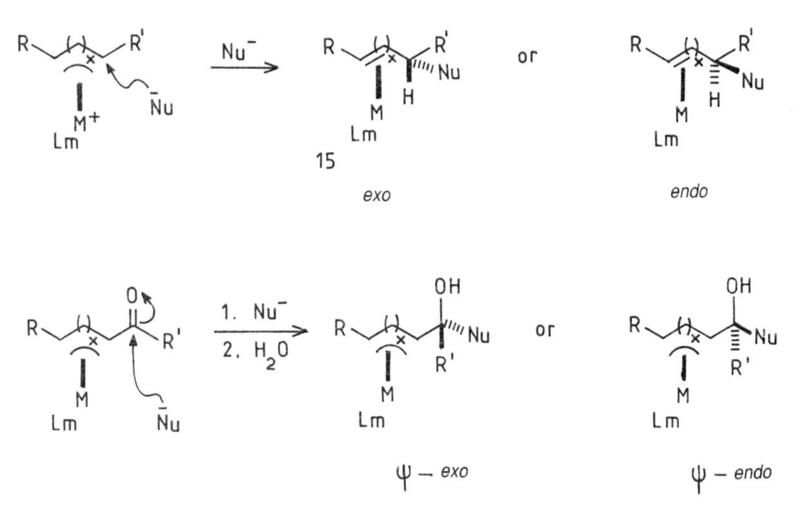

Figure 16.5 Products of nucleophile addition to η^n π-complexes have two stereogenic units: for $x \neq 0$, the product contains atom-centred chirality in addition to the planar chirality of the metal/ligand system; for $x = 0$, and for direct addition to the η^n unit, two positions with atom-centred chirality are present in the product.

additions to electrophilic π-complexes are completely controlled, and often only the *exo* diastereoisomer is formed. In these cases, if an enantiopure metal complex is used as the starting material, complete asymmetric induction at the carbon-centred chirality will be expected. There is no chiral auxiliary in this process; it is the chirality of attachment of the metal to the prochiral ligand system which is controlling the reaction. Stereocontrol in these reactions are examples of diastereoselectivity relative to the planar chirality of the metal-bound π-system.

16.2 Working ligands and their role in synthesis design

In target molecule synthesis, reactions take place at a metal-bound ligand that is destined to become incorporated into the target structure. Useful skeletal bond-formation reactions can be achieved by exploiting the reactivity properties of the metal/ligand assembly. It is helpful to refer to the ligand that is being elaborated as the 'working ligand'. There may be other ligands ('auxiliary ligands') on the metal, but these will not usually ultimately be built into the target molecule. Their role is to stabilise the transition metal complex and to adjust its reactivity properties. Therefore, in the selection of transition-metal complexed intermediates, substituent placement in the target determines the most suitable substitution pattern in the working ligand, and it is the partnership between the working ligand and the metal bearing its auxiliary ligands that determines the chances of success for the chosen route to the target molecule.

The formation of (**15**) by *exo*-selective nucleophile addition (Figure 16.5) illustrates an important additional requirement for control if these reactions are to operate in enantioselective synthesis. To be chiral, starting material must have an unsymmetrical substitution pattern, so even in the case of nucleophile addition to a terminus of the π-system, there are two possibilities; addition at the terminus carrying substituent R′ to form (**15**), or addition at

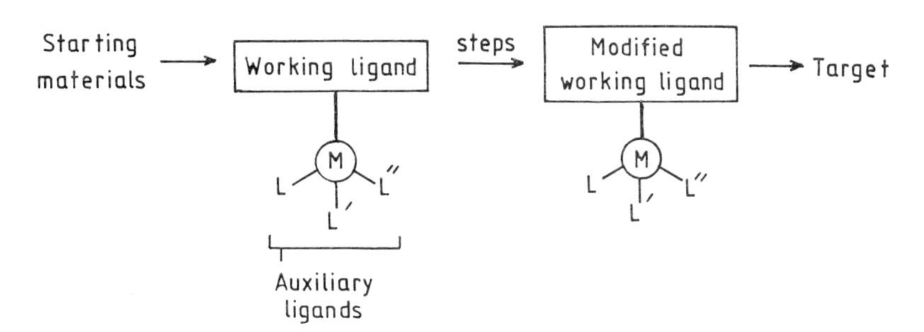

Figure 16.6 Organic starting materials and the atoms in the working ligand are combined stoichiometrically to form an organic product or target molecule in reactions that can be fine-tuned by the choice of the metal (M) and the auxilliary ligands L, L′ and L″.

the R-substituted end, which would form a regioisomeric product. Besides stereocontrol, which is often straightforward because of the potent influence of the metal, efficient regioselectivity is also required, and it is the unsymmetrically placed substituents that direct the regiocontrol in the reactions of the working ligand. Therefore, when choosing appropriate starting materials for a synthetic route, the extent of hapticity and the patterns of regiocontrol should both be considered. With reactions at organic functionality, fine-tuning must be performed by changing the substituents, and these changes can often carry unnecessary or even undesirable substitution through into later intermediates, requiring functional group deletion or interconversions to complete the target structure. In reactions of stoichiometric π-complexes (Figure 16.6), the choice of the metal, and the auxiliary ligands, provides a strategy for fine-tuning that leaves the substitution pattern on the working ligand unchanged. Ideally, substituents would be restricted to features that are needed either for direct incorporation into the target molecule or for an anticipated role in conventional (entirely organic) bond-formation steps.

For cationic η^5-metal complexes, the implications of these considerations in target synthesis were set out by Birch's group in the early 1980s,[3] where the role of the metal and its auxiliary ligands is illustrated by discussion of the strategy of the lateral control. This concept is applicable in general to all the η^n π-systems.

16.3 Stoichiometric metal complexes in skeletal bond-formation steps

To illustrate this chemistry, we will start with simple examples in which the metal participates in the control of the formation of only one skeletal bond. Later, larger ligand systems will be considered. It is here that the multiple use of the metal becomes most important. Since cationic π-complexes provide the best electrophiles, a sequence of reactions with nucleophiles also requires a sequence of activating steps to return to the cationic state after each nucleophilic addition. The design of these reactivation steps is often a crucial decision in the planning of the synthesis. Since electrophilicity is imparted on the working ligand by the transition metal and is not dependent on the usual activating effect of organic functional groups, nucleophile addition can be brought about at positions of either natural or umpolung reactivity, relative to the organic substitution pattern. The placement of the η^n-unit determines whether reactions correspond to natural or umpolung processes, and, in the case of sequential bond formations controlled by the same transition metal centre, the relative position of the new skeletal bonds formed during a reaction sequence is similarly a consequence of the choice of the size and position of the initial η^n-complex and whether iterative or linear reaction sequences[1b] are employed (Figure 16.7).

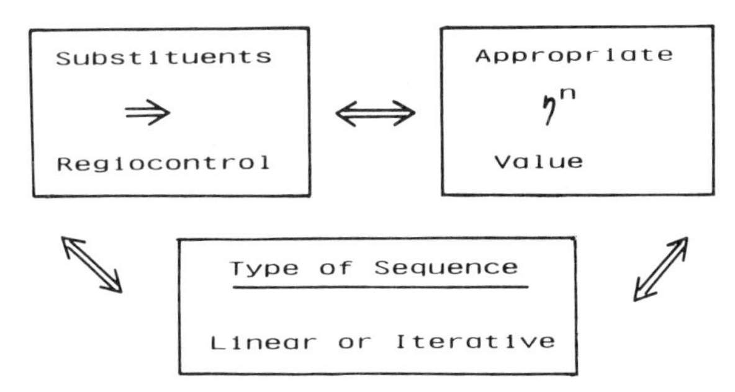

Figure 16.7 Both substitution pattern and the extent of hapticity are important factors in the choice of starting materials for regio- and stereoselective reaction sequences in target molecule synthesis.[4]

16.3.1 η^2-Complexes

The most commonly used cationic stoichiometric η^2-complexes are found in the dicarbonyl(cyclopentadienyl)iron series. The example (**16**) shows how attachment of the transition metal has imparted electrophilicity to the ethene working ligand (Scheme 16.1).

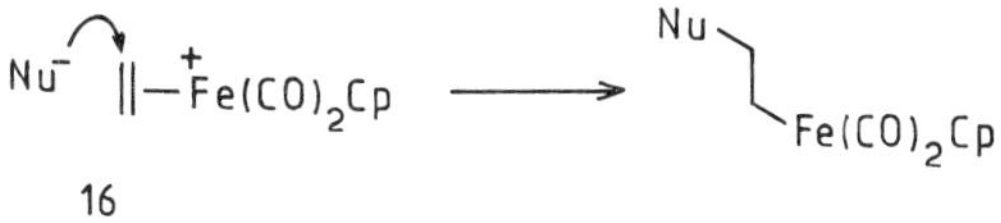

Nu^- ‖—$\overset{+}{Fe}(CO)_2Cp$ ⟶ Nu–CH_2CH_2–$Fe(CO)_2Cp$

16

Scheme 16.1

In the case shown in Scheme 16.2, reaction of the enolate of cyclohexenone with the vinylether complex (**17**) produced (**18**) as a single diastereoisomer. Hydride reduction of the ketone in (**18**) was successfully performed in the presence of the $Fe(CO)_2Cp$ unit, so preparing for lactonisation during carbonyl insertion upon oxidation of the metal complex with ceric ammonium nitrate (CAN). In this reaction sequence, the metal participated in two carbon–carbon bond formations, and one carbon–oxygen bond formation, and three new chiral centres were built along the way.[5] Racemic (**17**) was used, so a racemic product (**19**) was obtained.

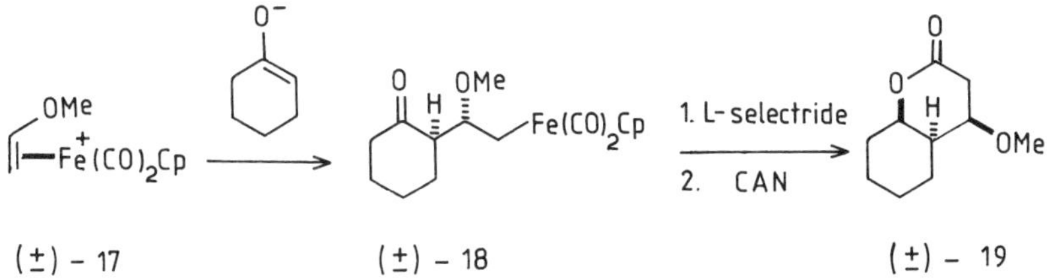

(±) - 17 ⟶ (±) - 18 ⟶ (1. L-selectride, 2. CAN) (±) - 19

Scheme 16.2

(a) *Synthesis of the insect pheromone sitophilate.* It is the unsymmetrical substitution pattern in (**17**) which renders it chiral. The dimethoxy-substituted alkene complex (**20**), by comparison is symmetrical and so is achiral. Acid-catalysed alkoxy group exchange with the chiral diol (**21**) affords a chiral structure (**22**) in which a chiral auxiliary has been covalently incorporated into the working ligand (Scheme 16.3). The methyl group on the face of the ring opposite to the metal blocks one end of the metal-bound alkene so nucleophile addition occurs at the unhindered end. Reaction with Me_2CuLi,

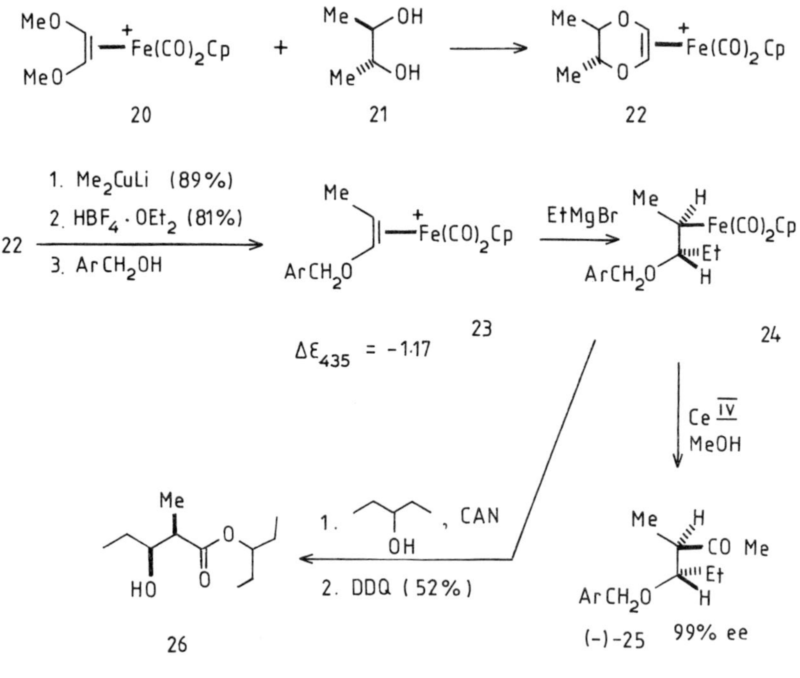

MeO, MeO, $\overset{+}{Fe}(CO)_2Cp$ + Me, OH, Me, OH → Me, O, Me, O, $\overset{+}{Fe}(CO)_2Cp$

20 21 22

22 —1. Me_2CuLi (89%) 2. $HBF_4 \cdot OEt_2$ (81%) 3. $ArCH_2OH$→ Me, $ArCH_2O$, $\overset{+}{Fe}(CO)_2Cp$ —EtMgBr→ Me, H, $Fe(CO)_2Cp$, Et, $ArCH_2O$, H

23 24

$\Delta\varepsilon_{435} = -1.17$

Ce^{IV} MeOH

Me, HO, O, O 1. OH, CAN 2. DDQ (52%) Me, H, CO Me, Et, $ArCH_2O$, H

26 (−)-25 99% ee

Scheme 16.3

followed by acid treatment effects the overall replacement of one of the carbon–oxygen bonds in (**22**) with a methyl group. A second alkoxy group exchange, using *p*-methoxybenzyl alcohol, detaches the chiral auxiliary to afford the propenyl ether complex (**23**). A further nucleophile addition can now be performed. Reaction with ethylmagnesium bromide occurs at the alkoxy-substituted end of the ligand to form (**24**). Oxidative removal of the metal in methanol extends the carbon chain by carbonyl insertion to produce the methyl ester ((−)-**25**), which was shown by means of a chiral shift reagent to have been formed in 99% ee. The iron complex (**24**) has been used as an intermediate for the synthesis of the insect pheromone sitophilate (**26**) in the unnatural enantiomer series. Instead of methanol, 3-pentanol was used in

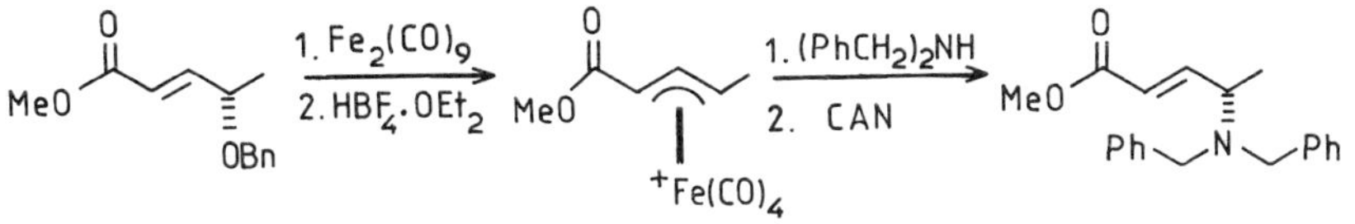

(−)-27 → 1. $Fe_2(CO)_9$, 2. $HBF_4 \cdot OEt_2$ → 28 (83% yld.) → 1. $(PhCH_2)_2NH$, 2. CAN → 29 (91% yld., 97% ee)

Scheme 16.4

the metal-removal step. Detachment of the benzyl protecting group completed the product structure. In this synthesis, three carbon–carbon bonds, a carbon–oxygen bond and two chiral centres have been formed by a sequence of nucleophile additions to the metal complex.[6]

16.3.2 *η^3-Complexes*

Stable stoichiometric cationic η^3-complexes are typically afforded by the tetracarbonyliron group. These metal complexes react well with a variety of nucleophiles. Complexation of ((−)-**27**) by reaction with $Fe_2(CO)_9$ affords an η^2-alkene complex which is converted into the cationic η^3 form by reaction with tetrafluoroboric acid. Only the *syn*-stereoisomer of the product (**28**) was formed. Nucleophile addition to this η^3-complex proceeds at the end of the allyl unit remote from the electron-withdrawing group. Reaction with dibenzylamine, for example, followed by oxidative removal of the metal gave the product (**29**) in high optical purity (Scheme 16.4).[7]

Similar complexation (Scheme 16.5) of the optically pure pyrrolidinone (**30**), which is readily obtained[8] from (*S*)-malic acid, afforded a mixture of two diastereoisomeric metal complexes ((+)-**31**) and ((−)-**32**). Stereocontrolled replacement of the allylic leaving group from ((+)-**31**) was achieved (Scheme 16.6) using a Lewis acid-promoted silylenol ether addition, to form the (*R*)-product (**33**) in >95% ee. Similar reaction with an allylsilane afforded (*R*-(**34**)).[8] The conversion of (**31**) into (**33**) or (**34**) involves displacement of the leaving group from the face of the ligand opposite to the metal and the introduction of an *exo*-substituent. Overall, the result is substitution with retention of configuration. In the *cis*-isomer (**32**), the leaving group must depart from the *endo*-face. The stereoselectivity of substitution in this case was considerably worse, and products were formed in 33–55% ee. A slow interconvertion of (**32**) into (**31**) has been proposed to account for the loss of stereoselectivity.

16.3.3 *η^4-Complexes*

(*a*) *Intermediates for macrolide antibiotics tylosin and carbomycin B_4.* In the η^4-series, dicarbonyl(cyclopentadienyl)molybdenum complexes have been

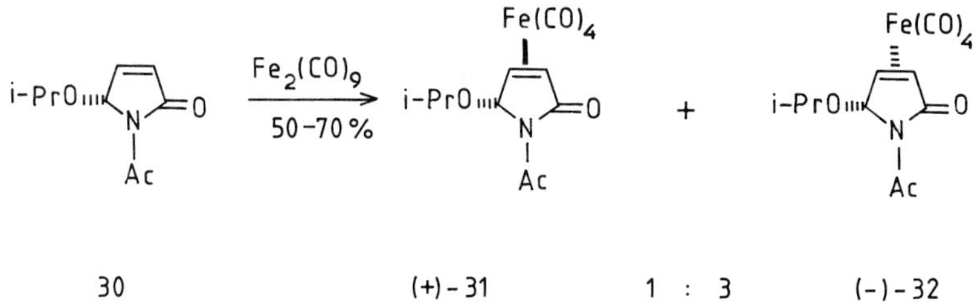

Scheme 16.5

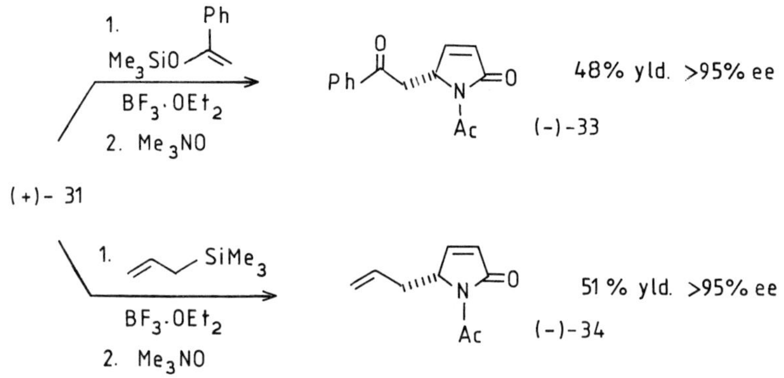

Scheme 16.6

much explored. A sequence of two nucleophile additions (Scheme 16.7) can be used to control relative stereochemistry. In a route to a macrolide building block, Pearson has first used a Grignard reagent and then an enolate to form two important chiral centres with *cis* relative stereochemistry in the elaboration of (**35**). Nucleophile addition to an η^4-cation affords a neutral η^3-product, which is reactivated by hydride abstraction from the less hindered end of the cyclohexenyl ligand to form the chiral η^4-complex (**36**). A second nucleophile addition returns the working ligand to the neutral η^3-form. When the metal is removed, further bond formation must occur. In this case, oxidation of the molybdenum with iodine effected a cyclisation to form a lactone with three stereodefined chiral centres. Oxidative cleavage of the remaining double bond produced[19] the dialdehyde (**37**) which has been taken on to a partially protected building block[10] which is suitable as a subunit in a synthetic route to the macrolide antibiotics tylosin and carbomycin B.

A similar reaction sequence (Scheme 16.8) employing two Grignard reagents as nucleophiles leads to the 1,5-disubstituted η^3-product (**38**). In this reaction sequence, as was the case in the sequence leading to the bicyclic lactone (**37**), hydride abstraction with the triphenylcarbenium ion is used to effect reactivation between the first two nucleophile addition steps. An

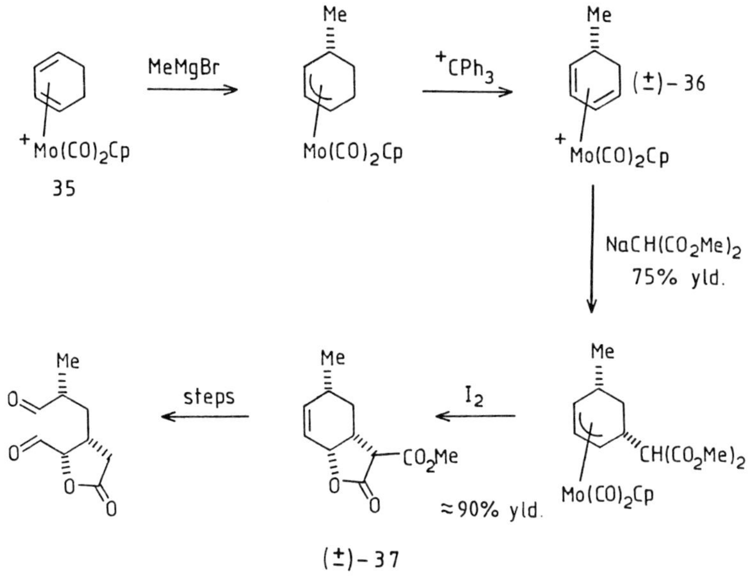

$^+Mo(CO)_2Cp$ **35** —MeMgBr→ Me, $Mo(CO)_2Cp$ —$^+CPh_3$→ Me, $^+Mo(CO)_2Cp$ (±)-**36**

$NaCH(CO_2Me)_2$ 75% yld.

Me, $Mo(CO)_2Cp$, $CH(CO_2Me)_2$ —I_2→ Me, CO_2Me, O, O (±)-**37** ≈90% yld. —steps→ Me, O, O, O, O

Scheme 16.7

alternative to the reformation of the η^4-cation is provided by exchange of ligands at the metal. NO^+ is isoelectronic with a carbonyl group, so exchange of these ligands in a neutral η^3-structure affords a modified η^3-cation with the same working ligand but different auxiliary ligands. In general, it has proved difficult to use these Mo(CO)NOCp cations in efficient carbon–carbon bond-forming reactions. In this case, the cationic product is employed with a hydride reducing agent as a nucleophile to afford the disubstituted cyclohexene (**39**).[11]

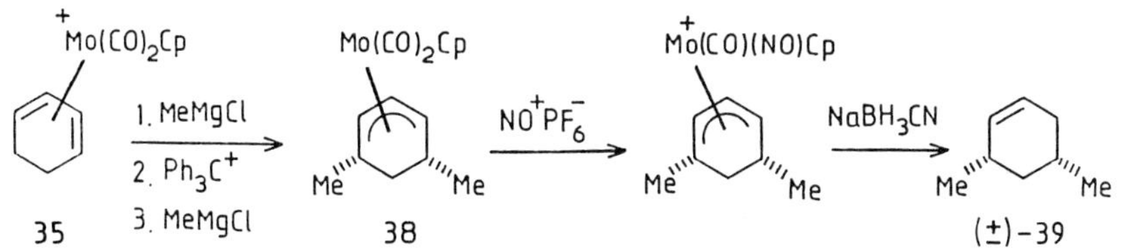

$^+Mo(CO)_2Cp$ **35** —1. MeMgCl 2. Ph_3C^+ 3. MeMgCl→ $Mo(CO)_2Cp$, Me, Me **38** —$NO^+PF_6^-$→ $^+Mo(CO)(NO)Cp$, Me, Me —$NaBH_3CN$→ Me, Me (±)-**39**

Scheme 16.8

(*b*) *Asymmetric induction in η^4-complexes.* A starting point for these reaction sequences is the prochiral metal complex (**35**), so, since no chiral auxiliaries have been employed, the products (**37**) and (**39**) are racemic. A strategy for the asymmetric modification of this chemistry has been described.[12] The two ends of the η^4-diene complex in (**35**) are enantiotopic, so in reactions with an enantiopure chiral nucleophile there is the possibility of differentiation leading to asymmetric induction in the product. The potassium enolate derived from an (*S*)-sulfoximine has been used to good effect in this way (Scheme 16.9). Addition of the enolate produced a mixture of diastereoisomers. These products were desulfonylated by a two-step procedure using first fluoride and then aluminium amalgam to afford ((−)-**40**) in 80% yield and 80% ee, and as a single diastereoisomer. This corresponds to exclusive *exo*-attack by the nucleophile combined with considerable enantiotopic group selectivity in the induction of asymmetry in the prochiral metal/ligand bonding system. The allyl complex (**40**) is chiral and can be further elaborated with complete control by hydride abstraction and nucleophile addition (of a phenyl group from diphenylcuprate in this instance) along the lines set out in Schemes 16.7 and 16.8.

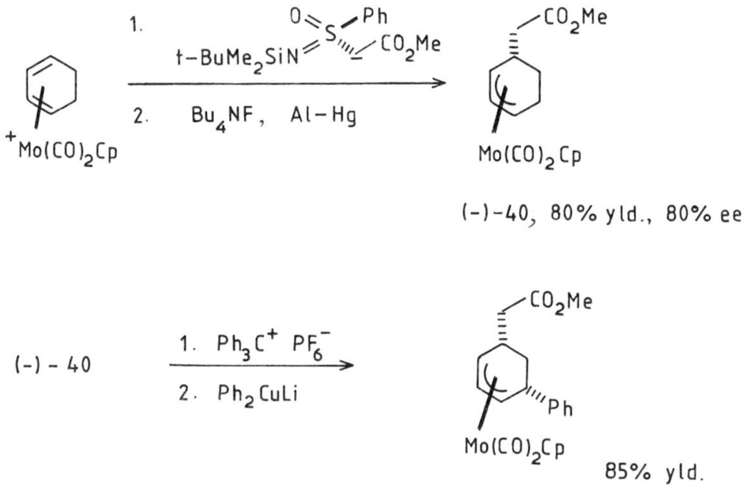

1. t-BuMe$_2$SiN=S(O)(Ph)CH$_2$CO$_2$Me

2. Bu$_4$NF, Al−Hg

$^+$Mo(CO)$_2$Cp

Mo(CO)$_2$Cp

CO$_2$Me

(−)-40, 80% yld., 80% ee

(−) - 40

1. Ph$_3$C$^+$ PF$_6^-$

2. Ph$_2$CuLi

CO$_2$Me

Ph

Mo(CO)$_2$Cp

85% yld.

Scheme 16.9

(*c*) *Transoid* η^4-*complexes.* Acyclic η^4-diene complexes can adopt a transoid structure (Scheme 16.10) and when these are involved in alkylation reactions, *syn*-η^3-complexes are formed. A recent example utilised low-temperature conditions for the formation of a racemic transoid η^4-dicarbonyl(cyclopentadienyl)molybdenum cation complex (**41**), which, if alkylated at a temperature below −40°C, could be trapped in the transoid form. At higher temperatures, equilibration to the more normal cisoid form (**42**) resulted in a different steroisomer of the neutral η^3-product.[13]

In cases where unsymmetrical substituents are present on the coordinated ligand, regioselectivity is of importance. The Liu group has examined the effect of an ester substituent at C-2 of an acyclic diene complex.[14] In the

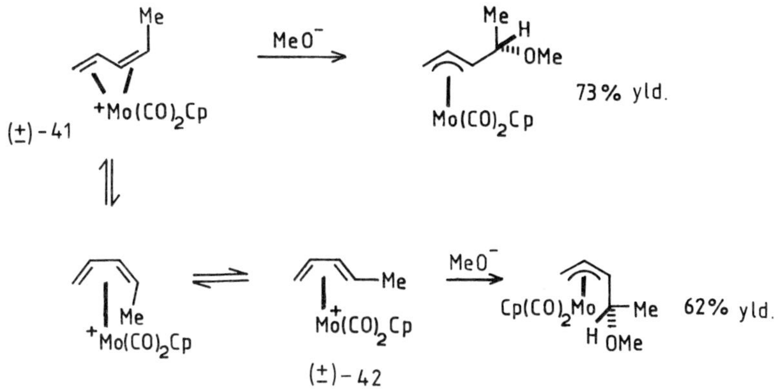

Scheme 16.10

case of many nucleophiles, preferential addition at the substituted terminus of structure (**43**) has been observed, though the product ratio depends on the type of nucleophile. Addition of benzylamine gave a single diastereoisomer (($\pm$)-**44**). In the case of the diphenylcuprate reagent, however, the major product was (($\pm$)-**45**) (Scheme 16.11).

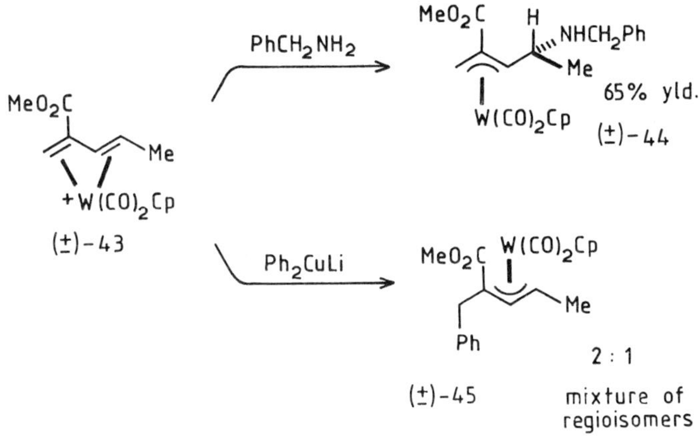

Scheme 16.11

16.3.4 η^5-Complexes

The most commonly used metal/auxiliary ligand unit to stabilise η^5-cations is the tricarbonyliron group. Related dicarbonyliron phosphine and phosphite complexes are also popular. The organoiron cyclohexadienyl complexes are optically stable and typically react with nucleophiles at the termini of the dienyl system. Larger ring systems, and acyclic complexes, behave in a more complicated fashion. In unsymmetrically substituted cyclohexadienyl ligands, substituents exert strong regiodirecting effects.[15]

Cationic tricarbonyl(cycohexadienyl)iron complexes are often formed by hydride abstraction from neutral η^4-precursors. Complexes of this type have been resolved[4,16] by separation of diastereoisomers using a chiral auxiliary. Alternatively, it is possible to effect an asymmetric transfer of the tricarbonyliron unit to a prochiral diene. Enantiopure enones derived from natural products have been used for this purpose, but the efficiency of chirality transfer is relatively poor and even in the best cases, products of only around 40% ee were obtained.[17] Nonetheless, by hydride abstraction to form the dienyl complex followed by precipitation of the enantiomerically enriched salt, it was possible to obtain (**46**) in optically pure form.[18] There has been much work on the use of 2-methoxycyclohexadienyl complexes as C-4 enone cation synthons (Figure 16.8).[19] The use of this approach to gain access to non-racemic 4,4-disubstituted cyclohexenones has been demonstrated.[20]

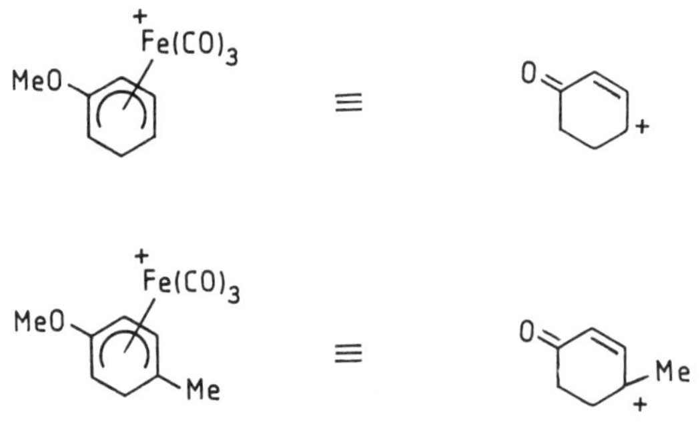

Figure 16.8 2-Methoxycyclohexadienyl complexes have been defined as equivalent of C-4 cyclohexenone cation synthons.

(*a*) *Trichodiene synthesis.* Routes towards trichothecenes developed in Pearson's laboratory illustrate the power of an OMe directing group to control regioselectivity in the formation of quaternary centres. In trichodiene (**49**), the two adjacent quaternary centres pose the main difficulty in total synthesis. Pearson planned his synthesis using a disconnection that combines a cyclohexadienyl complex with β-ketoester enolates and stannylenol ethers. It was the discovery, quite late in the research, that stannylenol ethers were exceptionally good reagents for this purpose, that opened up the best synthetic route.

In the synthesis of ($\pm$)-trichodiene (**49**), racemic **46** was employed. The key nucleophile addition step to form **47** and **48** showed 5:1 selectivity for the required stereoisomer. After separation, Wittig methylenation completed the right hand side of the molecule. Removal of the metal, with hydrolysis of the enol ether to reveal the cyclohexenone, was followed by a second methylenation and a 1,4 reduction. The target molecule was isolated after chromatographic removal of minor amounts of regioisomers. This synthesis (Scheme 16.12) indicates the power of the OMe substituent as an ω directing group (directing to the far end)[22] on the cyclohexadienyl ligand. Two adjacent quaternary centres have been built in the nucleophile addition step.

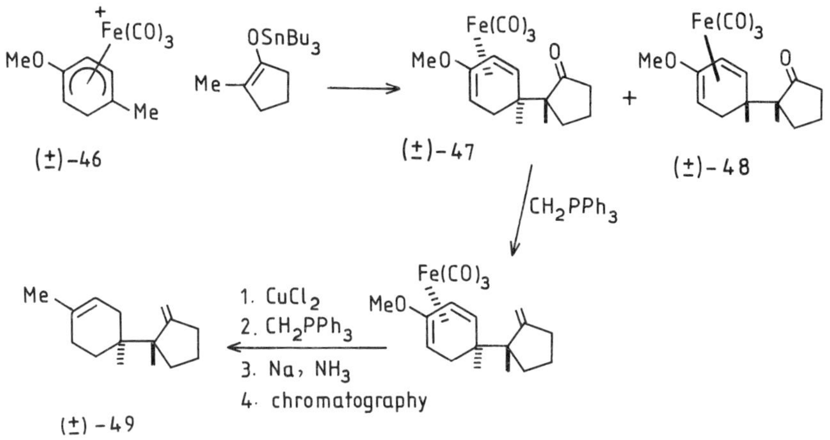

Scheme 16.12

The same approach has been used to complete ($\pm$)-trichodermol (**50**) in only 17 steps and in 5% overall yield from 4-methylanisole (Scheme 16.13). The use of a silyl substituent in the nucleophile sets the stereochemistry of a carbon–oxygen bond later in the synthesis.[22]

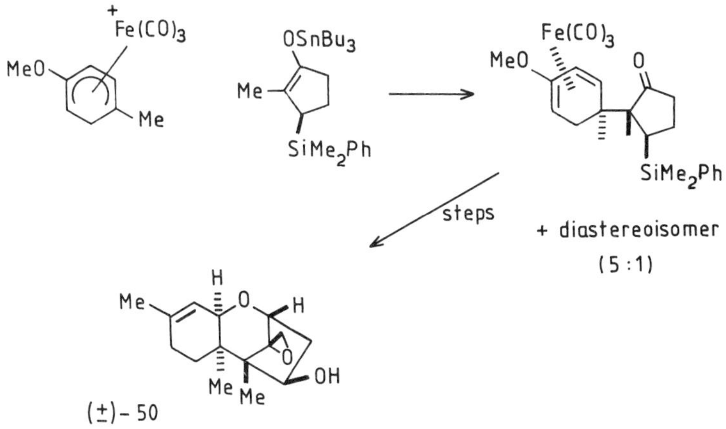

Scheme 16.13

(*b*) *Enantioselective synthesis of (−)-gabaculine.* Carbomethoxy substituents at C-1 of the dienyl system are also ω-directing. This provides the basis for Birch's approach[23] to an enantioselective synthesis of gabaculine (**53**), which served to prove the absolute configuration of this amino acid, since the configurations of the organometallic intermediates had previously been determined by chemical correlation with simple terpenes.[24]

The design of this synthesis uses natural regiocontrol in the nucleophile addition step (which conveniently requires a nucleophilic centre at nitrogen), and complete diastereoselectivity results from *exo*-addition to the metal π-complex. The BOC-protected amine H_2NCO_2-t-Bu was found to be the best source of the amino group. Metal removal from (**51**) used the trimethylamine *N*-oxide method to produce the free ligand ((−)-**52**). Two difficult deprotection steps followed to complete the synthesis (Scheme 16.14).

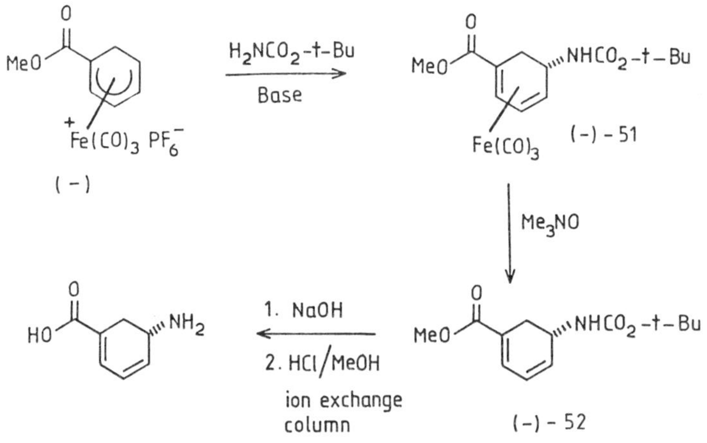

Scheme 16.14

The same approach has been used in an enantioselective synthesis of methyl shikimate, in which HO^- addition sets one of the three chiral centres in the target, the remaining two being formed by diastereoselective elaboration after removal of the metal.[25]

(*c*) *Stoichiometric control groups: the need for multiple use.* Tricarbonyliron complexes offer exceptionally powerful control, but in the cyclohexadienyl case, better methods are needed to enable the organometallic control effects to be brought into play at several key stages during the synthesis. Unlike the organomolybdenum examples (Schemes 16.7–16.9), once a nucleophile has introduced a chiral centre into the tricarbonyliron-complexed 6-membered ring, hydride abstraction to re-form the dienyl structure is blocked. Alternative methods using allylic leaving groups for the reactivation step, however, have

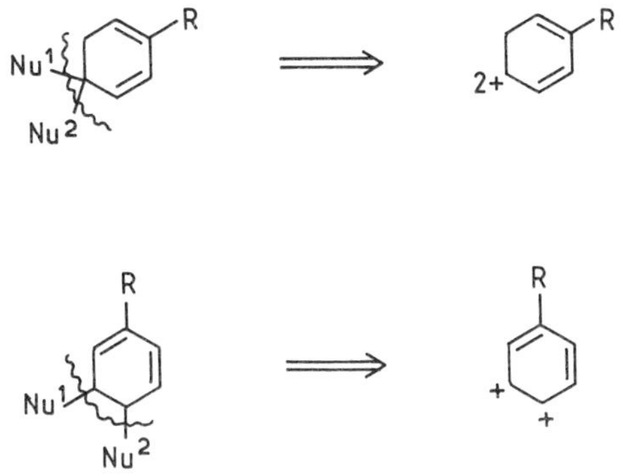

Figure 16.9 Multiple use of the organoiron control group allows the definition of dication synthons. In cyclohexadiene rings, two situations are important: 1,1- and 1,2-dications.

proved effective, and work in Norwich has explored these procedures in the context of alkaloid synthesis. In cases where sequential reactions can be performed, new dication equivalents can be defined (Figure 16.9).

(d) (±)-O-Methyljoubertiamine synthesis. *O*-methyljoubertiamine (**53**) provides a simple example of the asymmetric construction of quaternary centres in alkaloid synthesis. The introduction of both substituents as nucleophiles suggests a dication synthon and reactivation by a leaving-group strategy. The natural ω-directing effect of the aryl group[26] must be overcome by the inclusion of the methoxy substituent on the ring. The retrosynthetic analysis shown in Figure 16.10 suggests a 1,4-dimethoxy substitution pattern on the first dienyl complex. Both these substituents direct to the same pattern on the ring, so nucleophile addition at the substituted terminus of the dienyl system can be anticipated. This would leave the methoxy substituent as an allylic leaving group and so prepare the way for the reactivation step.

The synthesis starts (Scheme 16.15) from the prochiral 1,4-dimethoxy substituted complex (**54**).[27,28] Hydride abstraction produced a racemic sample of a single chiral cyclohexadienyl complex, and addition of the aryllithium reagent, followed by acid treatment gave access to the required 1-aryl salt on a multigram scale. Addition of the enolate of a cyanoacetate

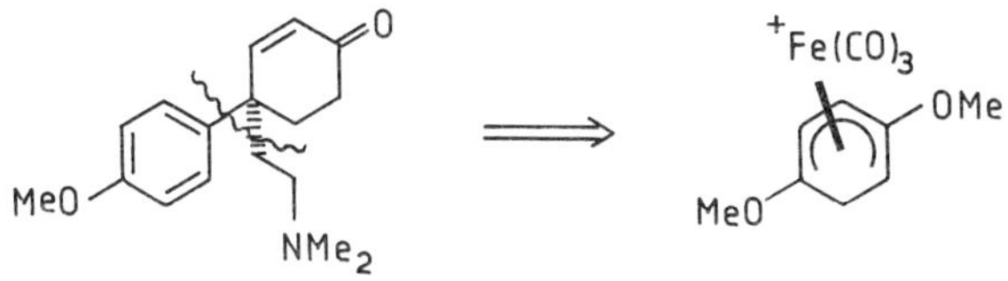

Figure 16.10 Retrosynthetic analysis indicates a 1,4-dimethoxy substitution pattern.

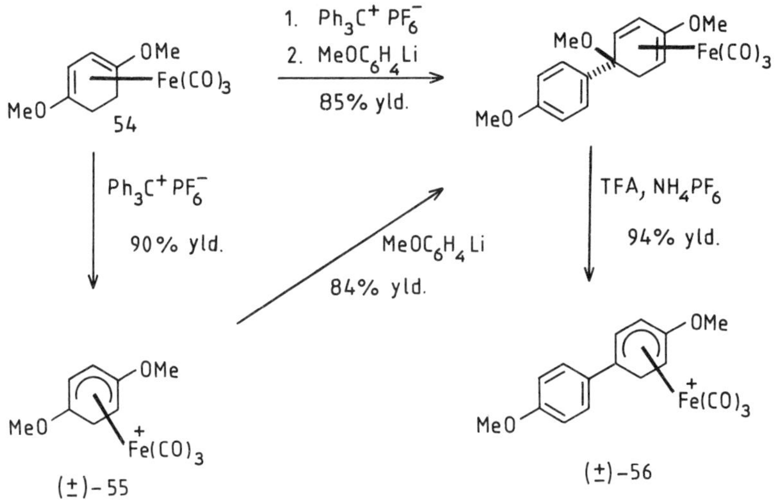

Scheme 16.15

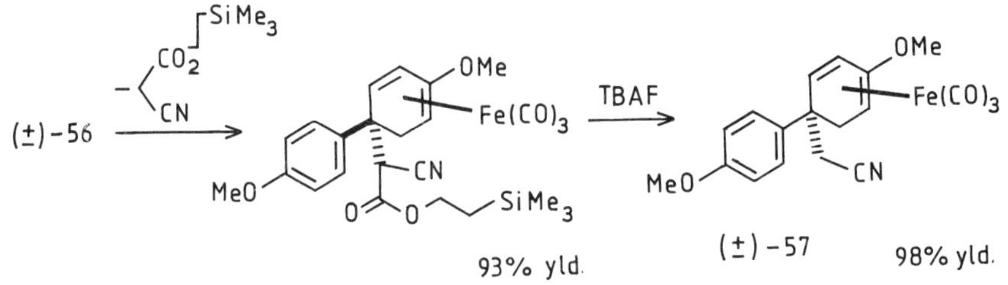

Scheme 16.16

silylethyl ester to (**56**) (Scheme 16.16) is the key step in the synthesis. Despite the presence of the *p*-methoxyaryl substituent, which directs strongly in the wrong direction, the methoxy group on the dienyl system dominated the regiocontrol, allowing the nucleophile to enter to form the required quaternary centre in the product. This nucleophile is prochiral, and the resulting chiral centre is not defined in this reaction. This constitutes no problem in the synthesis, since this chirality is removed in the next step to form (**57**). The crucial chiral centre in the metal-bound ring, however, is formed with complete diastereocontrol. Silylethyl malonate ester derivatives have been employed previously in combination with organoiron electrophiles.[29] Decarboxylation with fluoride was thus expected to proceed efficiently, and this proved to be the case, affording (**57**) in high yield. Hydrogenation of the nitrile must be performed before removing the metal, which is masking the enone functionality in thc target molecule. The construction of *O*-methyljoubertiamine in this way (Scheme 16.17) illustrates the use of tricarbonyliron complexes as 1,1 dication equivalents (Figure 16.9).[30]

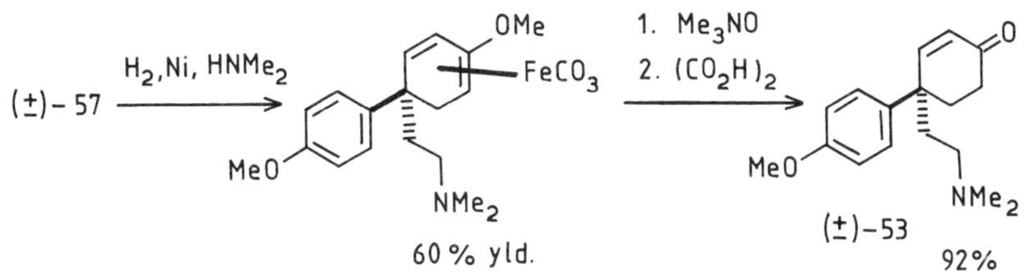

Scheme 16.17

(*e*) *Synthetic routes towards hippeastrine.* The target molecule hippeastrine (**58**) provides a case where 1,2 dication equivalence is required (Figure 16.11). Again, an allylic leaving group strategy is used, in this case one that is based on complexes of cyclohexadien-1,2-diol ligands, which are available[31] (see Chapter 13) from arenes by dioxgenation using the microorganism *Pseudomonas putida*. This can give convenient access to optically pure metal complexes. Enantiopure complexes related to (**59**, X = OEt) have been prepared in this way.[32] An approach to a racemic ABC-ring model (**60**) for hippeastrine can be based on the use of the prochiral 6-alkoxy cation complex (**63**) (Figure 16.12). A route combining dication and dianion synthons was adopted. The

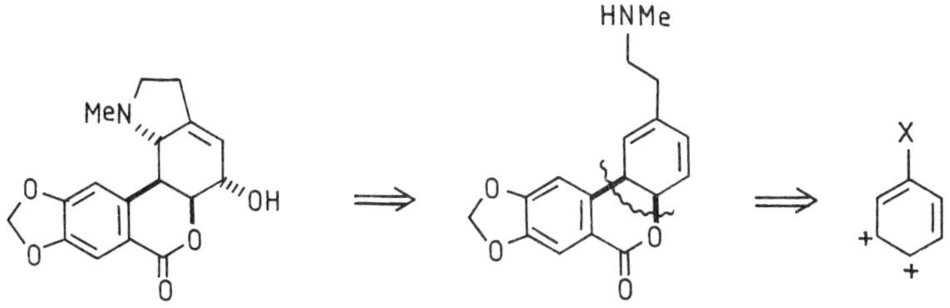

Figure 16.11 Retrosynthetic analysis indicates a 1,2 dication synthon for access to hippeastrine.

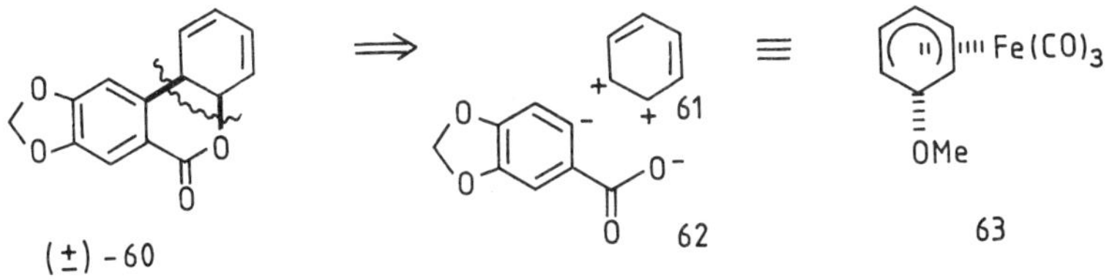

Figure 16.12 Dianion + dication approach in a model study for a synthesis of hippeastrine.

construction of the basic features of the lower four rings of hippeastrine illustrates the use of metal complexes as 1,2 dication equivalents.[33]

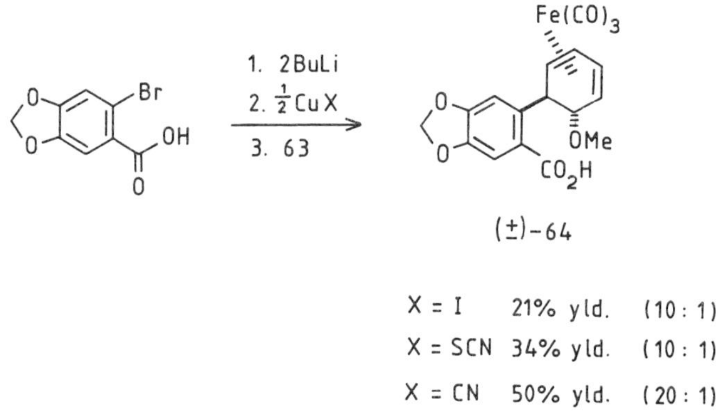

X = I 21% yld. (10 : 1)
X = SCN 34% yld. (10 : 1)
X = CN 50% yld. (20 : 1)

Scheme 16.18

Copper modification of a functionalised aryllithium dianion provided a starting point for the synthesis. Copper cyanide was found to be the best additive. The product ratios shown in Scheme 16.18 correspond to mixtures of regioisomers resulting from isomerisation of the position of metallation of the aromatic ring. By performing with the reaction at $-100°C$, the isomerisation was minimised. The intermediate (**64**) contains an allylic leaving group which can be removed by treatment with acid (Scheme 16.19) to afford

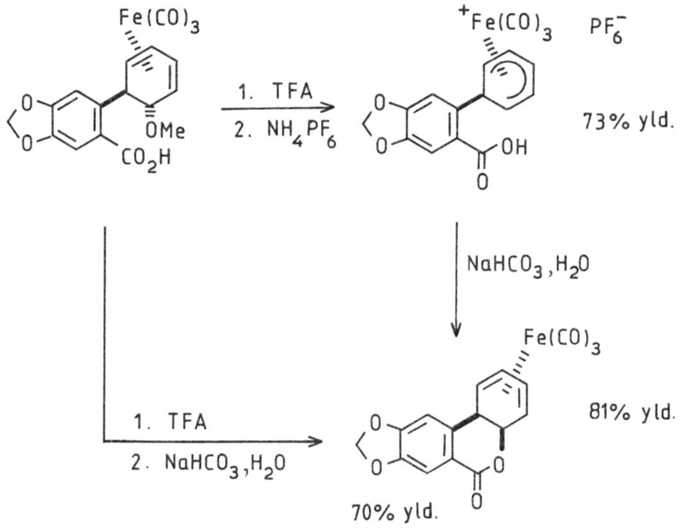

Scheme 16.19

a dienyl cation. Deprotonation of the carboxylic acid with mild base completed the lactone model. This reaction can be performed without isolation of the η^5-intermediate, gaining a slight improvement in yield.

(f) (+)-Prelog–Djerassi lactone synthesis. The leaving group methods discussed above are important in 6-membered rings because hydride abstraction cannot normally be performed after introduction of the first substituent. In 7-membered rings, steric hindrance at the methylene group is less, and hydride abstraction can proceed. Initial difficulties with inefficient and non-regioselective nucleophile addition have been overcome by use of $Fe(CO)_2P(OPh)_3$ complexes as electrophiles with organocuprate nucleophiles, to promote terminal attack on the dienyl complex.[34]

A stereoselective route to the Prelog–Djerassi lactone (**67**) was based on the metal-controlled introduction of the two methyl substituents. Control of relative stereochemistry in the metal-mediated steps formed a symmetrical intermediate (**65**), which was converted (via a *meso*-diacetate and lipase ester hydrolysis, see Chapter 13) into an unsymmetrical enantiopure enone (**66**) (Scheme 16.20). This was taken on to the target molecule by oxidative cleavage

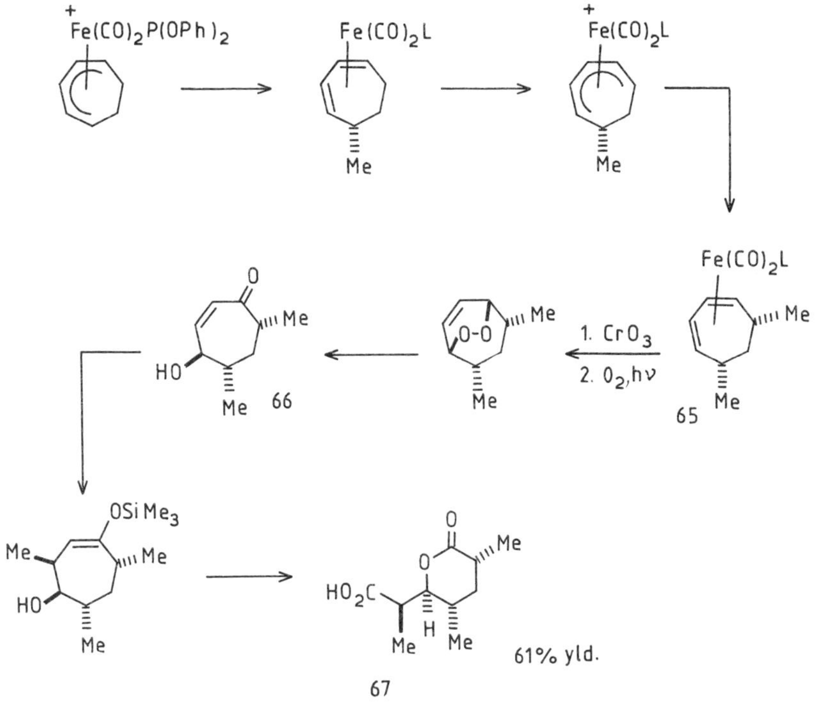

Scheme 16.20

of the 7-membered ring. Of the four chiral centres, two were controlled by the use of organoiron complexes.[35]

(*g*) *Other synthetic applications.* Similar methods have been used to produce a carbomycin B subunit,[34] in an alternative to the methods employing η^4-electrophiles discussed earlier. Attention has turned more recently to access to a fragment of the immunosuppressant FK-506, which requires the introduction of OMe in place of Me in the second nucleophile addition step.[36] Chiral nucleophiles have been used[12] to induce asymmetry in a prochiral dienyl complex, in a similar way to that illustrated for an η^4-electrophile in Scheme 12.9. The optical yield (50% ee) was somewhat lower in this case. These 1,5 difunctionalisation strategies have recently been extended to the cyclooctadiene series.[37]

Covalent introduction of achiral auxiliary via an ether link, as seen with η^2-complexes in Scheme 16.3, has also been used to obtain optically pure analogues of (**46**) (Figure 16.8). The separation of menthyl ethers is an effective procedure for this purpose.[38] The McCague group, working in the Drug Development Sector of the Cancer Research Campaign Laboratory, has used organoiron complexes in reaction with vinylcopper reagents to form highly substituted quaternary centres in a short synthesis of tamandron (**68**), a potentially antiandrogenic analogue of tamoxifen.[39] The use of isopropyl or menthyl ethers, and nucleophiles prepared from the copper(I) bromide dimethyl sulphide complex, proved the best conditions for quaternary centre formation. In other cases, considerable quantities of products arising from nucleophile addition at the wrong terminus of the dienyl system were observed. The synthesis of racemic tamandron has been completed in this way, and the key formation of the quaternary centre has been successfully achieved in 32% yield in the optically active menthyl ether series (Scheme 16.21).

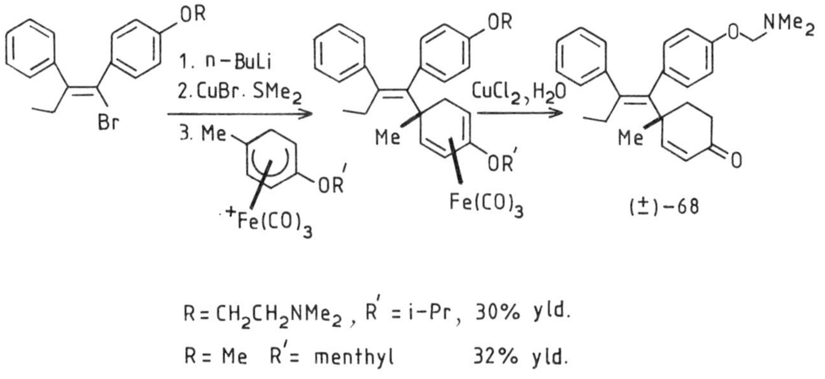

Scheme 16.21

16.3.5 η^6-Complexes

Cationic tricarbonylmanganese complexes of arenes provide typical examples of powerful η^6-electrophiles. The formation and alkylation of the η^6-complex (**69**) (Scheme 16.22) illustrates the use of an organometallic complex of this type in carbon–carbon bond formation. The cation (**69**) is prochiral so the product (**70**) is racemic.[40]

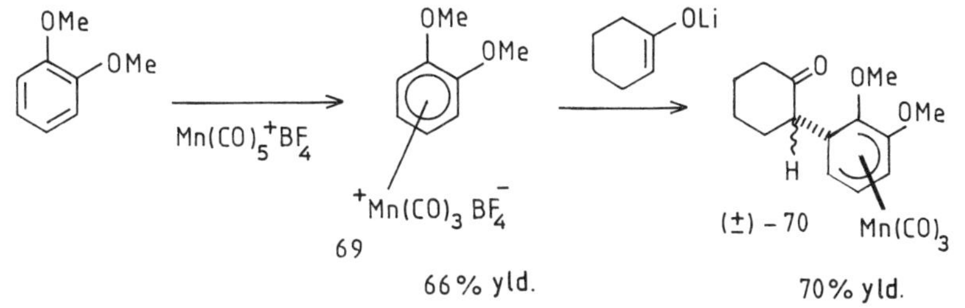

OMe OMe $Mn(CO)_5^+BF_4^-$ OMe OMe $^+Mn(CO)_3\ BF_4^-$ 69 66% yld. OLi O OMe OMe H (±) - 70 $Mn(CO)_3$ 70% yld.

Scheme 16.22

Cationic electrophile complexes often have the advantage of being compatible with a wide variety of nucleophile types. Even neutral complexes, however, are good electrophiles. In the η^6-series, the chemistry of tricarbonylchromium complexes has been very extensively explored. Nucleophile addition in this case (Scheme 16.23) affords an anionic η^5-intermediate. Protonation followed by oxidative removal of the metal can provide an elaborated product such as (**71**).[41]

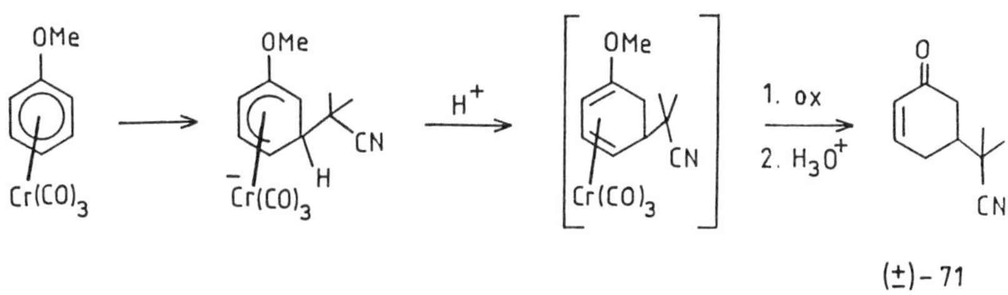

OMe $Cr(CO)_3$ OMe CN H $^-Cr(CO)_3$ H^+ OMe CN $Cr(CO)_3$ 1. ox 2. H_3O^+ O CN (±)- 71

Scheme 16.23

As with smaller π-complexes, sequences of repeated nucleophile additions can be performed. In the case of the route to (**72**) (Scheme 16.24), methyl group addition from a Grignard reagent is followed by NO^+ ligand exchange to effect reactivation. The result is an η^5-cation, with the working ligand bound to a modified manganese dicarbonylnitrosyl unit. Malonate enolate addition completed the formation of (**72**).[42] In this reaction sequence, both nucleophiles add *exo* to the metal, ensuring the *cis* relative stereochemistry at the two adjacent chiral centres. Transition metal η^6-complexes can also be contained in larger rings (Scheme 16.25), as is the case in the cycloheptatriene complex (**73**). Conversion into (**74**) again uses Grignard addition followed by NO^+ to produce an η^5-intermediate for the final nucleophile addition step.[43]

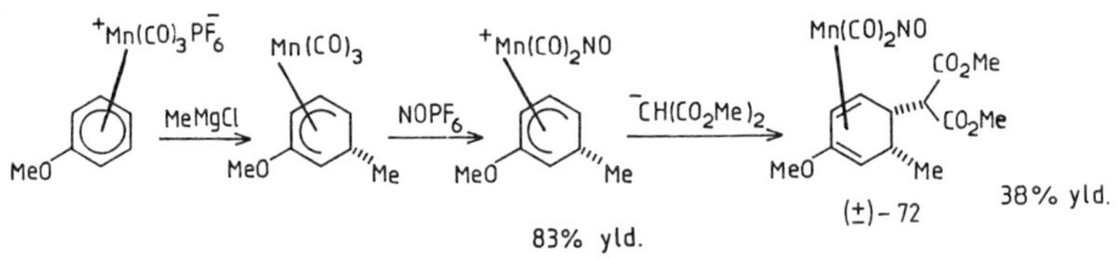

Scheme 16.24

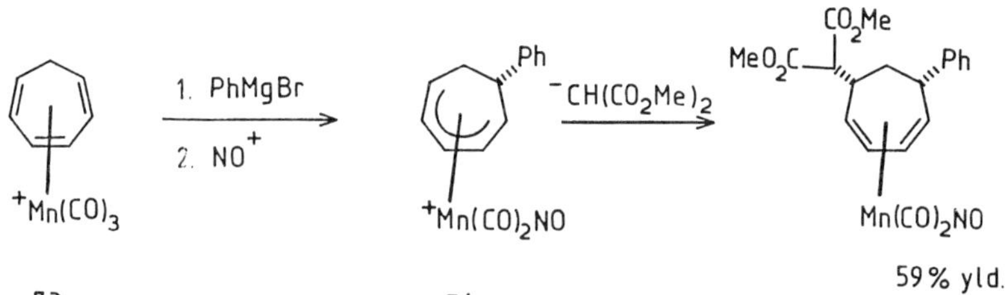

Scheme 16.25

(a) *Asymmetric induction in η^6-complexes.* In the case of η^6-chromium–arene complexes, considerable efforts have recently been made to develop easier methods to obtain these compounds in optically pure form. A typical approach is the covalent attachment of a chiral auxiliary within a substituent on the aromatic ring. Aminals (see Chapter 5) derived from *o*-methoxybenzaldehyde complexes have been used to effect the separation of diastereoisomers and hence, after hydrolytic removal of the aminal, to obtain optically pure tricarbonylchromium complexes of the 1,2-disubstituted arene.[44] In the course of this work, it was demonstrated that the two diastereoisomeric chromium complexes can be interconverted by heating, through epimerisation of the planar chirality within the arene complex. This complexation of the aminals gave roughly 10:1 mixtures of the two diastereoisomers, demonstrating a considerable asymmetric induction with the (2*R*) form of the metal complex as the major product. In contrast, delivery of the tricarbonylchromium moiety from tricarbonyl(naphthalene)chromium favoured the (*S*)-isomer, affording under kinetic control a product mixture in the ratio of 98:2 (Scheme 16.26). Selective hydrolysis during column chromatography of imines obtained from L-valinol has also proved a convenient way to obtain optically pure substituted benzaldehyde complexes.[45]

When the arene ligand is monosubstituted, its complex is prochiral. There are now a number of examples where asymmetry is induced into a prochiral

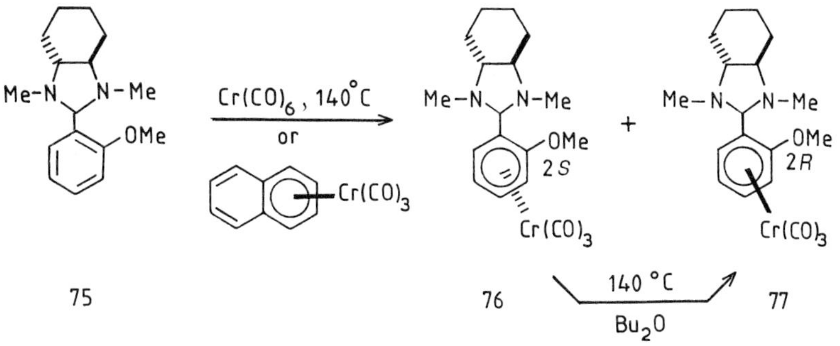

Scheme 16.26

complex of this type by directing metallation of the ring by means of a chiral auxiliary at a side-chain position. A recent example comes from the work of Kondo, Green and Ho with the tricarbonylchromium complex of tartrate-derived acetals of benzaldehyde (Scheme 16.27).[46] Metallation of (**78**) produced an ortholithiated intermediate which was alkylated with methyl iodide to afford a product in 92% de. Similar results were obtained with other electrophiles. The chiral auxiliary can be removed by hydrolysis. The sample of ((−)-**79**) obtained in this way contained no trace of the other enantiomer when examined in the presence of a chiral shift reagent by the method of Soladié-Cavallo *et al.*[47] Enantiopure chiral bases have been used to induce asymmetry (see Chapter 6).

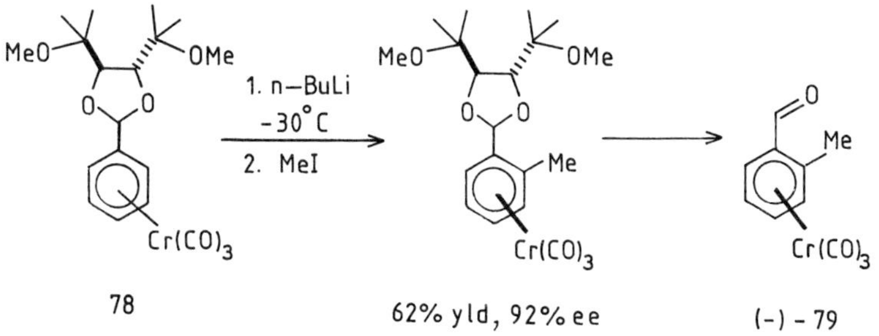

Scheme 16.27

When the metallated arene complex reacts with aldehydes as the electrophile, a new chiral centre is formed in the benzyl alcohol product. A considerable range of diastereoselectivity has been observed in reactions of this type, with the best results affording 94:6 mixtures of diastereoisomers.[48] In this case, the chiral auxiliary was an *N*,*N*-dimethylaminoethyl substituent. Microbial reduction[49] (see Chapter 13) has also been used for the preparation of optically active tricarbonylchromium complexes of benzyl alcohols. Organolithium reagents can add as nucleophiles to form anionic η^5-tricarbonylchromium intermediates. Reaction with the triphenylcarbenium ion returns these structures to the η^6-coordination mode. With SAMP as a chiral auxiliary attached by an imine, this nucleophile addition/hydride abstraction reaction can afford ((+)-**79**) in >97% ee (Scheme 16.28).[50]

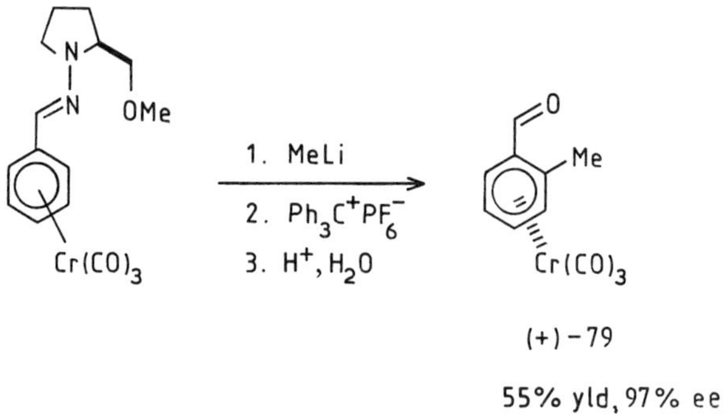

OMe
$Cr(CO)_3$
1. MeLi
2. $Ph_3C^+PF_6^-$
3. H^+, H_2O
Me
$Cr(CO)_3$
(+)-79
55% yld, 97% ee

Scheme 16.28

16.4 Reaching out beyond η^n

When working with stoichiometric π-complexes as intermediates in asymmetric synthesis, the greater the hapticity of the starting material (n in η^n), the greater are the possibilities for carbon atoms within the working ligand to come under the direct control of the metal through their bonding to the metal/ligand fragment. Even this, however, limits the applicability of metal-controlled bond-forming reactions. There are two strategies that can extend the range of the control effects. Either the metal must move its position of attachment within the working ligand, or stereocontrolled reactions must be developed at positions remote to the η^n portion. Examples of multiple chiral centre formation in which the metal moves have already been seen in Schemes 16.7, 16.8 and 16.18–16.20. Although the metal in these examples has only moved its point of attachment by one carbon atom (in these cases, one position round the ring), the reaction sequences serve to demonstrate the principle. In this section, we can progress to look at reactions where stereoselectivity is the result of the effects of the transition metal complex at reactive centres outside the η^n portion.

16.4.1 Metal-controlled reactions of metal-stabilised cations and conventional functionality, adjacent to the η^n unit

The η^2-alkyne complexes (e.g. **80**) provide the best example for $n = 2$. Many transition metal complexes stabilise positive charge adjacent to the site of metal attachment. In the η^2-alkyne series (Scheme 16.29), the generation and utilisation of metal-stabilised propargyl cations (**81**) has been the subject of much investigation and has recently become referred to as the Nicholas reaction. Hexacarbonyldicobalt alkyne complexes are typically employed in this procedure, and with chiral propargyl structures in the working ligand, considerable stereocontrol in leaving group replacement can be achieved.[51]

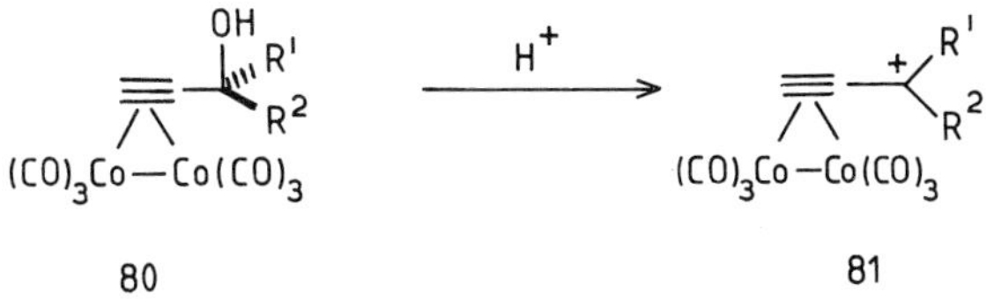

80 $\xrightarrow{H^+}$ 81

Scheme 16.29

Reactions of propargyl aldehydes are also possible in the presence of the dicobalt unit, and this has been recently extended to the use of enantiopure chiral allylborane nucleophiles to give enantioselective access to chirality in the working ligand. Metal removal from the product (**82**) could be achieved using ceric ammonium nitrate in acetone, but since the product has chirality adjacent to the dicobalt-complexed alkyne, there are further possibilities for stereocontrolled carbon–carbon bond formation (Scheme 16.30).[52] The use of chiral nucleophiles in conjunction with chiral propargyl complexes offers excellent stereocontrol. The complex (**83**) is easily obtained from (*R*)-BnOCHMeCHO by reaction with lithiotrimethylsilylethyne, methylation and complexation with dicobalt octacarbonyl. The complex (**83**) reacts with the chiral boron enolate (**84**) to produce (**85**) in optically pure form, as the major diastereoisomer (75:6).[53] Removal of the metal with ceric ammonium nitrate and desilylation and detachment of the chiral auxiliary (lithium peroxide) afforded ((+)-**86**) in 61% overall yield.

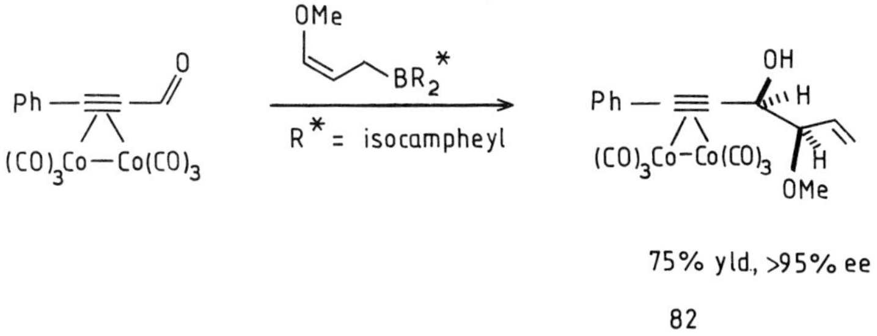

R^* = isocampheyl

75% yld., >95% ee

82

Scheme 16.30

Use of the boron enolate (**87**) allows this procedure to be employed in a formal total synthesis of (−)-thienamycin (Scheme 16.31).[53b]

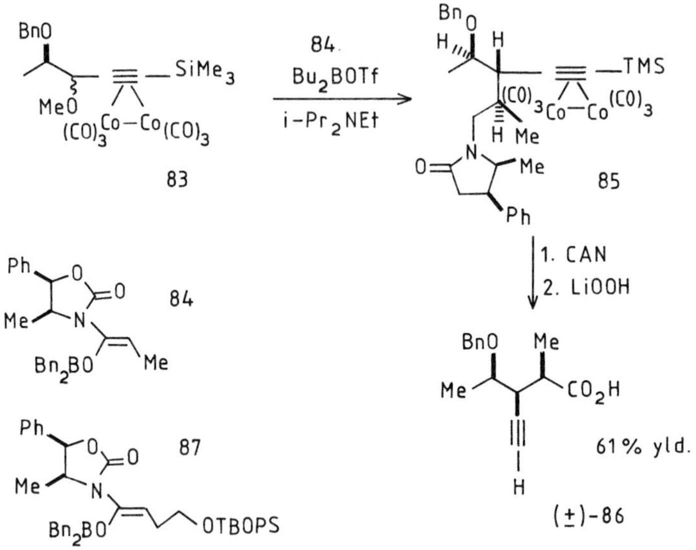

Scheme 16.31

Replacement of one carbon monoxide ligand on the hexacarbonyldicobalt fragment by the phosphine (**88**) renders the entire cobalt alkyne cluster chiral. Procedures for diastereoselective utilisation of this chiral dicobalt pentacarbonylphosphine cluster are not straightforward, since there are many potential isomerisation processes available to the cationic intermediates,[54] but recently, a stereoselective leaving group replacement has been described[55] in a reaction sequence that utilises a chiral cobalt cluster. Diastereoselectivity as high as 10:1 has been obtained (Scheme 16.32).

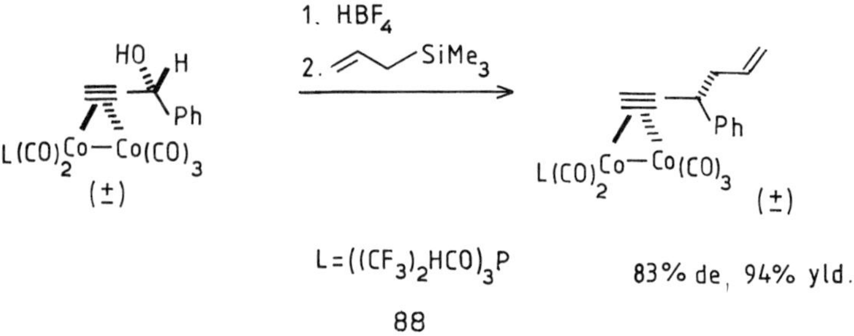

Scheme 16.32

With the surge of interest in the synthesis of enediyne natural products, the Nicholas reaction has found special utility because the metal-bound alkyne unit is bent, not linear like a free alkyne, and this facilitates the closure of alkyne-containing medium-sized rings. There have been many examples of such ring-closures, involving both propargylic alcohol[56] and aldehyde[57] dicobalt complexes. When an aldehyde is employed, a new chiral centre is generated adjacent to the alkyne metal complex. In the example shown in Scheme 16.33, the precursor (**89**) was available in enantiopure form through the use of a Sharpless epoxidation of an allylic alcohol, prior to attachment of the dicobalt unit. The right-hand side of this molecule was thus stereodefined, but the presence of the chiral centre in the 5-membered ring meant that (**89**) was used as a mixture of diastereoisomers. Cyclisation using the boron triflate procedure was diastereoselective with regard to the chiral centres built around the new carbon–carbon bond, though stereoambiguity at the top part of the molecule remained. The diyne, which contains the neocarzinostatin core, was obtained by removal of the metal (iodine oxidation) after separation of the diastereoisomer mixture.[58]

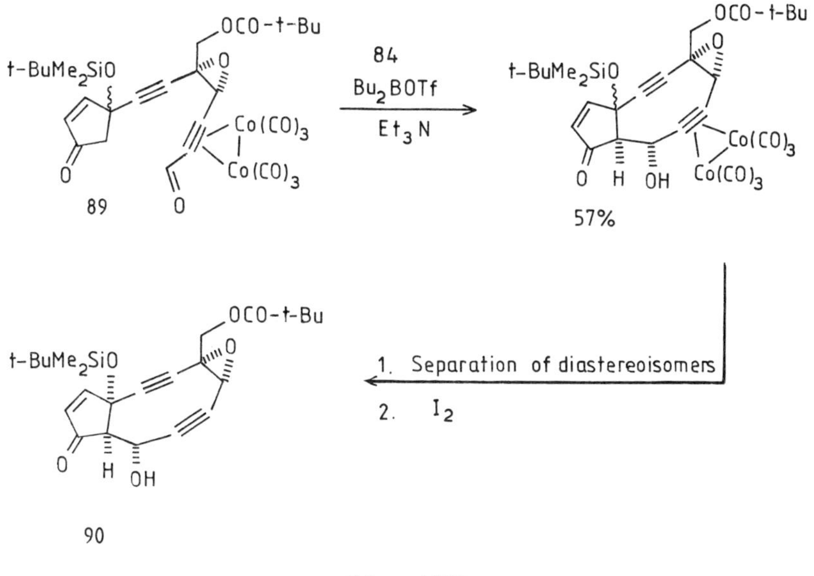

Scheme 16.33

The stereodirecting influence of the metal at conventional organic functionality can reach out a considerable distance from the site of complexation. In Scheme 16.34, an example using an η^3-dicarbonyl(cyclopentadienyl)molybdenum complex illustrates the principle. The cuprate addition to the enone proceeds in the normal manner, but since the organometallic allyl complex (**91**) is chiral, two possible diastereoisomers of the product (**92**) are possible, depending on the face selectivity of approach of a dimethylcuprate reagent.

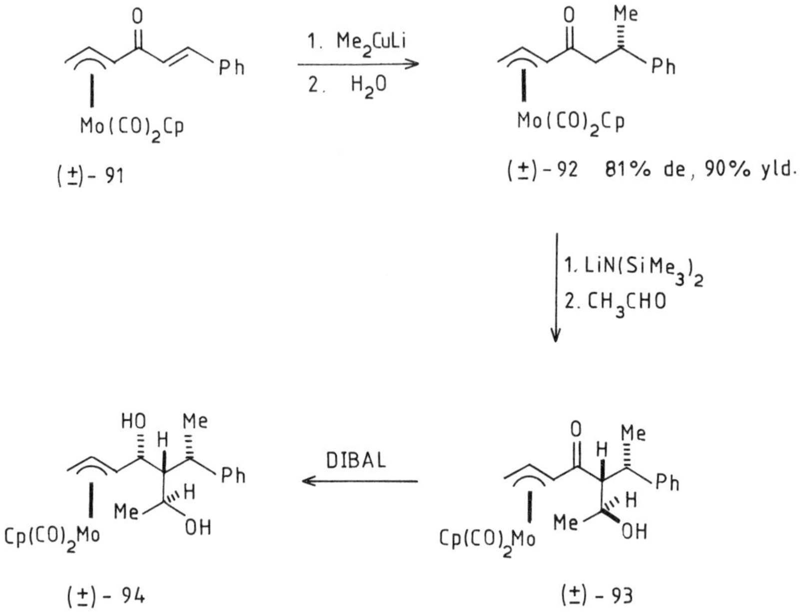

(±)-91 → 1. Me_2CuLi 2. H_2O → (±)-92 81% de, 90% yld.

(±)-92 → 1. $LiN(SiMe_3)_2$ 2. CH_3CHO → (±)-93

(±)-93 → DIBAL → (±)-94

Scheme 16.34

Diastereoface selectivity was high (14:1) and improved to 21:1 for ethyl group transfer from a magnesiocuprate reagent.[59] In the case of (**92**), a diastereoselective aldol reaction produced two more chiral centres in (**93**) and hydride reduction (DIBAL) afforded (**94**). In all, four chiral centres have been built at the positions α, β, and γ to the metal-complexed allyl unit. Apart from its stereodirecting influence, the cyclopentadienylmolybdenum complex has played little part in these reactions. However, by replacement of carbon monoxide by NO^+, an electrophilic allyl complex can be obtained. This undergoes a stereoselective intramolecular nucleophile addition by HO^- to form a tetrasubstituted tetrahydrofuran derivative with four contiguous chiral centres around the ring.

Chiral η^4-diene complexes can similarly exert a considerable stereocontrol influence in reactions at adjacent functionality. Tricarbonyliron complexes are commonly used in this way. An early application was directed towards pyrethroid synthesis,[60] as illustrated in Scheme 16.35, in which an optically pure metal complex was converted by a diastereoselective cycloaddition reaction into ((+)-**95**). The diene is not required in the final product but instead provides access to the aldehyde functionality by ozonolysis. Cycloaddition with a heterodiene has been used in a route to the methyl ester of 5-HETE (Scheme 16.36). The reaction sequence starts with an acetylide addition to an η^5-pentadienyl complex (**96**), affording the required (*E*, *Z*)-stereostructure for (**97**). Conversion of the ester into an aldehyde provided a substrate for

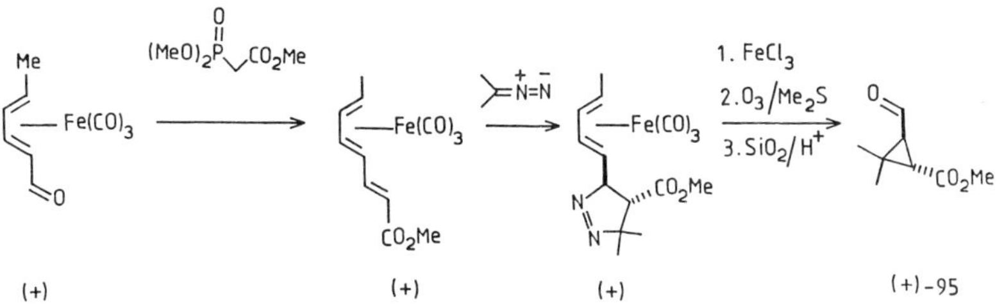

Scheme 16.35

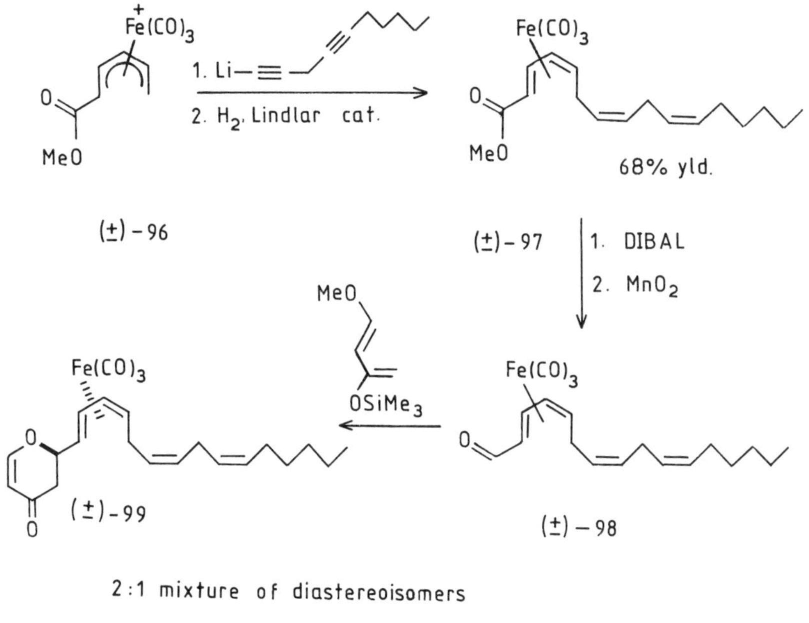

2:1 mixture of diastereoisomers

Scheme 16.36

the cycloaddition step. The diastereoisomer (**99**) was the major product. This was taken on to complete the synthesis at the target molecule.[61,62]

Donaldson's group have performed a second 5-HETE synthesis working in the enantiopure series. Resolution of the tricarbonyliron complex of methyl 6-oxo-2,4-hexadienoate (**100**),[63] followed by reduction of the aldehyde and treatment of the resulting alcohol with acid afforded ((2*R*)-**96**). This dienyl complex was taken through to ((2*R*)-**98**) following the reaction sequence shown in Scheme 16.36. Nucleophile addition at aldehydes next to η^4-tricarbonyliron complexes is rarely completely stereocontrolled (far better stereocontrol is available with ketone functionality) and in the case of the aldehyde (**98**), reaction with a nucleophile derived from (**101**) again afforded

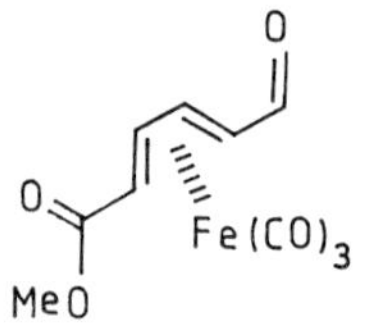

100

a 2:1 mixture of ψ-*exo* and ψ-*endo* products, which were isolated as the corresponding esters. The major product was converted into ((5*R*,6*R*)-**102**). Removal of the metal by oxidation with ceric ammonium nitrate afforded the natural product (5*R*)-(−)-HETE in >93% ee. The same reaction sequence starting from the (*S*)-isomer of the dienyl complex affording the corresponding (+)-isomer of (**103**) (>98% ee) (Scheme 16.37).[62]

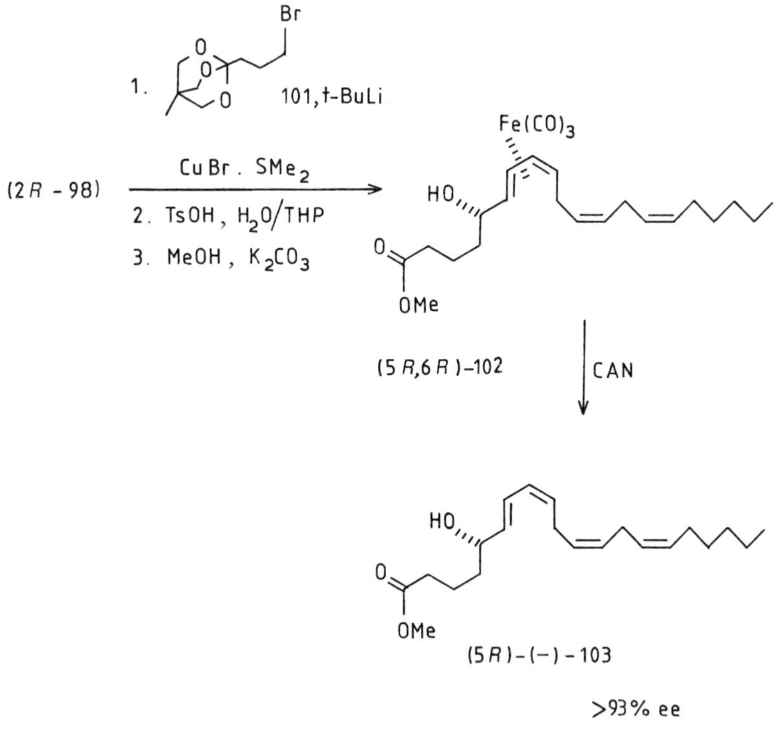

(2*R* - 98) → 1. 101, t-BuLi, CuBr . SMe_2; 2. TsOH, H_2O/THP; 3. MeOH, K_2CO_3 → (5*R*,6*R*)-102 → CAN → (5*R*)-(−)-103 >93% ee

Scheme 16.37

Stereocontrolled reduction of the ketone ((−)-**104**) (Scheme 16.38) provides an alternative access to an optically active derivative which played a key part in a synthesis of the methyl ester of (−)-(5*S*,6*S*)-LTA_4 (**107**).[64] This

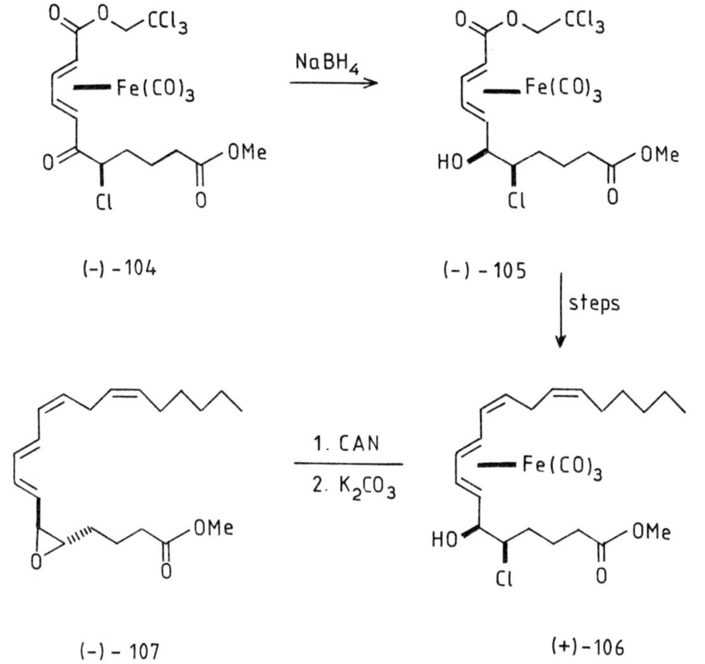

(−)-104 $\xrightarrow{NaBH_4}$ (−)-105 $\xrightarrow{steps}$ (+)-106 $\xrightarrow{1.\ CAN,\ 2.\ K_2CO_3}$ (−)-107

Scheme 16.38

chlorohydrin was a precursor for the epoxide in the lower portion of the target molecule. However, before the epoxide was introduced, extension of the upper part of the molecule was necessary. By a series of steps, aldehyde functionality replaced the trichloroethylester in (**105**), and the top chain was introduced using a Wittig alkenation. The metal was removed from (**106**) by

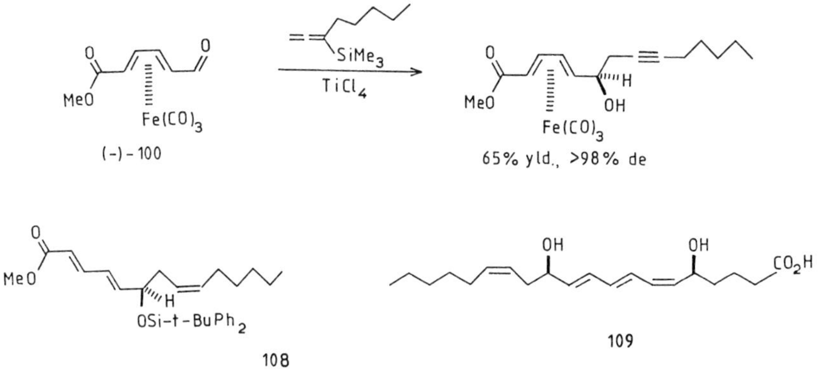

(−)-100 $\xrightarrow{TiCl_4}$ 65% yld., >98% de

108

109

Scheme 16.39

oxidation with ceric ammonium nitrate, and the epoxide closed with inversion, completing the enantiomer synthesis of the target molecule.

Lewis acid-mediated nucleophile addition to the aldehyde group of ((−)-**100**) affords a *ψ-endo* product. Although this is not the usual outcome for nucleophile addition at an aldehyde adjacent to the metal complex, it is the normal addition pathway in Lewis acid-mediated reactions, and in the case illustrated in Scheme 16.39, (**108**) was formed in greater than 98% de. This product was taken on to (**109**), which was required in an approach to the synthesis of leukotriene B_4 (LTB_4).[65] Working in the racemic series, 6,7-dihydro-LTB_4 has been prepared as its methyl ester.[66]

Hydroxylation of the metal-free alkene of an η^4-triene complex has also been shown to be stereocontrolled (Scheme 16.40). The triene was obtained by Wittig alkenation of a silyl ether derived from ((+)-**110**). *Cis*-hydroxylation of (**111**) with osmium tetroxide gave only the stereoisomer ((+)-**112**) in 93% yield. The intermediate was taken on through a series of steps to complete a synthesis of optically pure (+)-(11*R*,12*S*)-diHETE ((+)-**113**).[67]

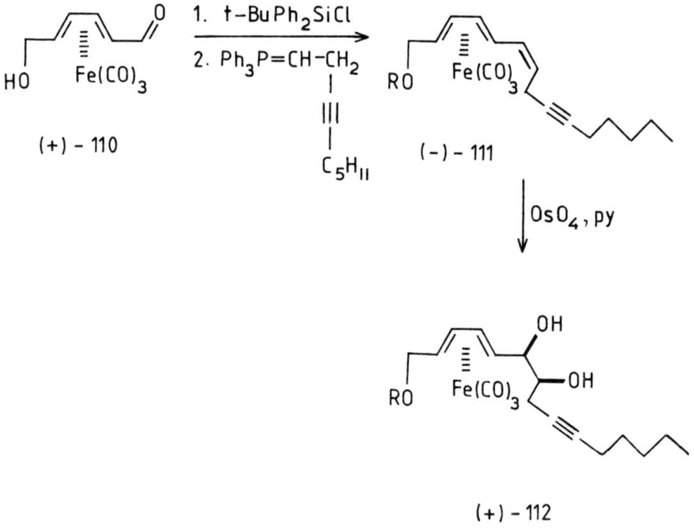

HO Fe(CO)3 O
(+) - 110
1. t-BuPh2SiCl
2. Ph3P=CH-CH2-C≡C-C5H11
RO Fe(CO)3
(-) - 111
OsO4, py
RO Fe(CO)3 OH OH
(+) - 112

Scheme 16.40

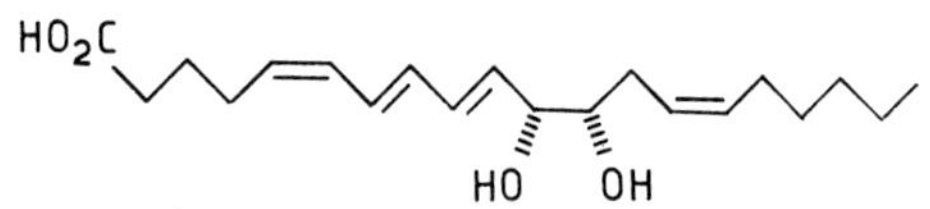

HO2C HO OH
(+) - 113

In the η^6-series, nucleophile addition to an optically pure tricarbonylchromium complex of *o*-methoxybenzaldehyde provides a nice example (Scheme 16.41) of an enantioselective synthesis of the amino acid analogue (4*S*,5*R*)-(**114**).[68]

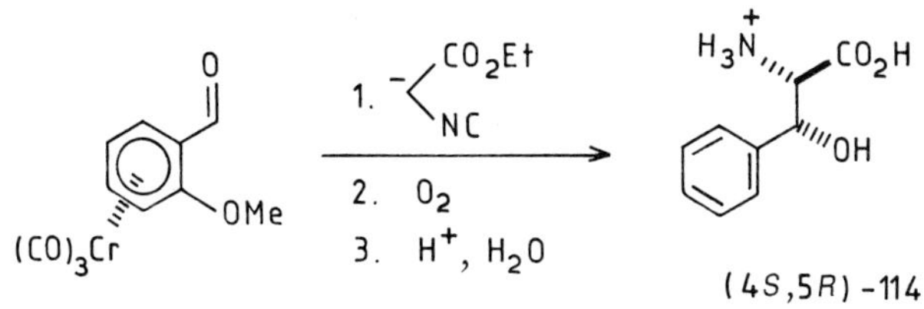

1. $^{-}$CH(CO$_2$Et)NC
2. O$_2$
3. H$^+$, H$_2$O

(CO)$_3$Cr OMe (4S,5R)-114

Scheme 16.41

Stereocontrolled nucleophile addition to benzaldehyde complexes is particularly effective when an *o*-substituent imposes a specific orientation on the aldehyde functional group. The OMe substituent in Scheme 16.41 and the Me_3Si group in Scheme 16.42 fulfil this role. Working in an enantiopure series, Hanaoka's group has used the aldehyde (**115**) as an electrophile with a titanium enolate prepared from (**116**) to achieve a high (95:5) enantioselectivity in the formation of (**117**).[69] The planar chirality of the metal-bound disubstituted aromatic ring has imposed diastereoface selectivity in the addition of the nucleophile to the aldehyde. The product (**116**) has been used

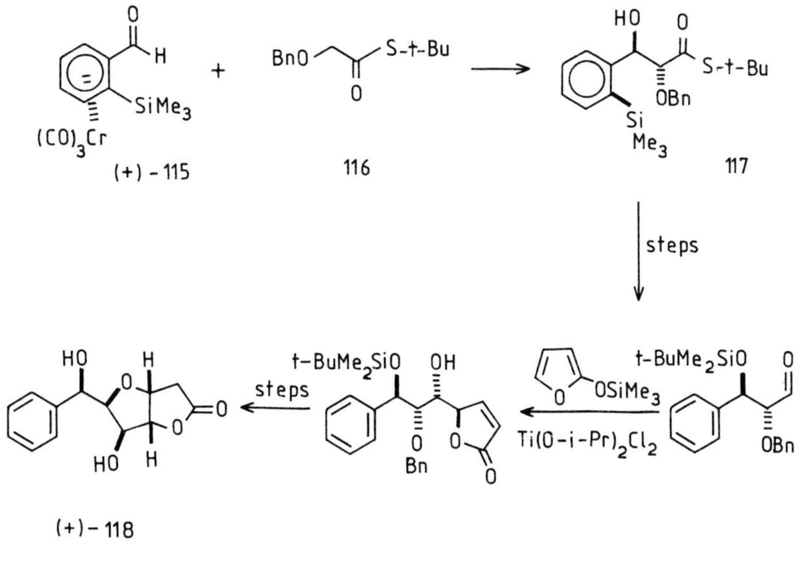

(+)-115 116 117

steps

Ti(O-i-Pr)$_2$Cl$_2$

steps

(+)-118

Scheme 16.42

as an intermediate in a synthesis of (+)-goniofufurone (**118**) (Scheme 16.42).[70] The target molecule contains a simple phenyl substituent, but to gain stereocontrol, the *ortho* silyl functionality in (**114**) was needed. This substituent (which becomes redundant once the asymmetric induction has been achieved), can easily be removed. As shown in Scheme 16.42, the side-chain functionality was further elaborated in a concise total synthesis of the target molecule in enantiopure form.

Chirality in tricarbonylchromium complexes can also influence the stereochemistry of enolates formed from acyl substituents on the aromatic ring. Diastereoselectivity as high as 90:10 can be achieved, as seen, for example, in the formation of (**120**) (Scheme 16.43).[71] Diastereoselectivity was

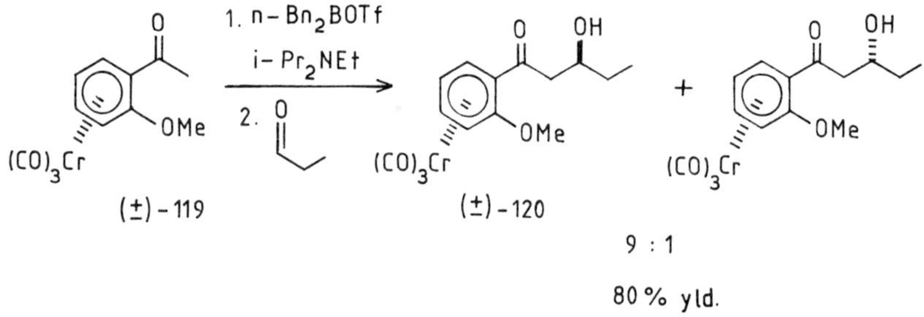

1. n-Bn_2BOTf, i-Pr_2NEt; 2. propanal

(±)-119 → (±)-120 + diastereomer, 9 : 1, 80 % yld.

Scheme 16.43

improved to 95:5 by the use of the more hindered aldehyde 2-methylpropanol. In Scheme 16.44, an alternative approach to the controlled construction of a chiral centre γ to the metal-bound ring proceeds by conjugate addition of a nucleophile to an enone (**121**). Free rotation of the side-chain is prevented by the tetralone structure, and addition of the nucleophile occurred exclusively from the face of the working ligand opposite to the chromium. Protonation of the resulting enolate, however, was not fully stereocontrolled and the major product (**122**) was obtained in 79% de.[72]

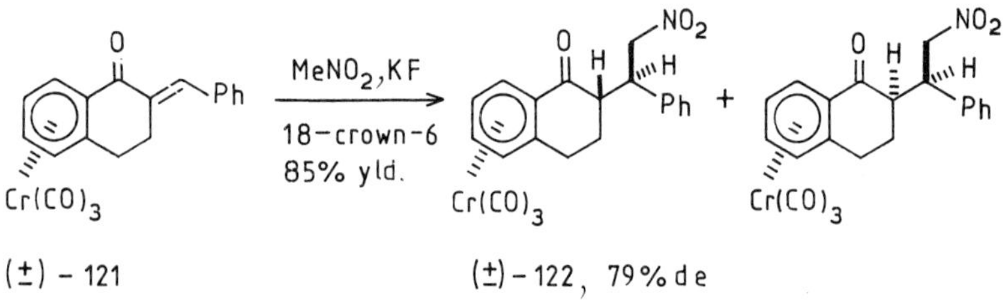

$MeNO_2$, KF, 18-crown-6, 85% yld.

(±) - 121 → (±)-122, 79% de + diastereomer

Scheme 16.44

16.4.2 Metal-derived activation remote from the η^n unit

With the exception of the Nicholas reaction, the examples discussed so far use conventional functionality adjacent to the tricarbonylchromium complex, with the metal serving only as a stereochemical blocking group forcing reagent approach to a particular face of the working ligand. Bond-forming processes exist, however, in which the transition metal can play a more significant role. This is well illustrated by the chemistry of η^6-arene complexes (Scheme 16.45).

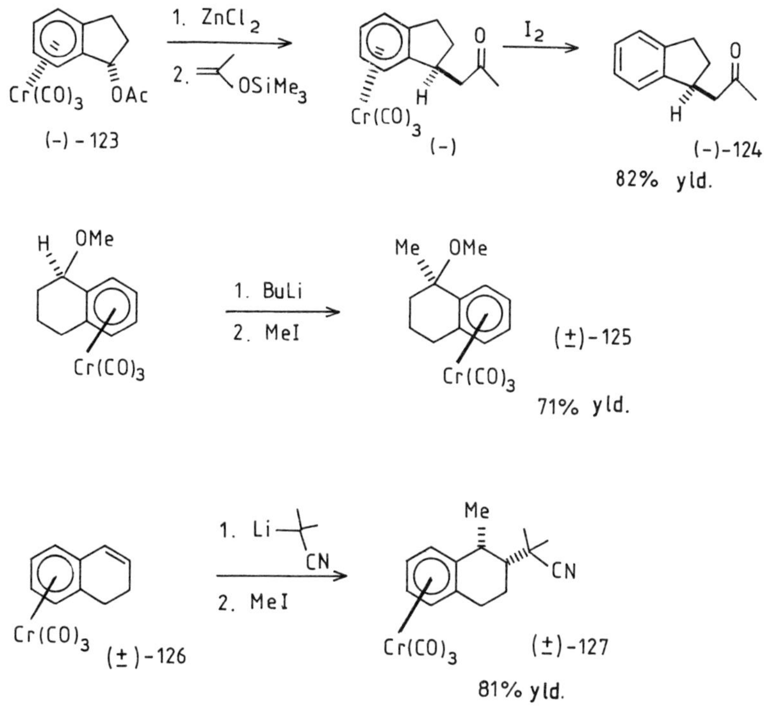

Cr(CO)3 OAc (−)-123 1. ZnCl2 2. OSiMe3 O H Cr(CO)3 (−) I2 O H (−)-124 82% yld.

H OMe Cr(CO)3 1. BuLi 2. MeI Me OMe Cr(CO)3 (±)-125 71% yld.

Cr(CO)3 (±)-126 1. Li CN 2. MeI Me CN Cr(CO)3 (±)-127 81% yld.

Scheme 16.45

Both benzyl cation and benzyl anion intermediates can be stabilised by the presence of a tricarbonylchromium moiety. The benzylic position can thus serve as either a nucleophilic or electrophilic partner in reactions which benefit from the stereodirecting influence of the chromium atom. The benzyl cation complex obtained by Lewis acid treatment of (**123**) is constrained within a ring. Nucleophile addition was completely diastereoselective, and removal of the chromium by oxidation with iodine afforded (**124**) as a single enantiomer.[73] The corresponding indane-derived anion, again stabilised by a tricarbonylchromium group, has been used as a nucleophile in the stereocontrolled reaction with formaldehyde.[74] Anions derived from A-ring aromatic steroid complexes offer similar stereocontrol.[75] Chromium-stabilised anions at alkoxy-substituted carbon atoms do not undergo a Wittig rearrangement and so can be used in reactions with electrophiles, as seen in the stereocontrolled formation of (**125**).[76] This capacity for chromium carbonyl complexes to stabilise anions can be used to impart electrophilicity to an alkenyl substituent. Conjugate addition to the metal-free double bond in (**126**) also proceeded exclusively *exo* to the tricarbonylchromium group. The resulting metal-stabilised anion could then be alkylated, again with complete stereocontrol, to produce the *cis* relative stereochemistry in (**127**).[77]

(a) *Synthesis of dihydroxyserrulatic acid and dihydroxycalamenone.* The use of tricarbonylchromium groups to stabilise positive and negatively charged intermediates adjacent to the metal-bound ring has been employed in a number of stereocontrolled target molecule syntheses. Two recent examples illustrate how well developed these methodologies now are. Electrophilic arene complexes, for example, hold the key to a synthesis of (±)-dihydroxyserrulatic acid (**134**).[78] Like the Donaldson synthesis of 5-HETE presented in Schemes 16.36 and 16.37, Uemura's synthesis of dihydroxyserrulatic acid (**134**) makes use of electrophilicity imposed on the metal-bound ring, as well as stereoselective functionalisation in side-chain positions, though in this synthesis, direct nucleophile addition to the metal complex comes at the end, after side-chain chirality has been established (Scheme 16.46). The starting material (**128**) was obtained by complexation of the corresponding aromatic ligand. Reduction of the ketone in the tetralone ring occurred with addition of hydride exclusively from the face opposite to the chromium. Allylsilane addition under Lewis acid catalysis replacing the acetate, afforded (**130**) in 72% yield. In this reaction, a second chiral centre is formed by asymmetric induction in the prochiral nucleophile, and the required stereoisomer (**130**) was obtained as the major component of a 3:1 mixture of diastereoisomers. The target molecule (**134**) has two chiral centres with *trans* relative stereochemistry. Introduction of the methyl group was performed by reaction with methyllithium to afford (**131**), followed by an ionic reduction in which hydrogen is transferred from the triethylsilane to a benzyl cation complex. In this reaction, it was the hydrogen, not the methyl group, that was placed opposite to the chromium on the stereochemically open face of the working ligand. This ensures the correct relative stereochemistry in intermediate (**132**). Nucleophilic substitution on the chromium-bound ring occurs *meta* to the OMe group, and oxidation of the resulting anionic metal complex affords an aromatic free ligand (**133**). This product was further elaborated to complete the synthesis of racemic dihydroxyserrulatic acid ((±)-**134**) (Scheme 16.46).

Calamenene synthesis provides a further example of this approach. Double use of chromium-stabilised anions resulting in a *cis* relative stereochemistry in (**135**). Completion of the substitution pattern on the aromatic ring was

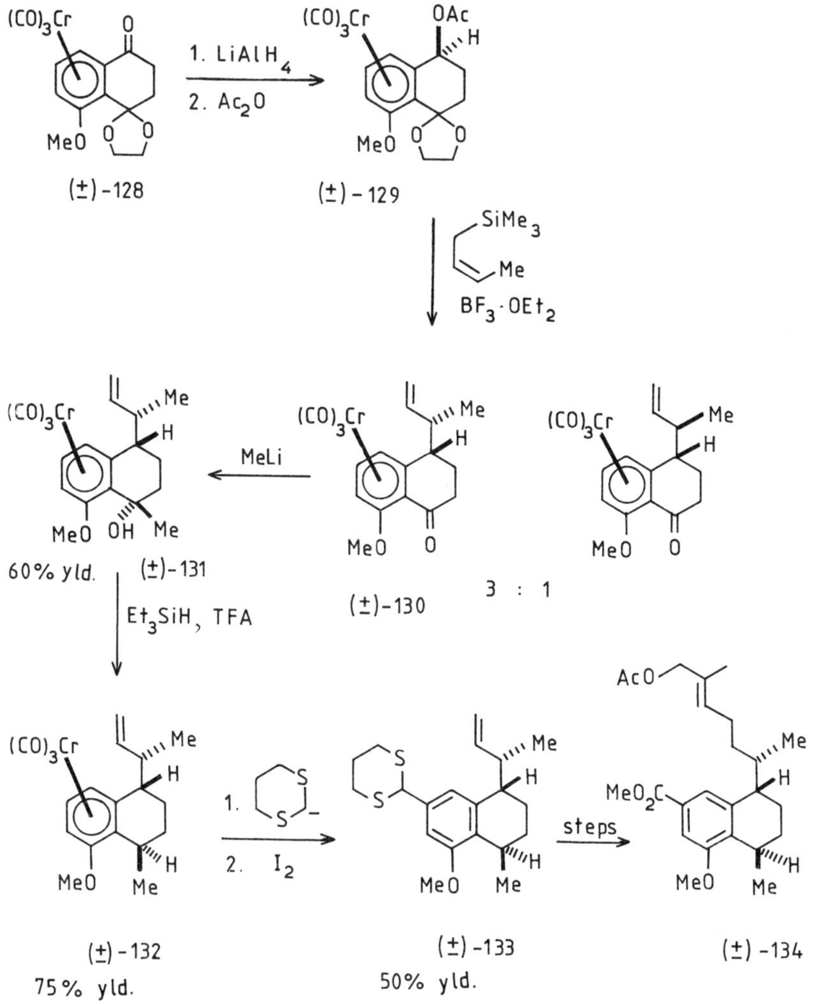

Scheme 16.46

affected by metallation and reaction with methyl iodide to introduce the final methyl group. Removal of the tricarbonylchromium moiety and ether cleavage completed an enantiomer synthesis of (−)-(1*S*,4*S*)-7,8-dihydroxy-calamenene ((−)-**136**) (Scheme 16.47).[79]

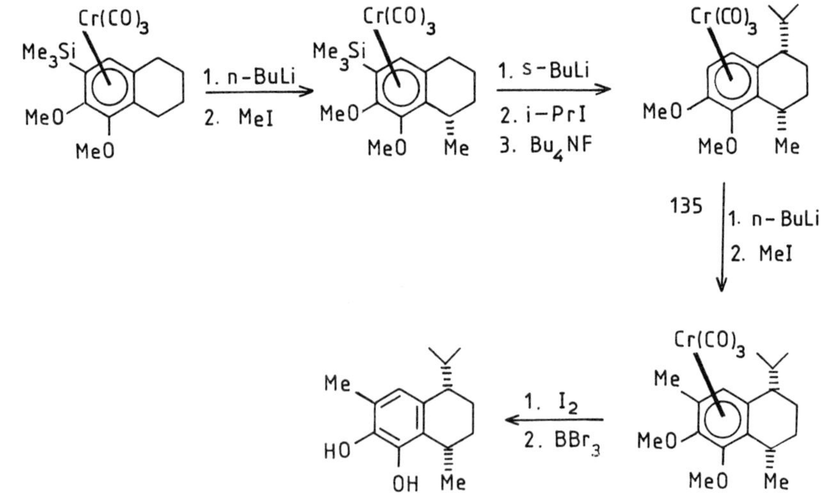

Scheme 16.47

16.5 Electrophilicity and nucleophilicity of transition metal π-complexes

The main focus of this review has been the electrophilicity of transition metal π-complexes and the utilisation of their reactivity in asymmetric synthesis. In the case of the tricarbonylchromium group at the end of the preceding section, other types of reactivity have also been encountered, illustrating how versatile transition metal complexes can be in imparting activation as well as stereocontrol to the working ligand. Indeed, nucleophilic metal complexes are now quite common, and recent examples can illustrate this in the chemistry of η^1-allyls,[80] η^3-allyls[81] and η^3-dienyls,[82] and η^4-trienes,[83] although the systems are yet to be used in target molecule synthesis.

16.6 Skeletal bond formation during the introduction and detachment of the metal/ligand fragment

16.6.1 Strategies to improve the efficiency of reaction sequences that make multiple use of the metal

In many cases, the working ligand serves as masked functionality or enables the protection of otherwise reactive groups. When masking or protecting strategies are used in synthesis, extra steps are typically introduced in the synthetic sequence to introduce and remove the protecting/masking group. So, too, with stoichiometric transition-metal π-complexes, separate complexation and decomplexation steps are the norm. In principle, shorter and hence more

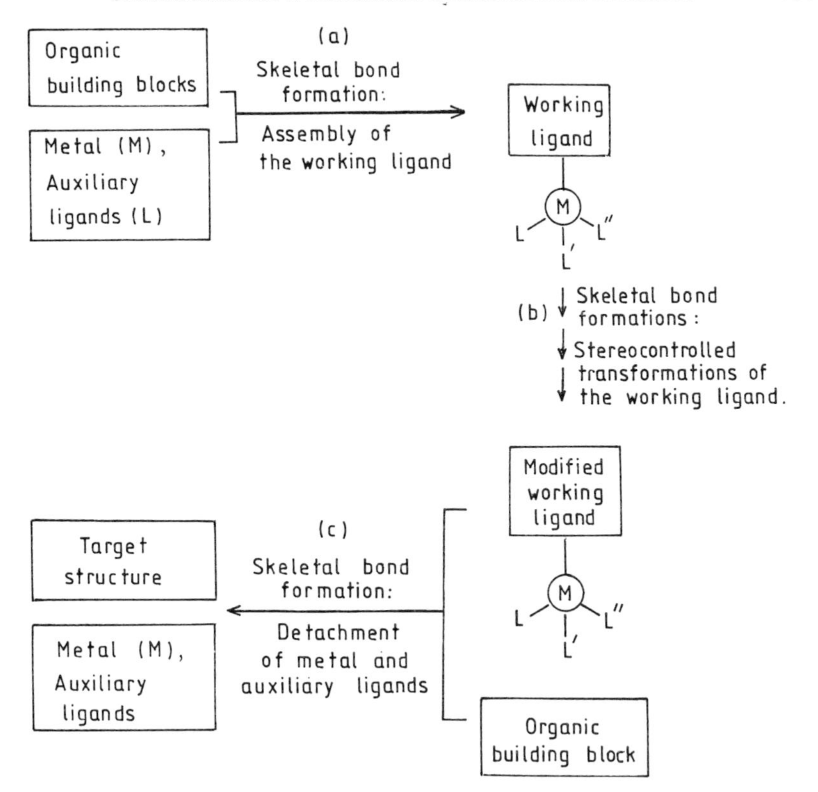

Figure 16.13 Idealised synthetic sequence converting organic building blocks and reagents into a target structure via a series of metal-mediated steps. (a) Construction of the working ligand in a process which also attaches the metal and auxiliary ligands. (b) Stereocontrolled transformations mediated by the activation and control effects imparted on the working ligand by the presence of the metal. (c) Final detachment of the modified working ligand in a reaction that achieves further elaboration to complete the target structure.

efficient synthetic routes would result if the step which attaches or detaches the metal also forms significant skeletal bonds and so directly advances the synthesis towards its final goal. The outline for such a synthetic methodology is illustrated in Figure 16.13. As yet, no single target molecule synthesis includes all these features, but stages (a) and (c) have each been combined with metal-mediated steps that transform the working ligand.

16.6.2 Reactions that form metal complexes with concomitant skeletal bond formation

Access to η^2-complexes can be achieved via an intermediate carbene complex. Nucleophile addition to a metal-bound carbon monoxide ligand and trapping of the resulting anion is a typical way to form Fischer carbenes. These can be converted into the corresponding aminocarbene derivatives, which are elaborated by using carbene-stabilised anions and then rearranged photolytically to cyclic amino acid derivatives in a process that is thought to involve intermediate η^2 aminoketene complexes (Scheme 16.48).[84] The product (**137**) (formed as a 1:1 mixture of diastereoisomers) has been taken on to complete a synthesis of (+)-bulgecenine by incorporation of a working ligand that has been constructed from the alkyllithium reagent, an aldehyde and two molecules of carbon monoxide, with formation of two carbon–carbon bonds along the way.

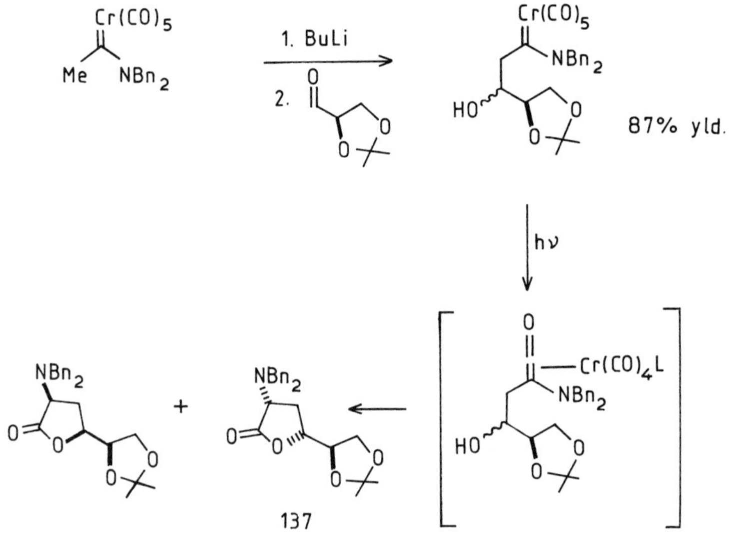

Cr(CO)5 Me NBn2 1. BuLi 2. Cr(CO)5 NBn2 HO 87% yld. hν Cr(CO)4L NBn2 HO NBn2 + NBn2 137

Scheme 16.48

The opening of vinyl cyclopropanes and epoxides by an iron carbonyl reagent illustrates a route to η^3-allyl complexes. In the cyclopropane case, as well as the allyl ligand, a metal–carbon σ-bond is formed through the cyclopropane ring opening, and σ-bond migration to insert a carbon monoxide ligand can produce a metal acyl structure.[85] This reaction, and the corresponding transformation of vinyl epoxides[86] have been known for some time, but it is the epoxide series that has been most extensively employed in organic synthesis, mainly by the work of the Ley group.[87] A recent example[88] that uses an enantiopure epoxide originating through a Sharpless asymmetric epoxidation illustrates this widely used carbonyl insertion chemistry (Scheme 16.49). During the formation of the η^3-allyl intermediate, a carbon–oxygen skeletal bond is formed to afford an intermediate in the preparation of a subunit (**138**) for the total synthesis of the ionophore antibiotic routiennocin.

Alkynes are valuable building blocks for the construction of η^4-working ligands. The formation of η^4-cyclopentadienone complexes from two alkynes and a carbon monoxide molecule has been investigated independently by Knölker[89] and Pearson.[90] The formation[90b] of (**139**) (Scheme 16.50) illustrates

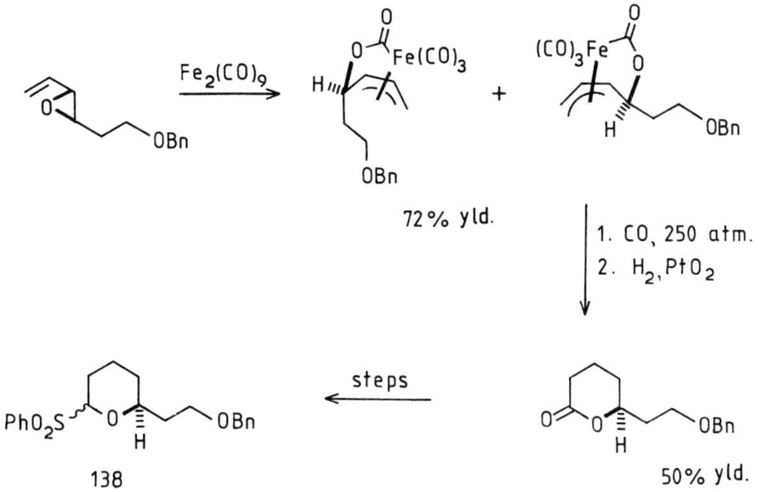

Scheme 16.49

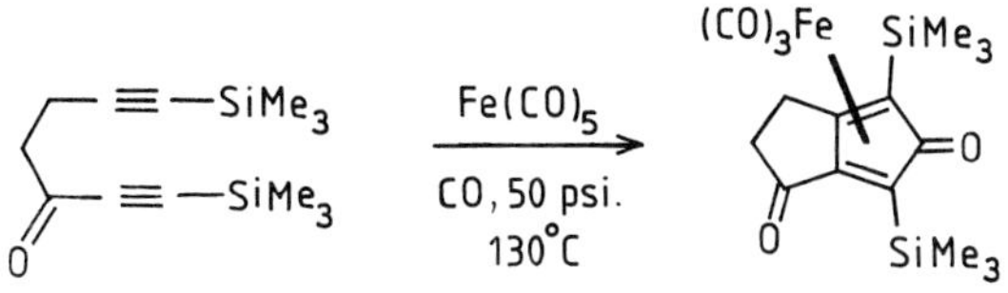

Scheme 16.50

this type of reaction. Cyclotrimerisation of alkenes and alkynes can also produce an η^4-diene complex. Cyclopentadienylcobalt complexes are typical of the metal complexes formed in this way.[91] In the example shown in Scheme 16.51, metal-mediated cycloaddition was followed by hydride abstraction

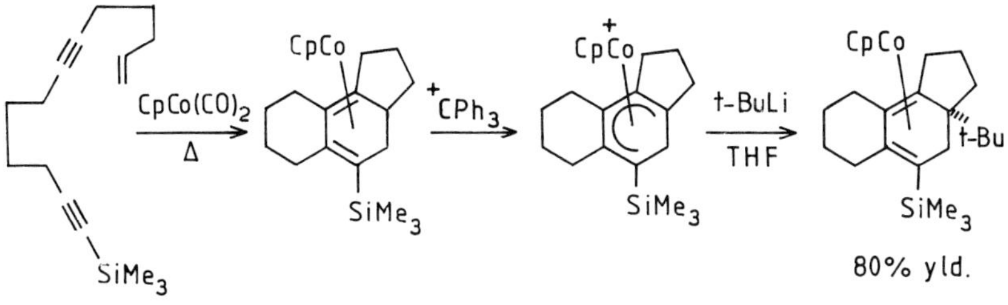

Scheme 16.51

from the resulting working ligand, and then reaction with a nucleophile.[91a] Iron carbene complexes react with alkynes and carbon monoxide to form η^4-pyrone complexes.[92] In a similar way, cationic η^5-dienyl complexes of the tricarbonyliron group have been obtained by the reaction of dienes with iron carbene complexes followed by removal of an allylic leaving group with acid (Scheme 16.52).[93] In the η^6-series, the widely used Dötz–Wulff cyclisation

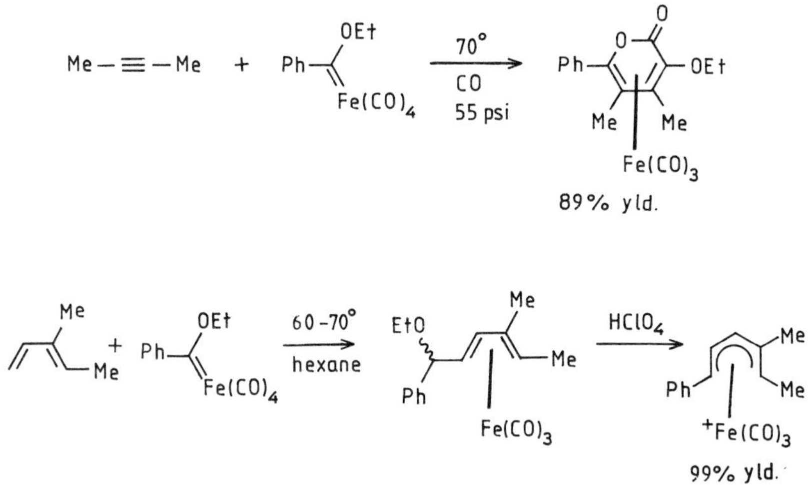

Me—≡—Me + Ph—C(OEt)=Fe(CO)$_4$ —(70°, CO, 55 psi)→ pyrone–Fe(CO)$_3$ complex (Ph, OEt, Me, Me) 89% yld.

diene (Me, Me) + Ph—C(OEt)=Fe(CO)$_4$ —(60–70°, hexane)→ EtO, Ph, Me, Me diene–Fe(CO)$_3$ —($HClO_4$)→ Ph, Me, Me dienyl $^+Fe(CO)_3$ 99% yld.

Scheme 16.52

provides a general way to form highly substituted arene complexes of tricarbonylchromium (Scheme 16.53). Most commonly, in synthetic applications, these complexes are not isolated but are oxidised to remove the metal. In a recent example,[94] the η^6-metal complex formed by the Dötz–Wulff step has been further elaborated using metal-mediated reactions. In the complexation procedure, three carbon–carbon bonds are made, but in this synthesis, the metal serves again to facilitate later skeletal bond-formation steps.

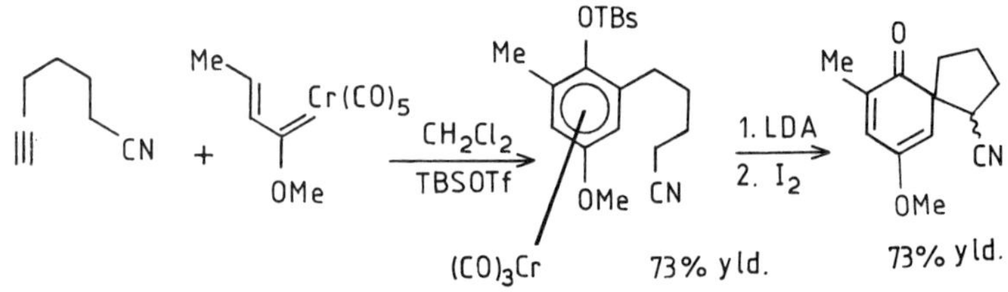

alkynyl nitrile (CN) + Me-vinyl C(OMe)=Cr(CO)$_5$ —(CH_2Cl_2, TBSOTf)→ arene (OTBs, Me, OMe, CN)–Cr(CO)$_3$ 73% yld. —(1. LDA, 2. I_2)→ spiro dienone (Me, OMe, CN) 73% yld.

Scheme 16.53

16.6.3 Decomplexation with concomitant carbon–carbon bond formation

(a) *Decomplexation by carbonylation.* η^2-Complexes of alkynes provide the most general illustration of a synthetically valuable carbon–carbon bond formation that occurs by carbonyl insertion during a decomplexation step. In the Pauson–Khand reaction, a dicobalt hexacarbonyl alkyne complex is heated with an alkyne to form a cyclopentenone. There have been several examples where the use of cobalt-stabilised propargyl cations (see Section 4.2) has been followed by a Pauson–Khand step to complete a substituted cyclopentenone. One of the best featured in Schreiber's synthesis of (**141**) (Scheme 16.54). Formation of an alkyne-containing 8-membered ring would not normally be possible, but as with the enediynes, the bent nature of the dicobalt complexes of alkynes assists the cyclisation process. The reaction of the cobalt-stabilised cation was diastereoselective, affording the *anti*-product (**140**), which was then converted in a stereocontrolled Pauson–Khand cyclisation into the target structure (**141**).[95]

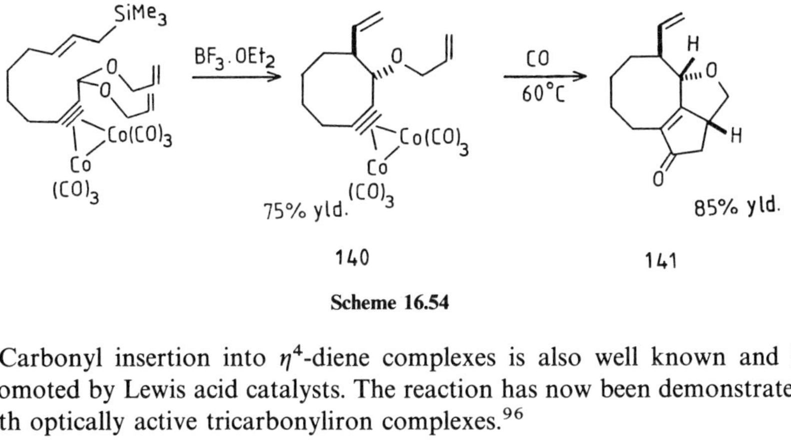

Scheme 16.54

Carbonyl insertion into η^4-diene complexes is also well known and is promoted by Lewis acid catalysts. The reaction has now been demonstrated with optically active tricarbonyliron complexes.[96]

(b) *Decomplexation by nucleophile addition.* Nucleophile addition is also used to effect combined decomplexation/skeletal bond-formation. Enolate addition to η^3-allyl structures can sometimes occur at the central carbon atom of the η^3-unit.[97] This forms an intermediate such as (**142**), which has two metal-carbon σ-bonds. Reductive elimination detaches the metal, forming a substituted cyclopropane (Scheme 16.55).[97a] In contrast to the examples employing cationic molybdenum complexes (Schemes 16.7–16.10), nucleophile addition to neutral η^4-complexes (Scheme 16.56) occurs at an internal carbon atom.[98] The resulting anionic intermediate is not isolated but can be reacted with electrophiles such as methyl iodide to affect carbonyl insertion. The overall result is the *trans*-addition of nucleophile and acyl group to the working ligand, in the reaction that detaches the metal. Reaction of the anion

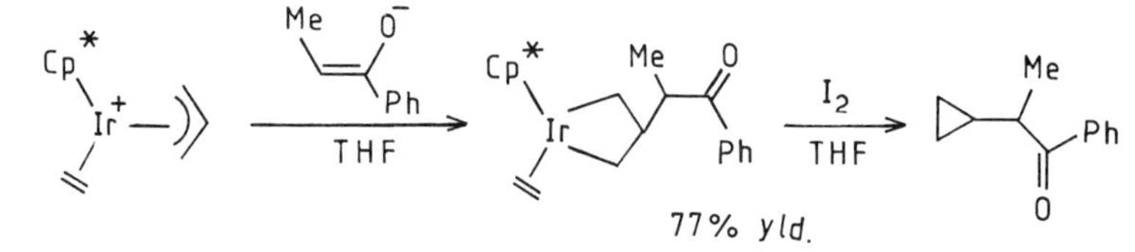
Cp*
Ir+
Me
O⁻
Ph
THF
Cp*
Ir
Me
O
Ph
77% yld.
I2
THF
Me
Ph
O

142

Scheme 16.55

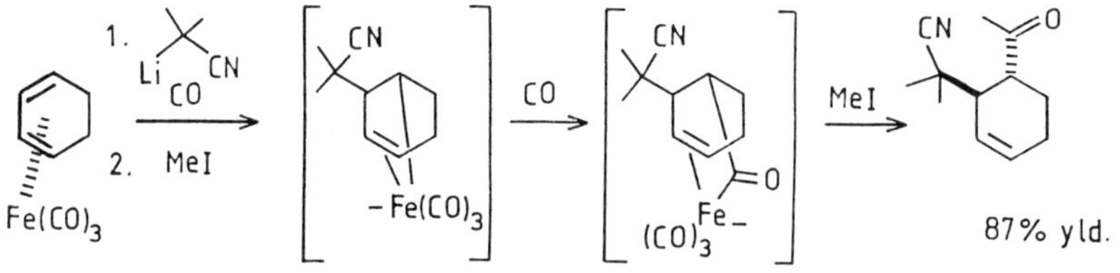
1.
Li
CN
CO
2. MeI
Fe(CO)3
CN
- Fe(CO)3
CO
CN
O
Fe
(CO)3
MeI
CN
O
87% yld.

Scheme 16.56

intermediate with other electrophiles such as Br_2 or H^+ has also been examined.[98c] Three new carbon–carbon bonds are formed in these reactions. The method has recently been extended to demonstrate an intramolecular version of the reaction (Scheme 16.57).[98d]

Organolithium reagents add to cationic η^5-pentadienyl complexes $Fe(CO)_2PPh_3$ to form η^1,η^3-structures which insert carbon monoxide upon air oxidation.[99] Cyclohexenones (**143**; R or R′ = Ph) can be obtained in this

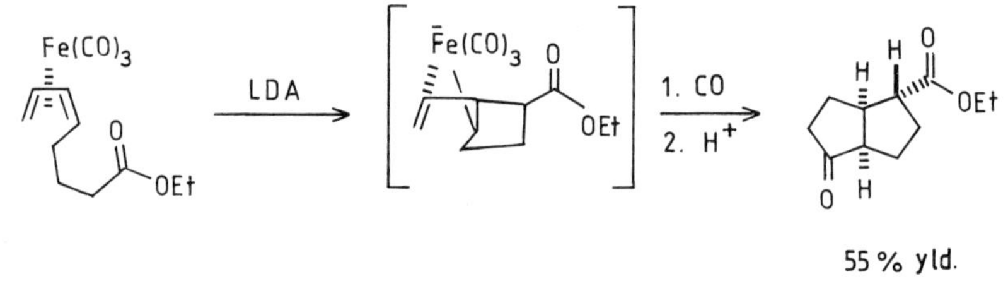
Fe(CO)3
1. Li
Ph
Ph
2. Br2,H2O
Ph
Ph
CO2H
58% yld.
Fe(CO)3
O
OEt
LDA
Fe(CO)3
O
OEt
1. CO
2. H+
H
H
O
OEt
O
H
55 % yld.

Scheme 16.57

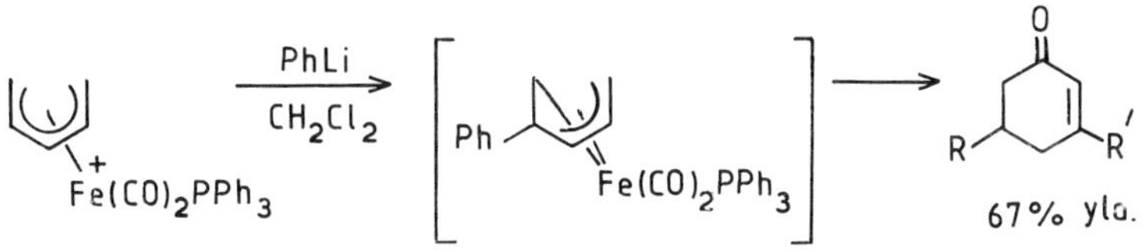

Scheme 16.58

way (Scheme 16.58). Nucleophile addition to neutral η^5-tricarbonylmanganese complexes produces a metal anion that can be oxidised to afford a 1,3-diene (Scheme 16.59). This reaction overcomes a long-standing limitation to the

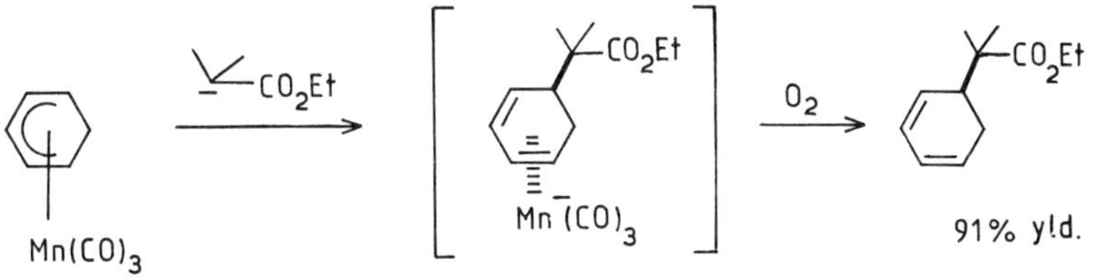

Scheme 16.59

multiple functionalisation of cationic manganese arene complexes. Earlier attempts to reactivate for a second nucleophile addition by exchanging CO with NO^+ met with little success, but through the nucleophile addition oxidation procedure, the sequence of two carbon–carbon bond formations can be completed, and metal removed in the final step (Scheme 16.60).

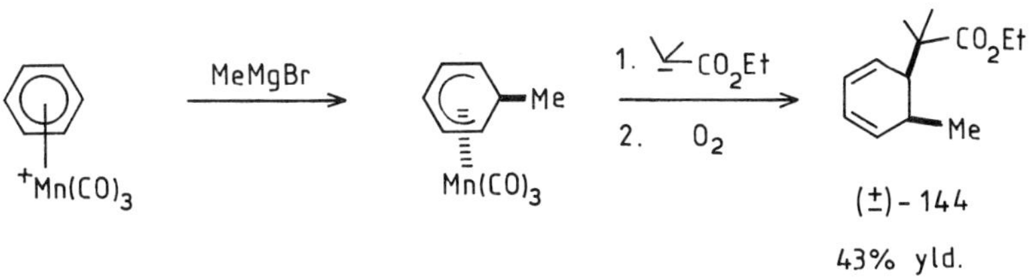

Scheme 16.60

Grignard addition[100] to the η^6-cation complex (**100**) afforded an *exo*-substituted dienyl complex which was taken on to the *cis* product (**144**) in 43% yield. Nucleophile addition to the related neutral η^6-tricarbonylchromium complexes can be combined with anion trapping and carbonyl insertion in a similar way to the reactions of the η^4-structures shown in Schemes 16.56 and 16.57. Examples of this type of reaction are shown in Scheme 16.61.[101]

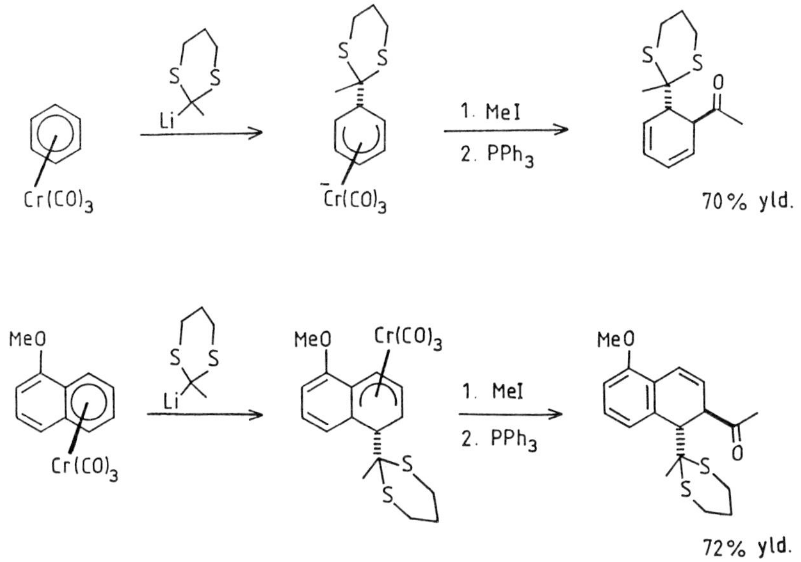

Scheme 16.61

The reaction has been applied in a formal total synthesis[102] of the aklavinone AB ring system, in which metal removal is combined with three carbon–carbon bond formations. A variant of the method is shown in Scheme 16.62. The nitrogen in the imine substituent can assist in stabilisation of negative charge in the anionic tricarbonylchromium complex. Reaction with methyl iodide under a slight pressure of carbon monoxide effects carbonyl insertion to afford an intermediate that rearranges by hydrogen shifts to form (**145**) as the major product.[103]

In the case of tricarbonylchromium complexes, the reactions shown in Schemes 16.53 and 16.61, combined with metal mediated elaboration of

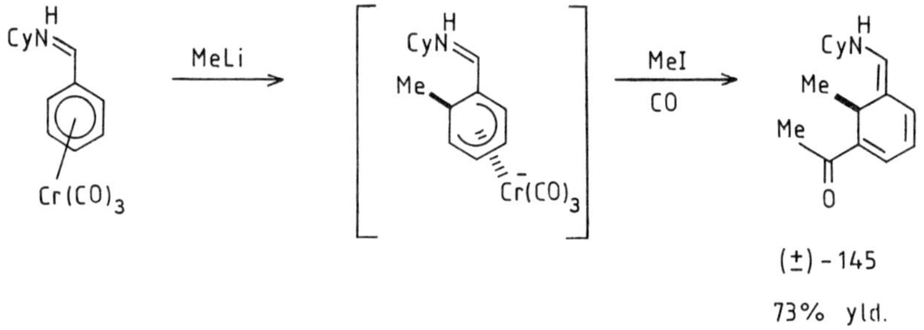

Scheme 16.62

side-chains in substituents to the aromatic ring (e.g. Schemes 16.41–16.47), could provide the means to achieve the objectives set out in Figure 16.13. Since, in terms of multiple utilisation of the transition metal centre, it is the chemistry of the tricarbonylchromium group that is most fully developed, it seems probable that the first examples of a fully integrated stoichiometric metal control strategy will emerge here. Other metal/ligand systems, however, are rapidly gaining ground, and the general realisation of this approach to asymmetric synthesis is not far off.

16.7 Conclusions

Transition metal π-complexes can participate in stereocontrolled skeletal-bond formations in a wide variety of ways. The use of several such reactions to effect key steps in a synthetic sequence is attractive because chirality of the metal complex can be carried through from one step to the next, so that the control of relative stereochemistry in the target structure is straightforward. In many cases in this review, optically pure, optically stable metal π-complexes have been used in enantioselective reactions. In others, chiral ligands have been elaborated in stereocontrolled fashion, or asymmetry has been induced into prochiral metal/ligand structures. A combination of these metal-mediated reactions with efficient procedures for the introduction and removal of the metal during significant bond-forming steps which advance the synthesis towards its final objective will be the goal of the future development of this subject.

References

1. (a) McQuillin, F.J., Parker, D.G. and Stephenson, G.R., *Transition Metal Organometallics for Organic Synthesis*, Chs 7 and 14, Cambridge University Press, Cambridge, 1991; (b) Stephenson, G.R., Alexander, R.P., Morley, C. and Howard, P.W., *Philos. Trans. R. Soc. Lond., A*, 1988, **326**, 545.
2. Davies, S.G., Green, M.L.H. and Mingos, D.M.P., *Tetrahedron*, 1978, **34**, 3047.
3. Birch, A.J., Ratnayake Bandara, B.M., Chamberlain, K., Chauncy, B., Dahler, P., Day, A.I., Jenkins, I.D., Kelly, L.F., Khor, T.-C., Kretschmer, G., Liepa, A.J., Narula, A.S., Raverty, W.D., Rizzardo, E., Sell, C., Stephenson, G.R., Thompson, D.J. and Williamson, D.H., *Tetrahedron*, 1981, **37**, 289.
4. Birch, A.J. and Bandara, B.M.R., *Tetrahedron Lett.*, 1980, 2981.
5. Chang, T.C.T., Coolbaugh, T.S., Foxman, B.M., Rosenblum, M., Simms, N. and Stockmann, C., *Organometallics*, 1987, **6**, 2394.
6. Chu, K.-H., Zhen, W., Zhu, X.-Y. and Rosenblum, M., *Tetrahedron Lett.*, 1992, **33**, 1173.
7. Enders, D. and Finkam, M., *Synlett*, 1993, 401.
8. Koot, W.-J., Hiemstra, H. and Speckamp, W.N., *J. Chem. Soc., Chem. Commun.*, 1993, 156.
9. Pearson, A.J. and Khan, M.N.I., *J. Am. Chem. Soc.*, 1984, **106**, 1872.
10. (a) Pearson, A.J., Khan, M.N.I., Clardy, J.C. and Cun-Heng, H., *J. Am. Chem. Soc.*, 1985, **107**, 2748; (b) Pearson, A.J. and Khan, M.N.I., *J. Org. Chem.*, 1985, **50**, 5276.

11. Faller, J.W., Murray, H.H., White, D.L. and Chao, K.H., *Organometallics*, 1983, **2**, 400.
12. Pearson, A.J., Blystone, S.L., Nar, H., Pinkerton, A.A., Roden, B.A. and Yoon, J., *J. Am. Chem. Soc.*, 1989, **111**, 134.
13. Vong, W.-J., Peng, S.-M., Lin, S.-H., Lin, W.-J. and Liu, R.-S., *J. Am. Chem. Soc.*, 1991, **113**, 573.
14. Cheng, M.-H., Ho, Y.-H., Lee, G.-H., Peng, S.-M. and Liu, R.-S., *J. Chem. Soc., Chem. Commun.*, 1991, 697.
15. Stephenson, G.R., Astley, S.T., Palotai, I.M., Howard, P.W., Owen, D.A. and Williams, S., *Organic Synthesis via Organometallics*, in K.H. Dötz and R.W. Hoffmann (eds), Vieweg, Braunschweig, 1991, p. 169.
16. Alton, J.G., Evans, D.J., Kane-Maguire, L.A.P. and Stephenson, G.R., *J. Chem. Soc., Chem. Commun.*, 1984, 1246.
17. Birch, A.J., Rafferty, W.D. and Stephenson, G.R., *Organometallics*, 1984, **3**, 1075.
18. Stephenson, G.R., *Aust. J. Chem.*, 1981, **34**, 2339.
19. (a) Kelly, L.F., Narula, A.S. and Birch, A.J., *Tetrahedron Lett.*, 1980, 2455; (b) Birch, A.J., Dahler, P., Narula, A.S. and Stephenson, G.R., *Tetrahedron Lett.*, 1980, 3817; (c) Pearson, A.J. and Raithby, P.R., *J. Chem. Soc., Perkin Trans. 1*, 1980, 395.
20. Birch, A.J. and Stephenson, G.R., *Tetrahedron Lett.*, 1981, **22**, 779.
21. For a discussion of α- and ω-directing effects, see Ref. 15 and Ref. 2, pp. 2–7. In discussion of regiocontrol, ipso (i) indicates addition at the atom carrying the substitutent, α indicates addition adjacent to this position, and ω indicates addition at the end of the π-system most distant from the substituent.
22. Pearson, A.J. and O'Brien, M.K., *J. Org. Chem.*, 1989, **54**, 4663.
23. Bandara, B.M.R., Birch, A.J. and Kelly, L.F., *J. Org. Chem.*, 1984, **49**, 2496.
24. Birch, A.J., Raverty, W.D. and Stephenson, G.R., *J. Chem. Soc., Chem. Commun.*, 1980, 857.
25. Birch, A.J., Kelly, L.F. and Weerasuria, D.V., *J. Org. Chem.*, 1988, **53**, 278.
26. Stephenson, G.R., Owen, D.A., Finch, H. and Swanson, S., *Tetrahedron Lett.*, 1990, **31**, 3401.
27. Birch, A.J., Cross, P.E., Lewis, J., White, D.A. and Wild, S.B., *J. Chem. Soc., A*, 1968, 332.
28. Birch, A.J., Kelly, L.F. and Thompson, D.J., *J. Chem. Soc., Perkin Trans. 1*, 1981, 1006.
29. Mincione, F., Bovicelli, P., Carrini, S. and Lamba, D., *Heterocycles*, 1985, **23**, 1607.
30. Stephenson, G.R., Owen, D.A., Finch, H. and Swanson, S., *Tetrahedron*, 1993, **25**, 5649.
31. Stephenson, G.R., Howard, P.W. and Taylor, S.C., *J. Organomet. Chem.*, 1989, **370**, 97.
32. Astley, S.T., Meyer, M. and Stephenson, G.R., *Tetrahedron Lett.*, 1993, **34**, 2035.
33. Astley, S.T. and Stephenson, G.R., *Synlett*, 1992, 507.
34. Pearson, A.J., Kole, S.J. and Ray, T., *J. Am. Chem. Soc.*, 1984, **106**, 6060.
35. Pearson, A.J., Lai, Y.-S., Lu, W. and Pinkerton, A.A., *J. Org. Chem.*, 1989, **54**, 3882.
36. Pearson, A.J., Lai, Y.-S. and Srinivasan, K., *Aust. J. Chem.*, 1992, **45**, 109.
37. Pearson, A.J., Balasubramaniam, S. and Srinivasan, K., *Tetrahedron*, 1993, **49**, 5663.
38. Potter, G.A. and McCague, R., *J. Chem. Soc., Chem. Commun.*, 1990, 1172.
39. Potter, G.A. and McCagne, R., *J. Chem. Soc., Chem. Commun.*, 1992, 635.
40. Pearson, A.J. and Richards, I.C., *J. Organometal. Chem.*, 1983, **258**, C41.
41. Semmelhack, M.F., Clark, G.R., Garcia, J.L., Harrison, J.J., Thebtaranonth, Y., Wulff, W. and Yamashita, A., *Tetrahedron*, 1981, **37**, 3957.
42. Chung, Y.K., Willard, P.G. and Sweigart, D.A., *Organometallics*, 1982, **1**, 1053; Honig, E.D. and Sweigart, D.A., *J. Chem. Soc., Chem. Commun.*, 1986, 691.
43. Honig, E.D. and Sweigart, D.A., *J. Chem. Soc., Chem. Commun.*, 1986, 691.
44. Alexakis, A., Mangeney, P., Marck, I., Rose-Munch, F., Rose, E., Semra, A. and Robert, F., *J. Am. Chem. Soc.*, 1992, **114**, 8289.
45. (a) Davies, S.G. and Goodfellow, C.L., *J. Chem. Soc., Perkin Trans. 1*, 1990, 393; (b) Bromley, L.A., Davies, S.G. and Goodfellow, C.L., *Tetrahedron: Asymmetry*, 1991, **2**, 139.
46. Kondo, Y., Green, J.R. and Ho, J., *J. Org. Chem.*, 1993, **58**, 6182.
47. (a) Solladié-Cavallo, A., Solladié, G. and Tsamo, E.J., *J. Org. Chem.*, 1979, **4**, 4189; (b) Solladié-Cavallo, A. and Suffert, J., *Magn. Reson. Chem.*, 1985, **23**, 739.
48. Uemura, M., Miyake, R., Nakayama, K, Shiro, M. and Hayashi, Y., *J. Org. Chem.*, 1993, **58**, 1238.
49. (a) Yamazaki, Y. and Kobayashi, H., *Tetrahedron: Asymmetry*, 1993, **4**, 1287; (b) Howell, J.A.S., Palin, M.G., Jaouen, G., Top, S., El Hafa, H. and Cense, J.M., *Tetrahedron: Asymmetry*, 1993, **4**, 1241.

50. Kündig, E.P., Liu, R. and Ripa, A., *Helv. Chim. Acta.*, 1992, **75**, 2657.
51. (a) Saha, M. and Nicholas, K.M., *J. Org. Chem.*, 1984, **49**, 47; (b) Nicholas, K.M., *Acc. Chem Res.*, 1987, **20**, 207.
52. Ganesh, P. and Nicholas, K.M., *J. Org. Chem.*, 1993, **58**, 5587.
53. (a) Jacobi, P.A. and Zheng, W., *Tetrahedron Lett.*, 1993, **34**, 2581; (b) Jacobi, P.A. and Zheng, W., *Tetrahedron Lett.*, 1993, **34**, 2585.
54. Bradley, D.H., Khan, M.A. and Nicholas, K.M., *Organometallics*, 1992, **11**, 2598.
55. Caffyn, A.J.M. and Nicholas, K.M., *J. Am. Chem. Soc.*, 1993, **115**, 6438.
56. (a) Magnus, P. and Fortt, S.M., *J. Chem. Soc., Chem. Commun.*, 1991, 544; (b) Maier, M.E. and Brandstetter, T., *Tetrahedron Lett.*, 1991, **32**, 3679.
57. Magnus, P. and Pitterna, T., *J. Chem. Soc., Chem. Commun.*, 1991, 541.
58. Magnus, P. and Davies, M., *J. Chem. Soc., Chem. Commun.*, 1991, 1522.
59. Lin, S.-H., Cheng, W.-J., Liao, Y.-L., Wang, S.-L., Lee, G.-H., Peng, S.-M. and Liu, R.-S., *J. Chem. Soc., Chem. Commun.*, 1993, 1391.
60. (a) Franck-Neumann, M., *Pure Appl. Chem.*, 1983, **55**, 1715; (b) Monpert, A., Martelli, J., Grée, R. and Carrie, R., *Nouveau J. Chim.*, 1983, **7**, 345; (c) Franck-Neumann, M., Martina, D. and Heitz, M.P., *Tetrahedron Lett.*, 1982, 3493; (c) Monpert, A., Martelli, J., Grée, R. and Carrie, R., *Tetrahedron Lett.*, 1981, **22**, 1961.
61. Donaldson, W.A. and Tao, C., *Synlett*, 1991, 895.
62. Tao, C. and Donaldson, W.A., *J. Org. Chem.*, 1993, **58**, 2134.
63. Monpert, A., Martelli, J., Grée, R. and Carrie, R., *Tetrahedron Lett.*, 1981, **22**, 1961.
64. Franck-Neumann, M. and Colson, P.-J., *Synlett*, 1991, 891.
65. Nunn, K., Mosset, P., Grée, R. and Saalfrank, R.W., *Angew. Chem., Int. Ed. Engl.*, 1988, **27**, 1188.
66. Nunn, K., Mosset, P., Grée, R. and Saalfrank, R.W., *J. Org. Chem.*, 1992, **57**, 3359.
67. Gigou, A., Beaucourt, J.-P., Lellouche, J.-P. and Grée, R., *Tetrahedron Lett.*, 1991, **32**, 635.
68. Colonna, S., Manfredi, A., Solladié-Cavallo, A. and Quazzotti, S., *Tetrahedron Lett.*, 1990, **31**, 6185.
69. Mukai, C., Kim, I.J. and Hanoaka, M., *Tetrahedron: Asymmetry*, 1992, **3**, 1007.
70. Mukai, C., Kim, I.J. and Hanoaka, M., *Tetrahedron Lett.*, 1993, **34**, 6081.
71. Uemura, M., Minami, T., Shiro, M. and Hayashi, Y., *J. Org. Chem.*, 1992, **57**, 5590.
72. Ganesh, S. and Sarkar, A., *Tetrahedron Lett.*, 1991, **8**, 1085.
73. Reetz, M.T. and Sauerwald, M., *Tetrahedron Lett.*, 1983, **24**, 2837.
74. Brocard, J., Lebibi, J. and Couturier, D., *J. Chem. Soc., Chem. Commun.*, 1981, 1264.
75. Joauen, G., Top, S., Laconi, A., Couturier, D. and Brocard, J., *J. Am. Chem. Soc.*, 1984, **106**, 2207.
76. Blagg, J., Davies, S.S., Holman, N.J., Laughton, C.A. and Mobbs, B.E., *J. Chem. Soc., Perkin Trans. 1*, 1986, 1581.
77. Moriarty, R.M., Engerer, S.G., Prakash, O., Prakash, I., Gill, U.S. and Freeman, W.A., *J. Chem. Soc., Chem. Commun.*, 1985, 1715.
78. Uemura, M., Nishimura, H., Minami, T. and Hayashi, Y., *J. Am. Chem. Soc.*, 1991, **113**, 5402.
79. Schmailz, H.-G., Hollander, J., Arnold, M. and Durner, G., *Tetrahedron Lett.*, 1993, **34**, 6259.
80. Agoston, G.E., Cabal, M.P. and Turos, E., *Tetrahedron Lett.*, 1991, **32**, 3001.
81. Faller, J.W., Nguyen, J.T., Ellis, W. and Mazzieri, M.R., *Organometallics*, 1993, **12**, 1434.
82. Cheng, M.-H., Ho, Y.-H., Wang, S.-L., Cheng, C.-Y., Peng, S.-M., Lee, G.-H. and Liu, R.-S., *J. Chem. Soc., Chem. Commun.*, 1992, 45.
83. Pearson, A.J. and Srinavasan, K., *Tetrahedron Lett.*, 1992, **33**, 7295.
84. Schmeck, C. and Hegedus, L.S., *J. Am. Chem. Soc.*, 1994, **116**, 9927.
85. Aumann, R., *J. Organometal. Chem.*, 1973, **47**, C29.
86. (a) Aumann, R., Fröhlich, K. and Ring, H., *Angew. Chem., Int. Ed. Engl.*, 1974, **13**, 275; (b) Aumann, R., Ring, H., Krüger, C. and Goddard, R., *Chem. Ber.*, 1979, **112**, 3644; (c) Annis, G.D. and Ley, S.V., *J. Chem. Soc., Chem. Commun.*, 1977, 581.
87. Ley, S.V., *Philos. Trans. R. Lond. A*, 1988, **326**, 633.
88. Kotecha, N.R., Ley, S.V. and Mantegani, S., *Synlett*, 1992, 395.
89. Knölker, H.-J., Heber, J. and Mahler, C.H., *Synlett*, 1992, 1002.
90. (a) Pearson, A.J., Shirley, R.J. Jr and Dubbert, R.A., *Organometallics*, 1992, **11**, 4096; (b) Pearson, A.J. and Shirley, R.J. Jr *Organometallics*, 1994, **13**, 578.
91. (a) Sternberg, E.D. and Vollhardt, K.P.C., *J. Org. Chem.*, 1984, **49**, 1564; (b) Boese, R., Rodriguez, J. and Vollhardt, K.P.C., *Angew. Chem., Int. Ed. Engl.*, 1991, **30**, 993.

92. Semmelhack, M.F., Tamura, R., Schnatter, W. and Springer, J., *J. Am. Chem. Soc.*, 1984, **106**, 5363.
93. Semmelhack, M.F. and Park, J., *J. Am. Chem. Soc.*, 1987, **109**, 935.
94. Chamberlain, S. and Wulff, W.D., *J. Am. Chem. Soc.*, 1992, **114**, 10667.
95. Schreiber, S.L., Sammakia, T. and Crowe, W.E., *J. Am. Chem. Soc.*, 1986, **108**, 3128.
96. Bernardes, V., Verdaguer, X, Kardos, N., Riera, A., Moyano, A., Pericàs, M.A. and Greene, A.E., *Tetrahedron Lett.*, 1994, **35**, 575.
97. (a) Wakefield, J.B. and Stryker, J.M., *J. Am. Chem. Soc.*, 1991, **113**, 7057; (b) Carfagna, C., Galarini, R., Musco, A. and Santi, R., *Organometallics*, 1991, **10**, 3956.
98. (a) Semmelhack, M.F., Herndon, J.W., *Organometallics*, 1983, **2**, 363; (b) Semmelhack, M.F., Herndon, J.W. and Springer, J.P., *J. Am. Chem. Soc.*, 1983, **105**, 2497; (c) Yeh, M.-C. and Hwu, C.-C., *J. Organometal. Chem.*, 1991, **419**, 341; (d) Yeh, M.-C.P., Sheu, B.-A., Fu, H.-W., Tau, S.-I. and Chuang, L.-W., *J. Am. Chem. Soc.*, 1993, **115**, 5941.
99. McDaniel, K.F., Kracker, L.R. II and Thamburaj, P.K., *Tetrahedron Lett.*, 1990, **31**, 2373.
100. Roell, B.C. Jr, McDaniel, K.F., Vaughan, W.S. and Macy, T.S., *Organometallics*, 1993, **12**, 224.
101. Kündig, E.P. and Simmons, D.P., *J. Chem. Soc., Chem. Commun.*, 1983, 1320.
102. Kündig, E.P., Inage, M. and Bernardinelli, G., *Organometallics*, 1991, **10**, 2921.
103. (a) Kündig, E.P., Bernardinelli, G., Liu, R. and Ripa, A., *J. Am. Chem. Soc.*, 1991, **113**, 9676; (b) Kündig, E.P., Ripa, A., Liu, R. and Bernardinelli, G., *J. Org. Chem.*, 1994, **59**, 4773.

17 Asymmetric oxidation

G.R. STEPHENSON

17.1 Introduction

Of all the recent developments in asymmetric synthesis, it is the asymmetric epoxidation process[1] developed in the laboratory of Professor Sharpless* that has had the greatest effect on the way organic synthesis is planned. This is because the titanium/tartrate catalyst system proved to be the first simple method to induce asymmetry that approached the goal of universal applicability and predictability. Already in this book, several examples of the Sharpless method have been discussed:

1. for kinetic resolution of allylic alcohols (Chapter 1);
2. as an example of a process in which non-linear effects are encountered (Chapter 2);
3. as entries to cobalt-mediated cyclisations employing the Nicholas reaction, and to iron-mediated methods of lactone synthesis (Chapter 16).

In the next chapter this topic will be encountered again:

4. as a symmetry-splitting reaction for asymmetric induction in extended *meso* structures.

The active catalyst species (see section 17.3) is prepared from titanium tetraisopropoxide and either diethyl tartrate (DET) or diisopropyl tartrate (DIPT), and the reaction is specific to allylic or homoallylic alcohols (as was its vanadium-catalysed precedent[2]). Though not fully universal in scope (for example, dienyl alcohols are unsatisfactory substrates and enantioselectivity with homoallylic alcohols is unreliable), the key to the wide applicability of the reaction is the role of the allylic OH group in coordinating titanium. To be truly catalytic in titanium, scrupulously dry (molecular sieves[3]) conditions are needed; the solvent of choice is dichloromethane. Auxiliary ligands derived from either (+)- or (−)-tartaric acid are available, so either enantiomer of the epoxide product is accessible. The overall process is set out in Scheme 17.1.

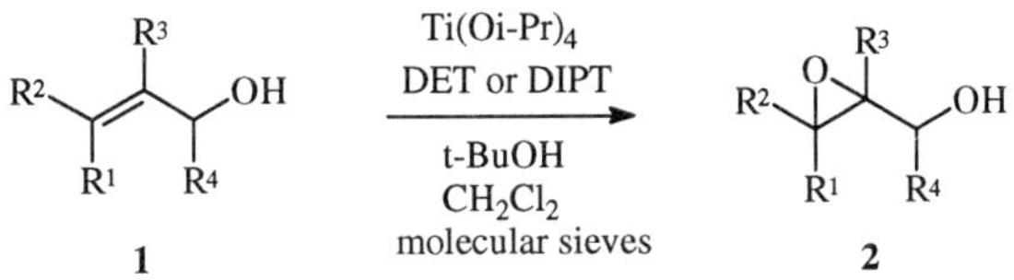

R3 R2 OH R1 R4 **1** — $Ti(Oi\text{-}Pr)_4$, DET or DIPT, t-BuOH, CH_2Cl_2, molecular sieves → R3 O R2 OH R1 R4 **2**

Scheme 17.1

* In fact this laboratory has produced three widely acclaimed oxidation methods. The first was an epoxidation procedure[2] employing an achiral vanadium catalyst to give excellent diastereoselectivity with chiral allylic alcohols. Subsequently the enantioselective epoxidation catalyst was developed, and most recently of all, attention focused on asymmetric dihydroxylation (see section 17.6).

17.2 Importance of the Sharpless asymmetric epoxidation

The impact of this reaction on synthesis design can be detected in the growing popularity of the use of epoxides as intermediates in target molecule synthesis. Since the advent of this chemistry, chiral epoxides are often among the first structures considered when enantioselective synthetic routes are contemplated. There has been a boom in the development of new techniques to manipulate epoxide groups, and many syntheses have been completed in which otherwise unnecessary alcohol functions have been introduced in intermediates solely to provide a handle for the Sharpless step. That this strategy should be considered worthwhile, despite the need for extra functional group interconversions, is truly a testament to the awe in which asymmetric epoxidation is held as a safe choice when planning to induce asymmetry.

A simple example is provided by consideration of amino acid synthesis (Scheme 17.2). Epoxidation of an allylic alcohol, followed by a stereocontrolled

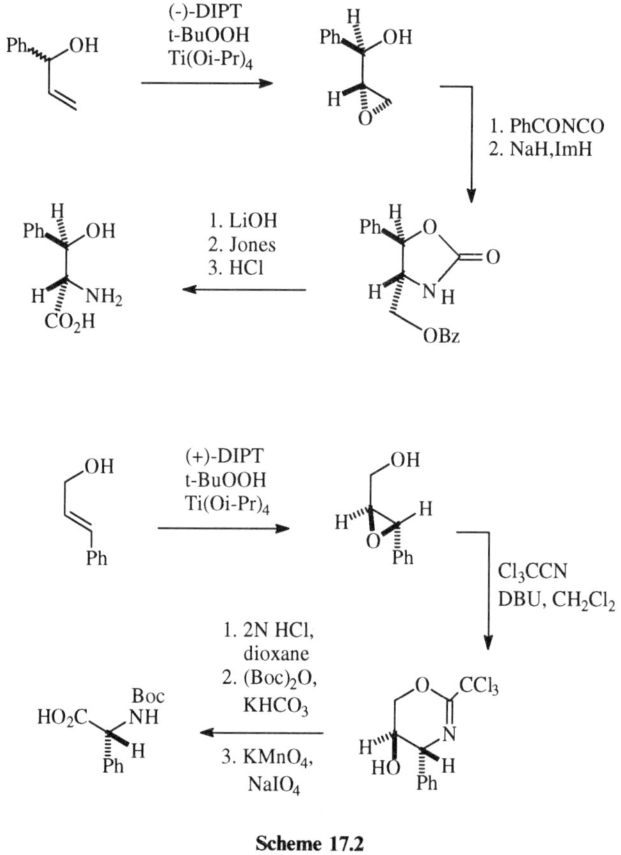

(-)-DIPT
t-BuOOH
$Ti(Oi\text{-}Pr)_4$

1. PhCONCO
2. NaH,ImH

1. LiOH
2. Jones
3. HCl

(+)-DIPT
t-BuOOH
$Ti(Oi\text{-}Pr)_4$

Cl_3CCN
DBU, CH_2Cl_2

1. 2N HCl, dioxane
2. $(Boc)_2O$, $KHCO_3$
3. $KMnO_4$, $NaIO_4$

Scheme 17.2

introduction of a nitrogen atom, and conversion of the alcohol into a carboxylic acid ester, can afford β-hydroxy-α-amino acids.[4] None of the functionality in the product was present in the epoxide, since the β-hydroxy group was introduced by opening the epoxide. The choice of epoxide-containing intermediates was motivated by the need for reliable asymmetric induction, which repaid the labour of several functional group changes and a special trick (intramolecular delivery of an isocyanate) to relay chirality during the introduction of the nitrogen atom. β-Hydroxyamino acids are rather specialised targets, but similar methods have been developed[5] for the general construction of α-alkylamino acids. Cyclisation relays chirality but in this case the original CH_2OH of the allylic alcohol is discarded by periodate cleavage to introduce the carboxylic acid.

Even more extreme (Scheme 17.3), is the design of the route[6] to a C/D-

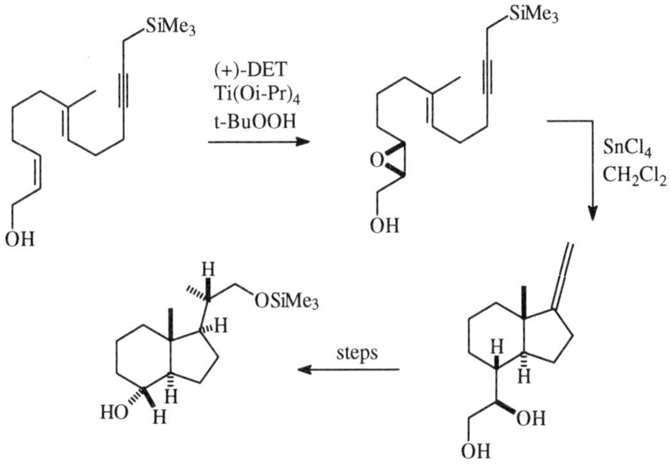

Scheme 17.3

ring synthon for vitamin D_3, in which two of the three carbons of the original allylic alcohol are discarded (at the cost of 4 steps). However, the epoxide was crucial for the formation of the two rings by a biomimetic Lewis acid induced cyclisation, and the allylic epoxide offered a strong binding site for the reagent (tin tetrachloride). In another case, the epoxide product was discarded altogether. The unusual substrate contained a thiophene not an alkene and the oxidation product spontaneously polymerised. This allowed kinetic resolution without difficult separation of residual alkene and epoxide.[7]

As in the amino acid case, oxidation of the alcohol after its role in ligation titanium in the epoxidation step is over, gives general access to epoxycarboxylic ester building blocks that are not available optically pure by epoxidation of acrylic esters.[8] Similarly, it can be advantageous to introduce arene sulfonates after the Sharpless step.[9] Tosylates, for example, have been used to introduce the lower chain of the pheromone (**1**). The Sharpless-based approach requires the carbon chain to be completed after the introduction of functionality (Scheme 17.4).[10]

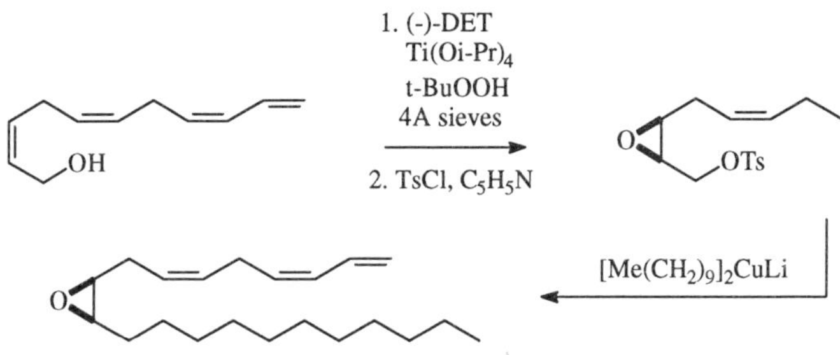

Scheme 17.4

The most natural use of the Sharpless epoxidation is found in cases where all the chirality is carried on into the target structure. Frequently, the destination contains no epoxide, but the reactivity of the strained ring has played a role in its construction. Intramolecular epoxide opening in a post-Sharpless step has been discussed already. Similar cyclisation to carboxylic acid is popular, as in a route to the C-10–C-24 fragment of FK-506, developed at the Merck Frosst Centre in Canada,[11] and in a C-9–C-15 fragment of (9*S*)-dehydroerythronolide A seco acid.[12] Both these cyclisations (Scheme 17.5) were performed with acid. A run of two epoxides (each

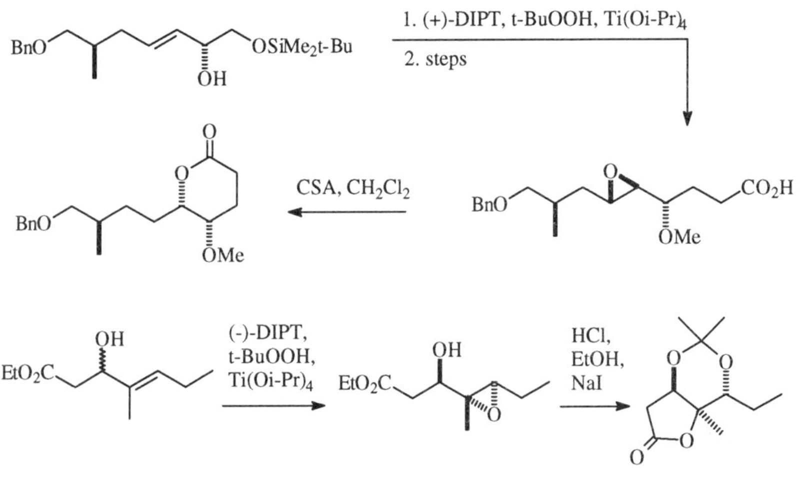

Scheme 17.5

introduced by a Sharpless step) has been zipped up to form (**2**) by the action of pig-liver esterase,[13] and three participate in the formation of (**3**)[14] (Scheme 17.6). Stereochemistry in each ring of the polyether is controlled by the choice of absolute configuration on the tartrate auxiliary for each epoxidation step.

These examples illustrate the popularity of the Sharpless epoxidation as a means to induce asymmetry. Looking back through the earlier chapters of this book, however, readers should recognise that in the future no one method will dominate asymmetric synthesis in the late 1990s as the Sharpless procedure did at the start of the decade. Indeed, our intention in this book on advanced asymmetric synthesis is to draw attention to the wide selection of general asymmetric methodologies that are now either already available or soon to be expected. Nonetheless, it is undoubtedly true that for the time being many synthetic chemists turn at once to the Sharpless epoxidation when planning a new route.

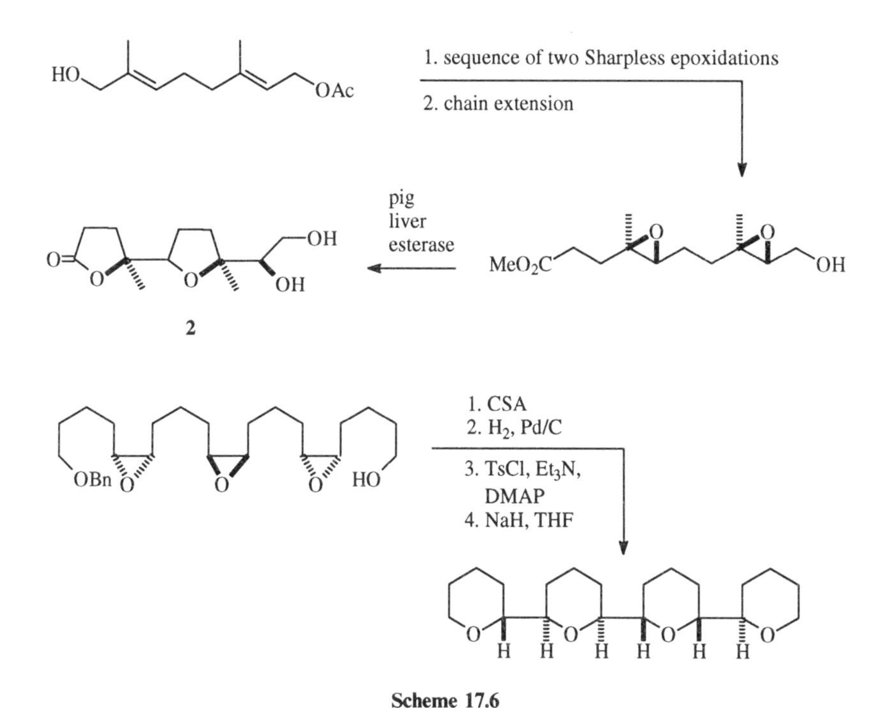

Scheme 17.6

17.3 Mechanism of the Sharpless asymmetric epoxidation

The role of titanium alkoxide complexes in the epoxidation process, and particularly the dimeric structure of the active form of the catalyst, has already been introduced in Chapter 3. The dimeric composition of the catalyst is crucial to the appreciation of the origin of non-linear effects, but to follow the molecular details of the catalytic cycle it is sufficient to consider events at one of the titanium centres. Indeed, although the cooperative involvement of both titanium atoms cannot formally be ruled out,[15] the best model available[15,16] for the reaction is based on the assumption that both allylic alcohol and t-butyl hydroperoxide (TBHP) ligands are bound to the same titanium for the oxygen transfer step. The *syn* epoxidation of cyclohexen-3-ols supports this assumption, but since chiral discrimination is poor in this case, it may not give an accurate guide to the exact details of highly enantioselective examples. Despite these difficulties, it is possible to present a simplified view of the mechanism (Scheme 17.7) by concentrating on events at just one of

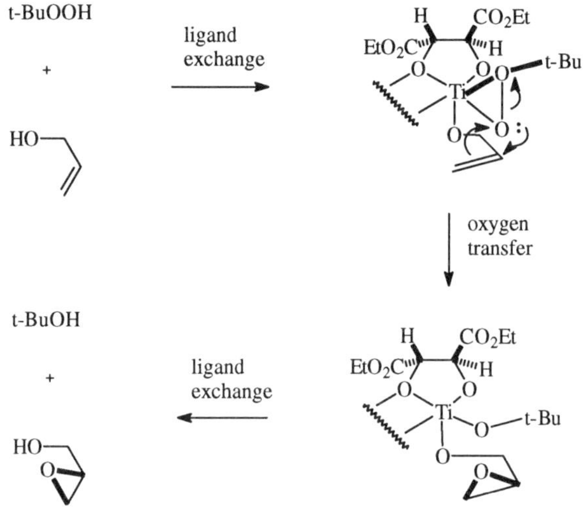

t-BuOOH + HO– (allyl alcohol) —ligand exchange→ Ti complex —oxygen transfer→ Ti complex —ligand exchange→ t-BuOH + HO– (epoxy alcohol)

Scheme 17.7

the two titanium atoms in the dimeric structure. The allylic alcohol substrate, the source of oxygen (t-butyl hydroperoxide) and the chiral auxiliary (a tartrate diester) are all coordinated to titanium by M–O bonds. Such titanium alkoxide species are labile and a selection of structures will be present in solution in catalyst preparations. However, one structure dominates the equilibrium mixture[15] and is also the most catalytically active. This fortunate property of the titanium/tartrate system endows it with high selectivity together with rapid substrate/product exchange. Comparison of around 50 ligands has indicated that titanium and tartrate are perfectly matched,[15] and the 1:1 stoichiometry of tartrate and titanium tetraisoperoxide is optimum.[17] In this mechanism, the alkene itself is never bound to the metal, and face selectivity of oxygen transfer is a consequence of the geometrical features of the complex and the tight association of reactants and chiral auxiliary in the coordination sphere of titanium. It has been shown that increasing the bulk of the reactants leads to an enhancement of both rate and enantioselectivity, suggesting the allylic alcohol and peroxy species become pressed together in a near perfect alignment, with the alkene approaching the coordinated epoxide along the O–O axis. Detailed kinetic studies have been reported;[15] it is the oxygen transfer step that is rate determining. Alkyl peroxides bind to early transition metals in a bidentate fashion, forming a triangular dioxametallacycle. This capacity to coordinate both oxygens activates the peroxide for oxygen transfer, and there is evidence that both C–O bonds in the epoxide are formed simultaneously (though secondary deuterium isotope effects are slightly larger at C-3 (1, $R^1 = D$, $R^2 = C_7H_{16}$). There is also evidence that the bidentate peroxy ligand occupies an equatorial site with the allylic alcohol bound axial, as shown in Scheme 17.7.

17.4 Jacobsen–Katsuki epoxidation

Following the success of the titanium-mediated asymmetric epoxidation, the search was intensified for similarly general methods which did not rely on allylic alcohols for substrate recognition. The main advances have been made in the independent laboratories of Jacobsen and Katsuki, both ex-coworkers from the Sharpless group. Salen-based chiral auxiliaries and molybdenum oxide complexes hold the key to the methods which emerged. Though not as reliable as a general source of high ee (the truly remarkable benefit from the titanium/allylic alcohol approach), after intense development over the last few years widely applicable methods are now available for the asymmetric epoxidation of some classes of unfunctionalised alkenes. Structures **5** and **6** (Scheme 17.8) are typical examples of currently popular catalysts.

The first cases of asymmetric epoxidation catalysed by manganese salen complexes appeared independently from Jacobsen[18] and Katsuki[19] in 1990. These catalysts are related to metalloporphyrin-based epoxidation catalysts (see section 17.5), and oxygen transfer to the alkene takes place from an oxomanganese intermediate. Most common oxidising agents are iodosylbenzene or aqueous solutions of sodium hypochlorite,[20–22] but direct utilisation of molecular oxygen has also been reported.[23] Hydrogen peroxide can be used as the oxidant, but is required in excess and the reaction is performed in the presence of an additional ligand.[24,25] The addition of donor ligands can also improve reactions in which iodosylbenzene is the oxidant.[26]

There has been much debate about the mechanism of this asymmetric epoxidation procedure. Originally, while good results were obtained only with *cis*-alkenes and bad results with *trans*-alkenes, it was felt that side-on approach of the alkene to the Mo=O unit held the key to efficient asymmetric induction and that this was blocked if the alkene was forced to approach from above. Since then, it has become clear that some *trans* disubstituted

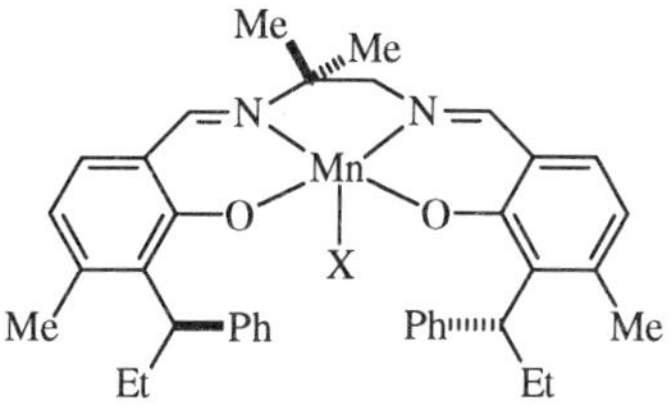

4

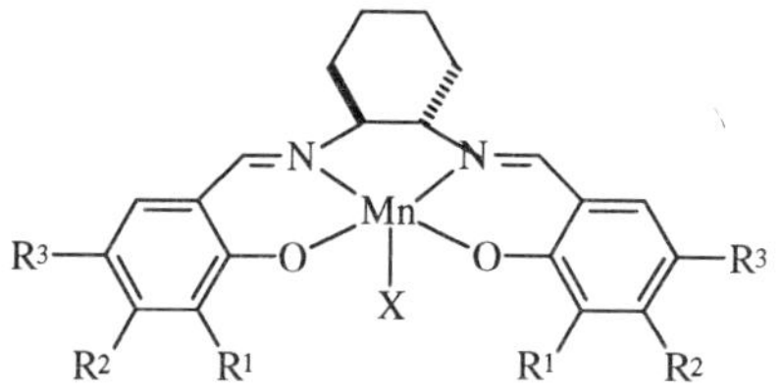

5 R^1 = H, R^2 = Me, R^3 = CH(Ph)Et

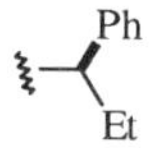

6 R^1 = H, R^2 = R^3 = t-Bu

cat. / PhIO or NaOCl

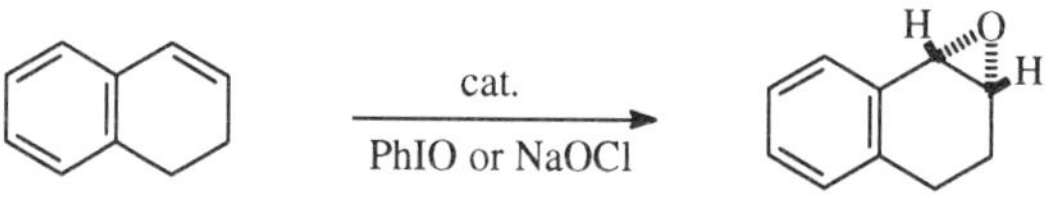

cat. = **4** ;
PhIO, CH_3CN: 87% yld, 36% ee

cat. = **5** / 4-Me_2N-py-N oxide;
PhIO, CH_2Cl_2: 38% yld, 91% ee

Scheme 17.8

alkenes,[27] trisubstituted alkenes[28] and even some tetrasubstituted structures[29] can be good substrates, and mechanistic theories have been adjusted. Since asymmetric substituents at the C–C link between the two imines are important for asymmetric induction, it is tempting to believe the proposal[30] that stereocontrol originates from the approach of the alkene parallel to the Mo–N bond furthest from the substituent that projects upwards (Figure 17.1). This model is being used as a rational basis for the design of new catalysts,[30] so it should soon be clear whether this offers a good empirical guide.

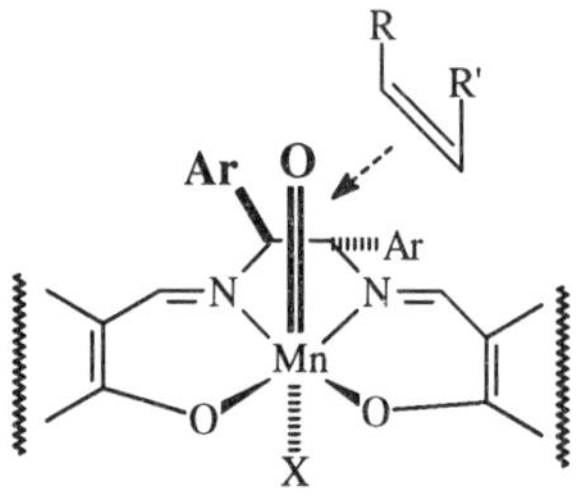

Figure 17.1 Approach of alkenes to the Mo=O centre is controlled by the Ar group that projects above the plane of the salen ligand.

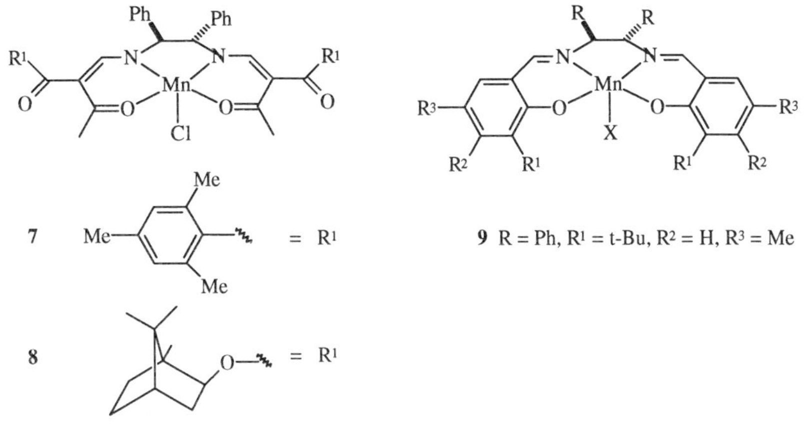

7 Me—(C6H2Me2)— = R1

8 = R1

9 R = Ph, R1 = t-Bu, R2 = H, R3 = Me

Figure 17.2 Comparison of ligands for manganese catalysed asymmetric epoxidation.

Already, there are clues that the full mechanistic picture will be more complicated. A comparison has been made of face selectivity for oxidation using oxygen (in the presence of an aldehyde and a structurally distinct class of ligands, Figure 17.2) and sodium hypochlorite. Scheme 17.9 shows two cases where the O_2-mediated oxidation not only proved more efficient, but also afforded the opposite absolute configuration of the product.[31,32]

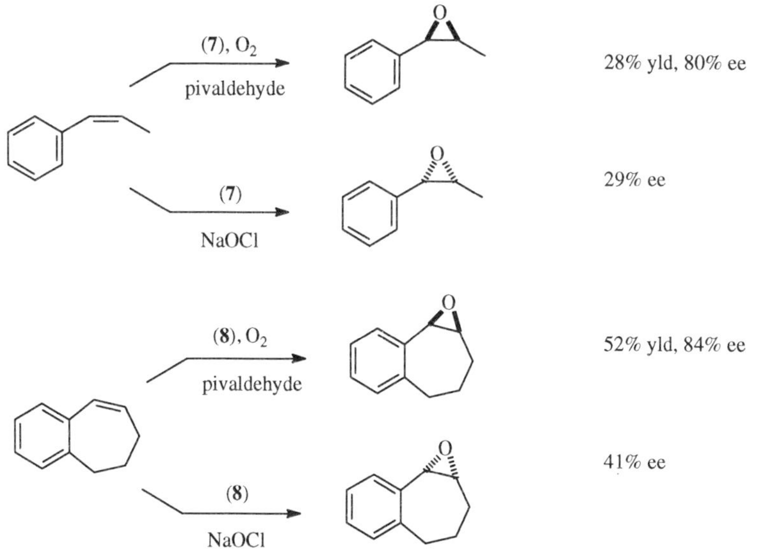

(7), O_2 pivaldehyde — 28% yld, 80% ee

(7) NaOCl — 29% ee

(8), O_2 pivaldehyde — 52% yld, 84% ee

(8) NaOCl — 41% ee

Scheme 17.9

A concensus seems to favour a stepwise mechanism for oxygen transfer, with free radicals involved as intermediates. The debate centres on the orientation of the alkene and the origin of enantioface selectivity. With more sterically hindered polysubstituted substrates, it seems clear that in some cases the alkene must be tilted as it approaches the oxygen atom. Based on this view, it has been possible to account for the reversal of face selectivity when changing from di- to trisubstituted alkenes.[28]

Scheme 17.10 shows examples of successful tri- and tetrasubstituted substrates.[28,29] Cycloalkenes have been popular and though initially optical yields were moderate,[33] considerable improvement has now been made with more elaborate catalysts (Scheme 17.11).[34,35]

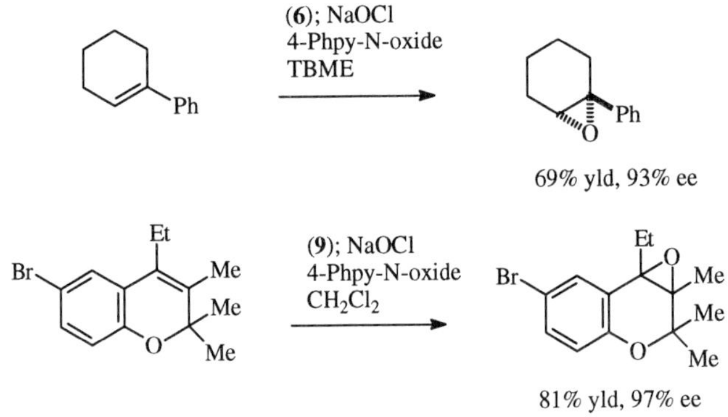

Ph
(**6**); NaOCl
4-Phpy-N-oxide
TBME
Ph
O
69% yld, 93% ee

Et
Br
Me
Me
O
Me
(**9**); NaOCl
4-Phpy-N-oxide
CH_2Cl_2
Et
O
Br
Me
Me
O
Me
81% yld, 97% ee

Scheme 17.10

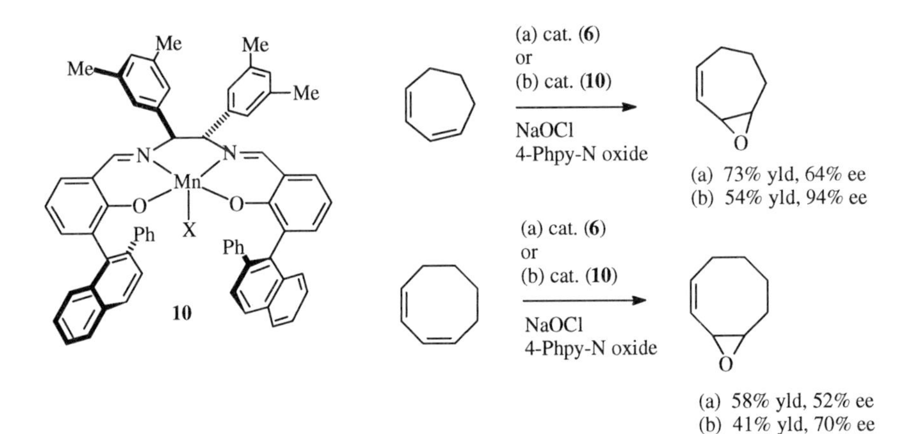

Scheme 17.11

17.5 Other methods of asymmetric epoxidation

Asymmetric epoxidation was born with the oxidation of styrene using peroxycamphoric acid in 1967[36] and has come of age in the last decade with the work of Sharpless, Jacobsen, and Katsuki. There are, however, many alternative procedures and even in the difficult field of epoxidation of unfunctionalised alkenes good results have been obtained. For example, an optically active oxaziridine has been recommended for the asymmetric epoxidation of simple alkenes.[37] Non-racemic molybdenum oxo-diperoxo complexes can similarly be used in stoichiometric procedures.[38] Metalloporphyrins have also proved effective.[39]

Enzymatic methods (see Chapter 13) are also effective in epoxidation of unfunctionalised alkenes. For example, *Pseudomonas oleovorans* has been studied in some detail[40] and affords epoxides from terminal alkenes (cyclic and internal alkenes are not substrates[41]). A virtue of the biotransformation approach is that it is suitable for use with small alkenes. Though *P. oleovorans* will not form epoxides from propene and 1-butene, many simple straight-chain 1-alkenes are substrates. 1-Octene, for example, can be converted into (*R*)-1,2-epoyxoctene by a biotransformation performed simply in 5 litre conical flasks. Often oxidation of methyl groups competes with epoxidation, and mixtures of epoxides and ω-epoxyalkan-1-ols can be encountered. Epoxidation of the α,ω-diene (**11**) (Scheme 17.12), which has no methyl group, provides an example that proceeds cleanly, and the product (**12**) can be isolated with ee >80%.[42] Alternatively, complete epoxidation can give the diepoxide in similar optical yield. Some functionalised alkenes [e.g. (**13**) can be excellent substrates, but the choice of *Penicillin spinulosum* in Scheme 17.12 was the result of a survey of many microorganisms.[43] This may be necessary to gain good results with a new substrate and, although time-consuming, is likely also to lead to the development of procedures for the use of relatively unstudied microorganisms.

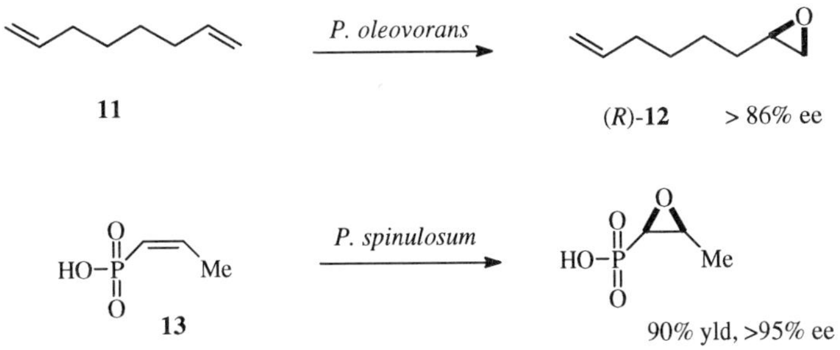

P. oleovorans
11
(*R*)-**12** >86% ee

HO–P(=O)(=O) Me
13
P. spinulosum
HO–P(=O)(=O) O Me
90% yld, >95% ee

Scheme 17.12

17.6 Sharpless asymmetric *cis*-dihydroxylation

The domination of the field of asymmetric oxidation by the Sharpless laboratory is also evident in the field of asymmetric induction during the introduction of a *cis* diol by the oxidation of an alkene. After a long and intense search simple procedures with opposite face-selectivities (summarised by the labels AD-mix-α and AD-mix-β) have emerged to universal acclaim. Once again the miracle of reliable widespread applicability has been achieved, and high optical yield and predictable stereoselectivity can be anticipated with a good selection of substrates in well-defined situations.[44]

The key to these reactions is the use of two epimeric look-alike chiral auxiliaries [dihydroquinine (DHQ) and dihydroquinidine (DHQD), Figure 17.3] to modify the classic osmium tetroxide *cis*-hydroxylation reaction. These two auxiliaries are not enantiomers, but have opposite absolute configurations at the two chiral centres α and β to the quinoline ring, the region of the auxiliary that is important for controlling the asymmetric induction.

Several features contribute to the popularity of this method. A catalytic quantity of osmium is introduced in a non-volatile form [$K_2OsO_2(OH)_4$] in combination with an inorganic oxidant ($K_3Fe(CN)_6$), and, by means of a two-phase reaction mixture (Scheme 17.13), a competing and virtually non-enantioselective second reaction pathway is blocked. This constitutes a very convenient method, and ensures the best prospects for asymmetric induction. Addition of one equivalent of $MeSO_2NH_2$ is also recommended to increase the rate of reaction, although with intrinsically reactive terminal alkenes this additive can be omitted. The sulfonamide accelerates hydrolysis of the glycolate Os(VI) ester (**15**), releasing osmium (VIII) to the aqueous phase to feed the reoxidation section of the catalytic cycle, ultimately ensuring satisfactory concentrations of the Os(VIII) oxidant.

Most undergraduates learn a [3 + 2] mechanism for *cis*-hydroxylation, but an alternative starts with [2 + 2] addition to an Os=O double bond to form an oxametallocyclobutane (**14**), followed by a σ-bond shift to complete the glycolate ester.[45] For the asymmetric version of the reaction the non-linear

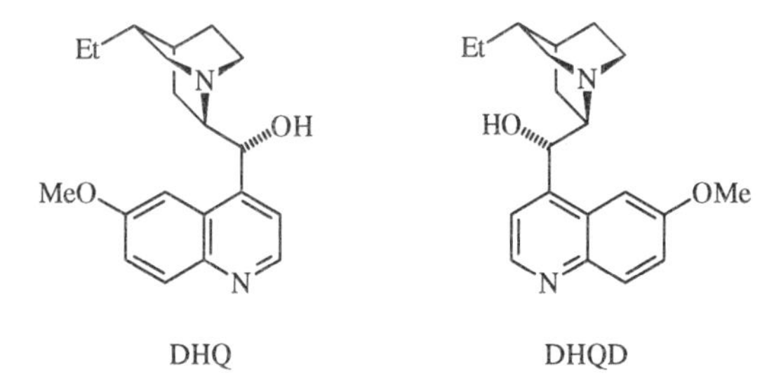

Figure 17.3 Chiral components of auxiliaries used in Sharpless asymmetric dihydroxylation.

temperature dependence of ee[46] points to a two-step process and so rules out the [3 + 2] mechanism, giving indirect support to the [2 + 2] alternative. Scheme 17.13 illustrates this mechanism and attributes asymmetric induction to the face selectivity of the [2 + 2] step.

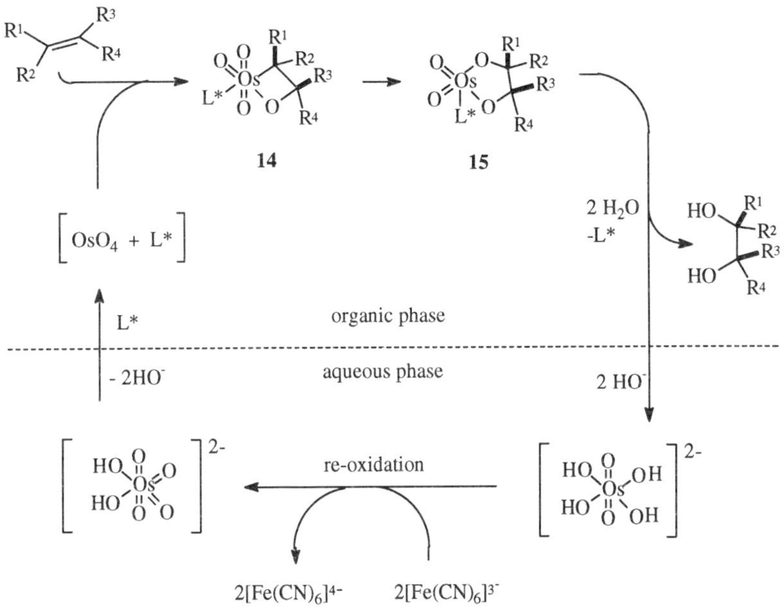

Scheme 17.13

There has been considerable progress in the design of the chiral auxiliary. Dimeric assemblies comprising two identical alkaloid building blocks linked by an aromatic spacer give improved results compared to ligands used in early studies.[47] There is evidence from modified MM2 force field calculations[48] that the linker is important in promoting π-stacking effects, so enhancing recognition and stabilising the transition state of the oxidation process. The second alkaloid does not bind a catalytically active centre, but still plays a part, shielding one side of the recognition cleft, and so contributing to the control of alkene orientation. The osmium atom binds to the bridge-head nitrogen (there is evidence for this from X-ray studies of osmate ester complexes of cinchona alkaloids[49]) and the quinoline nitrogen plays no part in coordination to osmium. The OMe substituent on the quinoline, however, is significant, and improves binding of the ligand to the metal. These considerations lead to the now familiar cartoon that guides the prediction of face selectivity of asymmetric dihydroxylation (the 'Sharpless mnemonic'[50]) and is illustrated in Figure 17.4 for the case of a trisubstituted alkene with large, medium and small substituents (R_L, R_M and R_S, respectively). Steric

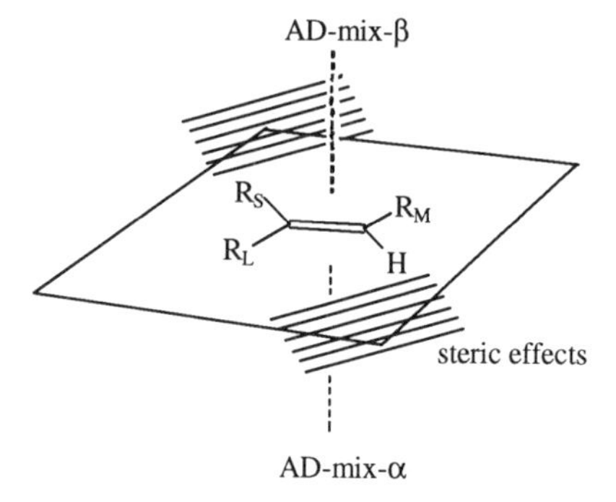

Figure 17.4 Steric blockade at two sides of the binding cleft controls dihydroxylation.

effects at two opposite corners of the plane drawn around the alkene leave a cleft in which the alkene lies with R_L to the left. Viewed from this angle, AD-mix-β (DHQD) introduces oxygen from above (β-face) and AD-mix-α (DHQ) reacts from below (α-face). Not all substrates fit this model, but it is popular as an empirical rule.

Some currently favoured chiral auxiliaries for the Sharpless asymmetric dihydroxylation are shown in Figure 17.5. Most widely applicable are the phthalazine-linked assemblies $(DHQ)_2$PHAL (**16**) and $(DHQD)_2$PHAL (**17**), which are the auxiliaries used in AD-mix-α and -β, respectively, and the diphenylpyrimidine-linked alternatives $(DHQ)_2$PYR (**18**) and $(DHQD)_2$PYR

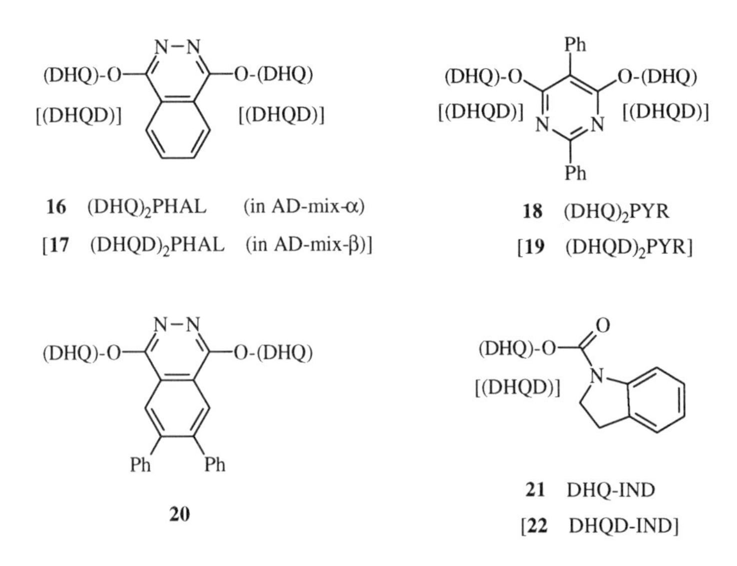

Figure 17.5 Most ligands for Sharpless asymmetric dihydroxylation are dimeric but some monomeric auxiliaries are still used.

(**19**). The less readily available $(Ph)_2(DHQ)_2$PHAL dimer (**20**) can give improved enantioselectivity in some cases,[51] and for most *cis*-alkenes the monomeric auxiliaries (**21**) and (**22**), carrying an indenoyl substituent, are superior to any of the dimeric structures (though there are exceptions where the PYR system is more suitable).

Unlike asymmetric epoxidation, where tartrate esters are used exclusively, a choice must be made from a range of auxiliaries for dihydroxylation. Even so not all classes of substrates can be accommodated and in his recent review[44] Sharpless predicts the development of further ligands with tighter binding clefts to address these limitations. Even now, however, the asymmetric dihydroxylation is impressive as a general method, though it works badly with cyclic ligands and is inexplicably ineffective in situations where kinetic resolution is required.

17.7 Choice of auxiliary for Sharpless asymmetric dihydroxylation

The PHAL-linked ligands are the best for use with 1,1-, and (*E*)-1,2-disubstituted alkenes (Scheme 17.14) and with trisubstituted compounds (Scheme 17.15), and in many cases >95% ee is obtained.[52,53]

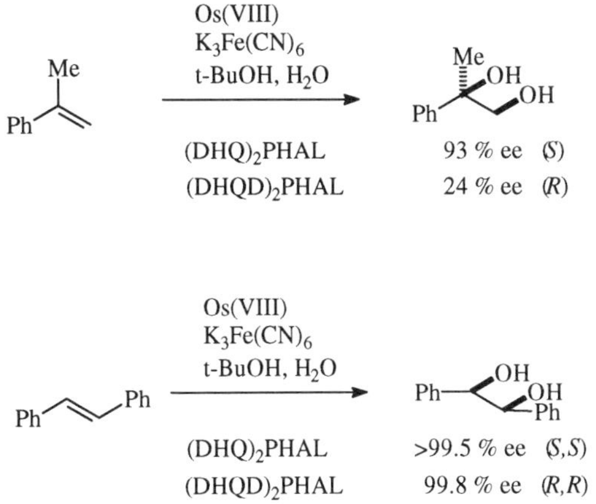

Os(VIII)
$K_3Fe(CN)_6$
t-BuOH, H_2O

Me, Ph → Me, OH, OH, Ph

(DHQ)$_2$PHAL 93 % ee (*S*)
(DHQD)$_2$PHAL 24 % ee (*R*)

Os(VIII)
$K_3Fe(CN)_6$
t-BuOH, H_2O

Ph, Ph → Ph, OH, OH, Ph

(DHQ)$_2$PHAL >99.5 % ee (*S*,*S*)
(DHQD)$_2$PHAL 99.8 % ee (*R*,*R*)

Scheme 17.14

At the two extremes, however, with mono-substituted[52,53] and tetrasubstituted[54] alkenes, the situation is less clear cut, and both PHAL- and PYR-linked ligands should be evaluated. As seen in scheme 17.16, there is often no outright winner. With some bulky enol ether substrates very good yields of enantiopure α-hydroxyketones are possible (e.g. Scheme 17.17).

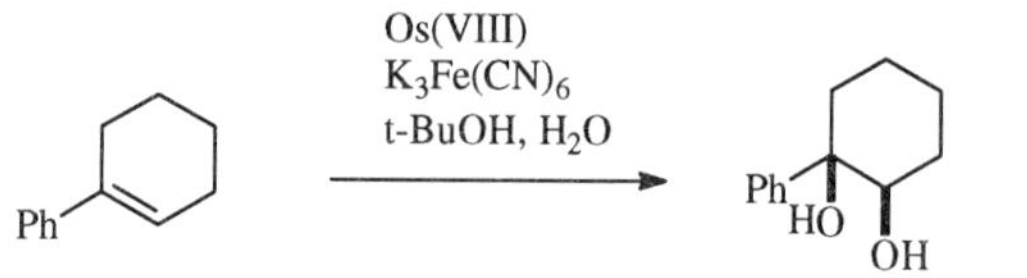

Ph

Os(VIII)
$K_3Fe(CN)_6$
t-BuOH, H_2O

Ph HO OH

$(DHQ)_2PHAL$	97 % ee (*S,S*)
$(DHQD)_2PHAL$	99 % ee (*R,R*)

Scheme 17.15

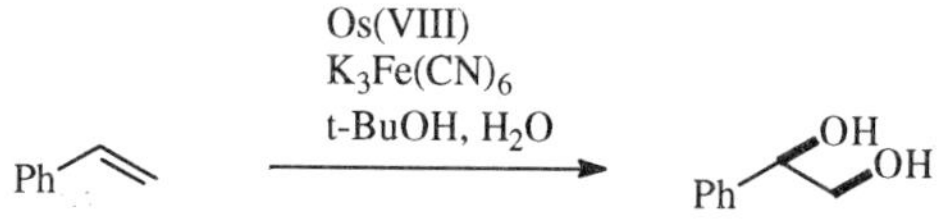

Ph

Os(VIII)
$K_3Fe(CN)_6$
t-BuOH, H_2O

Ph OH OH

$(DHQD)_2PHAL$	97 % ee (*R*)
$(DHQD)_2PYR$	80 % ee (*R*)

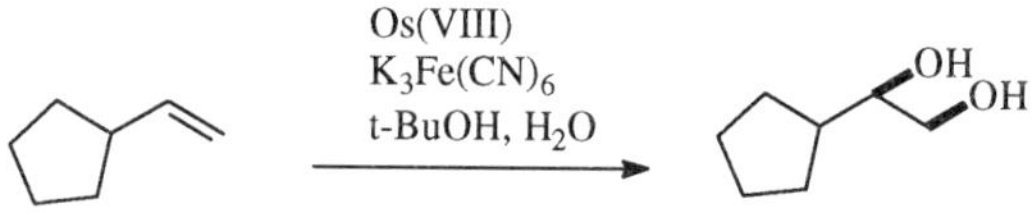

Os(VIII)
$K_3Fe(CN)_6$
t-BuOH, H_2O

OH OH

$(DHQD)_2PHAL$	80 % ee (*R*)
$(DHQD)_2PYR$	93 % ee (*R*)

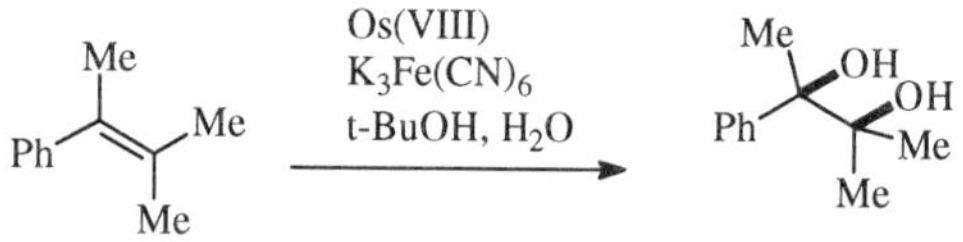

Me Ph Me Me

Os(VIII)
$K_3Fe(CN)_6$
t-BuOH, H_2O

Me Ph OH OH Me Me

$(DHQD)_2PHAL$	39 % ee (*R*)
$(DHQD)_2PYR$	47 % ee (*R*)

Scheme 17.16

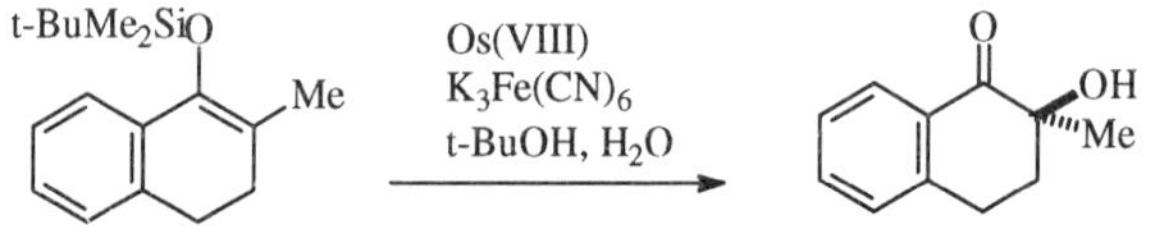

$t-BuMe_2SiO$ Me

Os(VIII)
$K_3Fe(CN)_6$
t-BuOH, H_2O

O OH Me

$(DHQD)_2PHAL$ 97 % yld, 93 % ee

Scheme 17.17

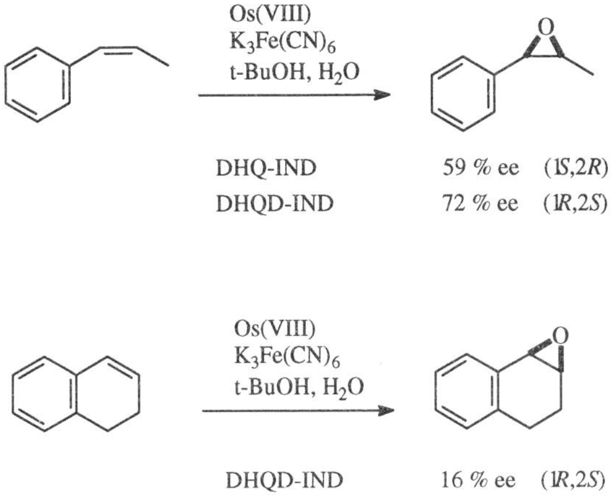

Os(VIII)
$K_3Fe(CN)_6$
t-BuOH, H_2O

DHQ-IND 59 % ee (1*S*,2*R*)
DHQD-IND 72 % ee (1*R*,2*S*)

Os(VIII)
$K_3Fe(CN)_6$
t-BuOH, H_2O

DHQD-IND 16 % ee (1*R*,2*S*)

Scheme 17.18

(*Z*)-alkenes fit the cleft in the dimeric catalysts less well than their *E* counterparts, and DHQ-IND and DHQD-IND are the best catalysts (Scheme 17.18) but optical yields are relatively poor (below 80% ee).[44,55] The most difficult substrates for efficient asymmetric induction are thus tetrasubstituted and (*Z*)-disubstituted alkenes.

In contrast to the titanium-catalysed epoxidation, 1,3-dienes can be used in dihydroxylation reactions affording attractive functionalised building blocks (Scheme 17.19).[44,56] A nice application of this type of product makes use of the opening of cyclic sulfates derived from the enantiopure diol, as in a route to the Carpenter bee pheromone.[57]

Some of the simple ligands used in early investigations gave excellent results and Scheme 17.20 shows an example in which dihydroxylation controlled

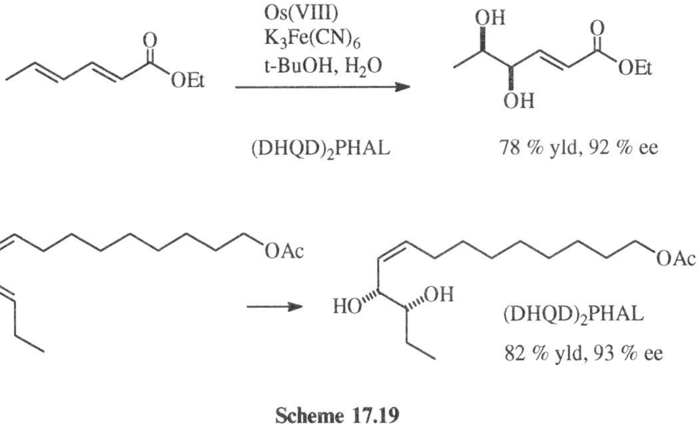

Os(VIII)
$K_3Fe(CN)_6$
t-BuOH, H_2O

O OEt → OH O OEt OH

$(DHQD)_2PHAL$ 78 % yld, 92 % ee

OAc → HO OH OAc

$(DHQD)_2PHAL$
82 % yld, 93 % ee

Scheme 17.19

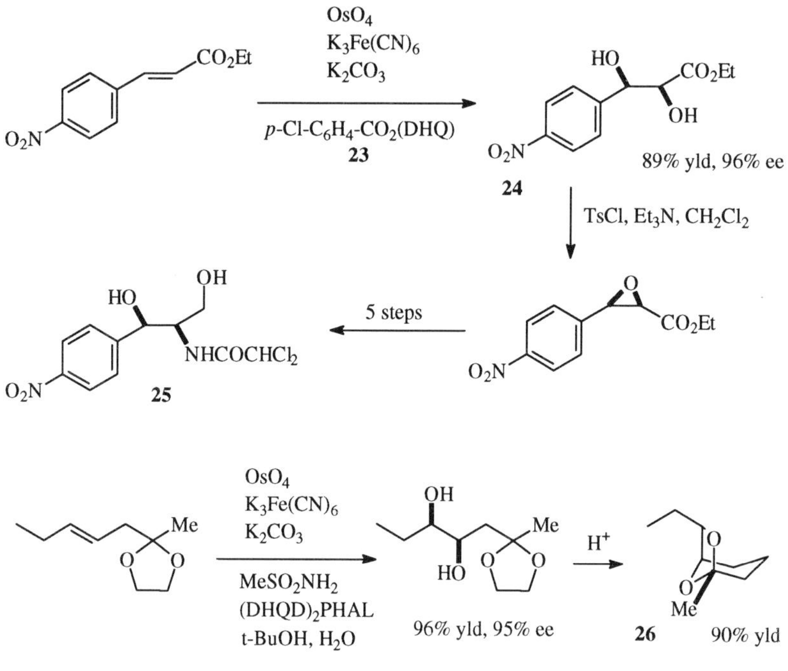

Scheme 17.20

by DHQD-CLB (**23**) afforded (**24**) in excellent optical purity for use in an enantioselective synthesis of chloramphenicol (**25**).[58] A similar intermediate provides a starting point for a synthesis of (*R*)-reticuline.[59] A short route to (+)-*exo*-brevicomin (**26**) starting with asymmetric dihydroxylation employing $(DHQD)_2PHAL$ illustrates an example of the use of $MeSO_2NH_2$.[60]

17.8 Limitations and alternative dihydroxylation procedures

Chemists turn to the Sharpless procedure, particularly the commercially available AD-mix variants, when first considering the induction of asymmetry in the preparation of a diol. Not all alkenes are suitable, however, and cyclic alkenes are usually poor substrates, though there are exceptions (Scheme 17.21). Good chemical yield, but poor induction of asymmetry, are common.[44] More unsymmetrical substrates, e.g. (**27**), sometimes give good results and occasionally high yields and enantioselectivities are encountered together, as in the case of (**28**) for which the dihydroxylation is also diastereoselective.[61]

No other ligand series has been studied in the detail possible in the Sharpless laboratory though other examples are present in the literature. Attempts at catalytic dihydroxylation with (**29**) by Hirama's team,[62] or (**30**) by Murahashi

OsO_4
$K_3Fe(CN)_6$
t-BuOH, H_2O

HO
HO

$(DHQD)_2PHAL$	97 % yld, 37 % ee
$(DHQD)_2PYR$	88 % yld, 23 % ee
DHQD-IND	88 % yld, 24 % ee

Ph
OsO_4
$K_3Fe(CN)_6$
t-BuOH, H_2O

HO Ph
HO

27 $(DHQD)_2PHAL$ 42 % yld, 91 % ee

AD-mix-β
$MeSO_2NH_2$
$[(DHQD)_2PHAL]$

HO
HO
O
O
H
Ph
O
O
Ph

28 85 % yld, 95 % ee

Scheme 17.21

et al.,[63] however, gave only moderate optical yields. In stoichiometric oxidations simple amine ligands (e.g. (**31**) and (**32**), Figure 17.6[64,65] can give

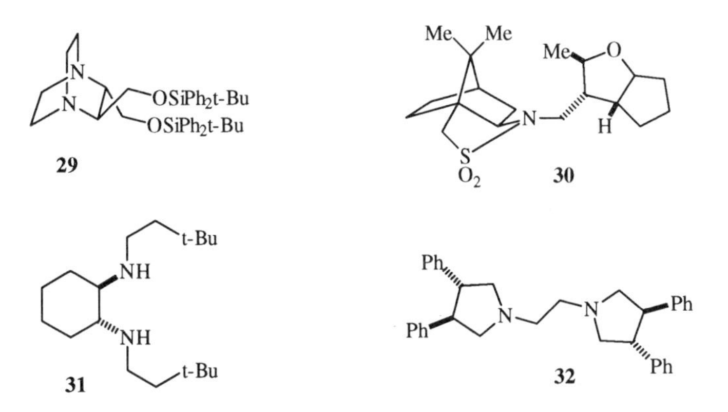

Figure 17.6 Alternative ligands for asymmetric dihydroxylation.

good induction of asymmetric, but slow hydrolysis of products prevents *in situ* recycling. In seeking alternatives to ligands employing cinchona alkaloids it is necessary to forgo the benefits of the catalytic process. With osmium as the oxidant this is literally a high price to pay.

No specific feature of the cinchona auxiliary renders it ideal, nor does the answer lie in a fortuitous balance of the kinetics of catalyst activity with equilibrium concentrations of active structures, as is the case with the tartrate/$Ti(O\text{-}i\text{-}Pr)_4$ systems for asymmetric epoxidation. Barring a chance discovery of a new system, it now seems likely that further fine tuning of ligands based on cinchona alkaloids will be the most probable way to overcome the remaining limitations. However, the best opportunities for a radically new approach lie in the development of new catalyst kinetic resolutions, where the AD-mix magic fails. Since the first demonstration using acetate esters of cinchona alkaloids by Hentges and Sharpless[66] great improvements have been made, though it remains to be seen whether a single auxiliary system can be developed to give successful enantioselective dihydroxylations in every situation.

17.9 Asymmetric oxidation of thioethers

Thioethers are rather specialised substrates for asymmetric oxidation but since the products (chiral sulphoxides) are important chiral auxiliaries (see Chapter 4), considerable efforts have been made to develop reliable procedures. The use of enzymatic methods and titanium/tartrate catalyst has already been mentioned in Chapter 4; an important point in the latter case, however, is the addition of water in the Kagan titanium alkoxide procedure, while the best results with the same catalyst system in the Sharpless asymmetric epoxidation require scrupulously dry conditions achieved by adding molecular sieves to the reaction mixture.

Salen-based manganese catalysts of the Jacobsen[67] and Katsuki[68] types have also been used to oxidise thioethers but optical yields rarely exceed 70%. Scheme 17.22 shows one of the best examples. This employs a salen complex with highly substituted phenolic rings. Salen vanadium complexes have been used.[69] Titanium/binaphthol[70] and manganese/salen catalysts[68] can be used to effect kinetic resolution of racemic sulphoxides.

1,3-Dithianes are interesting substrates for asymmetric oxidation because the products are mono-sulfoxides. The Kagan procedure can be used for this purpose.[71] When a prochiral centre is at C-2, chirality at sulfur and at carbon can be induced. This has been examined in 5-[72] and 6-membered[73] rings.

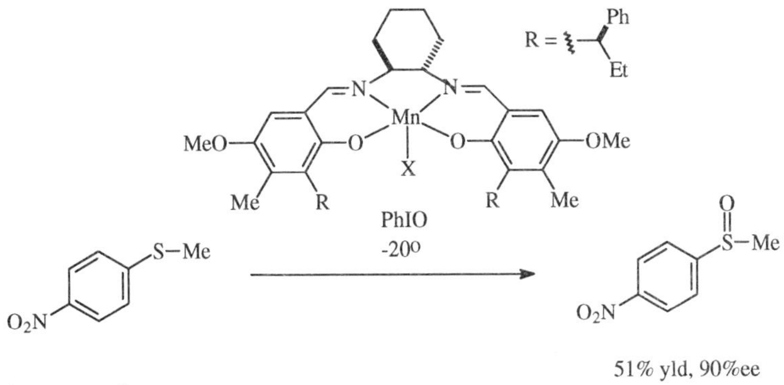

Scheme 17.22

17.10 Other asymmetric oxidations

The asymmetric modification of allylic oxidation is a useful goal because efficient methods are available for the stereocontrolled elaboration of the products by C–C bond formation (Chapter 15). Copper camphorate salts and amino acid complexes[74] have been examined. Feringa's group has described the use of chiral copper(II) bis-proline complexes in catalytic asymmetric allylic oxidation.[75] The oxidant is a perester. Cyclohexene can be converted into cyclohexyl proanoate but since the best ee so far is 61% the method is far from optimised. With the bisoxazoline complex (**29**), Andrus[76] has improved the best efficiency to about 80% ee, but results are still variable. Oxidation of cyclopentene and cyclohexene (Scheme 17.23) are

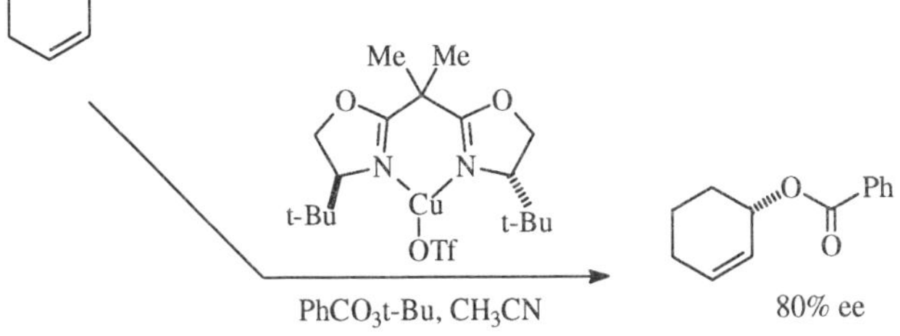

Scheme 17.23

the best cases but of the ligands examined, the best for cyclopentene gave only 47% ee with cyclohexene. Cyclooctene and 3-phenyl-1-propene were poor substrates, even with (**29**). Unfortunately, these methods lack the tolerance of different substrate types that have made the Sharpless asymmetric oxidations so popular. As shown in Chapter 1, the Sharpless epoxidation can be used for kinetic resolution of allylic alcohols, and a complementary simple procedure to rearrange Sharpless-derived epoxides to optically pure allylic alcohols has been described.[77]

Diastereoselective 1,4-difunctionalisation of dienes by the Backvall reaction also affords allylic alcohols. The reaction has been examined with an enantiomerically pure tolysulfoxide as a substituent of a chiral quinone co-oxidant, but although this influenced the stereochemistry of the reaction by improving *trans* selectivity, enantioselectivity was not achieved.[78]

At benzylic positions microbial oxidation allows enantioselective conversions but substrates are either rather simple, like ethyl benzene (oxidised to 1-phenylethanol by *Mortierella isabellina* in 33% ee[79]), or rather specialised (e.g. the oxidation of tetrahydroquinolines by *Cunninghamella elegans*[80]). Enzymes can also be used. Dopamine β-hydroxylase performs its natural function in Scheme 17.24 by forming (−)-noradrenaline (**30**) by benzylic oxidation of dopamine.[81] Optical yields are sometimes good, though in some cases with microbial methods high optical yield may be due to the selective removal of the minor enantiomer of the product by a second oxidoreductase. Oxidation of a *substituent* at the benzylic position can also give good results, and in the case of the anti-inflamatory drug (*S*)-naproxen (**31**), *C. militaris* effects its preparation in >97% ee by oxidation of an enantiotopic methyl group (Scheme 17.24).[82]

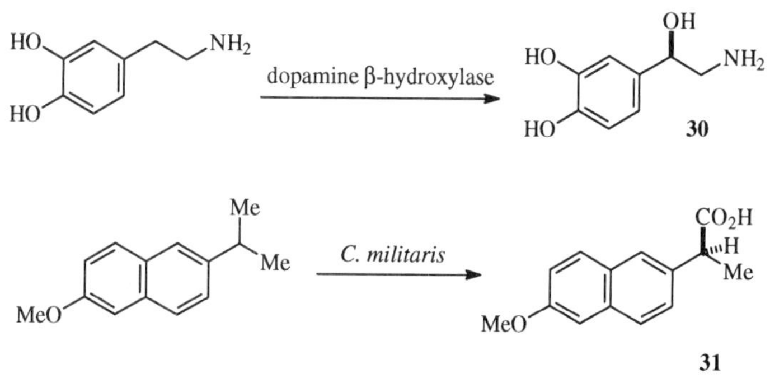

Scheme 17.24

17.11 Conclusions

Epoxidation and dihydroxylation of alkenes are so popular as procedures for asymmetric induction that many syntheses contrive the introduction of the epoxides or diols, or of reactive functionality obtained easily from them. These asymmetric conversions are functional group changes and do not participate directly in the assembly of the carbon skeleton. Carbon–carbon bond formations that make use of these functional groups, however, have enjoyed a surge of popularity as the general applicability, particularly of the Sharpless procedures, has become recognised. There is still scope for some improvement in methods for production of particular classes of epoxides and diols, and in other asymmetric oxidations, perhaps particularly at allylic and benzylic positions, but it can be argued that progress with highly general asymmetric C–C bond formation is now strategically more important. Many good methods exist and most have been surveyed in this book, but they still lack the extraordinary tolerance of substrate variation that is exemplified in the Sharpless asymmetric epoxidation.

In the next chapter of this book some design techniques are discussed to help the reader marshal ideas in this complex field, and even perhaps squeeze a little extra selectivity from reactions drawn from the now extensive list of highly developed methods for asymmetric synthesis.

Acknowledgement

The author thanks Professor Istvan E. Marko for detailed discussions of some of the asymmetric induction procedures reviewed in this chapter during his visit to Norwich to present a lecture on this topic at the UEA Asymmetric Synthesis Short Course.

References

1. Sharpless, K.B., Woodard, S.S. and Finn, M.G., *Pure Appl. Chem.*, 1983, **55**, 1823.
2. Sharpless, K.B. and Michaelson, R.C., *J. Am. Chem. Soc.*, 1973, **95**, 6136.
3. Hanson, R.M. and Sharpless, K.B., *J. Org. Chem.*, 1986, **51**, 1922.
4. Jung, M.E. and Jung, Y.H., *Tetrahedron Lett.*, 1989, **30**, 6637.
5. Schmidt, U., Respondek, M., Lieberknecht, A., Werner, J. and Fischer, P., *Synthesis*, 1989, 256.
6. Hatakeyama, S., Numata, H., Osanai, K. and Takano, S., *J. Chem. Soc., Chem. Commun.*, 1989, 1893.
7. Kitano, Y., Kusakabe, M., Kobayashi, Y. and Sato, F., *J. Org. Chem.*, 1989, **54**, 994.
8. Legters, J. and Zwanenburg, L.T.B., *Tetrahedron Lett.*, 1989, **30**, 4881.
9. Klunder, J.M., Onami, T. and Sharpless, K.B., *J. Org. Chem.*, 1989, **54**, 1295.
10. Mori, K. and Takeuchi, T., *Lieb. Ann. Chem.*, 1989, 453.
11. Wang, Z., *Tetrahedron Lett.*, 1989, **30**, 6611.
12. Chaimberlin, A.R., Dezube, M., Reich, S.H. and Sall, D.J., *J. Am. Chem. Soc.*, 1989, **111**, 6247.
13. Russel, S.T., Robinson, J.A. and Williams, D.J., *J. Chem. Soc., Chem. Commun.*, 1987, 351.

14. Iimori, T., Still, W.C., Rheingold, A.L. and Staley, D.L., *J. Am. Chem. Soc.*, 1989, **111**, 3439.
15. Finn, M.G. and Sharpless, K.B., *J. Am. Chem. Soc.*, 1991, **113**, 113.
16. Carlier, P.R. and Sharpless, K.B., *J. Org. Chem.*, 1989, **54**, 4016.
17. Woodard, S.S., Finn, M.G. and Sharpless, K.B., *J. Am. Chem. Soc.*, 1991, **113**, 106.
18. Zhang, W., Loebach, J.L., Wilson, S.R. and Jacobsen, E.N., *J. Am. Chem. Soc.*, 1990, **112**, 2801.
19. Irie, R., Noda, K., Ito, Y., Matsumoto, N. and Katsuki, T., *Tetrahedron Lett.*, 1990, **31**, 7345.
20. Hosoya, N., Irie, R., Ito, Y. and Katsuki, T., *Synlett*, 1991, 691.
21. Hosoya, N., Hatayama, A., Irie, R., Sasaki, H. and Katsuki, T., *Tetrahedron*, 1994, **50**, 4311.
22. Lee, N.H., Muci, A.R. and Jacobsen, E.N., *Tetrahedron. Lett.*, 1991, **32**, 5055 and 6533.
23. Yamada, T., Imagawa, K., Nagata, T. and Mukaiyama, T., *Chem. Lett.*, 1992, 2231.
24. Schwenkreis, T. and Berkessel, A., *Tetrahedron Lett.*, 1992, **34**, 4785.
25. Irie, R., Hosoya, N. and Katsuki, T., *Synlett*, 1994, 255.
26. Irie, R., Ito, Y. and Katsuki, T., *Synlett*, 1991, 265.
27. Chang, S.B., Lee, N.H., Jacobsen, E.N., *J. Org. Chem.*, 1993, **58**, 6939; Chang, S., Galvin, J.M. and Jacobsen, E.N., *J. Am. Chem. Soc.*, 1994, 6937.
28. Brandes, B.D. and Jacobsen, E.N., *J. Org. Chem.*, 1994, **59**, 4378.
29. Brandes, B.D. and Jacobsen, E.N., *Tetrahedron Lett.*, 1995, **36**, 5123.
30. Sasaki, H., Irie, R., Hamada, T., Suzuki, K. and Katsuki, T., *Tetrahedron*, 1994, **50**, 11827.
31. Nagata, T., Imagawa, K., Yamada, T. and Mukaiyama, T., *Chem. Lett.*, 1994, 1259.
32. Nagata, T., Imagawa, K., Yamada, T. and Mukaiyama, T., *Inorg. Chim. Acta*, 1994, **220**, 283.
33. Chang, S., Heid, R.M. and Jacobsen, E.N., *Tetrahedron Lett.*, 1994, **35**, 669.
34. Mikame, D., Hamada, T., Irie, R. and Katsuki, T., *Synlett*, 1995, 827.
35. Sasaki, H., Irie, R. and Katsuki, T., *Synlett*, 1994, 356.
36. Ewins, R.C., Henbest, H.B. and McKarbevey, M.A., *J. Chem. Soc., Chem. Commun.*, 1967, 1085.
37. Davis, F.A., Harabal, M.E. and Awad, S.B., *J. Am. Chem. Soc.*, 1983, **105**, 3123.
38. Schurig, V., Hintzer, K., Leyrer, U., Mark, C., Pitchen, P. and Kagan, H.B., *J. Organometal. Chem.*, 1989, **370**, 81.
39. Collman, J.P., Lee, V.J., Zhang, X., Ibers, J.A. and Brauman, J.I., *J. Am. Chem. Soc.*, 1993, **115**, 3834.
40. May, S.W., Gordon, S.L. and Steltenkamp, M.S., *J. Am. Chem. Soc.*, 1977, **99**, 2017.
41. May, S.W., Schwartz, R.D., Abbott, B.J. and Zaborsky, O.R., *Biochem. Biophys. Acta*, 1975, **403**, 245.
42. May, S.W. and Schwartz, R.D., *J. Am. Chem. Soc.*, 1974, **96**, 4031.
43. White, R.F., Birnbaum, J., Meyer, R.T., ten Broeke, J., Chemerda, J.M. and Demain, A.L., *Appl. Microbiol.*, 1971, **22**, 55.
44. Hartmuth, C.K., VanNeieuwenhze, M.S. and Sharpless, K.B., *Chem. Rev.*, 1994, **94**, 2483.
45. Sharpless, K.B., Teranishi, A.Y. and Backvall, J.-E., *J. Am. Chem. Soc.*, 1997, **99**, 3120; Jorgensen, K.A. and Schiot, B., *Chem. Rev.*, 1990, **90**, 1483.
46. Gobel, T. and Sharpless, K.B., *Angew. Chem., Int. Ed. Engl.*, 1993, **32**, 1329.
47. Sharpless, K.B., Amberg, W., Beller, M., Chen, H., Hartung, J., Kawanami, Y., Lubben, D., Manoury, E., Ogino, Y., Shibata, T. and Ukita, T., *J. Org. Chem.*, 1991, **56**, 4585.
48. Norby, P.-O., Kolb, H.C. and Sharpless, K.B., *J. Am. Chem. Soc.*, 1994, **116**, 8470.
49. Pearlstein, R.M., Blackburn, B.K., Davis, W.M. and Sharpless, K.B., *Angew. Chem., Int. Ed. Engl.*, 1990, **29**, 639.
50. Kolb, H.C., Andersson, P.G. and Sharpless, K.B., *J. Am. Chem. Soc.*, 1994, **116**, 1278.
51. Kolb, H.C., Bennani, Y.L. and Sharpless, K.B., *Tetrahedron, Asymmetry*, 1993, **4**, 133; Arrington, M.P., Bennani, Y.L., Gobel, T., Walsh, P.J., Zhao, S.H. and Sharpless, K.B., *Tetrahedron Lett.*, 1993, **34**, 7375.
52. Sharpless, K.B., Amberg, W., Bennani, Y.L., Crispino, G.A., Hartung, J., Jeong, K.-S., Kwong, H.-L., Morikawa, K., Wang, Z.-M., Xu, D. and Zhang, X.-L., *J. Org. Chem.*, 1992, **57**, 2768.
53. Crispino, G.A., Jeong, K.-S., Kolb, H.C., Wang, Z.-M., Xu, D. and Sharpless, K.B., *J. Org. Chem.*, 1993, **58**, 3785.
54. Morikawa, K., Parh, J., Andersson, P.G., Hashiyama, T. and Sharpless, K.B., *J. Am. Chem. Soc.*, 1993, **115**, 8463.
55. Wang, L. and Sharpless, K.B., *J. Am. Chem. Soc.*, 1992, **114**, 7566.
56. Xu, D., Crispino, G.A. and Sharpless, K.B., *J. Am. Chem. Soc.*, 1992, **114**, 7570.
57. Hang, S.-K., Park, Y.-W., Lee, D.-H., Sim, H.-S. and Jeon, J.-H., *Tetrahedron, Asymmetry*, 1992, **3**, 705.

58. Rao, A.V.R., Rao, S.P. and Bhanu, M.N., *J. Chem. Soc., Chem. Commun.*, 1992, 859.
59. Hirsenkorn, R., *Tetrahedron Lett.*, 1990, **31**, 7519.
60. Soderquist, J.A. and Rane, A.M., *Tetrahedron Lett.*, 1993, **34**, 5031.
61. Takano, S., Yoshimitsu, T. and Ogasawara, K., *J. Org. Chem.*, 1994, **59**, 54.
62. Oishi, T. and Hirama, M., *Tetrahedron Lett.*, 1992, **33**, 639.
63. Imada, Y., Saito, T., Kawakami, T. and Murahashi, S., *Tetrahedron Lett.*, 1992, **33**, 5081.
64. Hanessian, S., Meffre, P., Girard, M., Beaudoin, S., Sanceau, J.-Y. and Bennani, Y.L., *J. Org. Chem.*, 1993, **58**, 1991.
65. Tomioka, K., Nakajima, M. and Koga, K., *J. Am. Chem. Soc.*, 1987, **109**, 6213.
66. Hentges, S.G. and Sharpless, K.B., *J. Am. Chem. Soc.*, 1980, **102**, 4263.
67. Palucki, M., Hanson, P. and Jackobsen, E.N., *Tetrahedron Lett.*, 1993, **33**, 5055.
68. Noda, K., Hosoya, N., Irie, R., Yamashita, Y. and Katsuki, T., *Tetrahedron*, 1994, **50**, 9609.
69. Nakajima, K., Kojima, M. and Fujita, J., *Chem. Lett.*, 1986, 1483; Nakajima, K., Kojima, M. and Fujita, J., *Bull. Chim. Soc. Jap.*, 1991, **64**, 1318.
70. Kosatsu, N., Nishibayashi, Y., Sugita, T. and Uemura, S., *Tetrahedron Lett.*, 1992, **33**, 5319.
71. Samuel, O., Ronan, B. and Kagan, H.B., *J. Organometal. Chem.*, 1989, **370**, 43.
72. Di Furia, F., Licini, G. and Modena, G., *Gazz. Chim. Ital.*, 1990, **120**, 165.
73. Samuel, O., Ronan, B. and Kagan, H.B., *J. Organometal. Chem.*, 1989, **370**, 43.
74. Denney, D.B., Napier, R. and Cammarata, A., *J. Org. Chem.*, 1963, **30**, 3151; Muzart, J., *J. Mol. Catal.*, 1991, **64**, 381.
75. Rispens, M.T., Zondervan, C. and Feringa, B.L., *Tetrahedron, Asymmetry*, 1994, **6**, 661.
76. Andrus, M.B., Argade, A.B., Chen, X. and Pamment, M.G., *Tetrahedron Lett.*, 1995, **36**, 2945.
77. Discordia, R.P., Murphy, C.K. and Dittmer, D.C., *Tetrahedron Lett.*, 1990, **31**, 5603.
78. Grennberg, H., Gogoll, A. and Backvall, J.-E., *J. Org. Chem.*, 1991, **56**, 5808.
79. Holland, H.L., Carter, I.M., Chenchaiah, P.C., Khan, S.H., Munoz, B., Ninniss, R.W. and Richards, D., *Tetrahedron Lett.*, 1985, **26**, 6409.
80. Crabb, T.A. and Soilleux, S.L., *J. Chem. Soc., Perkin Trans. 1*, 1985, 1381.
81. Battersby, A.R., Sheldrake, P.W., Staunton, J. and Williams, D.C., *J. Chem. Soc., Perkin Trans. 1*, 1976, 1056.
82. Phillips, G.T., Matcham, G.W.J., Bertola, M.A., Marx, A.F. and Koger, H.S., *Eur. Pat. Appl.*, EP 20521517, Dec. 1986.

18 Worked examples in asymmetric synthesis design

G.R. STEPHENSON

18.1 Symmetry and synthesis design

The problem of the design of stereoselective approaches for the construction of complex molecules can be simplified by an analysis of the symmetry properties of potential intermediates and the consequent prospects for stereocontrol at each stage of the synthesis.[1] The following factors are important considerations in planning an asymmetric synthesis:

1. types of chiral element present
2. symmetry elements in partial structures
3. oportunities for stereodifferentiation.

There are a number of ways in which the presence of symmetry elements can help in synthesis design. In symmetrical structures, problems of regiocontrol are reduced. It can be worth while planning to introduce symmetry to achieve this.

18.1.1 C_2 axes in synthetic planning: simplification of synthetic routes

(a) *Example: aglycon of venturicidin macrolide antibiotics.* The synthesis of a potential intermediate for the C-15 to C-23 portion of the aglycon (**1**) (Scheme 18.1) illustrates the importance of the consideration of symmetry elements. The synthesis can be simplified by the removal of one chiral centre to introduce a C_2 axis. Synthetic precursors should, therefore, also contain a C_2 axis. Two methyl groups can now be introduced into (**2**) at a single step (Scheme 18.2). Alkylation of a spirocyclic dilactone proved a convenient way to obtain stereocontrol.[2]

18.1.2 S axes: prochiral intermediates in synthetic planning

It can be helpful to look deliberately for prochiral intermediates when planning a synthesis. This is particularly true in enantiomer synthesis because of the possibility of effecting asymmetric induction by removing the symmetry element as the synthesis progresses. Compounds containing arrays of oxygenated chiral centres such as (**3**) are of interest because of their relationship to degradation products from mycoticin A. Schreiber and Goulet[3] chose an approach (Scheme 18.3) in which symmetrical precursors

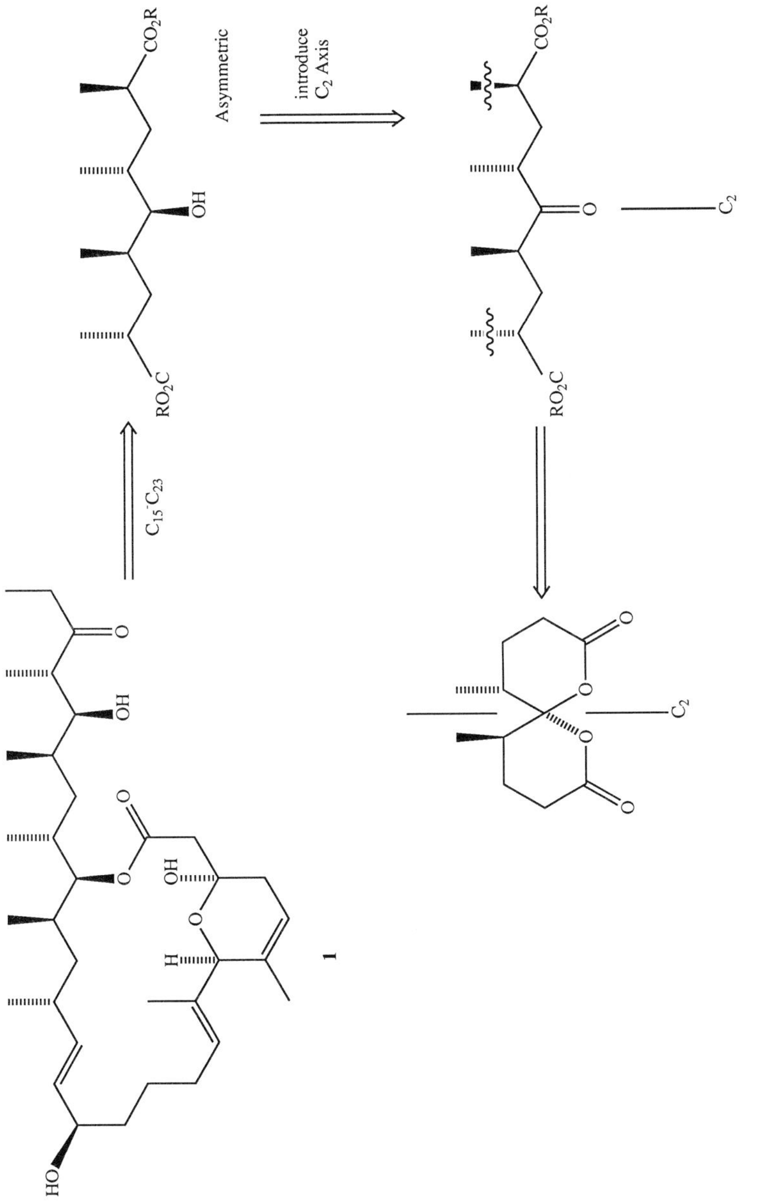

Scheme 18.1

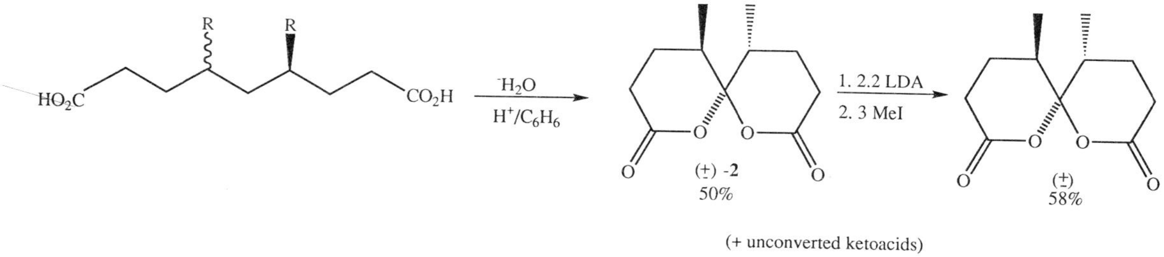

R
R
HO_2C
CO_2H
$-H_2O$
H^+/C_6H_6
O
O
O
O
(±) -2
50%
1. 2.2 LDA
2. 3 MeI
O
O
O
O
(±)
58%
(+ unconverted ketoacids)

Scheme 18.2

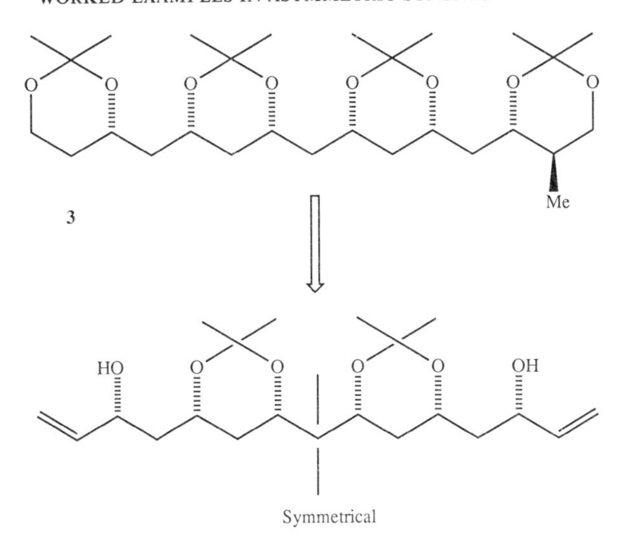

Scheme 18.3

simplify the construction of the chain. A *meso*-bisepoxide is elaborated at each and to extend the chain, first by a copper-catalysed vinyl Grignard addition and subsequently by a Wittig rearrangement. The enantiotopic allyl alcohols are differentiated by Sharpless epoxidation (see Chapter 17) in a reaction that also distinguishes the diastereofaces of the alkene (Scheme 18.4).

18.2 How to use symmetrical intermediates in enantioselective synthesis

Some of the best approaches to enantiomer synthesis start from symmetrical intermediates, so when learning to plan synthetic routes, it is important to look for structure simplifications which place symmetry elements in convenient places for efficient asymmetric induction when the synthetic sequence is performed. Two types of symmetrical intermediate are important: those chosen for enantioface differentiation and those chosen for enantiotopic group differentiation. In the first case, reactions take place at a plane of symmetry. In the second, reaction occurs remote from the symmetry plane. When planning a synthesis it is helpful to evaluate these possibilities separately. Guidelines for this process are given in Sections 18.2.1 and 18.2.2.

18.2.1 Enantioface differentiation

These reactions take place at a plane of symmetry (or S_n axis).

(*a*) *Guide for retrosynthetic analysis.* For enantioface differentiation, look

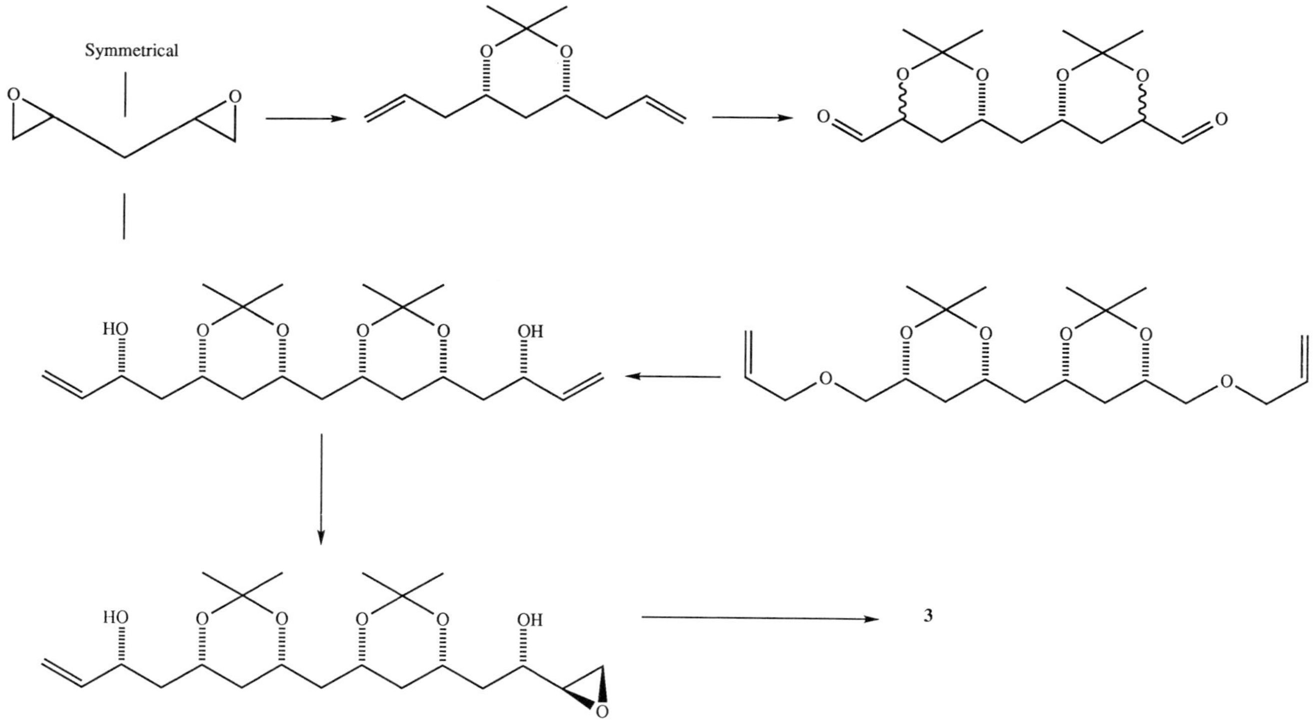

Symmetrical

3

Scheme 18.4

for a disconnection in which substituents around the chiral centre of the target molecule are located on a symmetry plane in the precursor.

(*b*) *Example: CD ring system for steroid synthesis.* The target structure (**4**) contains two chiral centres but can be simplified (Scheme 18.5) to a prochiral precursor for the D ring. The approach shown in Scheme 18.6 shows a

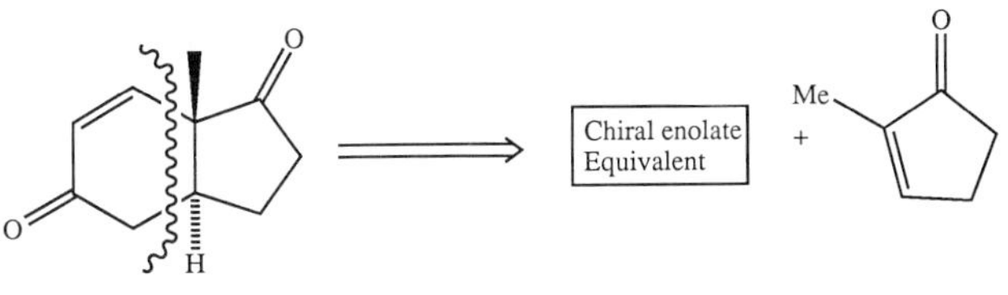

Chiral enolate Equivalent

Scheme 18.5

t-Bu

OMe

CuLi

1.

2. Me_3SiCl

3. NH_4OH/NH_4Cl

$OSiMe_3$

$(MeO)_2HC$

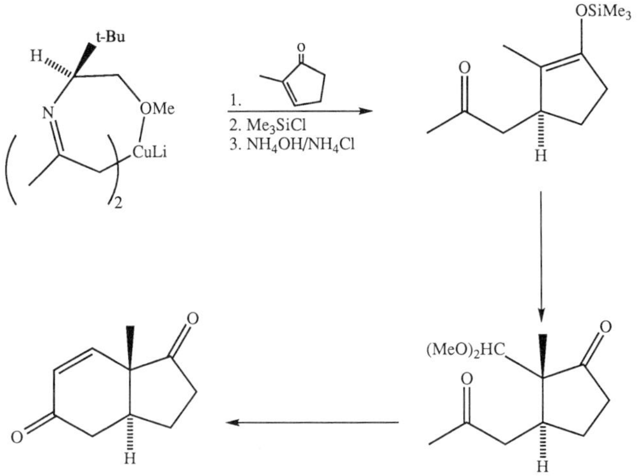

Scheme 18.6

third-generation asymmetric induction. A survey of Chapter 3 will reveal many possibilities for the chiral enolate equivalent. Alternatively, fourth-generation chiral catalysts (Chapter 12) could be considered. In Scheme 18.6, an anullulation reaction of 2-methylcyclopentenone can introduce both chiral centres by conjugate addition/enolate trapping.[4]

18.2.2 Enantiotopic group differentiation

These reactions take place remote from the symmetry plane.

(*a*) *Guide for retrosynthetic analysis.* For enantiotopic group differentiation:

1. if there is a single chiral centre in the target molecule, look for a disconnection or functional group interconversion such that substituents around that centre lie on the symmetry plane of a prochiral substrate. This rule will also apply when two centres can lie on the same symmetry plane; or
2. if the target molecule contains more than one chiral centre, then look for a functional group interconversion which involves a substrate with a symmetry plane between the positions of carbons destined to become the chiral centres.

(*b*) *Example: an alternative CD ring subunit.* Target (**5**) is a regioisomer of (**4**) and has a single chiral centre. Applying approach (**1**) with a Robinson annulation disconnection (Scheme 18.7) the symmetrical structure (**6**) can be

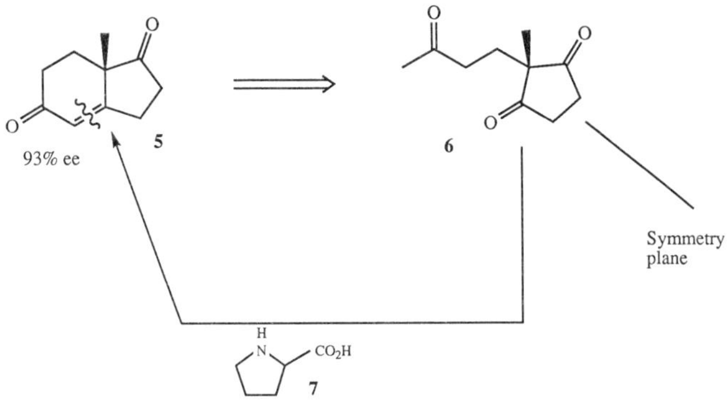

93% ee 5 6 Symmetry plane 7 CO_2H

Scheme 18.7

identified as an attractive precursor. Proline (**7**) catalyses the asymmetric cyclisation into form (**5**) in 93% ee.[5] Asymmetric amplification effects in this reaction have been discussed in Chapter 3.

(*c*) *Example: intermediate for carbacyclin synthesis.* Carbacyclin, a stable PGI_2 analogue has become important because of its inhibition of platelet aggregation. Structure (**8**) is an important intermediate in carbacyclin synthesis and can be obtained from (**9**) by employing a reconnection strategy. The β-keto ester (**9**) contains three chiral centres, but by employing strategy (2) and using the Claisen approach to form the 5-membered ring (and one at the chiral centres), the possibility of introducing a symmetry plane (Scheme 18.8) becomes clear.

The *meso*-structure (**11**) has two enantiotopic esters and is suitable for asymmetric induction employing biotechnology, as outlined in Chapter 13. Nagao and Fujita *et al.* based their enantioselective synthesis of carbacyclin on a chemical enantiotopic group differentiation.[6] The chirality at the ring

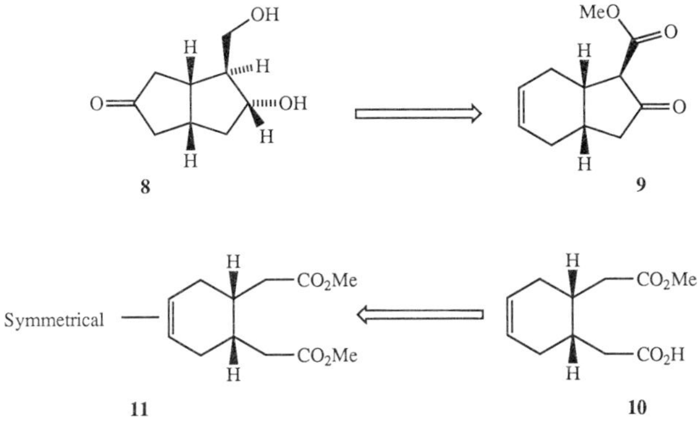

Scheme 18.8

junction introduced in this way subsequently determines (Scheme 18.9) the relative stereochemistry at other chiral centres. The thioester (**12**) was used in the Claisen step.

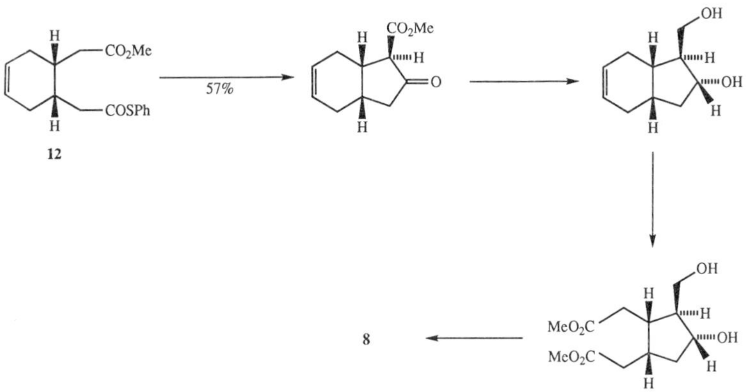

Scheme 18.9

18.2.3 Deleting a chiral centre to induce asymmetry

Approach (2) (Section 18.2.2) can be used to suggest a route to a target with only one chiral centre in a process that induces asymmetry and removes a chiral centre at the same time. This is worth while, if the relative stereochemistry in the precursor is easily established, as in the prochiral epoxide shown in Scheme 18.10. Biotechnology can be employed again (Scheme 18.11) to differentiate the two sides of the symmetry plane: selective hydrolysis using

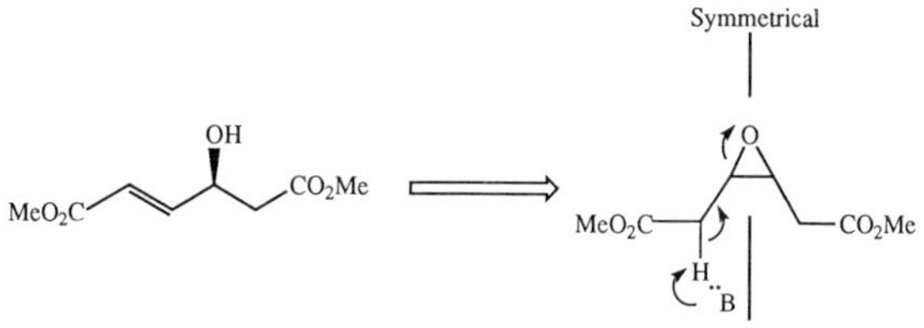

Scheme 18.10

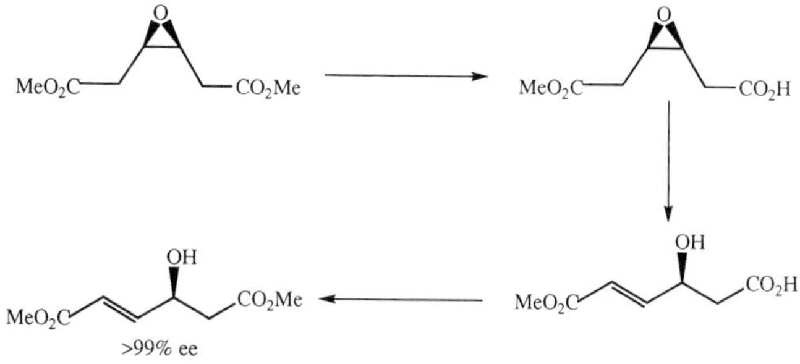

Scheme 18.11

pig-liver esterase was followed by deprotonation to form a dianion to effect epoxide ring opening.[7]

An alternative and more direct approach,[8] deprotonation by a chiral base (Scheme 18.12) (Chapter 6), cannot yet match the degree of enantioselectivity available from the use of esterases.

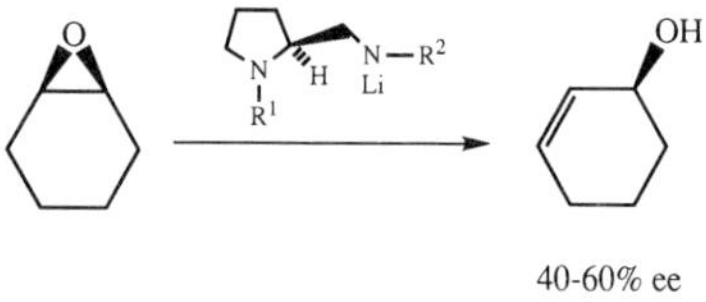

Scheme 18.12

18.3 Comparison of enantioface and enantiotopos differentiation

A simple example of the application of these guiding principles is that of the lactone (**13**), because there is only one chiral centre (*) involved. Enantioselective

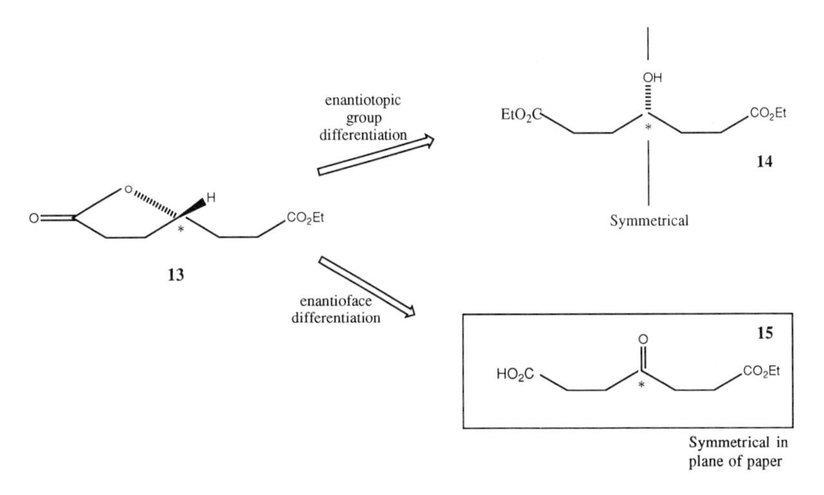

Scheme 18.13

synthesis can be approached in two ways. We could place the starred atom (*) on a symmetry plane of a prochiral intermediate identified by a functional group interconversion, or we could plan a disconnection by removing chirality to convert the starred atom into a prochiral group (Scheme 18.13).

Enantiotopic ester hydrolysis employing (**14**) looks attractive but is limited by the selectivity of the hydrolysis enzymes. Instead, a microbial half hydrolysis of the ketone (**16**) was used, followed by highly enantioselective reduction by yeast (Scheme 18.14).[9]

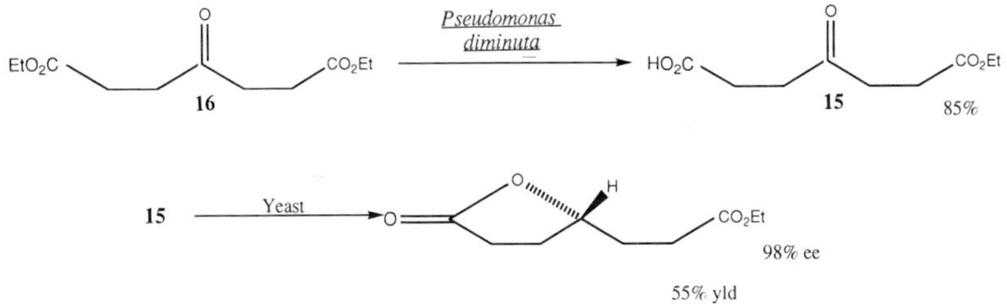

EtO2C, CO2Et **16** — *Pseudomonas diminuta* → HO2C, CO2Et **15** 85%

15 — Yeast → CO2Et, 98% ee, 55% yld

Scheme 18.14

18.4 Diastereoselectivity

Control relative to an existing chiral centre in the molecule (first-generation asymmetric synthesis) requires the consideration of diastereoselectivity in the design process. The same considerations apply in second-generation

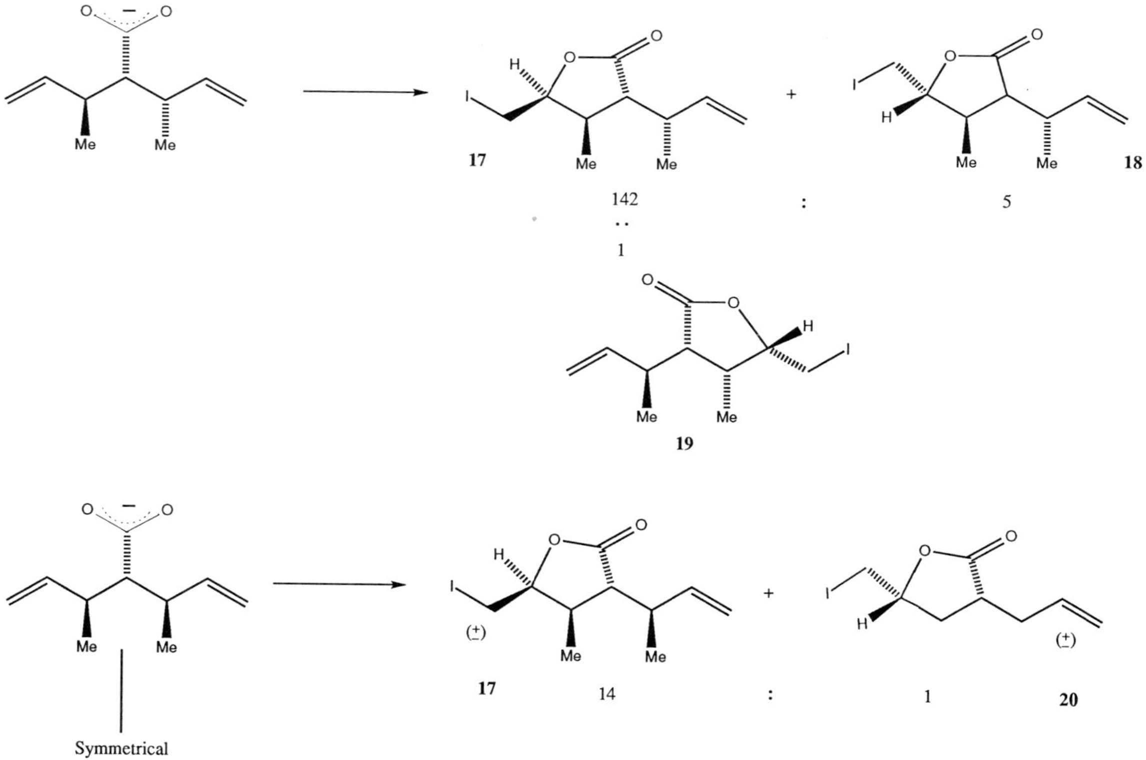

17 : 18 = 142 : 5

17 : 19 = 142 : 1

Symmetrical

(±) 17 : (±) 20 = 14 : 1

Scheme 18.15

approaches to asymmetric induction, in which a chiral auxiliary is covalently incorporated into the starting material.

Guide for retrosynthetic analysis. For diastereoface differentiation:

1. look for a disconnection adjacent to an existing chiral centre or in a position where special circumstances will lead to stereocontrol; or
2. look for a disconnection close to a substituent that can be modified by the covalent attachment of a chiral auxiliary prior to the synthetic reaction.

Diastereoface selectivity is often combined with enantioface and enantiotopos differentiation. Good diastereoface selectivity has been obtained (Scheme 18.15) in the iodolactonisation of symmetrical bisalkenes.[10] In one case, enantiotopic alkenes were differentiated. Even better stereocontrol was obtained in an example in which diastereotopic groups were used. The lactone (**17**) was by far the preferred product.

18.5 Planning for stereocontrol

The identification of symmetry elements in potential precursors, and in the target molecule, has been the main theme in the examples given in this chapter. It provides a useful technique for the assessment of possibilities arising from a retrosynthetic analysis and gives an indication of the type of stereocontrol that will be required at each step. This aids the design of the synthetic process. One final example (Scheme 18.16) from the work at the Hoffmann-LaRoche laboratory in New Jersey will draw these points together.

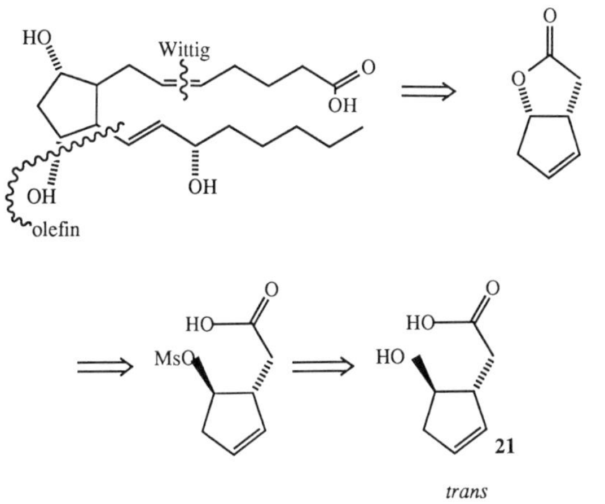

HO Wittig O OH OH OH olefin O O

HO O MsO HO O HO 21 trans

Scheme 18.16

Example: prostaglandin synthesis. Although containing two differently substituted chiral centres, the cyclopentenol intermediate (**21**) can be obtained from a symmetrical precursor if the alcohol is introduced by hydroboration of an alkene. The second substituted centre now lies on the symmetry plane of the cyclopentadiene. Control of relative stereochemistry by diastereoselective hydroboration on the less hindered face, combined with an asymmetric induction (Scheme 18.17), is required to achieve stereocontrol.[11]

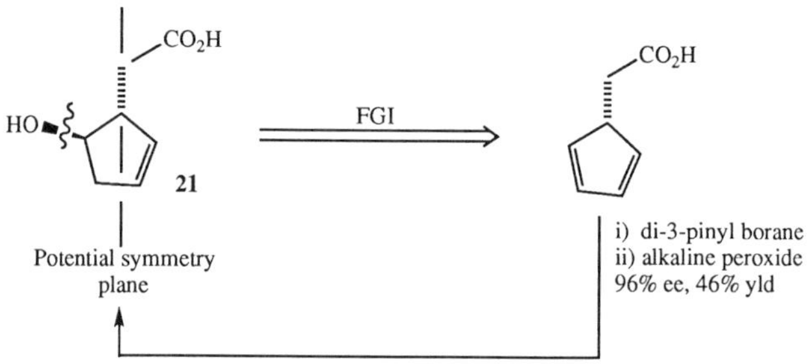

CO_2H
HO
21
Potential symmetry plane
FGI
CO_2H
i) di-3-pinyl borane
ii) alkaline peroxide
96% ee, 46% yld

Scheme 18.17

The stereochemistry at C-15 can be tackled in two ways (Scheme 18.18):

1. introduce resolved subunit
2. introduce prochiral subunit (e.g. a ketone) then reduce with diastereoface selectivity.

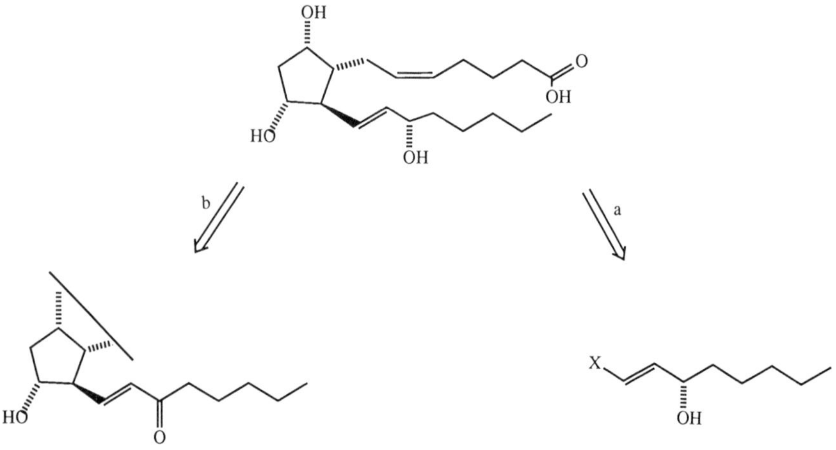

OH
O
OH
HO
OH
b
a
HO
O
X
OH

Scheme 18.18

18.6 Conclusions

Enantioselective synthesis is the last frontier in the development of the application of organic chemistry to the construction of target molecules. Designing efficient enantioselective routes is a far from routine task. At the present time, most of the main classes of reactions have been modified to asymmetric versions. The earlier chapters of this book lay out these possibilities. To put these reactions to work in an effective manner, however, requires new planning skills to guide key decisions about strategies for asymmetric induction, and the symmetry properties of intermediates and starting materials. This final chapter provides a guide to this process. As pharmaceutical and agrochemical products increasingly require enantiopure compounds, these skills will be at a growing premium in industrial and academic laboratories around the world.

References

1. Bertz, S.H., *J. Chem. Soc., Chem. Commun.*, 1984, 218.
2. Hoye, T.R., Peck, D.R. and Trumper, P.K., *J. Am. Chem. Soc.*, 1981, **103**, 5618.
3. Schreiber, S.L. and Goulet, M.T., *J. Am. Chem. Soc.*, 1987, **109**, 4718.
4. Yamamoto, K., Iijima, M., Ogimura, Y. and Tsuji, J., *Tetrahedron Lett.*, 1984, **25**, 2813.
5. Hajos, Z.G. and Parrish, D.R., *J. Org. Chem.*, 1974, **39**, 1615.
6. Nagao, Y., Nakamura, T., Ochiai, M., Fuji, K. and Fujita, E., *J. Chem. Soc., Chem. Commun.*, 1987, 267.
7. Mohr, P., Rösslein, L. and Tamm, C., *Helv. Chim. Acta*, 1987, **70**, 142.
8. Asami, M. and Kirihara, H., *Chem. Lett.*, 1987, 389.
9. Moriuchi, F., Muroi, H. and Aibe, H., *Chem. Lett.*, 1987, 1141.
10. Kurth, M.J. and Brown, E.G., *J. Am. Chem. Soc.*, 1987, **109**, 6844.
11. Partridge, J.J., Chadha, N.K. and Uskokovic, M.R., *J. Am. Chem. Soc.*, 1973, **95**, 7171.

Key to abbreviations for chiral auxiliaries

This is a list of chiral auxiliaries, mostly chiral ligands, which have been discussed in this book or have closely related structures. They are often referred to by acronyms, so their structures are given here. This is not intended to be a comprehensive glossary of acronyms, but serves as a guide to material presented in the book.

ALAPHOS

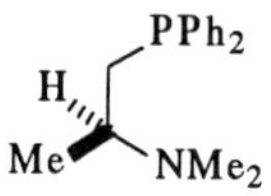

PPh_2 H Me NMe_2

APPM

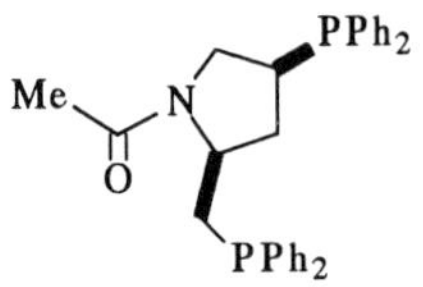

Me N O PPh_2 PPh_2

BCCP (page 168)

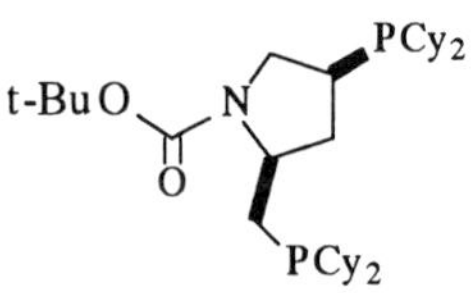

t-BuO N O PCy_2 PCy_2

BCPM (pages 168, 169)

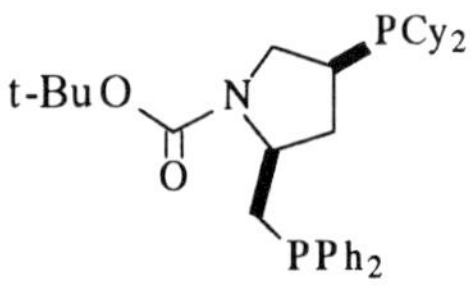

t-BuO N O PCy_2 PPh_2

BCPP

t-BuO N O PPh_2 PCy_2

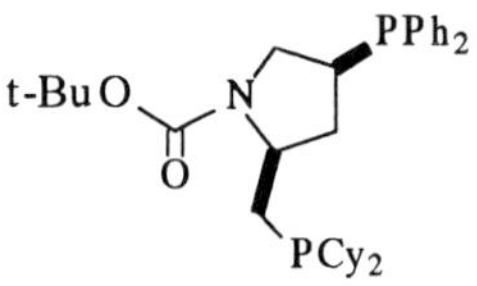

BDPP (pages 147, 153, 154, 175, 176)

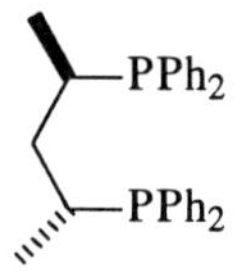

PPh_2 PPh_2

BINAP (pages 147–150, 153–155, 157–161, 164, 166–174, 176, 196–198, 279–285, 287, 289, 291, 293, 297)

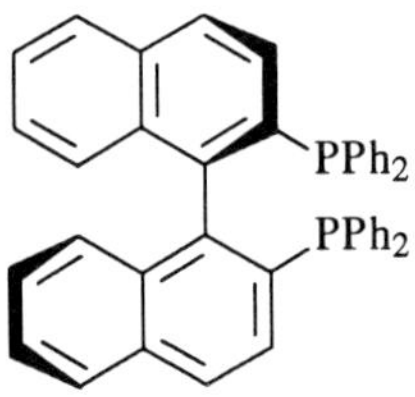

PPh_2 PPh_2

BINAPHOS (page 219)

BINOL (pages 16, 17, 18, 44, 133, 141, 248, 252, 388)

BIPHEMP (pages 147, 149, 150, 154, 158, 159, 162, 168, 170, 171, 172)

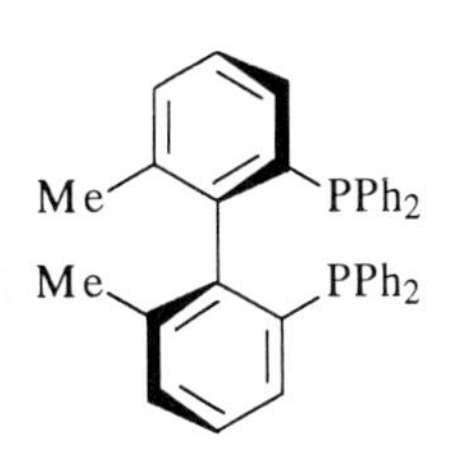

Bipymox-ip (page 214)

BITAP (page 173)
o-BITAP

m-BITAP

BPPFA (pages 147, 153, 154)

BPPFOH (pages 47, 164, 165, 284)

BPPM (pages 147, 149, 159, 167–169)

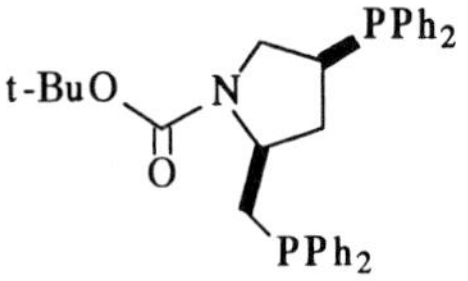

t-BuO, N, O, PPh_2, PPh_2

BuTRAP (page 214)

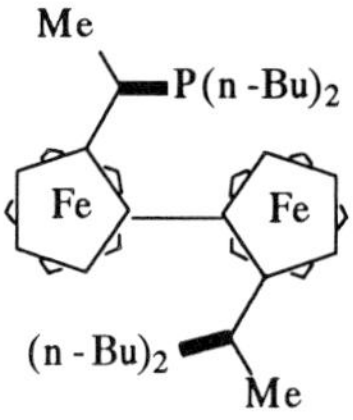

Me, $P(n\text{-}Bu)_2$, Fe, Fe, $(n\text{-}Bu)_2$, Me

BZPPM

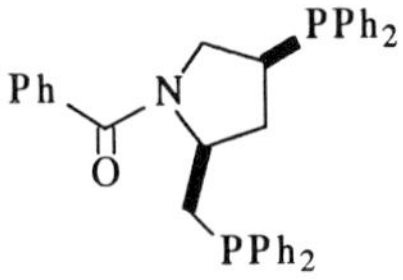

Ph, N, O, PPh_2, PPh_2

CAPP (pages 147, 157, 159)
Phenyl-CAPP

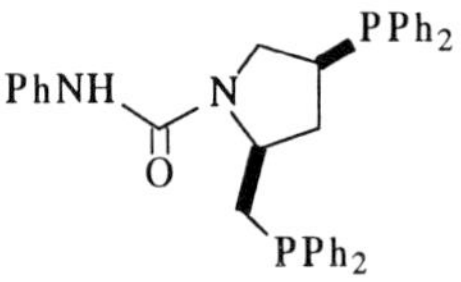

PhNH, N, O, PPh_2, PPh_2

CBD (pages 147, 149, 157, 159, 163, 174)

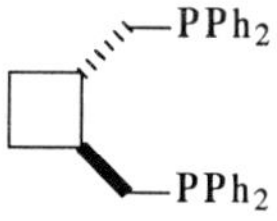

PPh_2, PPh_2

CHIRAPHOS (pages 147, 149, 154, 155, 157, 163, 173, 174, 219, 309)

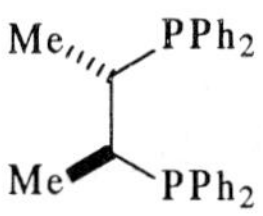

Me, PPh_2, Me, PPh_2

CYCPHOS (pages 175, 176)

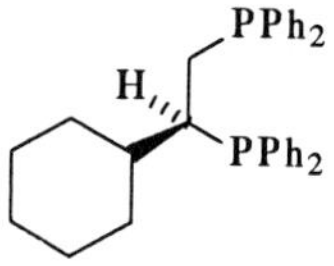

PPh_2, H, PPh_2

DAG (pages 35, 40, 241, 242, 244, 248)

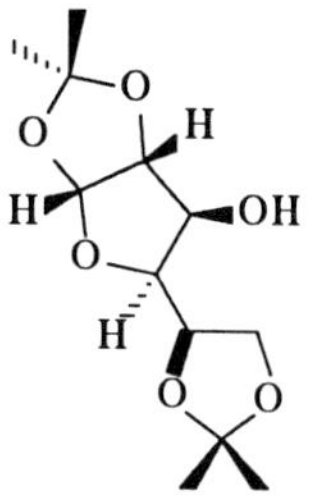

O, O, H, H, OH, O, H, O, O

DAIB (pages 11, 13)

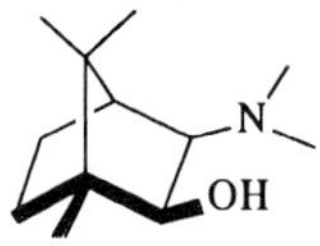

N, OH

DBP-DIOP (page 219)

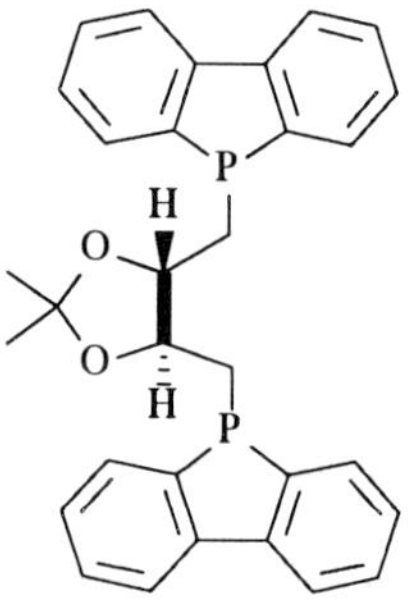

P, H, O, O, H, P

DEGUPHOS (pages 147, 149, 155)

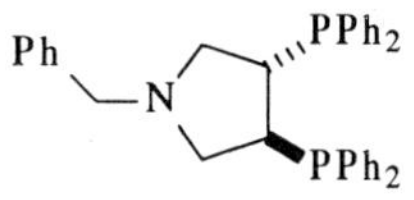

Ph, N, PPh_2, PPh_2

DET (pages 9, 10, 16, 367, 370)

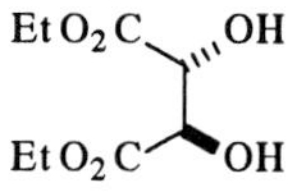

DHQ (pages 379, 381)

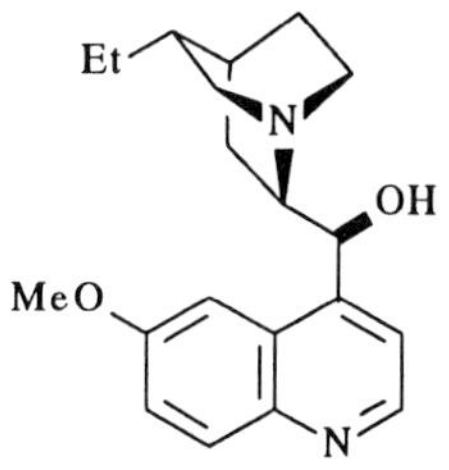

DHQD (pages 379, 381)

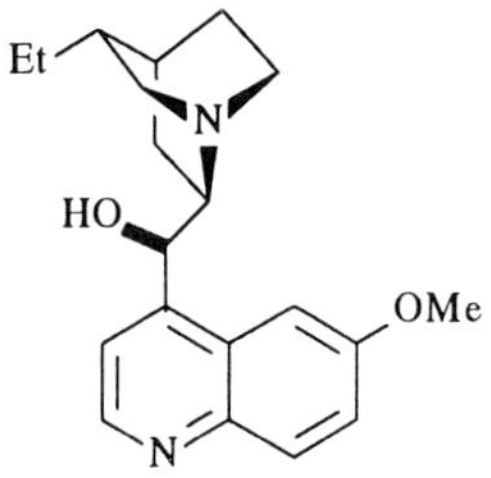

DHQD-CLB (page 384)

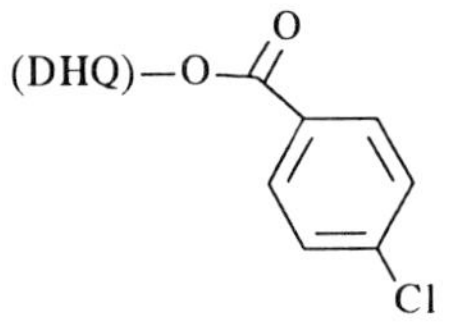

DHQD-IND (pages 382, 384, 386)

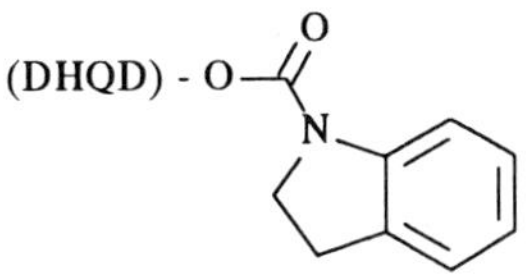

(DHQD)$_2$PHAL (pages 381, 382, 384–386)

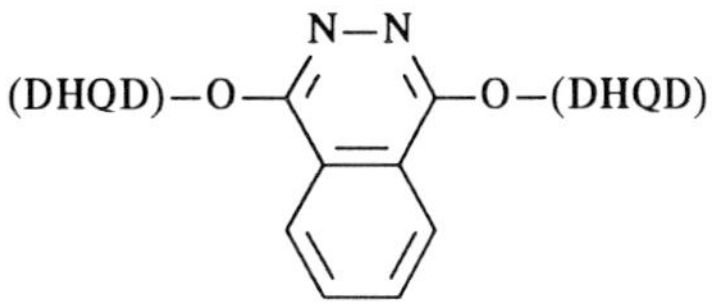

(DHQD)$_2$PYR (pages 381, 382, 386)

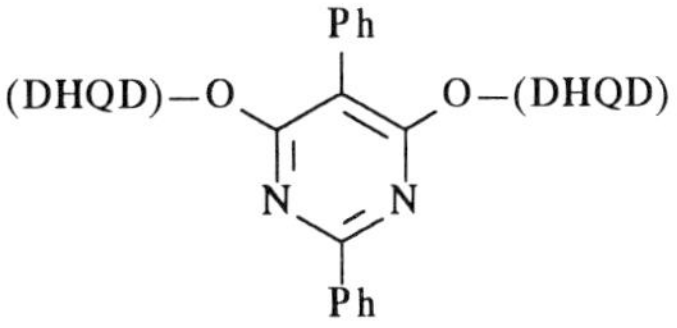

DHQ-IND (pages 382, 384)

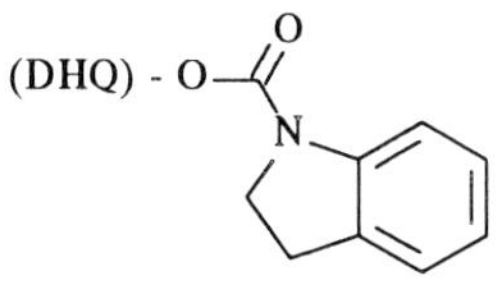

(DHQ)$_2$PHAL (pages 381, 382)

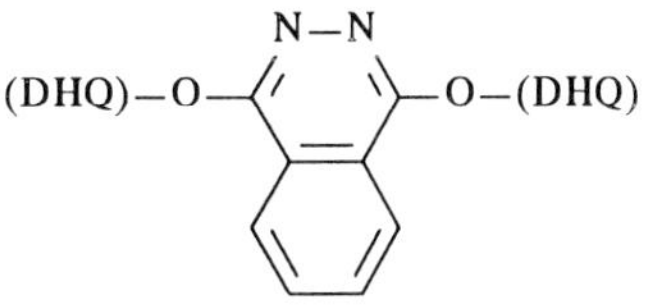

(DHQ)$_2$PYR (pages 381, 382)

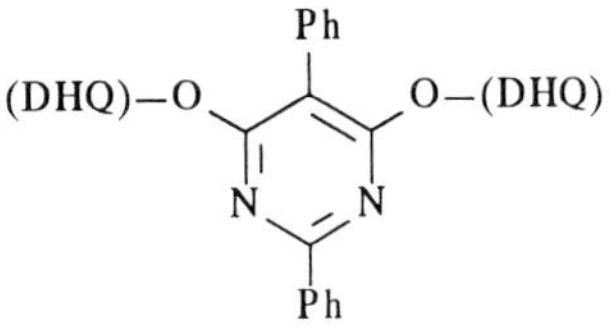

DIOP (pages 146, 147, 149, 153, 154, 157, 158, 160, 165, 173, 174, 195–197, 215, 216, 219)

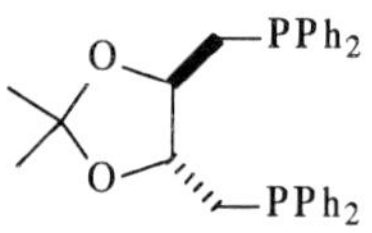

PPh_2

O

O

PPh_2

DIOP′

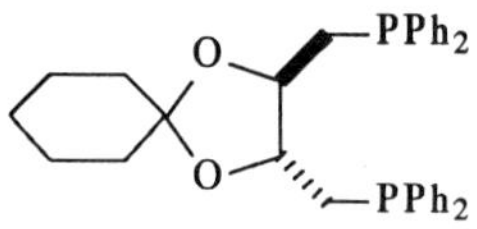

PPh_2

O

O

PPh_2

DIPAMP (pages 147–149, 153, 154, 157, 168, 169, 173, 174)

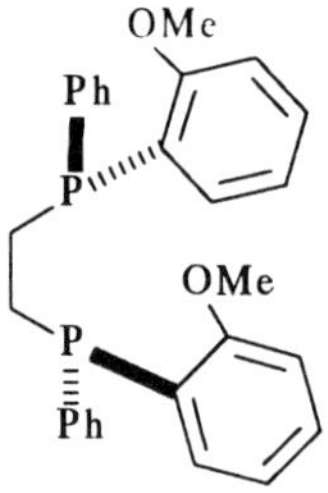

OMe

Ph

P

OMe

P

Ph

DIPT (pages 67, 371)

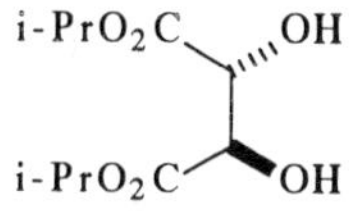

i-PrO_2C OH

i-PrO_2C OH

DUPHOS (pages 147, 149, 153, 175)

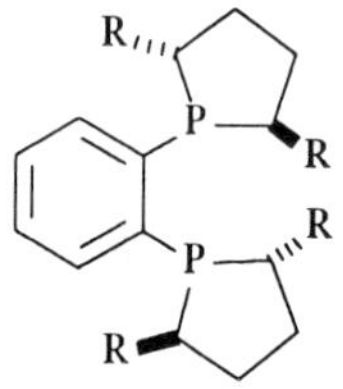

R

P R

P R

R

EapBH$_2$ (pages 187, 188)

Et

BH_2

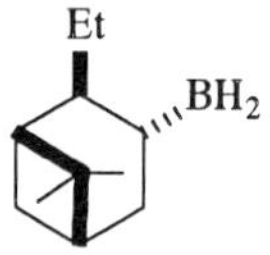

Et-DUPHOS (pages 154, 175, 176)

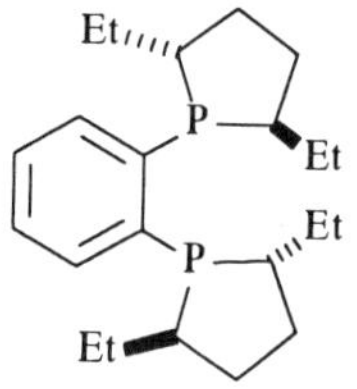

Et

P Et

P Et

Et

FPPM

PPh_2

H N

O

PPh_2

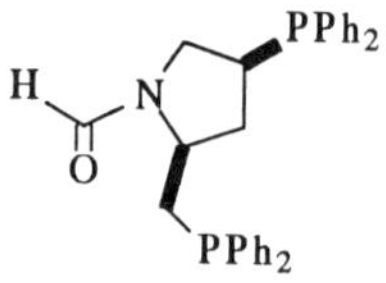

Icr$_2$BH (pages 184, 188)
2-^{d}Icr$_2$BH

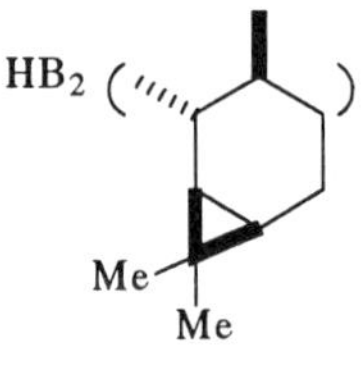

HB_2 ( )

Me

Me

3-^{d}Icr$_2$BH

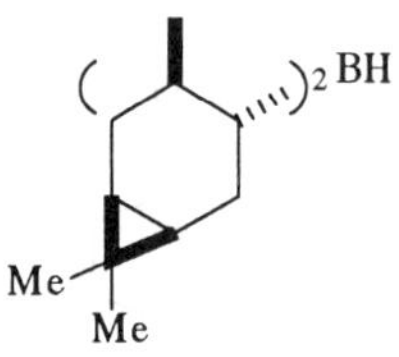

( )$_2$ BH

Me

Me

Ipc_2BH (pages 183, 184, 188)

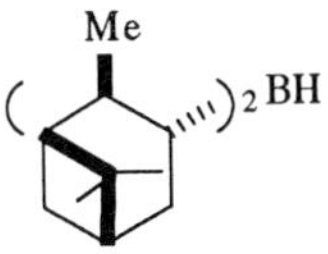

$IpcBH_2$ (pages 187, 189, 191, 205)

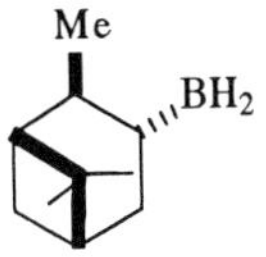

LEUPHOS (page 295)
t-LEUPHOS

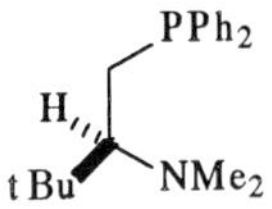

Lgf_2BH (pages 188, 189)

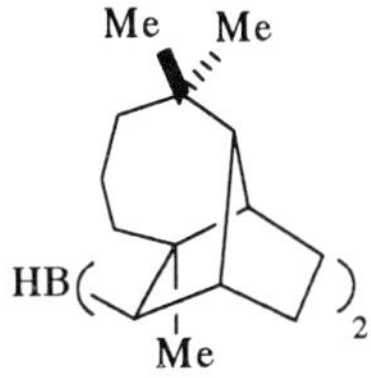

LimBH (pages 184, 185, 188)

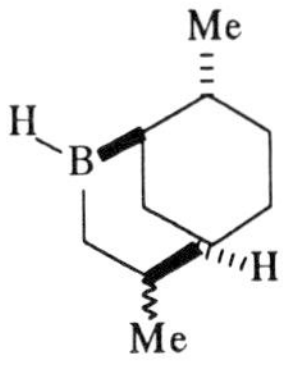

MAPP (page 19)

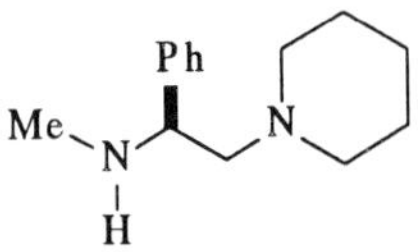

MCCPM (page 165)

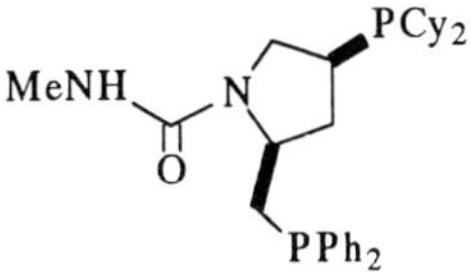

$MDBH_2$ (pages 187, 188)

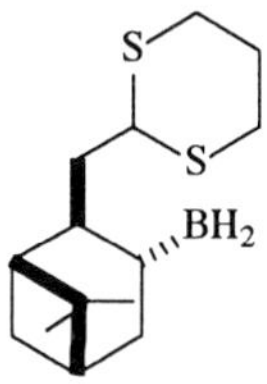

Me-DUPHOS (pages 168, 170)

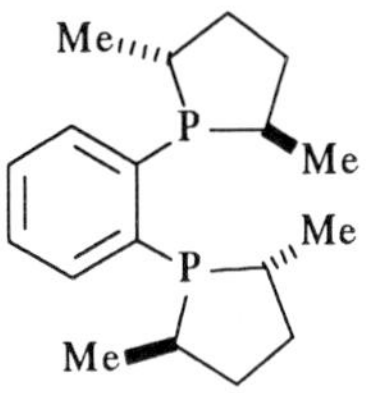

MeO-BIPHEP (pages 147, 149, 150, 159, 160, 162, 166–168, 170, 173)

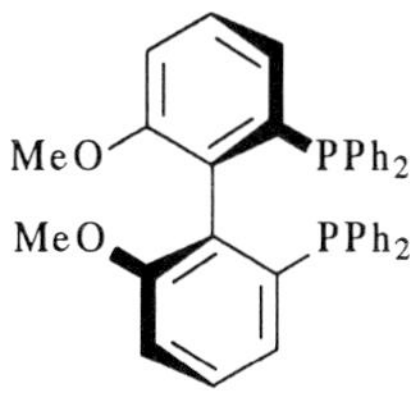

MOD-DIOP (pages 147, 159)

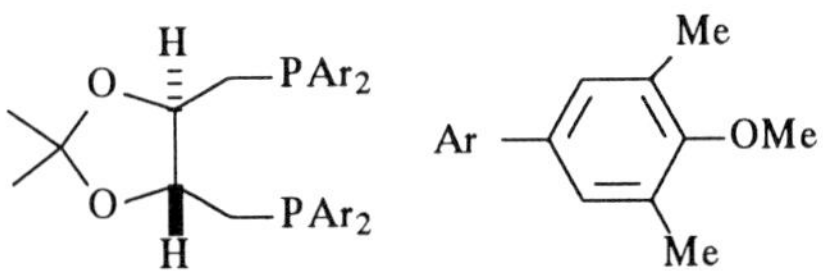

MOP (pages 215–217)

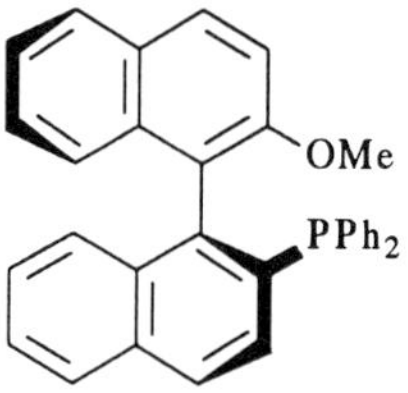

OMe
PPh_2

MSCPM

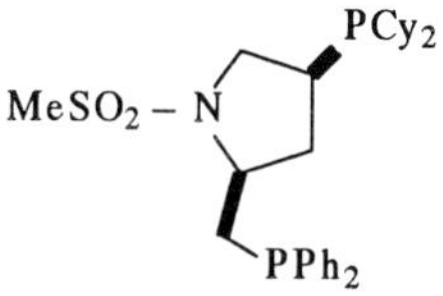

PCy_2
$MeSO_2$ – N
PPh_2

NORPHOS (pages 147, 149, 154, 173)

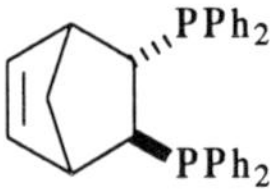

PPh_2
PPh_2

PapBH₂ (pages 187, 188, 190)

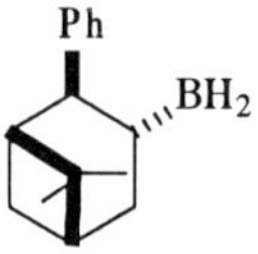

Ph
BH_2

PCPM

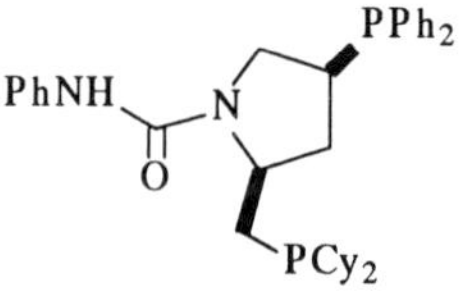

PPh_2
PhNH
N
O
PCy_2

(Ph)₂(DHQ)₂PHAL (page 381)

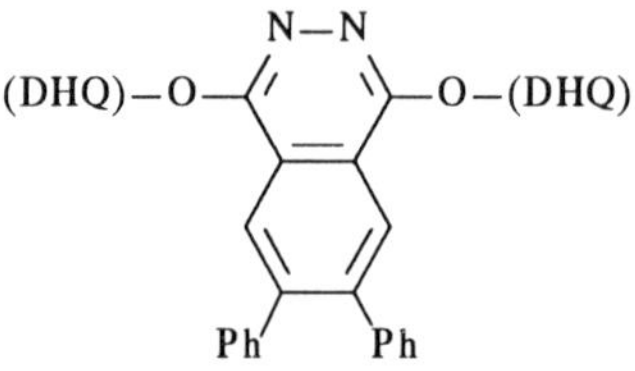

N–N
(DHQ)–O   O–(DHQ)
Ph   Ph

PHEGLYPHOS

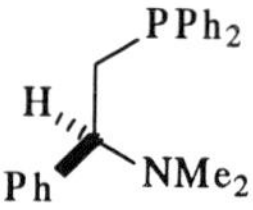

PPh_2
H
Ph NMe_2

PHEPHOS

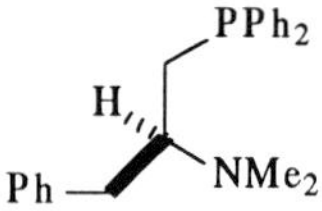

PPh_2
H
Ph NMe_2

Ph-TRAP

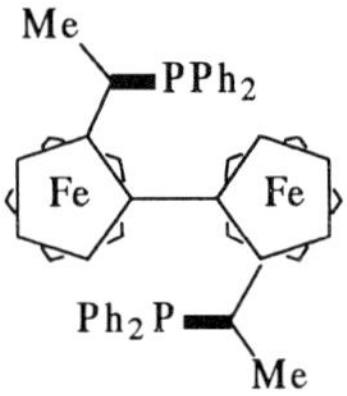

Me
PPh_2
Fe   Fe
Ph_2P
Me

PNNP (pages 147, 156)

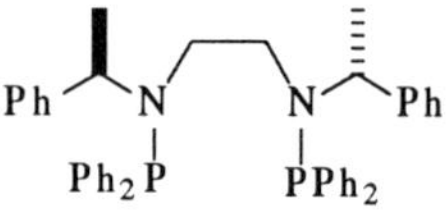

Ph   N   N   Ph
Ph_2P PPh_2

PPFA (pages 292, 293, 295, 298)

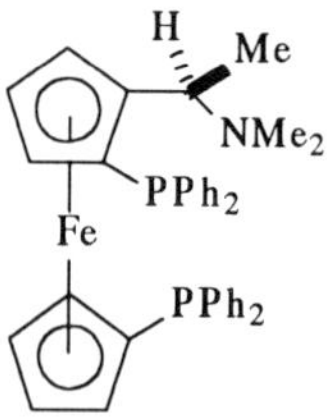

H   Me
NMe_2
PPh_2
Fe
PPh_2

PRONOP (pages 167, 168)

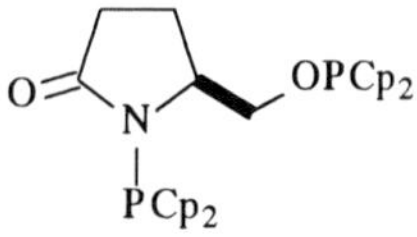

O   N $OPCp_2$
PCp_2

PROPHOS (pages 147, 149, 153, 154)

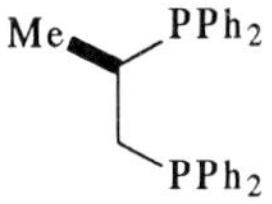

Me, PPh_2, PPh_2

RAMP

OMe, N, H, NH_2

Pybox-ip (page 214)

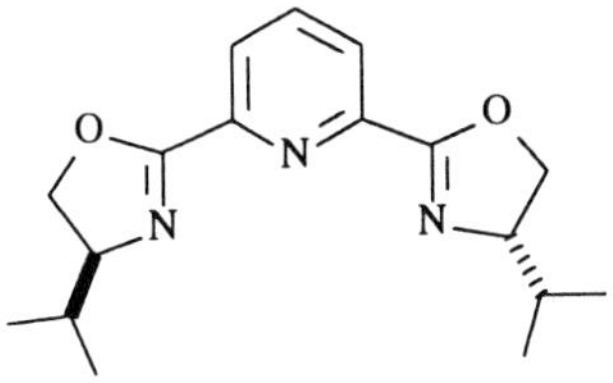

O, N, N, O, N

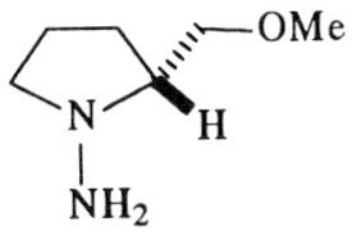

SAMP (page 340)

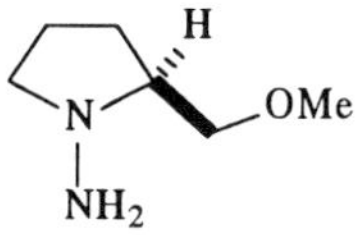

H, N, OMe, NH_2

TADDOL (pages 214, 247)

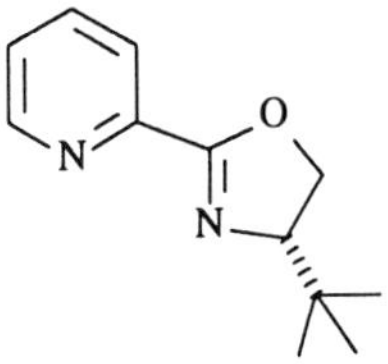

Nap Nap, H, O, O, P–Nap, O, O, H, Nap Nap

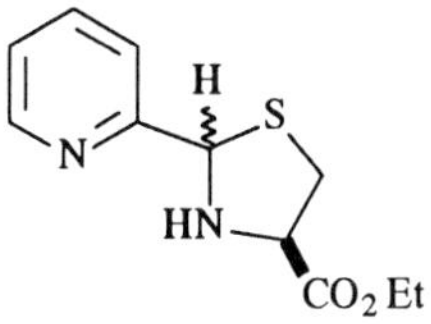

Pymox-tb (page 214)

N, O, N

Pythia (page 214)

N, H, S, HN, CO_2Et

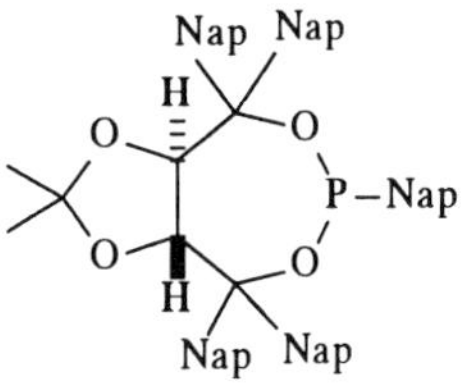

VALPHOS (page 288)

PPh_2, H, NMe_2, Me, Me

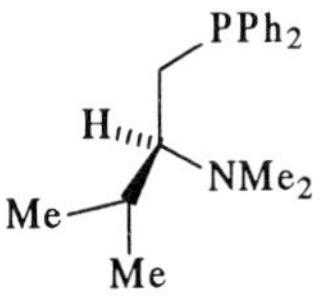

Index

ab initio calculations 201
acetophenone 213
acetylalanine 152
acetylphenylalanine 152
Achiwa 157, 166
acrylimide 134
acyl boronates 139
acyl complexes 49
acylases 262
acyloin condensations 271
AD-mix 384
  α 378, 380
  β 378, 380
adrenergic blocking agent 163
adrenoceptor agonist 164
aggregation 114, 118, 119
agrochemicals 7
Akermark 307
aklavinone 362
Alcalase 262
alcohol dehydrogenase 265, 266
aldol reaction 4, 9, 27, 239, 344
  aza- 33
  homo- 243
  nitro- 22
aldolase 271, 272
alkene epoxidation 15
allose 233
allothreonine 173
allylic acetates 301
allylic alcohols 159, 166, 367, 368, 369
allyltitanation 245
aluminium 135, 136
Amano P-30 lipase 269
amino acid dehydrogenase 265, 267
amino acids 7, 23, 49, 53, 101, 128, 132, 135, 141, 152, 172, 204, 241, 242, 263, 267, 330, 349, 355, 388
  α-alkyl 368
  β-hydroxy-α- 369
amino alcohols 9, 11, 12, 19, 34, 101, 226, 229, 253, 254
aminocarbene 355
amphetamine 8
amphotericin B 76
analytical methods
  CD 61
  GC 245
  HPLC 95
  molecular weight measurements 18
  NMR 14, 66, 76, 95, 115, 157, 188, 233, 237, 238, 244, 245, 248, 299, 321
  ORD 61
  X-ray structure determination 13, 61, 105, 106, 119, 128, 136, 247, 233, 234, 236, 238, 299, 306, 307, 308, 379
Andersen 61
Andrus 388
ANIC S.P.A. 156
Annunziata 64
anomeric effect 84, 131
anti-selectivity 37, 203, 204, 241, 247, 250
anti-substitution patterns 316
antiperiplanar 201
antisense oligonucleotides 249
arabino-configuration 249
arabinose 130, 251
aroylhydrazones 175
arylketimine 175
aspartam 7, 153, 263
Aspergillus acylase 262
Aspergillus niger 264, 265, 267
asymmetric
  amplification 4, 9
  conjugate addition 19
  deprotonation 111–125
  dihydroxylation 379, 383
  epoxidation 4, 9, 231, 343, 356, 368, 395
  hydroboration 181–211, 404
  hydrocyanation 23
  hydroformylation 217
  hydrogenation 146–180
  hydrosilylation 212
  induction, definition of 2
  oxidation 9, 367–394
    of thioethers 387
  reduction 21, 146–180, 265
  synthesis
    design of 1–6
    generations
      first 3, 240, 401
      fourth 3, 297
      second 3, 240, 401
      third 3, 242, 397
    importance of 7–8
  transfer of tricarbonyliron 328
  transformation 123, 295
ate-complexes 190, 251, 252

auxiliary ligands 318
axial chirality 279
aza-aldol 33
azanorbornene 289

Backvall 388
Baeyer–Villiger oxidation 268, 269
bakers' yeast 265–268
Balenovic 60
9-BBN 204, 205
BCCP 168
BCPM 168, 169
bda 292
BDPP 147, 153, 154, 175, 176
BDPP (sulfonated) 175, 176
BDPPH 176
BDPPM 157
Beauvaria sulfurescens 264, 265, 270
Beck 155
benzoquinolizine 104
benzoylhydrazines 175
bidentate phosphines 40, 146–151, 153–177, 196–198, 214, 218–220, 278, 279, 281–293, 295–297, 305, 306, 310
BINAP 147–150, 153–155, 157–161, 164, 166–174, 176, 196–198, 279–285, 287, 289, 291, 293, 297
BINAPHOS 219
binaphthol 16, 17, 18, 44, 133, 141, 248, 252, 386
binaphthols 138
binaphthyl 279
BINOL 16, 17, 18, 44, 133, 141, 248, 252, 386
biological activity 6
biotechnology 260–274, 398–399
biotoxication 8
BIPHEMP 147, 149, 150, 154, 158, 159, 162, 168, 170, 171, 172
biphenanthrol 133
biphenomicine A 173
Bipymox-ip 214
bis(dimethylamino)naphthalene 286, 288
bis(diphenylphosphino)pyrrolidine 163
bis(methoxymethoxymethyl)pyrrolidine 47
bisoxazoline 388
bisphenylethylamide 112, 119
BITAP 173
β-blocker 8
boat transition state 241
Bolm 20
borane
  -dimethylsulfide 183
  -THF 42
borinic esters 190, 207
bornane-10, 2-sultams 29, 31, 48, 54
borneol 229
bornyl-10, 2-sultam 49
borolanes 192, 193
boron 27, 38, 136
boronic esters 190, 209
Bosnich 309
BPPFA 147, 153, 154
BPPFOH 147, 164, 165, 284
BPPM 147, 149, 157, 167–169
BPPP 164
brasilenol 115
Brassard's diene 137, 140
Braun 32, 33
brevicomin 99, 100
Brown 156
Brunner 288
Buchwald 177
bulgecenine 356
Burk 147, 153, 175
butadiene 131, 132
butene 188, 189, 219
butenolides 68
BuTRAP 214
4-butylcyclohexanone 112
t-butyldimethylsilyl triflate 31
t-butyl hydroperoxide 373

C. militaris 388
C_2 axes 392
C_2 symmetric 13, 34, 47, 79, 85, 93, 94, 112, 133, 136, 142, 147, 151, 153, 164, 173, 193, 238, 279
Cahn–Ingold–Prelog nomenclature for sulfoxides 61
calamenene 352
calculations
  ab initio 201
  force field 237
  MM2 299, 379
  semiempirical 190
calyculin A 29
camphor 113
camphorate salts 387
camphorsulfonyl imines 140
Candida cylindracea 261, 262
cannabisativine 140
capnellenol 290
CAPP 147, 157, 159
carbacyclin 116, 398
carbapenem 173
carbidopa 155
carbocyclic nucleosides 263
carbohydrates 130
carbomycin B 323, 324, 336
carbonyl insertion 321, 322
carbovir 263
carboxypeptidase 262
carene 184, 186
carnitine 267

Carpenter bee pheromone 383
Carreira 44
castanospermine 284
catecholborane 196–198, 203, 204
CBD 147, 149, 157, 159, 163, 174
central nervous system (CNS) 8
chair transition state 35, 241, 243, 246
chalcone 226
chelation control 97, 109
chiral auxiliary, definition 3
chiral enolate equivalent 400
chiral pool 2, 126, 147
chiral reagents
  allylborane nucleophiles 341
  aminals 93–110
  bases 339
  boron enolates 341
  cobalt clusters 341
  dienes 131, 140
  dienophiles 126, 137, 139, 141
  enolates 27–59
  enones 328
  nucleophiles 293, 303, 326, 336, 341
  phosphines 40, 146–151, 153–177, 196–198, 214–220, 281–293, 295–297, 305–309
  sulfoxides 60–92, 123, 128, 131, 386, 388
chiral shift reagents 322, 339
chirality
  axial 279
  memory of 53
  planar 318, 338, 349
  relay 2, 3, 303, 369
CHIRAPHOS 147, 149, 154, 155, 157, 163, 173, 174, 219, 309
Chirasil–Val 245
chloramphenicol 383
chromium 108, 293, 298, 352
Chromobacterium viscosum 262
chymotrypsin 262
cinchonide 52
Cinquini 64
citronellol 161
cladospolide A 78, 83
Claisen 401
closed π-systems 317
cobalt 150, 161, 218
cod 148, 149, 151, 175, 176, 201, 215, 216
cofactor-recycling 265
conduritol C 269
conjugate addition 96, 103, 222–230, 344, 352, 400
contraceptive activity 7
control
  by *convex* asymmetry 248
  chelation 97, 109
  kinetic 108, 121, 304, 338
  lateral 320
  reagent 3, 182
  stereoelectronic 111
  steric 109
  substrate 2, 3, 182, 200
  thermodynamic 76, 105, 108, 121, 237, 244
copper 19, 73, 75, 96, 97, 99, 103–105, 109, 224–228, 336, 344
Corey 32, 34, 38, 44, 64, 135, 136
Cotton effect 61
$Cr(CO)_6$ 108
Crabtree 200
Cram's rule 246
cross-coupling 275, 292
18-crown-6 52, 350
cryptone 115
Cunninghamella elegans 388
cuprates
  higher-order 227
  medium-order 227
Curvularia falcata DAH 268
Cy_2BH 202, 205
cyanocuprates 73, 76
cyanohydrin 16, 23, 270
cyclobutadiene 317
cycloheptatriene 338
cycloheptenone 19, 224, 227
cyclohexadiene 141
cyclohexadienediols 269
cyclohexene 388
cyclohexenones 224, 360
cyclopentadecenone 226
cyclopentadiene 127–130, 133–137, 139
cyclopentadienone 356
cyclopentadienyl ligands 49, 233, 321, 323, 326, 343, 344
cyclopentadienylmolybdenum 344
cyclopentene 387
cyclopentenone 227
cyclooctadiene 148, 149, 151, 175, 176, 201, 215, 216
CYCPHOS 176, 175

d^1 nucleophiles 239, 251
d^3-nucleophiles 239
DAG 35, 40, 241, 242, 244, 248
DAIB 11, 13
Danishefsky 137, 138
Danishefsky's diene 131, 141
Davies 50, 55
Davies/Liebeskind enolate equivalent 49, 50, 55
Davis oxaziridine reagents 53
DBP-DIOP 219
de novo asymmetric synthesis 2
DEGUPHOS 147, 149, 155
Degussa 267
Degussa Co. 155

dehydroalanine 153
dehydroamino acids 153, 155, 157
dehydroerythronolide A seco acid 371
dehydrohalogenation 123
dehydropeptides 155
dehydrophenylglycine 153
deltamethrin 7
deoxynojirimycin 271
2-deoxy-D-ribose 78
2-deoxy-L-xylose 78
Deprenyl 8
Dess–Martin 56
desulfonylated 326
desulfurization 68, 74, 79, 84, 85
DET 9, 10, 16, 367, 370
deuterium isotope effects 373
DHQ 378, 380
DHQ-IND 382, 383
DHQD 378, 380
DHQD-CLB 384
DHQD-IND 383, 384, 386
$(DHQD)_2$PHAL 380, 382, 384–385
$(DHQD)_2$PYR 380, 382, 385
$(DHQ)_2$PHAL 380, 382
$(DHQ)_2$PYR 380, 382
1, 2-diamines 93
1, 2-diphosphines 147
1, 4-diphosphines 147
1, 3-dithianes 388
diacetone-D-glucose 35, 40, 241, 242, 244, 248
dialkoxycyclopentadienyltitanium chlorides 233
dialkylboron enolates 27
dialkylzinc reagents 11, 13, 21, 229, 253, 254
DIBAL 66, 73, 76, 84, 88
diborane 66
dicarbonyl(cyclopentadienyl)iron 321
dicarbonyl(cyclopentadienyl)molybdenum 323, 326, 343
Diels–Alder 16, 18, 22, 106, 126–145
  hetero- 137
  inverse electron demand in 106
dienophiles 126, 128, 130, 134
dienyl alcohols 367
diethyl tartrate (DET) 9, 10, 16, 367, 370
diethylaluminium chloride 50
diethylzinc 9, 11, 21
diHETE 348
dihydro-LTB_4 348
dihydroactinidiolide 115
dihydrocitronellol 162
dihydrofuran 285, 287
dihydroquinidine 378
dihydroquinine 378
dihydroxycalamenone 352
dihydroxyserrulatic acid 352
diisopinocampheylborane 183, 247
diisopinylcampheylboron triflate 33
diisopropyl tartrate (DIPT) 367, 371
diketodisulfoxides 85
dimeric catalysts 371
(2-methoxycarbonyl)cyclopentanone 173
2, 3-dimethylbuta-1, 3-diene 135
dimethylcyclohexanone 112, 113
dimethyl fumarate 135
dimethylcuprate reagents 343
(2, 5-dimethyl)methoxyborolanes 192
dimethylzinc 13
DIOP 146, 147, 149, 153, 154, 157, 158, 160, 165, 173, 174, 195–197, 215, 216, 219
dioxin 280
DIPAMP 147–149, 153, 154, 157, 168, 169, 173, 174
dipeptides 23, 155
diphenylsilane 212
diphenylzinc 228
DIPT 367, 371
Donaldson 345
dopa 155
dopamine 389
dopamine β-hydroxylase 389, 390
double asymmetric hydrogenation 159
double stereodifferentiation 4, 132, 139, 162, 197, 247
drug receptor 7
drugs 218
DSM-Toyo Soda 263
DUPHOS 147, 149, 153, 175
Duthaler 35
dynamic kinetic resolution 173
dynamic resolutions 262

$EapBH_2$ 187, 188
Electric eel choilinesterase 262
elimination, β- 276, 280
enamides 157, 158
enantiocontrol by *convex* asymmetry 248
enantioface differentiation 225, 278, 246, 279, 298, 403
enantiomer recognition 6
enantioselective autocatalysis 4, 15, 24
enantiotopic
  alkenes 289, 404
  group differentiation 288, 398, 400, 401
  leaving groups 309
  termini in π-ligands 325
enantiotopos 278, 288, 403
endo 318
  ψ- 318, 346, 348
ene reactions 17, 236
enediyne natural products 341
enzymatic reactions 121
  epoxidation 378
  hydrolysis 260, 261

oxidation 61, 268
reduction 265
enzymes 260–268, 270–274
epimerisation 62, 338
epoxides 6, 10, 75, 77, 78, 95, 260, 261, 264, 265, 343, 347, 357, 368–375, 383, 386, 397, 399, 400
eptazocine 282, 284
EQ 113, 117, 118
erythro 4, 42
Escherichia coli 272
ester hydrolysis 260, 399
esterase 260, 261
Et-DUPHOS 154, 175, 176
Et_2Zn 229
Et_2Zn 254
ethoxycarbonyltetralone 173
ethyl citronellate 160
ethyl *p*-tolyl sulfoxide 61
ethylapopinene 187
ethylenic esters 98
ethylmagnesium bromide 322
$Eu(fod)_3$ 137, 138
$Eu(hfc)_3$ 137
Evans 21, 27, 29, 31, 46, 53, 88, 240
exo 37, 304, 318, 352
ψ- 318, 346
addition 319, 326, 330
-brevicomin 385
external quench (EQ) 113, 117, 118

Faller 248
Fe (η-C_5H_5) (CO) (PPh_3) 49
Feringa 21, 388
ferrocene 283, 292, 295
ferrocenylphosphines 39
first-generation asymmetric synthesis 3, 240, 404
Fischer carbenes 355
FK-506 171, 336, 370
force field calculation 237
formate dehydrogenase 267
fourth-generation asymmetric synthesis 3, 297
formylpyridine 102
frontalin 98, 100
FruAld 272
fructose-1, 6-bisphosphate aldolase 271
FucAld 272
Fujita 398
fungi 265

gabaculine 330
GC 245
Genet 148
Geotrichum candidum 262
geraniol 161
Gerlach's method 74
Gilman 61, 228
gingerol 76
Gleave 113
glucose 35, 131, 233, 241
glyceraldehyde 249–251
glycine 35, 48, 52, 241
glycopeptides 264, 272
glycosides 260
glyoxal 99, 100
glyoxylate-ene reaction 17
gold 39, 40
goniofufurone 350
Grabowski 52
Green 339
Grignard reagents 61, 62, 64, 98, 99, 101, 192, 228, 244, 251–253, 298, 324, 337, 338, 361, 395

hafnium 253, 254
Hajos–Parrish reaction 9, 23, 401
Halpern 156, 160
Hanaoka 349
hapticity 314, 319, 340
Harpp 62
Hayashi 215, 285, 287, 295, 298, 305, 308
Heck 275
Heck coupling 288, 289
Hegedus 219
Helmchen 47, 307
HETE 344, 345
hetero-Diels–Alder reaction 137
heterochiral 13, 15, 18
heterodiene 344
hexafluoroacetylacetonate 150
higher-order cuprates 227
hippeastrine 333
Hirama 386
Hiroi 303
HMG-CoA synthase inhibitor 171
HMPA 46, 48, 114, 115, 117, 122, 227, 228
Ho 339
Hoffmann–LaRoche 168, 404
hog-kidney acylase 262
homoaldol-synthons 243
homoallylic alcohols 367, 368
homochiral 11, 13, 15, 18, 19, 20, 23, 111
homocuprates 226
Hoppe 234, 243
Horner 146
horse-liver alcohol dehydrogenase 266, 268
horse-liver esterase 262
Houk 201
HPLC 95
hydrazones 175
hydrindans 289
hydrobenzoin 135
hydroboration 181–211, 404

hydrocyanation 23
hydrogen peroxide 375
hydrogenolysis 53
hydrolase enzymes 261
hydroxycyclopentenone 162
hydroxyketones 381
hydroxylation, *cis*- 379
hydroxyproline 226
hydroxypyruvate 271
hydroxysteroid dehydrogenase 266, 268
hydroxyvitamin D_2 132

ibuprophen 8, 159
Icr_2BH 184, 188
Ikariya 159, 171
imidazolidine 93, 108
imines 175
immobilization 163
in situ quenching 113, 116–118, 123
indene 197
indoloquinolizine 104
Ingold–Thorpe effect 236
insecticides 7
inverse electron-demand 106, 132
inversion of stereochemistry 301
iodolactonisation 406
iodosylbenzene 375
ionic reduction 352
Ipc_2BH 183, 184, 188
$IpcBH_2$ 187, 189, 191, 205
iridium 150, 161, 175, 200
iron 49, 50, 55, 95, 296, 305, 321–324, 327–336, 345–348, 357, 358, 360, 361
Isicon 155
isoborneol 229
isoborneol-10-sulfinyl auxiliaries 128
isobutyraldehyde 243
isocyanate 368
isopropylidene glyceraldehyde 139
isopropylnoradrenaline 8
isotopic labelling 62, 373
ISQ 113, 116–118, 123
itaconic acid 157, 159
iterative reaction sequences 321

Jacobsen 374, 375, 388
Jacobsen–Katsuki epoxidation 373
James 148, 175
Jansen 21
Johnson 269
juglone 130

$K_2OsO_2(OH)^4$ 378
$K_3Fe(CN)_6$ 378, 382, 383, 384
Kagan 9, 16, 60, 146, 153, 388
Kanemasa 105
Katsuki 373, 387
Kende 29, 43
ketopantolactone 168, 169
3-ketosteroids 120
kinetic control 108, 121, 304, 338
kinetic resolution 4, 50, 120, 122, 161, 162, 173, 260–262, 285, 286, 386
  dynamic 173
kinetic studies 373
Kirby 228
Kishi 201
Klibanov 263
Knochel 255
Knolker 356
Knowles 146, 153
Kobayashi 22
Koga 112–114, 120
Kondo 339
Kunieda 64

LAC 15
lactam, β- 100
lactamases 262, 263
lactate dehydrogenase 265, 267
Lactobacillus kefir ADH 268
lanthanide catalysts 21
lasiodiplodin 74
lateral control 320
LDA 114
leucine 267
leucine dehydrogenase 267
leukotriene B_4 75, 348
LEUPHOS 295
Lewis acid 31, 33, 34, 38, 42–44, 50, 70, 97, 126–130, 133–135, 137, 138, 140, 141, 231, 234, 236, 239, 242, 243, 248, 253, 254, 322, 348, 351, 352, 359, 369
Ley 356
Lgf_2BF 184
Lgf_2BH 188, 189
$LiAlH_4$ 66, 67
LICA 48
$LiEt_3BH$ 67
ligand accelerated catalysis 15
ligand classification
  η^1 354
  η^2 276, 314–316, 321, 340, 355, 356, 359
  η^3 299–312, 314–317, 322, 324, 326, 354, 356, 359
  η^4 314–316, 323–325, 328, 354, 356, 359
  η^5 315–317, 327, 335, 337, 338, 358, 360, 361
  η^6 314, 315, 337, 338, 358, 361
  η^7 314
  η^8 314
  auxiliary 318
  odd and even 315
  open and closed 316
  working 318
LimBCl 185

LimBH 184, 185, 188
limonene 184, 185
limonylborane 185
linear reaction sequences 321
lipases 260, 261–264, 335
lipase M (*Mucor javanicus*) 264
lipase N (Amano) 264
Lipozyme 262
Lippard 227
Liu 326
longifolene 184, 186
LTA_4 346
LTB_4 348
Luche 228
lyxo-configuration 249, 250

m-CPBA 56, 115, 190
magnesiocuprate reagent 344
Majewski 113
maleic anhydride 132
malic acid 322
malyngolide 98, 100
Mander's reagent 116
manganese 373, 386
mannitol 238, 249
MAPP 19
Marko 164
Markovnikov hydroboration 197
Masamune 27, 34, 42, 43, 74
masked functionality 332, 354
McCague 336
MCCPMB 165
$MDBH_2$ 187, 188
Me-DUPHOS 167, 168, 170
Me_2CuLi 344
mechanism of
  cross-coupling 277
  Heck coupling 276
  Jacobsen–Katsuki asymmetric
    epoxidation 375
  Sharpless asymmetric epoxidation 372
medium-order cuprates 227
Meerwein–Ponndorf–Verley reduction 21
memory of chirality 53
menthol 61, 62, 127
menthyl acetate 27
menthyl *p*-toluene sulfinate 61
menthyl sulfinate 79
MeO-BIPHEP 147, 149, 150, 159, 160, 162,
  166–168, 170, 173
MeO-DUPHOS 167
Merck 159
Merck Frosst 371
meso
  intermediates 11, 13, 14, 18, 19, 261, 288,
    302, 304, 398
  bisepoxide 395
  complexes 304, 308
  diacetate 335
metalloprophyrins 377
methacrolein 133
methamphetamine 8
methoxyphenyl acetate 281
methylbenzylamine 141
2-methyl-2-butene 189
1-methylcyclopentene 189
3-methyl-3-phenylcyclobutanone 116
methyl lactate 76
methyl shikimate 330
methyl-*p*-tolyl sulfoxide 16, 73, 84, 85
methylcarbapenem 162
Meyers 33, 50, 51, 129
microbial oxidation 388
microbial reduction 339
Mikami 17, 18
Mikolajczyk 60, 62
Mislow 61, 147
mismatched reaction 247
Mitsui 27
mixed aggregates 112
MM2 calculations 299, 379
MOD-DIOP 147, 159
molecular biology 273
molecular sieves 17, 18, 233, 237
molecular weight measurements 18
molybdenum 377
monoalkylboranes 186
monoisopinocampheylborane 186
monomorine 1 117
Monsanto 155, 159
Montanari 60
MoOPH 53
MOP 215–217
Mortierella isabellina 388
Mortreux 168
Mucor javanicus lipase 262, 264
Mucor javonicus ADH 268
Mucor miehei lipase 261, 262
Mukaiyama 34, 40, 41, 93, 98
Mukaiyama reaction 242
Murahashi 384
Musco 297
muscone 19
mutarotation 62
mycoticin A 392
Myers 31
myrtanylborane 187
myrtenol 188

N-bromosuccinimide 50
n-butylboron triflate 28
N-tosyl-(*S*)-tryptophan 135
n.O.e. 136
$NaBH_4$ 66
NAD 265
NADH 267

NADP 265
NADPH 267
Nagao 398
Nagel 155
Nap(Ph)SiH_2 216
naproxen 159, 388
nbd 148, 198
neocarzinostatin 343
neooxazolomycin 29
nerol 161
neuraminic acid 272, 273
Ni$(acac)_2$ 229
Nicholas reaction 341
nickel 19, 20, 229, 341
nicotinamide adenine dinucleotide 265
nitrile oxides 106
nitrilease 264
nitro-aldol reactions 22
nitroalkenes 131, 132
NMR 14, 66, 76, 95, 115, 157, 188, 233, 237, 238, 244, 245, 248, 299, 321
NO^+ 325, 337, 338, 361
non-linear
  dependence of ee 9–26
  temperature dependence 378
nonactic acid 86, 88–90
nopol 187
noradrenaline 388
norbornadiene 148
norbornene 195, 197, 288
NORPHOS 147, 149, 154, 173
Noyori 10–12, 148, 150, 159
nucleic acids 7
nucleoside derivatives 127
Nugent 256

O'Donnell 52
O-methyljoubertiamine 331
O-methylmandeloxy dienes 132
octene 378
odd and even π-systems 315
Oguni 9, 16
oligonucleotides 249
oligopeptides 155
oligosaccharide 272
open π-systems 316
Oppolzer 29, 31, 32, 48, 49, 54, 128, 137, 141, 175
ORD 61
organolanthonide 198
organomagnesium reagents 227
organozinc 229, 298
organozinc reagents 228
Osborn 150
osmium 378, 379
osmium tetroxide 348
OsO_4 386
over-expression 273
Overman 275, 281
oxazaphospholidines 147
oxazolidine 129
oxazolidinones 27, 31, 47, 53, 54
oxazolines 33, 39, 136, 307
oxidation of benzene 269
oxidative addition 276, 277, 300
oxidative insertion 299
oxidoreductase 388
oxo process 217
5-oxo-ProNOP 169
oxygen 372
oxynitrilase 270
Ozawa 285, 287

π-allylpalladium 299
palladium 113, 115, 212, 275–312
pantolactone 129, 168
pantothenic acid 168
papain 262
PapBH_2 187, 188, 190
Parkinson's disease 155
Paterson 38
Pauson–Khand reaction 359
Pearson 324, 328, 356
PEG-NADH 267
Penicillin spinulosum 377
pentamethylpiperidine (PMP) 281
pentoses 249
peptidases 262
peptides 23, 241
periodinane 56
peroxy ligand 373
peroxycamphoric acid 374
Pfaltz 150, 227, 305
PGI_2 398
Ph_2CuLi 326
$(Ph)_2(DHQ)_2$PHAL 381
Ph_2SiH_2 213, 216
pharmaceutical products 6, 8, 164
phase-transfer catalyst 52
2-phenyl-1-butene 163
3-phenyl-1-butene 295
3-phenylcyclobutanone 116
phenyl glyoxal 98
phenylalanine 156
phenylapopinene 187
phenylbutyraldehyde 246
phenylethene 197
phenylethylamine 60, 112
phenylethylmagnesium chloride 295
phenylglycine 112–114
phenylglyoxylic methyl ester 166
phenylzinc chloride 301
Philipps 62
phosphoaoxazines 226
phyllanthocin 47
physostigmine 281, 282

phytenal 252
pinene 183, 187, 184, 186
pig-liver esterase 261, 262, 371, 400
piperazic acid 54
planar chirality 318, 338, 349
planar sterogenic unit 278
plane of symmetry 395, 397, 398, 399
platinium 60, 212, 219
PMP 281
PNNP 147, 156
polymer-bound catalyst 162
Porcine pancreas (ppl) 262
porcine pancreatic lipase 261
potential symmetry planes 315
PPFA 292, 293, 295, 298
PPTS 84, 88
$Pr(hfc)_3$ 132
Pr-DUPHOS 153, 154
Prelog's rule 267, 268
Prelog–Djerassi lactone 335
preparation of aminals 93
prochiral
  intermediates 403
  ligands 315
  nucleophiles 310
proline 23, 34, 37, 93, 112, 132, 141, 387, 401
prolinol 31, 46, 193, 252
PRONOP 167, 168
propanolol 8
PROPHOS 147, 149, 153, 154
prostaglandins 162, 404
proteases 262
proteins 7
pseudo-sugars 128
pseudoaxial 94
pseudoephedrine 50
pseudoequatorial 94
Pseudomonas
  diminuta 404
  fluorescens 262
  oleovorans 377
  putida 269, 333
  sp. ADH 268
  spp. 262
  spp. lipase 261
Pummerer rearrangement 75–81, 123
Pybox
  -ip 214
Pymox-tb 214
pyrethroids 7, 344
pyrone 358
pyrrolidines 31, 46, 130
pyrrolidinone 322
Pythia 214

quinones 135

racemisation 302
Raney nickel 53, 67, 68, 79, 85, 88, 101, 102
rat-liver microsomes 265
Re-face 46, 48, 54, 130, 225, 240, 244, 245
reagent-controlled diastereoselectivity 3, 182
Red-Al 51
reductive elimination 277, 297
Reformatsky reagent 29
resolving agents 94
resolution
  dynamic 173, 262
  kinetic 4, 50, 120, 122, 161, 162, 173, 260–262, 285, 286, 386
  regiodivergent 121, 122
retention of configuration 190, 301
reticuline 385
$Rh_2Cl_2(cod)$ 156
RhaAld 272
Rhizopus arrhizus lipase 262
Rhizopus japonicus lipase 262
Rhizopus niveus lipase 262
rhodium 148, 153–157, 159, 161, 163–169, 174, 175, 196, 212, 218
$[Rh(COD)Cl]_2$ 195
$[Ru(benzene)Cl_2]_2$ 156
Rhodococcus equi 263
Rhodococcus spp. 264
ribo-configuration 249, 250
ribonucleosides 249
ribose 7, 249, 251
RNA 7
Robinson annulation 9, 398
Rossiter 19, 224
routiennocin 356
roxaticin 171
$Ru(cod)(2\text{-methylallyl})_2$ 148
ruthenium 148, 153–155, 157, 159–163, 166, 167, 169–171, 174–175

σ-bond migration 276, 278, 279
S axes 395, 398
Saburi 148, 159, 171
Saccharomyces cerevisiae 266
salen 373, 386
salt effects 112
samarium 21, 175
SAMP 340
Santi 297
scandium 22
Schneider 64, 395
second-generation asymmetric synthesis 3, 240, 404
Seebach 254
selenium 307
semicorrin 151
(semicorrinato)cobalt 150

semiempirical calculation 190
serine 247
Sharpless 15, 385, 386
  asymmetric *cis*-dihydroxylation 378, 381
  epoxidation 4, 9, 343, 356, 395
  mnemonic 381
Shibasaki 22, 275, 281, 282
shikimic acid 330
showdomycin 116, 128
Si-face 32, 39–41, 46–129, 137, 244, 248, 249
$(Sia)_2BH$ 206
sialic acid aldolase 272, 273
$(Siam)_2BH$ 205
$SiCl_4$ 134
side-on alkene ligands 278, 314
silicon 31
silyl enol ethers 42
silyl ketene acetals 42, 44
sitophilate 322
sodium hypochlorite 373
Soladie–Cavallo 339
sparteine 234, 235
spinach 272
spiroacetals 84, 85
spirocycles 281
spirooxindoles 281
Springler 150
stannylenol eters 328
statine 171
stereoelectronic control 111
steric control 109
sterochemistry of β-elimination 280
steroids 200, 400
Still 205
Stille 219, 275
Stille coupling 293
stoichiometric π-complexes 313–366
strategy of lateral control 320
styrenes 218, 378
substrate-controlled diastereoselectivity 2, 3, 182, 200
subtilsin 262
sugar 131
sulcatol 266, 267
sulfates 384
sulfide oxidations 10, 16
sulfides 60
sulfinate esters 62
sulfinyl epoxides 73
sulfonamide 379
sulfonated diphosphines 163
sulfonates 370
sulfones 60
sulfonimines 175
sulfoxides 60, 386, 387
  Cahn–Ingold–Prelog nomenclature 61
  cyclic 122
  β-keto 60–92
  vinyl 128
sulfoximine 326
sulfur 307
sultams 29, 31, 48, 128, 137, 141, 175
Suzuki 275
Suzuki coupling 277, 292
Swern oxidation 79
sym-tetraisopinocampheyldiborane 183
symmetrical bisalkenes 403
symmetry plane 398, 399, 401
syn 37
  -diastereomers 250
  -elimination 201
  -periplanar 105, 280
  -selectivity 204, 241, 250, 292
Syntex 159
synthesis gas 218
synthesis of
  5-HETE 344, 345
  aspartame 263
  biphenomicine A 173
  brasilenol 115
  brevicomin 99, 100
  bulgecenine 356
  calamenene 352
  calyculin A 29
  capnellenol 290
  carbacyclin 116, 398
  carbapenem 173
  carbomycin B 323, 324, 336
  carbovir 263
  Carpenter bee pheromone 384
  castanospermine 284
  chloramphenicol 384
  cladospolide A 78, 83
  conduritol C 269
  dehydroerythronolide A seco acid 371
  deoxynojirimycin 271
  diHETE 348
  dihydroxycalamenone 352
  dihydroxyserrulatic acid 352
  dopa 155
  enediyne natural products 341
  eptazocine 282, 284
  exo-brevicomin 384
  frontalin 98, 100
  gabaculine 330
  gingerol 76
  goniofufurone 350
  hippeastrine 333
  HMG-CoA synthase inhibitor 171
  ibuprophen 159
  lasiodiplodin 74
  leukotriene B_4 348
  leukotriene B_4 75
  LTA_4 346

LTB$_4$ 348
malyngolide 98, 100
medium-sized rings 343
methylcarbapenem 162
monomorine I 117
muscone 19
naproxen 159, 388
neocarzinostatin 343
neooxazolomycin 29
nonactic acid 86, 88–90
nucleoside derivatives 127
O-methyljoubertiamine 331
phyllanthocin 47
physostigmine 281, 282
Prelog–Djerassi lactone 335
prostaglandins 404
pyrethroids 344
reticuline 383
routiennocin 356
roxaticin 171
showdomycin 116, 128
sitophilate 322
statine 171
steroids 397
sulcatol 266
thienamycin 341
trichodermol 329
trichodiene 328
trichothecene 328
tylosin 323
venturicidin 395
vernolepin 291
vitamin D$_3$ 369
yashabushiketol 76
zearalenone 74

TADDOL 214, 247
TagAld 272
Takaya 159, 160, 168
tamandron 336
Tamm 53
tamoxifen 336
tandem reaction 54
target molecule synthesis 368
tartaric acid 41, 136, 139, 367
tartrate 15, 244, 247, 339, 373
 -derived acetals 134
terpenes 184
tetrabutylammonium borohydride 67
tetrahydrofarnesol 162
tetralone 213, 352
tetramethylammonium triacetoxyborohydride 88
tetramethylthreitol 247
TFAE 245
thalidomide 8
thermitase 264
Thermoanaerobium brockii 266
Thermoanaerobium brokil ADH 268
thermodynamic control 76, 105, 108, 121, 237, 244
thermolysin 262, 263
thexylborane 201, 207
thienamycin 341
thioesters 34
thiophene 369
third-generation asymmetric synthesis 3, 242, 397
Thornton 132
threo-stereochemistry 242, 271
threonine 173
ThxBH$_2$ 202, 203, 206–208
tin 29, 34, 37, 40, 293
tin(II) triflate 34
titanium 4, 9, 15, 129, 135, 138, 177, 231–259, 367–369, 373
 enolates 240, 349
 tetrachloride 31, 32, 127, 128
 tetraisopropoxide 367
 isopropoxide 134
 /binapthol 386
titanocene 177, 231, 238
TMEDA 186
tol-BINAP 162
(*p*-tolylsulfinyl)acetophenone 64
transition metal catalysed hydrogenation 146–180
transketolase 271
transoid η^4-complexes 326
tributylvinyltin 293
tricarbonyl-
 (cyclohexadienyl)iron 328
 (η^6-arene)chromium 122, 123
 (naphthalene)chromium 108, 338
 chromium 108, 337, 338, 349, 350, 352, 358, 361, 362
 iron 327, 330, 333, 344, 346
 manganese 337, 361
trichodermol 329
trichodiene 328
trichothecene 328
2, 2-trifluoro-1-(9′-anthracenyl)ethanol 245
triisocampheyldiborane 183
triphenylethane diol 33
tris-s-butyl borohydride 68
tropinone 116, 118
Trost 299, 304
tryptophan 44
Tsuji 299, 302
tylosin 323, 324

UEB Isis Chemie 155
Uemura 16, 298, 352
umpolung reactivity 321

valinol 50, 338
VALPHOS 288
van Koten 228
vanadium 386
Vederas 53
venturicidin 392
vernolepin 291
vinyl sulfoxide 128
Vitagliano 307
vitamin
  B_3 168
  B_5 168
  D_3 369
  E 252

Warren 56
water soluable ligand 156, 163, 175
Weissensteiner 295
whole cells 263–271
Widdowson 31
Wilkinson's catalyst 146, 164
Wittig alkenation 78, 79, 328, 347, 348
Wittig rearrangement 123, 352, 395
Wittig–Horner alkenation 62, 99
working ligands 318

X-ray crystallography 13, 61, 105, 106, 119, 128, 136, 233, 234, 236, 238, 247, 299, 306, 307, 308, 380
xylose 233

Yamaguchi 79
Yamamoto 33, 42, 133, 136
yashabushiketol 76
yeast 271, 272, 401
yeast alcohol dehydrogenase 266, 268
Yoshioka 254
ytterbium 22

zearalenone 74
zeaxanthin 268
zinc 9, 11, 12, 19, 298
  borohydride 67, 68
  chloride 31
zirconium 32, 231, 253, 254
zirconocene 31, 32

Built to Move Millions

RAILROADS PAST AND PRESENT

George M. Smerk, editor

Built to Move Millions

STREETCAR BUILDING IN OHIO

CRAIG R. SEMSEL

Indiana University Press ◉ Bloomington & Indianapolis

This book is a publication of

Indiana University Press
601 North Morton Street
Bloomington, IN 47404-3797 USA

http://iupress.indiana.edu

Telephone orders 800-842-6796
Fax orders 812-855-7931
Orders by e-mail iuporder@indiana.edu

The paper used in this publication meets the minimum requirements of American National Standard for Information Sciences—Permanence of Paper for Printed Library Materials, ANSI Z39.48-1984.

Manufactured in the United States of America

Library of Congress Cataloging-in-Publication Data

Semsel, Craig R.
Built to move millions : streetcar building in Ohio / Craig R. Semsel.
p. cm.—(Railroads past and present)
Includes bibliographical references and index.
ISBN 978-0-253-34985-9 (cloth : alk. paper) 1. Street-railroads—Ohio—History. I. Title.
TF724.O3S46 2008
625.6'609771—dc22 2007031480

1 2 3 4 5 13 12 11 10 09 08

To Charles A. Knapp, for that first streetcar ride,
and to Violet E. Knapp, for all of the streetcar rides that followed.

WORKER SOCIETY
207
210
The WORLD'S FARE

CONTENTS

Acknowledgments ⊙ *ix*

1 *An Introduction to the Street Railway Industry* ⊙ *1*

2 *Car Builders of Ohio* ⊙ *11*

3 *Making the Cars Go: Components Essential for Operation* ⊙ *47*

4 *Couplers: When, Where, and Why They Were Used* ⊙ *71*

5 *Protecting the Public (and Themselves): Street Railways and the Manufacture of Safety Appliances* ⊙ *89*

6 *Fare Collection and Registration* ⊙ *115*

7 *Seldom Mentioned: Trimmings, Hardware, and Ventilation* ⊙ *137*

8 *The Decade of Transition, 1910–1919* ⊙ *153*

9 *Promise and Stagnation: Streetcar Technology during the 1920s* ⊙ *165*

10 *Parts of the Whole: Streetcar Component Manufacture during the 1920s* ⊙ *195*

11 *Streetcar Manufacture during the 1930s* ⊙ *213*

Afterword: 1938 and the End of an Era ⊙ *227*

Appendix: Tables ⊙ *233*

Notes ⊙ *257*

Index ⊙ *287*

ACKNOWLEDGMENTS

In a very real sense, this book is the product of many individuals and not solely the author whose name appears on the cover. Without their help, this book would never have happened. As I pore over the correspondence, e-mails, and reams of notes that were generated during this book's gestation, certain names and organizations stand out. For those I may have missed, I offer my sincerest apologies.

This book started and ended at the Cleveland Public Library. For anyone who has not had the privilege of conducting research there, I would encourage him or her to do so. Throughout the research, revision, and illustration stages of this project, the staff at CPL lived up to the term *professionalism.* A special thank-you should be extended to the staff of the Microfilm Department, which endured countless trips into the CPL basement to unearth the various reels of industry journals and transactions that are kept there. When hard copies of the trade literature were required, the Science and Technology and Business Departments never failed to answer the call. The Photograph Department must have set a record for the speed with which they located and reproduced the images of local streetcar activity.

At Case Western Reserve University in Cleveland, Dr. Kenneth Ledford was a patient and helpful mentor. On more than one occasion he settled my nerves and explained the mysteries of how books get published. His good humor, pithy insights, and general encouragement were invaluable.

Numerous museums contributed photographs, provided access to artifacts, and offered plenty of information. By far the greatest help came from the Branford Electric Railway Association in Connecticut, which operates the Shoreline Trolley Museum. Archivist and curator Michael Schreiber found plenty of excellent photographs of Ohio-built streetcars, while BERA president William Wall spent what had to be the hottest, most humid day of the year leading me and my wife through basements, barns, and yard trackage to find examples of Ohio-built streetcar components and other items needed to illustrate the various topics covered in this book. Fred Sherwood also helped us gain better access to some of the museum's exhibits.

The Indiana Historical Society, where most of the Cincinnati Car Company's photographs are preserved, rivaled the CPL in terms of its professionalism and general helpfulness. Susan Sutton, coordinator of visual references, was a pleasure to work with. She and her assistants came through in the eleventh hour to find the right photograph over the telephone when I had written down the wrong folder number in my notes. To Susan and all of the staff at IHS, thank you.

Surviving Barney & Smith interurbans are relatively rare, but a fine example is undergoing restoration in Coopersville, Michigan. James Budzinsky, president and curator of the Coopersville Area Historical Society and Museum, took me on a tour of this car. Housed in an old interurban electrical substation, the museum still captures the "feel" of the interurban era.

In West Henrietta, New York, the New York Museum of Transportation also has a number of useful cars and artifacts in its collection, as well as one of the most scenic settings for an electric railway museum. Charles Lowe and James Dierks, secretary of the board of trustees, were incredibly helpful in providing information on streetcar construction as well as providing access to areas of the museum that are not normally open to the public.

In Ohio, Dayton History, the organization that operates Carillon Park, went above and beyond the call to allow my wife and me to photograph some of their collection. Amanda Lakatos, communications manager, did a great job of arranging for our visit to Carillon, while education director Alex Heckman made sure that my wife and I had access to everything we needed. Docent Harold Boat was also a great help in an earlier visit to Carillon, especially in providing stories and information that otherwise would not have made it into the book.

In Columbus, the Ohio Railway Museum holds a number of rare artifacts. I thank museum president William Wahl for allowing me to use some of the museum's postcard images.

Closer to home, useful artifacts were discovered at the Northern Ohio Railway Museum at Chippewa Lake. Fund-raising director Steven Heister took time out of preparing for an open house event to take my wife and me through cars and storage areas to ensure we got the right photographs.

With the exception of photographs provided by CPL and IHS, many of the photographs appearing in this book required professional assistance in their development, as well as substantial coaching as to how to take them in the first place. Amanda Yeaton at Dodd Camera, Westlake, was "in" on the photographic portion of this project from the beginning, and she helped my wife and me avoid costly (not to mention embarrassing) mistakes. She performed wonders with a number of photographic mediums.

Edward Siplock was helpful in more ways than he will ever know. A gifted researcher and genealogist, Ed was working on his own project at the CPL while I was working on this one. Conversation over numerous lunches in downtown Cleveland made an arduous task more pleasant than it otherwise would have been. (For the record, Ed's book came out before this one.)

At Indiana University Press, series editor George Smerk has been a constant source of encouragement and advice. Without him, this book would never have happened.

Linda Oblack has also done much to demolish the stereotype of nasty editors and has been a pleasure to work with.

Last but certainly not least, there is my wife, Autumn. She has been a source of endless encouragement and so much more. She has read and reread all of the drafts, coordinated all of the research trips, formatted all of the text, and taken about half of the photographs. To say "thank-you" to her does not even begin to cover all that she has done.

Built to Move Millions

1

AN INTRODUCTION TO THE STREET RAILWAY INDUSTRY

PREDECESSORS TO MODERN STREET RAILWAYS

The modern street railway system was a late-nineteenth-century invention, evolving out of a desire to replace the horsecar. Horsecars first appeared on city streets in the 1830s and were common in most large cities by the 1860s. Essentially, a horsecar was a large carriage with metal wheels designed to run on metal rails laid in the middle of the street. Rails were used because they provided a smoother ride, enabling the horse to pull a much heavier load. The cars were not exceptionally fast, usually running at 4–6 miles per hour.

Although popular, the horsecar had numerous disadvantages. Horses moved slowly and typically could only work four to six hours per day, requiring a street railway to have three to five times as many horses as cars. Each horse consumed 30 pounds of feed per day.

A large workforce was required to care for the horses. In addition to blacksmiths and veterinarians (an outbreak of disease could ruin an operation), one stable hand was necessary for every 12 to 14 horses. Street crews were required to clean up after the horses, as most cities had strict regulations about the removal of manure from their streets.

The average car horse had a useful service life of only five years. They were expensive to replace—for example, in 1880 a new car horse cost $150. This cost might be recovered partially through the sale of retired car horses, but not all of it. Some operators attempted to economize by substituting mules for horses. Although less expensive initially, mules also had a lower resale value than retired car horses.[1]

It should come as no surprise that street railway operators sought mechanical alternatives to the horsecar. By the 1880s there was a plethora of alternatives, ranging from the conventional to the bizarre. During the 1889 American Street Railway Association convention, mechanical alternatives to the horsecar were discussed at length. Streetcars propelled by steam (produced both by conventional coal-fired boilers and by "fireless boilers" that generated heat using caustic soda), gasoline engines, ammonia, and compressed air were a few that were presented. A committee, assigned to the task of evaluating the alternatives, made the following cynical remark: "Of motors there are two kinds: motors and promoters; and of the two it is no small question in most cases to determine which is the more impractical."[2]

The three alternatives that appeared to be the most promising were systems dependent upon steam engines, mechanically driven cables, or electricity for their motive power. At first glance, steam-powered streetcars would seem to be the logical replacement of the horsecar. Steam power had been successfully applied to many industries and modes of transportation, including the street railway's distant cousin, the railroad.

Most steam-powered streetcars were not steam-powered at all; instead, they were unpowered trailers (often former horsecars) towed behind a small steam locomotive. Their noise and appearance tended to frighten horses and annoy pedestrians. One design solution was to build a shell resembling an ordinary horsecar around the locomotive. Called "steam dummies," these vehicles were used in a number of U.S. and European cities. However, steam dummies could not overcome public prejudice toward steam-powered vehicles running through city streets. Fearing boiler explosions, city ordinances either placed severe restrictions on steam-powered streetcar operations or banned them outright. Steam-powered street railway vehicles were virtually extinct by World War One.[3]

A more promising mechanical alternative to the horsecar was the cable car. This used a long loop of steel cable running through a trench, or "conduit," that was located between the rails. The trench's opening was kept as narrow as possible (for obvious reasons) and was referred to as a "slot." The car itself had a metal "grip" similar to pliers or a claw that hung from its underside and extended through the slot into the conduit. To move, the cable car's operator, called a "gripman," pulled a lever that caused the grip to grasp the moving cable. To stop, the gripman released the cable and manipulated a handbrake.

Cable car systems enjoyed a number of advantages over horsecars. They were at least twice as fast, could operate unimpeded in all types of weather, were cleaner and very quiet, could handle sudden crush loads of passengers without requiring additional power output, and were of particular advantage on hills. No fewer than 26 North American cities had cable car systems.

Despite the inherent advantages of cable railways over horsecars, there were numerous problems with cable railways that could not be overcome. The most obvious problem was cable breakage. Breaks were usually caused by premature wear, which indicated poorly aligned guide pulleys and sheaves in the conduit. This danger was minimal along straight routes, but increased considerably on routes containing hills and curves.

Worse than an outright break was the danger of frayed or kinked cables. Frays or kinks could easily snag a cable car's grip, making it impossible for the gripman to release the cable. Such cars were helpless, doomed to be dragged along their route until word was sent to the powerhouse or the car collided with something massive enough to stop it and allow the grip to be wrenched free. In light of the above, cable railways routinely stopped their cables (usually late each night) to inspect them for signs of damage.

Dangers aside, cable car routes were costly to construct, adding a level of complexity unheard of with horsecars or steam dummies. Conduits required excavation and an extraordinary amount of cast iron or steel. Their sides were reinforced with yokes spaced at regular intervals, usually 6 feet or less. Yokes often weighed 300–500 pounds apiece. Guide pulleys and sheaves (needed to direct the cable) as well as powerhouse machinery added to the already formidable cost of installation. Cost also depended upon the location of the powerhouse. Powerhouses needed to be sited along a line's immediate route or at a junction of lines, forcing cable railway owners to buy property at whatever price realtors demanded.

Naturally, once installed, cable car routes were inflexible. In order to ensure the long-term success of a prospective line (or at least to ensure that the line paid for itself), street railway owners had to be certain that the intended route would provide a high level of ridership. As a result, cable car lines tended to be constructed only in densely settled neighborhoods of populous cities.

An additional disadvantage was the cable car's lack of maneuverability. This term is admittedly a loose one, as any rail vehicle is limited by its inability to steer around obstacles. However, cable cars were limited further by their inability to travel in reverse. (This was also difficult to do with a horsecar, but not impossible.)

Finally, there was the rather disappointing running speed of the cable car. Cable speeds were limited by the power of the driving apparatus and by the amount of hardware in the conduits. The average running speed of cable railways in the United States was 10 miles per hour. The maximum speed for any cable car line was 14 miles per hour. This does not mean that 14 miles per hour was excessively slow, but it does mean that there were limits to the radius that cable car lines could cover.[4]

Although the theoretical radius of a cable railway was over twice that of an animal-powered one, the cable railway's running speed, combined with the necessity of locating lines in centers of high population density, meant they often did little to promote urban expansion. For example, in his study of transportation in Pittsburgh, historian Joel Tarr concluded that its three cable car lines had a minimal impact on the city's growth in comparison with the subsequent electric street railway and the automobile.[5]

DEVELOPMENT OF THE MODERN STREET RAILWAY SYSTEM

The mechanical alternative to the horsecar that proved to be the most successful was the electrically powered street railway system. Its precise origin is a matter of some dispute among historians, although it is generally acknowledged that the first commercial lines were designed and built in Europe by Dr. Ernst Werner von Siemens. Siemens successfully demonstrated a small, experimental electric railway at the 1879 Berlin Industrial Exhibition. He built his first full-sized commercial street railway system at Lichterfelde in Berlin in 1881. This was soon followed by additional lines in Charlottenburg (a Berlin suburb) in 1882 and in Potrush, Ireland, in 1883.[6]

In the United States, electric railway technology (either railroad or street railway) did not pass the experimental stage until the mid-1880s. The first commercial line was opened in Cleveland in 1884. It was not successful and ran only until the fall of 1885. However, the success in Europe, the enthusiasm with which electrical inventions were being received both in Europe and the United States, and the promise of a motive power that was less expensive to build than a cable railway and less expensive to operate than a horsecar line and was free of the stigma of the steam boiler proved irresistible for American inventors and entrepreneurs.[7]

The man who is credited with developing the first practical electric street railway system is Frank Julian Sprague. Born in Milford, Connecticut, in 1857, Sprague graduated from the United States Naval Academy in 1878. While serving as a naval officer, he traveled extensively and often reported on European developments in electrical science and technology.

In 1883, Sprague went to work for Thomas Edison at Menlo Park, New Jersey. His stay with Edison was brief, for in 1884 he formed the Sprague Electric Railway & Motor Company. He initially designed and installed electric motors for industrial plants, but he was also interested in railway applications. In the late 1880s Sprague designed and perfected his own electric street railway system, drawing on his experiences in Europe and those of inventors in the United States.[8]

Historian Clay McShane describes the Sprague system as not so much a new system as it was a synthesis of existing systems. To deliver electrical power, Sprague adopted the method of centering a copper wire directly over the rails. Also like some earlier systems, Sprague used an under-running pole (trolley) for current collection. Sprague's motors combined simplicity with sturdy, durable construction. He used conventional rheostat and resistor technology for motor control. Although not as efficient as some types of motor control, Sprague probably felt that the convenience of simplicity justified the loss in efficiency.[9]

Sprague employed a unique type of motor linkage. The "wheelbarrow" method of motor suspension (as Sprague called it) consisted of mounting half of the motor on the axle and half on the truck frame. The motor's pinion gear (the small gear attached to the motor's armature shaft) rested on larger gears mounted on the car's axle. The rest of the motor was attached to the truck frame by springs. This type of mounting enabled the motor to withstand any shocks associated with normal operation while keeping the gears constantly enmeshed.[10]

Sprague's first technically successful demonstration of the system was carried out in 1888 at Richmond, Virginia. His first commercial success followed in early 1889 between Boston and Brookline, Massachusetts.[11] Sprague's system proved that electric street railways could be the ideal successor to the horsecar. Like the cable car, electric streetcars were clean. Although expensive to install, they were less expensive than a cable car system. Since power was distributed through overhead wires, powerhouses did not need to be located along a railway's immediate route. Instead, they could be located where land was cheaper or where there was ready access to a supply of coal. Unlike cable car lines, power output could be altered to meet existing traffic demands. Electric streetcars also had the capability of operating in reverse, if necessary, and their ability to vary their speed enabled them to make up for lost time.[12]

Advances in motor controls enabled a degree of standardization for control systems and contributed to the rapid expansion of street railways. Popularly known as K-controllers, these motor control systems were first offered by General Electric and Westinghouse in 1893. The K-controller was essentially a rheostat that simplified the various electrical connections that cut out resistance from the motor circuits to accelerate the car. The K-controller was nearly universal on streetcars into the 1910s, and it remained the most common control system into the 1920s and 1930s.[13]

GROWTH OF ELECTRIC STREET RAILWAYS

Part of the appeal of the electrically powered streetcar was its potential operating radius. Historian Clay McShane once estimated that at an average speed of 3 miles per hour (allowing for stops), the effective area served by animal-powered railways could be 28 square miles. Cable-powered railways, at an average speed of 10 miles per hour, had the potential of serving 78 square miles.[14] As noted previously, this was rarely (if ever) accomplished.

The electric street railway, on the other hand, offered greater promise. Less expensive to build than a cable railway, less expensive to operate than an animal railway, and faster than both, the electric street railway could more easily cover the theoretical operating area of either. At an average speed of 15 miles per hour, the potential area covered could reach as high as 176 square miles.

Street railways tended to radiate outward from a city's center, usually terminating in areas that had yet to be developed. Their construction was generally supported by three groups of people: downtown real estate owners, businessmen, and executives who wished to draw more people into the city's center; real estate developers who wanted ready access to new development projects at a city's periphery, and those whose commute was already overcrowded or poorly served.[15]

Street railways allowed cities to grow. In the years before the automobile, they were the principal means of getting around (if one discounts walking). New residential neighborhoods were constructed at the peripheries of cities, allowing the middle class to move away from the city's center. Although wealthier than the working class, members of the middle class were still dependent upon working regular hours, usually at jobs in or close to downtown. The street railway made such outward movement possible.

Another form of electric railway was the interurban, which, as its name implies, ran either between cities or from a city deep into the hinterland. Distinguishing between an interurban and a street railway with extensive suburban operations is often a challenge to historians. George Hilton and John F. Due suggested the following set of characteristics as a rough guide: electric power, service based primarily upon passenger traffic (although some interurbans had significant freight operations as well), equipment that was both heavier and faster than that of street railways, and a mixture of running conditions (street railway trackage within cities and private rights-of-way outside city limits).

Interurbans filled a significant gap in urban and regional transport in the days before the automobile. They connected smaller cities and towns with larger ones, often serving areas that had been neglected by the steam railroads. They also ran frequent service, often on an hourly basis.[16]

The growth of the electric street railway industry was nothing short of explosive. The *Street Railway Journal* reported that by 1900 there were already 905 street railways of all types either in operation or in the planning stages in the United States. These railways had built over 20,400 miles of track, were operating nearly 63,000 cars, and represented a total investment of over $1 billion.[17]

A decade later, the number of street railway companies had increased to nearly 1,300. Over 40,000 miles of track had been built, and nearly 90,000 cars were in operation. It was determined that over 6 billion passenger fares were being collected annually, resulting in almost $500 million in gross revenue and netting nearly $200 million.[18]

TRADE ORGANIZATIONS AND TRADE PRESS

Naturally, an industry of such magnitude developed its own professional culture, complete with trade organizations and literature. Until the 1880s, the street railway industry was a highly localized affair. In 1882, as the industry became more national in scope, the American Street Railway Association was founded. ASRA's purpose was to provide a forum for railway owners and operators to discuss common business, legal, and technical issues and to develop committees to address those issues. As a result, membership was restricted to railway companies. ASRA grew in size and changed its name several times into the 1930s. The first time this occurred was in 1906, when ASRA became the American Street and Interurban Railway Association (ASIRA). This change was prompted by the proliferation of interurbans. ASIRA's name was simplified to the American Electric Railway Association (AERA) in 1910.

The focus of AERA was again challenged during the 1920s as greater numbers of bus operators joined its ranks (a number of existing members were either converting to buses or incorporating buses into their operation). In 1933 AERA adopted the generic name American Transit Association (ATA).[19]

Whether it was ASRA, ASIRA, AERA, or ATA, the association created a number of specialized "subassociations" as the need arose. In 1897, the Accountants' Association was founded to deal with methods and procedures of street railway accountancy. As technological issues grew more complex, the Engineering Association was organized in 1903. In 1906, the Claims Association was founded to address legal issues, and day-to-day operating issues became the focus of the Transportation and Traffic Association, which was established in 1908. The latter simplified its name to Operating Association in 1933.[20]

The association held annual conventions between the mid-1880s and 1930. Until 1908, the convention was held in a different North American city each year. From 1908 onward the convention was usually held in Atlantic City, New Jersey, although a few were held in New York City and Cleveland. Due to the Great Depression, few conventions were held during the 1930s. When no conventions were held, trade publications produced what were called "Conventions in Print." An early version of the Internet's "virtual reality," these consisted of attractively illustrated journal sections showing the latest products of car builders and component manufacturers.

Despite their importance to the industry, manufacturers were not allowed full membership within ASRA until the 1920s. Concerned that their issues might not receive the attention they deserved, manufacturers formed the American Electric Railway Manufacturers' Association in 1904. This organization did not intend to rival ASRA, but rather hoped to complement it by advancing "the interests of its members and of the American Street Railway Association by providing for and having custody of such exhibits of material as may be made at the annual [ASRA] conventions, and the establishment of friendly relations with each other and with the delegates of the railway companies."[21]

The industry spawned several trade publications. The oldest of these was the *Street Railway Journal,* which began late in 1884. Considered the best of the lot, the *Street Railway Journal* was based in New York City and published by McGraw-Hill. Another was the Chicago-based *Street Railway Review,* which began in 1891. It provided similar (though not as detailed) coverage as the *Journal.*

In 1908 the *Journal* and the *Review* merged to form the *Electric Railway Journal.* This new publication was the most thorough the industry had to offer, running feature articles that highlighted specific operating companies, provided overviews of transit operations in entire cities, and discussed key developments and trends within the industry. Regular departments reported on new products, finance, and other news pertinent to street railway men. Issues were often illustrated with numerous photographs and drawings. The *Electric Railway Journal* became the *Electric Traction and Bus Journal* in 1932 and *Mass Transportation* in 1935.

Electric Traction was another general trade publication. Independent of the *Electric Railway Journal, Electric Traction* ran as *Electric Traction Weekly* between 1906 and 1912. Based in Chicago from 1912 and published by Kenfield-Davis, this journal was similar to the *Journal* in scope and format.

The J. G. Brill Company published *Brill Magazine* between 1907 and 1927. Although devoted to its own manufacturing activity and that of its subsidiaries, the firm's prominence within the industry merits this magazine's inclusion.

TYPES OF STREET RAILWAY VEHICLES

The industry produced three basic car types: the streetcar, the interurban, and the rapid transit car. Streetcars operated on city streets. They were typically 40–50 feet long, sat 40–55 passengers, weighed 15–25 tons, and traveled at 25–40 miles per hour.

Between 1900 and 1914, streetcars came in a variety of design types. The most common was the "closed" car, in which the passenger compartment was fully enclosed. "Open" cars, sometimes referred to as "breezers," had passenger compartments that were open to the elements. Open cars were very popular with passengers during the summer months and year-round in southern regions of the country. Railway companies, however, were ambivalent toward them. Although open cars could handle large crowds, they were seasonal vehicles in much of the country and limited to use in fair weather.

A compromise between open and closed vehicles was the "convertible" car, so named because it had removable side panels. A variation of the convertible car was the "semiconvertible," in which only the window sashes were removable. The manner in which the sashes were removed on semiconvertibles varied. On some they were removed completely, while on others the sashes were dropped into pockets built into the side paneling. J. G. Brill had its own design in which the sashes were stored in two pieces in the car's roof.

The California car combined open and closed bodies on the same vehicle. These cars had an enclosed passenger compartment in the center of the car body, with open sections at each end. The cars received their name because they were developed in California and used most frequently on street railways in that state.

Another major car type was the "center-entrance" car, in which passengers entered and exited through doors mounted at the center of the car's side. Center-entrance cars were used by railways to keep the end platforms free of passengers, a plus on systems where crush loads were frequent and could hinder the motorman's ability to operate his car.

Interurbans were larger than streetcars, 45–60 feet in length and weighing 30–50 tons. Their passenger capacity was similar to streetcars, but car interiors contained larger, more comfortable seats, smoking compartments, baggage racks (sometimes baggage compartments), and restrooms. Interurbans were also faster than streetcars, capable of speeds up to 85 miles per hour.

There were fewer variations of interurban car types. Most interurban cars were closed and had separate passenger and smoking compartments. "Combines" included a separate baggage compartment that sometimes doubled as the smoking compartment. "Express" cars were used for baggage and light packages. Since express cars usually operated during the day, they were given a finished appearance and could run either by themselves or in train with passenger equipment. General freight equipment usually consisted of self-propelled boxcars (or small boxcars towed by electric locomotives) and were operated at night, when they would not interfere with daytime traffic on city streets.

Parlor cars were described in the *Electric Railway Journal* as "a type of Luxurious interurban car or special chartered car for city service fitted with individual seats or chairs."[22] Luxuries often included elaborate paneling, tile floors or wall-to-wall carpeting, fully equipped kitchens, and often dining and sleeping compartments. Observation platforms or oversized end windows were also common.

Another type of vehicle, called a "suburban car," was sometimes referred to in the trade literature. Suburban cars were basically a cross between a streetcar and an interurban. They were slightly larger and faster than streetcars and tended to run along suburban routes within the immediate vicinity of a city.

Rapid transit cars generally ran within city limits, occasionally reaching into neighboring suburbs. They usually ran in trains and traveled along grade-separated rights-of-way, such as an elevated structure or subway tunnel. These cars were generally designed to swallow crowds, with seating arranged to facilitate quick loading and unloading, and to accommodate large numbers of standing passengers. They typically operated at speeds between 25 and 50 miles per hour.

STREETCAR MANUFACTURE

An industry that experienced such incredible growth was bound to attract the attention of numerous manufacturing concerns. The industry averaged orders for nearly 3,000 cars per year between 1900 and 1910. The two largest car builders, the J. G. Brill Company and the St. Louis Car Company, served railways throughout North American and beyond. However, there were also numerous smaller car builders scattered about the country. One of the largest concentrations of these companies could be found within the state of Ohio.

Ohio was home to no fewer than five builders of streetcars, interurbans, and rapid transit vehicles. Their market was largely midwestern, but they served other states and even other countries as well.

The Midwest was a fertile market.[23] Returning to our statistics, we find that 341 of the nearly 1,300 railways in 1910 were located in the Midwest, with 91 in Ohio alone. Over 14,500 miles of track were located in the Midwest (over 4,000 in Ohio), and over 25,200 cars were in operation (over 5,700 of those in Ohio).

Ohio boasted the world's largest concentration of interurbans, with nearly 2,800 miles in service at the industry's peak. No Ohio town with a population of 10,000 or more was not served by at least one interurban. The second largest concentration was in Indiana, with over 1,800 miles of track. Upstate New York contributed an additional 1,129 miles.[24]

What follows is an examination of streetcar technology and manufacture between the years 1900 and 1940. This study is not intended to be a comprehensive history of the industry, nor does it cover all aspects of street railway manufacture (track work and power generation and delivery systems are not covered). Instead, this book will focus on the streetcars themselves, describing the various issues that affected their design and construction, as well as those for the components that went into them. It will also identify the major Ohio firms that built them. Occasional reference will be made to major firms outside of Ohio in order to place a particular type of component in its proper context.

The majority of the following chapters will concentrate on the period between 1900 and 1910, when the industry was at its peak. The remaining chapters will describe the challenges presented by World War One and the decline of Ohio streetcar manufacture in the decades that followed. In many ways the history of streetcar manufacture in Ohio reflects that of the industry itself. It is hoped that by the end of this study the reader will have gained a deeper understanding of the complexity of the streetcar and an appreciation for the many individuals and skills required to bring them into being.

Figure 2.1. Barney & Smith was primarily a builder of railroad rolling stock, making surviving electric railway vehicles by this car builder comparatively rare. Interurban no. 8 of the Grand Rapids, Grand Haven & Muskegon Railway is seen here undergoing restoration at the Coopersville Area Historical Society and Museum in Coopersville, Mich. *Author photo.*

2

CAR BUILDERS OF OHIO

All histories that address the street railway industry agree that its greatest period occurred between 1900 and 1914. More than 30 companies were devoted to the manufacture of streetcars, interurbans, and subway and elevated cars. Five of them were located in Ohio. Before addressing the car builders specifically, one should first take a glimpse at the cities in which they operated.

Ohio's streetcar builders were scattered throughout the state. To the north, in Cleveland, the G. C. Kuhlman Car Company produced thousands of streetcars and interurbans. To the southwest, in Cincinnati, the Cincinnati Car Company did more of the same, adding rapid transit cars to its repertoire. The progressive Niles Car & Manufacturing Company, which acquired a well-deserved reputation for building sturdy interurban equipment, was located to the east in Niles, and the massive works of the Barney & Smith Car Company were situated in Dayton to the west. Near the center of the state, the Jewett Car Company was located in Newark.

By 1900, most of Ohio had developed enough industrially to support heavy manufacture in a number of locations. Ohio was also sufficiently connected (mostly by rail) with other regions of the United States to encourage manufacturing on a national scale. Ohio's financial institutions were also strong enough to help finance manufacturers. Finally, local markets for street railway vehicles were strong and generated a large concentration of car building activity.

WHY LOCATE IN OHIO?

The early settlement of Ohio was one clear factor that made Ohio an integral part of the industrial Midwest. Legislation encouraging land purchase and settlement and improvements in communications and transport made the settlement of interior regions possible by the early 1800s. Initially, people moved to the Midwest mainly to acquire land, usually for agriculture. Later, as interior regions developed, emerging industries drew workers, enticed by economic opportunity.

Large numbers of European immigrants, many of whom were skilled laborers, came to the Midwest during the 1800s. The largest groups came from Germany, Ireland, Scandinavia, and the Slavic nations. They established themselves in the Ohio region fairly rapidly, with members of these ethnic groups holding political office in Ohio cities by the mid-1800s.

The growth of manufacturing in the Midwest was part of a broader national trend. Comparatively, before 1860 the United States' industrial output had been less than that of Great Britain, France, or the German states. In the 40 years that followed, U.S. industrial production grew to exceed the *combined* production of all three. Not only was there a shift in the size of U.S. industrial output, but there was also a shift in the emphasis of production from consumer goods (such as textiles, boots, shoes, and grain milling) to producer goods (such as iron and steel, machinery, and printing and publishing).

A shift also took place in the nature of manufacturing concerns in the United States. Firms in 1860 tended to be small, family-owned operations or partnerships. They were both specialized and labor-intensive. Goods were produced in small batches and were often intended for local or regional markets. This changed by 1900. Although small companies continued to exist, they were eclipsed by a new, larger type of company that was corporately owned and "bureaucratically managed." They produced goods in large quantities, often intended for national and international markets.[1]

During the late 1800s, Ohio became one of the nation's leading industrial states. By 1900, the state was ranked fifth in terms of manufactures, employing 345,869 people and turning out $832 million worth of goods annually. A number of factors contributed to Ohio's industrial growth, including natural resources, transport, and a large and growing population. The southeastern portion of the state was particularly rich in coal and iron ore, which encouraged the establishment and prosperity of many iron manufacturers within the state.

In addition to natural resources, such as coal and ore, water was useful both as a source of power and as a means of transport. The earliest metropolitan centers in the Midwest were located along major rivers and the Great Lakes. Ohio's location between Lake Erie to the north and the Ohio River to the south, along with its many rivers in between, made the state accessible both from within and without. This provided numerous attractive locations for mills and other manufacturers. Canal and later railroad construction further enhanced this natural advantage, ensuring that the state would be well connected with raw materials and markets elsewhere in the country.[2]

There were two major canals in Ohio, both of which connected Lake Erie with the Ohio River. The oldest of these was the Ohio & Erie, which connected Cleveland with Portsmouth and Marietta. It opened in 1835. In 1845, three other canals in the western half of the state were combined to form the Miami & Erie, which connected Toledo with Cincinnati.

A number of smaller canals were also opened in Ohio. Of these, one of the more prominent was the Pennsylvania and Ohio, which opened in 1840. This canal connected Pittsburgh and Akron and contributed to the development of the eastern portion of Ohio.

The significance of these canals should not be underestimated, as all five of Ohio's car builders would eventually locate in cities whose development and growth was made possible by canals. Both Cleveland and Newark were located along the Ohio & Erie Canal, Cincinnati and Dayton were served by the Miami & Erie Canal, and Niles was located on the Pennsylvania & Ohio Canal. However, only the Barney & Smith Car Company actually depended upon the canals, using the Miami & Erie Canal both to transport its goods and to supply power to its shops.

Each of the car builders' cities had established industrial infrastructures. The steady expansion of goods to markets in the Northeast, South, and Mid-Atlantic states contributed to the region's growth, and cities in the Ohio River valley became a viable market themselves. Pittsburgh, Cincinnati, Louisville, and St. Louis shipped goods to outside markets as well as to one another.[3]

Cincinnati was blessed with its location along the Ohio River. It was also at the mouth of numerous tributaries to the Ohio River, including the Little and Great Miami Rivers, Mill Creek, and the Licking River. Within 100 miles were the cities of Hamilton, Dayton, and Springfield. Initially settled during the 1780s, Cincinnati developed rapidly. By 1811, it was enjoying steamboat service. This was augmented by canal service in 1827. The railroad was quick to enter Cincinnati. By the time of the Civil War there were eight railroads and at least five locomotive manufacturers active in Cincinnati. The city soon became a major railroad center.

Cincinnati's economic influence was felt well beyond southern Ohio by 1850. It was a major gateway to the South. St. Louis, Chicago, New Orleans, and Louisville were all in intimate contact with the city. As canal and railroad linkages allowed Cleveland to develop, Cincinnati initially gained much from the northern city. In addition to another outflow for its manufactured goods, Cleveland's location on Lake Erie gave Cincinnati additional access to raw materials and markets in the Northeast.[4]

By the turn of the century, Cincinnati was the tenth largest city in the United States, known for a variety of products, especially tools and machinery. Other products included boilers and engines, carriages and wagons, copper and brass, tinware and sheet iron, and a variety of metal roofing materials.

Due to its close proximity to Cincinnati, Dayton grew quickly. Though already connected to Cincinnati via the Miami River, freight service between the two cities was improved considerably when a canal opened in 1829. The first of what would become nine railroads serving Dayton entered the city in 1851.[5]

Similar activity was taking place in northern Ohio, though slightly later. Cleveland's population more than doubled each decade between 1850 and 1870 and increased at a rate near or above 60 percent through 1930. Cleveland overtook Cincinnati as the state's most populous city in 1900. Cleveland did better in comparison with nearby Lake Erie ports, surpassing Detroit in 1870 and continuing to do so until 1920. Like Cincinnati, Cleveland acquired a reputation for diversity. In addition to steel, foundry, and machine shop production, Cleveland firms were producing electrical supplies, paints and varnishes, stamping products, wire products, and woolen goods. Practically every type of industrial city emerged in Ohio following the Civil War.[6]

Cleveland also became a major transport center in its own right. In addition to becoming a major lake port, Cleveland was well served by the railroads. Manufacturers took advantage of the city's access to outside markets and plentiful coal from southeastern Ohio and western Pennsylvania.

Ohio industry not only was prevalent in larger centers like Cleveland, Cincinnati, and Dayton but also emerged in smaller communities that boasted excellent railroad facilities, had been linked to canals before the railroads, and possessed a skilled labor force. Newark became known for its stoves, furniture, rubber products, furnaces, glassware, and oil refining. Niles became a major producer of sheet steel, boilers, metal lathes, and light bulbs.[7]

With the exception of Barney & Smith, Ohio's streetcar builders clearly did not come to dominate their respective cities, nor were they solely responsible for developing their cities' industrial base. Even Dayton, which was known as a center for car building during the nineteenth century, experienced a shift in its identity during the twentieth century. Once a city known for its railroad rolling stock, the city became a nationally recognized center for office and business equipment.[8]

THE CAR BUILDERS

Roughly one out of every five street railway vehicles produced between 1909 and 1913 was made in Ohio. By car type, Ohio car builders accounted for a fifth of all streetcars, 15 percent of all suburban cars, a third of all interurban cars (no doubt owing to the dense concentration of interurban railways in the state), and nearly a quarter of all rapid transit cars.

Each of Ohio's streetcar builders has a unique story. How they came into being, as well as how they came to their locations, is as diverse as their products. The companies will be examined in the order in which they were established.

Barney & Smith Car Company, Dayton

Barney & Smith was a builder of *railroad* rolling stock, something it had been doing since 1850. The company was founded by Eliam Barney and Ebenezer Thresher.

Born into a marginally successful farming family in 1807, Barney first worked as a schoolteacher in 1834. He left teaching and purchased a sawmill in 1840, which he ran successfully and sold at a profit in 1845. By the late 1840s he had met Thresher, a Baptist minister who had come to Ohio from Connecticut.

Deciding to start a business venture together, the two gentlemen purchased a sawmill in 1849, intending to use its lumber output in the manufacture of railroad cars. Located in Dayton, the new mill was sited near the Miami & Erie Canal and the confluence of Dayton's three rivers (the Mad, Stillwater, and Great Miami). The canal and rivers provided the mill with water power and convenient transport to Cincinnati and markets north and south. (Cars were shipped along the canal via barges until the mill was connected to the railroads when they entered Dayton.) By 1853, the firm employed 150 people and produced one freight car per day and one passenger car per week.

Over the next 15 years, Barney went through a number of business partners. Thresher sold his share of the business to Caleb Parker in 1854, who in turn sold his share to Preserved Smith following the Civil War. In 1867, Barney & Smith became the first company in Dayton to incorporate. The business continued to prosper through the late 1880s, employing just over 2,000 people in 1887.

In a manner similar to Chicago's Pullman Palace Car Company, Barney & Smith formed its own small village near Dayton to house the many immigrants in its workforce. Called Kossuth, this small village had its own bank, grocery and dry goods stores, and beer hall. Kossuth was surrounded by a high wall with only one point of entry. Much of Kossuth's success was due to J. D. Moskowitz, a successful Dayton immigrant labor contractor (and, conveniently, the Dayton agent for Cunard Steamship Lines).

In 1892, Barney & Smith was purchased by a group of Cincinnati-based investors. Their timing could not have been worse. The Panic of 1893 was particularly harsh on railroad rolling stock manufacturers, nearly a third of which were shut down due to lack of business. Ironically, the street railway industry was beginning to experience its greatest growth. Desperate to keep its plant operating (employment at Barney & Smith was down to 450 in 1894), the company's owners began canvassing street railway companies in hopes of securing orders.

Enough street railway car orders came in to sustain Barney & Smith until the railroad market rebounded in 1898. The car builder shifted its attention back to railroad rolling stock at this time. Of the 1,429 street railway vehicles known to have been produced by Barney & Smith, 1,082 were built during the nineteenth century. Barney & Smith also relied on local street railway markets more than Ohio's other car builders. Nearly two-thirds of its output went to Ohio railways (see table 2.1).

The Dayton plant was by far the largest among Ohio's car builders. At its height in 1900, the plant sprawled over 58 acres and required 8 miles of track to connect its buildings. At 3,500 workers, the car builder was the second largest employer in Dayton. By the turn of the century, most if not all of its machines and tools were powered by electricity or compressed air. Over a mile of electrical cable was needed to supply current to the 18 large electric motors that powered the company's machinery. Barney & Smith also possessed a compressor plant that was capable of delivering 500 cubic feet of air at a pressure of 80 pounds per square inch each minute. Illumination on the property was electrical, and the plant's power house was designed to burn wood shavings from the company's woodworking shops. To ensure a ready supply of lumber, Barney & Smith owned 86,000 acres of Georgia timberland.[9]

G. C. Kuhlman Car Company, Cleveland

The G. C. Kuhlman Car Company began as a cabinet-making and finished carpentry firm on Cleveland's east side in 1867. It was founded by Frederick Kuhlman and was located at 490 St. Clair Street. Fifteen years later, Kuhlman's two sons, Charles E. and Gustave C., joined their father. Hardwood furniture and interiors remained the company's principal business into the 1880s, when it began producing streetcars.

By 1888, the firm, now known as Kuhlman Brothers, was listed as a car builder in the *Street Railway Journal.* However, company advertisements during this same period suggest that it engaged in other street railway activity as well. Through these advertisements, Kuhlman claimed only to have provided "all the wood supplies" to three local street railway companies. An April advertisement announced, "All kinds of wood work for interiors a specialty. Street Railway Companies building their own Cars will do well to correspond with us."[10] In an 1888 international directory of street railway manufacturers published by the *Street Railway Journal,* Kuhlman was listed as a supplier of car ceilings, window sashes, seats, and panels. It was also the only Ohio firm identified as a car builder.[11]

During the early 1890s, Charles Kuhlman left the company and started a finished carpentry firm of his own. Gustave Kuhlman was building entire car bodies by this time (rather than just finishing them), although his market was still restricted to local customers. In 1892 he claimed his company would "guarantee the lowest possible price consistent with good material and workmanship."[12]

In November 1895, a fire swept through the Kuhlman works, destroying one-third of the property. Kuhlman rebuilt, but this proved to be a mistake. The St. Clair property was too small.

At some point during the mid-1890s, Kuhlman moved to a former horsecar yard at Broadway and Aetna. The yard was dominated by two long car barns. Consistent with the industrial practice of the day, Kuhlman organized his car building in a linear fashion.

Raw materials entered the foundry, cabinet, and subassembly shops at one end of the largest building. Completed streetcars emerged from the opposite end. The smaller building was used as a paint and varnishing shop.

Despite the more efficient layout of the Broadway property, it was still too small to keep up with Kuhlman's orders. For example, in early February 1900 the company received five orders totaling 41 cars ranging from small streetcars to large interurbans. In May the company received an additional seven orders totaling 61 cars. Also comprising city and interurban cars, these orders embraced no fewer than eight major car types.

During that year, Kuhlman incorporated his company and searched for a larger site. Locations under consideration were in the village of Collinwood to the east of Cleveland, Cleveland's west side, and the town of Elyria, located farther to the southwest. In May 1901 the company announced that it would relocate to Collinwood and reorganize its management.

Serving as president of the G. C. Kuhlman Car Company was Fayette Brown, a Cleveland industrialist best known as the president of the Brown Hoisting & Conveying Machinery Corporation. T. P. Howell and C. A. Ricks were vice president and secretary-treasurer, respectively. Gustave Kuhlman was general manager. The officers also served on the company's board of directors. Additional board members were Frank Rockefeller (of Standard Oil), I. H. Morley, C. C. Bolton, and R. A. Harriman.

Hailed by the *Street Railway Journal* as "one of the strongest manufacturing companies ever gotten together in Ohio," the company's new property was large, 31 acres, half of which were taken up by seven large buildings. It was well served by transport, located along the Lake Shore and Southern Michigan Railroad (later part of the New York Central), along a streetcar line on Adams Avenue (now East 140th Street), and on the "shoreline route" of the Cleveland, Painesville & Eastern interurban. The entire plant was designed by industrial architect J. Milton Dyer. Like its previous property, the Kuhlman works were laid out in linear fashion, with materials storage at the eastern end of the property (Adams Avenue) and paint and varnishing shops at the west end.

Kuhlman's growing position within the street railway industry did not escape the attention of other national car builders, and in 1904 Kuhlman was acquired by the J. G. Brill Company. Based in Philadelphia, Brill was both the largest streetcar builder in the United States and in the process of assembling a veritable car building empire. In addition to Kuhlman and its principal plant in Philadelphia, Brill came to control the American Car Company (St. Louis), the Wason Car & Manufacturing Company (Springfield, Mass.), the John Stephenson Works (Elizabeth, N.J.), the Danville Car Company (Danville, Ill.), and car building firms in France and Canada. Through these acquisitions, Brill was able to bid competitively on car orders anywhere in the country.

As part of the reorganization, most of Kuhlman's local leadership was replaced by Philadelphians. Most notable among the replaced locals were Fayette Brown (succeeded by Samuel Curwen) and Thomas Farmer, superintendent of the works. Secretary Ricks and board member Harriman were retained. New board members included C. E. Cowan, P. M. Hitchcock, and D. B. Dean.

Kuhlman gained a number of advantages from the takeover. Now part of a network of car builders, Kuhlman was assured competitive leverage in securing orders, either through direct bid or by handling the "overflow" of other builders in the network. The

Figure 2.2. An example of a streetcar at the Kuhlman works late in 1904. *Courtesy of the Cleveland Public Library.*

Figure 2.3. Convertible cars had sides that could be removed in warmer weather. This is a city convertible with its sides in place, sitting on a transfer table at the Kuhlman works. *Courtesy of the Cleveland Public Library.*

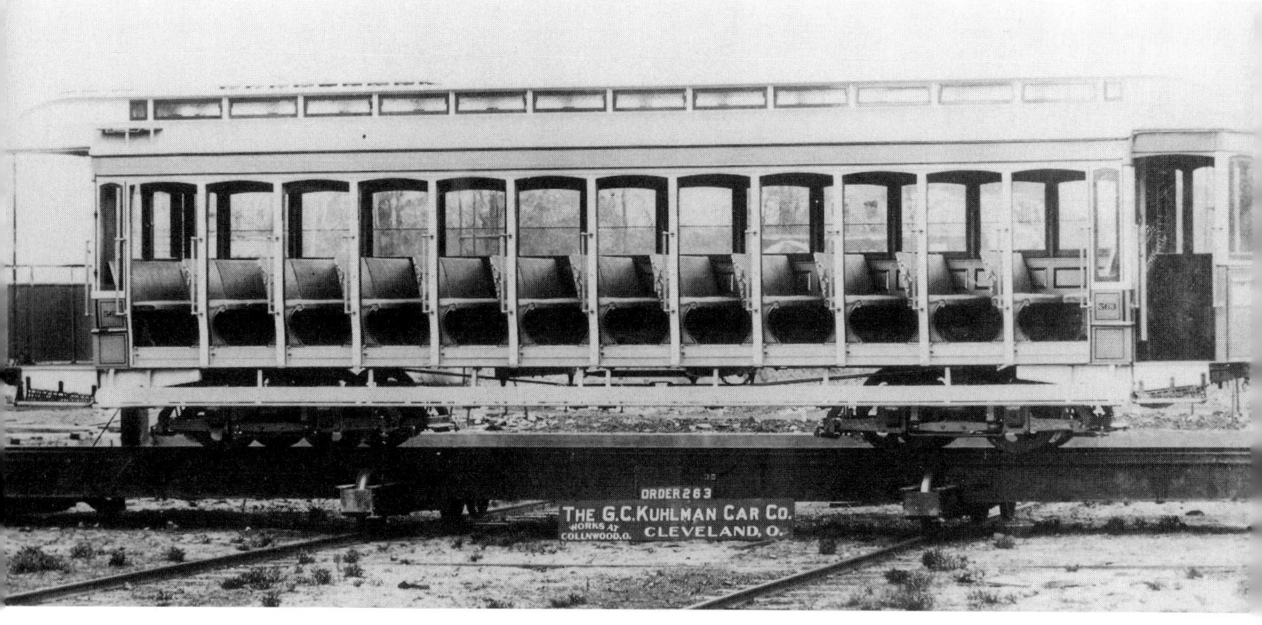

Figure 2.4. Here is the convertible car with one side removed. Note the running board and the grab handles along the side posts. *Courtesy of the Cleveland Public Library.*

Figure 2.5. A combination passenger and freight car (called a "combine") on the Kuhlman property. Note the protection given to the windows in the freight compartment on the left. *Courtesy of the Cleveland Public Library.*

Brill company supplied Kuhlman with seats and numerous trimmings, fittings, and components (especially trucks). In addition, Kuhlman was given access to Brill car designs and patterns, enabling it to partially streamline its production.

It did not take Kuhlman long to begin integrating Brill designs and components into its orders. During the summer and fall of 1905, 50 new closed cars for the Cleveland Electric Railway Company were built according to a Brill convertible design and using Brill trucks. An order of six 10-bench open cars for Knoxville, Tennessee, sported Brill handbrakes. Twenty open cars bound for Memphis, Tennessee, were also built according to a Brill design using Brill trucks.[13]

Although no longer active in the company's affairs, Gustave Kuhlman remained a presence at the Collinwood plant until his death in 1915. The company continued to bear his name into the next decade.

From the time Kuhlman started building streetcars in 1888 until the outbreak of World War One, the car builder produced over 2,500 street railway vehicles. Most of these were streetcars, though over 400 were interurbans. In keeping with Brill's practice of restricting its subsidiaries' car building activity to regional markets whenever possible, much of Kuhlman's output went to railways in Ohio, most of which, in turn, went to Cleveland (see table 2.2).

Figure 2.6. Following its absorption by Brill, Kuhlman began to use some of Brill's designs. One common design was the "semiconvertible," in which portions of the car's sides folded up into the car's roof. This view shows the side tracking on a car built by Kuhlman using Brill's design in 1904. *New York Museum of Transportation, author photo.*

Figure 2.7. Car builders used a variety of ways to advertise their workmanship. This elaborate Kuhlman sign is located on an interior bulkhead. *Courtesy of Dayton History.*

Figure 2.8. This casting is located on the step leading from the platform into a car's interior. *Courtesy of Dayton History.*

Figure 2.9. This builder's plate is located on the floor of a car. *Ohio Railway Museum, author photo.*

Looking specifically at the period between 1909 and 1913, when comparative data for the entire industry was published, Kuhlman produced 1,596 cars, or more than half its output between 1888 and 1914. The overwhelming majority of vehicles built between 1909 and 1913 were streetcars (84 percent). Eleven percent were interurbans, with the balance consisting of suburban and special types of vehicles. Nearly half of the cars of this period went to Ohio railways (45 percent), and 49 percent went to Illinois, Michigan, New York, Pennsylvania, and West Virginia.

Jewett Car Company, Newark

The origin of the Jewett Car Company has become clouded over the past century. Its first mention in the *Street Railway Journal* came in 1893, when it filled an order for the Sandusky, Milan & Norwalk Street Railway Company. The only information that addressed the car builder specifically stated that it was located in Jewett, Ohio. It also provided the following policy statement: "The Jewett Car Company believes in the employment of the best material and workmanship in all of its products, and the records made by its cars testify to the excellence turned out at the factory of this company."[14]

Aside from the rather generic nature of this statement, one can infer that the company had probably been in existence for at least a year, as the announcement referred (however vaguely) to other cars turned out by the company.

The early history of Jewett appears to have been plagued with financial difficulty. The company was reorganized in 1895. Two years later, the company was reorganized again as the Jewett Car & Planing Mill Company. At the time, the plant was located at the juncture of two railroads—the Pittsburgh, Cincinnati, Chicago, and St. Louis Railroad (a Pennsylvania Railroad subsidiary known as the "Panhandle Route") and the Wheeling & Lake Erie Railway. Aside from its milling operations, the plant's capacity was estimated at 300 cars per year. The company was run by manager A. H. Sisson, secretary C. E. Krebs, superintendent Neil Paulson, and W. H. Lorentz. Sisson, Krebs, and Lorentz were from Wheeling, West Virginia.

Around 1900 Jewett underwent yet another reorganization. This one provided the company with greater stability. The president of the "new" company was W. S. Wright, and the secretary was H. S. Sands, who left the company in 1901. As with previous management, both were from Wheeling. Sisson remained as manager, adding the duties of treasurer to his responsibilities. Paulson was kept on as superintendent.

Wright proved to be an important addition to the company. He took a keen interest in the design of streetcars and streetcar appliances. During his years with Jewett, he took out a number of patents that varied from onboard passenger gate mechanisms to window sashes and couplers. Jewett also made several modest contributions to methods of production, car interior layouts, and designs for entire cars. These included snow sweepers, automatic car couplings, and semiconvertible car designs.

On 1 March 1900, the company moved from Jewett to Newark. The new property sat on 10 acres of land, half of which was indoors. It bordered on two railroads, Pennsyl-

Figure 2.10. Fancy Jewett signage from 1904 on a car's interior bulkhead. *Branford Electric Railway Association, author photo.*

vania's "Panhandle Route" and the Baltimore & Ohio. By the summer of 1900, seven buildings were in service, with others under construction.

Too small for a linear orientation of its buildings, the Jewett plant made use of transfer tables to link the erecting, finishing, and machine shops. It employed 400 workers. Contemporary accounts have exaggerated Jewett's capacity, but it is safe to estimate that Jewett's shops contained at least 2,000 feet of track space.

Like Barney & Smith, Jewett's plant was steam- and air-powered initially. The plant was electrified over the winter of 1909–10. Work was slowed when the car builder experienced a fire on 27 December 1909. Although newspapers reported the worst, the only significant areas destroyed were a mill building and a lumber shed.

One aspect of its operations that made Jewett unlike most car builders was that it also engaged in car maintenance for local interurban railways. Its shops were not large enough for the car builder to handle all maintenance needs, but it could handle all of the woodwork maintenance.[15]

Looking at Jewett's output between the 1890s and 1914 and comparing it with Ohio car builders, the Newark car builder appears to have enjoyed the most well rounded car building activity. Of its output, 43 percent consisted of streetcars, 32 percent interurbans, 23 percent rapid transit cars, and 2 percent specialized vehicles. Jewett was one of two Ohio car builders in which streetcar output was outnumbered by the combined total of other types of railway vehicles (see table 2.3).

Jewett was also the only Ohio manufacturer dependent upon orders outside of Ohio. Through 1914, Ohio car orders accounted for only 12 percent of its output. This may have been due to its lack of an immediate market (such as Cleveland's street rail-

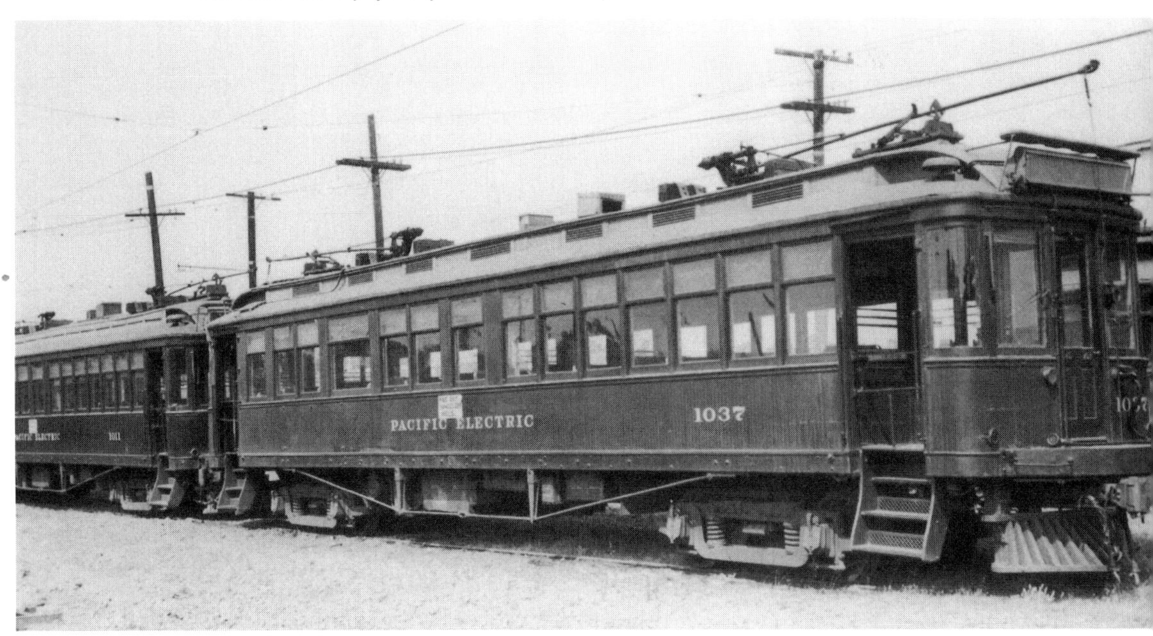

Figure 2.11. A Jewett interurban built for the Pacific Electric Railway. Note the curved glass at the car's end. *Courtesy of Branford Electric Railway Association.*

Figure 2.12. An interurban train being led by Jewett center-entrance car 7 on the Shoreline Electric Railway, Conn. *Courtesy of Branford Electric Railway Association.*

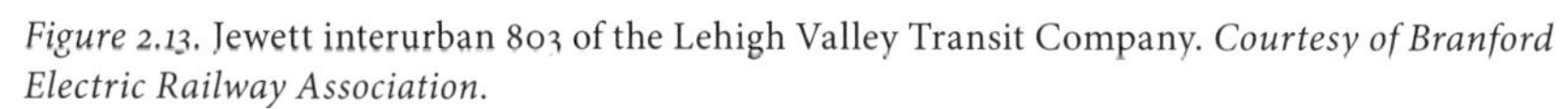
Figure 2.13. Jewett interurban 803 of the Lehigh Valley Transit Company. *Courtesy of Branford Electric Railway Association.*

Figure 2.14. Express cars were designed to haul packages and light freight. Here is a Jewett express from the Lehigh Valley Transit Company. *Courtesy of Branford Electric Railway Association.*

Figure 2.15. A Jewett passenger-baggage combine built for the Winona Interurban Railway, shown here on the successor, Winona Railroad. *Courtesy of Branford Electric Railway Association.*

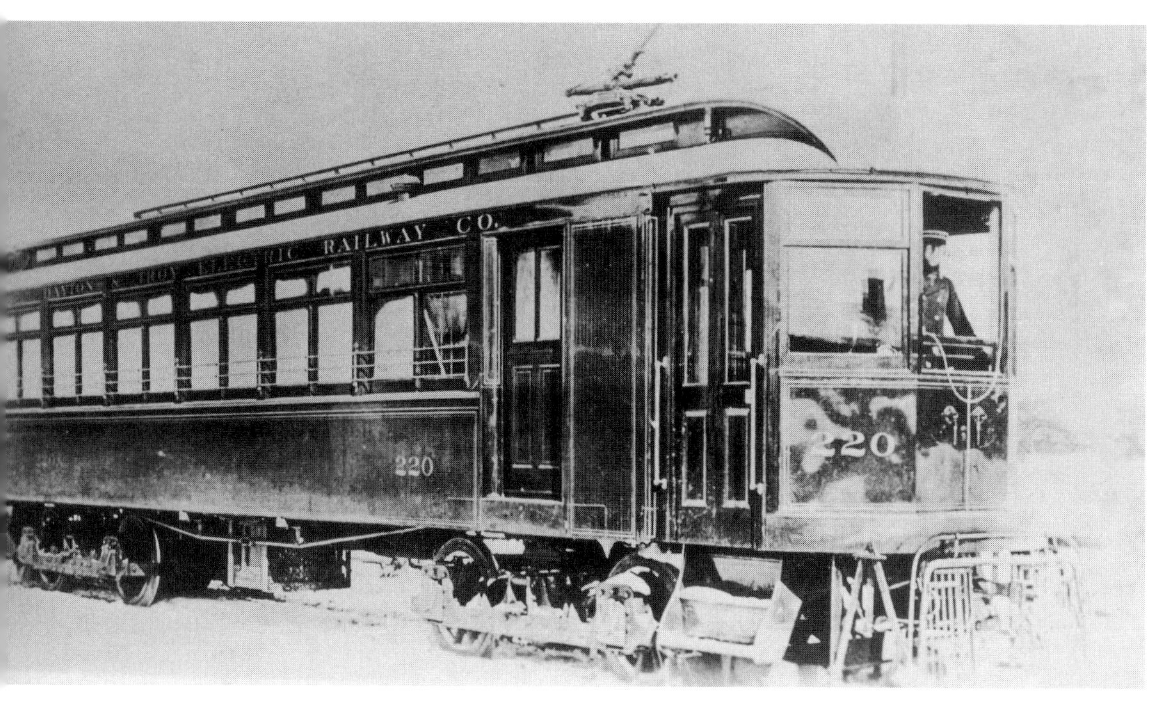

Figure 2.16. Another Jewett passenger-baggage combine operating along Dayton & Troy Electric Railway Company trackage in Ohio. *Courtesy of Branford Electric Railway Association.*

ways for Kuhlman or Dayton's street railways for Barney & Smith). Jewett's greatest customers were located in California, New York, and Illinois.

Between 1909 and 1913, however, we find a different story with car types. During this period, Jewett produced 913 street railway vehicles. Two-thirds of these (606) were streetcars. Seventeen percent of these cars were interurbans (155), and rapid transit cars and special vehicles (mostly combines) accounted for 7 percent each. The remainder were maintenance-of-way vehicles.

Jewett's geographic output remained the same for the five years. Jewett was also the only Ohio car builder not to depend upon local orders, with Ohio railways accounting for only 5 percent of its output. Nearly half of its output (46 percent) went to California and Washington, D.C. An additional 28 percent went to railways in New York and Massachusetts. The remainder of Jewett's cars went to railways in 13 states.

Cincinnati Car Company

Like Kuhlman and Barney & Smith, Cincinnati did not begin as a street railway manufacturer. It was not recognized as a car builder until 1903, although its existence can be traced to 1898. In February 1898 the *Street Railway Journal* described a very large car shop owned by the Cincinnati Street Railway Company.

Located in Chester Park on the site of a former racetrack (then on the outskirts of the city), the future car builder came into being as a centralized shop complex following the consolidation of Cincinnati's street railways (see table 2.4). The huge combine intended to economize not only by centralizing its major maintenance operations but also by building as many of its own cars as possible. Plans for such a facility were drawn up as early as 1895.

The 8-acre site had clear limitations that other car builders did not have, stemming from the original intent to devote the property to maintenance and "in-house" manufacturing. It had access to only one railroad, and a local railroad at that (the Cincinnati, Hamilton & Dayton). To complicate matters further, the property was bisected by a major street, Mitchell Avenue.

When completed, the property on one side of Mitchell Avenue contained the lumber shed, carpentry, mill, and paint shops. The property on the other side contained iron and brass foundries, electrical and machine shops, and a supply depot. All buildings were wood-framed and sided with quarry-faced limestone. For obvious reasons, the buildings were equipped with an elaborate sprinkler system. Water mains and 45 fire hydrants were scattered about the grounds.

The supply depot was a scene of constant activity, as it received supplies not only for the Chester Park complex but also for the yards and car barns of the entire Cincinnati street railway system (the street railway scheduled delivery runs during off-hours). The electrical shop could perform all types of repairs, saving the railway as much as 50 percent on its electrical maintenance budget. The brass foundry could turn out 7 tons of castings per month.

The painting and erecting shops each held six tracks, which were laid out in linear fashion, connected via a transfer table. A steam-heating system kept the paint shop at a constant 70 degrees Fahrenheit, and could do so even if the outside temperature dropped to –14 degrees.

The Chester Park facility's transition from a car maintenance department to an independent car building firm occurred quickly, taking just over a year. Between 1898 and

Figure 2.17. A streetcar built by the Cincinnati Car Company for the Georgia Railway & Power Company. *Courtesy of the Indiana Historical Society, Cincinnati Car Corporation Collection.*

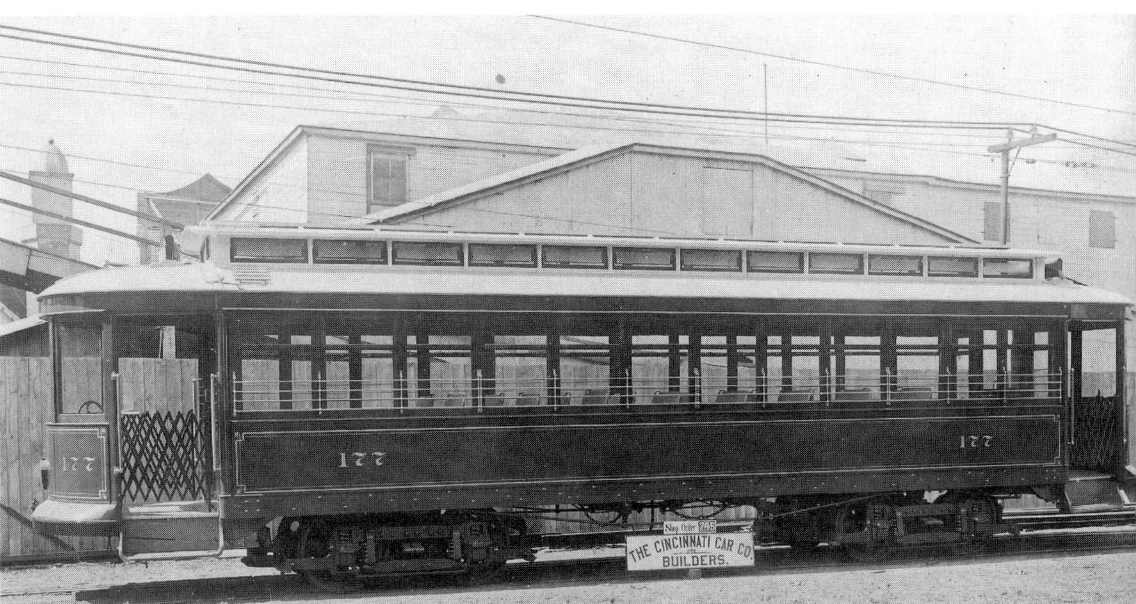

Figure 2.18. A rapid transit car built by Cincinnati for service on New York's Interborough Rapid Transit system. Note that the car is resting on "shop trucks." *Courtesy of the Indiana Historical Society, Cincinnati Car Corporation Collection.*

Figure 2.19. How Cincinnati loaded railroad flatcars to ship its vehicles. A flatcar was eased into a sloped pit, allowing completed streetcars to be pushed (carefully) from yard tracks onto the flatcar's deck. Here a single-trucked streetcar is being prepared for shipment. *Courtesy of the Indiana Historical Society, Cincinnati Car Corporation Collection.*

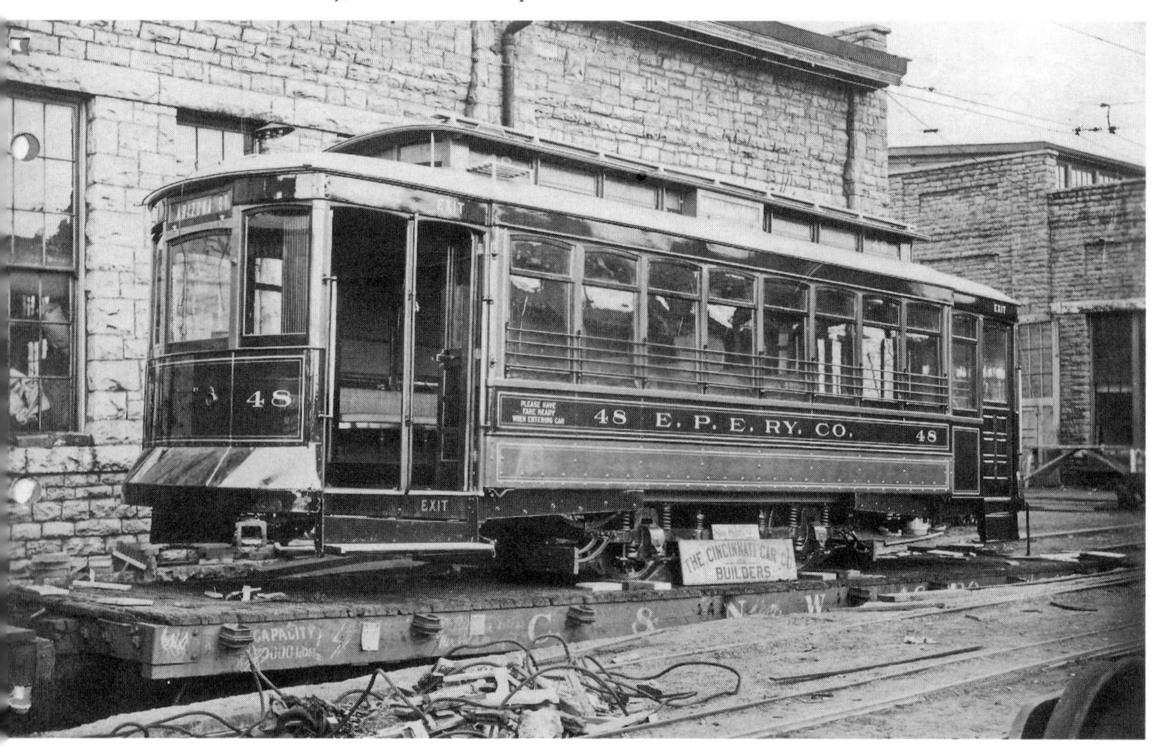

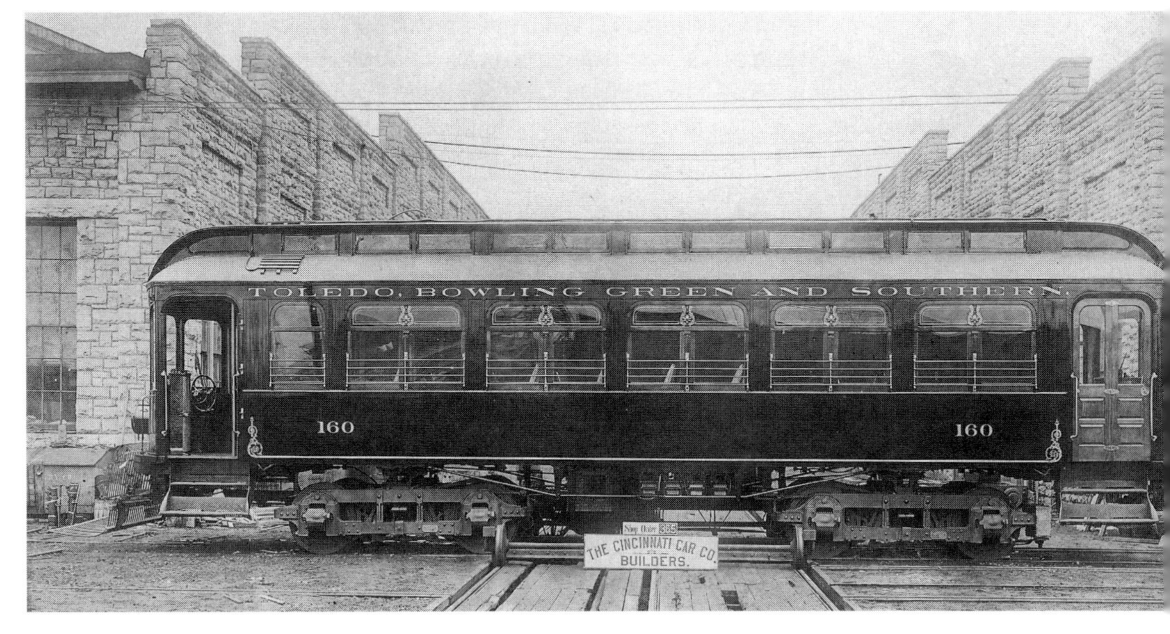

Figure 2.20. An interurban bound for the Toledo, Bowling Green, and Southern (Ohio) is being shifted into position on a transfer table at the Cincinnati Car Company. In a few moments it will be loaded onto a railroad flatcar. *Courtesy of the Indiana Historical Society, Cincinnati Car Corporation Collection.*

Figure 2.21. Another Cincinnati interurban ready for shipment. The Cincinnati Car Company owned at least a few of its own flatcars. Note the amusement park in the background. *Courtesy of the Indiana Historical Society, Cincinnati Car Corporation Collection.*

1902, the facility was dedicated to car maintenance. In 1902, a decision was made to replace all of Cincinnati's single-trucked streetcars with larger, double-trucked models, with all of the new cars produced at Chester Park. As this large project got underway, it became clear that despite the property's limitations, its facilities and staff were well suited to car building.

The Cincinnati Car Company was officially organized on 31 December 1902. It was a subsidiary of the Ohio Traction Company. Ohio Traction was a large corporation that controlled the interurban of the same name and Cincinnati's street railways. In January 1903 the new company built its first order apart from the Cincinnati street railway system, an undisclosed number of broad-gauged interurbans for the Cincinnati, Dayton & Toledo. City directories and the trade press recognized the new car builder the following month.

In 1906 Henry C. Ebert became the car builder's president. He was from Westinghouse originally, where he rose to the position of third vice president. During his years at Westinghouse, Ebert had been construction superintendent for the huge hydroelectric facility at Niagara Falls. He was also president of the Ohio Traction Company. In 1911, the Cincinnati Car Company expanded its shops. Two years later, it reorganized its personnel, with Ebert becoming manager of the company's sales department. He was replaced as president by W. Kelsey Schoepf, who was also president of the Cincinnati Traction Company (successor to Cincinnati Street Railway) and leader of a large syndicate that controlled street railway and interurban companies in Ohio, Indiana, and western Pennsylvania.[16]

Cincinnati enjoyed a broader geographic market than other Ohio car builders. Cincinnati built cars for 28 states plus railways in two Canadian provinces and Washington, D.C. Seventy percent of its output during this period was for streetcar lines, 20 percent for interurban railways, 6 percent for rapid transit lines, and the remainder for special vehicles (see table 2.5).

The Cincinnati Car Company was also the leader among Ohio car builders between 1909 and 1913, when it produced 1,821 street railway vehicles. Like Jewett and Kuhlman, most cars produced were streetcars (71 percent). The rest were fairly balanced among other car types, with 8 percent of its output comprising rapid transit cars and 6 percent each of interurbans, special cars (mostly combines), and maintenance-of-way equipment. Suburban cars made up the balance.

Ohio and Indiana accounted for 40 percent of Cincinnati's car orders during this period. However, no other state or Canadian province accounted for 10 percent or more of Cincinnati's orders. Although the Cincinnati Car Company already enjoyed a secure position in the industry in 1914, its most influential years were still in the future.

Niles Car & Manufacturing Company

The final streetcar builder to be organized in Ohio was the Niles Car & Manufacturing Company. It was also the shortest lived. Niles was incorporated with $250,000 of capital stock in 1901. George B. Robbins served as president, with A. G. McCorkle as vice president, C. P. Soulder as secretary, and William Herbert as treasurer.

The men given charge of day-to-day operations were all well acquainted with heavy industry. W. C. Allison, the general manager, was a veteran of mill and lumber opera-

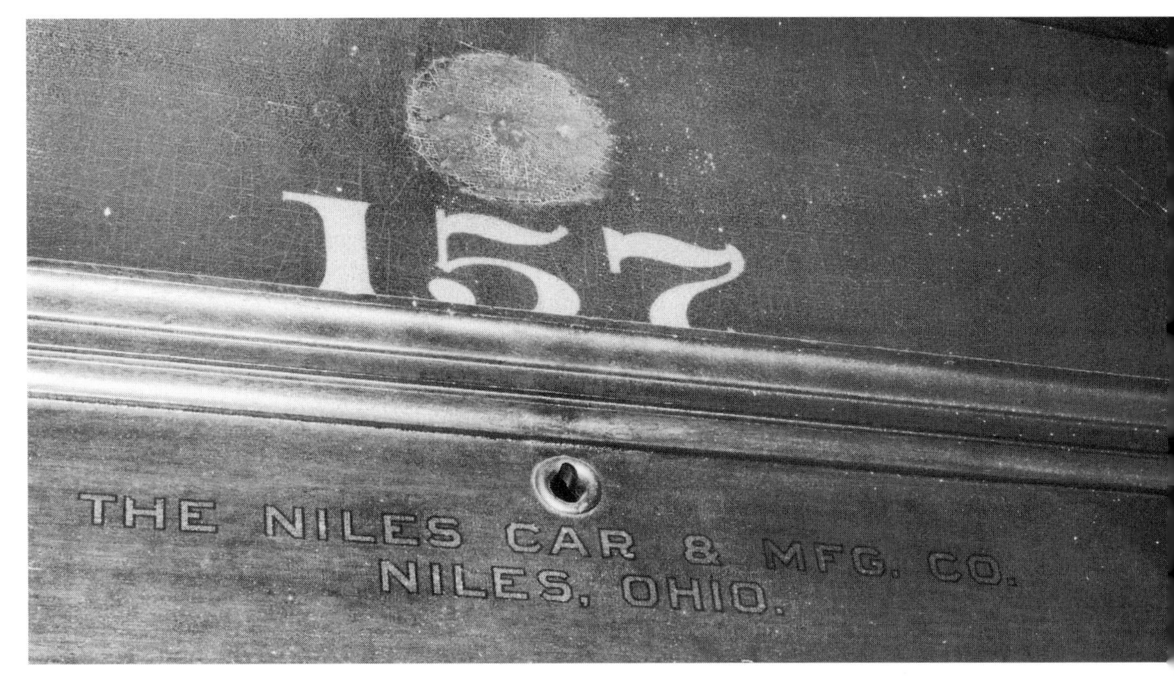

Figure 2.22. Niles's no-nonsense labeling on the interior bulkhead of an interurban. *Ohio Railway Museum, author photo.*

tions in the Niles area. His assistant, G. E. Pratt, had been a contracting agent for Jackson & Sharpe, a streetcar builder in Wilmington, Delaware. Before that, Pratt had been affiliated with the Star Brass works of Kalamazoo, Michigan. The company's general superintendent, A. L. Jacobs, was considered by the *Street Railway Journal* to be "one of the most modern car builders and designers in the industry." Jacobs' chief draftsman had 15 years of experience.

Niles intended to build passenger rolling stock of all descriptions to serve both the street railway and railroad markets. The car builder also intended to build bodies only, requiring customers to make arrangements for trucks and electrical equipment. Niles vowed not to become wedded to "antiquated horse-car construction," intending to take advantage of the latest advances in steel railroad car design and construction. For example, Niles used a novel method of car body construction. Ordinarily, cars were built starting with the floor framing. Niles decided to build both floor and body framing simultaneously. Once both were finished, they were joined. Although this took up more shop space (the frames and bodies were built adjacent to each other), the arrangement shortened production time.

The Niles plant enjoyed an ideal location for transporting raw materials and finished goods. The 7-acre site (smallest among Ohio's streetcar builders) was situated along the tracks of three major railroads (the Erie, the Baltimore & Ohio, and the Pennsylvania). The entire manufacturing complex was electrically illuminated and powered by both electricity and compressed air. All buildings were equipped with dust collection, steam heat, and fire protection systems.

On their own, the site's eight major buildings took up 4.5 acres. Two enormous erecting shops served as the heart of the complex, having a combined track capacity of

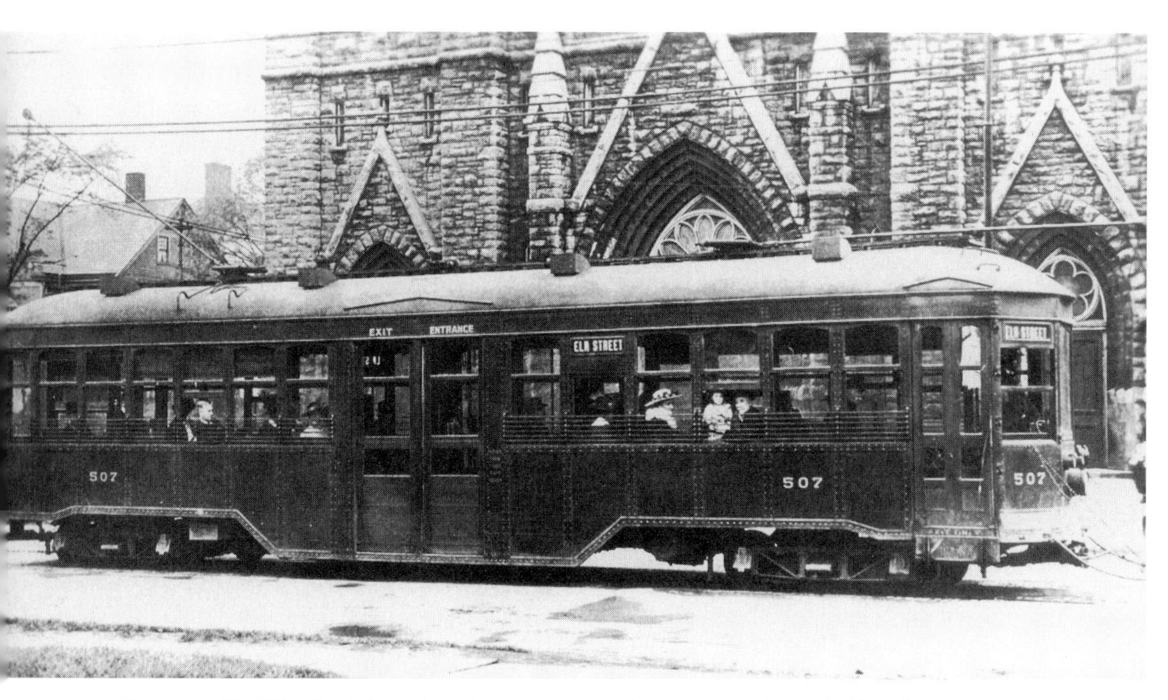

Figure 2.23. The Niles Car & Manufacturing Company was known primarily as a builder of interurban cars and was committed to building steel cars. Penn-Ohio car 507 was originally a center-entrance car built by Niles in 1915. *Courtesy of Branford Electric Railway Association.*

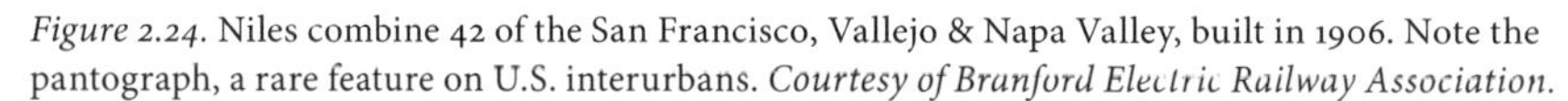

Figure 2.24. Niles combine 42 of the San Francisco, Vallejo & Napa Valley, built in 1906. Note the pantograph, a rare feature on U.S. interurbans. *Courtesy of Branford Electric Railway Association.*

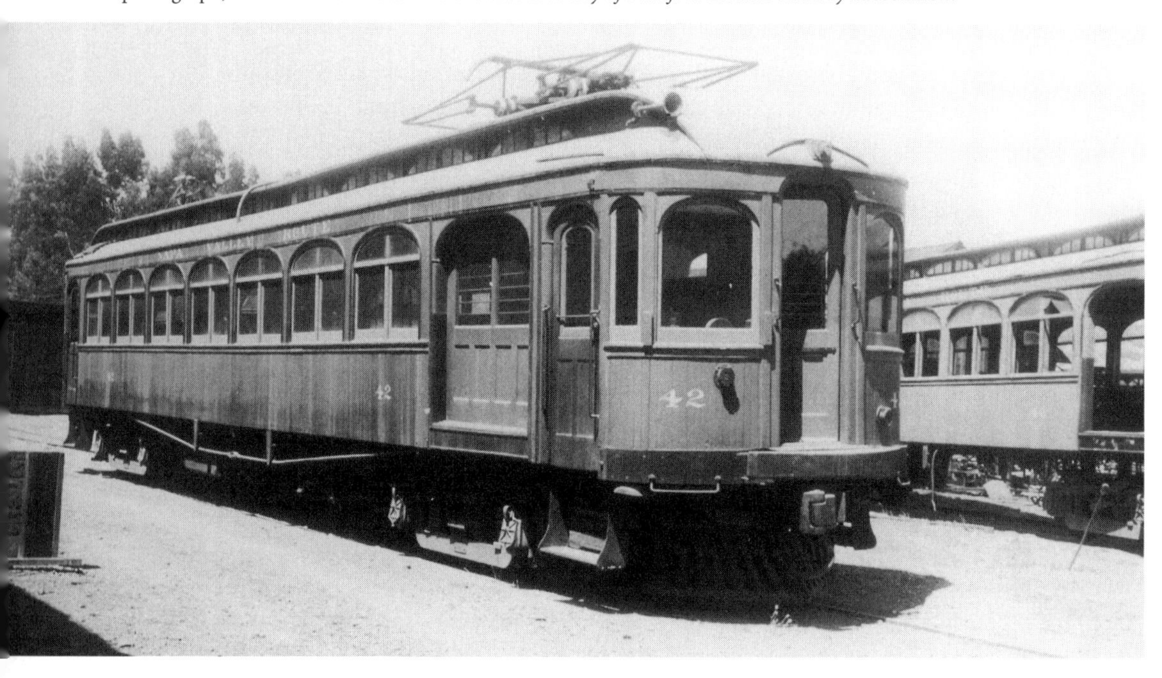

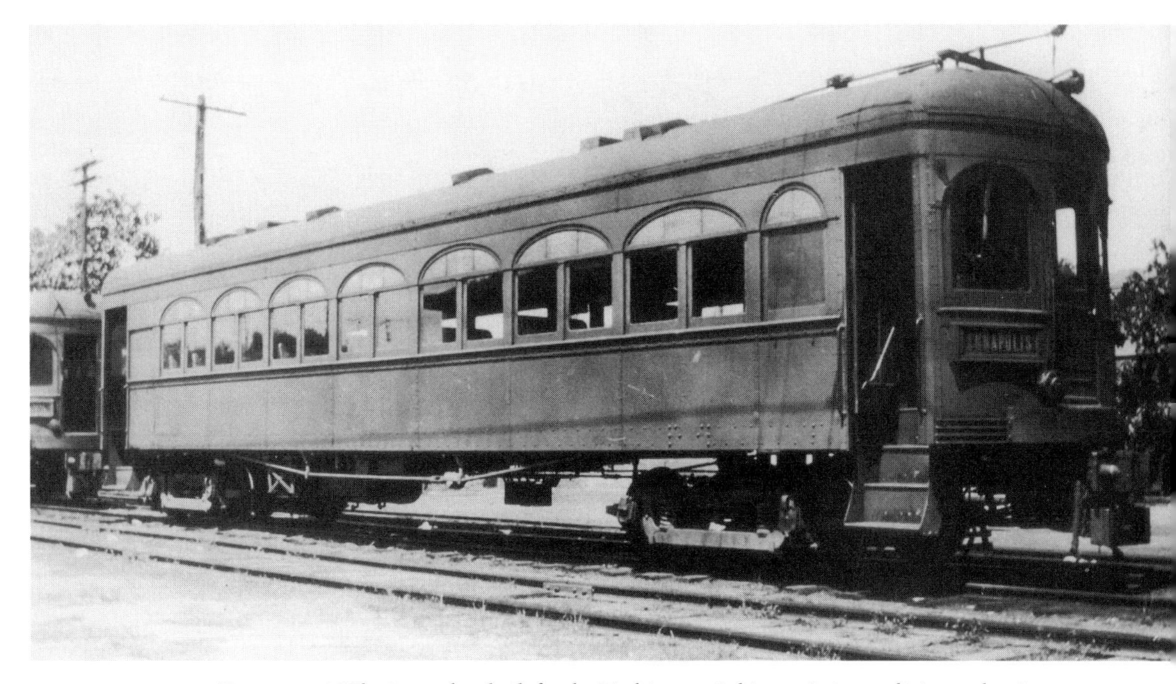

Figure 2.25. A Niles interurban built for the Washington, Baltimore & Annapolis interurban in 1907. This car originally ran on alternating current, but it was rebuilt to run on direct current in 1910. *Courtesy of Branford Electric Railway Association.*

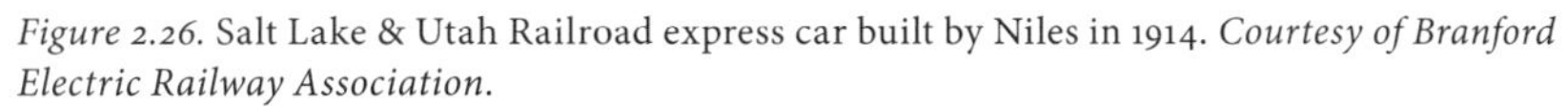

Figure 2.26. Salt Lake & Utah Railroad express car built by Niles in 1914. *Courtesy of Branford Electric Railway Association.*

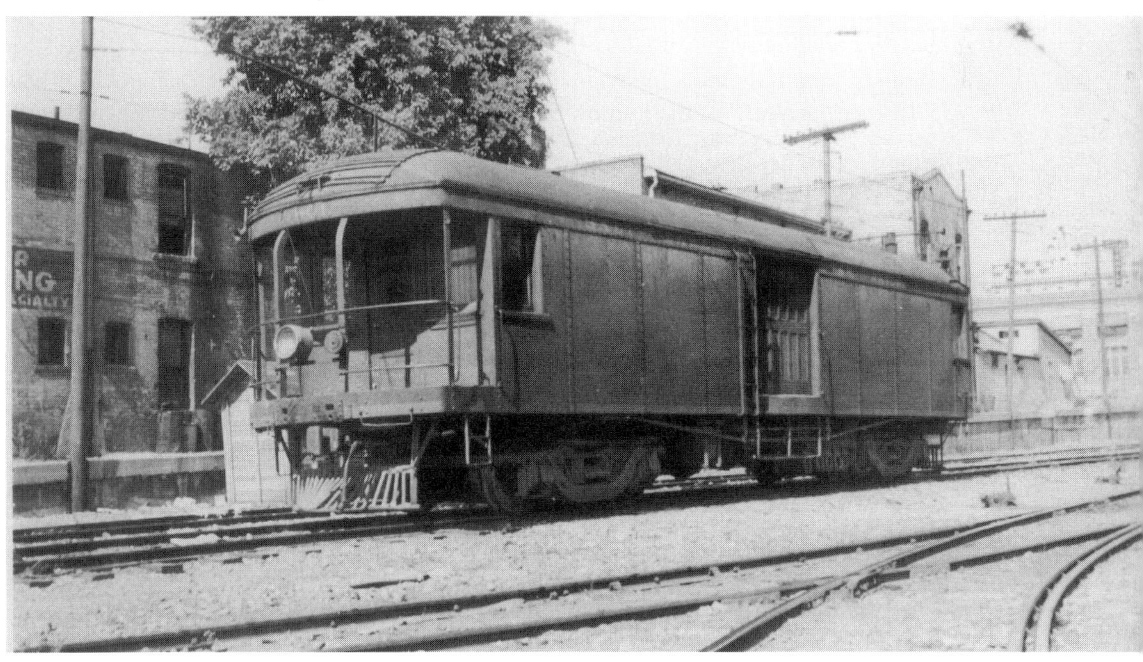

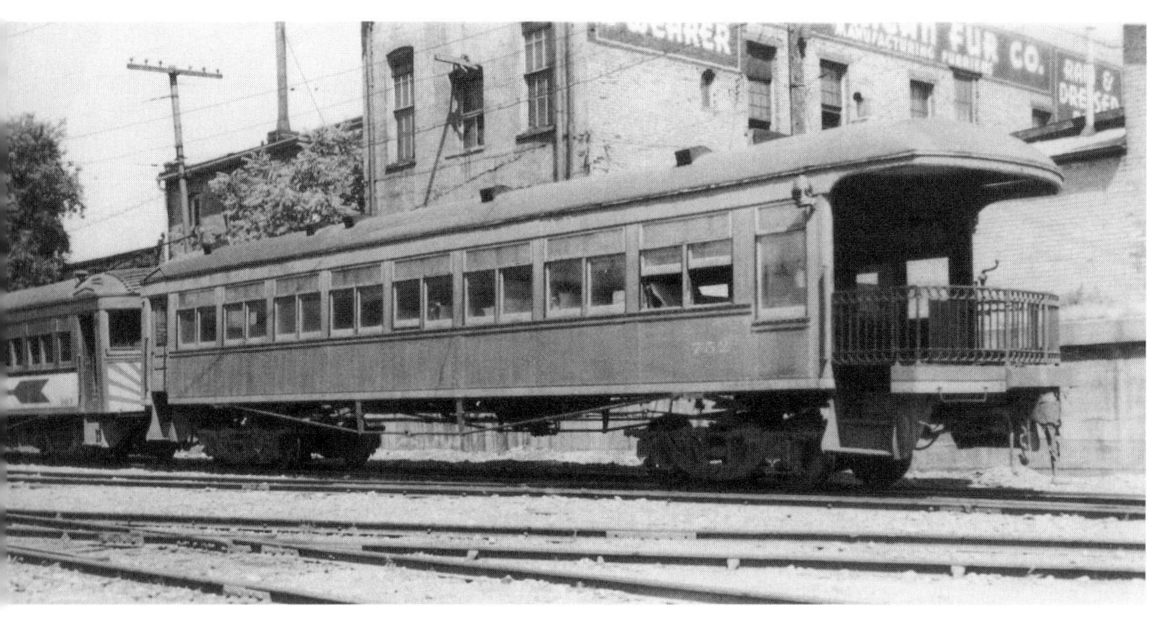

Figure 2.27. This Salt Lake & Utah Railroad observation car, built in 1916, was one of Niles's last cars. *Courtesy of Branford Electric Railway Association.*

83 streetcars. They were connected by an 80-foot transfer table. The paint shop, the only other building on the property capable of holding completed rolling stock, could accommodate as many as 25 streetcars at one time.

Despite its preference for steel car construction, Niles maintained a large carpentry shop. The car builder realized it was dealing with a conservative clientele, so it accepted orders for wooden cars to keep its shops full.

Early in 1904, the Niles Car & Manufacturing Company was reorganized, mostly to acquire additional capital to expand its production. A. W. Schall was retained, although a new individual, A. W. Ludlow, was engaged as general sales manager. Ludlow had been secretary of the Ludlow Supply Company, a Cleveland-based dealer in railroad appliances and maintenance equipment. The most immediate changes made to the plant after the reorganization were the rebuilding of the blacksmith and machine shops.[17]

Niles was the smallest of the Ohio car builders not only in physical size but also in output. Its orders accounted for only 10 percent of the five car builders' combined total. Niles gained its greatest reputation as an interurban manufacturer. Fully 520 of the 850 street railway vehicles it produced through 1914 (just over 60 percent) were interurbans (see table 2.6). Between 1909 and 1913 it produced only 379 cars: 54 percent were interurbans, 23 percent were streetcars, and 23 percent were special vehicles (mostly combines).

Niles shipped its cars to 18 states and to Canada. Michigan railways took the majority (28 percent), with an equal percentage going to railways in Ohio and Maryland (14 percent each). No other state or province accounted for 10 percent or more of Niles's output.

THE NATURE OF PRODUCTION, 1900–14

The Art of Car Building

Streetcars were never built entirely by one manufacturer. Car builders usually built the car bodies but relied on other manufacturers to supply electrical components (motors, controls, lights, wiring, etc.), trucks, brakes, and heating systems. Depending upon the breadth of a car builder's facilities, interior furnishings (such as seats and fixtures) might be manufactured in-house. Some large car builders had foundries that could produce trucks and subshops that could turn out a variety of furnishings and specialty items. However, even the largest car builders, such as Brill or the St. Louis Car Company, could build only about 75 percent of a streetcar.

The actual manufacturing process changed very slowly between the 1890s and 1914. The most noticeable changes in a streetcar built in 1914 from ones built years earlier were the car's size and materials used to build it. Initially, most car bodies were fairly short, usually around 20–30 feet maximum to accommodate a single truck beneath their bodies. Cars built after 1900 tended to have two trucks, thus enabling bodies to be

Figure 2.28. The milling shop at Kuhlman's Collinwood plant, located on Cleveland's East Side. One of the plant's features was the extensive use of electricity for illumination and machinery power. The arc light in the center of the photograph was probably made locally by the Adams-Bagnall Company. Note the precautions taken to collect sawdust. Kuhlman "recycled" its sawdust in the furnaces of its power plant. *G. C. Kuhlman Car Co.*

Figure 2.29. The Kuhlman cabinet shop in 1904. Stacks of window sashes are in the foreground, while the men appear to be assembling bulkheads. Hoods (or "bonnets") can be seen in various states of assembly in the upper right of the photograph. *G. C. Kuhlman Car Co.*

longer (usually 40 to 50 feet in length). Most streetcars were made completely from wood until about 1905. From then on, steel began to replace wood.

Before the introduction of steel, a variety of woods were used to build railway vehicles. Woods generally fell into "structural" and "finishing" categories. The two most common types, oak and ash, could be used as either. Common structural woods included white and yellow pine, poplar, hickory, and rock elm. Common finishing woods included cherry, maple, red birch, and mahogany.

Streetcar bodies consisted of four basic structures: the frame, the sides, the roof, and the end platforms. Frames consisted of a number of longitudinal beams called "sills." The most important were the two side sills, so named because they ran the length of the car body at its sides. "The side sills are the beginning of a car. To them are mortised the end sills and cross pieces, and upon this structure the body is built."[18]

Other sills included the center sill, which was located along the car's "centerline," or halfway between the two side sills. Intermediate sills were located between the center and side sills. Cross sills ran widthwise, joining the side, intermediate, and center sills. The cross sills at the ends of the frame were called "end sills." Sills were originally constructed of oak. When oak beams became scarce in the lengths required (a side sill consisted of a solid piece of wood measuring *at least* four inches thick, eight inches deep, and anywhere from 20 to 50 feet in length), yellow (also called "hard") pine was substituted.

Body framing was generally done with ash. Although oak was once popular in car framing, it had several disadvantages. In addition to increasing scarcity, oak took a con-

Figure 2.30. The Kuhlman erecting shop in 1904. The cars in the foreground are in various stages of frame assembly, although all appear to have their corner posts in place. Cars in the upper left and background are in various stages of having their bodies framed, and a hood is already receiving its canvas covering. Cars in the center and center-right are having their roofing installed. One car on the far right appears to have its end platforms and roofing completed and is awaiting a trip to the finishing shop. *G. C. Kuhlman Car Co.*

siderable amount of time to season. Also, oak contains a large amount of tannic acid, which gradually eats away at anything metal. Since streetcar bodies experienced a number of different twisting strains while in service, it was imperative that all of the car body's metal hardware (nails, screws, and fasteners) remain intact.

The lower third of most streetcar body frames curved inward on cars built before 1900. This was a holdover from carriage design, where bodies had to curve inward to clear large-diameter wheels. This practice was retained by streetcars long after their wheels were located entirely under the car, and it finally disappeared in the twentieth century.

Once the body was framed, the car builder could install the floor and roof. Flooring was usually made of hard pine boards, while roof framing was fashioned from any of a number of structural woods. After the roof was framed, large panels were attached to the car's sides. Streetcar sides were originally divided into four quadrants (upper and lower left and right). Additional panels were required once longer car bodies came into demand.

The most common paneling wood was poplar. Poplar could also be cut into narrow strips and laid over the roof framing to form the car's roof, and it was occasionally used in thin sheets for headlining (the "inside ceiling" of a car's roof). Once the roofing was

Figure 2.31. The Kuhlman finishing shop in 1904. Note the stacks of window sashes awaiting installation in front of the car to the right. *G. C. Kuhlman Car Co.*

in place, canvas was laid over the roof, tacked down, and given several coats of paint for waterproofing.[19]

End platforms could be added at any time during the construction process once the sills were in place. A common time for this work to begin was once the corner posts and/or end bulkheads of the body framing were in place. Short, heavy beams called "platform sills" or "knees" were attached underneath the ends of the side sills. Platform framing was built atop these sills.

Since most platforms were rounded at the front, a number of panels (as many as eight) were required to side the platform. The platform roof, called a "bonnet" or "hood," was usually preassembled in a car builder's cabinet shop and attached to the platform in one piece. The only different wood used in platform construction was hickory or elm, which was bent to form the rim of the hood. Before 1900, platforms were open, leaving motormen exposed to the elements. Enclosed platforms were adopted after 1900, largely due to Progressive-era legislation intended to protect car crews.

Interior finishes were installed according to the railway owner's taste and could involve a wide variety of finishing woods. Oak and ash were common choices, and oak was especially popular with owners desiring a plain, utilitarian appearance. Oak could be used as a plain finish (like ash), or it could be "quarter sawn" to produce striking diagonal patterns in the grain. Cherry and maple were other common finishing woods. Mahogany was highly prized for both its appearance and durability, but it was also used at great expense due to its scarcity.

After the basic streetcar body was completed, the car could be placed on its truck or trucks and fitted out with its electrical equipment, wiring, and plumbing for air brakes

Figure 2.32. A completed streetcar on the transfer table between Kuhlman's erecting and finishing shops in Collinwood. *Courtesy of Cleveland Public Library.*

and/or hot water heaters, as well as seats, windows, and the like. Some car builders did at least part of their painting on the shop floor, while others did all of their painting in a separate shop. Most car builders had a separate finishing shop where the final coats of paint were applied.

Car painting was an arduous task. During the days of wooden streetcars, up to 15 coats were required, not counting all of the striping, lettering, and other designs. All of this was applied by hand by skilled craftsmen. Even maintaining car bodies during this period was an ordeal. Anywhere from 9 to 14 days were required to apply six to nine additional coats once the body had been stripped.[20]

Unlike some types of manufacturing, such as automobile production, which lent themselves easily to standardized mass production techniques, there were clear limits as to how far streetcar production could be standardized. Car builders needed to be as flexible as possible.

Each street railway company had its own ideas regarding car design and appearance. Although the basic design and appearance of streetcars changed little from 1900 to 1914, car orders that looked similar outwardly were bound to have numerous variations. Added to this were changes in building materials as the new century progressed.

Because of this, car builders existed in a market that was based more on an economy of scope than one of scale. They often had to work on multiple orders that differed widely from each other simultaneously. Vehicles varying in size, materials required, and even type could be seen taking shape on the same shop floor on any given day.

The Business of Car Building

Car builders often enjoyed peculiar bonds with their customers, with some cities dealing almost exclusively with a given firm. It was common for this relationship to be based upon geographic convenience (such as Kuhlman and Cleveland or the Cincinnati Car Company and Cincinnati), but was not always limited to such. The Los Angeles Traction Company, for example, ordered most of its streetcars from the St. Louis Car Company in St. Louis, Missouri.

Car orders varied in size from as few as 1–5 cars to over 100. Custom orders of one or two cars, while not uncommon, were comparatively rare. Occasionally, small operators would "add onto" an existing order of a larger railway, merely having the car builder produce one or more identical cars beyond the larger customer's order. Rarer still, a particular car order could approach "standard" status, with numerous railways placing orders for identical cars.

Car builders usually had a representative in most major metropolitan areas. This could be either an individual agent or a brokerage firm. Builders also sent agents on sales trips, canvassing railways in a specific region in the hopes of securing orders.

When a street railway company was in the market for new equipment, it usually contacted a number of car builders, requesting bids. The railway specified the type of car, general dimensions, materials, components and furnishings desired from other manufacturers, and the delivery date. Also specified would be how much of the final assembly would have to be done by the builder and how much would be done by the railway (see table 2.7).

Once a bid was accepted, a formal contract was drawn up, specifying everything the railway required of the builder. If the railway owner provided no plans, the car builder drew up the required plans based on specifications in the contract. Working drawings were made from those plans, and once completed, materials were ordered and work began on the shop floor. When multiple orders occupied the shop floor simultaneously, the superintendent and foremen had to plan their activities carefully. The resultant coordination could resemble a chess match, especially when delivery dates coincided. Such a situation emerged at Jewett in spring 1906, when car orders were shipped to railways in Wilkes-Barre, Toledo, and Chicago in the same week.[21]

Although streetcar builders advertised, they often relied upon word-of-mouth to attract orders. Railway owners paid careful attention to each other, observing how well the vehicles of various car builders performed. Car builders realized that each order they filled could result in additional orders for this reason. In a 24 February 1917 bid solicitation from the Dayton, Covington & Piqua Traction Company, the Cincinnati Car Company was informed it was being approached by the interurban because it had "seen cars manufactured by you for the Dayton Street Railway Company in Dayton."[22]

With five streetcar builders in the same state, there was understandably some competition to secure orders. Few of the papers and correspondence from the car builders have survived, but those that have suggest the competition was stiff indeed. For example, on 3 May 1906, Jewett Car Company president W. S. Wright received the following letter from agent W. B. Wingertner:

> I enclose you herewith the contracts for two cars, lot 185, Camden Interstate Ry., to be changed to combination baggage and a few other changes which you will note by looking over the specification. Also a sheet showing how I arrived at my price quoted, and which was accepted, $2988.00 each.

As you know the contract price for Camden cars was $3200.00, deducting for trucks $570.00 gives cost of bodies $2630.00, after deductions and additions it amounted to $2819.00 and I put on another profit $169.00 per car, which I trust will meet with your approval.

The terms are the best I could get these people to make and they stated if we would not accept them, they would buy Niles Cars through Johnson as Mr. McKnight [a Camden official] had been to Cleveland Monday and Tuesday and had borrowed or bought three cars from the Cleveland Electric which were delivered last night and I saw them in Bowling Green [Ohio] myself this morning. He was at the Kuhlman [Car Company] plant and had seen Johnson and Hanna [Niles Car officials].

As the price is very good and since the Bowling Green Ry. Co. had signed the contract and not Mr. McKnight, I have reason to believe that you will accept this contract as I made inquiries and found they have never gone back on a contract. The person I talked to was Mr. Harding, Cashier of the First National Bank of Bowling Green, and he spoke very highly of all the men constituting the Company and stated that they represent the wealthiest men of Bowling Green.[21]

Participation in Conventions

Starting in 1893, exhibitions were staged at conventions by manufacturers. The halls at most American Street Railway Association conventions lent themselves more readily to component manufacturers, but the car builders did their best to maintain a presence. This was not always easy, as the conventions shifted from city to city during the first 25 years of the association's existence. If a particular car builder's vehicles were running in a convention city, that car builder usually sent only representatives with literature and photographs (the cars engaged in actual service could speak for themselves). If circumstances were favorable (or if their product was absent on local street railways), car builders would send a display car to the convention. If a hall could not hold them, temporary tracks were erected outdoors.

Transporting display cars to a convention was always an adventure. Component displays could be crated and shipped to a convention with relative ease. Whole cars, however, had to travel as part of a freight train, either atop a flat car or on their own wheels. Problems multiplied if a particular move required interchanges with several railroads. The Jewett Car Company was disappointed when its display car for the 1899 convention could not make all of the necessary interchanges in time.

When it first entered the street railway market in 1894, Barney & Smith brought two of its cars to the association convention in Atlanta. The Dayton car builder's motive was twofold. Not only did it wish to announce its entry into the field, but Barney & Smith also wished to use the cars to demonstrate the number of components it was capable of producing.

The cars were furnished with spring seats of its own patented design upholstered in mohair. The trucks had been produced by the car builder's own foundry. The cars' durability was soon apparent to all, as they withstood not only the convention's crowd but also throngs of visitors from the Barnum & Bailey Circus, which was performing in Atlanta at the same time.

In 1906, the Niles Car & Manufacturing Company brought two interurbans to the American Street Railway Association convention in Columbus, Ohio, one of which was a parlor car. The parlor car was immense, measuring 67 feet long and 8 feet, 8 inches wide. Despite its size, the car was astonishingly light, weighing only 17.5 tons (cars of this size could be expected to weigh over 50 tons).

Like many parlor cars, the seats on this car were not fixed to the floor, but could be moved about. The spacious passenger compartment was connected to a baggage room and men's and women's lavatories via a corridor that ran along one side of the car. A small smoking compartment was located on the opposite side of the baggage room in order to prevent tobacco smoke from wafting into the other passenger areas. The finish was described as being very attractive, consisting of mahogany inlaid with "rare colored woods."

The Cincinnati Car Company, also taking advantage of the 1906 convention's location, brought a parlor car of its own. Nearly as big as the Niles car, Cincinnati's parlor car was intended for the private use of Cincinnati Traction Company president W. Kelsey Schoepf. Unlike the Niles car, Schoepf confined smoking to an open, 9-foot observation platform at the car's rear. It was separated from the main passenger compartment by a buffet. Passenger seats were high-backed and plush-upholstered. The seats on the observation platform were leather.

So popular were parlor cars at conventions that the Jewett Car Company once sent a "faux parlor" to the 1902 convention in Detroit. The car was actually the completed body shell of a conventional interurban. Jewett officials used temporary furnishings to give the interior the appearance of a parlor car. Convention delegates were impressed with the car's permanent finish, which was mahogany with rosewood and holly inlay.[24]

Changes in Streetcar Production

In appearance, the most noticeable changes that took place in streetcar design between 1900 and 1914 were the enclosure of the end platforms to protect motormen and conductors from the elements and the introduction of the arched roof. Originally, most street railway vehicles were built with a raised or "clerestory" section in the roof that added natural light to the car's interior and provided ventilation. Clerestories gradually fell out of favor as the twentieth century progressed. A point of structural weakness in the car body (as well as an added expense), clerestories were replaced with stronger arched roofs on most cars around 1910. Clerestories also came to be shunned by the riding public, because they tended to develop leaks as they aged. "Aisle seats" on such cars were often empty during inclement weather.[25]

The greatest change in street railway vehicles before 1914 was not in how they appeared but in how they were made. A major change that affected all streetcar builders was the transition from wooden to steel car bodies. Safety was an obvious motivating factor. Wooden car bodies tended to shatter or "telescope" inside each other during collisions. Damage to steel cars tended to be concentrated at the end platforms, leaving the passenger compartments intact. Wooden cars were also highly combustible, especially as they accumulated numerous coats of paint and varnish. Although not as great a concern on cars that ran outdoors, the possibility of fire was a major concern for subways.

Steel also introduced greater structural strength to car bodies. It has been estimated that in some cases, wooden car bodies could only support loads amounting to about half their weight. Steel was quick to gain acceptance on cars that were large in both size and passenger capacity.[26]

In addition to safety and strength, economy was another major concern. The price of structural woods, particularly the long beams required for the side sills, increased

Figure 2.33. A car frame being assembled by the Cincinnati Car Company, illustrating the transition from wood to steel. Note that the end platform has been installed as an "appendage" to the main body framing. By the 1920s, car ends were an integral part of the car's frame. *Courtesy of the Indiana Historical Society, Cincinnati Car Corporation Collection.*

dramatically during the early 1900s. Finishing woods, however, continued to be used to "cap" exposed metal in car interiors into the 1920s.

The appearance of steel in railroad cars as early as 1896 led to a new trend in both the railroad and street railway industries. As steel cars became the new standard, new car builders, such as the Pressed Steel Car Company and the Standard Steel Car Company (both Pittsburgh firms) started up with plants designed specifically to build the new types of cars. Existing car builders either were forced to adapt or risked going out of business.[27]

The introduction of steel to streetcar body construction occurred gradually. Steel was first used in floor framing to replace the wooden sills. Car bodies that had wooden bodies and steel or partially steel frames were known as "composite cars."

Once all-steel frames became commonplace, the body framing became the next part of the car body to be replaced by steel. (As early as 1900 the *Street Railway Journal* was questioning how long ash would remain in plentiful supply.) Sheet steel made ideal flooring and side panels. The last part of a car's structure to be made of steel was the roof, and the use of steel roofs was by no means universal. (In some cases, wood was used in roofs into the mid-1930s.) Steel cars with wooden roofs or with steel reinforcement became known as "semi-steel" cars.

An interesting phenomenon of the transition period were "faced" wooden cars. As their name implies, "faced" cars were nothing more than wooden cars with thin sheets of steel fastened to their sides. While this might have given the appearance of greater safety and protected the wood underneath from the elements, the sheets did little or nothing to improve the structural integrity or safety of the vehicle.[28]

Ohio's car builders met the challenge of adapting to the new technology with varying degrees of success. G. C. Kuhlman made a relatively easy transition. By the time the car builder relocated to Collinwood, steel was becoming more common in railroad rolling stock. It is possible that Kuhlman officials anticipated a similar shift in the street railway industry when the new works were designed. Also, unlike other car builders, Kuhlman was a subsidiary of a larger corporation when it made the transition, roughly between 1908 and 1910. Brill was able to coordinate the conversion among its various plants, limiting the number of plants that were down at any one time and diverting orders to those that were kept open.

An examination of Jewett's orders indicates that it made the transition to steel around 1908. Like Kuhlman, since Jewett's Newark complex was relatively new, it is possible that the Jewett factory was designed with steel car construction in mind. Jewett had added incentive to make the change because rapid transit cars made up a significant percentage of its output.

Cincinnati took an unusual path toward converting its facilities to steel car production. In 1914 Cincinnati acquired the Armor Steel Foundry to aid in the manufacture of steel structural parts. Located near the Cincinnati Car Company, the foundry was remodeled and expanded. This proved to be a prudent move, as Cincinnati received an order for 128 rapid transit cars from the Chicago Elevated Railways shortly after this acquisition. The order specified all-steel construction.[29]

Niles intended to produce steel cars from the outset, and if anything its management was often frustrated with the street railway industry's slow acceptance of steel car design. As late as 1914, the following appeared in a 1914 Niles catalog:

> At this stage of the transition from wood to steel cars, we believe this to be the most practical construction. The entire and outside sheathing . . . [is made out] of standard steel shapes and plates which always can be obtained and which can be repaired or replaced in any railway shop. We do not recommend the use of parts pressed or forged from special dies which are liable to be out of existence when replacements are wanted.[30]

That same year, the *Electric Railway Journal* echoed these sentiments. Not only were some railways and manufacturers slow to adopt the new technology, but those that did merely replicated wooden car designs in steel. The *Journal* was critical of this practice.[31]

Barney & Smith was the last Ohio car builder to adopt steel car construction. This is somewhat puzzling, as the builder's principal product was railroad freight cars, which were the first type of rolling stock to make the transition from wood to steel. Since railroads made the transition *before* the electric railway industry, one might assume that Barney & Smith would have been the first to change.

Older companies that had made their reputation during the wooden rolling stock era must have recognized their need to adapt. Barney & Smith's reluctance appears even more surprising when one discovers that it probably built the first composite car order by Ohio car builders in 1895 (55 elevated cars for Chicago's Metropolitan Elevated).

There are numerous explanations behind Barney & Smith's reluctance to adapt. One of the firm's board members, W. H. Doane, was also president of a major producer of woodworking machinery, Cincinnati's Fay & Egan. Most of Barney & Smith's carpentry shops were equipped by Fay & Egan, and Doane may have been reluctant to leave a market that benefited his other interests.

One must also keep in mind that the Dayton car builder enjoyed a tremendous reputation for the quality of its woodwork. Arthur M. Kittredge, president of the company from 1906 until 1912, once stated that the public would not give up "the masterpieces of the woodworkers art." It is clear that the car builder was hoping to catch the few remaining wooden rolling stock orders that existed.[32]

Despite the conservative posture of its management, market trends, declining steel prices, and rising timber prices (despite its Georgia holdings) forced Barney & Smith's hand. In 1911, the company engaged in a huge retooling program, enabling the car builder to produce both wood and steel rolling stock. The lateness of the move would have negative consequences in the years ahead.

CONCLUSION

In his history of the St. Louis Car Company, Alan Lind divided the emergence of streetcar builders into two groups. The first were those founded in the early to mid-nineteenth century. These companies tended to be railroad coach builders that branched into streetcar manufacture to expand their markets. Of Ohio's car builders, only Barney & Smith fits this group. Ohio's remaining four car building firms were established between 1880 and 1910.

This tells us that Ohio's streetcar-building activity emerged and grew along with the street railway industry itself. By examining the output of these companies, we discover that at its peak, the industry provided enough business to support large firms with production on a large scale, as with Cincinnati and Kuhlman; midsized firms with midsized production, as with Jewett; and even small firms with relatively limited output, such as Niles. This was the industry and Ohio car building activity at its peak.

However, Ohio's car builders tell us only part of the story. Car building firms only represent a portion of Ohio's streetcar manufacturing activity. The next five chapters will examine component manufacture within Ohio and the influence they exerted upon the street railway industry.

Figure 3.1. The most common form of current collection on street interurban railways was the pole and trolley wheel. *Branford Electric Railway Association, author photo.*

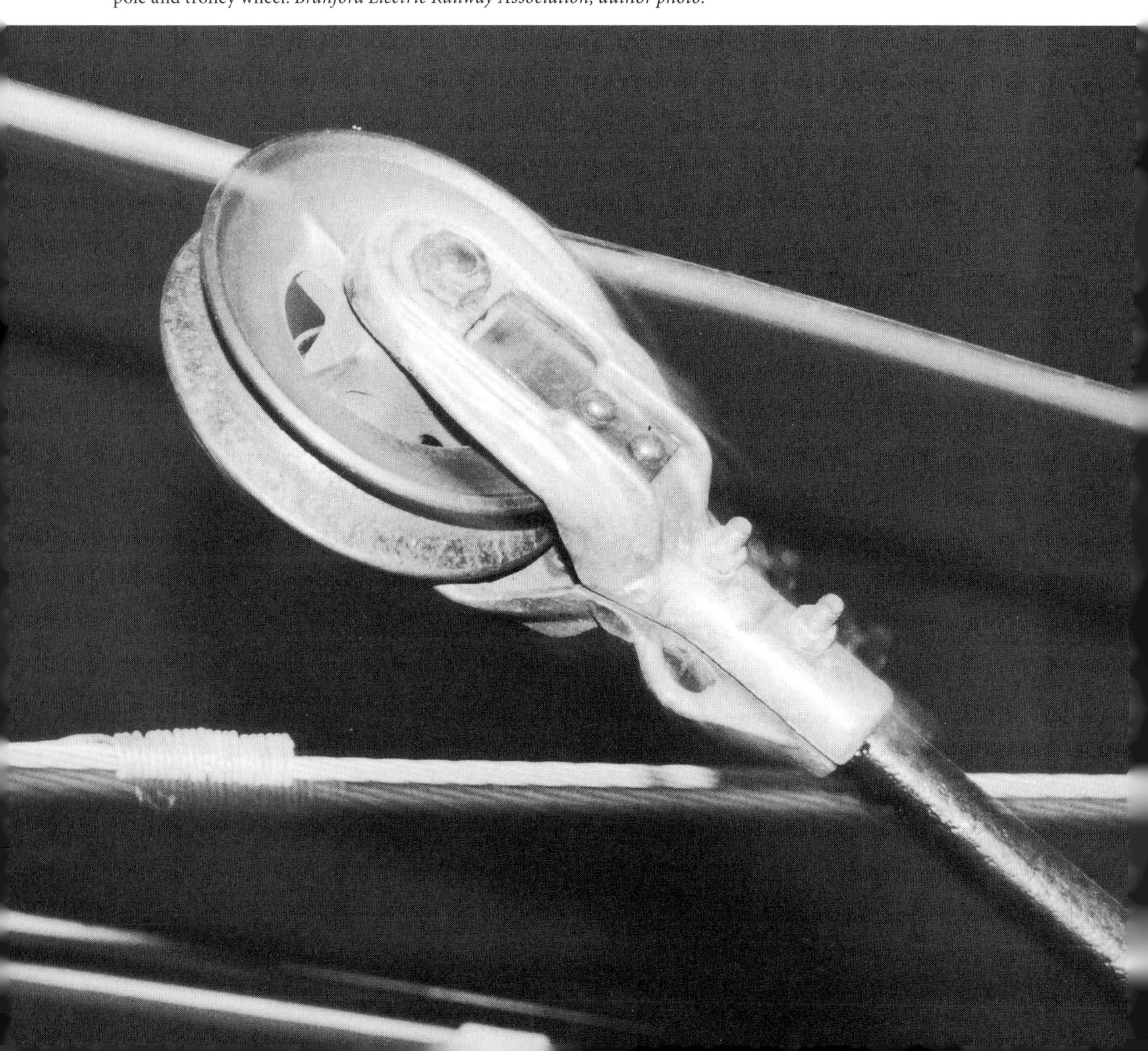

3

MAKING THE CARS GO: COMPONENTS ESSENTIAL FOR OPERATION

No single factory could build an entire streetcar. Scores of companies manufactured components for street railway operators and streetcar builders. Unlike car builders, who dealt directly with railway companies, many component manufacturers found themselves dealing with two types of customer: car builders and the railway companies themselves. Certain types of components were best installed by the car builder as the vehicle was being assembled. Others were best installed at the railway's own shops.

For component manufacturers in Ohio, two major patterns emerge. First, the presence of five active car builders in the state meant there was a considerable amount of local business. If a railway did not specify a particular manufacturer for a given component, the car builders did not have to look far to find a suitable supplier. Second, the state of Ohio was a significant market on its own. In addition to several large and numerous small and midsized street railway systems, Ohio contained more interurban mileage than any other comparable region in the world, making the state an ideal self-contained market for many Ohio-made components.

In addition to market patterns within the state of Ohio, several geographic trends also emerged. The presence of a major car builder in a community was no guarantee that component manufacturers would locate there. Of Ohio's five "car builder cities," three had numerous component-producing firms, but two did not. Other cities in Ohio became important centers of component manufacture without the presence of a major car builder.

Ohio-based manufacturers of streetcar components varied in their significance to the industry. In the case of motors, trucks, and control equipment, Ohio's greatest period as a manufacturing center was during the 1890s, while the industry was still maturing. By 1900, most motors and control equipment were manufactured by General Electric and Westinghouse, while trucks were made either by specialist firms located outside of the state or by very large streetcar builders, such as Brill or St. Louis. (Cincinnati was an exception.)[1]

This does not mean that Ohio firms did not make a significant contribution to the manufacture of essential operational components. In fact, several Ohio firms became industry leaders. There were three types of operational components through which Ohio firms distinguished themselves: couplers (which will be discussed in chapter 4), current collection apparatus, and gears and pinions.

CURRENT COLLECTION APPARATUS

One of the problems that plagued street railway pioneers was in finding a safe, reliable means of getting the electrical current to the streetcars. The most common method of current collection used a metal pole mounted on the car's roof. The pole itself was mounted on a base that was capable of swiveling in a complete circle (enabling the pole to negotiate sharp curves while traveling in either direction). The base contained strong springs, which were used to hold the pole against the wire. At the pole's base, power cables routed current through a main circuit breaker, a controller, a bank of resistors, and finally to the motors (the metal wheels and rails were used as a return to complete the circuit). The other end of the pole held a pronged device called a "harp." Attached to the harp was a grooved brass or bronze wheel that made contact with the wire.[2]

When not in use, trolley poles were secured under a hook on the streetcar's roof. A rope tied near the pole's harp enabled motormen to raise or lower the pole when necessary. Since trolley wheels occasionally left the wires, one of two devices was commonly used to keep the pole from whipping around. Trolley *catchers* were similar to seatbelts on today's automobiles. They had spools that payed out or reeled in rope to permit the pole to rise and fall gradually, but locked in place if the pole moved upward suddenly, as

Figure 3.2. Ohio Brass trolley wheel, side view. *Branford Electric Railway Association, author photo.*

it did when it left the wire. Thus the dewired pole was held in place until the motorman could stop and the conductor could pull the pole back under the wire.

Trolley *retrievers* were similar to trolley catchers. However, instead of its spool locking in place during a dewirement, the retriever reeled in the pole rope at high speed, slamming the pole onto the car's roof. Retrievers were usually found on interurban cars, where operating speeds were much higher than on city streets (a pole snagging a support wire at over 50 miles per hour could wreak tremendous havoc). Cars with retrievers often had protective structures at the ends of their roofs to prevent retrieved poles from damaging roofs and hoods.

Current collection equipment could be delivered either to car builders or to individual railways. Since poles and pole bases were vulnerable if left installed during transit, streetcar builders often installed them for testing, then removed them for shipping. Some railways simply ordered current collection devices and installed and tested the equipment themselves.

When one examines the photograph of a streetcar, the pole on the wire, if it is noticed at all, appears to be the simplest of devices. Few realize that they are actually looking at six components (wheel, harp, pole, pole base, rope, and catcher/retriever), which could have been made by as many as six manufacturers. Despite this complexity and the fragility of their appearance, current collecting devices proved to be quite durable.

For example, the *Electric Railway Journal* published a tally of all equipment failures in New York City for February 1911. Current collection devices accounted for less than 2.5 percent. The Hartford Division of the Connecticut Company published a similar

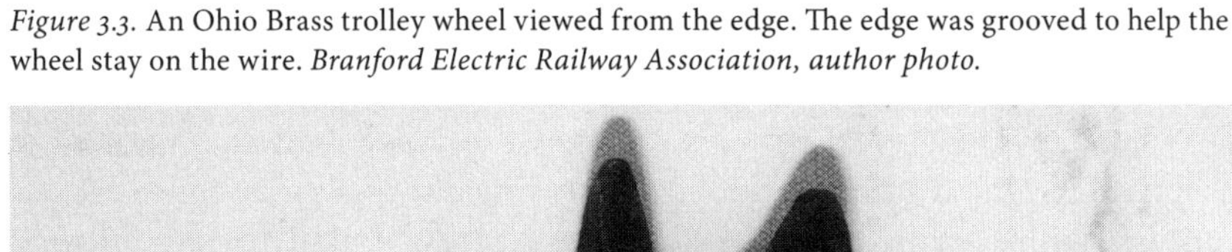

Figure 3.3. An Ohio Brass trolley wheel viewed from the edge. The edge was grooved to help the wheel stay on the wire. *Branford Electric Railway Association, author photo.*

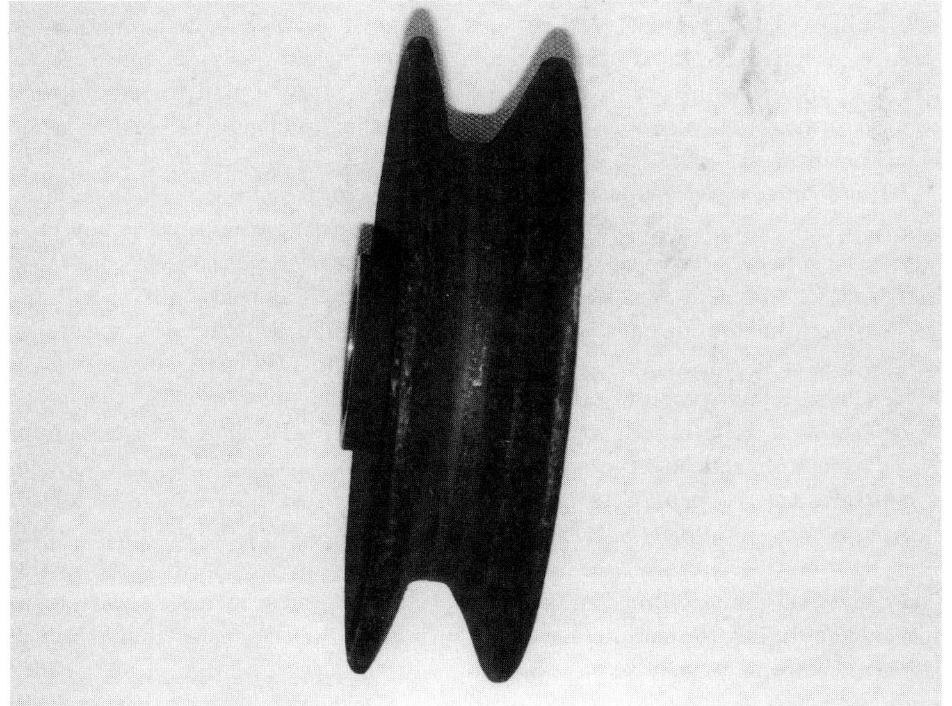

study for December 1911. Collection device failures there were under 1.5 percent. For the Middlesex & Boston Street Railway Company, failures of current collection devices accounted for 0.06 percent of all failures for the nine months between November 1909 and July 1910.[3]

Trolley Wheels

As little and unobtrusive as they might appear, trolley wheels had to be incredibly durable items, capable of withstanding the voltages they would encounter and the general wear to which they would be exposed. For instance, a trolley wheel 6 inches in diameter on a streetcar traveling at 30 miles per hour completed 1,740 rotations each minute. There were two points of friction: at the wire itself and where its rotating axle met the harp. Too much friction would result in premature wear and excessive arcing. Uneven wear might destroy the wheel and conceivably damage the overhead line if the wheel began to "bounce."

Bouncing wheels on an electric railway vehicle at speed were highly destructive. If the balance of the wheel was off by as little as 1 ounce, it could deliver hammer blows of 4.8 pounds to the wire at only 20 miles per hour. This force magnified itself greatly at 40 miles per hour and usually resulted in a dewirement at best and damaged trolley wires at worst.

Most wheels were made of bronze. Early wheels were made of steel, but it was discovered that steel was prone to excessive arcing and wear. Wheel sizes varied from 4.5 to 8.5 inches in diameter. Smaller wheels were preferred by companies whose vehicles traveled at low average speeds, usually on city lines. Larger wheels were used on higher speed lines, such as interurbans or rapid transit lines. Interurbans belonging to the Central Electric Railway Association (an association of midwestern railways) adopted the 6-inch wheel as their standard.

Trolley wheel manufacturers tried to strike a balance between maximum strength, minimal weight, durability, and low cost. This often led to designs that satisfied the criteria with varying degrees of success. At least initially, street railways purchased makes of wheels based as much on personal preference and whim as on actual performance.[4]

Two popular wheels made outside Ohio were the Ideal and the New Haven. Ideal trolley wheels were made in Buffalo by the Lumen Bearing Company. Unlike most wheels, Ideal flanges (the sides of the wheel) were made of "a special, soft, cold-rolled and pickled steel" that was supposed to minimize arcing. The tread of the wheel was a copper insert, and the hub on which the wheel rotated was made from a patented bronze alloy with graphite bushings. The two most common models were a 4.5-inch wheel intended for city service and a 6-inch wheel intended for high-speed service.[5]

New Haven trolley wheels were made by the Recording Fare Register Company of New Haven, Connecticut. A busy port city located on the Long Island Sound, New Haven was also a diverse manufacturing center, the home of Yale University, and the headquarters of the New York, New Haven & Hartford Railroad, the largest railroad in New England. New Haven was also the headquarters of the Connecticut Company, a subsidiary of the New Haven Railroad and operator of most of the state's street railways, totaling over 800 miles. It should come as no surprise that a national manufacturer of streetcar components would emerge in such a city. The New Haven wheel had a sealed bushing that required no servicing. Recording Fare Register claimed that its wheel

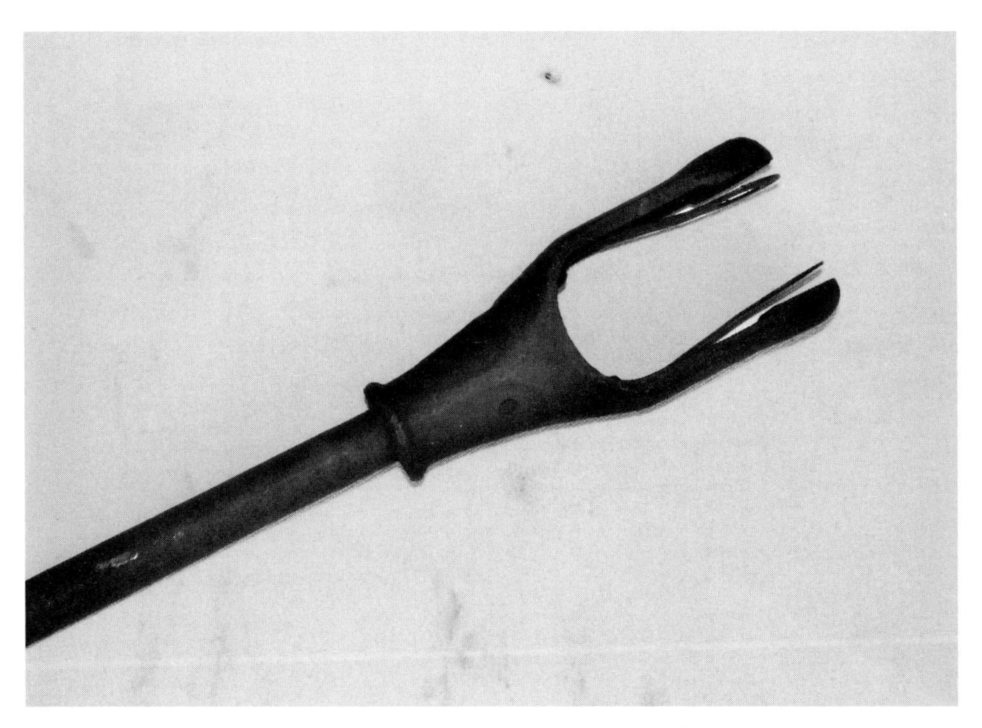

Figure 3.4. Ohio Brass trolley harp, without the wheel. *Branford Electric Railway Association, author photo.*

surfaces were known to wear out long before the bushings and that its wheels were 88 percent copper and contained no lead (a common wheel additive). New Haven wheels were available in three sizes: 4 inches, 4⅜ inches, and 5¾ inches. Among New Haven customers was the Schenectady Railway in New York.[6]

In Ohio, one of the more successful wheels was made by the Eureka Trolley Company of Ironton. Ironton is located along the Ohio River at the western edge of coal country, where its manufacturers could take advantage of both convenient river and railroad transport. Eureka drilled out one of the two prongs of its harp to act as a grease reservoir to ensure continual lubrication of the wheel. Its trolley wheels lasted an average of 20,000 miles. Ohio railways using Eureka wheels and harps included those in Delaware and Newark. Eureka's largest customer was probably the Brooklyn Rapid Transit Company, which ordered its wheels in lots of 10,000.[7]

The Standard Brass Foundry Company of Cleveland, one of the largest brass foundries in the United States, made its wheels entirely from virgin raw material rather than from scrap or material that was otherwise processed. Taking advantage of Cleveland's location on Lake Erie, the company stated that only "lake copper" was used in its products. Standard Brass wheels came in three sizes—4, 5, and 6 inches—and in two groove patterns–U or V.[8]

Other Ohio trolley wheel manufacturers included the Akron Trolley Wheel Company (which used ball bearings instead of graphite bushings), the Cleveland Trolley Wheel Company, and Ohio Brass.[9]

Finally, there were those large railways that preferred to make their own wheels. Chicago Railways made its own wheels from scrap material, casting them in six-wheel

Figure 3.5. Ohio Brass trolley harp, with wheel. *Branford Electric Railway Association, author photo.*

blocks. The Boston Elevated Railway made its own 4-inch wheels that lasted an average of 40,000 miles. Its casting formula was 90 percent copper, nearly 7 percent tin, and smaller amounts of zinc, phosphorus, and lead.[10]

Harps

The harp was a two-pronged device that connected the trolley wheel to the trolley pole. Some designs featured flexible spring-mounts for the wheel, which allowed it to flex inside the harp as the streetcar negotiated curves. This was often necessary, as a wheel held too rigidly would cut into the trolley wire, wearing out both wheel and wire.[11]

Harp springs were tricky affairs, as they needed to exert an even pressure on both sides of the wheel. If they did not, the wheel would engage the wire at an angle, even on straight sections of track, wearing out the wheel prematurely. The United Copper Foundry of Boston, for example, marketed harps with highly balanced springs.[12]

Other popular harp manufacturers located outside of Ohio included both the Lumen Bearing Company in Buffalo, New York, and the Recording Fare Register Com-

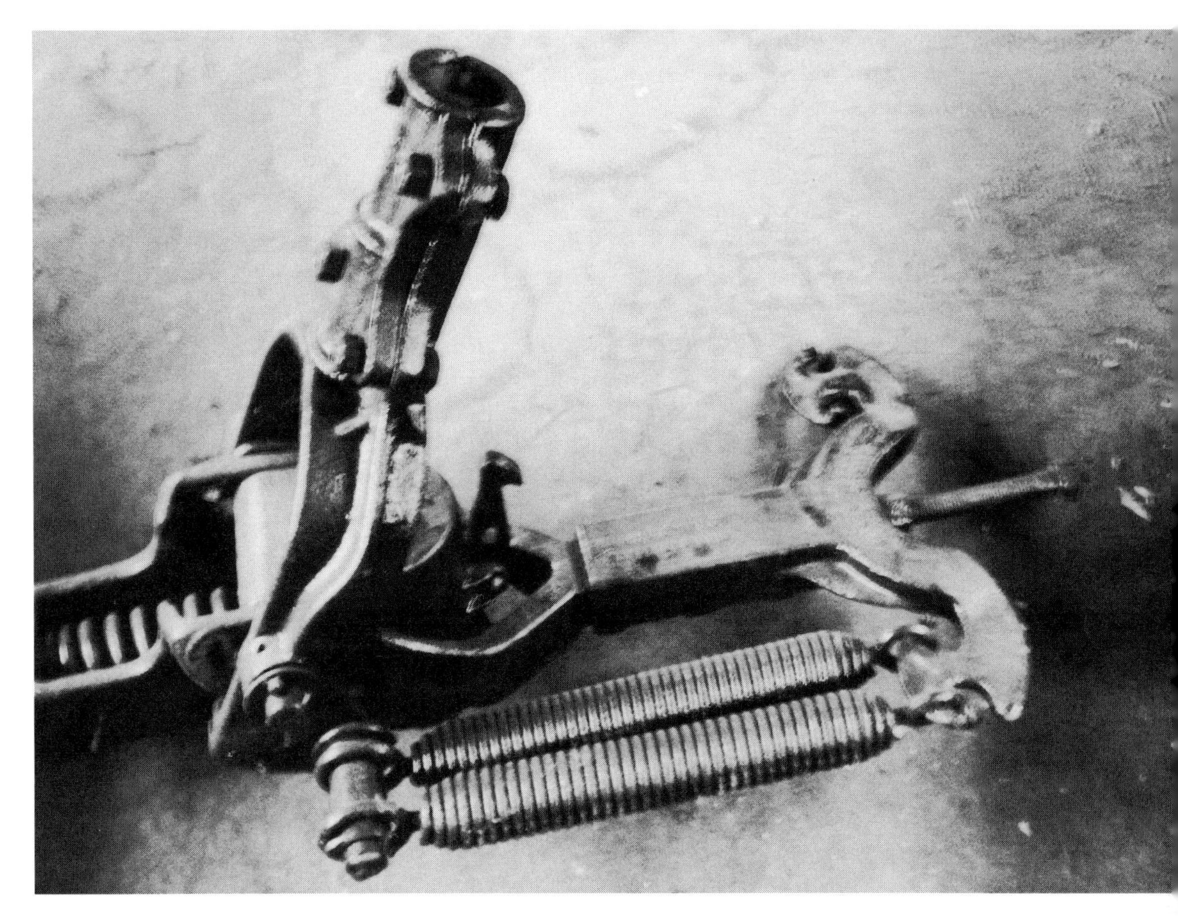

Figure 3.6. Ohio Brass trolley pole base. The pole fit into the socket in the left-center portion of the photo, where it was clamped in place. The pole was held against the wheel by two pairs of springs, one on each side of the base. In this view, the springs have been removed from one side. *Ohio Railway Museum, author photo.*

pany in New Haven, which made a harp with a hinge that reduced the amount of time required to change wheels. The Star Brass Works of Kalamazoo, Michigan, made both wheels and harps under the "Kalamazoo" label. Star did not have an easy time entering this portion of the street railway market, having to fend off legal challenges from General Electric for patent infringement of its harp designs.[13]

The Western Electric Company of Chicago included a harp in its railway product line that minimized the amount of bronze-to-metal contact. Critical of most harp designs, which featured both bronze and steel parts, Western Electric's only point of contact between the two metals was where a pair of steel dowel pins connected with the bronze hub of the trolley wheel. Lubrication was provided by wool packing saturated in oil.

The Western Electric harp was subjected to hundreds of tests and refinements over a period of three years before it was introduced to the market. It was designed for high-speed operation and had an estimated service life of 10,000 miles.[14]

Perhaps the best-known maker of harps among Midwest interurbans was the Bayonet Trolley Harp Company in Springfield, Ohio. Though only 20 miles from the larger city of Dayton, Springfield managed to retain its individuality even as its rival to the west grew at a faster rate. Agriculture was the city's primary economic driver, mostly due to the presence of four rivers in its immediate vicinity. The passage of the National Road through Springfield in 1832 and the subsequent addition of seven railroads ensured that Springfield would become an ideal location for industry.

However, the city remained true to its agricultural roots, and its first large-scale firms were flour and grist mills. The city later became known for its agricultural machinery, although by 1910, nearly half of the city's 10,000 industrial workers would be engaged in manufacturing activity other than agricultural. Prominent industries included publishing and medical supplies, Bayonet was another example of Springfield's diversifying economy.

Bayonet was concerned with two key design issues: ensuring the secure attachment of the harp to the pole, and allowing for quick and convenient wheel replacement. Bayonet harps were designed to be riveted to the pole, so there was little chance of the harp coming off (at least intact). Special springs allowed for quick wheel replacement, which the manufacturer claimed could be done in as little as 10 seconds.

Bayonet harps were common on interurbans in New York, Pennsylvania, Kentucky, Indiana, and especially Ohio. Bayonet also enjoyed a brisk foreign business. For example, during the fall of 1908 the company sent 200 detachable trolley harps to Lima, Peru. Two major Canadian railways (the Winnipeg Electric Railway Company and the Dominion Power & Transmission Company of Hamilton, Ontario) adopted Bayonet's harp as standard. In 1907 Bayonet built a new plant in Springfield. In addition to being larger and better laid out, the plant also featured "a full line of labor-saving machine tools."[15]

Another Springfield manufacturer of harps was the Economic Manufacturing Company. Economic's harp differed from conventional harps in that it had a spring-tensioned "wrist action" that allowed the wheel to handle abrupt changes in wire heights more smoothly. In Cleveland, the Holland Trolley Supply Company offered a harp that was so large that its prongs had flanges that extended beyond the wheel itself. Holland felt wheel life could be prolonged if the harp took the brunt of the punishment.[16]

Trolley Poles

When asked why the Boston Elevated Railway made its own current collecting devices, a maintenance official replied: "To find a trolley pole that will work without bending the trolley base, that will always be in good operating condition, and a trolley wheel that will stay on the wire and not pull the overhead down, is one of the hardest maintenance problems to solve."[17]

Clearly, the most obvious current collection component on a streetcar was the trolley pole itself. The pole is what most people noticed, and it is what gave the trolley its name in the first place (because the pole is dragged, or "trolled," along the wire by the streetcar). Trolley poles were also a major concern to street railway operators, as they were the most essential component used in current collection. This was not lost on manufacturers, but because of the pole's importance and inherent fragility, few firms dared to enter this part of the street railway market.

Trolley poles were basically long, narrow tubes of iron or steel. Lengths varied, depending on the type of car. Double-ended cars with a single pole mounted at the center of the roof tended to be longest, although railways using cars with low bodies and high wires also used long poles. Typical lengths ran from 12 to 15 feet. Pole diameters varied, tapering from a maximum of 2 inches at the base to less than 1 inch at the harp. Weight was kept down as much as possible, lest the poles overtax the car's roof. Weights averaged 18 to 30 pounds.

Good poles were supposed to be straight, flexible, and able to maintain good contact between the wheel and the wire. Flexibility proved most problematic, as poles could bend into any number of bizarre shapes if they dewired and caught a span wire while the streetcar was moving at speed (and sometimes even if the car was barely moving at all, depending upon the strength of the springs in the pole base). Backing maneuvers were probably the most common cause of pole damage, usually occurring when the wheel hit a snag in the wire and the motorman continued to back the streetcar, unaware of the problem. Conductors were often given instructions to hold onto the trolley rope whenever the streetcar passed through sections of known difficulty.

The *Electric Railway Journal* studied trolley poles in 1914. It was discovered that one of the most common problems was a "set," or permanent bend in the vertical plane that a pole took over time. It was also discovered that poles were most likely to take a set when they were being pulled off the wire and secured under the roof hook. This was partly the fault of the pole base. If base springs reached full compression before the pole reached the hook, trainmen were bending the pole to make it fit without realizing it. The *Journal* also discovered that heavier or reinforced poles were useless against this problem. Deflection-resistant poles simply broke apart when tasked too strongly.

In terms of elastic limits, most poles broke after bending a little over 24 inches. This usually occurred under 50 to 75 pounds of tension. Poles made of cold-drawn steel were stronger and could withstand as much as 110 pounds of tension and deflect as much as 32 inches.[18]

Most trolley poles were made by the National Tube Company. Based in Pittsburgh, National Tube made poles in all of the standard sizes, offering them in two models. "Standard A" poles were designed to withstand sets caused by compressed base springs. "Standard B" poles were reinforced throughout their entire length.

National Tube subjected *all* of its poles to stringent tests before they were shipped to customers. The thick end of the pole was secured in a restraining clamp with a graduated scale in the background. A reading was taken before the testing began. Weights were then added to the poles to measure their deflection (36–48 pounds for Standard A, 55–75 pounds for Standard B). If the pole did not return to its original position or "index" once the weight was removed, it was scrapped.[19]

The only Ohio city known to have pole manufacturing was Elyria. A small city located about 20 miles west of Cleveland and 3 miles inland from Lake Erie, Elyria was crisscrossed by a number of important railroad lines. Elyria's Garford Company subjected its poles to deflection tests before they were shipped. Garford was eventually acquired by National Tube.[20]

Pole Bases

Pole bases performed a couple of critical functions. They had to secure the trolley pole to the car's roof, and they served as the current's transfer point from the pole to the car's wiring. Bases needed to be strong enough to hold the pole to the wire, yet flexible enough to traverse curves without introducing excessive stress to the trolley wheel or overhead wire.

Among the most common problems experienced with pole bases were weak or broken springs, worn bearings, burned-out electrical contacts, and ordinary wear and tear. During a 1914 study of maintenance problems, the *Electric Railway Journal* discovered that most spring damage occurred during a base's installation. It was found that shop workers often inserted objects between the spring coils for leverage when hooking them into the base. Poor or nonexistent maintenance accounted for most of the other problems. Worn bearings occurred when bases were not properly lubricated, and burned-out electrical contacts occurred if the base was not cleaned regularly. Loose or missing bolts were also attributable to poor maintenance. In extreme situations, snagged poles could be ripped out of their bases entirely.[21]

Most makers of pole bases were careful to minimize the friction caused by the swiveling or rotative action of the base. General Electric bases used roller bearings to reduce friction. They came with two base springs that could be adjusted to vary the tension from 20 to 45 pounds. Railways of all sizes used General Electric pole bases, from large systems such as the Boston Elevated Railway Company and the Twin City Rapid Transit Company to smaller ones such as the Schenectady (N.Y.) Railway Company.

Another major manufacturer of pole bases was the R. D. Nuttall Company of Pittsburgh. Like General Electric, Nuttall used ball bearings to reduce rotative friction. Nuttall bases were manufactured under the "Union Standard" label. Models carried anywhere from one to five base springs.[22]

A number of Ohio manufacturers made pole bases. In Cleveland, Holland Trolley Supply came out with a base that used single springs of varying size and ball bearings. Unlike most pole bases, which placed their springs alongside the pole, the poles actually passed through the springs on Holland bases, taking up less space.[23]

Bases were also made in another of Ohio's major centers of component manufacture, Canton. Located roughly 55 miles southeast of Cleveland, Canton was a growing industrial center between 1900 and 1914. The seat of Stark County, Canton's population of 50,000 nearly doubled between 1880 and 1910. Its economy was driven by steel. Agricultural machinery, ceramic tile, and rubber products were also made there. The city became famous as the headquarters for Timken Roller Bearing, although this prestigious firm would not enter the railroad and street railway field until the mid-1920s.

Canton's Trolley Supply Company made four lines of pole bases under the National, Peerless, Simplex, and Star labels. All used ball bearings. The Simplex base featured a device that automatically adjusted the tension in the base springs, depending on the height of the wire. Normal operating tension was set at 32 pounds. It increased to as much as 40 pounds under high stretches of wire, and reduced the tension to as low as 15 pounds when the pole was being lowered by a trainman.[24]

In Bucyrus, the Milloy Electric Company departed from standard practice by making its bases out of malleable iron rather than steel. It used two tracks or "races" of ball bearings rather than the usual one. Tracks were kept sealed, eliminating (or so it claimed) the need to oil the base. Bases came with either two or four base springs, with an equal number of springs along either side of the pole.[25]

The Fixler Trolley Stand Company, located in the small southern Ohio town of Delta, was founded by Edgar R. Fixler, a former motorman. Frustrated with having poles come out of their bases, Fixler designed a base with a large, complex clamping mechanism that would not release the pole under any circumstances. Despite the clamp's impressive grip, Fixler claimed a pole could be released and changed out in as little as 20 seconds. The base also incorporated "holding clutches" that governed the vertical movement of the pole. Although the design was used successfully on the Toledo and Indiana Railroad (an interurban), the stand's complexity prevented it from coming into widespread use.[26]

Bayonet Trolley Harp also made bases and pole clamps. Unveiled in 1906, Bayonet claimed that its clamps enabled poles to be changed out rapidly, but advertised a more realistic time of 5 minutes to accomplish the task.[27]

Catchers and Retrievers

A popular maker of both catchers and retrievers was C. I. Earll of New York City, who began production in 1900. Earll catchers were designed for city service or lines with maximum speeds of 20–25 miles per hour. They were easily detached, allowing shop personnel to change the catchers out quickly for maintenance, or to allow crews to change ends if the streetcar was short enough to use the same pole in either direction. The first Earll catchers required trainmen to reset the catcher before rewiring the pole, which often required the car to come to a complete stop. Later Earll models, especially the retrievers, were designed to be reset while moving.

Ease of detachment, simple operation, and the fact that Earll catchers and retrievers had been around longer than most manufacturers, contributed to their popularity. Interurbans using Earll products included the Scranton Division of the Lackawanna & Wyoming Valley Railroad Company in Pennsylvania as well as portions of the Capital Traction Company's system in Washington, D.C.[28]

An unconventional model of catcher was produced by the Q-P Signal Company near Boston. Q-P catchers used a steel ball rolling around a ratchet. When the spool of trolley rope moved slowly, the ball was free to roll around. When the spool moved suddenly, the ball was caught in the teeth of the ratchet, locking the catcher.[29]

Another unconventional type of catcher was devised by W. H. Kilbourn of Greenfield, Massachusetts. Concerned with the safety of street railways, Kilbourn designed other safety devices as well, such as a rail sander. Unlike most catchers, which were housed in round metal housings, the Kilbourn catcher consisted of an iron block that slid up and down an iron rod. Poles were generally allowed 4 feet of play, but could be allowed to travel between 6 and 10 feet provided the pole moved smoothly. The device locked if a sudden tug on the rope exceeded 2 inches of travel.[30]

At least one manufacturer made conversion kits that allowed its catchers to be turned into retrievers. The Wilson Trolley Catcher Company of Boston entered the street railway market in the early 1900s. The conversion kits came out the same time that Wilson introduced its line of retrievers.

Some retrievers were air-powered. The Wason Engineering and Supply Company of Milwaukee (not to be confused with the Massachusetts-based car builder) made a retriever that used compressed air drawn from the car's braking system to pull the pole down to the approximate level of the roof-mounted hook. In order to prevent the pole from penetrating the car's roof, a small air cylinder was used to serve as a "shock absorber" to slow the pole's descent.[31]

Figure 3.7. Ohio Brass trolley catcher. *Branford Electric Railway Association, author photo.*

Perhaps the most complex and safety-conscious maker of air-powered retrievers was the International Trolley Controller Company of Buffalo. Sensitive to the damage a pole could cause if brought to the roof too quickly, International's retriever only dropped the pole 2 inches. Also mindful that trainmen were not always aware that a dewirement had occurred, International retrievers incorporated a brake interlock that threw the brakes into an emergency application the moment the pole left the wire. To alert others that the car was coming to an abrupt halt, the brake interlock also opened an air connection to the car's whistle, sounding it until the motorman was able to reset the equipment after the car had stopped.

In Ohio, the Trolley Supply Company of Canton made one of the best-known retrievers on the market. With the Knutson retriever, railways could not only adjust the distance it would pull down the trolley but they could also adjust how far the pole could be allowed off the wire before the device was triggered. Sudden rises could be set from 3 to 6 inches. Trolley Supply also made catchers under the Ideal label (see table 3.1).[32]

In 1913 Ohio Brass began marketing a special type of trolley catcher. Combining the best features of a catcher and a retriever, the catcher functioned conventionally, unless the pole came in contact with a span wire after the catcher locked. If this occurred, the catcher released, partially reeled in the pole, then locked again.[33]

Examining orders written up in *Electric Railway Journal* between 1909 and 1913, orders totaling 1,655 cars specified the type of catcher or retriever used. Trolley Supply accounted for nearly a quarter of these. Most of these orders were for Knutson retrievers, with over 40 percent going to the states of Ohio, Indiana, and New York.

Although one might be tempted to dismiss Trolley Supply as a local supplier of retrievers, two things belie the notion. First, of the next three states receiving Knutsons, only Illinois was close to Ohio; the other two were Florida and Texas. Also, Knutsons were very popular among interurbans, and the states of Illinois, Indiana, New York, and Ohio had the densest concentration of this type of railway.

A further argument against Trolley Supply being solely a local manufacturer is the market for its Ideal catchers. In terms of raw numbers, Ideals appeared on only 8 percent of the orders specified between 1909 and 1913, but none were located in Ohio. They went to railways in California, Michigan, Pennsylvania, Washington, and Mexico.

The Challenges of Winter

In his classic history of the street railway industry, Frank Rowsome related the numerous struggles experienced by Frank Sprague while building his historic system in Richmond, Virginia, in 1888. One of these was an unusually harsh winter storm that created huge icicles that hung precariously from the overhead wires. Taken aback by a hazard he

Figure 3.8. Knutson trolley retriever, made by the Trolley Supply Company. *Ohio Railway Museum, author photo.*

had not anticipated, Sprague was startled by the solution of Pat O'Shaughnessy, one of his mechanics. O'Shaughnessy stood on the roof of a streetcar using a broom to whack the ice off the wire before the streetcar's pole could reach it.

While O'Shaugnessy's method was effective, it was also dangerous, and for obvious reasons it did not become the standard operating procedure for any railway that had to deal with winter weather. For northern railways, winter provided a host of weather conditions that could shut them down.

While railways utilizing third rail systems had some success by spraying the conducting rail with solutions such as calcium chloride, this could not be used on overhead wires. Most railways resorted to contact devices that could scrape the ice off the wires. While there were a host of commercial products on the market, some railways opted to handle the problem in-house.[34]

Frustrated that none of the leading products seemed to work, railway operators in Newark, New Jersey, resorted to placing small squares of steel on the ends of their trolley poles. The squares did not work well either.

In Cincinnati, railway officials dealt with a considerable amount of sleet and ice each winter. The winters in southern Ohio are not as severe as they are in the northern portion of the state, but the region can be hit with snow. For the most part, however, winter temperatures in Cincinnati hover between the low and mid-30s. That and the fact that the city sits in a large river valley creates ideal conditions for wet, slushy weather.

The Cincinnati Traction Company decided to develop and manufacture its own sleet cutters, which it was doing by 1912. The railway's shops were extensive, so the task was not a daunting one. Furthermore, Cincinnati was unique among cities in that instead of using the rails as a ground, it utilized two parallel overhead wires (one supplying current, the other serving as the ground). As a result, the Cincinnati Traction Company had twice as much overhead wire and twice as many trolley poles as railways of identical size.[35]

Cincinnati Traction devised a sleet cutter that utilized a cast-steel harp and brass rotor. The cutter had a series of offset grooves that kept forcing the wire to each side of the harp in alternating fashion. Cincinnati Traction reasoned that if the grooves were not sufficient to break the sleet and ice from the wires, the percussive action of the harp and wire jumping about would be sufficient to shake the ice clear.

Most of the commercial products came in the form of wheels or devices that slid under the wire. One of the better-known makers of sleet-scraping devices was the Root Track Scraper Company of Kalamazoo, Michigan. Root's track scrapers kept rails clear from debris and light snow. For ice on trolley wires, Root developed a retractable device that was spring-mounted on the trolley pole just beneath the trolley wheel. The device consisted of three small scraping wheels that engaged the wire from both sides and underneath. A lanyard was used to deploy and retract the scraper. When deployed, the scraper came into contact with the wire in front of the wheel.[36]

To market this new device, Root formed a separate company, the Root Spring Scraper Company. In its first years, the device was received with enthusiasm, and 553 of the devices were ordered between January and May 1912. These orders are also useful to illustrate the dual nature of a component manufacturer's market. Of the 553, 144 went to car builders (123 to the St. Louis Car Company, the balance to Preston Car & Coach). The remainder went to individual railways, with the largest order (250) going to the Boston Elevated Railway. Other customers included railways in Virginia, Michigan, and Quebec.

Figure 3.9. Sleet cutters came in a variety of designs. This cutter was shaped like a trolley wheel, and it fit onto the harp in the usual manner. *Branford Electric Railway Association, author photo.*

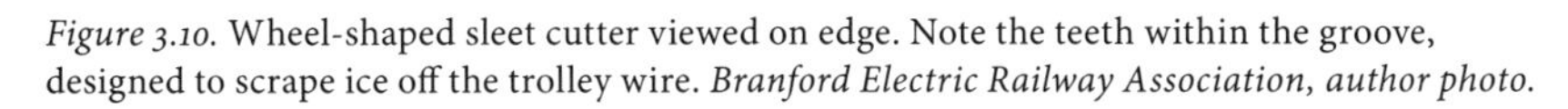

Figure 3.10. Wheel-shaped sleet cutter viewed on edge. Note the teeth within the groove, designed to scrape ice off the trolley wire. *Branford Electric Railway Association, author photo.*

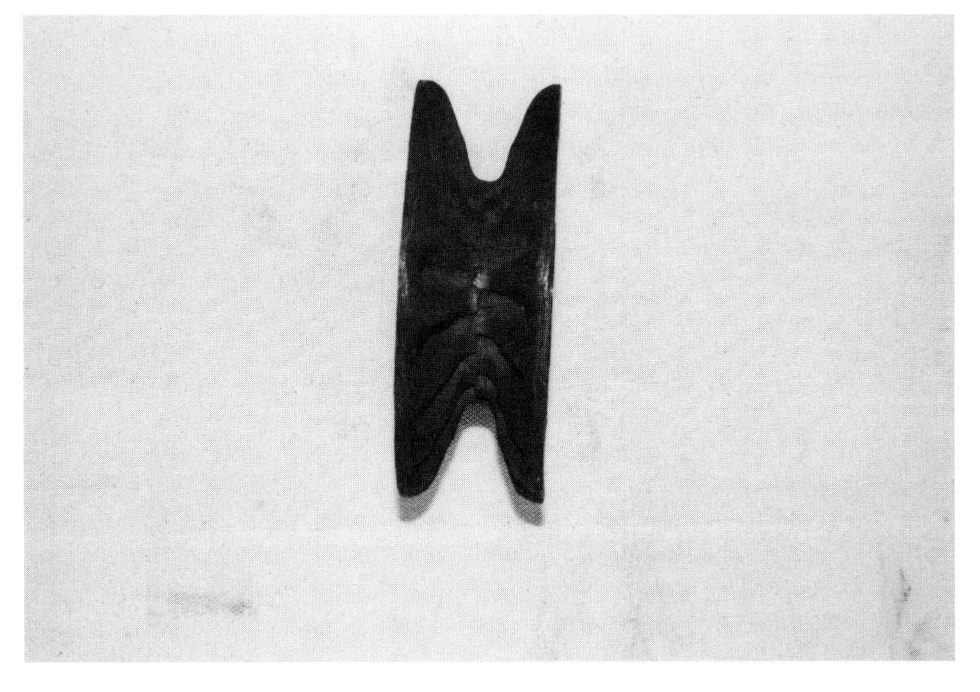

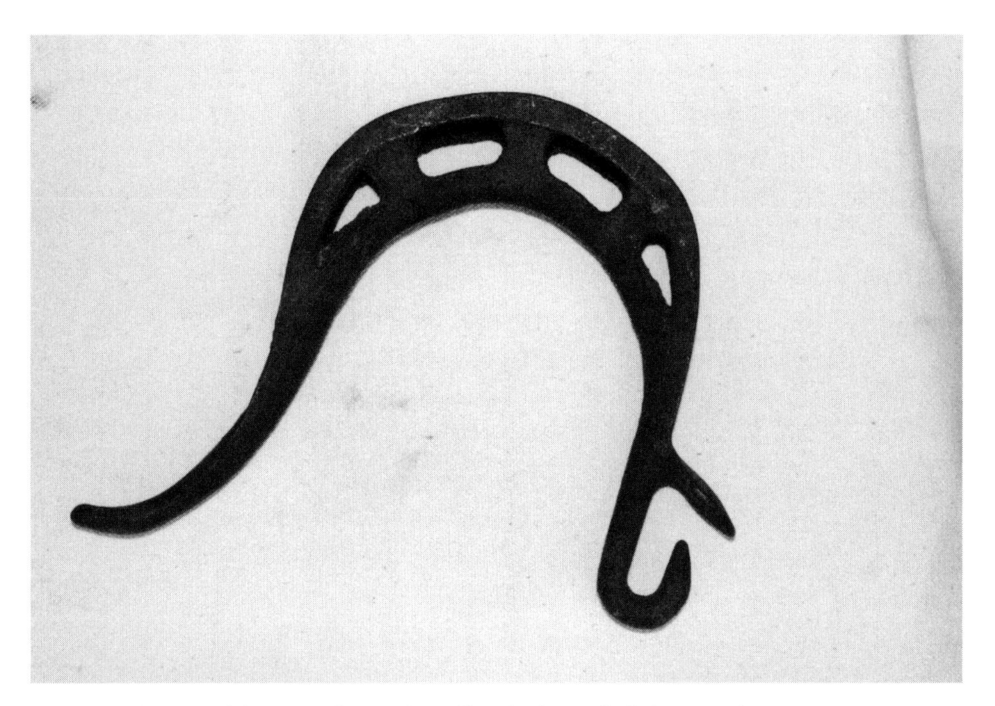

Figure 3.11. This type of sleet cutter fit over the trolley wheel. *Branford Electric Railway Association, author photo.*

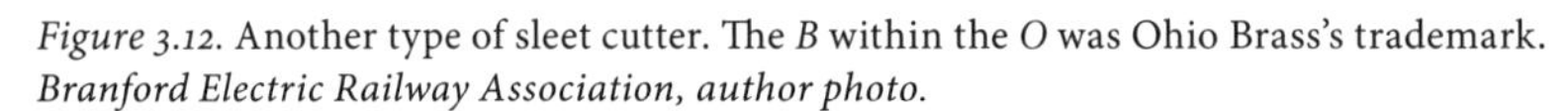

Figure 3.12. Another type of sleet cutter. The *B* within the *O* was Ohio Brass's trademark. *Branford Electric Railway Association, author photo.*

Of course there were also occasional novelty items that made the market. One inventor in Scranton, Pennsylvania, developed an odd Y-shaped pole that placed two contacts on the wire, one in front of the other to ensure a connection in icy weather. It did not catch on.

In Ohio, both Bayonet and Holland carried a line of sleet scrapers. Bayonet was known for lightweight scrapers that averaged 1–2 pounds instead of the usual 2–3 pounds. Holland offered both wheel-shaped and straight cutters to combat sleet and ice.[37]

GEARS AND PINIONS: THE SIGNIFICANCE OF THE UNSEEN

Not only were gears and pinions never seen, if properly designed, installed, and maintained; they were also seldom heard. Despite their lack of visibility, gears and pinions were absolutely essential to making the streetcars move. Once the current reached the motors, the motors' shafts would rotate. This motion was transferred to the streetcars' axles though gears and pinions.

A gear was a large, toothed metal disc mounted on the streetcar's axle. A pinion was a much smaller version of the gear, and it was mounted on the end of the motor shaft. As described in previous chapters, the motors were mounted with half of their weight supported by heavy springs attached to the truck frames, while the other half was supported by the pinions resting on the gears. This was called the "wheelbarrow" or "nose suspension" method of motor mounting, and it prevented the motor from shaking to pieces while ensuring that the gears and pinions stayed constantly enmeshed.

Gear and pinion life depended upon a railway's service conditions. If the tracks were standard gauge (the inside edges of the rails were 4 feet 8.5 inches apart), gear and pinion life was demonstrably longer than those on narrow gauge systems. On average, standard gauge gears lasted 135,000 miles while narrow gauge gears lasted 67,000 miles. Pinion wear was similar, with standard gauge lasting 42,000 miles and narrow gauge only 23,500 miles. Also, if a railway operated in cities with lots of hills, gear and pinion life was shorter than railways with mostly level tracks. On narrow gauge lines, the average was 104,000 miles versus 27,000 for gears and 43,000 miles versus 8,100 for pinions. Other factors affecting gear and pinion life included how well they were machined and bored at the time of manufacture, how well they were maintained in service, and whether or not they were made according to standard patterns or were customized.[38]

For the most part, railways inspected their gears and pinions regularly, lest they face the consequences. If the pinion bearings wore out prematurely, the pinion could start gouging portions of the motor itself. One New York manufacturer actually created a device that would automatically disengage the pinion if its bearings wore out.

A major concern among railways was gear ratios. The gear ratio is the number of complete pinion revolutions required to make one gear revolution, and it was a critical factor in determining how fast a streetcar could travel and how much power it would consume. In 1904, the *Street Railway Journal* lamented, "If there is any one sure way of wasting money in street railway operation, it is in operating cars geared for such a high maximum speed that they do not easily attain it between stops."[39] It was estimated that power savings of as much as 20 percent could be attained through the use of appropriate gear ratios and motors.[40]

Figure 3.13. A streetcar's main gear mounted on a wheel set. The motor's pinion gear rested on the large gear to the right of the photo. Half of the motor's weight was supported by the pinion, while the other half was supported by spring mounts attached to the truck frame. Thus supported, the motor's pinion remained constantly enmeshed with the main gear while the spring mounts absorbed any shocks as the streetcar traveled along its route. *Branford Electric Railway Association, author photo.*

The Brooklyn Rapid Transit Company discovered that due to the relatively close spacing of stations on some of its elevated lines, high acceleration with a lower top speed was much more desirable than lower acceleration with a higher top speed. Railways stopped short of adopting standard ratios, however, because they observed (correctly) that these were often dictated by the individual service characteristics of specific railways. Despite this, manufacturers tried to use standardized gear patterns whenever possible. General Electric and Westinghouse even went to the extent of having common-sized pinions on some motor models.

There was also considerable debate preceding World War One as to whether or not a gear should be solid (that is, a gear made in one piece) or split (a gear that came in two halves and was bolted together). Initially, all street railway gears were split. Solid gears were not used due to the difficulty in maintaining them. Like the wheels, solid gears had to be pressed onto their axles by force, but they had to be pressed deeper into the axle in order to leave room for the wheels. However, the advent of the interurban in the mid-1890s introduced heavier cars and higher speeds to the industry, which often proved too taxing for split gears. Solid gears were also preferred on rapid transit lines.

All manner of materials were tried in the manufacture of gears and pinions. Cloth and rawhide were used in an effort to reduce noise, but when these wore out too quickly, it became obvious that some sort of steel would be required. By 1914, at least four major types of steel were used.

One of the things that hindered the street railway industry's understanding of gears before World War One was the lack of any discernable standards, either in the types used or in the method of handling them. In a 1908 survey of railway companies, the *Electric Railway Journal* determined that only slightly more than a quarter (26.5 percent) of all railways used any sort of standards at all. About 10 percent were noncommittal, and slightly over half used none. The remaining 9.75 percent had developed standards that were unique to themselves.[41]

This problem was addressed on many fronts, notably by the AERA Engineering Association, which conducted a series of tests on various types of steel gears and pinions. The strongest type of gears was determined to be those of case-hardened, hard alloy steel. These were steel gears that had been treated either chemically or with heat to make the outer surface harder and more durable. For pinions, heat-treated forged steel yielded the greatest strength (see table 3.2).

The American Society of Mechanical Engineers (ASME) also addressed this concern in 1908 by investigating gear and pinion failures on Interborough Rapid Transit (IRT) subway cars in New York City. The ASME determined that the problem with most pinions came with handling procedures in the IRT's shops. The proper method of installing pinions on motor shafts was to "sweat" them on. In other words, a pinion was heated until the metal expanded enough for it to be slipped onto the motor shaft and then sprayed with water until the metal cooled and shrank enough to stay affixed to the shaft. In most cases of pinion failure, the ASME discovered that shop mechanics had abused the pinions by driving them onto the motor shafts with sledge hammers (one wonders how the motors survived).[42]

Gears and Pinions in Service: City and Interurban Lines

As mentioned above, streetcar gears were initially split, as the relatively small size and low speeds of early streetcars did not require anything stronger. As streetcars became larger, with longer and wider double-trucked bodies, and began to operate at higher speeds, solid gears began to make some inroads among city railways. One example of a city whose railways made the shift to solid gears was Birmingham, Alabama, in 1906.

To the north, the Rhode Island Company made the switch several years earlier, but not solely for reasons of gear strength. This railway discovered that unless they were machined and fitted precisely, split gears had a gap where the two halves met that did not match the teeth on the pinion. This gap became a point of premature wear on the gear and shortened the life of the pinion as well. If left too long, the gears became noisy and eventually delivered a rough ride for the passengers.

Other railways preferred to stay with split gears. In Kansas City, the Metropolitan Street Railway found that the split gears provided no trouble if the halves were attached securely.

On interurbans, most of the Ohio railways used solid gears, having had too many frustrations with split models. Both the Dayton & Troy Electric Railway and the Cleveland, Southwestern & Columbus received 300,000 miles of service from their solid gears. In Illinois, the Aurora, Elgin & Chicago (AEC) received well over 200,000 miles. It is interesting to note that the AEC also discovered it got longer service from smaller gears and pinions than it did from larger ones. By 1908 it had reduced its gear sizes from 58 to 48 teeth and its pinions from 36 teeth to 30.

Most if not all rapid transit lines used solid gears on their lines. Despite troubles such as those uncovered by the ASME in New York, gear and pinion failures were uncommon. Over one four-week period in 1908, of the 850 cars in service on the IRT, broken and loose gears accounted for 1.85 percent of all equipment defects. The failure rate for pinions was only slightly higher at 1.95 percent.[43]

Most of IRT's gears were solid steel, although it occasionally experimented with other compositions on its elevated division. Perhaps the most radical attempt at reducing wear and simplifying maintenance on gears and pinions was made by Chicago's Northwestern Elevated Railroad. In 1909, the road experimented with welding gears to the inside of the car wheels. It was hoped that the wheel would afford greater protection to the gear and would also simplify maintenance, as workers would not have to remove a wheel and then a gear.

Manufacturing Gears and Pinions

The Peerless Sectional Gear Company of New York made a complicated version of the split gear. It consisted of three subassemblies: the two gear halves, which were bolted together, and a "hub" to which the halves were mounted. Like solid gears, the hubs were pressed onto the axles. Peerless claimed that one man with a "safety key" could changes out gear halves in under a minute.[44]

One very popular supplier of solid gears was the Atha Steel Casting Company of Newark, New Jersey. Its gears were sold under the Titan label. Not surprisingly, it handled all of the gears and pinions for the Public Service Railway Corporation of New Jersey. Other major customers included the Rhode Island Company, the Rochester (New York) Railway Company, United Railways of Baltimore, the New York City Railway (a predecessor of New York Railways), and the Washington (D.C.) Railway & Electric Company. In tests run on the IRT's elevated division, it was estimated that Titan gears would average between 300,000 and 500,000 miles. The Pittsburgh-based R. D. Nuttall Company also distinguished itself in the field of gears and pinions.[45]

Among the Ohio firms, two early manufacturers were the Van Dorn & Dutton Company and the Whitmore Manufacturing Company, both of Cleveland. Van Dorn & Dutton made both solid and split gears. Among its customers were the lines of Detroit United Railways. By the turn of the century, Van Dorn & Dutton's orders had increased over 50 percent.

Whitmore made protective gear compounds for maintenance shops and did special repairs to damaged gears and pinions. Among the railways using its services were the Spokane & Inland Empire interurban, the Schenectady Street Railway Company in New York, and the Chicago Railways Company.[46]

Tool Steel

The most prominent Ohio company involved in gears and pinions and one of the industry's most significant was Cincinnati's Tool Steel Motor Gear & Pinion Company.[47] Initially, gears and pinions were made from untreated cast steel. However, it was discovered that treated gears and pinions lasted longer. There were three common treatment methods: hardening their surface, converting low carbon steel to high carbon steel, and by combining other elements (such as manganese) with the steel to harden it.

The strongest gears were determined to be those made from hard alloy steel, while the weakest were determined to be cast steel. The strongest pinions were determined to be those made from forged, heat-treated low carbon steel. The weakest pinions were determined to be made from forged, low carbon steel with no heat treatment.

Tool Steel used a surface hardening process to treat its gears and pinions. The process was developed by Russell Bloomfield, a Chicago bicycle shop owner, in 1900. Bloomfield happened onto the method while trying to find a way to harden the ball bearings used in his bicycles.

Bloomfield attempted to use the process to harden plow blades in 1905, and he established a small plant in Moline, Illinois. The venture failed, and he moved to Cincinnati, where he sought the aid of his father-in-law, Charles Sawtelle. The two perfected the hardening process and successfully tested it on a set of bevel gears for a shop crane at the U.S. Pipe & Foundry Company's plant in North Bend, Indiana. The manager of U.S. Pipe suggested that the process be applied to gears and pinions.

Bloomfield and Sawtelle's first major customer for treated gears and pinions was the Cincinnati Traction Company in 1907. Encouraged, the two founded the Tool Steel Motor Gear and Pinion Company on 16 August 1907 in Carthage (just north of Cincinnati, ultimately annexed by the city).

Tool Steel enjoyed wide popularity in the industry. For example, its gears and pinions were adopted as standard on all Stone & Webster properties. Stone & Webster was a large engineering firm that managed street railway and public utility companies nationwide. The firm also designed, built, and financed utility and street railway facilities.

Tool Steel was confident in the quality of its product, and by 1913 the company guaranteed its gears and pinions would last six times longer than untreated gears and pinions, four times longer than those that were oil-treated, and twice as long as case-hardened gears and pinions. To simplify maintenance, Tool Steel cut grooves into the sides of its gears and pinions. When wear reached the grooves, maintenance crews knew it was time for the gear or pinion to be replaced.[48]

Tool Steel's confidence in its product was well founded. By the early 1920s, the company cited the following percentages as an indicator of its success: in Illinois, 66 percent of all electric streetcars and interurbans were using Tool Steel gears and pinions (3,801 cars total). In Indiana, 8 of every 10 companies using 25 or more streetcars or interurbans used them. Eighty-nine percent of all cars (1,820) used Tool Steel gears and pinions in Texas, 88 percent (363 cars) in Alabama, and 74 percent (202 cars) in Utah.

Some railways opted to make or remake their own gears and pinions. For example, the Denver City Tramway Company found that it could recycle its pinions. Denver cars used 19-tooth pinions initially. When the teeth wore down too much, they were removed and machined smooth, in effect creating a "blank" for the railway's machinists. The "blanks" were then recut to form 15-tooth pinions. In 1909, the railway estimated that new pinions cost $2.24, while the remade pinion cost only 90 cents to produce.[49]

MOTOR COMPONENTS

Just because streetcar motors themselves were not made in Ohio does not mean that Ohio firms were not engaged in motor manufacture. Ohio served as the headquarters to one of the nation's largest (if not *the* largest) makers of carbon brushes. In direct-current (DC) motors, brushes provided the electrical contact between the motor's external circuit and the commutator, the latter of which was part of the armature shaft, or the part of the motor that rotated.

Brushes were made of copper originally, but copper's low resistance resulted in excessive sparking. By the late 1890s brushes in all but very low voltage motors were made of compressed carbon. Carbon conducts electricity, but its higher resistance minimizes sparking.[50]

The National Carbon Company could trace its existence to 1886. During most of the 1880s National Carbon focused upon the production of carbons for arc lamps. The main works, located on Cleveland's far west side, was producing 1.5 million carbons per year by 1888. This was a large complex, comprising several tall brick buildings along the Lake Shore & Michigan Southern (New York Central) Railroad. (This was also the same railroad line that passed next to the Kuhlman plant on Cleveland's east side.)[51]

In 1902, National Carbon expanded into the street railway market by purchasing the Partridge Carbon Company, located in Sandusky, a small city located along Ohio's western lakeshore. Formed during the late 1890s, little is known about Partridge's production except that its carbon brushes were used extensively by street railways operating in New York City. National Carbon did little to alter operation at Partridge, keeping the same management and even marketing Partridge brushes under their original name.

National Carbon's product lines were soon expanded. In 1909, its original Columbia and Partridge brush lines were augmented by the Laclede and Perfection labels. Laclede brushes were especially popular among street railways, as they could last as long as 500,000 miles. National Carbon came to own numerous plants nationwide. Its last expansion within the street railway market came in 1911, when a new plant opened at Niagara Falls, producing three new lines of carbon brushes. One of these was intended for railway use, and the others were for other direct current applications.[52]

One final manufacturer of motor components was the Homer Commutator Company of Cleveland. On streetcar motors, commutators consisted of individual copper bars mounted around one end of the armature shaft. The bars were separated and insulated from one another by mica sheets. Retaining rings held them in place.

Founded in 1891, Homer Commutator billed itself as the largest commutator manufacturer in the world (it qualified this statement by claiming to be the largest *exclusive* maker of commutators). Homer used a drop-forging process in the manufacture of its bars to ensure that no air gaps existed in the bars that might affect conductivity. Hard-drawn copper was used for custom patterns.[53]

CONCLUSION

In addition to providing a mind-numbing array of descriptions regarding components for current collection and propulsion, from the above we can deduce the following. Obviously, Ohio manufacturers were by no means the only ones in the field. In assessing the relative importance of Ohio as a center of essential operational components, we can see clear leadership in some areas and a relative lack of leadership in others. In terms of current collection, Ohio was clearly a major provider of trolley wheels, harps, and pole bases, but not a leader. It was only a minor supplier of poles. It did become a leader in catchers and retrievers, which grew in importance as car speeds increased.

Ohio was a consistent leader in the manufacture of gears and pinions, particularly between 1895 and 1905 and again after 1910 with the appearance of Tool Steel. Ohio was a consistent leader in the development of carbon brushes, although this leadership became diluted as National Carbon opened additional factories outside of the state. Centers of production could be found near major centers of car building activity, such as Cleveland, and in cities with no car building activity, such as Mansfield, Canton, and Springfield, so the presence or absence of a car builder does appear to have been a factor.

The factors that affected the design of the above components were dictated mostly by technical concerns: by the roles they were intended to perform and by the class service the railway intended to perform. What follows is a chapter that will focus on one specific component whose design was dictated primarily by service conditions.

Figure 4.1. A powered streetcar with trailer in downtown Cleveland. Note that the trailer is heated by a coal stove, a remarkably persistent feature on Cleveland streetcars. *Courtesy of Cleveland Public Library, James Thomas photo.*

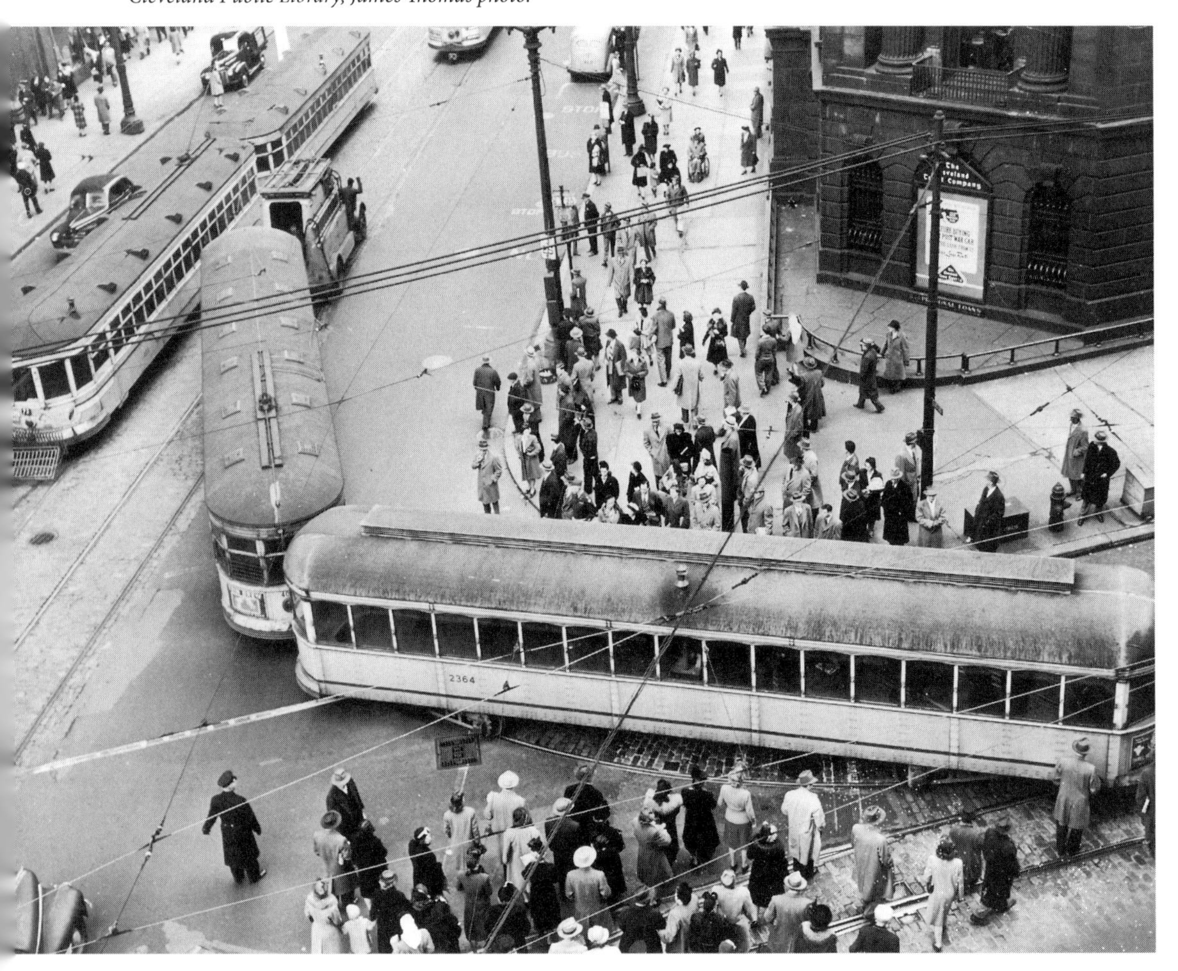

4

COUPLERS: WHEN, WHERE, AND WHY THEY WERE USED

THE CHALLENGES OF TRAIN OPERATION

In 1913 the Boston Elevated Railway Company conducted an internal study as to the causes of traffic delays on its surface lines. Records for October 1911 and October 1912 were scrutinized. It was estimated that service delays averaged 56.5 hours per month in 1911 and nearly 43 hours per month in 1912.[1]

While some delays could not be avoided (such as those caused by disasters like a bad fire along a particular line), many of the major causes *were* avoidable and spoke volumes to management. In some cases, accidents or breakdowns caused longer delays than necessary due to the inadequate dispatching and equipment of wrecking crews. Derailments were another culprit, especially around special work (complicated switches, junctions, and crossings) and on curves. Some delays were caused by lazy or incompetent railway employees.

Still another cause of delays was obsolete or underpowered equipment. Slow running and incompatible coupling equipment may not have been an issue on smaller systems or on lines with light passenger traffic, but they could become monumental on larger systems, especially on lines with a high passenger density.

Compatibility of coupling equipment or the question of whether or not to use coupling devices at all was an important issue for railway operators. While by no means universal, couplers occupied a number of niches within the street railway industry. For rapid transit, they were absolutely essential. For interurbans, their use depended upon the type of operation (traffic demand, frequency of service, the presence or absence of freight) envisioned by management. Although most street railway systems did not use couplers, there were a few that found them useful, especially those in larger cities.

Manufacturers of couplers had to take numerous factors into consideration when designing their products. What sort of railway did a prospective customer operate? What sort of service conditions did the railway experience? Did it interchange equipment with other railways or with steam railroads? Additional factors included safety, simplicity and ease of use, and speed.

This chapter will explore the use of couplers during the peak of the street railway industry, taking into consideration the reasons for their use, factors affecting their design, and specific examples of couplers used from 1900 to 1914. It will conclude with specific examples of coupler usage on city, interurban, and rapid transit systems. While Ohioans were certainly not alone in the design and manufacture of couplers, the firm Ohio Brass developed the most sophisticated coupler of the era, a coupler that would meet with increasing favor among all branches of the street railway industry.

Types of Service

Rapid transit vehicles (subway and elevated lines) were totally dependent upon couplers, since they nearly always ran in trains. Couplers with automatic air and electrical connections were favored, as they simplified train makeup and did away with complex arrays of cabling and hoses. As with street railways, couplers on rapid transit cars were radial, allowing for sharp curves in tunnels or on elevated structures.

Interurbans often used couplers, as their private rights-of-way enabled them to run long trains. Compatibility of coupling equipment was often an essential issue for interurban railways, as some engaged in through passenger and freight service with neighboring lines. More common, however, was freight interchange. Despite the greater length of their lines than those found on street railways or with rapid transit, most interurbans did not cover enough ground to make them self-sufficient as freight railways. Thus interchangeability and compatibility of equipment with other interurbans and (especially) steam railroads was essential. For example, the Pacific Electric Railway conducted interchange operations with both the Southern Pacific and Santa Fe railroads, and the Cincinnati & Columbus Traction Company interchanged freight with the Baltimore & Ohio. An example of interurbans interchanging freight with each other would be the upstate New York interurbans Hudson Valley Railway and Greenwich & Johnsonville Railroad.[2]

A variety of couplers could be found on interurbans. Although couplers designed with automatic air and electrical connections were preferred initially, many interurbans shifted to standard railroad couplers that required separate air and electrical connections. This was especially true if the interurbans hauled freight. Even though this meant more work for the interurban crew, it enabled cars to be exchanged with railroads more readily, which meant additional revenue, as well as an impartial standard with which to comply.

For street railways, the question of whether or not to use couplers depended upon the density of a given railway's passenger traffic. If the demand was heavy enough, railway managers considered operating cars in trains of two or three. There were multiple ways to go about this, depending upon how heavy the "heavy demand" actually was and how much a railway was willing to invest in equipment.

Trailers: The Simple Solution

Trailers were the simplest type of train used. The lead car was powered, and the cars it towed were usually unpowered (and thus lighter). There were several reasons for a street railway to consider using trailers.

First of all, they increased the carrying capacity of individual runs. Two-car trains were often slower than single units, owing in part to the added burden on the motorized car and to the confusion created by larger crowds getting on and off at certain stops. It was felt the delays were offset by reducing street congestion, as trains permitted fewer individual runs to operate on the streets.

Second, trailers saved on labor costs. Before 1914, most streetcars had two crew members: a motorman and a conductor. So two cars required four employees (two motormen plus two conductors). By having a trailer towed by a motor car, one crew member (a motorman) could be eliminated. Also, trailers cost less than motorized equipment.

Trailers were usually connected by a device called a drawbar, which was simply a long steel bar with holes at each end.[3] They were usually attached permanently to a car's ends and resembled couplers. Even railways that did not use trailers equipped their cars with removable drawbars. In this case, the drawbars were usually hung underneath the car's frame to enable towing in an emergency.

Drawbars were often accompanied by large hoses and cables to allow electrical and air connections to be made between cars. This enabled the motorman on the lead car to control motor and braking functions on all cars. Due to the tightness of many street railway curves, most streetcar drawbars were radial. In other words, they could swing from side to side. By swinging, the drawbars did not wrench themselves free from the underside of the car on sharp curves. Most drawbars were linked by overlapping holes at their ends and inserting steel pins. Others were more complex and had coupling heads at their ends.[4]

A more complex option for trailers than the drawbar was the coupler. Couplers differed from drawbars in their intent. Whereas drawbars were often used by railways regardless of their use of trailers, couplers were designed specifically for train operation.[5]

Trailers were not without their faults, however. As Boston and other railways discovered, they did not work if the lead car was not sufficiently powered. Being lighter than motorized equipment, trailers had an annoying tendency to derail at switches and special work. Some railways converted obsolete equipment into trailers only to find they were not as efficient at moving people as cars with better interior designs. However, if properly constructed for their role, teamed with adequately powered cars, and operated with care, trailers could be both convenient for the public and economic for the railway.[6]

Successful Operations

Traffic along Montreal Tramways' St. Catherine Street line was particularly dense. By 1913, ridership along the line was heavy all day, with peak schedules calling for 90-second headways (the time interval between runs) or an incredible 40 individual runs per hour. Montreal looked to other railways for examples. Although it had a large stock of old equipment available, it was decided that the St. Catherine line was best served by trailers built as such from the wheels up.

Montreal's new trailers were equipped with devices that would ensure the rapid and safe making and breaking up of trains. Automatic couplers and jumpers with three-way cables (incorporating braking, lighting, heating, and door signals) were to be used. Since Montreal only ran single-ended equipment, car design was simplified somewhat. Twenty-five trailers were ordered initially, along with 25 motorized cars of similar design to ensure operational compatibility.

One of the cities Montreal Tramways looked to as an example of trailer operation was Pittsburgh. Pittsburgh Railways had experienced a 12 percent growth in ridership between 1905 and 1906. Initially, the railway attempted to use retired equipment as trailers, but discovered this did not work well. The railway experimented with an initial order of 50 double-trucked steel trailers and determined this was the way to go. The trailers boasted the latest technology, including couplers that made all the necessary air and electrical connections in addition to the physical connection of the cars.

The brakes on the Pittsburgh trailers were of a semiautomatic type. All brakes were applied if the air connection failed for any reason, although crew members were given a nine-second delay to correct any problems. Entrance and exit doors slid open and closed (sliding doors interfered less with crowds on either side of the doors) and were air-powered. An electric signal system kept all crew members in contact with each other. By the beginning of 1912, Pittsburgh had 300 trailers in operation, 180 of which were of the new design and 120 of which were still of the older, single-trucked wooden design. Pittsburgh Railways had plans to replace all of its outmoded trailers.

Also, in 1912, Cleveland Transit Commissioner Peter Witt (one of the industry's true visionaries) was concerned that motorized cars would be too expensive to meet Cleveland's growing transit needs. He recommended that a recent Cleveland Railway order for 50 motorized streetcars be changed to 100 trailers. The railway followed his advice and began trailer operations on 1 August. Like the Pittsburgh trailers, the Cleveland trailers boasted many of the industry's most advanced features. Automatic couplers were used with air and electrical connections, air-powered sliding doors, electrical signaling systems, semiautomatic brakes, and water heaters.[7]

MULTIPLE-UNIT CONTROL: THE SOPHISTICATED SOLUTION

A more complex means of dealing with trains of two or more cars was multiple-unit control, in which at least two cars in a train were motorized and controlled by a single motorman from a single set of controls. Multiple-unit was an extremely flexible system, allowing trains of any number of cars to be controlled from a single control station anywhere on any of the cars (similar controls could be put on trailers as well).

In addition to flexibility, there were several other advantages enjoyed by multiple-unit systems. Perhaps the greatest advantage next to flexibility was safety. Non-multiple-unit controllers were large (from the floor to about the waist of an average-sized adult), drum-shaped cabinets that had the full operating voltage coursing through them. The only thing between the motorman and the full voltage of the trolley wire were the thin, asbestos-lined wooden sides of the controller. Should this insulation fail, a fire and/or electrocution could result.

The electrical contacts within a non-multiple-unit controller were small copper tabs called "fingers." If a motorman did not stop the controller handle at exactly the correct position, arcing could result that would burn out the fingers and create gaps in the control circuitry. This created extremely uneven rates of acceleration that were hard on motors and passengers alike.

Multiple-unit control required large, heavy switching devices known as "master controllers" that operated synchronously with one another. These were usually suspended underneath the cars. To operate them, motormen used smaller, low-voltage controllers on the car platforms, thus keeping the main voltage off of the platforms and as far from human contact as possible.

For interurbans, multiple-unit control offered the advantage of not having to divide large runs into separate sections. While some critics objected to long trains of interurban cars traveling down city streets, proponents countered that there was no technical reason why interurbans could not, so long as clearance issues were addressed. Furthermore, although long trains snake-dancing down the middle of city streets were annoying when they went through, fewer trains running at greater intervals disrupted street traffic far less than multiple sections. By making fewer but larger runs, multiple-unit trains reduced the chance of accidents, sped train movement, increased a line's carrying capacity, reduced traffic density, and lowered operating expenses.[8]

FACTORS AFFECTING COUPLER DESIGN: MULTIPLE-UNIT SYSTEMS

Several types of multiple-unit (m.u.) systems were available between 1900 and 1914. Which system a railway chose affected how it could run its railway, as the different systems required different types of hose and cable connections that had to be made when the car was coupled. Some systems allowed trains to be shortened or lengthened using only the train crew(s) involved. Others required personnel on the ground to make the necessary connections.

For the railway, the best system was the one that not only performed well on the road but also worked well in the yards where the trains were assembled. For the coupler manufacturer, the challenge was to design a coupler that would meet these requirements, ideally reducing the amount of manpower required in the yard and enabling a railway to add or reduce the number of cars on a run as swiftly as possible.[9]

The Sprague System

The idea of multiple-unit control was first conceived by Frank Sprague in 1885, three years before he installed his celebrated electric street railway system in Richmond, Virginia. He installed his first m.u. system on the South Side Elevated in Chicago in 1898. He then introduced refinements to the system for the Boston Elevated's elevated-subway rapid transit lines in 1901.

Sprague's system made use of a small "pilot motor" that operated master controllers on each car. Rather than have direct control over the master controller, the motorman's controls manipulated the pilot motor. The pilot motors on all of the cars were linked through cables connecting each of the cars in the train. This system proved rather cumbersome and was regarded as inferior to later systems of General Electric and Westinghouse. Sprague sold his company to General Electric in 1902.[10]

General Electric Systems

General Electric brought out its first m.u. system in 1902, marketing it as "Type M." This system did away with the drum-type master controller, replacing it with a set of electrical "contactors" or switches that cut resistance out of the motor circuit. The contactors were controlled by the motorman using a small, low-voltage controller on the car platform. The system was introduced on New York City's Manhattan Elevated Railway.

Following its acquisition of Sprague Electric, General Electric came out with a refined version of Type M in 1904. First used on New York's first subway (operated by the Interborough Rapid Transit Company, or IRT), the newer version used a "limiting relay" (a feature introduced by Sprague) that delayed the sequence by which the contactors closed, thus creating automatic acceleration. Although it was possible to accelerate trains at rates as high as 3 to 5 miles-per-hour-per-second, acceleration was usually kept at a conservative 1.5 miles-per-hour-per-second.

General Electric incorporated numerous safety devices into its m.u. system. Automatic switches cut all power from the train cables whenever the motorman placed his control handle in the "off" position or whenever a car parted from the train accidentally. Under normal conditions, isolation switches were used to cut power from a specific set of cables before decoupling. In addition, the manufacturer offered its own air brakes, which were patterned after those found on steam railroads (a lack of air pressure

Figure 4.2. Westinghouse electro-pneumatic multiple-unit system. It was called a "turret controller" due to its cylindrical design, which resembled a turret lathe. *Branford Electric Railway Association, author photo.*

resulted in a full brake application). General Electric's system allowed for rapid recharging should the brakes go into an emergency application.

A version of Type M control called "Type MK" was made available for streetcars and smaller interurbans in 1912. It offered the same general advantages as the Type M, though MK was intended for cars operating singly or in two-car trains. Also, the various components in the MK system were more compact, weighed less, and were designed for service conditions that were less severe than those encountered on rapid transit lines or large interurbans.[11]

Westinghouse Systems

The Westinghouse system was similar to the one developed by General Electric, only it used something called "unit switches" in place of "contactors," and operated the unit switches using compressed air instead of electricity. Before he became a prominent electrical manufacturer, George Westinghouse made his fortune by designing automatic air brakes for railroads. Thus it should come as no surprise that the Westinghouse m.u. system made use of air-powered as well as electrical components.

The original Westinghouse system, which came out about the same time as General Electric's Type M, used a small controller on the car platform to operate electrical valves that were connected to an air cylinder. The air cylinder controlled a drum-type master controller. Criticism of the air system's reliability was soon put to rest when it was discovered that the system could remain fully pressurized for several days with no power.

Westinghouse's original system was modified in 1903, introducing a squat, cylindrical master controller. Because it resembled a turret lathe, Westinghouse called the device a "turret controller," and the system in general was soon referred to as "turret control." As on the Sprague-General Electric version of Type M, a limit relay controlled the timing of the contact closures so that turret control could offer automatic acceleration as well. The first major user of turret control was the Brooklyn Rapid Transit Company.

Westinghouse subsequently condensed its turret control system into a more compact version suitable for use on streetcars or smaller interurbans. In an attempt to save space and weight, some of the features of turret control, such as automatic acceleration, had to be eliminated. For motormen, the controller looked and operated just like a conventional K-type, only it was much smaller. "Type HL" control held two great advantages over conventional controllers: it kept the main voltage off of the car platforms, and it was m.u. compatible.

A subsequent version of HL was offered in 1909 with automatic acceleration. Storage batteries were incorporated into the control circuits, further automating the operating sequence of electrical contacts as the car accelerated. One Westinghouse engineer remarked that all a motorman had to do was place the controller handle into the "full series" or "full parallel" position and the control system would take care of the rest.[12]

Wire Systems

Despite the broad influence General Electric and Westinghouse enjoyed within the street railway industry, other multiple-unit systems emerged from the period. One, called the "two-wire system," was developed by the Cutler-Hammer Manufacturing Company in 1904. This system required only a single small cable containing two control wires connecting the cars in the train.

The system was remarkably simple in conception. The direction of travel was determined by the direction of current in the two wires. Speed was determined by the amount of voltage in the two wires. Like the GE and later Westinghouse systems, the motorman could advance his controller all the way to the maximum position and allow the system to accelerate the train automatically. A variation called a "single wire" system was developed two years later. Perhaps because many felt neither system was rugged enough, they did not catch on.[13]

FACTORS AFFECTING COUPLER DESIGN: STANDARDIZATION

Another serious consideration for manufacturers and users of couplers was standardization. Standards tended to come into play under two circumstances. The first involved large systems, such as those located in heavily populated cities. Whether by default or design, large railways usually adopted some sort of "in-house" standard and made certain it applied to as many classes of equipment as possible. Couplers were no exception. The other case involved car interchange. Naturally, railways involved in equipment interchange needed to make sure their equipment was compatible with other railways.

Standardization at the National Level: The Master Car Builders' Association

By the twentieth century, the railroad industry was already operating on a partial set of national standards for rolling stock construction and components. Introduced partly due to federal legislation regarding safety and partly due to operational convenience and common sense, these standards included the structural design and strength of car bodies, braking systems, and couplers. The principal trade organization that presided over the development and adoption of these standards was the Master Car Builders' Association (MCB).

The establishment of standards within the street railway industry was a bit different from the railroads, however, and it did not attract the degree of federal attention that the railroads did for several reasons. Most important was the fact that relatively few street railways and interurbans engaged in interstate commerce. Even among those that did, the nature of the traffic was more local in nature than regional or national.

This does not mean that street railways were ignored, nor does it mean they were ignorant of or indifferent toward developments in the railroad industry. Some federal regulations that applied to railroad couplers also applied to the street railway industry. Additionally, MCB activities (especially its national conventions) were reported regularly in the street railway trade press, and its standards and policies were often debated at American Street Railway Association conventions and meetings. By the 1900s, the coupler issue had come and gone for the railroaders, and it is only natural that street railway manufacturers at least considered the railroads' experience when developing their products.

Figure 4.3. An MCB or "knuckle" coupler, modified for electric railway service. *Branford Electric Railway Association, author photo.*

Most people are familiar with the MCB coupler. Also called a "knuckle" coupler, its design became the railroad standard during the late 1800s. MCB couplers were referred to as "automatic" because they automatically engaged and held when brought together.

Earlier coupler designs had been based upon "link-and-pin" systems, where two steel bars or rods with holes in their ends were brought together. When the holes lined up, a brakeman standing between the cars dropped a heavy steel pin through the holes, thus linking the cars. Uncoupling often presented greater challenges for the brakeman. To uncouple, the cars often had to be pushed into one another slightly to allow enough slack in the couplers to permit the brakeman to remove the pin. This was a dangerous practice, and serious injuries and even fatalities were common.

Automatic couplers did away with the necessity of placing men in harm's way to assemble and separate trains. Not only was coupling accomplished automatically, but uncoupling was made easier by providing each coupler with its own locking pin. To uncouple, the pin needed to be pulled on one coupler only, and this was accomplished by raising a lever at the car's side. Once uncoupled, the brakeman released the lever and the pin dropped back, allowing for automatic coupling.

As early as 1901, MCB couplers were available for streetcars. Philadelphia's McLaughlin Car Coupler Company based all of its couplers on MCB standards, citing their convenience, safety record, and compliance with safety legislation. The Illinois Traction System, which had both statewide passenger and freight service, used modified MCBs. In a 1914 editorial, the *Electric Railway Journal* supported MCB couplers, citing many of the same reasons as McLaughlin. However, the *Journal* also acknowledged that MCBs would probably be essential only for those interurbans engaged in freight interchange with railroads. For systems that operated independently, coupler designs that met their individual operating characteristics might be more appropriate.[14]

Standardization at the National Level: The American Electric Railway Association

As mentioned earlier, the street railway industry had a national organization of its own, the American Electric Railway Association (AERA). Among the "sub-associations" that dealt with various industry topics was the American Electric Railway Engineering Association (AEREA). Both AERA and AEREA debated coupler development and the possibility of creating industry standards for them.

Unfortunately, the various types of electric railway construction and operational practices inhibited the development of coupler standards. For instance, when addressing MCB couplers, AERA committees were concerned that, if adopted, the MCBs would require suitable modification to meet conditions of streetcar and interurban operation that were absent on railroads. Curves and grades were chief among these concerns. Curves tended to be sharper and grades were more severe and changed more abruptly on streetcar and interurban lines than they did on railroads. While MCBs could be modified to allow greater radial and vertical play, such modifications would render them incompatible with railroad MCBs, which was the principal reason for considering adoption of MCBs in the first place.

Not only did AERA address issues regarding coupler design, but it also addressed concerns pertaining to multiple-unit operation. While it realized both electrical and electro-pneumatic systems would always have their adherents, there could be a standardized system of m.u. connections. Standard locations for such things as electrical and air receptacles, hoses, mounts for trolley catchers and retrievers, headlights, and marker lights were suggested. However, while standards of this type were proposed from time to time, they were rarely adopted, largely due to concerns that the standards would not be adequate for an individual railway's circumstances. Only the standards applying to the location of air hoses gained any headway within the industry.[15]

Standardization at the Regional Level: Central Electric Railway Association

Standardization was more successful at the regional level, mainly because railways within a specific geographic region tended to hold more in common with each other than they did with other areas of the country. One of the more successful regional efforts was that of the Central Electric Railway Association.[16] CERA membership was extended to interurbans of the Midwest, which held the world's greatest concentration of interurban systems. Similarities in terrain, types of service, and even equipment made it easier for CERA to develop standards that its members would accept.

During a 1906 meeting in Dayton, it was decided that CERA engineering committees should focus on three basic coupler design issues: length, height, and the profile or shape of the coupler head. MCBs soon fell out of favor with most CERA members due to their locking pins. Automatic couplers that did not rely on pins were felt to allow closer spacing of cars, reduce the play between cars, and further reduce danger to employees on the ground.

CERA's attitude softened toward MCBs in subsequent years, largely due to the fact that many CERA members dealt with freight interchange with railroads (more than 12 large CERA roads were involved in railroad interchange). Various MCB standards covering coupler subassemblies were adopted, and it was ultimately recommended that couplers on CERA roads at least be compatible with MCBs. Further recommendations went to the extent of stating that in an emergency, MCB couplers could be substituted for whatever coupler a CERA member car was using.[17]

The MCB in the Street Railway Industry

Pittsburgh's McConway & Torley Company made a number of different MCB couplers for the street railway industry, but specialized in those used by roads with railroad freight interchange. As a test of its couplers' strength, a freight train with 18 23-ton cars was fully equipped with McConway & Torley couplers and sent from St. Louis to Los Angeles via New Orleans. The 4,724-mile trip passed without incident. Pacific Electric, a Los Angeles-based interurban that enjoyed heavy freight interchange with the Southern Pacific Railroad, was a major customer.

McConway & Torley's other MCB products included a special three-quarter-sized MCB coupler for use on narrow-gauged lines. Its "Janney" coupler added a lug cast onto the knuckle of an MCB to handle the strain of tight curves.

Some interurbans felt compelled to use MCB couplers due to their extensive freight operations. The Illinois Traction System, for example, exchanged freight cars with three railroads: the Chicago & Eastern Illinois, the Chicago, Rock Island & Pacific Railway, and the St. Louis–San Francisco. Perhaps the most unusual adoption of the MCB coupler was the Inter-Urban Railway of Des Moines, Iowa. A large interurban that handled both passenger traffic and freight interchange, Inter-Urban was perhaps the only railway to use *two* MCB couplers at each end of its cars. The couplers were arranged with one over the other, with the lower one used only along a particular section of track that contained a series of very tight curves.[18]

The Van Dorn Coupler

The most widely used couplers before World War One were those manufactured by the W. T. Van Dorn Company of Chicago.[19] Van Dorn made a wide range of couplers (13 models, all automatic), draft gear, coupler heads, and even a few MCBs. Van Dorn held two crucial design patents that enabled it to gain a large share of the coupler market. The first allowed for vertical play, and the second allowed for anchoring the draft gear underneath the car body.

Figure 4.4. A portion of a Van Dorn coupler, one of the more popular couplers before World War One. *Branford Electric Railway Association, author photo.*

Van Dorn's products were used mostly by rapid transit lines and interurbans, but could also be found on street railways. They were used worldwide. In the United States, early customers included two of Chicago's four elevated lines (Northwestern and Lake Street), the Boston Elevated Railway's main line elevated, and the Manhattan Elevated in New York City. Later customers included the Hudson & Manhattan (locally referred to as the "tubes"). Interurbans and street railways included the Chicago & Southern Traction Company, the Winona Interurban Railway, and the New York & Queens County Railway. Van Dorn couplers could also be seen as far abroad as Scotland and Yokohama, Japan.

Essentially, Van Dorn couplers incorporated both male and female links in the same coupler head. The heads were blunted in a pattern that allowed proper alignment and contained notches several inches back that interlocked. They were well regarded for their simplicity and automatic features.[20]

Other Couplers

Some companies specialized in equipment for streetcars and interurbans. The Automatic Car Coupler Company of Los Angeles once produced 270 couplers for the Terre Haute, Indianapolis & Eastern, one of the biggest interurbans in Indiana.

One of the problems with previous coupler designs was that they all required a man to be stationed on the ground to make the necessary air and electrical connections. One manufacturer that arrived at a partial solution was the Westinghouse Traction Air Brake Company. Westinghouse's coupler not only coupled cars together, but also incorporated the necessary pneumatic connections for air brakes into the coupler heads. This coupler soon found its way into rapid transit service and was in use in Boston and New York City by 1907.[21]

Figure 4.5. A Tomlinson coupler. This was designed to couple two cars, as well as make all of the air and electrical connections automatically. The coupling and air connections are made at the center of the coupler, and the electrical contacts are made at both sides. *Northern Ohio Railway Museum, author photo.*

Figure 4.6. A Tomlinson coupler designed to make coupling and air connections only. *Branford Electric Railway Association, author photo.*

In Ohio, MCB-type couplers were available from Cleveland's National Malleable Castings Company. Marketed under the "Vulcan" and "Sharon" labels, National couplers combined the standard MCB coupler head with the pivoting radial bar essential to street railway interurban, and rapid transit operation. National pointed out that its couplers were automatic, fully compatible with railroad couplers, and would allow 50-foot cars to negotiate curves down to a 33-foot radius.

The Ohio Brass Company

By far the most significant Ohio coupler manufacturer was Ohio Brass, located in Mansfield, a small city 66 miles north of Columbus. Mansfield's industrial district produced sheet steel, brass, and rubber.

The Ohio Brass Company was founded in the 1890s to produce components for the street railway, electric railroad, and electric utility industries. The plant was situated on a 5-acre triangular plot surrounded by all three of the railroads serving

Figure 4.7. A Tomlinson coupler in multiple-unit service. Note that the couplers are designed to make coupling and air connections only. Electrical connections were made via "jumper cables," the sockets for which can be seen on both sides of the headlight. The lead car of this interurban train is a 1920 Kuhlman product, operated by the Northern Ohio Traction & Light Company. *Courtesy of Branford Electric Railway Association.*

Mansfield: Baltimore & Ohio, the Erie, and Pennsylvania. Its buildings enclosed 200,000 square feet of space. The company's foundry could produce 8–10 tons of brass per day.

In 1906, the company purchased the rights to a coupler patented by Charles H. Tomlinson, who enjoyed a lengthy career at Ohio Brass. Like the Van Dorn coupler, the Tomlinson was an automatic coupler that used male/female connectors. Like the Westinghouse coupler, the Tomlinson made all of the necessary air connections at the same time. Tomlinson continued to improve his product, so that by 1914 his coupler could also make all of the necessary electrical connections at the same time, completely eliminating the need for anyone to be stationed on the ground when cars were being coupled and uncoupled.[22]

In its fullest state of development, the Tomlinson coupler was compatible with all forms of m.u. control, whether they were purely electrical or electro-pneumatic. Tomlinsons could also handle auxiliary electrical circuits for such things as heat, lighting, and on-board signal systems. Models were manufactured for use on rapid transit, interurban, and streetcar lines.

At first, Ohio Brass offered MCB adaptors that would allow Tomlinsons to be used in conjunction with MCB equipment. Starting in 1907, the manufacturer offered a line on Tomlinsons that were MCB-compatible without the use of adaptors. Clearly, the Tomlinson was the most sophisticated coupler on the market before World War One.

Looking at Ohio Brass's manufacturing activity within the street railway system overall, one can see that it shipped products to car builders and railways. Orders published in the trade literature identified 765 cars that were built using Ohio Brass components. Two-thirds of these cars were built by Ohio car builders, and half of all cars built were made at the Kuhlman and Cincinnati plants.

The trade literature also includes products shipped directly to the railways themselves. Orders are listed for 1,220 Tomlinson couplers. Of these, half went to the Boston Elevated Railway. Boston Elevated had extensive subway, elevated, and m.u. streetcar operations, so the convenience and flexibility afforded by the Tomlinson made it a very attractive coupler (see tables 4.1 and 4.2).[23]

Looking specifically at comparative data published in the *Electric Railway Journal*, of the 2,783 cars reported to have couplers, we find that Ohio manufacturers accounted for 29 percent. One in five of these cars were equipped with Tomlinsons, with 40 percent of these cars operating in Massachusetts, Ohio, and Colorado. The remainder were scattered among 14 states and two Canadian provinces.

Roughly 8 percent of all cars were equipped with couplers manufactured and installed by the Cincinnati Car Company. Coupler manufacture was rare but not unheard of among streetcar builders. If a car builder could make its own couplers, it usually would use them on car orders being completed in its own shops. What is a bit perplexing in Cincinnati's case is that none of these orders were made for Ohio railways. Nearly half of Cincinnati coupler-equipped cars ran in New Jersey, with the remainder spread over nine states (see table 4.3).

CONCLUSION

Clearly, couplers were not for every railway. Railways that operated their units singly had no need for couplers, and even some railways that used trailers could get by without them. For railways that required couplers, however, there was a variety from which to choose. Railways that interchanged cars with other railways, and especially with railroads, were best off using MCB or MCB-compatible equipment. For railways that did little or no interchanging, the choice was more difficult.

For the coupler manufacturer, there were numerous factors to consider when marketing its product. Were its couplers of an MCB or MCB-compatible design? Did they conform to any additional standards or prevailing schools of thought within the industry? Did the couplers allow for sufficient radial or vertical clearances?

For those not opting for MCBs, simplicity seemed to be the rule of the day. Van Dorn couplers were ideal in this vein, although other models such as those by Automatic Car Coupler proved effective as well. More technologically sophisticated couplers were also available that either reduced the amount of work that ground employees needed to perform, as in the case of Westinghouse, or eliminated it entirely, as in the case of Ohio Brass and the Tomlinson.

An overriding concern among railways was the safe operation of whatever coupling device it decided to use. In this they were not completely alone, as there was federal legislation that placed certain minimum standards on coupler operation. However, safety was a major concern among operators—so much so that an entire segment of street railway component manufacture was devoted to the production of safety appliances. These devices will be discussed in the following chapter.

Figure 5.1. Braking controls came in a variety of forms. The mass of piping to the left is an air brake stand. The handbrake is the large handle to the right. In operation, it was swung up 90 degrees and then worked back and forth like the tiller on a boat. The three lines running from the brake stand off to the right of the photo connect the air brakes to the streetcar's doors. The doors will not open unless the brakes are fully applied. *Branford Electric Railway Association, author photo.*

5

PROTECTING THE PUBLIC (AND THEMSELVES): STREET RAILWAYS AND THE MANUFACTURE OF SAFETY APPLIANCES

An accident damaging a street railway's property or injuring an employee or passengers was obviously something to avoid. Accidents had all sorts of repercussions, none of which were good. For the railway, a streetcar in the shop was not on its route earning the company money. If the accident was serious enough, damage might also have occurred to the tracks or power wires, causing portions of the railway to be shut down until repairs were completed. The human cost of accidents could be far greater. The permanent or temporary loss of an experienced employee affected service and possibly company morale.

The greatest concern with accidents was, of course, the public. Damage to others' property could be expensive. Even worse, cases of personal injury could prove costly in numerous ways. In addition to the actual cost of a settlement or a court's judgment, such accidents could mean an unfavorable reputation and/or critical scrutiny from local, state, or even federal agencies.

Ironically, the street railways' concern over personal injury lawsuits, while stemming from a dramatic increase in the number of such suits, may not have been based upon actual increases in accident frequency. In his 1992 study of personal injury cases in New York City from 1870 to 1910, historian Randolph Bergstrom argues that the number of accidents involving street railways and individuals did not rise disproportionately to the increase in population. Furthermore, the number of judgments against street railways and the amount of compensation awarded the victims actually decreased. However, the fear of such lawsuits persisted.

Bergstrom attributes this fear to a rise in what he calls "assertion individualism," or a shift in people's conception of individualism from "self-help and personal responsibility" to "an aggressive, rights-oriented individualism." The causes of this shift are numerous. Bergstrom contends they can be traced to the social and economic changes wrought in the local environment by industrialization, mechanization, and the growth and spread of capitalism.[1]

The foregoing, when combined with the optimism of reformers in their attempts to identify and correct social problems, placed a greater burden on the street railways than had been present in the nineteenth century.

> Growth in the knowledge of how to prevent the problems of accidents and injuries and confidence in society's ability to do so converged. . . . The solutions contained in this expanding knowledge, the means to prevent injuries, required no more than ordinary care to implement. . . . If the preventative measures had required extraordinary effort, popular thinking as well as the New York courts' rules would have exempted injurers from having to employ them.[2]

Much of what street railways experienced in New York City was being felt by street railways in other parts of the country. This is what the railways faced, and their concerns and needs shaped the market for safety appliances.

ACCIDENT REDUCTION STRATEGIES

It should be noted that there were a number of strategies a street railway could employ to minimize the risk of accidents, aside from appliances. The most obvious were internal, such as adopting operating procedures that emphasized safety above all else. These procedures embraced not only the physical operation of the railway but also penalties and positive incentives for how well the procedures were followed. Thorough training regimens were adopted to ensure that all employees were familiar with the railway's equipment and its operation.

Another strategy was to create a claims department to handle cases of alleged damage or injury resulting from accidents involving the railway. It made good administrative sense to maintain a claims department. It ensured not only that all claims would be handled by individuals familiar with all of the procedures a claim entailed, but also that investigations would be carried out by specialists who were trained and determined to uncover all of the possible facts of a case, including physical evidence, witness accounts, and expert testimony when needed. Claims departments could also prove invaluable in cases where the railway was at fault. A prompt settlement could prevent future litigation and even place the railway in a positive light.

Claims departments also provided protection against fraud. Human nature being what it is, railways were often the victims of false accident claims. Street railways made convenient targets. They were in the public eye, a part of everyday life.

Fraudulent claims were perpetrated by amateurs as well as professionals. Amateurs were easier to deal with, as careful investigation often led to conflicting witness accounts, questionable medical diagnoses, or stories that simply were not plausible given the circumstances. The professional injury victim, however, was more difficult to expose.

Professional victims choreographed "accidents" (in at least one instance a retired circus contortionist was used) to ensure maximum effect upon an unsuspecting car crew and passengers. The appearance of "casual" or "disinterested" bystanders was carefully orchestrated to make certain there were reliable "witnesses" to the accident.

Unscrupulous medical doctors with impressive-sounding credentials were secured and coached in how to handle any queries. Furthermore, the professional victims were constantly on the move–it was not unheard of for one individual or group of individuals to have several cases pending in different cities at the same time.

To combat the professionals, claims agents from different railways kept in contact with each other and even had their own organization, the American Electric Railway Claims Association, under the American Electric Railway Association's umbrella. Some railways sponsored agencies that maintained master databases on accidents occurring within a prescribed multistate area. Databases were assembled from information provided by claims departments and cross-indexed by name, alias(es), physical description, and modus operandi. In some instances, these regional databases could grow to several thousand entries.[3]

Externally, public education campaigns were another strategy to help reduce accidents. Railways employed a number of tactics, including posters in public locations, advertisements and articles in local newspapers, and lectures to various civic groups and (especially) schools. A pioneer in the creation of safety presentations for school children was the Cleveland, Southwestern & Columbus interurban in Ohio.[4]

SAFETY APPLIANCES

Despite the thoroughness of training regimens, the money invested in public safety campaigns, and the diligence of claim departments to protect the company, accidents still happened. Even the best-trained crews could be careless for a few moments, or equipment could fail unexpectedly. Even worse, all of the caution in the world could not protect a streetcar from the foolish mistakes and actions of others—especially in the crowded, bustling streets of a city.

Street railway companies realized this and tried to equip their cars for every reasonable contingency. In most cases this was voluntary, the railways being motivated by conscience or common sense. In some cases they were required by law to provide certain appliances. In rare cases, court decisions had an impact on the devices a railway had to carry on its cars.

Between 1900 and the summer of 1914, the *Street Railway Journal* and its successor, *Electric Railway Journal,* published nearly 3,000 legal cases involving streetcar accidents. The overwhelming majority of these cases involved negligence. Cases involving safety devices were comparatively rare. Of the cases published, only about 200 involved safety devices, and with the exception of a dozen, the cases focused on the failure of car crews to use the safety devices rather than on the devices themselves.

Of the twelve cases involving injuries sustained due to the absence of safety devices, three are instructive. In a 1901 New Hampshire case, a street railway was cleared of any wrongdoing when a child was run over by one of its streetcars and killed. Although there were no protective devices on the front of the railway's cars, it was demonstrated that at the time of the accident, there were no devices on the market that would have saved the child given the circumstances. However, the presiding judge issued an unmistakable warning to the railway: "The duty of a street railway company to equip its cars with safety appliances is not limited by their convenience, but includes the adoption of such as men of average prudence would use under the same circumstances."[5]

The next case involved a passenger who fell between two cars on an elevated train running in Chicago. The plaintiff won a judgment against the railway. The ruling stated in part that "it was the defendant's duty to use due care in providing suitable gates or other safeguards to prevent persons from falling between the cars."[6]

The third case involved a class action lawsuit against a Utah railway in 1908. The judge's opinion doubtlessly attracted the attention of street railways across the country:

> The rule that it is necessary to prove that certain appliances are in general use by street railway companies before negligence can be predicated on the omission to supply them does not apply to appliances, the use of which is a matter of common knowledge.
>
> As applied to street railway cars, a fender is a guard or protection against danger to pedestrians coming in contact with a car.
>
> Where negligence, in an action for death of a pedestrian by being struck by a street car, was predicated entirely on the omission to provide the car with any fender or guard whatever, the court was entitled to take judicial notice of the purpose of fenders, as applied to streetcars; such appliances being in common and general use as street cars.[7]

Safety devices can be grouped into two general categories: those intended to protect the public at large, and those intended to protect passengers. Devices intended to protect the public can be further divided into preventive devices, or devices that would prevent an accident from occurring in the first place, and "actuative" devices, or devices that would come into play the moment an accident occurred.

There were four general groups of preventive safety devices: brakes (which may also be considered to be a component essential to operation), sanders (which could assist in the braking and starting of a car), headlights (to help the motorman see and be seen by objects in the distance), and gongs, bells, and horns (to alert the unwary of an approaching car's presence). The major actuative devices consisted of fenders, wheel guards, bumpers, and anti-climbers.

DEVICES TO PROTECT THE PUBLIC

Brakes

Streetcars built between 1900 and 1914 were well served by braking technology, generally possessing three semi-independent braking systems. The first was an air-powered brake, which constituted the primary braking system. Air-powered brakes could function in one of two ways. For single cars, the most common was a system that increased the braking power as the amount of compressed air was increased in the brake lines. For a train of cars (most common on rapid transit and interurban lines), the brakes were actuated by *reducing* the pressure in the brake lines (similar to a tractor trailer on today's highways, or a modern freight or passenger railroad train). The most popular air braking systems were made by Westinghouse, with other systems being manufactured by General Electric, National Air Brake, and Christensen (which was absorbed by National around 1910). None of these were based in Ohio.

Compressed air was commonly obtained from air compressors and reservoirs located on the individual cars. However, some cities utilized a "storage air" system in which compressed air was obtained from stationary compressors located at the ends of the streetcar lines. Storage air systems enjoyed their greatest popularity when compressors that were small and reliable enough to be carried on cars were still in their infancy (roughly 1900–1905), but were never used on more than a handful of railways.

The second means of stopping a car was by a hand brake. Hand brakes were the original type of braking device used on streetcars, dating back to the horsecars of the early to mid-1800s. They were fairly effective on smaller, single-trucked streetcars, but began to require too much effort from the motorman as the cars grew in size, weight, and speed. Hand brakes were retained, however, and served as an emergency alternative in case the air brakes failed and as a "parking brake" when the trolley poles were removed from the wire for long periods, such as when a car was stored in a yard or barn. The handbrake was only semi-independent from the air brake, as both systems used the same rigging to actuate the car's brakes.

The final means of stopping the cars were the motors themselves. By shutting off the controller and reversing the motor circuits, the motors in effect turned into generators and could be used to stop the car in an emergency. Though usually quite effective, this practice was discouraged for two reasons. First, although the motors could stop the

Figure 5.2. A popular form of handbrake control was the vertical wheel. An air brake stand appears on the left. The handle on the right is a switch iron, which the motorman could use to manipulate switches along a given street railway route. *Branford Electric Railway Association, author photo.*

Figure 5.3. The earliest form of handbrake control was the "goose-necked" handle. *Branford Electric Railway Association, author photo.*

car, they could not "hold" it. The motorman had to come up with a means of blocking the car wheels soon after stopping the car, especially if the car was on a hill.

The most important reason for not using the motors to stop the cars, however, was that the motors often sustained heavy damage in the process. Most streetcar motors of this period were designed to operate on 500–600 volts direct current and were insulated for up to 1,000 volts. When a motor's circuit was reversed, it generally developed twice its operating voltage, which meant a total of 1,000 to 1,200 volts. Still, this was an option for a motorman to employ in an emergency, although the best railway training programs emphasized *in an emergency.*

Sanders

Sanders were devices that spread sand on the rails in front of a streetcar's wheels to gain better traction. Sanders were used most often when the rails were slippery, but could also be used if the car needed to be stopped in a short distance (although too much sand on a rail could also cause a streetcar to skid). Slippery rail conditions occurred under circumstances similar to bad road conditions for automobiles. The most frequent time for rails to grow slippery was at the beginning of a rain shower, when oily deposits and residue created a thin film along the rail surface. In northern climates, snow, ice, and sleet also made it difficult to gain traction. In areas with large numbers of deciduous trees, wet fallen leaves were an additional hazard.

Sanders consisted of a sandbox, or hopper, where the sand was stored (usually under passenger seats near the streetcar's wheels), hoses leading from the sandbox

Figure 5.4. Hoppers for car sanders were usually located inside the streetcar. When the covers for this hopper were in place, it doubled as a seat. *Branford Electric Railway Association, author photo.*

to the wheels, and piping (if the sanders were air-powered) or mechanical linkages (if the sander was operated manually). If air-powered, a control valve was located near the motorman's brake valve and could be either hand- or foot-actuated. The reason hoppers were kept inside the car bodies was to ensure the sand was exposed to a minimum of moisture and other atmospheric conditions that might result in "clumping."

Sanders were made by a variety of manufacturers, from small specialist firms to the streetcar builders themselves. The sand itself was generally obtained locally. "Lake sand" was preferred where possible, as it had a "fine, even grain" and was easily dried and screened.[8]

Maintaining an adequate supply of usable sand could be quite an operation for a street railway. To be most effective, sand had to be dry and free of foreign objects, such as pebbles or small stones. Sand drying and screening operations varied from railway to railway, ranging from small, self-contained units on small railways to elaborate plants on larger systems comprising multistoried buildings, furnaces, conveyors, screens, and storage hoppers.

Of the Ohio streetcar builders, Kuhlman, Jewett, and Cincinnati made sanders. Not surprisingly, Kuhlman's sanders were Brill-patented (although it did make its own version before the Brill takeover), and could either be made at the Cleveland plant or obtained from Brill's Philadelphia plant. Jewett not only made its own sanders but also helped to create a small specialty firm, the Newark Air Sand Box Company, in the same city. Newark Air's sanders were based on the work of a Jewett foreman. While Cincinnati also made sanders, it only made them for specific cars it was building, and it did not market its sanders separately (see table 5.1).[9]

One of the more popular Ohio makers of sanders was the Nichols-Lintern Company. Located on Cleveland's West Side, it began as the Lintern Car Signal Company. The company changed its name and expanded its product line in 1904, when William Lintern, former master mechanic of the Cleveland, Southwestern & Columbus interurban, joined the firm.

Nichols-Lintern sanders could work independently of, or in conjunction with, a car's braking system. The company made a point of stressing its sanders' compatibility with any type of air braking system on the market. Among the exhibits at the 1906 American Street Railway Association convention were Nichols-Lintern sanders installed with Christensen, Westinghouse, and General Electric air brake systems (see table 5.2).[10]

In addition to complete systems, Nichols-Lintern also made sander components. Among these were sand traps that helped form the linkage between the sand hoppers and the discharge hoses. The traps could be used to clear out sand lines.[11]

Nichols-Lintern made its own sanders and licensed other manufacturers to make them as well. For example, its sanders could also be purchased through Cleveland's Pneumatic Railway Equipment Company. Ohio Brass created a new department within its factory to build and market a host of new railway appliances, including such Nichols-Lintern products as sanders, signal lights, and hose bridges.[12]

Ohio Brass was attracted to the Nichols-Lintern sander for several reasons. The discharge hoses were mounted so that they always remained in front of the car's wheels, even on sharp curves. Also, the sand hoppers were convenient to mount—they could be placed anywhere on the streetcar body. Finally, by tying the sanders into the braking system, Ohio Brass felt the Nichols-Lintern sanders would reduce wear on a streetcar's air system.[13]

Of the orders specified in the trade literature, most of Nichols-Lintern's products went to Ohio car builders. However, most did not go to Kuhlman, which was also located in Cleveland. Instead, most Nichols-Lintern sanders went to the farthest builder, the Cincinnati Car Company. Two-thirds of Lintern components produced before World War One were sanders, and the rest were marker lights. Most railways using Lintern's components were located in Ohio or Indiana.

By 1908 Ohio Brass was manufacturing sander components of its own design. Marketed under the "Universal Sander" name, Ohio Brass made both control valves and traps. It started making complete sanders in 1910.[14] (See table 5.3.)

In Dayton, the Dayton Manufacturing Company also made sanders. Called the Simmons-Moore sander, it could be operated either by compressed air or by hand or foot-powered mechanisms. Among the notable features of the Simmons-Moore sander was a set of revolving blades in the sand hopper, which ensured the sand would not "clump" or otherwise clog the sand hoses. This was not a common feature among sanders, although the St. Louis Car Company also used this arrangement. Simmons-Moore sanders were especially popular among Ohio interurbans. Examples could be seen on the Columbus, Urbana & Western; Scioto Valley Traction; the Columbus, New Albany & Johnstown; and the Columbus, Delaware and Marion.[15]

Of the car orders specified in the *Electric Railway Journal* between 1909 and 1913, sanders are mentioned on 2,538 cars. Ohio companies figured prominently in sander manufacture, accounting for 40 percent of all cars. Most sanders were manufactured by Ohio Brass (nearly 22 percent), with Nichols-Lintern a distant second (13 percent), and the balance made by Ohio car builders.

Of the cars equipped with Ohio Brass sanders, two-thirds went to New York railways, the remainder being divided among nine states plus Washington, D.C. Most cars equipped with Nichols-Lintern sanders (56 percent) ran on railways in Wisconsin (32 percent), Ohio (18 percent) and Illinois (16 percent). The remainder were divided among railways in sixteen states.

Headlights

Headlights were essential safety devices that allowed a motorman to see the track ahead during night operations and allowed people and animals on the tracks to notice the approaching car. Most headlights on the market were electrically powered by 1900 and used either arc or incandescent technology. Arc headlights were much brighter than incandescent ones and tended to appear on interurban systems only, as they were much too bright for city use (their intense light could stun or "freeze" a person or animal on the tracks). Rapid transit lines could use either type of headlight, but they tended toward incandescent ones, as their segregated rights-of-way made spontaneous encounters with people or animals unlikely. The brightness of arc headlights made them questionable even for interurban service at first. Tests conducted with arc headlights on railroad locomotives in 1890 showed that they not only had a "dazzling" effect upon people on the track, but also tended to blind crews on trains traveling in the opposite direction. For this reason, oil headlights persisted on steam railroads for a while longer, although the *Electric Railway Journal* commented in 1910 that the 1890 tests "cannot be considered at all conclusive against the use of electric headlights on suburban and interurban electric lines."[16]

"Dimming" features on arc headlights were first introduced either by adding resistance into the circuit or by using a combination of arc and incandescent lighting (arc lighting for the "high" beam, incandescent for the "low" beam). By 1910, arc lights could also be dimmed by reversing their polarity. Headlights of this type were introduced by General Electric.[17]

In addition to the reason stated above, most streetcars used only incandescent headlights because city streets tended to be better lit than those in suburban or rural areas. Streetcar headlights were kept lit night and day to aid pedestrians in determining whether or not a distant streetcar was approaching.

In Cleveland, the Multiplex Reflector Company enjoyed considerable success in the headlight market during the early 1900s. For example, the St. Louis Transit Company placed an order for 800 headlights in 1900. International orders included headlights for street railways in Barcelona, Berlin, London, and Melbourne.

A number of smaller Cleveland firms were active for part of the 1900–1914 period. The Duplex Headlight Company existed around 1904 and made what were known as "combination headlights" (headlights that used both arc and incandescent lighting). In addition, Cleveland's Globe Machinery & Stampings Company made cases or shells for headlights between 1901 and 1905.[18]

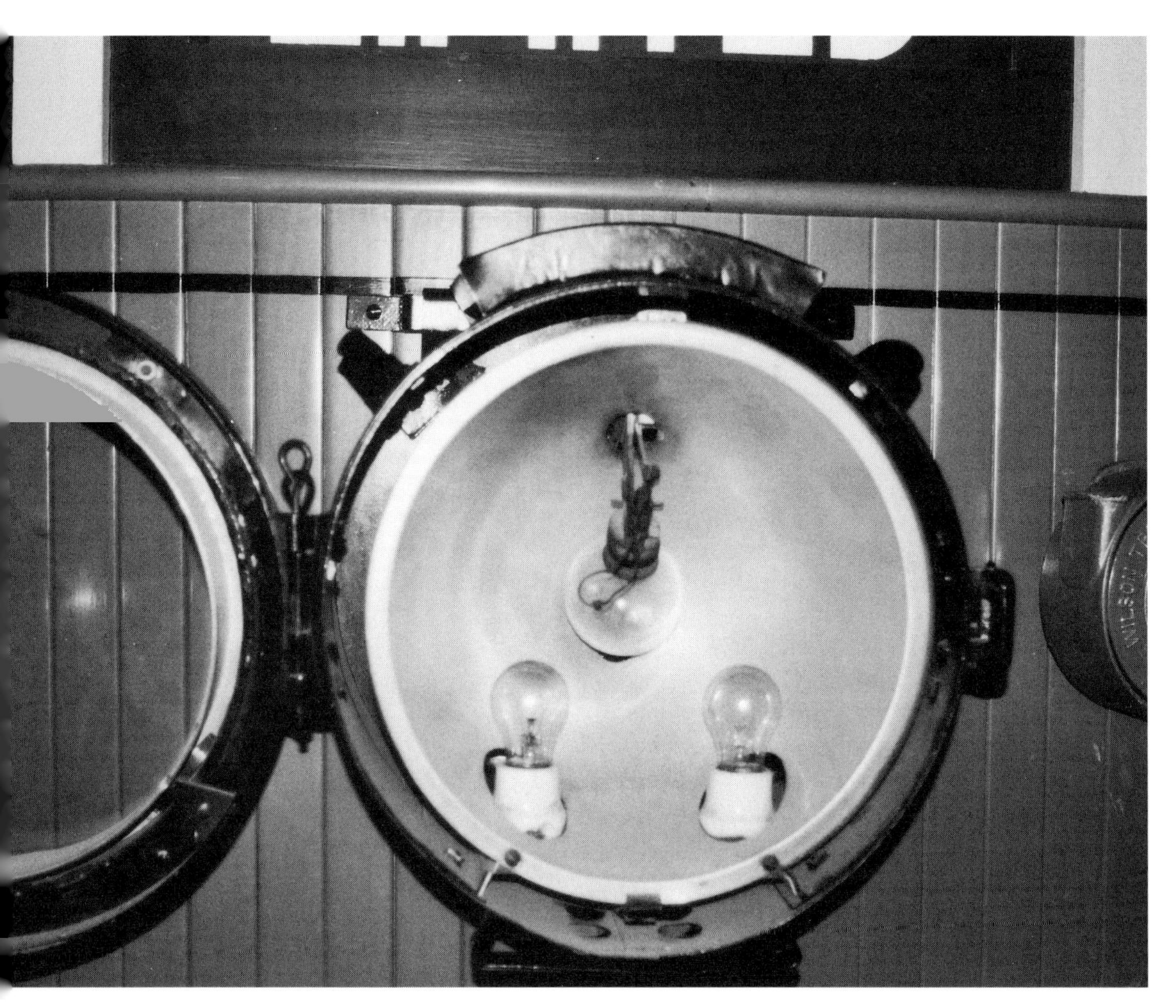

Figure 5.5. An interurban headlight. The two light bulbs (called "lamps") at the bottom of the headlight reflector are incandescent, and they were used within cities and towns (corresponding to the "low beam" of an automobile headlight). The larger lamp near the reflector's center is actually an arc light, which is much brighter than incandescent lighting, and this would be used on the interurban's private rights-of-way (corresponding to the "high beam" of an automobile headlight). *Branford Electric Railway Association, author photo.*

The Trolley Supply Company made headlights, which it marketed under the Climax and Star labels. Climax headlights first appeared in 1906. Like Duplex headlights, Climax headlights combined arc and incandescent technology.[19]

Trolley Supply unveiled its "Star" headlights in 1908, which were also combination headlights. However, unlike conventional headlights, which tended to use polished aluminum reflectors, Star headlights used nickel-plated brass. The manufacturer had discovered that polished aluminum had an annoying tendency to grow opaque over time. In addition to combination headlights, Trolley Supply also made straight incandescent and arc headlights under the Globe label.[20]

In 1909, Trolley Supply expanded into semaphore lights. Semaphore lights had several different colored lenses that could be changed by the motorman. They were used as on-board signaling devices that could be used to communicate with other cars or dispatchers encountered along a route. Nine inches in diameter, the semaphore lenses were available in white, blue, green, and red. Like the Star headlights, Trolley Supply made the semaphore reflectors out of nickel-plated brass.[21]

Another important Ohio-based maker of headlights that deserves mention is the Dayton Manufacturing Company. Dayton Manufacturing was originally a supplier of *railroad* components and a major supplier of oil headlights. Dayton made the transition to electric headlights in the early 1900s.

According to *Electric Railway Journal* orders published between 1909 and 1913, 3,520 cars had the type of headlight specified. Most of the cars (27 percent) had headlights manufactured by New York–based Crouse-Hinds, a company that is still in existence (among other things, it is a well-known maker of traffic lights). Dayton Manufacturing ran a close second, accounting for 22 percent of all orders. Dayton's market was mostly local, with Ohio and New York accounting for half its orders. An additional 29 percent went to railways in New Jersey and Indiana, and the rest went to railways in seven states.

Marker Lights

A different type of exterior light, known as a "marker light," was sometimes placed at the rear end of a car for following vehicles. These devices were most common on interurbans and rapid transit vehicles. Manufacturers of exterior lighting systems were scattered throughout Ohio, but the most prominent were located in Cleveland, Dayton, and Canton.

Before its reorganization as the Nichols-Lintern Company, Cleveland's Lintern Car Signal Company specialized in the manufacture of marker lights. The early history of the company is obscure, though it is possible the firm was once affiliated with Ohio Brass. Lintern marker lights were a common feature on Ohio-built streetcars, and they were sent to all Ohio car builders except Barney & Smith.

A Niles order provides the most detail. In 1905, Nichols-Lintern marker lights were used on a 10-car order for the Lake Shore Electric Railway, an interurban that ran along Lake Erie between Cleveland and Toledo. Rear markers included two red lenses in the panel above the vestibule windows. Front-end markers were green and white. The lights were controlled by individual switches, allowing the proper lights to be used depending upon the car's direction of travel.[22]

Gongs, Bells, and Horns

Last but certainly not least among a streetcar's arsenal of preventive safety devices were those that made noise, alerting those around the car of its presence. Street railways imitated the steam railroads and developed signals for their motormen to announce what the car was about to do. In most cases two reports meant that the car was about to proceed forward, while four meant the car was about to back up.

The most common form of audible warning device was the gong (not to be confused with bells or buzzers, which were internal devices used by conductors or passengers to signal the motorman). Most gongs were mounted on the underside of the car platform, beneath the motorman's station. To ring it, he struck a foot knob. Some railways mounted their gongs on the roof near the ends of the cars, requiring the motorman to pull a rope to ring the gong.

Figure 5.6. A Nichols-Lintern marker light. Marker lights were often used on the ends of interurban cars. *Courtesy of Dayton History.*

Horns could be mounted anywhere and were tied into the car's braking system to make use of the compressed air. A rope was usually tied to the actuating valve, and a practiced motorman could raise or lower the horn's tone and volume, depending on how he manipulated the rope. These devices were made by specialty manufacturers and by large car builders.

Fenders and Wheel Guards

In her study of personal injury cases involving railroads and street railways, Barbara Welke argued that the two greatest sources of street railway injury cases involved people being struck by cars and boarding and alighting accidents.[23] With the exception of powered brakes, no safety device received the greatest amount of attention in the trade press before World War One than fenders and wheel guards. These devices were intended to be the last line of defense for the unwary pedestrian who crossed in front of a moving streetcar. Their basic purpose was to prevent people or objects from getting crushed underneath the car wheels.

Some safety devices were designed to protect the cars in the event of a collision. Pilots, also known as cow catchers, were V-shaped slanting fenders mounted under the end of a car body in front of the trucks. They were heavily built, designed for knocking objects clear of the tracks when hit at high speed. Thus pilots were more common on interurbans than on street railways. Pilots were often made by the indi-

Figure 5.7. An Eclipse fender, mounted on the front of a Cleveland streetcar. *Courtesy of Cleveland Public Library, Byron Filkins photo.*

vidual streetcar builder, who could tailor them to a specific order. For example, Niles made and fitted its own pilots for four interurbans for the South Cambria Railway of Johnstown, Pennsylvania.[24]

Bumpers were another safety feature used on the ends of cars, although they served more to protect car ends from routine bumps and scrapes. Like pilots, bumpers were often made by the car builder. An example of an Ohio-built bumper order could be found on two cars Niles built for the Twin City Light Company of Minnesota.

For streetcars, the most important collision devices were the fender and wheel guard. Like pilots, their purpose was to prevent people or large objects from getting caught underneath a streetcar's wheels. Fenders consisted of wire baskets mounted rigidly in front of a streetcar, while wheel guards were protective aprons that dropped from the underside of a car's vestibule when triggered. Triggering mechanisms usually consisted of small fences or gates suspended under the front of a streetcar, just above the rails. When the trigger was pushed backward, the apron dropped.[25]

Fenders and wheel guards were made both by specialist firms and by large component manufacturers that had diverse product lines. On occasion, streetcar builders themselves made fenders or wheel guards of their own design or to the custom specifications of an individual railway. This practice was by no means universal, and the only Ohio streetcar builders to do so consistently were Kuhlman and Cincinnati. Niles made "pick-up fenders," but only one order featuring these was reported in the trade literature.[26]

Cleveland's Eclipse Car Fender Company was a leading national manufacturer of car fenders. Unlike most fenders, which consisted of rigid screens held in place by a framework of piping and rods, the Eclipse fender consisted of two framed screens. The first was held rigidly in front of the streetcar. The second was suspended in front of the first and was pivoted forward. The second screen would pivot backward upon striking an object. By pivoting backward, the fender absorbed some of the backward momentum of whatever was struck, lessening the impact. The frames were covered with rubber sleeves to provide a cushioning effect.[27]

Developed by Benjamin Lev, the Eclipse Fender was first unveiled locally on the Brooklyn Heights Railroad Company's lines in 1903. Barbara Welke notes that during the early 1890s at least one manufacturer was careful to advertise that its fenders were tested on live human beings. Lev took this a step further. First, he allowed *himself* to be run down by a streetcar traveling at 6 miles per hour and then at 21 miles per hour. Both times he smoked a cigar to show onlookers that not only was he unharmed but that his cigar was undamaged and had stayed lit both times. Although he used a 3-foot-tall, 25-pound bag filled with sand and wood shavings to simulate a child being struck, he made a point of using live subjects for demonstrations whenever possible. At the 1903 ASRA convention he did so repeatedly, although six other company officials spelled him from time to time.

The fender's design, combined with Lev's personal demonstrations, made the Eclipse Fender an immediate success. By the end of 1903, the Cleveland Electric Railway had installed 225 of them, and in New York the Brooklyn Rapid Transit Company made them standard on its vast system, starting with an order then under construction at the Kuhlman plant.

Eclipse not only made fenders for streetcars but also developed devices appropriate for the larger and faster interurbans. For example, Eclipse fenders were used on cars for Northern Ohio Traction & Light. By 1911, Eclipse Car Fender had expanded its product line to include not only Eclipse fenders but also wheel guards in a separate ACME line (see table 5.4).[28]

One final group of front end devices deserves mention: anti-climbers. These were similar to bumpers, only they had external ribs that helped prevent one car from overriding the other car and crashing through it (if one car was to climb over the platform of another, the anti-climbers would have to rub past each other, and the ribs would slow or stop them). Anti-climbers were common to all forms of railed transit vehicles.

As one might well imagine, there were numerous designs of fenders and wheel guards for a street railway to choose from, and the inventors and manufacturers of each made bold and extravagant claims as to how wonderful their devices were and how they were better than anything else on the market. While some of these devices were, in fact, very effective at doing what they were designed to do, there were many others that were not. Concerned with the number of faulty "safety devices" that were appearing on the market, large street railway systems, municipal governments, and even some state railroad commissions conducted tests using either manikins or sandbags of varying sizes and weights to determine how effective these devices were.[29]

Figure 5.8. The actuating mechanism of a wheel guard. If the gate is pushed back by an obstacle, a protective gate dropped in front of the car's wheels. *Branford Electric Railway Association, author photo.*

There were at least eleven tests reported in the trade literature. Of these, the one that received the most attention was conducted in 1908. This was actually a pair of tests conducted at roughly the same time under the auspices of the Public Service Commission of New York. One was carried out by General Electric at its Schenectady complex, and the other was run by Westinghouse at its East Pittsburgh facility. Between the two, 38 fenders and 29 wheel guards were examined.

A thorough set of 12 tests were run for each of the safety devices. The devices were evaluated not only on their ability to protect the unwary pedestrian but also on their likelihood of overrunning objects that might be encountered lying across the tracks. Three dummies were used, representing a 5 foot 9 inch man of 170 pounds; a 5 foot 3 inch woman weighing 120 pounds; and a 4 foot 6 inch child weighing 50 pounds. Two types of pavement were simulated: paving block and a surface simulating asphalt or macadam. Tests were run at both 6 and 15 miles per hour.

Of the 67 devices tested, 23 achieved a rating of 75 percent or higher (15 fenders and 8 wheel guards). The highest ratings were attained by wheel guards. The very highest was achieved by the Hudson & Bowring wheel guard (commonly referred to as the "H-B wheel guard"), which scored an impressive 86.9 percent. The second highest rating (83.6 percent) was achieved by the Parmenter wheel guard. Benjamin Lev's Eclipse Fender received the third highest rating (80.6 percent), tied with the Watson wheel guard.[30]

Among the *Electric Railway Journal* car orders between 1909 and 1913, 911 cars are mentioned with the type of fender specified. The Eclipse fender did well, accounting for 16 percent of all cars, more than any other fender manufacturer.

DEVICES TO PROTECT THE PASSENGERS

The other major category of safety devices used on streetcars comprises those that were used to protect the passengers themselves. Although this is by no means a comprehensive list, the most common were gates and steps, stanchions, railings, and straps, and interior lighting.

Gates and Steps

The second type of accident that Welke discussed was one that occurred while a passenger was either boarding or alighting from the streetcar. Since many of these accidents occurred when people attempted to board a vehicle that was either moving or just starting to move, some railways used gates that effectively prevented unnoticed boardings and also helped keep passengers from falling off. The most prominent type of gate used by the industry was the Minneapolis gate, developed by the Twin City Rapid Transit Company.[31]

For all of the current scholarship directed toward gates, they were only a small part of the effort to eliminate boarding and alighting accidents. The industry directed far more attention toward the cars' steps. Unlike gates, which involved appliance manufacturers and the railways, steps also brought car builders into the debate.

Between 1900 and 1910, the *Street Railway Journal* and *Electric Railway Journal* published a number of statistical compilations put together by large railways on the types of accidents they encountered over a given period (usually at least one year). Of those accidents involving passengers, the most common was passengers falling from a car's steps. Naturally, many of these accidents were the direct result of operational issues—usually the conductor signaling the motorman to proceed without checking to see that the steps were clear.

In addition to more careful operating habits, however, railways and (to an increasing degree) regulatory bodies began to take a closer look at the steps themselves. At issue were two features of car steps—their height from the pavement and from each other, and their surface. The latter problem was usually an issue between railways and makers of step treads, but the former was more complex.

Most streetcars required passengers to negotiate three or four steps to enter the car's interior. There were usually one or two steps leading to the end platform (or vestibule), and then an additional step from the platform into the car body proper. If the steps were not spaced evenly, the first step was usually the highest and often exceeded 16 inches in height from the railhead. On streetcars with only one step between the ground

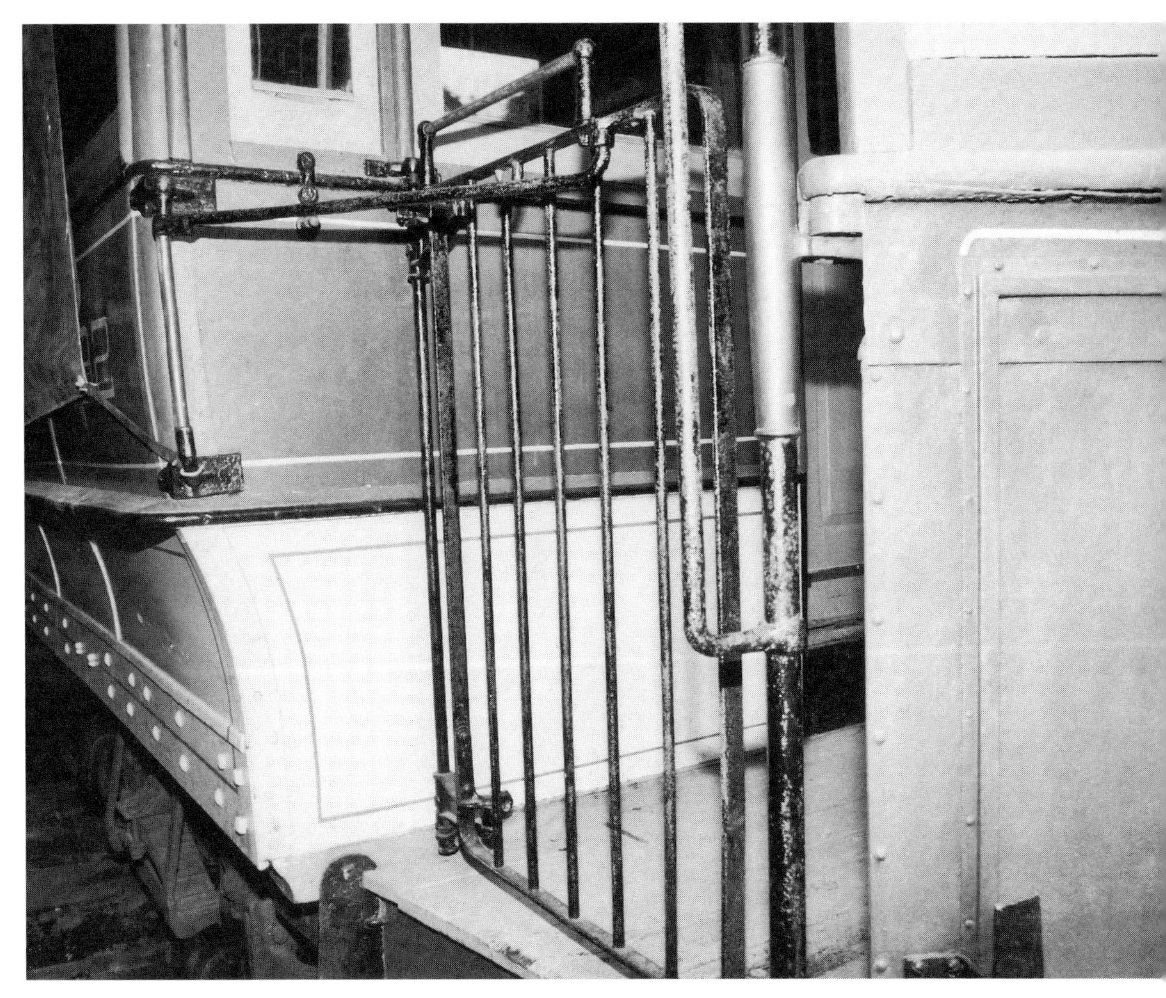

Figure 5.9. Like many other streetcar components, platform gates came in a variety of forms. This is a swinging gate. *Branford Electric Railway Association, author photo.*

and platform, the step onto the platform could be even greater, especially if the car was intended for suburban and not solely for city service.

Step *height,* on the other hand, was a far more serious issue. The *Electric Railway Journal* once went so far as to say, "Perhaps no single detail of electric railway car construction affects the comfort, convenience, and safety of passengers to such an extent as the dimensions of the platform steps."[32] For a typical double-trucked city streetcar, the two major design constraints affecting step height were car width and truck design. In most cases with car width, steps could not project beyond the sides of the car. With trucks, the floor had to be high enough to provide sufficient clearance for the trucks to swivel. This in turn was dependent in part on the size and shape of the truck's frame as well as the diameter of its wheels.[33]

As might be expected, there were varying opinions as to how high an individual step could be. Basing its recommendations on an overall floor height of 41 inches, in 1908 the American Electric Railway Engineering Association (AEREA) recommended three steps with maximum heights of 17, 14, and 10 inches, respectively. The Public

Figure 5.10. A folding gate. *Branford Electric Railway Association, author photo.*

Service Commission of New York based its recommendations on an overall floor height of 40 inches and specified its maximums as 15.25, 13.25, and 11.5 inches. In Canada, the Ontario Railway & Municipal Board agreed with a 40-inch floor height, but only specified the height of the first step as 15 to 16 inches.[34]

For some major cities, such as Chicago, AEREA's recommended standards presented little problem. Chicago streetcars had a *maximum* floor height of 41 inches and maximum step heights of 16, 14, and 11 inches. For other cities, such as New York, tougher regulations, such as those of the Public Service Commission, could mean trouble. On average, New York streetcars conformed to AEREA recommendations but fell wide of the commission's mark.[35]

There was also the problem of existing car fleets that were around before the regulations were put in place. One solution was to modify car entrances and exits using folding doors and steps with a brake interlock. This not only promised to solve the step issue, but also made the gate issue irrelevant. Folding steps had been around as early as 1901, and they became a common means of making steps easier on the riding public. For railways that did not wish to adopt folding steps or could not use them, the solution could entail some radical car surgery. Platform supports had to be relocated, redesigned, or both. Brake rigging might also require relocation. For car builders and components manufacturers, stricter regulations that were either in place or coming into place by 1910 meant redesigning bodies, end platforms, trucks, and motors.

Starting in 1910, "low floor" car designs began to appear. The first "low-floor" car design was introduced in Pittsburgh. Built by the Standard Steel Company, the design was used on 50 lightweight center-entrance trailers that were built almost entirely of steel (wood was retained for the seats and floors). Unlike most center-entrance cars, which separated the entrance and exit area from the rest of the car via interior steps and heavy partitions, the floor was kept at a uniform level throughout the car, and no partitions were used. The car's trucks used 22-inch wheels, which brought the car floor to within 30 inches of the rails (the car entrances had one step built into the car floor just inside the doors). Longitudinal seats provided a seating capacity of 67.

It was originally intended to have these cars motorized, but motor technology had not advanced to the state where motors small enough for 22-inch wheels existed. A compromise was reached two years later, when Westinghouse developed motors that could be used with 24-inch wheels. The motors were longer than conventional ones but smaller in diameter.

Figure 5.11. This Washington, D.C., streetcar displays several safety devices. The front platform is protected by a folding gate, and the car's front end is protected by both a fender and a wheel guard. Note that both the wheel guard's actuating bar and protective gate are visible. *Courtesy of Indiana Historical Society, Cincinnati Car Corporation Collection.*

Figure 5.12. Perhaps the best way to prevent unanticipated attempts at platform boarding was to use doors. The folding type shown here was the most common. Another refinement was the folding step, also shown here. Steps and doors were interlocked, so that when the doors were closed, the steps were up. *Branford Electric Railway Association, author photo.*

The next advances in low-floor design came in 1912, with designs by the New York Railways and Brooklyn Rapid Transit systems. Both companies were able to eliminate steps entirely by using "maximum traction" trucks with their small wheels facing inside.

These trucks were common before World War One. Like most trucks, maximum traction trucks had two axles, one with a motor and one without. The motored axle had standard-diameter wheels, while the unmotored axle had smaller diameter wheels. The idea was to concentrate most of the car's weight in the powered axle, thus increasing the car's tractive effort. Usually facing outward, the smaller wheels ensured the axles remained perpendicular to the rails.

The New York Railways design was more radical in appearance (railway historian Andrew Young once described it as "ferocious"), but less daring than the Pittsburgh cars technologically, using conventional motors mounted on axles with 30-inch wheels. The smaller wheels were only 19 inches in diameter, bringing the floors of the New York cars to an astonishing 10 inches above the rails. Like the Pittsburgh cars, the floor height was uniform throughout, with no partitions separating the entrance/exit area. Transverse seats (with semicircular seating at the ends) gave a seating capacity of 51.

Figure 5.13. With interlocked steps and doors, when the doors were in the open position, the steps were down. *Branford Electric Railway Association, author photo.*

The car's unusually low floors required a rather unorthodox equipment arrangement. The compressors and other components had to be clustered at the car's ends, and the motorman was isolated from the car's interior (access to his compartment was gained through an outside door). Despite the provision of large windows at the rear of the motorman's compartments, electrically operated signals were necessary for communication between the motorman and conductor.

The prototype's body and trucks were designed in-house, although they were built partly by Brill and finished in the railway's own shops. A production order of 175 cars was placed with the St. Louis Car Company the following year. They were similar to the prototype, only the motorman's access door was moved from the car's nose to one side of the car's ends. Brill made a number of production models of these cars for other railways throughout the 1910s.

The Brooklyn design was conventional in appearance on the outside, but embodied some unusual features of its own. Like the New York Railways design, the Brooklyn car used conventional motors, although it used smaller (28-inch) wheels and large (21-inch) "pilot wheels," bringing the floor to 16 inches above the rails. It embodied a combination of transverse, longitudinal, and semicircular seats to provide a seating capacity of 58. Also like the previous designs, it eliminated the use of interior partitions.

Unlike the previous designs, the floor on the Brooklyn car sloped upward toward the ends, allowing the car's operating equipment to be mounted in the usual fashion (the Pittsburgh cars also did not require an unconventional equipment arrangement, as its floors were 30 inches above the rails). The car was double-ended, and the motorman's cabs were arranged to be folded down and replaced with semicircular passenger seats when not in use.[36]

Stanchions, Railings, and Straps

The purpose of stanchions and railings is fairly obvious. They were there to assist passengers in boarding and negotiating their way through the car to and from their seats (especially when the car was in motion). If all of the seats were taken, the stanchions and railings provided standing passengers with a convenient handhold to maintain their balance. It was common for streetcars to have grab handles mounted on the aisle ends of seat backs as well.

Straps were intended solely for the use of standing passengers once a car got underway. Initially, leather straps were fastened in loop fashion to horizontal bars mounted near the car's ceiling. There were usually two bars per car, one on each side of the aisle. It was the leather strap that created the familiar term "strap hanger."

Although beyond the scope of this study, advertisements suggest straps attracted more than a little concern from public health officials. Makers of leather straps went to great lengths to emphasize the sanitary nature of their products, and some even incorporated the term *sanitary* into their company or product name. Sanitary or not, leather straps had the annoying tendency to pull apart over time. This was a never-ending bother for maintenance crews.

Straps began to fall into disfavor by the early 1910s, and it became more common for railways (especially rapid transit systems) to make greater use of stanchions. Writing in 1914, one New York consulting engineer stated, "Hand straps . . . do not afford the same degree of confidence and sense of security that is inherent in the vertical stanchion, which offers a firm hand hold to invite the passenger to approach the exit door while the car is still in motion."[37]

Interior Lighting

The final safety item to be considered is often the most overlooked: interior lighting. Streetcar lighting changed tremendously over time. The first most common form was the oil lamp. By 1910, all streetcar interiors were lit with incandescent lights. However, there were several types of incandescent light bulbs (called "lamps" before World War One) available. The most common had carbon filaments. While they gave off acceptable amounts of light, they were highly fragile and had relatively short life spans. A more durable type of filament was tantalum, which, like carbon, gave off an acceptable amount of light. However, they were also more expensive. It was not until about 1910 that tungsten lamps came into common use. Tungsten lamps gave off acceptable light and were more durable and less expensive to produce than tantalum lamps.

Figure 5.14. Some streetcars were designed to swallow crowds by accommodating large numbers of standees. Here the seats were placed along the car's sides. Note the large number of leather straps for standees, hence the term *strap hangers.* A tall passenger could hold onto the bars used to secure the straps. Note also the clerestory roof and the rows of lights on either side of the clerestory. *Branford Electric Railway Association, author photo.*

Streetcars and interurbans usually had two independent types of interior lighting. One illuminated the passenger seating area, while the other illuminated the vestibules. Vestibule lighting was necessary to assist passengers in boarding and alighting. Vestibule lights were kept independent of the interior lights so that motormen could extinguish the light in the particular vestibule from which they were operating the car. This helped diminish the glare created from interior lighting on the vestibule windows, especially at night. Vestibules were often equipped with curtains that motormen could draw behind themselves to reduce glare from lighting within the passenger area as well.[38]

It is important not to confuse street railway lamps with the light bulbs we are familiar with today. Even though the two look similar, street railway lamps were (and still are on some transit systems) designed for direct current, not alternating current. Lamps

Figure 5.15. A Nichols-Lintern light selector switch. This device prevented a single burned-out lamp from extinguishing all of the lamps wired in a series by isolating that particular lamp. Conductors manipulated the device by twisting the knob in the center of the device. *Northern Ohio Railway Museum, author photo.*

used for interior lighting were normally wired in series, five or six lamps to a circuit, with the number of circuits depending on the size of the streetcar and the preference of the railway.[39]

In addition to the type of lamps used by street railways, there was also the question of how many to use and how to arrange them. There was no consensus among street railways on this subject, and even investigations by the American Electric Railway Association's Engineering Association proved inconclusive. As a result, there were numerous lighting fixtures and configurations for a street railway to choose from.

At one point Ohio ranked third in the nation for the production of all types of lighting equipment. Most of this activity was centered in the Cleveland area, where General Electric had a number of plants. However, lighting equipment, particularly lamps, were made elsewhere in Ohio, notably in Warren and Niles in Trumbull County.[40]

Like other communities in the Connecticut Western Reserve, Warren was founded late in the eighteenth century. The Pennsylvania and Ohio Canal opened in Warren in 1839, and the town began to grow soon afterward.[41] Warren first made a name for itself as a cheese producer. The peak year for cheese was 1880, when over 112 tons were made. Even as late as 1938, dairy products accounted for over half of the area's agricultural output. Industry came to Warren during the late nineteenth century, and for a time the city was the nation's second largest producer of lamps. Other products that came out of this city were pipe-threading equipment, steel ranges, and railroad tank cars.[42]

J. Ward Packard and his brother, William D., began manufacturing lamps in Warren in 1880. They eventually founded two major companies: the Ohio Lamp Works in 1891 and the Packard Electric Company in 1893. The Packards held patents on lamp designs, lamp sockets, and vacuum systems for use in lamp production. Most of their production took place at the Ohio Lamp Works. This company was eventually absorbed by General Electric and converted to produce high-intensity lamps. During the 1890s, the Packard brothers founded an additional company, the New York and Ohio Company, and located it in part of the Packard Electric Works. New York & Ohio specialized in incandescent lamps, transformers, dynamos, and motors. By 1916, New York & Ohio had become the Packard Lamp Division of General Electric and was also beginning to specialize in automotive ignition cables.[43]

Of all specialty manufacturers, electric lamp factories were probably the most given to mass production. Even if they received a small number of orders, they were usually large enough to demand such organization. For example, the Warren Electric & Specialty Company was another manufacturer involved with street railways. In 1911 the company received an order from New York City's Interborough Rapid Transit Company for 41,000 incandescent lamps. Of these, 26,000 were intended for interior car lighting and 15,000 for station lighting.[44]

Since interior lamps were divided into banks of five or six, each of which was wired in series, a burned-out lamp could plunge entire sections of a car's interior into darkness. To remedy the problem, Nichols-Lintern developed a selector switch that enabled burned-out lamps to be isolated without affecting the rest of the lamps in the circuit. Kuhlman orders sometimes featured this device. Examples of Nichols-Lintern lighting controls could be seen on New York State Railways and Cleveland Railway equipment.[45]

CONCLUSION

Safety devices came in different sizes and shapes and had many functions to perform. Because safety was such a complex issue, railways could not anticipate everything. Thus there were numerous manufacturers specializing in all manner of safety equipment, each claiming to have "the solution" to a particular problem.

Ohio manufacturers did not encompass the whole of this issue, nor did they make devices that were necessarily the best or most common. When put to the test, however, Ohio safety devices performed well. Hopefully what this chapter has demonstrated is that Ohio safety devices were a fairly common feature on street railway vehicles before World War One. While they may not have engaged all of the safety issues of concern to the railways, Ohio manufacturers certainly grappled with many of them and were able to make their presence felt within the street railway industry.

Figure 6.1. A car crew for a North Jersey Rapid Transit Company interurban. The large man in the striped uniform on the left is the motorman. The man in the bow tie is the conductor. The conductor was responsible for the collection and registration of fares, as the change dispenser worn at his waist attests. *Courtesy of Branford Electric Railway Association.*

6

FARE COLLECTION AND REGISTRATION

> Conductor, when you receive a fare,
> Punch in the presence of the passenjare.
> A blue trip slip for an eight-cent fare,
> A buff trip slip for a six-cent fare,
> A pink trip slip for a five-cent fare,
> Punch in the presence of the passenjare!
> Punch, brother, punch with care,
> Punch in the presence of the passenjare!
>
> ISAAC BROMLEY, NOAH BROOKS, W. C. WYCKOFF, AND MOSES W. HANDY

In today's world of subsidized, government-run transit agencies, it is easy to forget that in the early twentieth century, urban transit was a private industry. It derived most of its income from fares–what passengers were charged for riding the streetcars. The business of collecting fares and accurate recordkeeping were essential to the industry's success. As the above poem (once derided in a Mark Twain short story) suggests, the job could be complex, confusing, and tedious. Manufacturers had to cut through the complexity, confusion, and tedium to design equipment that was accurate, durable, and simple to use. What follows is an overview of the issues involved and the types of devices that were developed to satisfy them.

FARES AND TRANSFERS ON STREETCAR LINES

The most common fare for street railways between 1900 and 1914 was 5 cents. It dated back to the early days of mechanized street railways, when a nickel was less than what omnibuses and horse car lines were charging (often 6 to 9 cents). Based on the assumption that a flat rate would automatically yield a good return, the five-cent fare was created in an era when investment in physical plants was limited and the competition was more expensive.

As electrical technology developed and matured during the early 1900s, substantial sums were poured into upgrading physical plants and rolling stock. As investment increased and the rate of return from the nickel fare dropped, there was understandable movement among street railways to increase their fares. Here they ran afoul of the Progressive Movement.

Fearing that railway ownership might try to increase earnings beyond a reasonable level and sensitive to anything unpopular to their constituents, politicians at state and local levels fought to keep fares fixed at a nickel. Many state governments established regulatory bodies to ensure riders were not being exploited by nefarious businessmen. However, despite the unpopularity of fare increases, state regulators usually allowed them if a railway could demonstrate the need. Although this could (and often did) lead to months or even years of squabbles between the railway, local municipalities, state governments, and the courts, a railway generally prevailed if its case was sound.

Concerned over the legal expense and length of time involved in a proposed fare increase's approval or rejection, the American Electric Railway Association created the Fare Research Bureau in 1914. One of the Bureau's principal tasks was to investigate factors affecting the cost of passenger service. Association members were encouraged to discuss these factors freely with Bureau personnel, who compiled data on investment, depreciation and related charges, the actual cost of operations, and the nature of railway operations (length of haul, traffic density, etc.).

Another major task given to the Bureau was the compilation of statistical data both on railway operations and finances. All findings were summarized in reports that were made available to all Association members. Ideally, the Fare Research Bureau was to become *the* authoritative source on the economics of railway fares.[1]

Closely related to the issue of fares was the issue of transfers. Even in a small or midsized city, a passenger's journey often required travel on more than one line. Most railways allowed for the free transfer from one line to another, and a few even permitted the free transfer to two. Other railways charged an additional cent for a transfer.

Legally, street railways were only required to issue transfers at the time passengers paid their fares, and then only upon request. They also did not have to issue transfers that would allow the passenger to travel in the opposite direction. By 1912, most railways had added specific time limits to their transfers, ranging from two hours to as little as 15 minutes. To reduce transfer abuse, tickets often included such additional information as the current date, direction of travel, and sometimes even the intended final destination of the passenger.[2]

Most railways experienced an increase in transfer activity over time. Nationally, the trend was toward a slight increase, from 18.2 percent of all passengers in 1902 to 20.9 percent in 1907. Locally, the upward trend could be more dramatic. In Philadelphia, for example, transferring was almost unheard of in 1900 (less than 2 percent of its passengers changed lines). By 1907, this figure had jumped to nearly 16 percent. Similar increases were experienced in other cities. In a 1908 survey conducted by AERA, for instance, it was determined that over 60 percent of all Chicago passengers transferred to at least one additional line in the course of their journeys. The figure was over 50 percent in Nashville, Tennessee, and over 40 percent in Kansas City, Missouri.[3]

Other cities were less prone to generate a large amount of transfer traffic. In New York City, surprisingly, only a little over a third of its passengers transferred between lines. In 1908 the least amount of transfer activity seemed to take place in Spokane, Washington, where it was estimated that only 10 percent of its passengers transferred from one line to another.

For large, complex street railway systems, transfers evolved from a public courtesy into a colossal headache. For instance, the Brooklyn Rapid Transit Company (BRT) issued transfers at only six locations before 1895. By 1908, BRT was issuing them systemwide. To do so, 11 types of transfer tickets were required, and in some cases two separate transfers had to be issued to a single passenger. Although the types of transfer tickets were reduced to seven by 1910, transfers remained a perpetual source of difficulty for BRT.[4]

The transfer of passengers between railways owned by different companies depended in part upon the attitude of the companies and local government(s) involved. In most cases (such as in Boston, Detroit, or Chicago), railways combined into single, comprehensive systems, making the issue irrelevant. However, some cities (such as New York) were served by multiple railways throughout the 1900–1914 period. Ideally, neighboring railways worked out agreements among themselves, but there were cases of prolonged feuds between municipal governments and street railways over inter-railway transfers. One of the more notable disputes occurred in New York City over railways operating within the borough of Manhattan.

For the passenger, however, the nickel fare, especially when accompanied by the free transfer, was a bargain by any measure. In 1907, the average longest ride for 5 cents was just under 11 miles. Combined with a transfer, the average longest ride extended to just over 20.5 miles.[5]

FARES AND TRANSFERS ON INTERURBANS

Unlike street railways, interurbans could not charge systemwide flat rates due to the different nature of their service, which featured fewer stops and much greater distances than city lines. Passenger fares were calculated either on a per-mile or zone basis. As their names suggest, a per-mile fare was calculated by charging a specific rate per mile of travel. With zone fares, a line was divided into a number of sections called zones, with a flat rate established for each zone. A passenger's fare was determined by multiplying this rate by the number of zones through which the passenger traveled.

Most interurbans charged a minimum fare of 5 cents, with per-mile rates ranging from as little as 1.25 cents to as much as 8 cents depending upon the nature of the service. Obviously, a fare calculated on 8 cents per mile would be considered extravagant by most if not all passengers of the day. Rates this high were reserved for specialized lines, or those offering higher than average levels of service. An example of an 8 cent rate was Pacific Electric's excursion line that climbed Mount Lowe.

Concerned that the interurbans might turn exploitive, Progressive reformers did not limit their attention to street railway fares alone. By 1907, 20 states had either established regulatory bodies or passed legislation that limited interurban passenger rates. Fourteen of these states capped interurban rates at 2 cents per mile, which often corresponded to rate limits imposed on steam railroads.

Such regulatory action provoked the ire of the interurbans, which argued that consideration should be given to the nature of a particular railway's service. Steam railroads, for instance, offered service that covered even greater distances. There was also a big difference between an interurban line that ran between two large cities and a line of the same length that ran from a large city into the hinterland. However, some interurbans derived additional income from freight operations, giving them a source of revenue that was neither available to nor convenient for most street railways.

While interurbans did not have to deal with transfers to the same extent as street railways, they often made connections with other interurbans and steam railroads. Interurbans entered into agreements with each other to allow passengers to change from one railway to another, and in some instances they even had reciprocal agreements with steam railroads. As mentioned in previous chapters, equipment interchange was another possibility.

A fairly common practice in areas of dense interurban concentration was the use of interchangeable coupon books. These books contained between 200 and 300 nickel coupons that were valid on participating interurbans. Interurbans honoring these books were usually members of a regional association or owned by a large syndicate, such as those found in the states of Indiana, Ohio, New York, and Illinois.[6]

RAPID TRANSIT OPERATIONS

As with most street railways, the usual fare on subway and elevated lines was 5 cents. However, rapid transit differed from street railways and interurbans in that fares were rarely, if ever, collected on board the trains. Rapid transit fares were usually sold and collected by agents on duty at each station to speed the process of passengers boarding the trains. A limited amount of free transferring was permitted between rapid transit and street railway lines, and it seldom took place between rapid transit lines and interurbans.

DISCOUNTS AND FREE RIDES

In an attempt to speed the process of fare payment aboard the cars, many street railways sold tickets that passengers could purchase ahead of time in lieu of having to come up with exact change upon boarding (or delaying the boarding process by having the conductor make change). To encourage ticket purchases, railways offered quantities of tickets at a reduced rate. A common practice was to charge passengers a nickel cash fare, although they could purchase six tickets, each good for a single fare, for 25 cents. Tickets were available for purchase at the railway's offices or sometimes at designated stores or other locations along the railway's lines.

The policy of extending free rides to passengers varied greatly among electric railways. The practice was never widespread. One estimate placed the percentage of free rides nationally at about 1 percent of total ridership. To say the least, they were popular with those that were offered them. One railway official quipped, "The grant of a pass [often seems] the easiest way to gain a man's friendship, because nothing delights the average American so much as the thought of getting something for nothing." But free rides were not so popular with the accounting office.

For those railways offering free rides, such as the Philadelphia Rapid Transit Company, most limited them to employees or officials of the railway. Often those employees had to be in uniform. If the free ride was extended beyond the railway personnel, it was usually offered to policemen and firemen. As with railway employees, police and fire personnel had to be in uniform.[7]

FARE MEDIA

Passengers had several options to choose from when it came to paying their fares. Tickets and coupons were preferred, as they sped up the collection process and were the best way for the railway to compile accurate statistical information. Tickets came in a number of shapes, sizes, colors, and even materials. Most tickets were made of paper of varying thickness, although a few were celluloid.

If a railway was large enough, it could print its own tickets and transfers. Such was the case with the BRT. Housed in a former cable car powerhouse, the BRT printing department was established in 1903. It consisted of five presses, three stitching machines, and one cutter. The plant required 16 employees to produce an average of 35 million transfers and tickets per month.[8]

Railways like BRT were more the exception than the rule. Most railways purchased their paper fare media from outside firms. Of those printers specialized in railway tickets and transfers, the largest firm was probably the Globe Ticket Company. Based in Philadelphia, Chicago, and San Francisco, Globe offered numerous types of standard ticket and transfer formats, as well as the ability to produce custom orders.[9]

In Ohio, Cleveland's MacDonald Ticket and Ticket Box Company devised a system that used two-part tickets. The first part went to the passenger, while a stub remained with the ticket agent or conductor. A metal box, with eight slots that each held up to 100 tickets at a time, kept the ticket portion in the open, leaving the stubs concealed. To issue a ticket, the conductor pulled a ticket from the box, leaving the stub behind. To audit a run, accounting officials unlocked the box, retrieved the stubs, and checked them against the money turned in. An early user of this system was the Toledo & Indiana Railway Company.

In 1907, MacDonald began to branch into other markets by modifying its tickets for use as checks in coat rooms. MacDonald did not abandon the transit market, however, and in 1908 produced a new register with 40 individual slots, yet weighed only half as much as its original units.[10]

Another large printer was the National Ticket Company of Cleveland. As its name suggests, National Ticket was initially a printer that produced various kinds of tickets—interurban tickets among them. Recognizing the value of combining recordkeeping with ticket issuance, National Ticket began making two-part tickets with carbon sheets in 1905. The system did away with strong boxes or other items that might clutter a ticket counter or conductor's station. The company developed a further system that used a combination turnstile and canceling box in 1906. Passengers entering a station platform or an interurban car placed their ticket in the canceling box, which accepted and cut up the ticket once the passenger passed through the turnstile.[11]

"Metal tickets" began to appear on some systems by 1910. Also known as "tokens," examples of this fare medium were the same general size and shape as ordinary coins, with the name of the railway and sometimes the words "one fare" stamped on their surfaces. Some systems used different styles of token for different classes of fare.

Of course, there was always cash, but cash presented difficulties. Passengers paying with exact change were no problem. However, passengers did not always have exact change. It was easy to make change for dimes, quarters, or 50-cent pieces, but problems arose for passengers attempting to pay their fare with large denomination bills.

Legally, conductors were not required "to furnish change for anything more than a reasonable amount." The definition of "reasonable amount," however, differed from state to state. For example, in New York and Georgia, conductors were only required to change amounts up to $2, while in California and Tennessee the amount was $5. Traveling salesmen with large territories chafed at such disparities.

Some railways started their conductors off with enough change to last them through their first few runs (most providing $25 in change), with change for later runs coming from the till. There were also railways that required conductors to provide their own change. While this tended to keep conductors honest, it also required conductors to go through complicated administrative processes to get reimbursed at the ends of their shifts and tended to make them visibly and audibly surly toward passengers requiring them to make change.[12]

"CONDUCTING" BUSINESS: THE PROCESS OF FARE COLLECTION AND REGISTRATION

The member of the car crew responsible for the collection and registration of fares was the conductor. Since conductors were also responsible for the safety and well-being of the passengers, they usually assumed responsibility for the actual run. Motormen were required to follow the conductor's signals exactly.

At one time streetcar conductors were considered to be models of the community. According to Alan Lind, they were expected to be "neat and of pleasing appearance; must be courteous, agreeable, and even tempered; must be clean in his work habits; must be honest, accurate, careful and known to be reliable and dependable; . . . must grasp situations quickly and be able to make a correct statement of any unusual occurrence."[13]

In the early years, passengers merely boarded the streetcar at the rear, entered the passenger compartment, and made their way to the nearest available seat or clear space in the aisle. Once all of the passengers were aboard and all exiting passengers had departed, the conductor signaled the motorman to proceed. Once underway, the conductor entered the passenger compartment and collected the fares from each of the individuals who had just boarded. If possible, he would try to make a mental note of where they wished to get off.

This procedure was not a problem with a small to moderate number of passengers on board, but it became a daunting task in large cities during peak traffic hours. If the conductor was a big man, he risked antagonizing passengers as he bulled his way into the throng. If he was small in stature, the conductor rarely got far enough into the passenger area to collect all of the fares before the next stop. Either way, the railway ran the risk of missing fares, irritating passengers, or both.

Between 1905 and 1910, the industry experimented with different ways of improving fare collection and passenger flow in an attempt to collect as many fares in as little time as possible. The most ambitious schemes involved the design and arrangement of car interiors. For example, Brill introduced the "nearside car" in 1911, in which the car was designed to stop at the near side of intersections only, requiring passengers to board the cars at the front platform instead of at the rear. This required some redesigning of the end platforms (mostly increasing their length and rearranging the motormen's controls) to better facilitate passenger movement. Nearside car interiors were arranged with seats running along the car's sides in the front half (thus giving passengers more floor space) and cross seats with a center aisle in the rear half.

One highly influential car design was the Pay-As-You-Enter type, or "PAYE" car, in which passengers paid the conductor immediately upon boarding the streetcar. Although versions of this type of car had been around as early as 1905, the design was formalized and patented by the Pay-As-You-Enter Car Corporation in 1907. Railways wishing to purchase cars conforming to PAYE's designs had to obtain licenses from the corporation.

PAYE cars could be based either on conventional or nearside designs. Passengers entered at one end only, but were allowed to exit from either end. Mechanical doors were often a feature of such cars. The conductor usually controlled the rear door, while the motorman controlled the front. The first of these streetcars appeared on the streets of Chicago in 1907. Within two years, there were over 5,000 PAYE types operating in 42 cities. Another variant of the prepayment car required passengers to pay their fares upon exiting the car.[14]

In Cleveland, a different type of prepayment car was designed. Peter Witt, who had served as Cleveland's transit commissioner, realized that if one could increase a car's scheduled speed (the time required to make one run, including stops), it would reduce traffic congestion. Witt observed that most delays occurred when passengers boarded, fumbled for change, and requested transfers at the streetcar's entrances.

Witt reasoned that if conductors could be moved away from the car entrances, passengers could board more swiftly, and the car could proceed with minimal delay. He accomplished this by designing a single-ended car with a front entrance and center exit, placing the conductor at the exit doors. Like the nearside car, the seats in front of the conductor ran along the car's sides. The seats in the rear half of the car ran crosswise with a center aisle. Upon boarding, passengers had the option of either paying the conductor immediately and then passing to the rear seats, or remaining at the front of the car and paying the conductor upon alighting.

The front section, with its longitudinal seats, enabled passengers to board speedily and allowed the car to swallow large crowds. Witt developed his prototype in 1914, and "Peter Witt" cars quickly caught on in other cities.[15]

It was soon discovered that prepayment arrangements carried a number of benefits. In addition to increasing the number of receipts (fewer passengers could dodge the conductor), improved passenger flow resulted in significant time savings. The International Railway in Buffalo discovered that cars could be scheduled 6 minutes faster with prepayment operation than without, and the Chicago City Railway found it could save between 5 and 10 minutes. Another benefit was a dramatic reduction in the number of boarding and alighting accidents.[16]

While the above may have assisted the conductor in collecting the fares, the railway still needed an accurate gauge of how many fares were being collected. This was necessary for keeping records, but it also served to keep conductors honest. Called "fare registration," this was another vital function of the conductor.

Most railways were resigned to the fact that they would not be able to collect every fare. They could collect most of the fares through such methods as PAYE, but there would always be situations in which collection would be impossible, and there were always a few individuals willing to take risks to get a free ride. This was especially true

Figure 6.2. The interior of a Kuhlman-built Peter Witt streetcar, from the front of the car looking toward the rear. Passengers boarded the front of the car and could either loiter there and pay their fare to the conductor as they exited at the car's center or pay their fare and pass to the cross seats at the car's rear. Note the front of the car was designed to handle large crowds, as attested by the longitudinal seats in the foreground and the grab bars above the windows. The conductor's station was just beyond the longitudinal seat on the left. *Courtesy of Branford Electric Railway Association.*

when the cars were carrying crush loads during peak hours. Dishonest passengers waited for moments when the conductor was distracted, or they created plausible-sounding stories that might confuse a new conductor facing his first shifts of peak hour service. Occasionally, passengers resorted to counterfeit tickets.

The most common type of "fare beating" involved transfer abuse. Passengers sometimes engaged in trading, selling, or giving away transfers. Newsboys were a frequent source of the problem, as they would beg transfers from exiting passengers and then sell them to individuals buying newspapers.

Dishonest employees developed numerous schemes enabling them to pocket large amounts of cash fares, while appearing to have collected and turned in all of the day's fares. One common method was to replace cash fares with a corresponding number of unused transfers. Conductors working different lines would exchange unused transfers

Figure 6.3. The rear of a Kuhlman-built Peter Witt streetcar, standing at the rear looking toward the front. Passengers who paid their fares to the conductor at the car's center passed to the car's rear, where they sat in the cross seats in the foreground. *Courtesy of Branford Electric Railway Association.*

to make them appear legitimate. Such "conductor rings" were usually easy to expose once auditors at the accounting office detected a pattern among the cash fares and receipts being turned in.

Some enterprising conductors went to the extreme of counterfeiting tickets and using those to substitute for the cash fares. At times, stories involving the detection and capture of counterfeiters assumed the character of a detective novel. In Cleveland, a counterfeit ring was uncovered in 1911 when the Cleveland Railway hired private investigators. The detectives followed a suspicious conductor to Akron during the conductor's day off and found him purchasing several lots of phony transfers. The detectives then followed the conductor's contact and discovered the counterfeiter working out of a print shop located in a private home in the Akron area. In 1901, a ring was uncovered in Washington, D.C., in which the culprits actually imported their counterfeit transfers from Scotland (one of the ring's spouses crossed the Atlantic to get them).[17]

The skills required to catch dishonest employees varied according to the ability and imagination of the culprit. As noted above, most cases of fare pilfering left discernable patterns that could be spotted in the auditor's office. There were occasions, however, when auditors suspected something was amiss but could not prove it. To act as an additional check on conductor behavior, railways often hired on-board inspectors. These employees traveled disguised as ordinary passengers, paying their fares and looking as inconspicuous as possible.

If the perpetrator was exceptionally clever, railways had to go to even greater lengths to uncover them. One such instance occurred in New Orleans in 1901, when counterfeit tickets were entering the street railway system despite conventional methods of detection. The railway resorted to chemically treating all of its tickets. Following the treatment, all of the tickets collected afterward were exposed to a second chemical. The resulting reaction helped the railway to determine which runs were receiving genuine tickets and which were not. Once auditors were able to pinpoint the line and specific runs that yielded the most counterfeits, it was easy to identify the culprits.[18]

TOOLS OF THE TRADE: FARE COLLECTION AND REGISTRATION DEVICES

The devices and tools used by railways to collect fares and record the transactions fell into two general categories: those used in the actual collection of fares, and those used to record them. Collection devices were normally called "fare boxes," while registration devices were called "fare registers."

Fare Boxes

Fare collecting devices varied according to the method of collection then in vogue. As methods changed, manufacturers were careful to follow the trend. For example, in the days of the conductor passing through the car to collect fares, an unconventional approach to the problem of fare collection was tried by Cleveland's Bellamy Vestlette Manufacturing Company. The Bellamy Vestlette was a type of vest worn by the conductor. Its pockets were claimed to be secure, preventing change from being lost and resistant to pickpockets. The vest had pockets for various denominations of coins, as well as pockets for tickets, transfers, and paper money. The vest was designed to be worn all year, replacing conventional vests in the summer and fitting over an overcoat in the winter. The company claimed to be servicing 200 street railways in 1906.[19]

A similar version of fare collection was developed by the Rooke Automatic Register Company. The system consisted of a small box the conductor kept in his hand as he passed through the car. Passengers were instructed to place their fares in a slot at the box's front. A bell would ring once the fare was properly deposited. Some Rooke boxes had counters that kept track of the number of coins collected.

The Rooke system was similar to an earlier method used on Canadian street railways. Nicknamed the "coffee pot" due to its appearance, conductors passed a small metal fare box through the car to passengers who had just boarded the car. Once they paid their fare, they either passed the box back to the conductor or on to the next passenger.

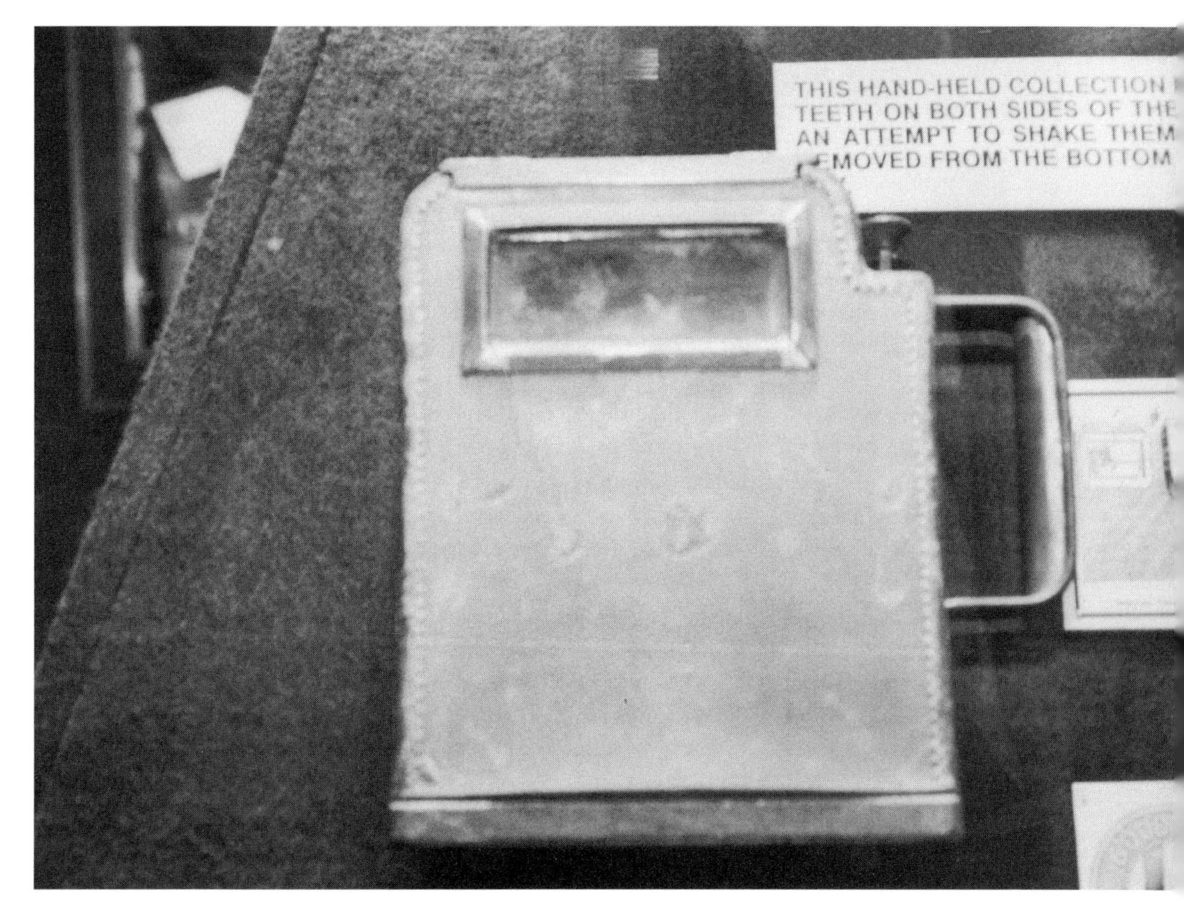

Figure 6.4. A simple, hand-held Coleman fare collection device. *Branford Electric Railway Association, author photo.*

Although the Rooke system had its adherents (particularly in Providence, Rhode Island, and on several smaller railways in New York), it had serious disadvantages. Its greatest problem was that the conductor was still required to circulate among the passengers. Furthermore, the small hand box could only accept one type of coin (usually a nickel) and then only a small number of them. Conductors using the Rooke system had to empty their boxes into a pouch they wore.[20]

Prepayment arrangements enabled conductors to remain at a specific location on the cars, and fare boxes usually wound up on the platforms. There were two types of fare box, registering and nonregistering. In its simplest form, a nonregistering fare box was merely a strong metal lock box into which passengers placed their fares. More sophisticated nonregistering fare boxes were fitted with an upper compartment featuring glass windows and a false bottom, so that the conductor could inspect the fare before opening a small door on the bottom of the upper compartment and dropping the fare into the lower compartment, which was the lock box. Unless access to this compartment was required for the conductor to make change, only railway managers or accounting office representatives had access to the second compartment.

Registering fare boxes varied in complexity from simple counting machines to those that sorted the fares as they were deposited. Most railways required the passenger to deposit the fare in the box, prohibiting conductors from handling the actual fare.

In 1911, the American Electric Railway Association's Traffic and Transportation Association (TTA) conducted a survey of railways using some form of prepayment system. The results illustrate how fare boxes were used in the years leading up to World War One. Of the 60 railways that responded, 26 used fare boxes, with an additional railway making the decision to use them just as the survey went out. Fifteen railways used boxes that could take any coin or ticket. Five used boxes that would accept coins of any denomination from pennies to half-dollars. One used boxes that accepted any coins except pennies, while another used boxes that accepted any coins except half-dollars. Three railways used boxes that accepted only pennies, nickels, and dimes, and two used boxes that accepted only nickels and dimes. Most of the railways (21) forbade their conductors to handle the fares directly, requiring passengers to place their fares into the boxes. Only five railways used registering fare boxes.

Some railways used more than one type of fare box. The Public Service Corporation of New Jersey, for instance, used 1,125 fare boxes. Most (750) accepted multiple types of fare, while 125 accepted only one. The remaining 300 fare boxes were of the registering type.[21]

The first registering fare boxes appeared on horsecars during the early 1870s. The most popular dated from 1872 and was designed by Louisville street railway official Tom L. Johnson (later of Indianapolis and still later of Cleveland, when he became the city's most legendary mayor). Its appearance and operation have been described as "a small cylindrical box much like a large alarm clock or gas meter, the face of which consisted of five dials. Both the one large dial and . . . four small dials located within the large one were divided into equals. Each time a coin or token of proper size was deposited, the larger dial moved one place. A complete revolution of the large dial moved one of the smaller dials one place, a complete revolution of the smaller dial moved the second small dial one place, and so on for the other dials."[22]

The Johnson Fare Box Company, established in Louisville and eventually based in Chicago and New York City, became one of the industry's leaders. Later models of the Johnson fare box also provided such data as passenger counts, individual trip numbers, and direction of travel. Examples of customers included the Third Avenue Railway system in New York City, the United Railways of St. Louis, the United Railroads of San Francisco, the Oakland Traction Company of California, and the Omaha & Council Bluffs Railway in Nebraska. Although most recording fare registers were operated by a hand crank manipulated by the conductor before World War One, Johnson Fare Box was offering electrically powered models in early 1914.

Another industry giant was the Recording Fare Register & Fare Box Company of New Haven, Connecticut, which made both recording and nonrecording types of fare box. Recording Fare products were known for their simplicity and durability. The company continually simplified its various models of fare box, reducing the number of springs and moving parts to a bare minimum. Although it initially used gears and pinions made from sheet-metal stampings, by 1910 all of the inner gears and wheels were machine cut.

Fare boxes of both types were also available from the streetcar builders themselves. Brill, its subsidiaries, and the St. Louis Car Company all made fare boxes. Such was the competition faced by Ohio manufacturers.

Figure 6.5. A Cleveland Fare Box Company box, mounted at the front of a car designed for one-man operation. *Courtesy of Indiana Historical Society, Cincinnati Car Corporation Collection.*

Figure 6.6. Another Cleveland-based manufacturer was the MacDonald Fare Box Company. The slides inside the glass-enclosed upper compartment served two purposes: they allowed the conductor to inspect the fares as they slid to the bottom of the compartment, and they prevented someone from reaching into the fare box if the conductor was distracted. The lever about halfway down the fare box's side opened a trap door at the bottom of the upper compartment, allowing the fares to drop into a strong box in the fare box's lower compartment. *Courtesy of Dayton History.*

Among the Ohioans, the products of the Cleveland Fare Box Company probably best embodied the registering fare box concept. Its fare boxes consisted of an upper glass-enclosed portion and a lower lock box. An added feature was a "slide" down which coins or tokens slid before dropping to the false bottom. This allowed the conductor to inspect the fares before they reached the bottom; it also served as a baffle to prevent people from reaching into the box opening. Cleveland fare boxes were highly regarded for their simplicity and sturdy construction.[23]

Fare Registers

One thing railways were extremely vigilant against was chronic carelessness or deliberate dishonesty on the part of the conductor. Theoretically, the conductor was required to account for every fare he collected. To assist conductors toward this end, railways equipped their cars with a device known as a "fare register," which was little more than a giant counting machine. The conductor recorded the register's reading on his day sheet at the beginning and end of each run or each shift. The resulting number of passengers was expected to equal the numbers of fares and transfers he had collected.

Returning to the 1911 TTA survey of "prepayment railways," 17 of the 60 respondents used fare registers in conjunction with fare boxes, while 11 did not. Thirty-two railways used registers only. As each passenger paid his or her fare, the conductor manipulated a pull-cord, lever, or pedal in order to register the number of fares that were paid. The register mechanisms were locked so that only senior railway personnel could manipulate or reset the numbers. Fare registers were used in part to assure the riding public that the conductors were performing their duties honestly.[24]

The simplest type of fare register merely counted the number of times its mechanism was actuated. For obvious reasons, this type of register could only be used on systems that either used a single type of fare or were concerned with obtaining an accurate passenger count. For systems with more than one fare, more sophisticated registers were needed.

Major producers of fare registers included the International Register Company in Chicago, the Recording Fare Register Company in New Haven, and the Sterling-Meaker Company of Newark, New Jersey. International and Sterling products came in "single" models that counted overall fares only and "double" models that distinguished between fares and transfers. They also sold register mountings and actuating hardware. Recording Fare registers could perform basically the same tasks as its registering fare boxes.

International products were in such demand that the company moved its operations to a larger Chicago facility in 1913. Not only did the new plant triple the company's floor space, but it also enabled the company to build its own brass foundry on the premises. Once the move was complete, International made much of its own brass stock.[25]

Sterling registers grew in popularity during the first decades of the 1900s, as evidenced by a business increase of nearly 50 percent every year from 1907 through 1912. In 1912, Sterling-Meaker was bought out by the Bonney-Vehslage Tool Company (also of Newark). The Sterling plant was closed, and all of its operations were moved to the Bonney-Vehslage plant, where registers continued to be made under the Sterling name.[26]

The city of Dayton became Ohio's (and eventually the industry's) greatest center for fare registration equipment. Home of the legendary National Cash Register, Dayton was already known as a center for business machine production when fare collection and registration equipment started to be produced.

The Dayton Fare Recorder Company produced recording fare boxes that also performed many of the functions of a fare register. In addition to counting fares, these boxes also kept track of information such as run data and subtotals for different fare types, as with a fare register. Similar to Cleveland Fare Box products, Dayton Fare Recorder boxes had slides or baffles enabling conductors to inspect fares while preventing wayward hands from intercepting them. The register provided totals for 5 cent fares,

3 cent fares, transfers, tickets (counts for the last two were entered by the conductor), and the total number of passengers.

Dayton Fare Recorder also produced fare registers, the latter for both street railway and interurban service. Its registers were sophisticated machines, capable of recording a considerable amount of information. Although small (10.5 inches by 15.25 inches), they kept track of the following information: arrival time, running cash totals, subtotals for each type of fare collected (including separate totals each for tickets, transfers, and passes), running passenger counts, and the identity number of the conductor. The registers were tamper-resistant, accessible only to an auditor with a key. Additionally, the register was designed to recognize up to 9,999 different conductors and 99 auditors.

Among Dayton Fare's largest orders were a trio of orders received from the Chicago Traction Company and street railways in Detroit and Buffalo. The Chicago Traction order alone totaled 2,500 registers, while the Detroit and Buffalo orders totaled 1,500 registers.[27]

The greatest register manufacturer in Ohio (and ultimately the largest in the world) was Dayton's Ohmer Fare Register Company, founded in 1898 by John F. Ohmer. Its fare registers were, in the words of their inventor, "a marvel of ingenuity and mechanical skill," combining the work of "an adding machine, a printing press, a time clock, and a cash register."[28]

Like Dayton Fare products, Ohmer registers tabulated a wide variety of data, including individual records of each fare collected and the specific fare type (not only the cash amount, but other types of fare as well, including tickets, transfers, passes, etc.). The registers also ran subtotals for each conductor using the register and totals for each fare type (including different cash fares such as 5, 3, or 2.5 cents). Additional trip information, such as the direction the car was traveling, the trip number, the date, and the time the conductor went on and off duty was also recorded. Ohmer registers, however, printed this information on multiple slips of paper, so that passengers could be given individual receipts and the conductors and auditing departments could have their own slips.[29]

Unlike most manufacturers, John Ohmer did not sell his registers. Instead, he rented them. Ohmer justified his position on the grounds that his registers would be prohibitively expensive if sold outright. In an address to the Pennsylvania Street Railway Association he remarked, "The rental charge we make is but a fraction of a cent per hour for work you could not possibly buy in any other way for 20 times its cost."[30]

In 1911, the going rate was 12 cents per register per day, whether a particular register was used or not. For registers on seasonal equipment, such as open cars in northern cities, the charge was only applied to the months the car was in service—but the rental fee was higher, 16 cents per register per day. Ohmer representatives were responsible for the inspection and maintenance of the registers throughout the rental period.[31]

John Ohmer believed not only in renting what he considered to be the finest fare registers available, but also in promoting sound business practice and moral accountability. By accurately tracking every transaction on every streetcar, Ohmer felt that it would become easier to identify dishonest or incompetent conductors. He attacked conventional fare registers on these grounds: "Our system controls the trip sheet, and

Figure 6.7. A self-explanatory example of a fare register. *Courtesy of Dayton History.*

nails the liar. With your system passengers have no interest in the register. With our system the passenger gets his receipt from the register indicator. Your system makes conflict between the company and conductor–on the part of the company to see how many fares it can make him register, on the part of the conductor to omit registration. Your system leaves both conductors and company in doubt. Our system proves the work of the conductor, and backs up his integrity. Your system works against nature, and breeds immorality. Our system, working with nature, teaches morality." It was imperative to Ohmer that railways renting his registers also accept his "system," because "without proper knowledge of how the machines should be operated, mechanically and systematically, there would be little hope for success. We consider the mechanical operation of our registers of second importance to the system which precedes it."[32]

Correspondence between John F. Ohmer and at least one of his contracting railways has survived among the papers of the Dayton, Covington & Piqua Traction Company (DC&P). Notorious for delaying or refusing to pay its bills, the DC&P papers contain a considerable amount of acrimonious correspondence with manufacturers. The correspondence provides an interesting study about how Ohmer the moralist dealt with customers who were perhaps less than honest, yet whose business

Figure 6.8. An Ohmer fare register. Fare amounts or messages such as "not registered" appeared in the center of the register. *Branford Electric Railway Association, author photo.*

he felt was still worth keeping. It also sheds further detail about how Ohmer's rental policy functioned.

On 15 March 1915, Ohmer sent a renewal contract to the DC&P, reminding them that their existing contract expired on 5 April. The term for renewal was five years. The DC&P replied on 27 March that it objected to the five-year term, proposing instead a one-year contract with an indefinite extension beyond the one year—terminable after three months' notice from either party. A hard bargainer, Ohmer replied on 30 March that he would be willing to extend the DC&P a *six-year contract* with a 5 percent discount on all rental charges. This contract could be terminated following one year, provided three months' notice was given and the discount was forfeited. The DC&P replied on 1 April that it wanted the 5 percent discount regardless, along with any other discounts Ohmer extended to his other customers.

Ohmer replied on 3 April that he could not accept those conditions, as they would not be fair to his other customers. He noted that he would, on occasion, issue

Figure 6.9. An installed Ohmer fare register. The dial to the right assisted the conductor in determining which fares to register. *Courtesy of Indiana Historical Society, Cincinnati Car Corporation Collection.*

a three-year contract with the 5 percent discount. Ohmer was quick to point out that regardless of the contract term, all roads were required to forfeit the 5 percent discount if they broke the contract.

The DC&P then proposed a "fair and equitable compromise" by accepting the six-year contract with the 5 percent discount, only it would forfeit 2.5 percent should the contract be terminated prematurely. Ohmer accepted this on 9 April.[33]

Ohmer, who initially made interurban registers only, started producing street railway registers in 1904. The first of these, model numbers Five and Six, not only kept track of the usual run information but also registered highly descriptive information about conditions for specific runs. Of particular note were the categories it used to describe any unusual occurrences that might have a bearing on schedule or fare discrepancies. For example, the registers recorded the following directions of travel: north, south, east, west, up, down, and extra. They recorded the following weather conditions: hot, cold, fair, rain, snow, and sleet. They also recorded the following special types of service: baseball, work train, circus, and fair. Finally, the two registers recorded the following explanations for any delays in service: late, accident, collision, off track, motor impaired, fuse out, wires down, washout, railroad blockade, railroad crossing, and passenger put off. The Number Five kept track of four types of fare, the Number Six, two.[34]

In 1905 Ohmer developed a smaller version of his register that could be "worn" by a streetcar or interurban conductor. Called the "Ohmergraph," the compact device was designed for a conductor to wear either at his chest or side. By 1906, the company was producing five models of the Ohmergraph, at least some of which were capable of issuing as well as registering transfers.[35]

Ohmer also made control equipment for fare registers. In 1907 the company developed a means for controlling two registers simultaneously. This "tandem arrangement" was displayed at the 1907 ASRA convention, along with a new type of 3 cent fare register for streetcars, a register capable of handling up to 30 types of fare, and a register run by electricity. By 1910, Ohmer was making registers that could handle up to 60 types of fare. Another Ohmer product was called the "turn-in car." The "turn-in car" was actually a turnstile designed to fit on any PAYE-type streetcar platform. The turnstile was equipped with a lock that prevented it from turning unless the conductor depressed a foot pedal (thus preventing people from entering the seating area without paying). It was designed to operate in conjunction with fare registers of any make.[36]

John Ohmer was a talented inventor, and many of his company's fare registers and register mechanisms were of his own design. In early 1909 the U.S. Patent Office reported that Ohmer, with 16 patents to his name, had received more patents than any other Dayton resident, ranking him above either of the Wright brothers (if only in this category).

Ohmer was quick to recognize the contributions made by others in his company as well. He instituted an award program at an annual banquet in which employees were given cash for ideas and devices the company adopted over the previous year. For example, on 9 September 1908, $148 was distributed among 14 employees for their efforts. As the company prospered, the value of the awards appreciated. In 1911 alone, Ohmer gave out 17 awards of $100.[37]

Ohmer also offered cash incentives to large railway systems using its registers. For systems in Denver, Colorado, and Portland, Oregon, the company offered cash prizes of $100 to the railway division with the highest percentage of accuracy (fares submitted with no overages or shortages). The second highest division received $75 and the third highest $25.[38] John Ohmer also kept abreast of recent patents issued in his field. If he noticed patents he felt would enhance existing products or help develop new product lines, he purchased them.[39]

Ohmer employees sometimes struck out on their own to form companies devoted to fare collection and registration systems. Two of these were the MacDonald Ticket and Ticket Box Company and Dayton Fare Recorder.[40]

John Ohmer was not only given to making fare registers and preaching morals to the industry. He also promoted the street railway manufacturers' cause in various trade organizations. He chaired the "supply committee" of the Central Electric Railway Association in 1909, serving at least through 1914. Other Ohio manufacturers represented on the committee were W. H. Bloss of Ohio Brass and L. G. Parker of Cleveland Frog & Crossing, the latter a maker of specialty rail switches and rail junctions for railroads and street railways.[41]

Ohmer received a considerable amount of attention in the trade literature, when some 2,123 registers were mentioned. Over 70 railways renting Ohmer registers were mentioned. Only a third were Ohio and Indiana railways.

Returning to the orders reported in the *Electric Railway Journal* between 1909 and 1913, 2,671 cars had fare registers specified (fewer than 500 cars with fare boxes were mentioned). The overwhelming majority of them (64 percent) were manufactured by International, with 19 percent manufactured by Dayton and 11 percent by Ohmer (see table 6.1).

Punches

As the poem that opened this chapter suggests, conductors often needed to mark transfers to indicate where the passenger was going as well as the time of issuance to assist their counterparts along other lines. Tickets were also marked to indicate that they had been used. The most common form of mark was the "punch mark," or a hole punched into the ticket or transfer.

Naturally, the devices used to make punch marks were called "ticket punches," "conductors' punches," or just "punches." They resembled the single hole punches found in most offices today, only they were heavier and more durable and usually made with distinctive punch dies or patterns. These patterns varied from conductor to conductor and were an easy way to determine which conductor punched which ticket. Some manufacturers offered hundreds of different dies.

Conductors favored punches that fit easily into their hand, were well balanced, had a long reach, and had clear, open sights so that they could see what they were punching. Punches were available from many manufacturers, including Sterling-Meaker/Bonney-Vehslage, the J. H. Stedman Printing Company of Rochester, New York, Woodman of Boston, and the American Railway Supply Company of New York City.

Many punch manufacturers were based in Newark, New Jersey. In addition to Sterling, Newark was also home to L. A. Sayre & Company, which advertised 1,065 patterns, and Kraeuter & Company, which had been making punches since the mid-1860s. Ohio punch makers included the Cincinnati Manufacturing Company, which produced 25 models of punch, and the Fred J. Meyers Company.[42]

CONCLUSION

As with other types of components described in these chapters, Ohio manufacturers were by no means the sole producers of fare collection and registration equipment for the street railway industry. However, Ohio companies were engaged in making virtually any type of device a railway might require, some of which (such as Ohmer and Dayton Fare Recorder) were among the most successful in the industry. The reputation gained by Ohio manufacturers would serve them well beyond 1914 and would only increase as the industry matured. Dayton-based manufacturers would become especially successful, eventually making this southwestern Ohio city the largest center of fare registration equipment in the world.

Figure 7.1. One of the Dayton Manufacturing's product lines was lavatory fixtures. This is the company's "Eckert Car Sanitary Closet." *Courtesy of Dayton History.*

7

SELDOM MENTIONED: TRIMMINGS, HARDWARE, AND VENTILATION

A number of items used on streetcars have tended to escape general descriptions of car building. Classified under such generic terms as *trimmings, furnishings,* or *specialties,* these items were either purchased from specialist firms or made by the car builders themselves. All five Ohio car builders made at least some of these items. There were also some specialist firms in the state, one of which was among the largest in the industry.

Another topic that is seldom addressed is ventilation. This is partly due to the limited amount of activity that went into ventilation research before 1914 and partly due to the types of cars requiring ventilation systems. While all streetcars could have benefited from some type of system, it would be the elimination of the clerestory from car roofs that forced railways and manufacturers to pay closer attention to ventilation.

TRIMMINGS AND HARDWARE

Trimmings are those items used in a streetcar's interior that are more decorative than functional in nature, while hardware is defined as anything that contributed to the structural integrity of the vehicle or to its operation. Hardware was usually fashioned from metal, with brass and bronze favored.

One thing that makes the terms *trimmings* and *hardware* hard to define is that the manufacturers were not consistent in using the terms among themselves. For example, in 1919 Dayton Manufacturing listed such items as lighting fixtures and door locks and hinges as "trimmings." The St. Louis Car Company called many of these same items "specialties."[1]

The city of Dayton grew fairly early during the nineteenth century, due in part to two transportation developments that occurred before 1840. The first segment of what would ultimately become the Miami and Erie Canal was opened between Dayton and Cincinnati in 1829. This gave Dayton access to the Ohio River and major river ports

Figure 7.2. Dayton Manufacturing's elaborate builder's plate was installed beneath the toilet seat on lavatory fixtures. *Courtesy of Dayton History.*

from Pittsburgh to New Orleans. Also, the National Road passed within 8 miles of Dayton during the early 1830s, making it relatively accessible to overland travelers.[2] Early companies that grew into major employers, such as Barney & Smith, and the arrival of the railroad in the early 1850s ensured Dayton of growth and prosperity.

During the late nineteenth and early twentieth centuries, Dayton came to be known for products pertaining to railroads, business machines, and agricultural machinery. Dayton also had numerous foundries, machine shops, and printing and publishing firms. Its population rose sharply from 85,333 in 1900 to 116,577 in 1910.[3] Although Dayton was similar to Cleveland and Cincinnati as a major industrial center, it was also significantly smaller than those cities, and thus its industries were not quite as diverse as those of the other two.

The Dayton Manufacturing Company was the greatest streetcar component manufacturer in Ohio and one of the largest in the nation. The company built components, appliances, and furnishings for both street railways and railroads. The company's founder, John Kirby Jr., possessed the potent combination of shrewd business sense and commitment to customer service. Born in upstate New York in 1850, he held a number of odd jobs during the late 1860s, working for a stove works, a photography studio, and a dry goods store. In 1869 Kirby joined an uncle who worked for the Illinois Manufacturing Company, a maker of railroad hardware. In 1870, Kirby moved with the company to Adrian, Michigan. By the following year he headed both the lamp department of Illinois Manufacturing and the brass department of the Lake Shore & Michigan Southern Railway.

Kirby moved to Cincinnati in 1875, where he became the superintendent of Post & Company, another railroad hardware manufacturer. Post did not enjoy a good reputation when Kirby joined the firm. It employed only 25 people, and its products were regarded with suspicion by the railroads. When Kirby left almost eight years later, Post & Company was held in high regard by the industry and had 300 employees.[4]

In 1883 Kirby moved to Dayton, where he was "instrumental in organizing the Dayton Manufacturing Company." Dayton Manufacturing had actually been established a year earlier by the Barney & Smith Car Company.

> Like other car builders, Barney came to depend on car accessory manufacturers in the 1870s for lamps, locks, hinges, toilet fixtures, baggage racks, and other trimmings for use in coaches. Rather than continue this dependence, E. J. Barney and several officers organized the Dayton Manufacturing Company in 1882 to produce hardware for railroad cars.[5]

E. J. Barney became president of the new company by 1889, and Kirby became general manager.[6]

Dayton Manufacturing developed a broad line of products by designing its own and buying out other firms. Cincinnati's Post & Company was acquired in 1892, and the Covington Brass Manufacturing Company of Covington, Kentucky, was acquired in 1894. Dayton also entered into agreements with other manufacturers to pool patents on certain items to control prices and production rates. One pool concerned railroad oil lamps for illuminating car interiors. In 1891 Dayton formed the Car Lamp Manufacturers Association with the firms Adams & Westlake (Chicago), James L. Howard & Company (Stafford, Connecticut), and Post Company (the latter just before Dayton's takeover).[7]

Unlike most component manufacturers, the trade literature published numerous car orders specifying Dayton Manufacturing components. Over 1,200 cars with Dayton products can be identified, embracing 10 types of components. Over half of all cars were equipped with Dayton headlights. The next most common components were sash fixtures (latches, locks, handles, etc.), which were installed on nearly one-quarter of the cars (see table 7.1).

To get a picture of how Dayton Manufacturing figured comparatively, orders published in *Electric Railway Journal* from 1909 to 1913 specified trimmings manufactures for a total of 1,157 cars. Dayton trimmings appear on 40 percent of these cars, with Kuhlman, Cincinnati, and Niles contributing an additional 6 percent. Taken together, Ohio accounted for nearly half of all the cars specified.

Aside from suggesting how large Dayton's output was, the figures make some additional assertions. For instance, it appears that Dayton's street railway market was largely

Figure 7.3. The Dayton Manufacturing Company also made car seats. These seats are on a railroad parlor car built by the Dayton-based Barney & Smith Car Company. *Courtesy of Dayton History.*

regional. Only 5 percent of its trimmings were used by railways outside of Indiana, Kentucky, Michigan, New York, and Ohio. This also demonstrates how great Ohio and its neighboring states were as markets in and of themselves (see table 7.1).

The nearest rivals to Dayton Manufacturing were the two largest car builders in the United States, Brill and the St. Louis Car Company. Brill came closest to Dayton's output, accounting for 30.5 percent of the cars. St. Louis produced 16.5 percent. Combined, these two car builders edged out Ohio manufacturing by a slight margin. This was probably due to the extensive list of specialty items both firms produced for their own orders.

Of the cars Dayton Manufacturing equipped, nearly 1,000 were built by Ohio car builders. The largest number of these (823) were turned out by the Cincinnati Car Company. One might question why Barney & Smith street railway orders did not account for

Figure 7.4. Small labels on Dayton Manufacturing's seats on a Barney & Smith parlor car. *Courtesy of Dayton History.*

most of Dayton Manufacturing's business, but this was due to the relatively few transit vehicles that Barney & Smith produced after 1900. Most of its street car building activity took place during the 1890s, when the market for railroad rolling stock was limited. One must also keep in mind that Dayton Manufacturing was itself primarily a supplier of components to *railroad* manufacturers and not the street railway industry. When one looks at equipment Dayton furnished to the railroads, Barney & Smith figures prominently (see table 7.1).

Dayton Manufacturing produced lighting fixtures from numerous materials and finishes. Three basic materials were used: bronze, brass, and iron. Each of these materials had finishes that were unique to them. Bronze finishes included real, polished, satin, fine wheel, scroll, antique, statuary, black ground, and oxidized gold. Brass finishes included yellow, satin, scroll, polished, red, "second art wheel finish," oxidized gold, "Persian," and "one wheel finish." Iron finishes were either japanned or plated—the latter available in brass, bronze, silver, gold, or nickel.[8]

Although many of Dayton's street railway fixtures were interchangeable with railroad coaches, Dayton was not ignorant of the specific needs of the transit market. For example, Dayton developed a high-efficiency system of lighting for street railways and interurbans. The system called for a series of powerful fixtures (five 90–100 watt lamps) running down the centerline of a car's ceiling.[9]

Figure 7.5. Barney & Smith also made a number of car furnishings, as this label for car lavatory fixtures attests. *Courtesy of Dayton History.*

A 1916 order for five double-ended suburban cars for the Steubenville & East Liverpool Light Company used these lighting fixtures as well as Dayton headlights. This order is also highly illustrative of how extensive Ohio transit vehicle manufacture was during the 1910s. The cars were built by the Cincinnati Car Company, which also furnished their wheel guards, gongs, seats, and ventilators. The cars also sported Lintern sanders from Cleveland, Knutson catchers from Canton, and Ohmer fare registers from Dayton.[10]

Most if not all of Ohio's streetcar window production was undertaken by the streetcar builders themselves. They usually made the window sash (the frames that held the glass) and often assumed responsibility for fitting the glass (called "glazing") to the frames. However, none of Ohio's streetcar builders manufactured complete windows.

Window sash fixtures were specified on 2,093 new cars reported in *Electric Railway Journal* between 1909 and 1913. Dayton was the leader among Ohio manufacturers, accounting for 12 percent of this market. Three firms outside Ohio produced more. The industry leader (with 38 percent) was the O. M. Edwards Company, which specialized in these fixtures. Both Brill and St. Louis sold 16 percent apiece.[11]

VENTILATION

The proper ventilation of streetcar interiors was poorly understood before World War One. A 1908 editorial in the *Electric Railway Journal* lamented that inventors were not doing enough to improve streetcar ventilation, but also admitted it was a difficult issue to address.

Figure 7.6. A Barney & Smith corner-mounted sink fixture. *Courtesy of Dayton History.*

Figure 7.7. A straight Barney & Smith sink fixture. *Courtesy of Dayton History.*

When empty, a streetcar's interior contained anywhere between 2,000 and 3,000 cubic feet of air. The number of passengers they carried depended upon their seating capacity and their arrangement for standees. Although many cars were equipped with movable window sashes, windows could not be kept open in cooler or inclement weather.

Doors provided a generous opening for fresh air to enter the car, but cars were usually stationary when the doors were opened, and during poor weather doors were kept open as little as possible. The issue, then, was twofold: how to change the air within a streetcar without relying upon doors or windows, and how frequently to change this air.

This issue was clouded by trying to define "good" and "bad" air and figuring out how to distinguish between the two. There was general agreement within the industry that air purity should be judged by the quantity of carbon dioxide in the air. For instance, air started to grow foul once carbon dioxide levels reached 15 to 20 parts per 10,000, and it could enter the lethal range at 50. Some cities had regulations specifying maximum levels for certain interiors. The Detroit Board of Health, for instance, decreed that maximum levels for Detroit school buildings should be 9 parts per 10,000.[12]

Types of Ventilation Systems

The most common type of ventilation before 1908 was the clerestory or "monitor deck" roof. Many clerestories were equipped with movable sashes, which could be arranged to open individually or in groups using rods. While they allowed fresh air to enter the streetcar's interior, they shared a similar problem with movable windows—they could not be used in inclement weather and could create disagreeable drafts, especially during colder months.

There was even some question over how efficient clerestories were. In 1905, the *Street Railway Journal* cited tests conducted on 19 types of streetcar. Only five types had carbon dioxide levels below 20 parts per 10,000, while five had levels between 20 and 30, and nine had levels over 30. It was clear that something had to be done, and the arrival of the arched-roof forced the issue, as they eliminated the deck sash.[13]

While there were numerous descriptive terms for ventilation systems and products between 1900 and 1914, they could all be classified into the general categories of "active" or "passive." Active ventilation systems used some sort of mechanical means to introduce fresh air to the streetcar's interior and to exhaust stale air into the atmosphere. Most used cone-shaped fans mounted in one of the streetcar's hoods. These fans were electrically powered and could draw up to 33,000 cubic feet of air per hour.

A 4-inch gap between the car's roof and headlining served as an exhaust duct, with adjustable vent openings in the headlining (up to 14), spaced along the car's length (up to 7 per side). Four air intakes were cut into each side of the car, near the floor. They were usually positioned so that incoming air would have to pass over a heating surface. This was done so that during the winter, the systems could also help heat the car's interior. Active systems were usually adjusted to change the air at a rate of 400 cubic feet per minute, which enabled them to do their job yet be "hardly perceptible to the passengers."[14]

Passive ventilation systems were more common and were dependent upon the speed and direction of the streetcar. They consisted of boxlike metal "exhausters" mounted either into the sides of a monitor deck or atop an arched roof. They usually contained baffles to prevent drafts, breezes, or rain from entering the streetcar's body. These baffles were also designed to "split" the air flow, allowing fresh air to enter the car's interior while aspirating stale air out.

Neither system was ideal. Although capable of operating at higher rates than 400 cubic feet per minute, active systems did so at the risk of creating drafts. They were also more expensive to install than passive systems and more costly to maintain. While passive systems were less expensive, they were also dependent upon the car's motion in order to work. They did little, if anything, to provide ventilation when the car was standing still.[15]

The 1908 Chicago Tests

Similar to devices providing front-end protection, there was a dazzling array of ventilation products for railways to choose from. While smaller railways tended to make their

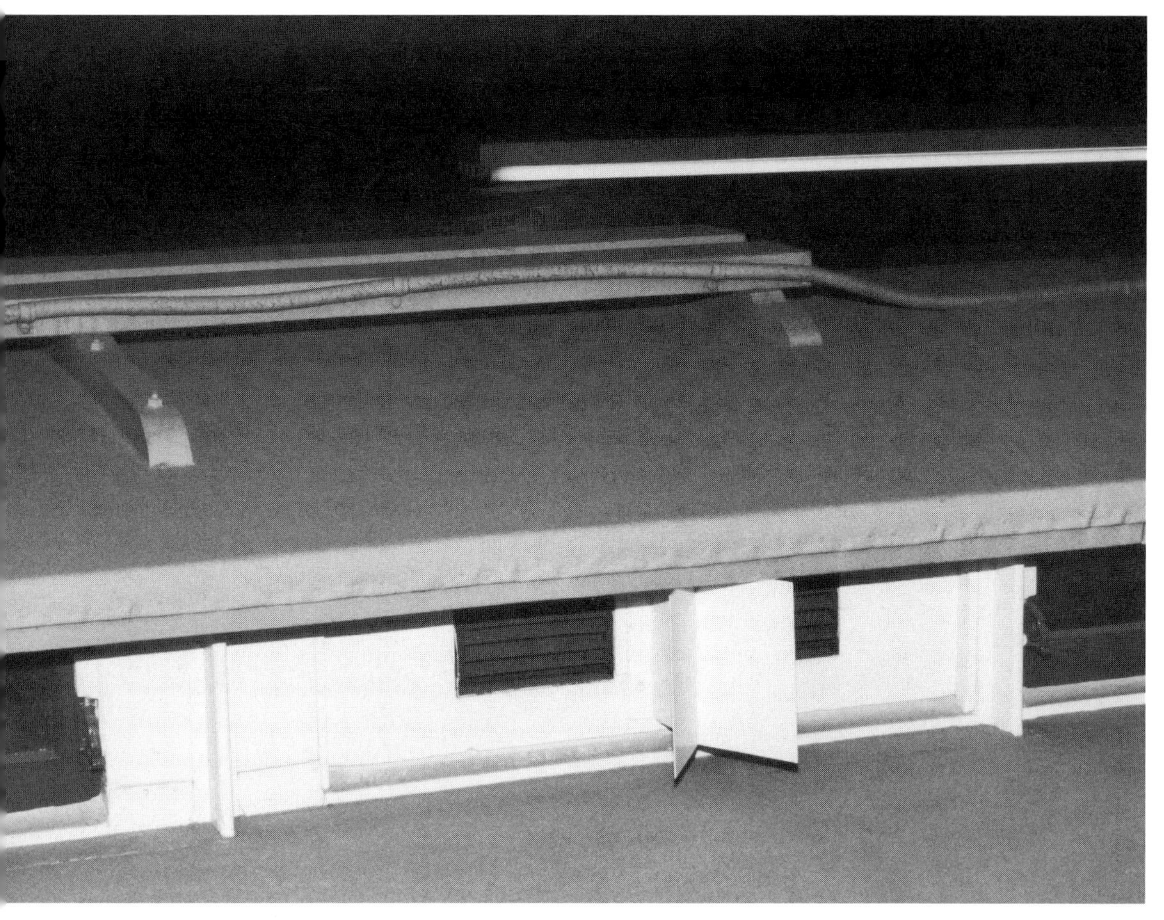

Figure 7.8. A clerestory-mounted ventilator. The fins in the center of the photo directed fresh air into the car while also exhausting stale air. This type of ventilator was dependent on the car's motion. *Branford Electric Railway Association, author photo.*

decisions based on least cost or the experience of other railways in their area, larger systems could afford to equip individual cars with samples of different products to judge their effectiveness before placing any large orders.

One of the more thorough tests of the era was conducted by the Chicago City Railway in 1908. This test involved five popular manufacturers and was covered extensively in the trade literature. Chicago City Railway's motivation for conducting these tests was to comply with the stringent regulations imposed by the city's board of health (350 cubic feet of air per hour per passenger). Representatives of the board assisted railway officials throughout the testing period. It was agreed that the minimum requirement would be to change 28,000 cubic feet of air each hour. Although all of the systems were tried on cars with monitor deck roofs, test officials were careful to keep the deck sashes closed.[16]

Two of the five ventilators were active, three were passive. The active ventilators were the Cooke and McGerry systems. The Cooke system was similar to the general description of active systems given above. The McGerry system was also similar,

Figure 7.9. The trailer in this 1915 Kuhlman builder's photo has an unusual "monitor" type of ventilator running nearly the length of the car. *Courtesy of Branford Electric Railway Association.*

Figure 7.10. Some streetcars, especially those built nearer the turn of the century, were heated by coal stoves. This stove was installed on a Toronto streetcar. *Branford Electric Railway Association, author photo.*

Figure 7.11. Some streetcars used hot water heat. This heater was a coal-fired model built by the Peter Smith Heater Company in Detroit. *New York Museum of Transportation, author photo.*

Figure 7.12. The pipes along the sides of a car that used hot water heat. *New York Museum of Transportation, author photo.*

although it used two ventilation fans powered by a common motor through belts or cords (see table 7.2).

The passive ventilators were the Garland, Perry, and Taylor systems. The Garland system was a local product, manufactured by the General Railway Supply Company of Chicago. It was developed by T. H. Garland, a general agent for the Chicago, Burlington & Quincy Railroad's refrigerator service. T-shaped ducts that conformed to the car's roof curvature were fitted to the deck sash, while a pair of funnel-shaped ducts was fitted to the inside of the sash to ensure proper air distribution.

The Perry system was a New England product, manufactured by the Perry Ventilator Company of New Bedford, Massachusetts. This system used a set of baffles placed in the deck sashes. The baffles could move freely, enabling them to self-adjust to the direction of the wind. Perry ventilators were extremely popular among street railways during the prewar years and could be seen on such major systems as the Boston Elevated Railway, Montreal Tramways, Pittsburgh Railways, and in New Orleans. The Taylor system was similar to the Perry, although it used stationary baffles.

Despite its superior performance, the Taylor system was not chosen by the Chicago City Railway. Officials from the company decided the Cooke system was a better choice partly due to its higher exhaust rate, because it functioned independent of the car's

Figure 7.13. Other cars used electric heaters. Many had a heater underneath each seat. *Branford Electric Railway Association, author photo.*

motion and because it was manufactured locally (it was made in Chicago by the Vacuum Car Ventilating Company).[17]

One other ventilating system worth mentioning was the experimental system installed by the Brill Company on its low-floor cars. Although technically a "passive system," it used a set of automatically actuated external louvers. The extent to which these louvers opened or closed was dependent upon an interlock with the cars' suspension. The greater the weight being carried, the farther the louvers opened.[18]

The *Electric Railway Journal* made note of ventilators on 2,561 cars built between 1909 and 1913. Most of the cars (25 percent) used the Cooke System, while 14 percent used the Taylor System and 12 percent used the Perry System. Among the car builders, the St. Louis Car Company was the leader with 18 percent, with Brill a distant 7 percent. Ohio manufacturers combined to provide ventilators for only 2 percent of the cars. These were Cincinnati, Niles, and Dayton Manufacturing.

CONCLUSION

The foregoing chapters have reviewed the street railway industry at its peak from a manufacturer's perspective. Starting at approximately 1914, the industry experienced a number of challenges. New railways and extensions to existing railways were no longer being built at the rate of the boom years, and railway failures were slowly on the rise. Orders for new cars began to drop slightly. None of this was cause for immediate concern, although it suggested the industry was entering a maturing stage.

The onset of war in Europe was an additional factor that would bring change to the industry. What is clear now is that the late 1910s was a period of transition for street railways, requiring manufacturers to either adapt to the new market conditions or fail. The next chapter will examine how the transition was handled by Ohio firms.

Figure 8.1. Shortages of materials and labor made it difficult for streetcar builders to turn out their products. Car 741 is a rare Kuhlman product from 1918–19, built for the Washington Railway & Electric Company.

8

THE DECADE OF TRANSITION, 1910–1919

World War One affected U.S. industry by creating a shortage of labor and materials and thereby forcing the cost of each to rise. This impacted street railway manufacturers and operators alike. Operators responded by instituting economies wherever possible and by raising fares (when they could overcome the opposition). This often meant delaying purchases of new equipment and purchasing only those supplies that were absolutely necessary.

Understandably, this had a negative impact on street railway manufacturers. Of the nearly 38,000 street railway, interurban, and rapid transit vehicles built over the decade, two-thirds were built between 1910 and 1914. In most cases, manufacturers survived if they enjoyed a broad market, secured a particular market niche, or managed to adapt to the new market conditions (by producing war materiel, for instance). Manufacturers who could not do any of the above, or who were already experiencing problems when they entered the 1910s, perished.

The war was also partly responsible for the cessation of the industry's growth. In 1910, the street railway industry expanded by 1,200 miles of streetcar, interurban, and rapid transit track. By 1919, expansion was less than 200 miles. Whether this decline in growth was attributable to the war or to the maturing of the industry would become apparent in the 1920s.[1]

THE SHORTAGE AND RISING COST OF MATERIALS

The shortage of raw materials began in 1915, soon after war erupted in Europe. In some cases, the war closed off access to certain products and raw materials. In others, raw material producers were overwhelmed by large orders placed by European governments.

The British blockade of German ports forced U.S. manufacturers either to seek new sources of materials or to develop their own. Before the war, for example, Germany had been a leading supplier of carbolineum, an oil used as a wood preservative. U.S. manufacturers were eventually able to develop their own preservatives using creosote oils, but there was a severe shortage in the interim.

Another critical shortage came in dyestuffs and pigments, especially German dyes used to produce red, yellow, and green paints. U.S. manufacturers had a more difficult time here, as they were only able to replicate yellow and green pigments by the decade's end. Potash, a key ingredient in soaps, was another European material denied to manufacturers.[2]

In addition to war-induced shortages, European orders for U.S. raw materials and manufactured goods placed a considerable strain on manufacturers and suppliers. One notably large order was placed for 224,000 tons of copper ore. To put this order into perspective, the *Electric Railway Journal* observed that when the Chicago, Milwaukee, St. Paul and Pacific Railroad electrified 440 miles of its main line, it required only 4,000 tons of copper. The European order was so large that its price was not made public, although speculation placed it in the neighborhood of $125 million.[3]

Some people were concerned about the long-term impact of these large orders. Commenting in 1916, American Electric Railway Association (AERA) president Charles L. Henry remarked, "The profits arising from this export business . . . have been such as to intoxicate the American people, and spreading out from the business centers . . . into every community throughout the land, there has developed a feverish, unnatural and unhealthy condition, until the American people have almost lost their moorings."[4]

Large European orders also created a serious residual effect. Shipment of vast quantities of raw materials to ports along the east coast disrupted normal freight traffic along U.S. railroads. On average, freight traffic in 1916 was 40 percent greater than it had been in 1915. Obviously, the bottleneck of freight cars awaiting unloading at ports along the eastern seaboard created freight car shortages everywhere else. This problem was exacerbated by shipping losses due to German submarine attacks, both in ships destroyed and a reluctance among shipping companies to ply European waters.[5]

During the fall of 1916, the shortage reached an estimated 115,000 freight cars. The ensuing delays were hard on manufacturers, both in the receipt of raw materials and in the shipment of finished goods. Coordination between the railroads, shippers, and the Interstate Commerce Commission managed to reduce the shortage to just under 60,000 by year's end, but the situation worsened again with the United States' entry into the war. Shortages in 1917 would climb to nearly 150,000.[6]

All of the forgoing led to dramatic price increases, both for raw materials and for manufactured goods. Most, if not all, prices on materials used by car builders and street railways increased 50 percent. A few cleared 150 percent (see table 8.1).

GOVERNMENT REGULATION

Before 1917, the federal government was largely content to let the business community handle its own affairs. Once the United States became involved in the war, however, the government played an increasingly larger part in regulating industrial production. At first, the War Department merely specified the type and quantity of materials required for the war effort, leaving manufacturers outside of the war effort to make do with what was left. AERA did its part by voluntarily suspending all exhibits at its 1918 convention in Atlantic City. The convention was subsequently canceled.[7]

As the war continued, government intervention increased gradually and took several forms, including production limits and quotas, the prioritization of the allocation and shipment of raw materials, price regulation, and the direct management of the railroads in an effort to reduce car shortages. In February 1918, the federal government adopted a four-tiered classification scheme grouping industries, by priority, for the purpose of allocating raw materials. The scheme was designed to be subdivided further if necessary. Affected companies included all users of copper, iron, steel, chemicals, cotton duck, and woolen cloth.

In September 1918, the government, acting through the War Industries Board, reclassified U.S. industries in the following manner. Highest priority was given to those whose products were of "intrinsic importance . . . for use during the war," followed by those necessary "for maintaining or stimulating and increasing the total quantity of production." A further category employed a relative measure of an industry's potential or capacity to produce "essential products." Using this scheme, street railways and nearly all manufacturing activity related to them were grouped under the second tier of industries. Curiously, electrical manufacturers were grouped under the third tier.[8]

Government regulation of commodity shipment took place in 1917. Highest priority was given to those commodities necessary for the railroads themselves (mostly coal). Food came next, followed by military supplies. Coal for use in coking plants fell beneath military supplies, while fuel and raw materials intended for public utilities (which included street railways) came fifth. Utilities were regulated further by having fuel quantities restricted to current needs only. Utilities were not permitted to stockpile.[9]

The government then tried to reduce the amount of fuel that utilities consumed. During the first three months of 1918, railways operating in all states east of the Mississippi River (including Louisiana and Minnesota) were required to limit their fuel consumption each Monday to the amount consumed the previous Sunday. The measure met with mixed results.

In a survey of selected major cities conducted by the *Electric Railway Journal,* only a handful of railways were able to comply. Cities whose railways were able to implement some service cuts included Buffalo, Chicago, Trenton (New Jersey), Louisville, and New York City (the latter succeeded by reducing most of its streetcar service by 20 to 30 percent). Cities that could not comply owing to large factories engaged in war production were declared exempt. These included St. Louis, Indianapolis, and Cleveland. Boston also merited an exemption owing to federally mandated service cuts in local railroad passenger service.[10]

In some cases, railways resorted to purchasing their own coal mines to ensure their energy needs would be met. For instance, Buffalo's International Traction Company purchased a coal mine in the Pittsburgh mining district. It intended to use only a third of the mine's production and sell the rest. The Commonwealth Power, Railway & Light Company of Grand Rapids, Michigan, purchased a three-quarter interest in a West Virginia mine, and the remainder was bought by a local power company.[11]

In rare instances, the government regulated production levels of particular industries or groups of manufacturers in an effort to conserve fuel. During the winter of 1918, manufacturers of window glass were ordered to cut their production by half. It had been determined that production within this group of manufacturers exceeded demand by a fair margin.[12]

The government also regulated production levels, prices, and the import and export of selected commodities. Coal was the first commodity affected, with prices set by state or region. Regional regulation was done in cases where the nature of mining or the type of coal varied dramatically within a state.[13]

Another important commodity the government regulated was copper, a vital material used by street railways in overhead line, motors, and control equipment, as well as in the production of brass and bronze trimmings. The War Industries Board fixed the price of raw copper ore at 23.5 cents per ton in early fall 1917. Copper producers were not permitted to cut wages as long as the price of copper was fixed, and production had to remain at maximum levels.

Steel prices were fixed anywhere from 47 to 70 percent of the highest levels reached in 1917. Prices were raised further in June 1918. Highest priority for steel deliveries was given to the War and Navy Departments and the U.S. Shipping Board's Emergency Fleet Corporation. The next highest priority was given to governments allied with the United States. "Private industries not engaged in war work must wait until the last before obtaining supplies."[14]

In early 1918 the federal government, acting through the War Trade Board, banned the export of all steel scrap, regardless of destination. Other bans included the importation of lead and all rattans and related products. These remained in place into 1919.[15]

A final means the government used to regulate industrial activity during World War One was in managing the railroads. Realizing the railroads' effort to coordinate their freight traffic would not solve the car shortage problem, the federal government took over all traffic management through the United States Railroad Administration. The USRA immediately ordered 100,000 new boxcars and hoppers from rolling stock manufacturers. The largest single order was for 30,000 cars placed with the American Car & Foundry Company. The remaining 70,000 cars were divided among 16 firms.[16]

RISING WAGES AND LABOR SHORTAGES

Shortages of both skilled and unskilled labor were first noticed late in the summer of 1916. One street railway manager from a "North Central" state lamented, "We raised the wages several times . . . in order to hold the old-time experienced mechanics, but were unsuccessful in many instances."[17]

At first, the labor shortage could be attributed to the need for more workers brought on by the large European orders. Beyond that, the labor shortage's most obvious cause was the war. War-related production siphoned off workers. For example, the U.S. munitions industry, which had been relatively docile during the early 1910s, was employing half a million men and women by 1917. Voluntary enlistment and conscription further reduced the nation's labor supply.

Wages rose. Within the steel and textile industries, for example, wages increased about 10 percent during 1916 alone. Some U.S. industrial sectors raised their wages by as much as 50 percent.

In 1917 labor legislation extended the eight-hour day to railroad workers. This created an "artificial" labor shortage, as a slightly larger number of workers were now required to maintain production levels over shorter shifts. Labor "pirating," in which firms deliberately outbid their competitors' wages, caused the federal government to require all industrial firms employing over 100 workers to hire new workers through the Federal Employment Service.[18]

Just as shortages of materials elevated prices, labor commanded a higher wage as the war progressed. By 1917, wages within the metal industries were climbing at a rapid rate. J. G. Buehler, president of the Columbia Machine Works & Malleable Iron Company in Brooklyn, New York, observed that his company's wages had increased anywhere from 15 to 25 percent. Columbia also paid year-end bonuses equivalent to a full month's pay, in addition to giving out five dollar gold pieces at Christmas.

In a survey of railways in 18 states, the *Electric Railway Journal* discovered wage increases averaged approximately 25 percent. Increases appeared to be the most dramatic along the east coast, gradually tapering off as one went farther west (California being an exception at 25 to 45 percent). The most dramatic increases took place among track workers, where wages had increased by 60 or even 100 percent.

A final labor trend worth noting is the changing composition of the industrial workforce. On 1 September 1917 a new child labor law went into effect, banning the employment of anyone under the age of 14. The law also restricted the employment of any individual between the ages of 14 and 16: eight hours maximum per day, no more than six days per week, and never between the hours of 7 p.m. and 6 a.m.

This legislation hit electrical manufacturers particularly hard, as children's smaller hands were ideal for winding coils of wire on generators and motors. Manufacturers resorted to hiring women, as their hands, though larger than children's, were small enough on average to do the work.

Within the street railway industry specifically, female workers were used the most in Philadelphia. The Electric Service Supply Company employed women in its machinery areas where their "natural deftness and intelligence is peculiarly fitted." Pleased with their performance, company officials remarked that the only changes Electric Service had to make to its facilities were creating dedicated restrooms and hiring a "matron" to "look after the physical well-being and comfort of the women."[19]

Comparatively few car builders devoted their entire facilities to war production. This was due in part to the scarcity of labor and higher wages commanded by production workers, but was also due to the high cost and scarcity of fuel. Most car builders that went into full-time war production were located in New England, where hydroelectric power was in ready supply. For example, the Osgood-Bradley Company of Worcester, Massachusetts, shifted its production from streetcars to gun carriages.[20]

OHIO MANUFACTURERS DURING THE 1910S

The war in Europe was met with mixed emotions by Ohio's street railway manufacturers. Some, like John F. Ohmer, saw no reason to be concerned, as they neither depended on materials from foreign countries nor had markets anywhere aside from Canada. Others, like Tool Steel Gear & Pinion, were very concerned, as they had large foreign markets. Still others, like Niles Car & Manufacturing, were wary. Niles did not use foreign materials or sell its vehicles in foreign markets, but feared the war might destabilize the banks, making it harder to finance new orders.

Ohio streetcar builders continued working, although longer delivery periods appeared in their contracts. They varied in their ability to secure new orders during the 1910s, particularly after 1917. Some had no difficulty finding new orders, while others struggled. Of the latter, some interpreted their difficulty as a sign to pursue war contracts or to leave the railway field altogether. Four of Ohio's five car builders participated in war production, turning out munitions and automotive vehicle bodies and components. In writing up contracts, car builders quoted delivery periods that were contingent on the receipt of materials. In more radical cases, orders were accepted without specifying a delivery date.

The J. G. Brill Organization (G. C. Kuhlman Car Company)

Despite its dominant presence in the industry, the Brill organization suffered from a dearth of orders once war broke out in Europe. In 1915, the main plant in Philadelphia built only 388 vehicles. Its workforce fell from 3,000 in 1912 to only 1,200 three years later. The situation among Brill's subsidiaries was not much better. Collectively, they produced fewer than 1,000 cars in 1915.

Sensing the transit market was not going to improve in the immediate future, Brill secured as many war contracts as possible. In Philadelphia, the company manufactured artillery shells, caissons, communications equipment, carts, and automotive bodies for ambulances, mobile kitchens, machine shops, and printing presses. By the war's end in November 1918, the Philadelphia workforce surpassed prewar levels, with nearly 4,000 at its main plant, including 300 women.

Although most streetcar orders placed during the war were a lesser priority than orders for war materiel, some car orders were placed by the U.S. Shipping Board's Emergency Fleet Corporation (EFC) or by the U.S. Housing Corporation (USHC) to help workers commute to defense plants. Brill did its best to fill these emergency orders, most of which were sent to its American Car plant in St. Louis. American Car also built a limited number of automotive vehicles for the army.

Brill's Wason plant in Springfield, Massachusetts, was turned over completely to war production. It produced the well-known "Jenny," a training airplane, for the army. Kuhlman continued to produce streetcars, although a portion of its facilities were devoted to war production. Kuhlman produced automotive parts and truck bodies, taking in as much as $1 million in war-related contracts.[21]

Cincinnati Car Company

Unlike most car builders during the war, the Cincinnati Car Company had relatively little difficulty in securing car orders. Between 1914 and 1915 it received orders for 250 all-steel rapid transit cars for elevated service in Chicago. In 1915, the company also received an order for 100 low-floored streetcars from the Pittsburgh Railways. No fewer than 300 streetcars were ordered by railways in Rochester, New York, the Public Service Company of New Jersey, and the Cincinnati Traction Company in 1916 and 1917. The Cincinnati Car Company also turned out streetcar components during the war. Orders for hand brakes, sash fixtures, and heaters were delivered to Charleston, South Carolina, and Cleveland.

Cincinnati also received EFC orders, such as one in 1918 for 20 streetcars for Staten Island in New York. This particular order is also illustrative of some of the delivery problems experienced by car builders, even when an order was considered a war priority. The Staten Island order was placed in August, with a delivery date of mid-October. None had been delivered by December, and Cincinnati officials were uncertain as to when the cars could be completed, much less shipped.

The Cincinnati Car Company probably felt the war a bit more personally than Ohio's other car builders. During the hostilities, the company lost its general sales manager, D. Hayward Ackerson. He had started at Cincinnati as a student-apprentice through the University of Cincinnati's cooperative education program in 1905. After graduation, he worked in the car builder's wiring department and at Westinghouse. Ackerson came to Cincinnati Car in 1914. He went to Europe in early 1918 as a first lieutenant in the army's ordnance department. There he suffered an attack of appendicitis and died on the operating table.

Jewett Car Company

The Jewett Car Company found itself caught between two extremes in the war years. Materials were increasingly difficult to obtain. When they became available, Jewett could scarcely afford the increased cost. On the other hand, Jewett managed to secure a number of very lucrative car orders (estimates placed the value of these orders near $400,000) and several munitions contracts. The car builder had neither a sufficient amount of materials for those orders nor enough cash to purchase them. Jewett declared bankruptcy on 12 December 1918.

The war may also have contributed to Jewett's difficulties by precipitating a crisis in leadership. Among surviving company correspondence is a letter dated 7 July 1915 to President Wright from Paul Reymann, a board member and minority stockholder. A German nationalist, Reymann was concerned over a military contract to produce shells and munitions that the company was considering. Reymann cautioned Wright not to accept it unless he received the consent of a majority of stockholders. If such consent was obtained, Reymann wrote, "please have the Board of Directors accept my resignation and appoint a successor. . . . Personally, I do not want to be in a position of accepting profits accruing from the manufacture of war materiel." Although Reymann emphasized these were his personal thoughts, it is possible that his actions (and possibly those of other board members) distracted management from running the company's affairs.[22]

Another factor behind Jewett's cash-flow problem may have been shoddy bookkeeping. A surviving account book reveals significant errors on several car orders, making certain expenses appear to be lower than they actually were.[23]

The car builder's location also served as a liability. Larger firms such as Kuhlman and Cincinnati benefited from being in large cities with extensive street railway systems, giving them a competitive edge when these systems solicited bids for new cars. Although neither firm could survive solely on local orders, they gave Kuhlman and Cincinnati a level of security that Jewett did not enjoy.[24]

Although Jewett entered into receivership, there was no immediate cause for alarm. President Wright stated that "some of [Jewett's] largest creditors . . . have thought it advisable to have a receivership to preserve the goodwill, property and assets of the company and to complete contracts on hand and to continue business."[25]

Unfortunately, Jewett's troubles grew. Early in 1919, sales manager Edwin Besuden resigned. Within a month of his resignation, the Brooklyn Rapid Transit Company, itself in dire financial straits, petitioned the courts to cancel a recent order for 50 rapid transit cars from Jewett. This cancellation represented a significant loss for Jewett and crippled the receiver's attempt to restore the car builder's finances.[26]

Barney & Smith Car Company

The twentieth century was anything but kind to the Barney & Smith Car Company. The railroads' shift to steel rolling stock held profound consequences for the car builder.

First, new car building firms dedicated to the new technology secured most new car orders. Two of the largest were the Pressed Steel Car Company (formed through a combination of two Pennsylvania rolling stock manufacturers) and the American Car & Foundry Company. ACF was a veritable empire, formed out of a consolidation of 13 rolling stock producers. Barney & Smith found it increasingly difficult to compete against such firms.[27]

The transition to steel railroad cars also meant the transition to steel components, notably trucks and couplers. This doomed Barney & Smith's foundry operation, which was based on cast iron. Faced with the loss of its primary market and a key secondary market, it became clear that Barney & Smith was in peril.[28]

Natural catastrophe made Barney & Smith's already precarious situation practically irreversible. During the spring of 1913, much of Ohio experienced severe flooding. Dayton was especially hard hit, and the Barney & Smith plant was no exception. The cost of recovery from the flood of 1913 added to the tremendous debt the company incurred while making a belated switch to steel car technology in 1911. The car builder passed into receivership that September.[29]

Barney & Smith managed to emerge from receivership in 1915 and survive into World War One. Since its primary market was railroad rolling stock, Barney & Smith was able to bid successfully on USRA freight car contracts, securing an order for 3,000 cars. Unfortunately, the car builder's "shortage of working capital delayed the placement of materials contracts till inflationary price movements increased costs to the point where completion of the contract proved ruinous."[30]

Barney & Smith struggled in the inflationary postwar market and reentered receivership in 1921. An attempt to restore the company failed when it received "an insufficient amount of subscriptions" toward a needed $2 million worth of mortgage bonds.[31] Barney & Smith kept its shops open through the spring of 1921. Among the orders it was able to fill was a contract to rebuild a number of 20-year-old open cars for the Indianapolis Street Railway Company.[32] Barney & Smith also produced four interurbans (two passenger/baggage combinations, two passenger coaches). Contrary to standards of the day and in keeping with Barney & Smith's tenacious conservatism, all four vehicles were made of wood.[33]

Business worsened for Barney & Smith, and the works were closed in June 1921. It was hoped that the shutdown would be temporary, as company attorney Lee Warren James and receiver Valentine Winters were negotiating with the Mexican government for car orders. The works were reopened briefly during the summer, but Winters closed them permanently in September when operating losses exceeded $150,000.[34]

The Dayton plant proved to be stubborn in death. In 1922, Judge Edward Snediker of the Montgomery County Common Pleas Court ordered the Barney & Smith property to be sold at auction. Bids for the sale of the property needed to be at least two-thirds of the property's assessed value. The first assessed value, which was $3,257,759, was apparently too high, as no bids were received. Despite this setback, Snediker ordered Barney & Smith to use its remaining cash to partially repay its creditors. Late in 1923 the assessed value of the property was cut nearly in half, to just under $1.5 million.[35]

It ultimately took four attempts to sell the property. Despite wishes to sell it intact, the Barney & Smith plant could only be sold piecemeal. According to a description by the *Electric Railway Journal*, "The plant comprises 47 acres of land and 76 buildings of various sizes. Most of them are said to be in a neglected condition. The plant has been . . . considered to be obsolete as a car-building establishment." All remaining assets were liquidated to pay off remaining creditors in July 1925.[36]

Niles Car & Manufacturing Company

Concerned with a drop in railway orders in 1915 and recognizing the potential of emerging automotive technology, Niles management decided to take advantage of the new market and shifted its production away from rail vehicles. Jewett had also experimented briefly in the automotive field by offering a "one-ton truck attachment for the Ford chassis."[37] Perhaps encouraged by Jewett's example, Niles began producing a line of automotive trucks early in 1916. Niles management also envisioned a scheme in which it would operate a chain of service stations in various cities.

Two models of trucks were produced, one with a maximum cargo capacity of one ton, the other with a capacity of two and a half tons. Many parts on the two models were interchangeable. The transition to automotive manufacture was made over the course of 1916, during which time Niles experimented briefly with interurbans powered by internal-combustion engines before leaving the railway field altogether. The only internal-combustion railway order Niles built was a "kerosene-electric" interurban car for a Tennessee railway. Niles stopped accepting railway orders late in 1916.[38]

ASSESSING THE FAILURES

The disappearance of Barney & Smith and Jewett resulted from inefficiencies, a failure to recognize industry and market trends, and poor business methods. Barney & Smith's plant was woefully outdated, and the belated attempt and cost of modernization (coupled with a natural disaster) overwhelmed the firm. Jewett cannot be regarded as an efficient firm either, as its demise was triggered by an inability to maintain a sufficient cash reserve.

Niles's disappearance from the street railway field, by contrast, was voluntary. It came out of the recognition of automotive technology's potential and a hastened effort to adopt it. In doing so, Niles left the transit market altogether, opting to enter the automotive truck market instead.

One thing that should be kept in mind about the decline of streetcar manufacture in Ohio is that it was never an essential part of the state's economy. None of the car builders were critical to their cities' economies. For example, Dayton was a city of sufficient size to withstand the loss of Barney & Smith. The loss was further offset by the fact that it was not abrupt. The car builder declined gradually over a period of nearly 10 years.

The departure of Niles Car & Manufacturing from the street railway industry was a market shift rather than a business failure. Furthermore, Niles's ultimate fate did not hold its local economy in balance. Niles's principal industries were large and labor-intensive, including sheet steel, boilers, metal lathes, and electric lamps (light bulbs). While the eventual loss of Niles Car & Manufacturing was regrettable, it did not spell the doom of Niles's economy.

A similar observation can be made about Jewett's relationship with Newark, Ohio. Newark businesses embraced a broad spectrum of manufacture, from stoves and rubber products to oil refining. In terms of the number of people Jewett employed, however, the loss of a car builder may have been harsher on Newark than it was on Dayton or Niles.

COMPONENT MANUFACTURERS

For the manufacturers of components, the war held mixed blessings. Unlike the car builders, who had large plants that could be modified at least in part to take on government orders, the smaller plants of the component manufacturers made it difficult to secure large war-related contracts. Their traditional markets were restricted by the high cost and unavailability of materials. The largest component manufacturers managed to survive. For them, the war meant only a temporary loss of business, something to ride out. The war provided additional opportunities through military contracts or by creating favorable trends among street railways.

Tool Steel Gear & Pinion did not depend on Europe for raw materials, but it did have markets in Great Britain, France, and Germany. E. S. Sawtelle, Tool Steel's assistant general manager, felt the German market was lost, but he had hope that the French and British markets would recover. In the meantime, the company pursued sales among South American countries to compensate for its loss of European business.[39]

Ohio Brass was another manufacturer that found the war years to be both a challenge and a blessing. In an effort to maintain its workforce, by 1917 Ohio Brass was forced to hire unskilled immigrants at 30 cents an hour. Less than a year earlier they had been hired at 17 cents. Inflated prices and shortages of materials were another problem. The various insulating compounds Ohio Brass used in its overhead line hardware jumped from 14 to 40 cents per pound, when they were available.

In other respects, the war years were very good to Ohio Brass. In particular, the success of its Tomlinson couplers suggested the demand for its products would remain strong. By 1916, the firm boasted that nearly 1,000 Tomlinsons were in service on interurbans alone.

On the eve of the United States' entry into the war, Ohio Brass was overwhelmed by coupler orders. The largest order came from the Boston Elevated Railway, which ordered 684 Tomlinsons divided into three groups: 200 for streetcars, 84 for trains on its elevated line, and 400 for streetcars using the East Boston Tunnel. An additional order for 120 couplers from Toledo Railways kept the Mansfield plant busy well into 1917.[40]

The war years were also challenging yet generous to John F. Ohmer. (Given his bright, optimistic nature, one wonders if he could have perceived the war years as anything less.) On the surface, Ohmer Fare Register appeared to be exceptionally vulnerable. Its average wage increased 50 percent by late 1916. Employees engaged in "special development work" were making as much as $80 per week.

Ohmer also suffered incredible increases in the prices of materials. The price of sheet steel jumped between 200 and 500 percent. Brass increased from 13 cents to 40–45 cents per pound. Light malleable castings rose from 6 to 18 cents per pound.

Although the above was typical of what other manufacturers were facing, Ohmer's business methods created a serious problem. Unlike most firms, which sold their products outright and could thus recover by raising the prices of their products, Ohmer leased his registers to railways at fixed rates for terms of five or six years. Thus Ohmer had to absorb a greater proportion of higher material costs than other Ohio firms.

Curiously, Ohmer chose not to recover costs in areas where he could have. For example, the price of his company's register paper, which he required his customers to purchase, was kept low, even though the cost of producing it continued to rise.

Ohmer was not blind to the problems introduced by the European war, however. In an August 1916 employee bulletin, he recalled the Panic of 1873, cautioning his employees to look and plan ahead: "It . . . behooves us to lay aside some of our earnings to carry us over the stormy days which may follow the close of the European conflict. Save now while the saving is good." Ohmer offered his factory to the federal government in support of the war effort. He was awarded several military contracts to produce naval gun sights. A portion of the plant continued to produce fare registers and register parts. Ohmer Fare Register advertisements adopted war-related themes: "The Ohmer System stands for economy. It is a proper wartime measure for the successful and profitable operation of electric railways."[41]

Other advertisements assumed a more militant air:

> Enlist today in the training camp for efficient business service. You are needed as an officer in the fight for better business. Your industry needs your example. Your men need your leadership. Will you do your bit? . . . Have you a system which has stood the annual test of business battles through years of aggressive campaigns? Have you a system that will ensure your complete victory?
>
> After 16 years of competitive warfare the flag of the *Ohmer Fare Register and System* is impregnably mounted on the pinnacle of business success in every state in the Union.[42]

Following the war, John Ohmer predicted that prices for raw materials and finished products would not go down for the street railway industry. Due to "depleted world stocks" of raw materials and manufactured goods in general, he expected manufacturing costs to hold for at least a year. Ohmer saw this as a help rather than a hindrance for register manufacturers.

Although the higher prices would mean higher production costs for manufacturers, they would also mean rising operating costs for railways. Rising operating costs meant that railway would probably attempt to raise fares. Before the cost increases, the fare structure of much of the street railway industry was based on the 5 cent fare, and registers and fare boxes had been designed accordingly. An increase in the number of railways charging different fares coupled with future improvements in fare box and register technology suggested a future demand for new box and register models.[43]

CONCLUSION

By 1920, the street railway industry appeared to be entering a period of maturity. The drop in new railway mileage suggested the years of growth were over. This transition, coupled with higher operating costs, presented railway and manufacturer alike with challenges that were unheard of before the war.

Already in Ohio, the impact was being felt. The state tended to run counter to the prevailing trend reported in the trade press. According to the *Electric Railway Journal,* most car builders endured the war by accepting large war contracts. Component manufacturers had a rougher time of it. In Ohio, exactly the opposite was true.

Of the five car builders present in Ohio in 1910, only two were prepared to enter the new decade by 1919. Ohio's component manufacturers, particularly large firms such as Ohio Brass and Ohmer Fare Register, did better, due in part to changing market conditions and their participation in the war effort. However, it will be recalled that the 1910s also witnessed a slackening of the street railway industry's physical growth. What the 1920s held for the industry and its manufacturers will be addressed in the final chapters.

Figure 9.1. An experiment in lightweight car construction. Cleveland Railway car 1376 was built entirely out of aluminum. *Courtesy of the Cleveland Public Library.*

9

PROMISE AND STAGNATION: STREETCAR TECHNOLOGY DURING THE 1920s

> The question of modernization of our equipment admits of no discussion. The electric railway industry must keep pace with the times or it will pass out just as surely as the horse car. It must constantly improve its electric cars and supplement its service with buses and such other forms of conveyances as are useful and in keeping with the public demand.
>
> L. S. STORRS, MANAGING DIRECTOR, AMERICAN STREET RAILWAY ASSOCIATION, IN THE *ELECTRIC RAILWAY JOURNAL*, 4 JULY 1925

The 1920 U.S. Census indicated that for the first time, the nation's population was more urban than rural. The street railways were a contributing factor. As cities grew in size, so did the street railway. In an address to the American Electric Railway Association at Atlantic City in 1921, Commerce Secretary Herbert Hoover outlined the vastness of the street railway industry. It carried 12 to 15 billion passengers annually. In 1920 there were just over 47,700 miles of street railway trackage in the United States. Over 100,500 passenger cars, freight cars, and locomotives were operated by electric railways.

In 1920, the industry took in nearly $1 billion in revenue, of which $60 million was "repaid to the community in taxes." Hoover observed that the street railway industry had $6 billion in "fixed investments." It employed 300,000 people. It paid out $300 million in wages and spent the same amount on supplies and materials. An additional $200 million was also spent on extensions and replacements. Each year, street railway power houses consumed 16 million tons of coal.[1]

An industry of such dimensions might appear to be quite solid, yet less than 20 years after Hoover's address, the industry was a shadow of its former self, with much of what remained in danger of closure. Signs of decline were apparent by 1929. Track mileage had shrunk by 19 percent to 38,470 miles. The number of cars in service had decreased by

29 percent, to 71,219. By 1937, these figures had plummeted to 26,187 miles of track (a 45 percent decline from 1920) and 49,147 cars (a nearly 50 percent decline from 1920).

Historians agree that a combination of factors working against the street railway industry during the 1920s caused the decline. Clearly, overly optimistic and/or fraudulent speculation led to the construction of many street and interurban railways that could never have operated profitably. Furthermore, many railways that could have shown a profit did not as a result of inept management.

Regulation created another obstacle for street railways. In some cases, fares were held unrealistically low by local governments out of concern that railways might try to increase their profits at the riders' expense, despite rising material and labor costs. Railways that were owned by utility companies were considered to be monopolistic in some states, and the railways were forced to exist independently—even though they could not meet expenses otherwise. A large number of railways fell victim to indifference as state and local governments began favoring other modes of transport over the streetcar.[2]

Speaking before the American Electric Railway Association in 1928, P. S. Arkwright, president of the both the National Electric Light Association and the Georgia Railway & Power Company, remarked, "I remember the time very well when our street railroad carried our electric light and power business. I can also remember the time when our electric light and power business carried our street railroad."

All historians agree that the single greatest cause of the downfall of the street railway industry was the rise of automotive technology. The principal vehicle emerging from this new technology was, of course, the automobile, but the technology also gave rise to a new form of public transport, the bus. The public's embrace of the new technology and the street railway industry's lackadaisical response to it hastened the industry's decline. Rather than improving and redefining their product, streetcar builders and operators exhibited great conservatism toward car design, opting to refine existing designs instead of developing new ones. Despite the availability of new technologies and new trends in vehicle design introduced by the automotive industry, street railway operators were reluctant to take advantage of them. A few railways ordered experimental vehicles embodying some of what was available, but larger orders representing "advanced" streetcar design were not forthcoming. This conservatism presented the automotive industry with a window of opportunity.

In accounting for the streetcar builders' lack of vision, it is important to see how the streetcar's environment changed. One must remember that the street railway industry bridged three eras of transport. The first, running into the 1880s, was the era of the pedestrian and horse. Mechanically powered streetcars emerged during the 1880s, ushering cities into the streetcar era, which ran until about World War One. Following the war, automotive technology had developed (both the vehicles themselves and the methods required to produce them) to the point that cities entered the "recreational auto era," which lasted until World War Two (the freeway era followed).[3]

With the rise of the automobile and related conveyances like the bus, spatial relationships within cities and towns began to change. The automobile and bus made it possible for real estate development to go wherever demand permitted. It was no longer necessary for a developer to work in concert with a street railway company to open up new sections of land. Buses could be routed wherever there was a demand, or people could use their automobiles to arrive at their destinations. The impact of automotive technology did not end there. Not only were streetcars losing influence in the expansion of cities during the 1920s but they were also losing ground to buses and automobiles where they already existed.

A paper delivered at the 1924 American Electric Railway Association convention claimed real estate developers and street railways would always be closely linked. The permanence of tracks suggested that transport service would always be available, while bus routes might easily be diverted elsewhere. Such short-sightedness was characteristic of many in the street railway industry.[4]

Evidence presented by economists Leon Moses and Harold Williamson during the 1960s suggests this relationship was changing as early as the 1910s. Examining Chicago between 1910 and 1920 they discovered two trends. First, the number of wagon registrations declined from 58,000 to 31,000, while truck registrations increased from 800 to 23,000. Second, manufacturers were moving outward, *away* from downtown. Four industrial groups (publishing, chemicals, nonelectric machinery and electric machinery) totaling 955 companies were compared between 1908 and 1920, and 473 companies were found to be common to both eras. Of these, 285 (roughly 60 percent) *increased* their average distance from downtown Chicago from 0.92 miles to 1.46. Clearly industry was on the move in Chicago, changing not only freight patterns within the city but workers' journeys to work as well.[5]

During most of the first three decades of the twentieth century, streetcar builders and operators exhibited a contradictory approach to the design of their vehicles. While all adhered to the same basic plan, they nonetheless failed to develop industry standards. During the mid-1920s George Kippenberger, an official with the St. Louis Car Company, was highly critical of the industry. While the automotive industry spent millions of dollars on research into improved designs, the street railway industry did not. "If there is a place for the radically changed street car or 'electric rail coach' today, why has not the car building industry initiated its use by offering it to the trade?" Part of the problem, in Kippenberger's view, was that there was no standardization within the industry:

> Were it possible for the car builder to merchandise his car—his *standard* car—if we dare mention the word standard in connection with cars before railway men—the car builder would have opportunity and incentive constantly to improve and develop style and design.[6]

The American Electric Railway Association (AERA) formed a design committee to address the matter of standardized car design in 1921, but the only car order resulting from this effort was for a dozen units for the street railway in Berkshire, Massachusetts.[7]

Even if most street railways had taken advantage of all the technology available to them, it is doubtful that new streetcar designs would have overcome all of the problems experienced by the industry during the 1920s. However, it is conceivable that updated equipment may have helped preserve large and midsized railways that were still viable during the 1920s. What follows is an overview of developments in streetcar technology during the 1920s and how Ohio's remaining car builders participated in those developments.

ISSUES AND TRENDS IN 1920s CAR DESIGN

Body Design and General Appearance

Car appearance changed little during the 1920s. Appearance depended largely upon the taste of the individual railway but generally embraced such features as clean, uninterrupted lines and the concealment of "unsightly apparatus." Charles Gordon, western editor of the *Electric Railway Journal,* noted, "Comfort alone is not sufficient. . . . Merchandising attractiveness must please the eye. An awkward looking piece of machinery is hard to sell."[8] Gordon also observed that the concept of "ride merchandising" was new to the industry. Previously, cars were designed from the perspective of the mechanic, but with the advent of the automobile, car designers were compelled to respond to the rider's perspective as well. In terms of general appearance, Gordon felt cars should look "low [to the ground], graceful and dignified." In other words, body lines should flow without being broken up by ventilators or hardware. He advocated tapered body ends, sloped windshields, and sun visors.

Sloped windshields could alter and improve the exterior appearance of the car, while eliminating "windshield glare" at night (caused when interior lighting shined back into the eyes of the car's operator, blinding him). If windshields were angled enough, operator curtains, which some designers felt cluttered cars' interiors, could be eliminated (although many systems retained them as an extra safety measure). Gordon also suggested the use of skirting to hide the trucks and equipment suspended under the car body.

Other body features were addressed by Gordon as well. He felt windows should be at least as wide as their height (preferably wider), as this accentuated body lines and offered passengers a better view. Window guards should be discarded. (AERA's Committee on Equipment advocated using automotive-style drop windows, as they afforded a similar degree of passenger protection.) Lower cars could do away with folding steps, which Gordon considered "noisy, unsightly, and difficult to maintain." Doors should be of the folding type. End vestibules should be part of the body, not added on, as he believed this improved the body's appearance and eliminated the claustrophobic effect of thick bulkheads at the ends of the seating compartment. Cincinnati built an early version of a "bulkheadless" car late in 1917 for service in Westmoreland County, Pennsylvania.

The AERA Equipment Committee advocated clean, arched headlining with no molding. It was recommended that colors above the window sills should be bright but not glaring and darker beneath the windows. The Committee also suggested that there should be no steps inside and that steps should not be built of wood. There should be no inspection hatches, and floors in general should be no more than 32 inches from the pavement. The first outside step was no more than 15 inches, the second no more than 14, with a slight ramp inside the car at the end(s) to make up the difference.

Grab handles were gradually (though not entirely) replaced with horizontal bars. Passenger window curtains, though still of use on interurbans, disappeared on most city cars. Signage presented a peculiar dilemma. On the one hand, essential signs (entrance and exit markings, safety messages, etc.) needed to be "large, clear, visible, and explicit" but kept to a minimum. Advertisements, on the other hand, were encouraged in greater numbers, as the distraction they offered was believed to make a passenger's trip pass faster.[9]

Materials

From the early 1910s into the 1920s, steel streetcars were made from a "mild carbon steel." Although strong, easy to work, and relatively inexpensive, carbon steel had a tendency to rust. During the 1920s, a steel alloy containing a small portion of copper came into use, but even this succumbed to corrosion eventually.

Stainless steel was rarely (if ever) used in street railway applications. Even though its strength, attractive appearance, and noncorrosive properties were ideal, its high cost discouraged its use. Other anticorrosive metals found use in street railway application, although they were not used for structural purposes.

Aluminum was first used in stanchions and seating fixtures. Alloys such as Duralumin allowed the lightweight metal to be used structurally during the 1920s. Although aluminum and aluminum alloys tended to be used on experimental equipment during the late 1920s, the Cleveland Railway demonstrated that it could also be applied to conventional designs. In 1926 the railway built the first all-aluminum streetcar, a Peter Witt. The car was significantly lighter than other Cleveland Witts, weighing only 15 tons as opposed to 22.5 tons.

Metal-faced plywood, or "plymetl," was used on a few orders with limited success. Although this material was used successfully by bus manufacturers, streetcar builders had difficulty attaching it to the steel structural members of the car frame.[10]

Lightweight and Safety Cars

Following World War One, operating and material costs did not return to prewar levels but continued to climb. Beyond maintenance, operating costs came from two principal sources: labor and power consumption. One tactic to save power was to save weight. Less material made for a lighter car, and a lighter car was easier and less expensive to move (see table 9.1).

Labor costs could be reduced by eliminating the conductor. Such a crew reduction introduced safety concerns, and new designs and devices were introduced by manufacturers to permit streetcars and interurbans to be operated safely by one man.

In 1922, *Electric Traction* identified lightweight car designs and the reduction of car crews as two of the most significant events for the street railway industry. In 1914, a typical four-motored, double-trucked streetcar averaged 28 tons. By 1923, streetcars weighed as little as 17 tons. Single-trucked streetcars weighed even less, averaging 8–10 tons.

Much of this weight reduction came from eliminating structural features that were necessary for wooden streetcars but were needlessly carried over into steel car design. Once railway owners discovered lightweight bodies required no more maintenance than heavier ones, they began to specify weight-saving features on their orders.

For example, most of the steel used in car bodies was concentrated at the side sills, which were the principal structural members of the car. Originally, wooden sills were not strong enough to carry the weight of the side posts, sheathing, and roof, so they were reinforced with truss rods. Early steel side sills continued to be excessively large. It was eventually realized that smaller plate girders would suffice.

The introduction of center-entrance and Peter Witt car designs created problems initially, as one or both side sills were effectively cut in half. Two approaches remedied the situation. Although some builders reverted to massive sills, others created a network of lightweight trusses out of smaller steel girders. Both designs worked, but the latter saved weight.

Figure 9.2. The operator's control station for a car built for one-man operation. A K-type controller is located on the left. The large handle atop the controller was used to accelerate the car (it is not a deadman handle), and the smaller lever to the right was the reverser key. The pedal near the floor to the right of the controller drum may have been a sander valve. The piping and handle in the center is the air brake stand, which controlled both the brakes and the doors. The large wheel on the right is the handbrake. The fare box (along with an electric light for night operations) is located next to the operator's seat, designed long before the age of ergonomics. The device resembling a fire extinguisher on the window post is a door indicator, telling the motorman whether the car's rear door (probably treadle-operated) was open or closed. Note the light mounted near the door step to assist passengers boarding at night. *Courtesy of the Indiana Historical Society, Cincinnati Car Corporation Collection.*

One-man cars introduced a number of new safety devices, and cars featuring them were sometimes referred to as "safety cars." Among them was the door-brake interlock, which prevented doors from opening unless the brakes were applied. Another safety device was the door treadle. Used at exit doors, the treadle was a floor-mounted control valve activated by the passenger's weight. A common safety feature on air-operated doors was the "fold-down step." This was a step located beneath a set of doors that folded downward when the doors were opened. They folded up along the car's sides when the doors were closed.

An additional layer of safety was added with the line switch, which automatically cut power from the motor circuit if the motorman accelerated too quickly. When tripped, they were less noisy and easier to reset than circuit breakers. Line switches were also more sensitive than traditional circuit breakers. In some cases, all that an inattentive operator had to do was "jiggle" the controller handle to trip the device.

Still another safety device was the "deadman" throttle, which used a spring-loaded controller handle. The operator (as motormen on one-man cars were called) had to hold the handle down whenever the brakes were released or even partially applied. If the controller handle was released prior to a full brake application, power was cut to the motors, the brakes went into an emergency application, and the door engines were "cut out," enabling passengers to push the doors open.

Safety cars came in a variety of configurations. Cars did not always have the same devices (deadman throttles, door treadles, etc.), and there were different versions of these devices. Safety cars began to appear during World War One. Cincinnati was building them in 1918.[11]

Combining advances in lightweight construction, one-man operation, and efficient interior arrangements, such as the Peter Witt, often resulted in streetcars that were considerably more economical than conventional equipment (see table 9.2). Unfortunately, even when combined to their greatest advantage, these advances did not produce a car design that could be considered "new." AERA's Equipment Committee was outright defeatist on this point. In 1926 it lamented "Car outlines have changed but little and for practical reasons there is little that can be done to improve them that has not been already tried."[12]

The Birney Car

One particular safety car design that came fairly close to being considered "standardized" was the "Birney car." The car was named after its designer, Charles O. Birney. Birney was an engineer with the national engineering firm Stone & Webster. Based in New York City, Boston, and Chicago, Stone & Webster designed public utility facilities (including power and substations, power transmission and gas lines, street and interurban railways, and buildings), supervised the construction of these facilities (whether designed by its staff or through another firm), managed public utility systems, reported on utilities, financed utility development, and brokered the sale of securities.

Birney's design called for a lightweight single-trucked car body, one-man operation, and the extensive use of air-operated appliances. "Birney" passengers boarded at the front (paying the operator) and exited at the rear.

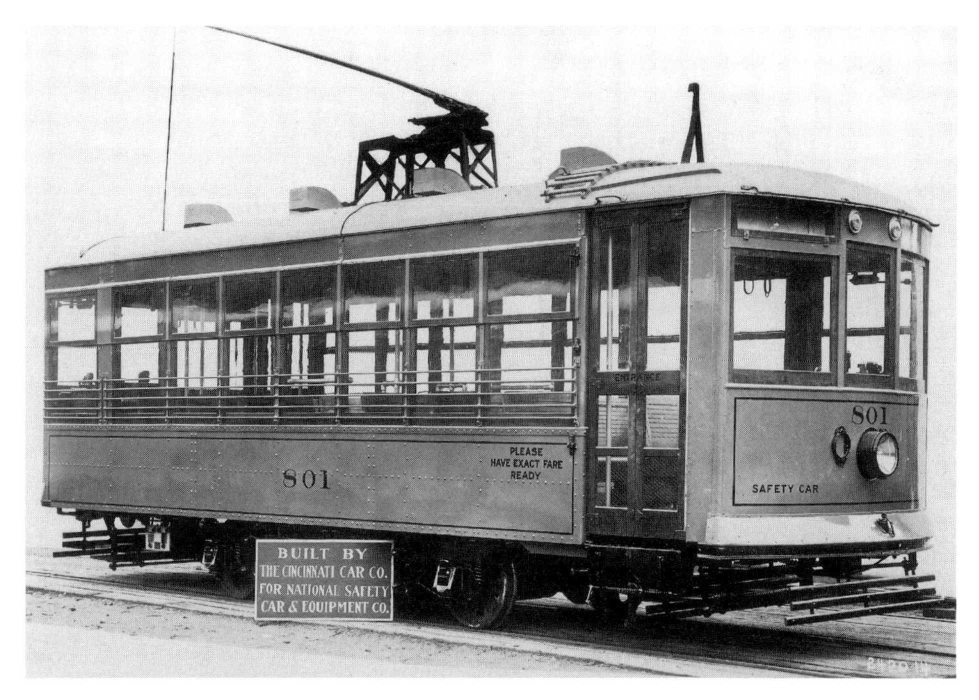

Figure 9.3. Although true Birney cars were built by Brill and Brill subsidiaries, other companies built cars that were similar in appearance. This was a Birney look-alike built by the Cincinnati Car Company for the National Safety Car & Equipment Company, a major manufacturer of safety car components. *Courtesy of the Indiana Historical Society, Cincinnati Car Corporation Collection.*

The following is a typical example of a Birney safety car, taken from an order of 20 built by the American Car Company for the Trenton & Mercer County (New Jersey) Traction Company in 1919. The cars were double-ended. They were 27 feet 9.5 inches long, 8 feet wide, and just under 9 feet 10 inches tall. They sat 32 passengers and were nearly all-steel (the roof was wood). The truck was of Brill manufacture, and the deadman controller had a foot pedal override (in lieu of holding the controller handle down, the operator could keep his foot on this pedal). General Electric motors and Westinghouse brakes were used. All interlocks were provided by a new firm specializing in these appliances, the Safety Car Devices Company (a St. Louis-based subsidiary of Westinghouse Air Brake).

Other features on the Trenton & Mercer County cars included Brill exhaust-type ventilators, American cross-over seats, hot water heat, Hunter illuminated destination signs, H.B. lifeguards, and air-actuated gongs. The finish varied: 14 cars were finished in cherry (a relatively inexpensive wood in 1919), and 6 were finished in mahogany.

One of the features that permitted Birneys to be such a compact design was the motors, which were rated at only 18 horsepower. The two motors had a slight advantage over conventional designs in that their armatures used ball bearings rather than friction bearings. Brill was supremely confident in the Birney. Its vice president, W. H. Heulings Jr., went so far as to exclaim, "What Ford did in the automobile field Birney has done in the electric car field."[13]

True Birneys were only built by the Brill organization (especially by its St. Louis subsidiary, the American Car Company). However, this was not the only single-truck streetcar design to emerge during the 1910s. As early as 1913 the Laconia (NH) Car Company produced a center-entrance design that used two independent (truckless) axles. Independent-axled streetcars rarely went past the experimental stage, although more cars of this type were built in 1916 for Tucson, Arizona. The Tucson cars also had such experimental features as one motor per wheel (two-axled streetcars rarely had four motors). Niles and Cincinnati built similar cars.

Despite its initial acclaim, the Birney proved to be a short-lived phenomenon. Systems in big cities soon discovered that a large number of small capacity, single-truck cars could not handle peak crowds as efficiently as a smaller number of large capacity, double-truck cars. The lightweight, single-trucked vehicles also provided "bouncy" rides that were extremely uncomfortable and occasionally led to derailments. A double-trucked version of the Birney was produced, although few orders were built.

Between 1914 and 1921, single-trucked car orders went from less than 20 percent to nearly 90 percent of all car orders, eventually totaling 5,500 cars. The Birney's reign was clearly over by 1923, when single-trucked cars accounted for only 10 percent of new orders.

Despite the near elimination of single-truck car construction during the 1920s, Birneys did not disappear from the streets. Although they failed in their intended market of large metropolises, their lower operating and maintenance costs *did* succeed in smaller cities (under 100,000 population).[14]

Unusual Equipment: Articulated Streetcars

Articulated streetcars enjoyed limited popularity during the 1920s. They consisted of several car bodies permanently connected by a flexible joint through which passengers could pass. The car bodies could be supported on their own trucks or by "sharing" trucks to save weight (in which case a truck was placed under the joint between two bodies and under the outside ends of the outer bodies).

In most cases, articulated cars were built in-house by reusing older equipment. A small group of cars of this type were built in Baltimore. Single experimental units were also built in Brooklyn and Milwaukee.

A few new articulated cars were created during the 1920s. Brill built a small number of cars for the Washington, Baltimore & Annapolis interurban, and Chicago Surface Lines built a single new experimental unit in its own shops. The Cincinnati Car Company built a trial articulated car for Detroit's Department of Street Railways. It was the first three-bodied articulated streetcar in the United States. Its center door served as the entrance and the end doors as exits.

One interesting aspect of this order was that the car traveled from the Cincinnati Car Company's plant to Detroit under its own power, illustrating the extent of street railway and interurban trackage in the Midwest. Although considerable attention was given to the car in the trade press, Detroit did not place any subsequent orders, nor did any other railway. Cincinnati also provided an articulated "drum" (the connection between the car bodies) for Chicago's experimental unit.

In 1928 Kuhlman built an entire fleet of 25 articulated streetcars for the Cleveland Railway. Consisting of two bodies on three trucks, the cars were over 100 feet long, yet surprisingly light (just over 38 tons). The cars were crowd swallowers, seating 104 passengers.[15]

Unusual Equipment: Specialized Cars

One of the most unusual orders of the 1920s was placed by the Chicago, North Shore and Milwaukee interurban with the Cincinnati Car Company in 1926. It called for five refrigerated freight cars to haul meat and dairy products between Chicago and Milwaukee. Unlike conventional railroad refrigerated cars, which used ice bins to cool the cars, the North Shore cars used self-contained, electrically powered refrigeration units.

The cars were considered "powered trailers." They had no trolley poles, motors, or air compressors. Instead, the cars used "jumper cables" to draw their power from other cars in the train for the refrigeration units. Ammonia was used as a refrigerant, and it kept the inside temperature at 38 degrees Fahrenheit. The cars were made of wood to improve their insulating capabilities and had interiors lined with 18 inches of cork.

Cincinnati received another odd North Shore order that same year that proved to have lasting significance. The railway requested a few flatcars that could carry two loaded wagons. Known today as "intermodal" or "piggyback" cars, they were intended to reduce door-to-door delivery time for light freight between Chicago and Milwaukee. The wagons used on these cars were specially designed for the flatcars, measuring 17 by 7 feet. Each wagon had a cargo capacity of 8 tons. These flatcars were among the first of their kind.[16]

PROTOTYPE VEHICLES

The bus's rise over the streetcar during the 1920s was not lost on the business community. Late in 1926 *Forbes* vice president and general manager Walter Dray commented on the situation:

> To my mind, the hardest pill that any group of men had to swallow was that which transportation took when they accepted the bus. When a man has been trained for years to believe in the supremacy of electric power to meet any and all conditions under any and all circumstances, by the time he reaches middle age it is difficult suddenly to realize that mechanical development has created a situation where the thing which he thought was omnipotent has limitations.[17]

The difficulty in recognizing their own vulnerability created an air of inflexibility among street railway men.

The issue of modernizing car fleets was of particular concern during the 1925 AERA convention. It was reported that some 25,000 cars in service nationwide were in need of replacement. As reported in the *Electric Railway Journal,* "These 25,000 cars parade daily before the public eyes . . . [giving] the impression of an industry down at the heel. These cars actually repel passengers. They are noisy, inefficient and expensive to operate."[18] It was held that operators had three choices: do nothing (and run the risk of further declines in ridership and perhaps even equipment failure), rebuild existing rolling stock (incorporating newer features wherever possible), or replace any equipment that could not compete effectively with buses or privately owned automobiles. Manufacturers naturally favored the last option:

> It is false economy to spend money trying to rehabilitate heavy, awkward cars. Such action postpones the time when new cars can be purchased. Frequently, money spent this way could have been used for financing new rolling stock. . . . Manufacturers are more than willing to help with such financing. . . . Experience on many properties shows that the new cars literally pay for themselves through increased earnings and reduced operating costs.[19]

Operating data from the city of Atlanta supported this view. Between 1921 and 1926, 178 streetcars were replaced with new equipment. Scheduled speeds increased from 9.28 miles per hour in 1923 to 9.81 miles per hour in 1926. Ridership experienced a similar improvement, from 91,358,379 in 1921 to 96,794,273 in 1926. Operators in other cities took more drastic action, forsaking car builders and building and rebuilding all of their own equipment. This practice continues today.[20]

By the late 1920s, some streetcar builders and operators recognized the need for a new design that broke with established practice. They realized it would not do well to replace 25,000 obsolete cars with "improved" equipment that resembled the vehicles they were replacing. What follows is a review of some of the more prominent attempts the industry made to redesign the streetcar before 1930.

The "New Birney"

During the late 1920s the St. Louis Car Company attempted to revive the old single-truck Birney concept. The new car featured rounded edges and limited protrusions from the car's surface. Much was borrowed from automotive styling, including bumpers and low, bus-type roof ventilators.[21]

The new Birney was larger than the old, measuring nearly 32 feet in length, but was about as light, weighing only 9 tons. The car also had a larger capacity, seating 39 versus 32 passengers. Experimental features included roller bearings and automotive-styled right-angle drive. These refinements were a considerable improvement over the original Birney, but they did not overcome the Birney's most serious problem, its bouncy ride. The new Birney never caught on. Only nine cars were ever built, with most operating in Jamestown, New York.

Springfield, Joliet, and Pittsburgh

In 1927 Clark Wood, president of the Springfield (Massachusetts) Street Railway, was concerned that "the tremendous increase in the use of automobiles had accustomed people to many refinements not incorporated in the usual type of street car." Wood's concern resulted in a joint effort between the railway, Brill subsidiary Wason (located in Springfield), and the Detroit-Timken Axle Company. An experimental car was ready by spring.

In appearance, the car made extensive use of flat surfaces and sharp angles, particularly at the ends. The result was a car that had a very "geometrical" or "angular" appearance. The motors were connected to the axles via worm drives, similar to those found on buses. The drives were sealed and completely immersed in oil. Automotive drum-styled brakes were used to stop the car. Rubber shock absorbers deadened any jolts resulting from uneven track. Tapered roller bearings were used instead of conventional friction bearings, which added to the smoothness and quietness of the ride. All acceleration and braking was controlled indirectly.

Use of aluminum alloys kept the car's weight under 12 tons. Low to the roof, bus-type exhaust ventilators preserved clean roof lines. All clutter was eliminated, and control equipment at the operator's station was enclosed in cabinets. Two rows of leather, double-bucket seats ensured passengers a comfortable ride.

The Chicago & Joliet (IL) Electric Railway Company took the Springfield concept a few steps further. The Joliet car was made entirely from aluminum (the Springfield car used steel frame assemblies) and included dynamic braking. Although the car was successful, it did not become as influential as hoped, largely because the Chicago & Joliet abandoned all rail service seven years later due to the Great Depression and competition from automobiles.

Encouraged by efforts in Springfield and Joliet, Pittsburgh Railways approached Westinghouse Electric, Westinghouse Air Brake, and the Osgood-Bradley Car Company to prepare an experimental car for Pittsburgh. Two cars were actually built. Each was single-ended, with front-entrance and treadle-activated rear-exit doors. The Pittsburgh cars had an angular appearance that resembled the Springfield body. However, their ends were more rounded beneath the windows than the Springfield car.

Despite the visual similarity, the Pittsburgh cars incorporated different design features. One car used Detroit-Timken trucks nearly identical to those used in Springfield, while the other used experimental trucks of Osgood-Bradley's own design, along with a new type of drive developed by Westinghouse. Called "W-N" drive, it used conventional helical gears in a completely sealed gear case. The brake design used conventional shoe-to-tread brakes, but broke from traditional practice by utilizing automotive-type diaphragms to actuate them.

Both cars featured "cabinetized" controls. The dynamic and air brake systems were operated by a common handle so that both systems came into play with each brake application. Warm air drawn off the dynamic brakes could be used to heat the cars' interiors in winter. Thirty-five rattan, single-bucket seats were situated along the sides of the car, angled forward and toward the center aisle.

The cars included numerous safety features, such as stoplights at the cars' rear, air-powered windshield wipers, and windshield defrosters. A battery-powered backup system actuated emergency lights inside the car in case of a dewirement.

Pittsburgh differed from Springfield and Joliet in not being concerned with "extreme light weight." Its cars used more steel than other experimental designs of the 1920s. Despite this, the cars were lighter than expected, totaling just over 18 tons for the car with Osgood-Bradley trucks and 17.5 for the car with Detroit-Timken trucks.

Rather than place orders based upon these experiments, Pittsburgh railways ordered a third experimental vehicle in 1929. Unlike the previous two, this emphasized saving weight. Much of the car was aluminum, resulting in a car weighing only 13.5 tons. Osgood-Bradley trucks were used.

The operator's controls were redesigned for foot operation. Foot pedals were arranged in similar fashion to those on an automobile or bus. The right pedal controlled acceleration, the left, braking. As a safety measure, the operator needed to place pressure on at least one pedal at all times, or a deadman feature would take over.

Despite the success of these three units, Pittsburgh did not place any new orders. Concerns over the Great Depression or a "wait and see" attitude toward the Electric Railway Presidents Conference Committee effort may have contributed to Pittsburgh Railways' reluctance.[22]

The Master Unit

Aside from the Birney, the Master Unit represented the closest the industry got to a truly standardized car before the PCC. Like the Birney, the Master Unit was built by J. G. Brill and its subsidiaries. Despite the size of the cars (most were over 40 feet long, with a seating capacity of 45), the use of aluminum alloys kept their weight down to 15 tons. Although their edges and corners were slightly rounded, Master Units made use of geometric shapes, flat surfaces, and angled ends. They rode extremely low to the ground (they were less than 10 feet tall, while typical low-floor cars of the 1920s were 12 feet or taller).

The low height was achieved partly from using 22-inch instead of 26-inch wheels. To accommodate them, Brill created a special truck. Despite the smaller wheels, the new trucks kept the Master Unit's motors further from the ground than those on conventional Brill trucks with 26-inch wheels.

Master Units used roller bearings and a special type of gearing to provide a smooth, quiet ride. Automatic acceleration (a version of HL) was provided by Westinghouse, assuring riders of consistent, even acceleration. Automotive-style drum brakes provided quick, quiet, and powerful deceleration.

The gearing used a helical reduction arrangement that was totally immersed in oil. The motors were connected to the gearing via "universal fabric joints." The controller had only three positions: slow speed (intended for yards, etc.), moderate speed (about 15 to 25 miles per hour), and top speed (approaching 60 miles per hour on some models).

Brill took the Master Unit to the next level in 1928 by developing eight specific models based on the design. Six were intended for city or suburban service and two for interurban service. One of the interurban models included a freight compartment for light merchandise. The city models included single- and double-trucked cars (available with single- or double-ended bodies), a single-ended front entrance/center exit car, and a double-bodied articulated car. The interurban version came in both single- and double-ended models.

The Master Unit enjoyed a fair amount of success, although only three of the eight types of carbody were ever built (mostly front-entrance, center-exit bodies and bodies with end doors only). Cities that used Master Units included Wilmington (Delaware), Indianapolis, Louisville, Lynchburg (Virginia), Portland (Oregon), and Youngstown (Ohio). Additional models could be seen on the Philadelphia & Westchester interurban, the lines of the Yakima Valley Transportation Company (Washington state), and in Brazil and Peru.[23]

The Electromobile

Based on its experience in Pittsburgh, Osgood-Bradley developed a double-ended streetcar design called the "Electromobile." Like Master Units, they had balanced, geometric proportions, were lightweight, and rode easily. Surprisingly, Osgood-Bradley reverted to wood for much of the Electromobile's lesser structural members. Operator's stations were similar to those on the first two Pittsburgh cars, although seating was similar to the Springfield car and Master Units, (two rows of leather, double-bucket seats). Westinghouse HL control was used, as was W-N drive.

Diaphragm-actuated railroad brakes were used, though without the dynamic feature. Electromobiles were 42 feet long and sat 54 passengers. Complete, single-ended Electromobiles weighed 17 tons.

Although the Electromobiles boasted fewer sharp angles at their ends than did the Master Units, their body surfaces were marred by countless rivets, making them similar in appearance to the conventional equipment they were trying to displace. Only a handful of companies purchased Electromobiles: the Altoona & Logan Valley Electric Railway (5), the Scranton Railway (10), and the Union Street Railway in New Bedford, Massachusetts (12).[24]

Louisville

In 1929, Louisville, Kentucky, accepted delivery of three experimental vehicles ordered from three car builders. Of particular significance to this study is that two of the three were Ohio products. The car builders were Kuhlman, Cincinnati, and the St. Louis Car Company.

The St. Louis car was the first to arrive, going into operation on Sunday, 7 April 1929. The car was single-ended (though it did have a trolley pole at the front end for reverse moves), and arranged for front-entrance/rear-exit passenger movement. It looked similar to a Master Unit, only with taller windows to improve visibility. It employed a mixture of single and double seats, seating a total of 51 passengers. The car weighed 18 tons. Control and braking were conventional, although the gears and pinions had hollow webbing filled with "noise absorbing material."

Kuhlman's contribution was the front entrance/center exit version of the Master Unit. It sat 50 passengers, and weighed 17 tons. Unlike a true Master Unit, however, its drive and braking apparatus were conventional.

Kuhlman had also produced a small number of similar cars in 1925 for use in Grand Rapids, Michigan. The cars were lightweight and sat 43 passengers in a 37.5 foot body. They were utilitarian in appearance (there was no headlining), with standard safety car features and an automotive-styled braking system (drum brakes on all four axles).

Cincinnati's car was made entirely from aluminum. Its trucks featured roller bearings. It sat 53 using leather bucket seats. There were 14 ventilators: 10 exhaust ventilators in the roof, and 4 louvered ventilators located in pairs at each end.

Despite the promise each of these cars offered, Louisville placed no orders for additional cars. It would not be until after World War Two that Louisville would purchase new PCC streetcars, and these would be diverted to Cleveland in 1946–47 without seeing service, following a decision to convert Louisville's transit system entirely to buses.[25]

ASSESSING THE LACK OF ORDERS FOR NEW EQUIPMENT

Why none of these designs "caught on," even when their performance was clearly better than conventional equipment, may be attributed to several factors. First of all, despite the number of railways that tried out new car designs, they were by far in the minority. The industry's conservative nature made it extremely difficult for new railway designs to catch on. In Atlanta, for example, a 1926 Cincinnati Car Company safety car order still featured clerestory roofs. While this case was a bit extreme, it nevertheless is illustrative of railway operators' reluctance to try anything truly innovative.

Most experimental equipment was ordered late in the decade, when new car orders were generally scarce. The financial condition of most railways by the late 1920s discouraged new cars purchases, much less those embodying new technologies. The Great Depression and business loss due to the automobile effectively destroyed the ability of most street railways and manufacturers to sustain themselves.

Still another factor in the industry's reluctance to embrace new car designs came late in 1929. Concerned with the industry's rapid decline and confident that a concerted effort by railway operators and manufacturers might yet develop a vehicle that could arrest the decline, a new and comprehensive effort was mounted. Charged with developing a streetcar that embodied all of the promises of standardization, advanced technology, and stylish design, this effort would have a retarding effect on new orders until the project produced a viable streetcar.

OHIO CAR BUILDING ACTIVITY DURING THE 1920S

Efforts of the G. C. Kuhlman Car Company

Despite the fact that only two car builders remained active in Ohio during most of the 1920s, Ohio manufacturers continued to play a significant role in the industry. Kuhlman carried on as part of the Brill organization. Despite a temporary lull in car orders in 1921 and 1922, sales by the Philadelphia plant and its three major subsidiaries (American, Kuhlman, and Wason) exceeded World War One levels in 1923 ($18,167,486 versus $16,761,155 in 1918). However, Kuhlman began to lose much of its local identity following Gustav Kuhlman's death on 4 September 1915. Within half a year, Brill added nearly 36,000 additional square feet of factory space to the original plant.[26] Between July 1914 and December 1929, 90 Kuhlman orders were published by the trade press, representing 2,510 cars. Of these, 87 percent were streetcars and 12 percent were interurbans. The balance were special vehicles, such as work, freight, or parlor cars. Kuhlman built no rapid transit cars during this period.

These orders reflect the decline of the street railway industry. Kuhlman's most productive period was 1914–19, in which nearly 1,500 cars were produced (almost 60 percent). Next came 1920–24, with 736 cars (nearly 30 percent). Only a little more than 10 percent of the builder's output came during the late 1920s.

Although Kuhlman's orders encompassed 15 states plus Washington, D.C., Brill clearly tried to keep Kuhlman's car building activity regional—over half of the orders were for railways in Ohio, Michigan, and upstate New York. Understandably, Kuhlman's largest single customer was the city of Cleveland. The Cleveland Railway Company accounted for 569 cars, comprising 43 percent of all Ohio orders and 23 percent of orders overall. The other states receiving Kuhlman car orders were Colorado, Illinois,

Figure 9.4. Car framing for a lightweight safety car at the Kuhlman plant. *Courtesy of the Branford Electric Railway Association.*

Figure 9.5. A 1920s safety car built by Kuhlman for service in Columbus, Ohio. *Courtesy of the Ohio Railway Museum.*

Indiana, Kentucky, Massachusetts, Minnesota, Missouri, Oklahoma, Pennsylvania, Texas, Washington, and West Virginia.[27]

In addition to their car building activity, many surviving car builders expanded into the production of bus bodies during the 1920s. Kuhlman offered at least five body styles, seating 21 to 29 passengers. The interiors varied in degrees of comfort, ranging from utilitarian city versions to a more comfortable intercity body.

Kuhlman's efforts were encouraged by the presence of a major local manufacturer of bus chassis, Cleveland's White Motor Company. Many of White's chassis were sent to Kuhlman for completion (the two manufacturers were located 3.5 miles apart on Cleveland's east side).

Kuhlman was also one of the first manufacturers to construct its bus bodies entirely of steel. Industry analysts noted that before the entry of streetcar builders into the bus market, bodies tended to be made entirely of wood or a wood-steel combination. Streetcar builders such as Kuhlman had already gone through the transition from wood to steel body construction and were quick to apply what they learned to this new form of transport.[28]

A 1912 Kuhlman advertisement had the following to say about steel bus construction:

Figure 9.6. Interior of a 1920s safety car built by Kuhlman. Note the arched roof, leather seats, and bars for standees. All are typical features of streetcars built during the 1920s. *Ohio Railway Museum, author photo.*

> By the use of steel, now predominately used in electric car construction, there is very little danger of injury to passengers from sharp, jagged pieces of splintered wood in collision, or from fire, and experience has taught the industry the economy of protection in public transportation. . . . Other advantages obtained by the use of steel-constructed motorbuses are longer life and low maintenance, and the familiarity which operating and maintenance departments of electric railways have with this type of construction.[29]

Efforts of the Cincinnati Car Company

The trade press published considerably more orders for the Cincinnati Car Company than for Kuhlman (150). However, Cincinnati's individual orders tended to be smaller than Kuhlman's, totaling 2,490 cars (20 fewer than Kuhlman). Of these, 1,531 were streetcars (just over 60 percent), 456 were interurbans, which included Chicago Rapid

Figure 9.7. A Kuhlman streetcar built for service on the West Penn Railways system. The car embodies a number of 1920s car features, including an arched roof with low ventilators and a body that is low to the ground. Note how easily passengers could step up into the car. *Courtesy of the Branford Electric Railway Association.*

Figure 9.8. A Kuhlman Peter Witt streetcar built for the New Albany & Louisville Railway in Indiana. *Courtesy of the Branford Electric Railway Association.*

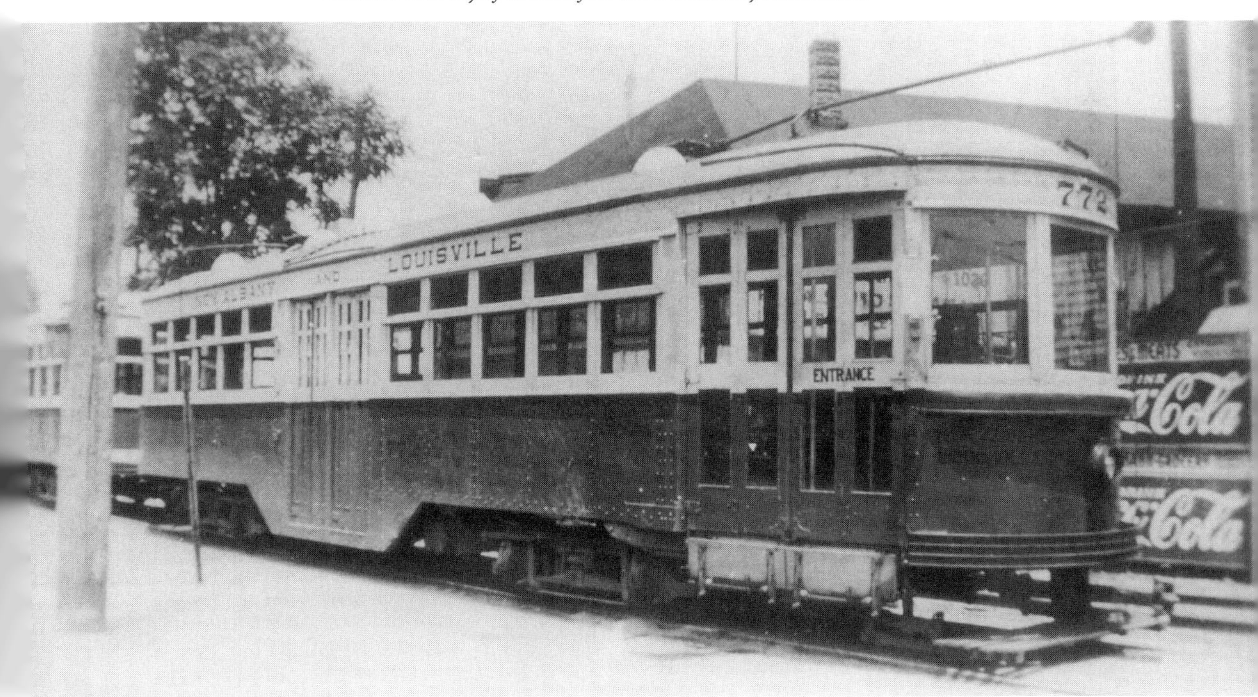

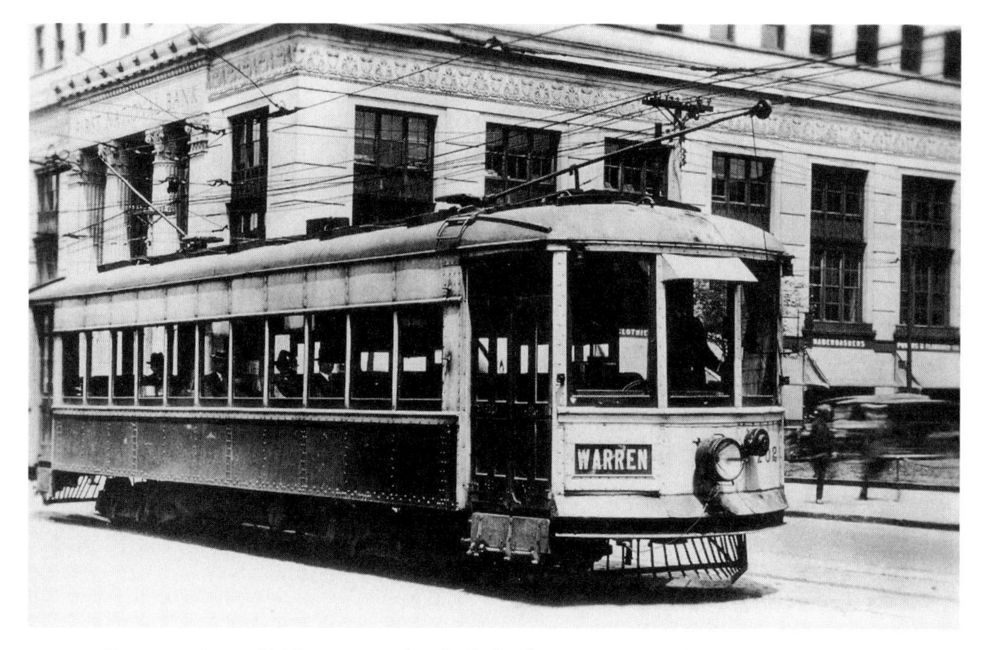

Figure 9.9. A 1924 Kuhlman interurban built for the West Penn Railways system. *Courtesy of the Branford Electric Railway Association.*

Transit's 4000 class (18 percent), and 422 were rapid transit cars (17 percent). The balance were specialized vehicles.

Like Kuhlman, Cincinnati's orders reflect the decline of the industry. The greatest concentration of orders came in 1914–19, when 1,273 cars (51 percent) were built; 720 cars were built over 1920–24 (29 percent), and 497 (20 percent) were built during the late 1920s.

Cincinnati's market of 21 states was a little broader and more evenly balanced. (Ohio and Illinois each made 20 percent of its orders, slightly over 500 cars apiece). Georgia, Pennsylvania, New Jersey, and New York each ordered more than 5 percent. Like Kuhlman, the local street railway operator (the Cincinnati Traction Company) was a significant customer (375 cars, or 73 percent of all Ohio orders and 15 percent of Cincinnati's orders overall). The other states receiving Cincinnati Car Company orders were Alabama, Colorado, Indiana, Kansas, Kentucky, Maryland, Michigan, Missouri, Oklahoma, South Carolina, Tennessee, Texas, Virginia, West Virginia, and Wisconsin.

The Cincinnati Car Company also enjoyed a closer relationship with its community than did other Ohio car builders. It benefited from local orders and participated in a cooperative education program with the University of Cincinnati. Thomas Elliott, vice president and general manager of the car builder and a prominent streetcar designer, encouraged the practice. He felt that students needed to balance their theoretical training with practical experience: "The clever manipulator of the slide rule should not operate to the disadvantage of the man who is equally good with the folding type, because the latter is far more useful and valuable in the shop."[30]

Figure 9.10. A lightweight Kuhlman streetcar built for the Interstate Public Service Company in Indiana. *Courtesy of the Branford Electric Railway Association.*

The Cincinnati Car Company also contributed much to the industry, developing new designs and construction methods that became virtually universal for lightweight cars. One of these was the practice of using continuous pieces of T-bar steel to form both the side posts and carlines. This both simplified a car's framing and increased its structural strength without adding weight. The *Electric Railway Journal* credited Cincinnati for popularizing the design, claiming it was "practically standardized by the Cincinnati Car Company." An early example of this type of construction was on an order for the Utica Division of the New York State Railways in 1917.[31]

Cincinnati was also the first car builder in the United States to start combining lightweight designs with safety car features on interurbans. The *Electric Railway Journal* credited Cincinnati with the nation's first lightweight interurban. Built in 1917 for the Princeton Power Company (West Virginia), the cars were made of steel, yet weighed only 10.5 tons. It was still running in 1928: "Its low energy consumption, averaging 0.8 kw per car mile, has effected a great saving. Its high rate of speed, 40 m.p.h., is especially appreciated by the patrons."[32]

Elliott introduced another, even greater innovation in streetcar design and construction. Conventional car building practice called for assembling the floor framing first (side, center, intermediate, end, and cross sills). Once completed, side and corner posts were attached to the frame, then the carlines, followed by the side sheathing.

Figure 9.11. A 50-ton interurban built for the Chicago, North Shore & Milwaukee by the Cincinnati Car Company in 1924. *Courtesy of the Indiana Historical Society, Cincinnati Car Corporation Collection.*

Figure 9.12. A passenger-baggage combine built by the Cincinnati Car Company during the 1920s for the Union Traction Company of Indiana. *Courtesy of the Indiana Historical Society, Cincinnati Car Corporation Collection.*

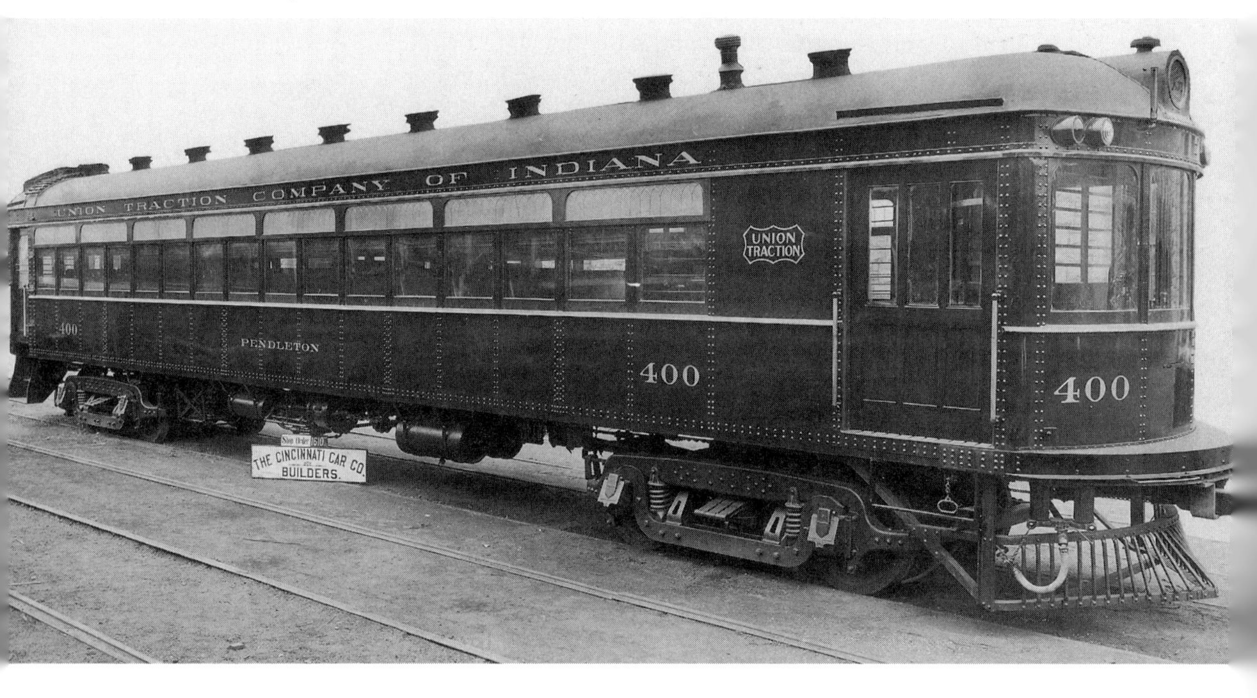

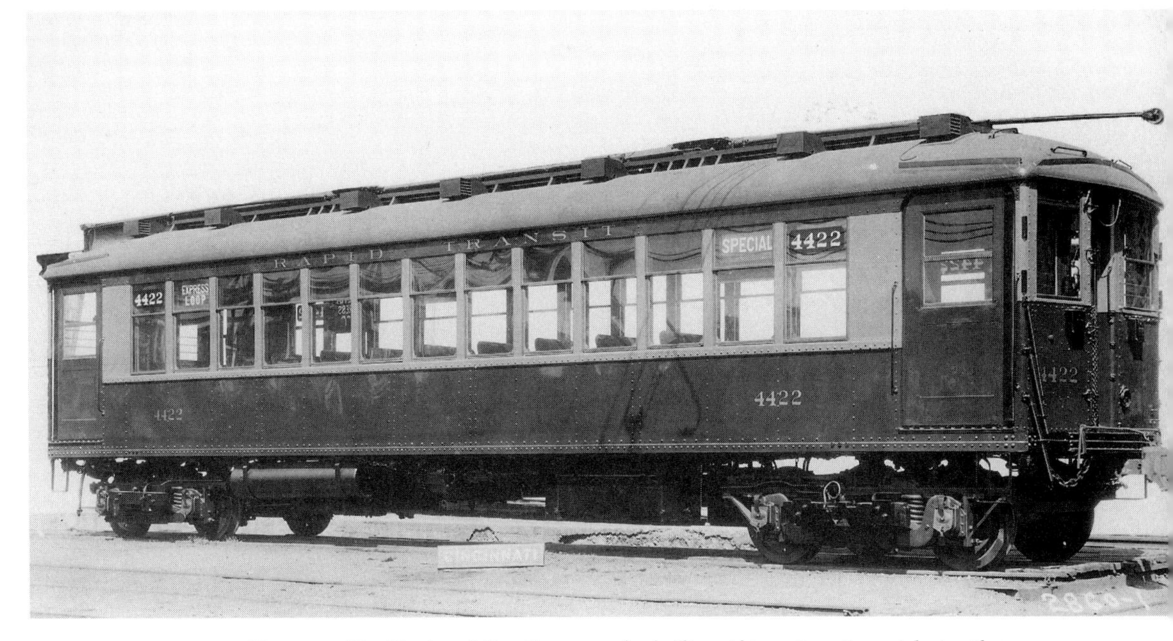

Figure 9.13. The Cincinnati Car Company also built rapid transit equipment during the 1920s. This is one of Chicago Rapid Transit's 4000 class of elevated cars. *Courtesy of the Indiana Historical Society, Cincinnati Car Corporation Collection.*

The practice of fabricating carlines and side posts from a continuous piece of T-bar simplified this somewhat, but Elliott went a step further. Rather than fashion his side sills from thick, narrow steel girders, Elliott substituted wide, thin plate girders instead. The width of the plate accomplished two things: it gave the girder tremendous strength while saving weight, and it was also wide enough so that when set on edge (thus changing the girder's width to its height), the girder actually formed the side of the car body, eliminating the need for floor-to-ceiling side posts and sheathing.

This design was first used on Birney streetcars, but was not used widely on other car types until Elliott adapted it to interurbans. Elliott's adaptation made its first appearance on five interurban cars delivered to the Cincinnati, Milford & Blanchester Traction Company in 1921. It was estimated that this design alone had the potential of reducing power consumption by as much as 50 percent. Kuhlman began making similar lightweight interurbans the following year. However, Kuhlman's cars were only composite designs. Everything above the plate girders was made of wood.

Called "preeminent" in the design of lightweight interurbans by historian George Hilton, the Cincinnati Car Company later refined Elliott's design by curving the car sides outward. Known in the industry as the "curve-sided car," the body's frame and roof were made several inches narrower than the car's maximum width. The sides of the car curved outward from the floor until they reached the windows, then started to curve inward again to meet the narrowed roof. By narrowing the floor and roof framing, less material was used, resulting in lighter weight and less cost. By allowing the sides to bulge below the windows, full-sized seats and aisles could be retained.

Figure 9.14. A group of Chicago 4000s being prepared for shipment in the Cincinnati Car Company's yard. The rail running perpendicular to the yard rails in the immediate foreground is connected to a transfer table. On the far right one can see a New York Central boxcar sitting on the car builder's loading track. Note the dual trolley wires above each of the yard tracks. This type of overhead was unique to Cincinnati. Also note the MCB coupler adapter in front of the rapid transit car in the right foreground. *Courtesy of the Indiana Historical Society, Cincinnati Car Corporation Collection.*

The Dayton Street Railway, which purchased an early order of one-man curved-sided cars from Cincinnati, discovered an additional advantage to narrower car framing: they could be run on lines with narrow streets without being obstructed by parked automobiles. Approximately 400 curve-sided cars were built by Cincinnati throughout the 1920s.

Some car builders experimented with narrower, straight-sided bodies. These cars realized the first two advantages of the curve-sided car, but did so at the expense of narrower aisles and/or smaller seats.

By the early 1920s, the Cincinnati Car Company had also become a major producer of its own components. Cincinnati developed three truck designs for use with its curve-sided cars. For example, a 1922 order for the Kentucky Traction & Terminal Company (Lexington) called for 10 convertible all-steel streetcars. Cincinnati not only built the cars but it also supplied the axles and bearings, hand brakes, couplers, pilots, bumpers, sanders, window sash fixtures, step treads, ventilators, and all of the brass trimmings.

Figure 9.15. The Cincinnati Car Company's greatest contribution to lightweight car design was the development of the curve-sided car. This photo shows one side of a curve-sided car being assembled on the Cincinnati erecting shop floor. *Courtesy of the Indiana Historical Society, Cincinnati Car Corporation Collection.*

Cincinnati received a blow similar to Kuhlman's in the wake of Gustav Kuhlman's death when Thomas Elliott stepped down as vice president and general manager in 1917. He retained some ties with Cincinnati as a consulting engineer, but moved to Birmingham, Alabama, as president of the Continental Gin Company, a manufacturer of cotton gins.

Another leadership crisis came in 1927 when W. Kelsey Schoepf died. Schoepf was a major figure in interurban railways who, along with Indiana traction magnate Hugh J. McGowan, controlled numerous interurbans in both Ohio and Indiana. In addition to his interurban holdings, Schoepf also controlled the Cincinnati Traction Company and its subsidiary, the Cincinnati Car Company. He had been an officer of the car builder since 1905. Schoepf was succeeded as president by Henry Sanders, who had been his assistant since 1905. Sanders resigned two years later (possibly influenced by the stock market crash) and was succeeded by Clinton Morgan. Morgan, who had been president of several railway companies (notably the large West Penn system), was openly critical of the inward focus of Cincinnati. He vowed to bring "the attitude of the buyer" to the company.[33]

Figure 9.16. During the 1920s, the Cincinnati Car Company simplified the construction of its arched-roof assemblies. This photo shows a completed roof being lifted off the car builder's roof assembly jig. *Courtesy of the Indiana Historical Society, Cincinnati Car Corporation Collection.*

Late in 1928, Cincinnati expanded into locomotive production with the "Cincinnati Line." The locomotives, which ranged from 2- to 50-ton models, were intended for industrial use or for switching cars in railway yards. A variety of power options were available, including electric (trolley wire or battery), internal combustion, and gasoline-electric. Two gasoline-powered locomotives were displayed at the 1929 American Road Builders' Association convention at Cleveland. Production was managed by H. R. Sykes, "one of the pioneers in the development, manufacture and sale of industrial locomotives, and who is well known to many of the locomotive using public."[34]

Cincinnati also participated in the manufacture of bus bodies during the 1920s. Although most of its bodies were of steel construction, Cincinnati also used ash carlines and wood body sheathing.

Cincinnati went a step further in the bus market by merging with the Versare Corporation in 1928. Versare was headquartered in Watervliet, New York, with an addi-

Figure 9.17. This photo shows a completed Cincinnati Car Company curve-sided car built for West Virginia City Lines. *Courtesy of the Branford Electric Railway Association.*

Figure 9.18. In addition to complete railway vehicles, the Cincinnati Car Company made a number of vehicle components as well. This view shows a truck made by the car builder. *Courtesy of the Indiana Historical Society, Cincinnati Car Corporation Collection.*

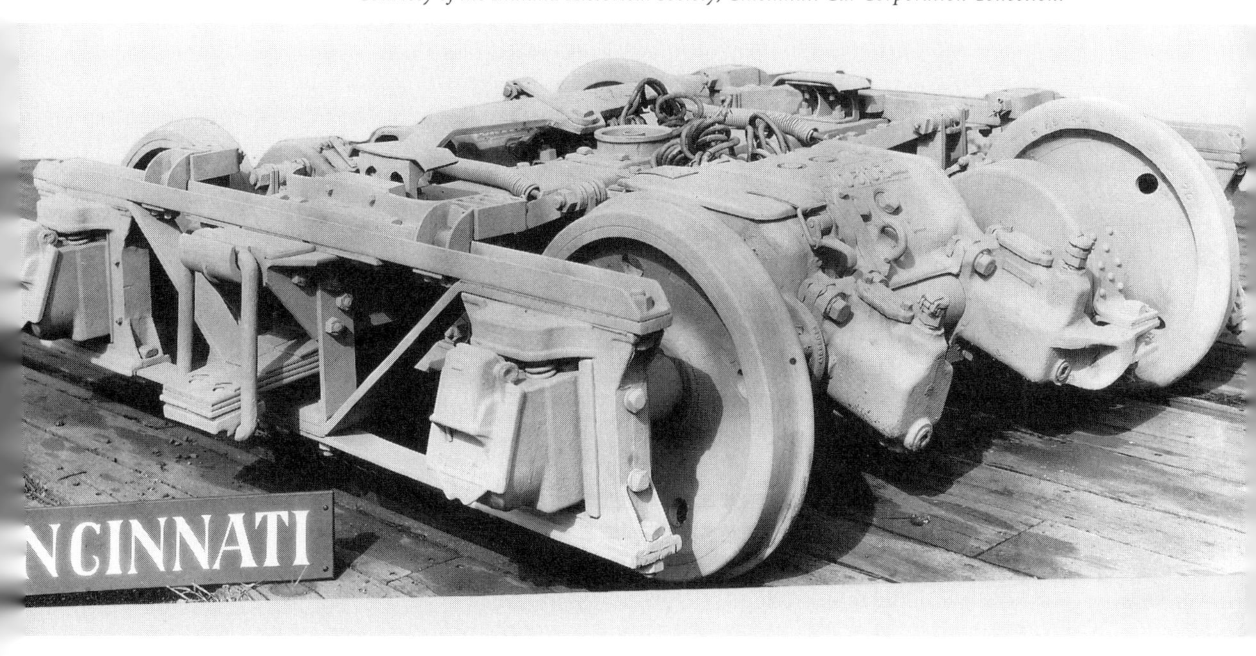

Figure 9.19. Another Cincinnati Car Company truck. *Courtesy of the Indiana Historical Society, Cincinnati Car Corporation Collection.*

tional production plant located in Buffalo. The *Electric Railway Journal* reported, "The two plants will be merged in the formation of the Cincinnati Car Corporation, with headquarters in Cincinnati. The engineering facilities and offices of the Albany corporation will be brought to the Ohio city, but physical property will remain in Buffalo." An example of a Cincinnati/Versare order was 10 trolley buses (electric buses that obtained their power from overhead trolley wires) for the Utah Light & Traction Company of Salt Lake City.[35]

CONCLUSION

The 1920s were disappointing years for the street railway industry in many respects. Decline in ridership led to the closure or contraction of many railways, accompanied by a corresponding drop in new car orders. Much of this decline can be laid at the feet of competing technologies, but not all of it. The industry's conservative preferences regarding car design did little to change public perception of the streetcars, particularly when compared to the dynamic automotive industry.

The industry's one attempt at a standardized design embodying many of the latest advances in streetcar technology, the Birney, met with disappointing results. This may have discouraged car builders and railways from developing further streetcar models that broke from convention until late in the decade. Of the few new designs to emerge, only one, the Master Unit, met with limited success.

The experience of Ohio's car builders during the 1920s reflected much of what was experienced by the industry in general. The sharp decline in their orders as the decade progressed is illustrative of what other car builders encountered. Both Kuhlman and Cincinnati exerted some influence in the development of streetcar technology—Cincinnati in its lightweight car designs, Kuhlman in participating in Brill's Master Unit program, and *both* car builders in the creation of experimental vehicles, the construction of articulated cars, and the manufacture of bus bodies.

While Ohio's car builders can be viewed as illustrating larger trends within the industry during the 1920s (most of which were negative), component manufacturers found themselves facing similar challenges and surprising opportunities. What the decade meant to the component manufacturers will be the subject of the next chapter.

Figure 10.1. Different types of streetcar gears. The large gear in the center is a spur gear (sometimes referred to as a "straight gear"): the teeth on this gear go straight across the gear's edge. The gear on the right is a helical gear, which became popular during the 1910s. The teeth on helical gears go diagonally across the gear's edge. They are quieter than spur gears. The gear on the left is a herringbone gear, so-called because of the tooth pattern. Although rare on street railways, herringbone gears were extremely quiet. *Branford Electric Railway Association, author photo.*

10

PARTS OF THE WHOLE: STREETCAR COMPONENT MANUFACTURE DURING THE 1920s

The previous chapter reviewed how car builders and their customers stubbornly adhered to an increasingly obsolescent car design. A double-trucked streetcar of 1925 looked remarkably similar to a streetcar built in 1905. At the same time, the industry failed to develop any coherent standardization scheme. This contrasts sharply with the automotive industry, where automobiles and buses of 1920 contrasted sharply with most models of 1929. What follows is an overview of major technological developments in streetcars during the 1920s and the role that component manufacturers played in their development.

Component manufacture remained strong in Ohio during the 1910s and 1920s. In most cases, companies that were firmly established before World War One continued to prosper. In some cases, stronger firms absorbed weaker ones. In other instances, changes in the industry led small manufacturers to gain added significance within the industry.

ESSENTIAL COMPONENTS

Motors

As mentioned previously, economies in car operation were sought around World War One. In addition to lightweight designs, energy saving came in the form of new electric motors. As cars weighed less, smaller and less powerful motors were built to propel them (the average horsepower per motor declined from 55 in 1914 to less than 40 in 1923). Older motors had "large empty spaces" within their casings. With the armature removed, a 6-foot-tall man could fit inside the casings of some rapid transit motors. In addition to wasted space, older motors had moving parts that were unsupported and did not permit sufficient lubrication. Worst of all, they were heavy (one motor usually exceeded one ton).

By rearranging the motors' components and using new materials, General Electric and Westinghouse were able to reduce motor weight by over 50 percent. The new motors were also more compact, allowing them to be used on trucks with smaller wheels.[1]

Final Drive

Another issue that confronted designers was noise. Cars often rattled and made grinding noises, all of which disturbed passengers, pedestrians, and anyone living or doing business along a car's route. Most railway officials agreed noise could result from bad operating habits (such as hard braking, which led to flat wheels) and from poor maintenance of both rolling stock and track. However, there were additional problems that were not so obvious.

Wheel noise was common on curves and switches. Early attempts to reduce wheel noise were made by drilling holes into the wheel's side (web) and bolting wood blocks to them in order to absorb vibrations. Bearing noise resulted from poor lubrication. By the late 1920s, this was eliminated entirely through the use of roller bearings. Welded trucks were quieter than trucks with bolted or riveted frames, as rivets could pop out over time and bolts could loosen.

Gear noise, the single greatest contributor to car noise, could be reduced both through better lubrication and new types of gears and pinions. Until World War One, the most common type of railway gearing was the spur gear. Spur gears (also called "straight gears") have teeth that run straight across their edges. They were simple to make, install, and maintain. However, they were not very efficient and tended to grow noisier as they wore. Gear and pinion wear was normally caused by both the motor's weight and the torque it exerted. Premature wear resulted from poor lubrication, the accumulation of grit in the gear casing, and misalignment, which resulted from poorly maintained track.

By the time the United States entered World War One, helical gears had come into use. These gears have teeth that run diagonally across their edges. In operation, the full width of the teeth on spur gears met at once, creating a jarring impact that could be felt throughout the car body. Helical teeth initially met at one end only, gradually transferring their load across their width. This gradual transference resulted in much smoother, quieter operation. Helical gears also lasted longer, in some cases reaching 300,000 miles before requiring replacement.

A variant was the double helical or herringbone gear. Its teeth form a "V" pattern across the gear's edge. Herringbone gears and pinions were virtually noiseless and even more efficient than helical gears. However, they were more complex to manufacture (and thus more costly to purchase) and maintain.

A further development was the use of automotive-type "right-angle" drives. Right-angle drive was quiet and permitted the motors to be rearranged so that most of the motors' mass was "placed . . . near the center of the truck."[2]

In addition to new types of gearing, roller bearings came into use late in the 1920s. Ideally, they minimized the friction and wear associated with journals and other friction-type bearings. Street railways and interurbans found two major applications for them: in motor armatures and axle bearings. Roller bearings resulted in power savings (as motors encountered less resistance), smoother acceleration, lower maintenance and lubrication costs, and better riding quality.

Smaller wheels also improved streetcar performance. They saved weight and brought the car bodies lower to the ground, making it easier for passengers to board and alight. Wheels that were 24–28 inches in diameter increased from just over 20 percent of new car orders in 1914 to 80 percent by 1923. During the same period, wheels that were 30–34 inches in diameter declined from 75 to 15 percent of new orders.

Another means of reducing floor height was to lower the end platforms/vestibules. Trucks were brought closer together, enabling the platform "knees" to drop lower. This gave the passenger a lower step-up into the vestibule, although it also created an additional step inside the car body.[3]

Tool Steel Gear & Pinion and Timken Roller Bearing

As railway operators of the 1910s and 1920s attempted to reduce operating costs, rolling stock maintenance received great attention. Cincinnati's Tool Steel Gear & Pinion Company benefited from this attention due to the quality of its gears and pinions. The firm had been in existence before 1910, but it was during the mid-1910s and 1920s that it attained prominence within the industry.

Tool Steel's gears and pinions were available in split and solid models. At one point in 1916 the company had a backlog of orders for nearly 3,000 gears and 5,000 pinions. The company was regarded as a good employer, as evidenced during a machinists' strike over the summer of 1916. Tool Steel was one of the few manufacturers that remained open.

Tool Steel gained a further foothold in the street railway market by developing the "wisdom tooth" pinion in 1922. The teeth on these pinions were thicker (and thus stronger) than those on other pinions. As a result, Tool Steel pinions needed only 12 teeth instead of the conventional 13.

By the mid-1920s, Tool Steel enjoyed a national and international market. Major offices were located in 12 North American cities and eight European countries. The Ackley Company represented Tool Steel in London, Paris, Berlin, Belgium, the Netherlands, Denmark, Norway, and Sweden. The U.S. Metal & Manufacturing Company represented Tool Steel in New York City, Chicago, and Atlanta. Separate companies represented Tool Steel in Portland (Oregon), Boston, Toronto, Pittsburgh, Los Angeles, San Francisco, Calgary, and Huntington (West Virginia). In the United States, it was estimated that 65 percent of all electric railway vehicles in New York used Tool Steel gears and pinions. Nearly 90 percent used them in Texas, 80 percent in Indiana, 74 percent in Utah, and 65 percent in Illinois.[4]

In some cases, firms not previously involved with the street railway industry entered the market successfully in the 1920s. The Timken Roller Bearing Company is an example. Henry Timken's experiments with wagon wheels led to the tapered roller bearing, which could withstand both radial (vehicle weight) and thrust (side) loads. His invention proved ideally suited for the automotive field, and during the 1920s his company created a small, vertically integrated manufacturing empire in Ohio, comprising steel mills in addition to bearing manufacturing plants.

Timken first entered the street railway market by supplying Ohio Brass with roller bearings for a new type of trolley pole base in 1924. However, Timken's greatest contribution to railroads and street railways was in its axle bearings. Even though the journal (a friction bearing) was a tried and true method of supporting a car's weight on the axles, journals required constant attention, lest their lubricating oil run dry. "Hot boxes" (resulting from excessive friction between the axle and bearing caused by insufficient lubrication) often led to fires, locked wheels, and derailments if not spotted quickly.

Tapered roller bearings solved this problem by providing a minimal and variable contact surface. They also withstood the sideward thrust of cars on curves. They were first tested on a Northern Ohio Traction & Light interurban line between Canton and Cleveland in 1923. Three years later a major railroad, the Milwaukee Road, adopted the bearings for its *Pioneer Limited* and *Olympian* passenger trains. Successful, they led to orders from numerous railroads, starting with "named" trains. By the mid-1930s, examples included Chicago, Burlington & Quincy's *Zephyrs,* Santa Fe's *Super Chief,* Rock Island's *Rocket,* and Seaboard's *Silver Meteor.*

Roller bearings also proved useful in motor armatures. Previously, armatures depended upon friction bearings that diminished motor efficiency. They could also destroy the motor if not lubricated properly. Roller bearings both improved motor efficiency and further minimized the risk of bearing and armature failure.

Despite the promise of its product, Timken was slow to enter the railroad and street railway markets, due in part to its senior management, which was not interested in railway applications for their product. Once the market was breeched, however, it was the railroads, not the street railway industry, that comprised Timken's principal rail vehicle market. Only eight orders for roller bearings (representing 136 cars) were reported in street railway trade literature, all during the late 1920s. The street railway industry's inherent conservatism may also have slowed Timken's entry.[5]

Control Systems

Before World War One, the most common method of controlling streetcars was the "direct method," as evidenced in the K-type of controller, mentioned in chapter 1. With direct systems, acceleration was entirely dependent upon the operator's skill. "Indirect" motor control, or a system in which the car's operator manipulated banks of switches that controlled the amount of resistance in the motor circuit via remote control, had been available since the late 1890s. These systems could provide smoother, more consistent acceleration (as well as keeping the full operating voltage out of the car's interior). However, due to the great size, weight, and complexity of early indirect units, indirect control systems were generally restricted to rapid transit and some interurban railways.

Shortly before World War One, General Electric and Westinghouse developed smaller indirect control systems that could be used on any railway vehicle, including streetcars. Westinghouse's system, known as "HL," was controlled electro-pneumatically. The operator manipulated a low-voltage controller that opened and closed valves on a set of air cylinders. Compressed air within the cylinders was used to open and close the resistor bank switches that accelerated the car. General Electric devised a system known as "MK" (a later refinement was known as "PC"), in which the low-voltage controller activated solenoids that opened and closed the switches in the motor circuit. Both systems accelerated cars at a smooth, uniform rate regardless of the operator's skill.

Although more complex electrically, indirect control systems greatly improved the performance of the streetcars. They eliminated the hazard of melting resistor grids or burning contact fingers by leaving conventional controllers in nonrunning positions. Rates of acceleration could be smoother and more consistent, something the passengers appreciated. Finally, indirect control made it easier for streetcars to operate in multiple-unit trains, something that was difficult to do previously. Unfortunately, their complexity resulted in higher purchase and maintenance costs. Thus, the K-controller never disappeared.

Another problem with indirect control systems such as HL, MK, and PC was that their rate of acceleration was constant (not flexible). By the late 1920s, both Westinghouse and General Electric developed systems that allowed operators to select the *rate* of acceleration. However, they were even more complex, which further magnified maintenance problems. These systems did not attain wide use.

Control equipment also benefited from cosmetic changes during the 1920s. Controls were concealed in cabinetry, with only the handles, switch knobs, and instruments necessary for operation remaining visible. For example, the Chicago & Joliet Electric Company cut away the flooring at the front of its cars to create a well for the controllers, thus allowing the cabinet top to be flush with the bottom of the windshields. "By having the airbrake piping and controller within cabinets, the front end of the car presents a very neat appearance, dust and dirt cannot accumulate in inaccessible corners, and the motorman can sit down in comfort."[6] Rearview and side mirrors gained favor in some cities, as did windshield wipers.[7]

Current Collection

Despite such alternatives as the pantograph and third rail shoe, the most common type of current collector remained the trolley pole. It was universal for street railways and common on interurbans. Pantographs and third rail were only used on a few interurbans. The only real development in current collection was the trolley "shoe." Shoes were alternatives to the copper or bronze trolley wheels used to contact the overhead wire. Rather than rotate under the wire, shoes "slid." Shoes increased the life of the wire, as copper wheels tended to wear the overhead wire more rapidly than steel shoes. They later used carbon inserts to improve conductivity and reduce arcing. While carbon inserts improved shoe performance, they also had the annoying tendency to create dark streaks on car roofs from falling carbon residue.[8]

There were at least two significant mergers among Ohio manufacturers in the current collection field. The Cleveland Trolley Supply Company was purchased in 1919 by the Advance Manufacturing & Tool Company, another Cleveland firm. Advance continued the product lines of Cleveland Trolley Supply, which included not only current collection devices (harps, wheels, sleet cutters, and pole bases), but also conductors' seats, window catches, ticket holders, drinking fountains (for use on interurbans) and brush holders. Of even greater significance was a merger between Ohio Brass and the Trolley Supply Company (before the merger, Trolley Supply had moved from Canton to Massillon around 1920). From 1927 onward, Ohio Brass assumed all manufacture of Knutson, Simplex, and Ideal catchers and retrievers. The former Trolley Supply plant was retooled to produce gas heaters.[9]

Ohio Brass remained a leading manufacturer of current collection apparatus. The company led all Ohio manufacturers in orders reported in the trade press with 349, embracing 10,233 cars and 17,079 sets of components. Of the 10,233 cars equipped with Ohio Brass components, over 75 percent used its current collection apparatus. Ohio Brass was successful with other components as well. For example, 40 percent of these vehicles (mostly streetcars) were equipped with couplers and multiple-unit connections.

The prominence of Ohio Brass in so many car orders was due to several factors. The sheer variety of Ohio Brass's repertoire gave the manufacturer more opportunities to furnish components for new rolling stock. Ohio Brass's other products included couplers and multiple unit connections, headlights, lighting systems, marker lights, sanders and sander parts, and on-board signals for communication between passengers and car operators.

Despite its prominence in the field, however, Ohio Brass's orders diminished between 1914 and 1929. Its greatest period was 1920–24, in which Ohio Brass equipped nearly half of the 10,233 cars reported in the trade literature (4,657). Slightly under a third (3,182) were equipped during the late 1920s.[10]

Ohio Brass improved its product line between 1914 and 1929. One example was the "self-lubricating" catcher. Virtually identical to standard trolley catchers, self-lubricating models included a small compartment for oil-soaked cotton or other waste. In 1926 Ohio Brass purchased rights to a trolley wheel that used graphite plugs. The wheel had been designed by Charles Feist, an employee of the Sioux City Service Company. Similar wheels had been produced by the Dayton Manufacturing Company since 1921.

Sensitive to the industry's concern over saving weight, Ohio Brass developed a lightweight trolley pole base during the late 1920s. It weighed just over 70 pounds (conventional bases exceeded 100 pounds). The company also produced a headlight with a prismatic covering that illuminated the car's dash panel as well as the road ahead, making approaching cars more visible to pedestrians and motorists.

Not all of Ohio Brass's products were successful, however. During the late 1920s it developed a "trolley silencer" that was supposed to deaden the noise created by the current collector at the end of the trolley pole. Unfortunately, the silencers added nearly nine inches of length to their trolley poles, requiring existing poles to be cut down or replaced by shorter ones.

Ohio Brass also worked to reduce its production costs. Concerned about the rising expense and long delivery time for malleable iron stock, Ohio Brass built its own malleable iron foundry in 1922. The foundry building covered 100,000 square feet.[11]

In Springfield, Bayonet Trolley Harp developed a "semi-rotary" sleet cutter that pivoted both forward and backward. The cutter supposedly removed sleet from the trolley wires while minimizing friction. Despite wartime shortages (notably in springs and brass washers), Bayonet experienced a 30 percent increase in its business in 1916. It remained active through the 1920s.[12]

SAFETY DEVICES

Brakes

Air brakes remained the industry standard. One of the more significant developments in braking technology during the late 1920s and early 1930s was the "self-lapping" brake valve. Unlike conventional "straight air" brakes, in which car operators had to admit air into the brake cylinder from the "apply" position, hold it in the cylinder using the "lap" position, and gradually release it by alternating between the "lap" and "release" positions, self-lapping valves did this automatically. Self-lapping brake valves had a "release" position and a range of "apply" positions. The farther the operator moved the brake handle to the right, the greater the braking effect. This simplified braking.[13]

Figure 10.2. During the 1920s, some interurban railways took to creating dazzling paint schemes for their cars to increase their visibility on city streets and at rural grade crossings. This example is a 1914 Niles combine running on the Salt Lake & Utah Railroad. *Courtesy of the Branford Electric Railway Association.*

In addition to self-lapping brakes, several electrical braking systems were tried during the 1920s. One was the magnetic brake, which used an electromagnet suspended by springs over the rail, usually between the wheels on a truck. When activated, the magnets clamped down on the rail, bringing the car to a rapid stop. Tests conducted on the Buffalo & Lake Erie Railway provided convincing evidence that magnetic brakes dramatically shortened a car's stopping distance. At 30 miles per hour the time and distance required for air brakes acting alone was 10 seconds and just over 259 feet. A combination of air and magnetic brakes required only 6.8 seconds and 186.8 feet at 32.5 miles per hour. Magnetic brakes were used on the Buffalo & Lake Erie, the Kentucky Traction & Terminal Company, and the West Penn Railway interurbans.[14]

Although effective, magnetic brakes were hard on rails and not widely used in the 1920s. In 1930, only 75 cars were equipped with a combination of air and magnetic track brakes. Forty-one were operated by the Kentucky Traction & Terminal Company, 18 by the Buffalo & Erie Railway, 10 by the Indianapolis & Southeastern Railway, 3 by the Jamestown Street Railway, and 1 each by the United Traction Company (Albany, New York), United Railways & Electric Company (Baltimore), and the Third Avenue Railway (in Manhattan and in the Bronx).[15]

Two other experimental braking systems were dynamic and regenerative. Both used the streetcars' motors. When in use, the motor circuits were reversed, which in effect turned the motors into generators, bringing the car to a stop. The current produced was dissipated as heat in dynamic systems and directed back into the trolley wire in regenerative systems.

The Chicago Rapid Transit Company tried regenerative braking equipment along its elevated trackage in 1926. Although the results were favorable both in terms of braking performance and power savings, neither regenerative nor dynamic brakes were adopted widely by the industry. Specially insulated motors were required, and their cost and complexity were deterrents.[16]

Brake diaphragms appeared on a few experimental applications. They were "borrowed" from automotive technology. Brake diaphragms were small, self-contained, air-powered mechanisms that could be mounted directly over a car's truck(s). They saved weight by eliminating main brake cylinders and cumbersome rigging.[17]

Although Ohio manufacturers contributed little to braking technology during the 1910s and 1920s, some conventional components were produced in the state. For example, the Dayton Manufacturing Company produced handbrake handles throughout the 1910s and 1920s. This was never a major part of Dayton's output, and published orders only indicated 145 of 1,497 cars equipped with Dayton components had Dayton handbrake handles. Handbrake handles were installed on one-quarter of all Dayton-equipped cars between 1914 and 1919, but this percentage dropped rapidly: only 16 percent of orders were so equipped between 1920 and 1924, and only 1 percent between 1925 and 1929.

Lighting

Lighting developments during the 1920s were guided by a desire to improve the amount and quality of interior light while reducing power consumption. The ability of lamps to properly diffuse light to the passengers' reading plane was enhanced through the use of reflective fixtures (such as prismatic glass) and lightly colored ceilings. The best interior lighting was felt to be produced by 56-watt lamps using prismatic reflectors.

To reduce power consumption, fewer interior lamps were used at a higher wattage. More efficient headlights were also adopted. For example, the original lighting system on Washington-Virginia Railway cars used 20 23-watt lamps in the interior and a 2,700-watt headlight. Improved headlights cut the wattage back to 150, while interior lamps were replaced by seven 72-watt lamps. Even though the interior wattage increased from 460 to 576, the new headlight enabled an overall wattage reduction from 3,160 to 726.

One of the greatest changes in street railway lighting technology was the shift from carbon to tungsten lamp filaments about 1916. Tungsten filaments extended the life of lamps while allowing them to operate on slightly less than their rated voltage. Current regulating devices further increased the reliability of car lighting. The *Electric Railway Journal* stated the preferred number and strength of lamps was a single 56-watt lamp for cars 25 feet long and less; 5 94-watt lamps for cars 25–35 feet long; 10 56-watt lamps for cars 35–45 feet long; and 10 94-lamps for cars over 45 feet long.[18]

The Trolley Supply Company developed several new headlight designs before selling its streetcar component business to Ohio Brass. It came out with an incandescent headlight, the "Perfect" model, in 1915. "Perfect" headlights used a single 100-watt lamp that was magnified by a nickel-plated brass reflector. Their effectiveness helped persuade street railway men of the superiority of incandescent headlights over arc headlights. Perfect headlights could illuminate greater distances while consuming only a third of an arc light's power. The entire assembly weighed 25 pounds.[19]

Over half of the cars Trolley Supply equipped in the early to mid-1920s had either the manufacturer's headlights, marker lights, or both. However, this figure pales in comparison with Trolley Supply's current collection business. Nearly 90 percent of all cars it equipped had either trolley poles, catchers, retrievers, or bases.

Trolley Supply remained a local supplier, with Ohio car builders serving as its principal market. Kuhlman and Cincinnati accounted for 36 percent of all published orders, with Brill's Philadelphia plant and the St. Louis Car Company running a distant second with 18. Ohio orders accounted for 41 percent of all cars (to Brill and Saint Louis's 12) and 35 percent of all components (to Brill and St. Louis's 14).

The Dayton Manufacturing Company remained a popular supplier of headlights and lighting fixtures. When General Electric developed a new incandescent railway lamp as part of its Mazda series, Dayton Manufacturing created a new line of lighting fixtures to accommodate it. The Mazda series represented a breakthrough in lighting for General Electric and for the street railway industry in particular. Mazda lamps provided 75 percent greater illumination than carbon filament lamps while consuming only 45 percent of the power.

Dayton's Mazda-compatible fixtures featured receptacles that could either protrude from or remain flush with the car's ceiling. Dayton also produced lighting fixtures that incorporated openings for ventilation systems. However, lighting fixtures were probably only a small portion of Dayton's street railway market, as only 18 percent of cars reportedly equipped with Dayton components used its lighting fixtures. Dayton's greatest period for lighting fixtures was 1920–24, when 44 percent of its cars were so equipped (208 cars).[20]

Like Ohio Brass, Dayton Manufacturing remained prominent in this industry between 1914 and 1929. Unlike other Ohio manufacturers discussed so far, Dayton actually experienced an overall increase in business during this period. Over half of its orders came in between 1925 and 1929, when nearly 800 cars were equipped with Dayton components.

Like Ohio Brass, streetcar components were only a part of Dayton Manufacturing's business. Very few account books have survived from this period (only 1919 and October 1924 through September 1925). Although too limited to form any solid conclusions, street railway manufacturers represented only 37 percent of Dayton's revenue from these periods. Of this revenue, only 6 percent came from Ohio firms.

Cleveland's Nichols-Lintern Company also continued to make the lighting fixtures, selector switches, and marker lights that it developed before World War One. However, lighting represented only a small percentage of its orders. Of the 6,034 cars represented in published orders for Nichols-Lintern, only 12 percent had the company's lighting equipment.

Other Safety Devices

Nichols-Lintern experienced modest success with its line of sanders during the 1910s and 1920s. Concerned about the industry's growing reliance upon air-powered appliances, Nichols-Lintern developed a foot-powered sander. The company feared that in the event of an air system failure, operators would not be able to use their cars' sanders to assist in stopping. The system was not failure-proof, as it required careful maintenance as well (to say nothing of the dexterity required of the operator). The manual sander did have some faithful adherents, as evidenced by an order for 600 mechanical sanders by New Orleans Railway & Light and nearly 200 by Georgia Railway & Power.

Despite the initial acclaim its mechanical sanders received, they only represented one-quarter of Nichols-Lintern's sander business. The remaining three-quarters were air-powered. Taken together, sanders were installed on 29 percent of all cars using

Figure 10.3. As a concession to increased automotive traffic on city streets, Cleveland's Nichols-Lintern Company created a device that told motorists when the streetcar was receiving power. This device was mounted on the car's ends. One light was lit when the streetcar's motors were drawing power. When the streetcar's operator shut the controller off, the power light went out and the other light lit up to warn motorists that the streetcar was either coasting or braking. *Branford Electric Railway Association, author photo.*

Nichols-Lintern components (1,756). The number of cars peaked between 1920 and 1924, although sanders accounted for an increased percentage of Nichols-Lintern orders through 1929. Sanders rose from 4 percent (212 cars) over 1914–19 to 26 percent (1,198 cars) over 1920–24 and then to 38 percent over 1925–29 (346 cars). Ohio Brass also manufactured sanders. Nearly 40 percent of the cars it equipped were given sanders.

In 1924, Nichols-Lintern developed what it called a "brake light." As with automobile brake lights, the Nichols-Lintern light was mounted on the rear of the car body. Better termed a "power off" light, the light was triggered not by a brake application but by the car's controller. A red light was illuminated whenever the controller was shut off. One can only imagine the confusion these devices caused as automobilists grew in number on city streets. The device was used in Cleveland and Toronto.[21]

Fenders and wheel guards also remained in demand throughout the 1910s and 1920s. Late in 1914 Trolley Supply came out with a fender that was more easily detached than other models. Cleveland Fare Box branched into fender production with a similar fender that could be folded against a car's dash panel when not in use.[22]

Cleveland's Eclipse Fender was active throughout the 1920s. In 1928 it was still offering its original safety fender, although it added a conventional wheel guard and a trolley pole catcher to its product line after 1924. Only 31 Eclipse orders were reported in the trade literature between 1914 and 1929, representing 1,208 cars, and 497 (41 percent) of these were equipped in the 1914–19 period. Eclipse experienced a decline in orders between 1920 and 1924 (283 cars, or 23 percent), although it recovered over 1925–29 with 428. Fenders comprised all of its equipment orders through 1924 and for 78 percent (946) of all 1925–29 cars as well. The remaining 22 percent (262 cars) were equipped with either trolley pole catchers (170) or wheel guards (92).[23]

FARE COLLECTION AND REGISTRATION

Corporate mergers among makers of fare boxes and registers ultimately made Ohio the world's largest center of fare box and register production. Dayton Fare Recorder began the process in 1917 by taking over the New Haven Trolley Supply Company (a large New England–based register manufacturer), the Recording Register & Fare Box Company (also in New Haven), and the Sterling Fare Register Company (Newark, New Jersey). Dayton's mergers were made easier by its sales manager, Frank B. Kennedy. Kennedy was engaged by Dayton in 1917, but had previously worked for the Recording Register & Fare Box Company as secretary and manager and for New Haven Car Register as assistant secretary. In 1923, Dayton itself was bought out by Ohmer Fare Register.[24]

In 1924, Ohmer broadened its market further by acquiring the American Taximeter Company, a New York City–based manufacturer of taxi meters. American Taximeter was founded in 1910 through the consolidation of the Jones Taximeter Company and the Franco-America Company. In addition to taxi meters, American Taximeter also made odometers that mounted on the hubs of automotive axles and dash-mounted schedule monitors.

Ohmer continued to consolidate its hold on the register market by forming a Canadian subsidiary (Canadian Ohmer) in order to buy out Canadian Taximeters, Ltd., in 1928. The *Electric Railway Journal* noted that "the combined company will be in a position to fill every recording device need and requirement of the transportation industry." Ohmer had experimented with taxi meters previously and had even developed a model in 1915. By 1929, Ohmer controlled 400 patents (100 by John F. Ohmer personally) and the product lines of five companies.[25]

Makers of fare boxes and registers followed industry trends and designed lightweight fare boxes. For example, the Cleveland Fare Box Company began making its fare boxes out of aluminum during the late 1910s. In addition to fare registers, the company also made portable change holders that conductors could wear on their belts.

In 1928 Cleveland Fare Box produced a fare register that could be worn. It was intended for fare collectors stationed along the street at heavily patronized stops during peak hours. By asking passengers to pay fares before boarding, collectors could save precious time.[26]

A total of 77 Cleveland Fare Box orders totaling 2,257 cars appeared in the trade literature. Despite selling its fare boxes in 15 states plus Washington, D.C., Canada, and Brazil, Cleveland Fare Box was primarily a local manufacturer. Ohio led all other states with 1,292 cars (70 percent). The city of Cleveland represented 30 percent of all Ohio orders

Figure 10.4. During the 1920s, Ohmer Fare Register simplified the appearance of its fare registers. *Branford Electric Railway Association, author photo.*

(375 cars) and 20 percent of orders overall. Indiana and Pennsylvania came next, although their orders comprised only 150 and 110 cars, respectively (13 percent combined).

In addition to published orders featuring new cars, the trade press also reported Cleveland Fare Box orders sent directly to railways for use on existing equipment. Of these, New York led with 375 fare boxes, Pennsylvania came next with 265 fare boxes, and Michigan followed with 235. Ohio was a distant fourth, totaling 95 fare boxes.

Before its acquisition by Ohmer, Dayton Fare Recorder produced many models of fare register, including a line of eight computing registers in 1915. They could handle 4 to 24 different fares. Unlike other register manufacturers, Dayton included the option of printing totals on cards rather than on rolls of paper tape. It was felt that register cards would simplify recordkeeping, as they were easier to file and could be changed after individual runs. With cards, audits of individual runs could be made faster than by searching a journal roll.

Dayton also came out with electrically powered registers in 1915. They used small motors to replace the conductor's strength and dexterity. Still other registers consisted of two machines, one which ran daily totals and the other which ran totals for individual trips.

Dayton Fare Register may have grown a little too aggressive in its product development, which possibly led to its undoing. In 1916, Ohmer sued Dayton for patent infringement (another register manufacturer, the Recording & Computing Machines Company, was sued as well). Ohmer won the case on appeal. An injunction was issued to prevent further infringement, and Dayton had to pay damages to Ohmer. The costs from these legal battles and its recent corporate acquisitions weakened Dayton's finances.[27]

Ohmer also introduced a number of new products to its line during the 1920s. The hub-mounted odometer was adapted to streetcars. Ohmer added ticketing machines in 1925. Two were offered, including a lightweight (71-pound) model. Electrically powered ticket-printing registers became available in 1928.

Like its fare registers, Ohmer ticketing machines printed a considerable amount of information. Included were the machine number, travel distance (origin and destination, with a capacity for as many as 1,000 stations), the fare amount (up to $9.99), the date, and the identity number of the ticket agent.

January 1929 was reported to be Ohmer's "best month in the entire history of the company." Sales in its California division numbered 778, while Pennsylvania sales nearly hit 1,400. Since the war, Ohmer had begun leasing its registers overseas and could boast of orders from Great Britain, Norway, Italy, Spain, Russia, Argentina, Australia, New Zealand, Columbia, and Venezuela.[28]

Ohmer installed registers on cars in at least 19 states plus Washington, D.C. Sixty-seven orders were published. Unlike Cleveland Fare Box, the orders suggest Ohmer did not rely upon the Ohio market for its success. For new equipment installations, the leading three states were Georgia with 31 percent (227 cars), New York with 21 percent (184 cars), and California with 12 percent (105 cars). Ohio represented only 5 percent, with 42 cars.

Ohio also ranked a remote fourth among Ohmer installations on existing equipment with only 13 registers (1 percent). New York shipments accounted for just over half with 566 (53 percent). Georgia ranked second with 418 (39 percent), and Louisiana ran a distant third with 44 (4 percent).

During the 1910s, another fare register company emerged in Dayton. The American Railways Equipment Company marketed a combination fare box and register in 1916 that could take both tickets and coins through a common opening. The device separated the coins and tickets, then registered their value. Coins went to a change drawer, and tickets were automatically canceled. Cleveland Fare Box came out with a similar fare box/register in 1922.[29]

The Bonham Fare Recording Company (Hamilton) developed rather sophisticated fare registers in 1915. This register recorded the boarding and exit times for individual passengers. The total figures could be broken down into as many as eight subtotals. The firm reorganized as the Bonham Recorder Company in 1917. Its president was E. E. Dwight, who was also president of Tool Steel Gear & Pinion. In 1919, Bonham developed a register that identified conductors by providing a facsimile of their signatures.[30]

Heating and Ventilation

Although forced-air and water heater systems persisted into the 1920s, electric radiant heat was the most common type. Power consumption was a particularly serious problem for railways that used electricity to heat their cars. For the few systems that used cars with dynamic braking, forced-air could be drawn into the car's interior to cool the motors. Another method of saving power was to include the heaters in the motor circuits, making the heating elements part of the electrical resistance used to accelerate the cars.

The single greatest advance made during the 1920s was the adoption of thermostatic control. Thermostats enabled railways to save on power consumption and to provide a consistent temperature inside their cars. Before thermostats, the car's conductor (or operator) had to continually turn the electric heaters on and off to regulate on the car's temperature.[31] Heater switches were not always conveniently located or easy to operate.

Passive, roof-mounted ventilators continued to be the principal method of car ventilation throughout the 1920s. Nichols-Lintern continued to make roof-mounted exhaust ventilators. Of the 6,034 cars that were reported built with Nichols-Lintern components between 1914 and 1929, 3,849 (64 percent) used the company's ventilators. Nearly one-quarter of the 507 cars Nichols-Lintern equipped in 1914–19 had ventilators (121 cars). This figure jumped to 3,007 cars for 1920–24 (65 percent). Although the overall number of cars equipped by Nichols-Lintern dropped dramatically over 1925–29 to 906 cars, 80 percent (721 cars) had ventilators.

TRIMMINGS AND MISCELLANEOUS ITEMS

Doors and Seating

Mechanically operated doors became nearly universal during the 1920s. The most common was the air-operated folding door, which used compressed air from the streetcar's air system. Sliding doors were also used, mostly on rapid transit vehicles.[32]

Seating improved markedly during the 1920s. Automotive "bucket" seats began appearing early in the decade. They did much to improve passenger comfort and were ideal for single-ended cars or for cars where the seats did not need to be reversed. Reversible bucket seats became available by the end of the decade. The American Electric Railway Association's Equipment Committee recommended a minimum spacing of 31 inches (a 1.5 -inch increase over 1926), with aisles made as wide as possible.[33]

One of the requirements for seats was that they be easy to clean. Although rattan remained the most common type of upholstery, seats with leather upholstery appeared during the 1910s and 1920s. Leather enjoyed several advantages. Like rattan, it had tremendous wearing qualities. In addition, leather was much easier to clean. While street railway companies regularly applied thin layers of varnish to rattan seats to protect them, they grew sticky in warm, humid weather. Railways in southern regions often resorted to hardwood seats because of this. Leather upholstery offered an ideal alternative.[34]

Imitation leathers also appeared during the 1920s. One such product was "Fabrikoid," a cotton fabric that received numerous chemical treatments and embossings to give it the toughness and consistency of leather.[35]

As leather gained acceptance by bus and streetcar operators alike, several manufacturers of leather upholstery emerged in Ohio. The most significant was the Cleveland Tanning Company, located in Cleveland's near west side. Nearly the entire process required to transform cow hides into leather upholstery was performed at the Cleveland plant. The hides themselves arrived salted from slaughterhouses located in the southern United States.

"Hyaline" and "alpha" leathers were Cleveland Tanning specialties. Alpha leathers were waterproofed using linseed oil. Hyaline leathers were finished with colored pigments and treated so they would not rub off. Hyaline leathers were preferred by transit operators and the railroads, the latter of which used them on such famous passenger trains as the *Twentieth Century Limited,* the *Broadway Limited,* and the *Capitol Limited.*[36]

Other upholstery manufacturers included Toledo's Textileather and the Leon L. Wolf Waterproof Fabric Company of Cincinnati. Wolf was established in 1926 and made a waterproof fabric called "kemi-suede." Kemi-suede was available in blue, green, black, nut brown, dark and light fawn, buckskin, and three shades of gray (blue, mixed, and neutral). Although Wolf's company was based in Cincinnati, the fabric was actually manufactured at Kemitex Products in Barberton, Ohio. Wolf's sales promotion manager, Edwin Besuden, had at one time been the sales manager for the Jewett Car Company.[37]

Painting

As railways grew more concerned over lowering operating costs, an obvious place to save on maintenance was car painting. Although steel car bodies simplified the task somewhat, many railways and car builders carried wooden car painting methods into the steel car era. Earlier painting methods, some of which consumed 18 days, were simplified into six basic coats: primer, body color, surfacer and putty, additional body color, and two coats of sealing varnish. Due to the high price of car paints, many railways deferred this type of body maintenance until it became absolutely necessary. Articles in the trade press were rather forceful in advocating simplified painting procedures. One declared, "To produce on a commercial vehicle traveling through the dust and grime of our cities a highly polished surface of a character which we would not think of attempting on the most expensive private house . . . seems ridiculous."[38]

Enamels were welcomed by the street railway industry. A relatively new product in 1910, enamels were in common use by 1916. They were less expensive and easier to apply than conventional paints and oils. Damaged surfaces were easier to repair, reducing maintenance costs. Enamels also came in a wider selection of colors and shades.[39]

Lacquer paints grew in favor during the mid-1920s. They were made from nitrocellulose, a "clean cotton fibre chemically treated so that it readily dissolves in certain liquids to form a viscous solution." Lacquers had the advantage of drying faster than traditional paints. They usually dried dust-free within a few minutes and could be handled easily within 30 minutes.

It was soon discovered that even wooden streetcars could be painted with lacquer. Lacquers did, however, require extensive preparation if they were to be used on wood, not only because of the different surface texture but also because of the numerous coats of paints and oils that had been absorbed by the wood over the years.[40]

The trade literature advocated paint schemes limited to a maximum of three colors. When using more than one color, the sides should contrast above and below the window bottoms. Bands of color should be narrow, as in pinstriping. The exterior should be illuminated, particularly the front end.[41] In Ohio, paints were manufactured by Cleveland's Sherwin-Williams (although its railway paints were made out of state), and varnishes (still used on interior woodwork) were manufactured by Glidden, also located in Cleveland.

Trimmings and Furnishings

A variety of trimmings and furnishings were produced by Ohio firms from 1914 to 1929. Dayton Manufacturing introduced a new type of curtain fixture in 1915 that prevented curtains and shades from swinging away from their windows. Dayton used spring-loaded "shoes" at each side of the bottom edge of the curtain. The shoes could be slid up and down the sides of the window frame, allowing passengers to adjust them at will.

Dayton's postwar product line was extensive, as advertisements from the 1920s suggest. Products covered everything from doors to lighting fixtures and hand brakes to vestibule trapdoors (the latter for high and low station platforms). Dayton also made a variety of water coolers and lavatories.[42]

Dayton Manufacturing's most successful street railway product was window sash hardware. As Dayton's business increased from 1914 to 1929, its sash business increased exponentially. Only 16 cars were so equipped over 1914–19 (7 percent). This figure increased to 68 cars (15 percent) over 1920–24. In 1925–29, 536 cars (68 percent) were equipped with Dayton sash hardware.

CONCLUSION

The close of the 1920s witnessed a street railway industry that was smaller in size though similar to 1914 in technology. It also witnessed an industry that was based on a technology that was remarkably similar to the years before 1914. With few exceptions, streetcars of the late 1920s continued to bear a striking resemblance to their 1914 forebears in terms of appearance, performance, and comfort. Direct motor control was still the rule. Even though automatic, variable-rate acceleration was available, its complexity placed it out of the reach of most railways. Despite new forms of braking technology, straight air systems prevailed.

Although there were advances that improved performance, reliability, and comfort such as improved gears and pinions, tapered roller bearings, and leather seats, other advances such as the Cincinnati Car Company's curve-sided design actually served to perpetuate earlier car designs. Both Kuhlman and Cincinnati participated in the industry's limited efforts to produce new car designs. Although there were fewer streetcar builders and component manufacturers active in Ohio in 1929 than there were in 1914, their continued success and influence allows the historian to view Ohio as a microcosm of the industry in general.

During the 1910s and 1920s, Ohio's car builders and manufacturers were subject to the same limitations and conservative attitudes as the rest of the street railway industry. There were no standardized designs. Early refinements to streetcar design were made when the streetcar had no viable competition. When automotive technology produced viable competition in the form of buses and private automobiles, the practice of refining existing designs persisted. Despite the number of experimental streetcars built toward the end of the 1920s, none gained industrywide acceptance.

Through all of this, Ohio remained an active producer of streetcars and streetcar components during the 1910s and 1920s. Despite the decline of the market, Kuhlman and Cincinnati continued producing streetcars. Ohio Brass and Nichols-Lintern continued furnishing a wide variety of components. Dayton became the leading city in fare registers. The Dayton Manufacturing Company continued producing trimmings and numerous examples of streetcar hardware.

Companies that either had not participated in the industry or only participated to a limited extent gained new importance during the 1920s. Tool Steel's gears and pinions gained tremendous acceptance within the industry, and Timken Roller Bearing introduced a simple yet effective means of minimizing friction in streetcar motor and axle bearings.

The following decade would prove to be a significant one for Ohio and the transit industry. Although Ohio manufacturing still had a viable presence in the industry entering the 1930s, the Great Depression and the industry's belated attempt to produce a standard state-of-the-art streetcar would push Ohio's manufacturers into the periphery.

Figure 11.1. An example of a modern interurban ordered from the Cincinnati Car Company by Dr. Thomas Conway for the Cincinnati & Lake Erie interurban railway. Due to their color and great speed, these cars were known as "Red Devils." *Courtesy of the Ohio Railway Museum.*

11

STREETCAR MANUFACTURE DURING THE 1930s

The 1930s were anything but kind to the street railway industry. Railway systems were closed or shortened, car fleets shrank in size, and most railways that remained in service made do with cars they already had. In a 1935 address to the American Transit Association, Malcolm Muir, president of McGraw-Hill (publisher of *Electric Railway Journal*), argued that one of the transit industry's greatest problems was the obsolescence of its equipment.[1] He estimated that within five years, of the 62,000 streetcars and buses projected to be in existence, as many as 47,000 would be obsolete. By "obsolete," Muir meant not only vehicles that were old and worn out but also those that were still capable of running but had been superseded by more efficient vehicle designs or newer styles. He implored the industry to recognize the importance of upgrading its equipment.

This does not mean that the industry accepted its decline passively. For the first time, the industry pooled its resources and embarked on a comprehensive program to develop a truly standardized, state-of-the-art railway vehicle. While successful, this effort would not be sufficient to save most of the industry's manufacturers. Some would remain, others would turn to different pursuits, and a number of companies would pass out of existence. What follows is a summary of the industry's major developments and how they impacted Ohio streetcar manufacture.

STREETCAR DEVELOPMENTS

By 1930, street railway mileage had declined to 38,470 (a 19 percent drop from 1920), and the number of streetcars had declined to 71,219 (a 29 percent drop). By 1937, the figures had plummeted further to 26,187 miles (a 45 percent decline from 1920) and 49,147 cars (a nearly 50 percent decline).

Car building also diminished. Between 1907 and June 1914, orders for new streetcars and interurbans averaged 4,079 cars per year. The impact of World War One cut this figure to 2,479 cars per year between July 1914 and December 1919. The industry recovered slightly during the first half of the 1920s to 2,653 new cars per year. During the late 1920s, an average of only 1,123 cars were built each year. As paltry as the latter figure may seem, it looked generous in comparison with streetcar orders between 1930 and 1937, which averaged only 210 cars per year (see tables 11.1 and 11.2).

Most of the nation's streetcars were concentrated in large cities, where ridership was still relatively strong. Of the nearly 38,000 streetcars and interurbans in service, 77 percent were in cities with populations over 100,000. All rapid transit cars (11,283 cars along 1,326 miles) ran in cities with populations over 500,000. Nearly half of all cars were in cities with populations over 500,000 (49 percent), and 28 percent were in cities with populations between 100,000 and 500,000 (see table 11.3).

Route mileage trends were not quite as severe. Fifty-two percent of all route mileage was in cities with populations over 100,000 (27 percent in cities over 500,000, and 25 percent in cities between 100,000 and 500,000). Thirty-one percent of all mileage was located in areas deemed by the *Transit Journal* as "interurban areas" (see table 11.4).[2]

Perhaps the most significant streetcar order of the early 1930s was for 150 Peter Witts for the United Railways and Electric Company of Baltimore in 1930. Brill built 100, and the Cincinnati Car Company built 50. In appearance, they closely resembled Master Units.

The cars used a unique control system that gave them the high acceleration rate of 3.2 miles per hour per second (conventional rates were 2 or 2.5 miles per hour per second). Self-lapping brakes were used. Most of the cars had conventional gearing, but 30 cars were given W-N drive.

In 1930, Westinghouse engineer J. G. Inglis conducted a survey of 11 recent interuban car orders and reported his findings in *Electric Traction*. All of the cars were lightweight, averaging just under 21.5 tons. The combined horsepower of their motors averaged just over 203. Wheel diameters, which had averaged 30 to 36 inches before 1914, averaged just over 26 in Inglis's survey, bringing the bodies closer to the ground. The cars were not especially fast, averaging nearly 50 miles per hour (only about 10–15 miles per hour faster than a streetcar), and six orders did not have automated control systems.[3]

Although it had been around since the late 1800s, it was not until World War One that safety glass received widespread use (it was used in eye pieces for goggles and gas masks). As late as the early 1930s, manufacturers of all types of vehicles preferred plate glass for "its superior beauty and clearer vision."

Safety glass consisted of two layers of plate glass sandwiching a thin, "transparent cellulose derivative." As late as 1928, only two automobile manufacturers produced models with safety glass. This changed during the mid-1930s, as states began to demand that safety glass be installed. In 1934, New York became the first state to pass comprehensive regulations concerning the use of safety glass. By the mid-1930s, 34 states had safety glass laws, 28 of which were "general."

Michigan serves as an example of a state with safety glass laws. All passenger vehicles operating in Michigan that were made after June 1932 were required to have safety glass windshields. All vehicles were required to have them after June 1934. In California, school buses were required to have safety glass for all of their windows.[4]

During the late 1920s, Dr. Thomas Conway Jr., who served as general manager for a number of interurban properties, had a reputation for rehabilitating endangered railways (in part) using lightweight equipment. In 1929, he ordered a series of lightweight, high-speed interurbans from the Cincinnati Car Company for the Cincinnati

& Lake Erie. The cars came to be known as "Red Devils," owing to their paint scheme and speeds that exceeded 70 miles per hour. These cars influenced future interurban designs.

In 1931, J. G. Brill produced an interurban car that embodied lightweight, high-speed, and many experimental components from the 1920s as well as successful features found on the latest bus models. The cars were streamlined, with the operator's compartment bearing a closer resemblance to the cockpit of a commercial airliner rather than a motorman's cab. Through the use of aluminum, Brill was able to keep the cars' weight to 21 tons. A combination of air and magnetic braking was used. The passenger compartment featured such comforts as forced-air ventilation, indirect lighting, and air-cushioned seats. The cars were exceptionally fast, designed for 90 miles per hour.

Due to their streamlined appearance and great speed, the cars were nicknamed "Bullets." A number of interurbans acquired them, including the Fonda, Johnstown & Gloversville and the Philadelphia and Western in Pennsylvania. A small number remained in commuter rail service in the Philadelphia area into 1995.[5]

ELECTRIC RAILWAY PRESIDENTS CONFERENCE COMMITTEE

Concerned with the large number of obsolete streetcars operating on city streets (as well as declining ridership and route mileage), several street railways and manufacturers, with the consent of AERA's Advisory Council and Executive Board, created the Electric Railway Presidents Conference Committee (ERPCC) in December 1929 "to expedite the development of the urban railway car."[6] Thomas Conway was elected chairman. ERPCC membership was said to represent approximately 60 percent of the industry's "potential car buying power" (see tables 11.5 and 11.6).

Despite the interest AERA had in the ERPCC, the ERPCC's activities remained "outside of the Association." Members contributed $252,000 to the project.

In May 1930, the ERPCC hired Dr. Clarence Hirshfeld to be chief engineer of the project. Hirshfeld had an impressive research background. He had served on the engineering faculty at Cornell University and chaired numerous committees on prime movers and power generation for the American Society of Mechanical Engineers, the Association of Edison Illuminating Companies, and the National Electric Light Association. He authored five books and had worked in Detroit Edison's research department since 1913, where he rose to department head. Hirshfeld also served in the Ordnance Department of the U.S. Army during World War One, attaining the rank of lieutenant colonel.[7]

Although Hirshfeld lacked railway experience, this was viewed as an asset, ensuring he would approach the problem without the inbred bias and conservatism of most street railway men. Hirshfeld was enthusiastic about the project: "It is a glorious adventure to those engaged in it, almost like an old-fashioned voyage of discovery."[8]

The ERPCC research team was given a set of objectives. Generally, the team was to develop "a modern, light weight car of attractive appearance and greatly improved performance" that could be purchased, operated, and maintained economically. Although standard dimensions and components were to be recommended, the overall design was to remain flexible enough to be adapted to any operating environment. Quiet, smooth, and powerful acceleration and deceleration were to be obtained. Shocks and jarring transmitted from the rails were to be minimized.[9]

Hirshfeld was given sound advice on approaching the issue of standardization. W. T. Rossell, general manager of the Pittsburgh Railways Company, felt that much of the industry's reluctance to abandon custom manufacture stemmed from a misunderstanding of what "standardization" meant:

> As in other industries, the standardization of the street car does not mean that one unchanging model must be used at all places and at all times. It does mean that the present method of each property ordering custom-built cars for its own use should give way to the adoption of a reasonable number of standards, to be brought up to date and redesigned periodically.[10]

Hirshfeld began by examining sample cars the ERPCC felt represented the industry's state of the art and identifying the problems that needed to be corrected. Next, he set up numerous experiments to probe these problems and establish a set of standards to guide efforts to produce a new streetcar. Most of Hirshfeld's work was based in New York City, using a Brooklyn car shop belonging to the Brooklyn & Queens Transit Corporation.

Hirshfeld's research team consisted of 30 men, including himself. In addition to shop facilities, the team was given a 1,550-foot test track along a protected right of way. Four streetcars were selected for testing: one of Baltimore's new Peter Witts (the dimensions of this car became the basis for ERPCC specifications), an experimental double-trucked streetcar built by Ohio bus manufacturer Twin Coach (it had been in service in Brooklyn), an additional Brooklyn car, and a car from Chicago.

THE PCC CAR

The production model based on the ERPCC's final specifications was called the "PCC" car. The PCC fulfilled its designers' goals for flexibility. Although intended to be single-ended, double-ended models were possible. Trucks could be adapted to any gauge, from narrow to broad. Basic seating could vary from 54 to 59, although other capacities were possible. The PCC weighed 25 percent less than a similarly equipped conventional streetcar.[11]

The PCC car body was streamlined. Hirshfeld encouraged his team to emulate "good automotive practice" when designing the production body. All "major lines and . . . obvious fittings and decorations" were given an automotive appearance. The letterboard above the windows was fluted to give the impression of speed. The height of the car was also kept to a minimum to further enhance this impression.[12]

The body was made of copper-bearing steel, with Cor-Ten steel framing. Despite the emphasis on modern design and materials, the roofs on the first PCC orders featured plywood center sections. All windows were made of safety glass, and window and body posts contained duct work for ventilation systems.

The PCC used right-angle drive, employing hypoid gearing. Hypoid gearing is commonly used on rear-wheel drive automotive vehicles. A ring gear, in which the teeth are located along one side rather than the outside or inside edges, is driven by a pinion resembling a worm gear. The pinion engages the ring gear at its middle. Hypoid gearing is quiet and durable.

Braking on the PCC was achieved through a combination of independent dynamic, air, and magnetic systems. Normal deceleration was between 4.25 and 4.75 miler per hour per second, although an emergency application increased this rate to between 8 and 9, or twice the rate of a conventional streetcar braking system.

The PCC was not fast, only reaching 42 miles per hour. However, this was offset by a high initial rate of acceleration. At 4.75 miles per hour per second, the PCC could accelerate to 15 miles per hour twice as fast as a conventional streetcar.

Acceleration on the first production orders was controlled indirectly by a 12-volt variable speed motor that cut out 66 resistance settings (a vast improvement over K-controllers, 8 or 9 settings that were cut out manually). The traction motors themselves were remarkably powerful for their compact size and weight. Both General Electric and Westinghouse produced 55 horsepower motors that weighed only 700 pounds.

The 66-setting accelerator designed by Westinghouse came out of experiments with the Model A. General Electric's accelerator was more complex, having 127 resistance settings both for series and parallel operation (254 settings total). Called "floating control," this accelerator was initially air-powered but later used an electric motor like Westinghouse.

PCC trucks consisted of two large steel tubes connected by cross members that also served as cradles for the motors. Rubber was used whenever possible to reduce steel-on-steel contact. All springs for mounting the motors and supporting the car body were made of rubber. Resilient wheels were used. This, combined with hypoid gearing, nearly eliminated running noise.

Chicago figured the average life of its first resilient wheels to be slightly over 65,000 miles. Baltimore figured the average life of its resilient wheels at 78,000 miles. Due to a resilient wheel shortage, Baltimore substituted solid steel wheels, which lasted 10,000 miles *less* than the resilient ones and had higher noise levels, especially on curves. Some of Washington's resilient wheels lasted over 100,000 miles.

PCCs used forced air for heating and ventilation. Air was drawn through openings at the trolley pole base and distributed throughout the car's interior via duct work. PCC ventilators could change up to 1,200 cubic feet of air per minute. A separate ventilation system cooled the motors, as the cars' dynamic braking developed considerable heat. During cold weather, the heated air drawn off of the motors was recirculated inside the cars. In general, PCC power consumption was 5–25 percent greater than on conventional equipment, although this was offset by lower power consumption during the winter, the PCC's higher scheduled speeds, and quieter, smoother operation.

Interior lighting for the PCC was based on compromise. Although experiments suggested that indirect lighting reflected off the headlining was the best method, it was rejected as too expensive to manufacture and maintain. Research determined typical streetcar illumination to be 2 to 5 foot candles at the "reading plane" as compared with 10 foot candles at a typical office desk. A system delivering soft light at about 15 foot candles to the passengers' reading plane was used. PCC wiring was designed to raise this level to 20 foot candles if needed, but Dr. Hirshfeld felt limits had to be placed on interior lighting, lest an alighting passenger step "into a poorly illuminated street from a brightly illuminated car . . . [and thus be rendered] temporarily almost incapable of seeing sufficiently clearly to guard against accident." Five types of lighting fixtures were used on PCCs during the 1930s. Interior lights on Chicago's first PCCs delivered 9 foot candles, but this level was upgraded 20 percent soon after the cars went into service.

Headlights, the controller motor, track brake, and other PCC auxiliaries were powered from a 32-volt battery. The battery was charged by a small motor-generator set. This set could be deactivated from the operator's station if necessary. Operator's controls were designed both for ease of operation and comfort. All acceleration and braking was controlled by foot pedals, arranged like the pedals on an automobile or bus.[13]

A third pedal was placed where the clutch pedal would normally be. This was the PCC version of a "deadman" interlocking. Should the operator's left foot slip off the pedal while the brakes were released, power would be cut from the motors and an emergency braking application would be made. Hirshfeld observed that the similarities between PCC and automobile pedals would make the PCC an easy streetcar for operators to master: "The operator need not know how the pedals operate to control the equipment. He need only know that further depression of either one of them produces more of the effect."[14]

All switches for lights, heaters, ventilation fans, gongs, etc., were arranged in a neat row on the dash panel in front of the operator and were labeled clearly. The operator's seat was large and comfortable, including a cushioned back, which helped to lessen fatigue.

Perhaps one of the more surprising aspects of the PCC car was that it embodied some features that had been developed but not embraced by the industry during the 1920s. Resilient wheels, dynamic braking (and its use to heat car interiors), and lighter motors were all available during the decade but were confined to experimental demonstrations. Hirshfeld must have hoped the work of his design team would encourage the street railway industry to be less reluctant about adopting promising innovations in the future.

In its first year of production, orders for nearly 400 cars were placed with the St. Louis Car Company. Initial orders came from Brooklyn (the first), Baltimore, and Chicago. The flexibility of the PCC's design was much in evidence, as each city gave different seating capacities and body dimensions. The dimensions on the Brooklyn and Baltimore cars were identical, but the former sat 59 passengers while the latter sat 54. The Chicago cars sat 58 and were larger.

Initial operations involving PCC cars were promising. By 1938 the streetcar market had improved, as street railways in Baltimore, Boston, Brooklyn, Chicago, Los Angeles, Philadelphia, Pittsburgh, San Diego, Toronto, and Washington, D.C., either had PCCs in service, had them on order, or were testing demonstrator models. Public reception of the new streetcar was also positive. Streetcar lines in Brooklyn experienced 12–33 percent increases in their revenue. Chicago and Pittsburgh PCCs reduced running time by 10 percent on some lines. Los Angeles experienced cost reductions of as much as 50 percent due to improved scheduling.[15]

Upon completion of its design work, the ERPCC's research team was reorganized into the Transit Research Corporation (TRC), which was intended to refine the original ERPCC specifications as experience was gained in PCC car operation. Hirshfeld was proud of the ERPCC's work and hoped that future car orders would be based upon its specifications. However, he cautioned industry officials not to become too rigid in their adoption:

> It is not intended that the industry shall be shackled by hard and fast standards, that improvement shall be checked and the streetcar crystallized in its present state of development. . . . It is intended that the specifications shall be modified as increasing experience, maturing judgment, and possibly renewed research may indicate to be desirable and profitable.[16]

ALTERNATIVES TO THE PCC

Despite the promise of the PCC car, not every street railway operator bought PCCs. For numerous reasons (particularly financial), many operators either opted to build their own cars or purchase cars developed independently of the ERPCC effort.

The Brilliner

Although a charter member of the ERPCC, the J. G. Brill Company withdrew from the committee about 1934. According to Debra Brill, the Philadelphia car builder withdrew for a number of reasons. The company had recently closed its subsidiaries, had experienced several changes in executive leadership within three years (one president died in office and another was recruited by Franklin Roosevelt to assist with the New Deal), and they were experiencing a shift in market emphasis through an agreement with ACF Motors, a major bus producer. Brill made its final break with the ERPCC over a dispute concerning Brill patents on resilient wheels.

Although several cities considered ordering PCC cars from Brill, no orders were forthcoming. Brill decided to develop its own version of a redefined streetcar, which it did several years after the PCC appeared. It was called the "Brilliner."

The Brilliner was very similar in appearance to the PCC, although its ends were slightly more "squarish" or "boxy." The two cars were roughly the same size. Brilliners averaged just over 45 feet long, with a maximum width of 8.5 feet. They sat 52 passengers.

Rather than build its body from all-welded alloy panels, Brill fashioned the Brilliner body from corrosion-resistant rolled steel plate. The plates were attached to a steel frame and to each other by a combination of welding and riveting. Like the PCC, other body components were fashioned from copper-bearing steel.

Brilliner trucks were made from steel beams and angle irons (as opposed to tubes on PCC trucks), and used conventional motor mounting with helical gears instead of right-angle drive. Cincinnati's Tool Steel supplied the gears and pinions. As with the PCC, rubber was used extensively for motor suspension, body support, and shock absorption. Acceleration and braking was comparable to the PCC, with a similar combination of air, dynamic, and magnetic braking systems. However, PCCs used automotive-styled drum brakes, while Brilliners used conventional tread brakes. Brilliners used resilient wheels.

Heating and ventilation were also similar to that found on PCCs, although Brilliners drew their air through four intakes located on the roof, while PCCs drew their air from a single large intake at the trolley pole base. Ohio Brass's PCC-styled pole bases were compatible with the Brilliner and were used by Brill.

The first Brilliner was a 1938 demonstrator built for the Atlantic City & Shore Railroad. Although it received favorable reviews in the trade press, the Brilliner was not a success. Between 1938 and 1941, only 40 were built. Brill did not receive any new streetcar orders after 1941.[17]

Remanufacturing

One of the problems posed by PCC cars were that they contrasted dramatically with conventional streetcars in terms of performance and maintenance requirements. Short of replacing all of their equipment with PCCs, many railways chose to find ways to make their older equipment conform to ERPCC standards as much as possible. For example, as mentioned in chapter 10, acceleration could be improved by rewinding armatures or by field-shunting. Rewinding was the most efficient, as it increased the motor's horsepower. Field-shunting did not increase horsepower, but it did increase the maximum or "balancing" speed of the motors.

Noise could be reduced by adopting helical gears, replacing worn brake rigging, rewelding or reboring truck frames, or using smaller axle bearings. Although it would be too expensive to replace entire control systems, new, multistep controllers could be retrofitted to most K-controlled systems by the mid-1930s.

Repainting could brighten streetcar exteriors and interiors dramatically. Controls could be cabinetized, and new lighting fixtures could improve the diffusion of interior light. Old types of flooring (especially flooring with wood slats) could be replaced with rubber matting.

The greatest example of car remanufacturing was New York City's Third Avenue system, which ultimately remanufactured over 600 cars during the 1930s. In some cases, new bodies were built out of aluminum or steel, but some were fabricated by splicing together single-truck cars. As a rule, anything that could be salvaged from older equipment or acquired secondhand from other railways was put into the cars. This included motors, brakes, trucks, seats, controllers, and current collection apparatus. All parts were thoroughly reconditioned before installation, and new gears and pinions (herringbone) were used.[18]

OHIO CAR BUILDERS

As the Great Depression reduced the number of car orders, it became increasingly difficult for car builders to secure new business. By the time the ERPCC completed specifications for the first PCCs, many car builders had gone out of business. Such was the case with Ohio's remaining car building firms, which only managed to secure orders during the early 1930s. Orders later went to larger firms such as the Saint Louis Car Company, Pullman, and Brill.

Looking at orders reported in the trade press, the Cincinnati Car Company built only 126 rail vehicles during the 1930s. Nearly half (60) were streetcars, just over a quarter (33) were interurbans, and the remainder were other types of rail vehicles, such as freight cars or locomotives. The geographic area of Cincinnati's market shrank as well, with nearly half of all cars (55) going to railways in Ohio and the neighboring states of Pennsylvania and Indiana. Although the remainder slightly outnumbered the "local" cars (61), they went to only two states: 50 to Maryland (Baltimore was Cincinnati's largest customer during the 1930s) and 11 to Louisiana.[19]

The most notable orders were the Baltimore Peter Witts and Thomas Conway's 1930 order for 20 high-speed passenger cars for the Cincinnati & Lake Erie interurban railway. Cincinnati continued to manufacture other types of equipment in addition to streetcars. In 1930 the Cincinnati Car Company developed a new streetcar truck called the "Type 101." The truck used standard journal bearings, yet mounted them so that the axles had considerable vertical play to improve riding quality. In 1931 Cincinnati began manufacturing steel containers for intermodal freight transport (the Cincinnati & Lake Erie Railroad was a customer).[20]

Despite building high-profile orders, new products, and its charter membership in the ERPCC, the Cincinnati Car Company did not escape the Great Depression, going into receivership in September 1931. In July 1933 the following statement was issued:

> Copies of [financial reports] are being furnished to all large creditors with the request that after the statements have been considered, the creditors recommend whether the receiver should continue operation of the plant or close it down and liquidate the company's affairs.[21]

The company continued to exist on paper into the late 1930s, but did not receive any new orders. By the time the final ERPCC specifications were announced, it was clear Cincinnati would not be building any PCC cars. The car builder ceased to exist in 1938.

Orders reported in the trade press suggest Kuhlman had an even greater struggle during the 1930s. Only 45 cars were built by Kuhlman. One-third were streetcars, and two-thirds were interurbans. All of the streetcars stayed in Cleveland, and all of the interurbans went to Santiago, Chile.

For Kuhlman, the first hint of trouble came in 1931 with a reorganization of Brill. Concerned with the declining number of streetcar orders, Brill divided its sales territories into eastern and western regions. Each division was then subdivided into districts (the east into five, the west, three). Brill claimed "With this arrangement, the Brill organization will possess close contact with every electric Railway in the United States and will be equipped to make shipments economically to all points." In 1930, Brill attempted to increase its position in the market by affiliating with Cummings Car & Coach.

Additionally, each of Brill's three major subsidiaries (American, Kuhlman, and Wason) were renamed. The American Car Company became the J. G. Brill Company of Missouri. The G. C. Kuhlman Car Company became the J. G. Brill Company of Ohio, and the Wason Car & Manufacturing Company became the J. G. Brill Company of Massachusetts (see table 11.7).

Brill also attempted to increase business for Brill of Ohio by having it build diners. This aspect of the car builder's business is somewhat confusing for historians because another company with a similar name, Michigan-based Kullman Industries, is still making diners.[22]

A further sign of trouble came halfway through 1931, when Brill closed Brill of Missouri. Brill announced that it was "following the trend toward the elimination of uneconomical operation by concentrating its manufacturing activities in the West and Middle West at the plant of the J. G. Brill Company of Ohio in Cleveland." Despite the promise of an expanded sales territory, Brill of Ohio was still vulnerable. Brill gave an ominous hint as to its fate in the same statement: "It was also felt that the facilities at the Cleveland plant were more adequately suited to *and possessing even a larger capacity than required* to meet the needs of electric railways in the middle and western territories" (emphasis mine).[23]

Brill continued to restructure its operations during the 1930s. Although the former Kuhlman and Wason plants remained open, Brill consolidated all sales operations in Philadelphia in 1932. All manufacturing at the Cleveland facility was stopped after 1932.[24]

COMPONENTS

Despite the cessation of car building activity in Ohio, component manufacturers continued to supply a variety of products to active car builders and street railways. Cincinnati's Tool Steel Gear and Pinion continued to make gears and pinions for streetcars and interurbans, but as the market shrank, the firm gradually shifted its emphasis to industrial machinery and heavy equipment.

Tool Steel did not ignore the street railway market completely, however. For example, it furnished all of the herringbone gears and pinions used by the Third Avenue Railway in its massive remanufacturing program. Tool Steel also supplied gears and pinions for Brilliner trucks and Brill's 1933 ERPCC demonstrator in Chicago and steel wheel rims for resilient wheels on PCC cars.[25]

Ohio Brass also remained active in the street railway market during the 1930s, furnishing components for most PCC and conventional streetcars built during the decade. Forty-three Ohio Brass orders representing 970 pre-PCC streetcars were reported in the trade literature: 83 percent (808) were streetcars and interurbans, and 13 percent (130) were trolley buses. The remainder were for buses or other vehicles.

Most of the components Ohio Brass furnished during the 1930s were for current collection. Pole bases were supplied for 85 percent of all vehicles. Three-quarters of all vehicles received trolley pole catchers or retrievers. One-quarter were given trolley poles, wheels, or shoes.

Of Ohio Brass's non-current-collecting components, the most common were headlights, which could be found on 42 percent of all vehicles. Twenty-nine percent used Ohio Brass sanders. Six percent sported Tomlinson couplers.

Ohio Brass's orders illustrate how the Great Depression affected the 1930s transit market. Of the 970 vehicles equipped with Ohio Brass components, 715 were built by only four car builders: Brill (mostly at the Philadelphia plant), the Saint Louis Car Company, Cincinnati, and Osgood-Bradley (Worcester, Mass.). The remaining 255 were divided among nine customers, 28 were remanufactured in-house by Los Angeles Railways.

Ohio Brass was able to continue its lead in the industry for several reasons. One was the quality of its products, but a more important factor was that it did not rely upon the street railway market alone. Ohio Brass was also active in the utilities market, furnishing hardware for power transmission lines. The firm also acquired the product lines of other manufacturers, such as General Electric's entire trolley wire hardware catalogue in 1931.

Ohio Brass not only made quality products, but continually improved them. In 1930, for example, it began making headlights with nickel-chromium wiring, which resisted corrosion and brittleness. Resistance coils were left exposed, which allowed them to operate at higher temperatures. All mounts were threaded, which lessened the possibility of the light working itself out of its mount. In 1931 a recessed, automotive-type version of Ohio Brass headlights became available.

Ohio Brass also participated in the ERPCC program. The challenges presented by the PCC car were twofold for component manufacturers. First, ERPCC specifications called for the elimination of all excess weight, which required a complete reworking of fundamental design concepts. Another challenge was to meet the aesthetic introduced by PCC body styling, which emphasized a clean, streamlined appearance.

For Ohio Brass, trolley pole bases were redesigned in order to meet both challenges. The new base was lightweight, weighing only 50 pounds. Furthermore, it was small enough to fit inside the streamlined cowling on the car's roof.

Headlights introduced a different challenge. After redesigning its headlights along automotive lines in 1931, Ohio Brass further refined its casings so that their casings became shallower, ensuring the headlight would be flush with the streetcar's dash. They were also redesigned to operate off of the 32-volt electrical system used by the PCC's auxiliaries.

Finally, the housings for Ohio Brass's trolley catchers were streamlined to complement PCC body styling. Other manufacturers of catchers, such as Earll (in Pennsylvania), had to redesign their products, too.[26]

Ohio Brass components were used on a group of "demonstrator streetcars" in Chicago in 1933, and they appeared on three of the initial PCC orders. Brooklyn PCCs featured Ohio Brass headlights, trolley pole bases, trolley wheels, and catchers. Chicago PCCs featured the same plus Ohio Brass trolley poles. Baltimore PCCs cars used Ohio Brass pole bases.

Ohmer Fare Register continued to produce its fare registers, ticketing machines, and recording devices during the 1930s. Although reported orders indicate only 160 new cars received its equipment, it should be remembered that Ohmer operated primarily through service contracts for existing fleets of vehicles and was not dependent on new car orders. Of those new cars fitted with Ohmer equipment, 151 were rail vehicles and 9 were trolley buses; 141 were equipped with registers, while 19 were given fare boxes. Seventy percent (112) of these vehicles ran in Indiana, Iowa, and Pennsylvania. The remainder ran in California, Ohio, Oregon, and Tennessee.[27]

John F. Ohmer, well into his seventies, remained optimistic and driven despite the Great Depression. He expanded into the rapid transit market in 1930 with a registering turnstile that could accept nickels, dimes, and tokens from a common slot. Another product was a register that featured an extra-large readout. Registers were also made available to operators in foreign countries, adaptable to whatever currency was required.

When asked about the Depression, Ohmer replied:

> I take no stock in people who are continually harping about hard times. . . . I have no patience with those who give us every imaginable cause for bad times. . . . Business depression is a necessary aftermath of the World War and . . . there is no effective panacea or antidote. . . . We must deal with the situation as it is, make the best of it, work cheerfully and a bright future will reward us.

To better handle his business during the Depression, Ohmer reorganized his company by forming a subsidiary, the Ohmer Register Company, to handle all of his sales.[28]

A new manufacturer whose product became increasingly important on both streetcars and buses during the 1930s was Toledo's Libby-Owens-Ford, a maker of safety glass. L-O-F made its safety glass by using a layer of clear cellulose glued between two glass plates. L-O-F's safety glass was popular because the cellulose did not "cloud up" or discolor (a problem that plagued other types of safety glass). In 1938 the company began offering ultra-violet-absorbent safety glass (plate glass absorbed only 8 percent of ultraviolet rays, whereas L-O-F 's new safety glass absorbed 22). In addition to PCC cars, L-O-F furnished safety glass for Brill's 1933 Chicago demonstrator and 10 "pre-PCCs" that Brill built for Washington, D.C. L-O-F emphasized the low cost of safety glass over

Figure 11.2. The ERPCC research effort redefined streetcar appearance during the 1930s. Many of the surviving component manufacturers redesigned their components to conform with the new car styles. This is a streamlined trolley catcher made by Ohio Brass. *Branford Electric Railway Association, author photo.*

plate glass. One of its advertisements claimed that "one cent added to every dollar of the total cost of new equipment will provide Safety Glass All-Around."[29]

Despite the survival of several Ohio component firms, others fell victim to the Great Depression. The Cleveland Fare Box Company was taken over in 1934 by the Johnson Fare Box Company of Chicago. The Cleveland facilities were retained by Johnson, and various additional products such as locks and turnstiles were made at the Cleveland plant. This takeover was particularly ironic, as one of the founders of Johnson Fare Box was Tom L. Johnson, a former Cleveland mayor and street railway owner. By 1938, the only purely Ohio-based fare box producer left in Cleveland was MacDonald Manufacturing.[30]

Like Ohio Brass, Nichols-Lintern manufactured components for rail vehicles, buses, and trolley buses. During the 1930s, Nichols-Lintern made components for 223 rail vehicles: 125 of these were streetcars or interurbans (56 percent), 48 were trolley buses (22 percent), and 50 were buses.

Nichols-Lintern's most popular products were its ventilators. Eighty-seven percent of the vehicles Nichols-Lintern equipped (195) used them. Twenty-three vehicles used Nichols-Lintern trolley pole catchers and lighting selector switches. Twenty were equipped with the company's sanders, and 15 used Nichols-Lintern tail lights.

Three manufacturers accounted for 10 percent or more of Nichols-Lintern's business (plus an additional 10 percent that went to in-house orders). They were Twin Coach (50 vehicles, 22 percent), Cincinnati (42 vehicles, 19 percent), and Brill (32 vehicles, 14 percent). The remaining 99 vehicles were spread among four streetcar manufacturers.

Nichols-Lintern developed a number of products during the 1930s, including forced-air ventilators intended for low-speed service with frequent stops. The ventilators could be operated off either 6 or 12 volt systems.

Although intended for urban service with frequent stops, Nichols-Lintern ventilators performed better at higher speed. While stationary, the system could change at least 193 cubic feet per minute. At 25 miles per hour, this improved to 258 cubic feet, and at 35 miles per hour this improved to 314. Despite its many products for both streetcars and buses, Nichols-Lintern was unable to remain intact. Detroit's Evans Products Company bought out Nichols-Lintern's heating and ventilation product lines around 1935 and turned it into the "Nichols-Lintern Division." The remainder of the company (sander and lighting manufacture) stayed in Cleveland and was reorganized into the Nichols-Lintern Corporation. The corporation's railway products were acquired by Bendix-Westinghouse during the late 1930s.[31] The corporation itself survived by entering the industrial air-conditioning market. It is still active today.

In addition to fare registers and boxes, Ohmer remained active in the bus field by making on-board monitors to record bus performance. One monitor recorded information on distance traveled, maximum speed, and running and idle periods on paper tape. Another device used 5-inch paper disks to record idle periods over a seven-day span.[32]

CONCLUSION

The 1930s were difficult for many industries, and the street railway industry was no exception. The interurban industry died out almost entirely. Unfortunately, it sometimes takes a major crisis to force people into action. The ERPCC effort produced a magnificent streetcar, but the irony is that much of the technology that came to be used in PCCs had been available during the previous decade, when there might have been a broader, more viable market. While some Ohio component manufacturers lasted long enough to take advantage of the PCC market, the market proved too small for medium-sized car builders like Kuhlman and Cincinnati. It even proved to be a challenge for large car builders such as Brill. Unless a company diversified into other product lines or served a highly specialized niche in the industry, the Great Depression was the end of the line for most of Ohio's street railway manufacturers.

AFTERWORD: 1938 AND THE END OF AN ERA

With the permanent closure of the Cincinnati Car Company in 1938, streetcar building in Ohio came to an end. This did not, however, mark the end of streetcar manufacture in the United States. To an extent, the PCC car gave new life to the streetcar market and enabled surviving streetcar and component manufacturers to continue.

The Cincinnati Car Company's closure also did not spell the end of Ohio transit vehicle manufacture, as the state had become a key center of bus production during the 1920s and 1930s. Manufacturers such as Twin Coach, which in 1927 introduced the modern "transit body" so familiar to today's riders, Flxible, and White were all producing buses for the transit, interurban, and long-distance markets. Jobbers such as Cleveland's Bender Body were producing streamlined bodies for chassis manufacturers, and firms such as Hercules Motor were producing gasoline and diesel engines. Akron, the nation's rubber-producing capital, became a new manufacturing center for the transit market, as numerous tire companies produced tires for buses.

The contrast between Ohio's street railway and bus industries is striking. The street railway industry was characterized by complacency. This stemmed from a faith in the quality of its product and that product's position in the transit industry at large. With a vehicle as sound as the streetcar and the security it gained in the urban infrastructure during the late 1800s and early 1900s, the industry felt no compulsion to reconsider the design or otherwise improve upon its product.

The bus industry, on the other hand, was characterized by a drive to continually improve the quality and design of its product. In reviewing the 50-plus years of Ohio transit vehicle manufacture that preceded the closure of the Cincinnati Car Company, several factors emerge that account for the contrast between the two industries.

In the first place, the street railway industry came into being in an era where there was no practical alternative transportation within cities. The 1880s and 1890s were exceptionally dynamic years for the industry, as different mechanical alternatives to the horse-drawn streetcar were devised, tried, and revised or rejected.

Until serious competition in the form of automotive vehicles began to appear around World War One, the street railway industry went unchallenged. When the bus began its ascendancy within the transit field during the early 1920s, the street railway industry's lackadaisical response could be attributed to a number of influences. In addition to its unchallenged emergence in the transit field at the turn of the century, the streetcar thwarted an automotive threat that preceded the bus.

Automobile owners were using their personal vehicles to compete directly with the streetcar lines during the 1910s and early 1920s. Called "jitneys," these automobilists would provide service along the same routes for less money or with the promise of faster service. They provided service that was erratic, and they often drove their cars irresponsibly. Street railway owners waged a successful legal battle to eliminate the jitneys, and this may have given them a false sense of security, leading industry leaders to feel they could deal with any subsequent threat.

The bus of 1920 was also clearly inferior to the streetcar of 1920. Early buses were a combination of assemblies that were neither intended for each other nor designed for transit service. They often consisted of small passenger bodies modified to fit onto truck chassis. They had smaller carrying capacities than streetcars and had poorer operating performance.

Finally, street railways had invested heavily in their physical plants (track, power distribution, etc.). Despite the theoretical promise of bus technology, street railways were probably reluctant to write off such a substantial investment.

The bus industry was not standing still during the 1920s. The bus's swift evolution over the decade from a combination of assemblies to an integral, transit-oriented design demonstrates the seriousness of the bus manufacturers. Integral design (chassis designed from the wheels up for bus service with bodies intended for specific chassis) was in place by the mid-1920s. This enhanced and promoted the bus's inherent advantages over the streetcar and allowed it to compete with the streetcar on all but the most heavily patronized routes.

In addition to its underestimation of the bus, the street railway industry was hampered by the nature of streetcar production. Car orders were highly individualized. Although certain common materials, methods, and features could be seen among all streetcar builders, car orders tended to be built to specifications dictated by the characteristics, preferences, and types of operation of individual railways.

Bus manufacturers, however, did not customize their orders. Throughout the 1920s, manufacturers only made specific bus models. This benefited the manufacturers by introducing a certain degree of standardization to their work. This enabled parts and subassemblies to be made in larger quantity, thus saving both production time and cost.

It was not until late 1929, when the street railway industry was losing ground at a rapid pace, that industry leaders made a decisive move toward improving the streetcar by forming the Electric Railway Presidents Conference Committee (ERPCC). Earlier efforts directed toward improving the streetcar and standardization had either met with failure (as with the Birney) or had not gained widespread acceptance (as with the Master Unit or the experimental cars in Joliet, Pittsburgh, and Louisville). By concentrating on performance specifications and individual components and subassemblies (rather than an entire streetcar), the ERPCC succeeded in producing a vehicle that both benefited from standardization and remained flexible enough to accommodate the needs of individual railways.

Unfortunately, by the time the first production PCCs entered service, streetcar building had ended in Ohio. Despite the shift to bus production, no former center of streetcar production was thrown into economic collapse as a result of the street railway industry's demise. The greatest reason for this was that despite the passing of such firms as the Cincinnati Car Company, Kuhlman, Jewett, Niles, Barney & Smith, Bayonet Trolley Harp, or Eclipse Fender, none of the cities in which these companies were located depended upon them for their economic well-being.

Ohio bus manufacturers continued to reap the benefits of a dynamic, progressive industry during the 1930s, despite the Great Depression.[1] Improvements to the bus did not end with the transit design. During the 1930s, bus bodies were streamlined. Although this did little (if anything) to improve performance, streamlining succeeded in altering the bus's appearance, making it more attractive and appealing to the prospective rider (the ERPCC followed the bus manufacturers' example in developing the PCC's body specifications).

Bus performance improved with the adoption of more powerful engines (especially the diesel engine), the availability of synchronous and automatic transmissions, and the relocation of the engine to the bus's rear. Clearly, Ohio's bus manufacturers were able to sustain the momentum they had developed during the 1920s, ensuring the continuance of bus manufacture in Ohio for many decades to come.

In addition to the closure of Ohio's last streetcar builder, 1938 was also marked by a series of deaths of individuals who had played key roles in the evolution of transit vehicle manufacture in Ohio. On one level, each of these men made significant contributions to the street railway industry, to the bus industry, or to both. On a deeper level, the companies each of these men represented serve to illustrate the contrast between the two industries.

On 25 September Thomas Elliott died in Cleveland at the age of 75. Elliott was an early advocate of lightweight streetcar and interurban design, and he helped to introduce a method of body and frame construction that became standard before the ERPCC program of the 1930s. He was also responsible for a method of pressing axle gears onto their axles without having to key the axle surface. The *Transit Journal* made the following comments about Elliott:

> For years he refused to publicize his many inventions, although later a number of patents were issued in his name. Cars which he built to his own design were of unusual stability, light in weight, and simple and inexpensive to maintain. He developed his designs from fundamental considerations without leaning upon previous practice.[2]

Despite the tremendous success of Elliott's basic design and the curve-sided variation the Cincinnati Car Company derived from it, the design embodied much of what went wrong with streetcar building. Rather than update the streetcar by developing a new carbody with improved comfort, performance, and appearance, Elliott merely simplified the construction of existing designs. He made them lighter and less expensive to fabricate. Elliott may not have "leaned upon previous practice," but his work reinforced it.

We can only speculate as to what Elliott's motives were for leaving Cincinnati in 1917. Perhaps he was seeking new challenges, perhaps it was money, or perhaps Elliott sensed where the street railway industry was headed. Whatever his motives, the car builder he left behind did not survive him. Neither did any car building activity in Ohio.

On 10 August Charles H. Tomlinson died.[3] Tomlinson's coupler was a significant streetcar component. It not only coupled cars together but also made all of the air and electrical connections necessary for multiple-unit operation. Some models were even compatible with the MCB knuckle couplers, which were standard on railroads.

Ohio Brass, the firm that made Tomlinson couplers, was one of the handful of streetcar component manufacturers to survive this period of study. That it survived when streetcar building, along with many other component makers, did not compels one to discover why.

The most important reason that Ohio Brass survived was that it was not solely a manufacturer of streetcar components. The company was founded during the 1890s as a maker of power line and power transmission hardware. From there the company branched into hardware for streetcar wires and finally to streetcar components. Even with its success in the component field, Ohio Brass never left the power transmission market.

Another reason Ohio Brass survived was the quality and versatility of the products it produced, of which the Tomlinson coupler is an excellent example. Tomlinsons were compatible not only with street railways but also with interurban and rapid transit operations. The simplicity and ruggedness of the coupler testified to the soundness of Tomlinson's original design, versions of which are still in use today.

The final reason Ohio Brass survived was that it made a point of acquiring product lines or even entire companies that rivaled its own products. Such was the case with General Electric's line of trolley wire hardware and Ohio Brass's merger with Trolley Supply. Charles Tomlinson may have left Ohio Brass with only one component among many in its product lines when he died, but the very fact that the company recognized it as one component among many in one market among many assured Ohio Brass's survival.

The last significant Ohio personage to pass was John F. Ohmer, who died on 4 November.[4] The Ohmer Fare Register Company was another rare survivor of this period of study. Although engaged in a different line of manufacture than Ohio Brass, Ohmer Fare Register had attributes that were similar to the Mansfield firm. Ohmer did not view his company as a component manufacturer/service provider just for street railways and interurbans. Instead, he viewed his company as serving the transit industry as a whole. As such, he modified his fare registers so that they could be used either on buses or streetcars.

Through the acquisition of other firms, Ohmer was able to eliminate competition, expand his product line, and broaden his market. His acquisition of American Taximeter enabled him to enter the taxi market. By acquiring Dayton Fare Register, he eliminated numerous rivals and gained access to their patents (Dayton Fare Register had itself bought out a number of register makers).

Furthermore, Ohmer continued to improve his products. At one point he was said to have held more patents in his name than any other Dayton resident. This persistent inventive activity led not only to new and improved registers but also to new products. The "hubodometer" stands as an example of this and as an example of Ohmer's broad market focus. Hubodometers could be used either on streetcars or on buses.

At the time of Ohmer's death, there were few (if any) urban transit forms that did not use some sort of Ohmer device. Streetcars, interurbans, rapid transit cars, buses, taxis, and even some railroads benefited from his equipment. Although he started as one register manufacturer among many in 1898, he left behind the largest company of its kind in the world four decades later.

A major lesson to be gleaned from Ohio's transit vehicle manufacturers is the danger of becoming too specialized in too narrow a field. Once the streetcar was eclipsed by the bus and automobile (more versatile vehicles), streetcar manufacturers were in trouble. The only firms to survive were those that were not dependent upon street railway orders or those that had the foresight to change their market focus.

By 1938 the clatter of Ohio-built streetcars was replaced by the hum and growl of Ohio-built buses along city streets. Vehicle and component manufacture, which had originally been located at the four points of Ohio's compass plus the center, was now concentrated in the state's northern half. Names such as Barney & Smith, G. C. Kuhlman, and Niles Car & Manufacturing had been replaced by names such as Twin Coach, Flxible, and White. The swiftness in which one type of urban transport manufacture supplanted another in Ohio explains, at least in part, why Ohio bus manufacture survived into the 1990s while streetcar manufacture has been largely forgotten.

APPENDIX: TABLES

Table 2.1. Profile of Barney & Smith's Car Building Activity, 1894–June 1914

Streetcars	861	60%	1800s Streetcars	739	1900s Streetcars	122
Interurbans	415	29%	1800s Interurbans	248	1900s Interurbans	167
Rapid Transit	103	7%	1800s Rapid Transit	63	1900s Rapid Transit	40
Special	54	4%	1800s Special	32	1900s Special	22
Total	1,429		1800s Total	1,082	1900s Total	347

Number of Cars by State

Ohio	882 (Dayton 159)	unknown	29
New York	107	Indiana	28
Michigan	101	Pennsylvania	20
Massachusetts	90	Colorado	4
Illinois	67	California	2
Maryland	62	Oregon	2
Wisconsin	34	Kentucky	1

Sources: Electric Railway Journal, Electric Traction Weekly, Street Railway Journal, and *Street Railway Review.* See also Scott D. Trostel, *The Barney & Smith Car Company: Car Builders* (Fletcher, Ohio: Cam-Tech, 1993), 222–25.

Table 2.2. Profile of G. C. Kuhlman Car Company's Car Building Activity, 1888–June 1914

Streetcars	2,081	(81%)
Interurbans	451	(17%)
Rapid Transit	—	
Special	46	(2%)
Total	2,578	

Number of Cars by State

Ohio	873 (540 Cleveland)	Wisconsin	21
New York	530	Indiana	12
Michigan	473	Maryland	7
Illinois	348	CANADA	6
Pennsylvania	74	Kentucky	5
Tennessee	38	Oregon	4
unknown	34	South Carolina	3
West Virginia	33	Massachusetts	1
Texas	25		

Sources: Electric Railway Journal, Electric Traction, Electric Traction Weekly, Street Railway Journal, and *Street Railway Review.*

Table 2.3. Profile of Jewett Car Company's Car Building Activity, 1894–June 1914

Streetcars	604	(43%)
Interurbans	453	(32%)
Rapid Transit	324	(23%)
Special	18	(1%)
Total	1,399	

Number of Cars by State

New York	295	Nebraska	27
California	274	North Carolina	23
Illinois	238	West Virginia	16
Ohio	171	New Jersey	12
Washington, D.C.	150	Michigan	10
Indiana	84	Utah	10
Connecticut	41	Maryland	3
Pennsylvania	38	Iowa	1

Sources: Electric Railway Journal, Electric Traction Weekly, Street Railway Journal, and *Street Railway Review.*

Table 2.4. Initial Equipment in Cincinnati Street Railway Shops (Forerunner of the Cincinnati Car Company)

Blacksmith Shop

12 forges
2 punches
1 steam hammer

Machine Shop

3 drill presses
1 bolt cutter
8 engine lathes (largest is 42 inches)
1 grindstone
1 planer
1 150-ton wheel press
2 boring machines
1 shaper
2 milling machines

Carpentry Shop and Mill Room

1 large double cylinder planer
1 small pony planer
1 cut-off saw
1 24-inch hand planer
1 heavy surfacer
1 self-feed rip saw
2 variety saws
1 double circular saw
1 large band saw
1 small band saw
1 scroll saw
1 double-headed friezing machine
1 single-headed friezing machine
2 graduated-stroke mortising machines
1 triple boring machine
1 triple drum sander
2 tenoning machines
1 variety wood worker
1 knife grinding machine
1 heavy molding machine
1 light molding machine
1 grindstone

Source: Street Railway Journal 14, no. 2 (Feb. 1898): 78–79.

Table 2.5. Profile of Cincinnati Car Company's Car Building Activity, 1903–June 1914

Streetcars	1,697 (70%)
Interurbans	498 (20%)
Rapid Transit	148 (6%)
Special	93 (4%)
Total	2,436

Number of Cars by State

Ohio	640 (425 Cincinnati)	South Carolina	12
New Jersey	360	Delaware	11
Indiana	292	Tennessee	10
Kentucky	189	California	9
New York	150	Colorado	7
Illinois	138	Oklahoma	6
Missouri	120	Connecticut	5
Texas	96	Maryland	4
Washington	95	Utah	4
Michigan	85	Wisconsin	4
Virginia	63	West Virginia	3
Pennsylvania	58	CANADA	3
unknown	18	Montana	2
Washington, D.C.	17	Georgia	1
Florida	15	North Carolina	1

Sources: Electric Railway Journal, Electric Traction, Electric Traction Weekly, and *Street Railway Journal.*

Table 2.6. Profile of Niles Car & Manufacturing Company's Car Building Activity, 1901–June 1914

Streetcars	297 (35%)
Interurbans	520 (61%)
Rapid Transit	—
Special	33 (4%)
Total	850

Ohio	200	New York	25
Michigan	166	Wisconsin	23
Illinois	76	Oklahoma	17
Maryland	62	Kentucky	15
Pennsylvania	37	California	12
Minnesota	36	Iowa	12
CANADA	35	Oregon	10
CUBA	35	Idaho	5
Utah	33	Montana	4
Indiana	30	unknown	2

Sources: Electric Railway Journal, Electric Traction, Electric Traction Weekly, Street Railway Journal, and *Street Railway Review.*

Table 2.7. Example of Formal Specifications for a Streetcar Order

The following is an example of formal specifications for a streetcar order between the Jewett Car Company and the Owosso & Corunna Electric Company (Owosso, Mich.). Anything appearing in normal print was a default specification that appeared in Jewett's specification forms, indicating material that Jewett would use unless directed otherwise. Anything appearing in **bold print** was specified by the railway.

Dimensions

Length—body **37 feet**
Width—over sills **8 feet 2.5 inches**
overall **47 feet 8 inches**

Lumber, Etc.

Side, center, intermediate, and cross sills to be made from long leaf yellow pine.
End sills, platform and buffer timbers, and needle beams (when used) to be made from well seasoned tough white oak.
Side posts of either ash or yellow pine.
Corner, door, and vestibule posts to be made from ash.
Ribbing, belt rail, side and deck plates, and deck sills to be made from long leaf yellow pine.
End plates to be made from either ash or oak.
Roof framing (carlines, etc.) to be made from either ash or oak.
Roofing to be made from half-inch yellow pine, **covered with no. 8 cotton duck.**
Filling in the side of car below sash rest to be made from white oak.
Sash rest capping to be made from ash or oak.
Letter boards, battens, sash rest apron, outside sheathing to be made from yellow poplar.
Flooring to be made of ⅞-inch yellow pine laid in **single** thickness and well nailed.
Floor framing as per plan no. **267** arranged to suit **Standard Steel C-60** trucks and **(unspecified)** motor equipment, with **(unspecified)** controller of the **(unspecified)** type.

Trucks

Jewett Car Company to ensure that the car body/bodies to be able to accept **motor equipment as may be selected by Purchaser.**

Motors

Railway company to supply, equip, and install motors.

Body Framing

(unspecified)

Interior Finish

Interior finish of main compartment and end doors to be made from either **oak or ash.**
Interior finish of vestibule (including side doors and panels beneath windows) to be made from **ash.**

Ceiling

Three-ply **veneer birch or maple.**
Ceiling back to receive two coats of paint.

Windows

Sashes to be made from either **oak or ash.**
Two sashes (top and bottom). Top sash to **remain stationary,** bottom sash to **raise**—with all necessary hardware and/or fittings, all according to specifications on plan no. **267**, hung on **hinges with the Jewett Car Company's catch.**
Sash for vestibule windows to incorporate two features: easy rain runoff, and will not blow in when car is at speed against the wind.

(continued)

Table 2.7. *(continued)*

Top sash to be **cathedral glass**, deck sash to be **chipped glass**.
All Windows, except top sash, to be glazed with best **D.S.A.** glass, set in putty and held in place by neat molding and bronze screws.

Curtains

Side windows to be equipped with **pantasote** curtains on Hartshorn Spring Rollers and **Curtain Supply Co.** fixtures.
Front doors as the same, accounting for the roll of the doors.

Wiring

Railway company to supply all wire, lamps, sockets, switches, electroliers, etc.

Seats

Eight walkover and longitudinal corner seats upholstered in **rattan** with **20**-inch backs.

Parcel Racks

None.

Vestibules

Each end of the car to be enclosed—the outside of each vestibule below the windows to be covered with **poplar sheathing.**

Brake Staff

1¾-inch round, forged iron—the top to have an **18-inch wheel.** Rigging to have the best quality ½-inch twisted chain link.

Steps

Each to be Jewett Standard Steps, **single** steps at vestibules.

Bumpers and Draw Bars

Each end of the car to be equipped with **Jewett Car Company radiating** drawbars.

Smoking Compartment

Space of the compartment to take up space equivalent to **two single windows.**
Smoking compartment to be separated from main compartment by a **swinging** panel and glass partition.
Interior finish to be made from **oak or ash.**
Seats to be **longitudinal** and upholstered with **rattan.**

Baggage Compartment

Separated from the rest of the car by a **swinging** panel partition.
Interior finish to be made from **ash,** though side next to adjacent compartment **(smoking)** to match that compartment.

Note: The remaining headings all have **none** specified by either Jewett or the railway—heaters, headlights, fenders or pilots, fare registers, and push buttons.

Source: Folder 11, Jewett Car Company Records, MSS 971, Ohio Historical Society, Columbus.

Table 3.1. Trolley Supply Company: Examples of Current Collection Orders from 1909

1) Buffalo, Lockport & Rochester Railway (New York)
   Knutson retrievers for 15 interurbans (cars built by the Cincinnati Car Co.)

2) Buffalo & Lake Erie Traction Co. (New York)
   Knutson retrievers for 8 interurbans (cars built by the Cincinnati Car Co.)
   Peerless trolley pole bases for same car order

3) Chicago, Lake Shore & South Bend Railway (Illinois)
   Knutson retrievers for 24 interurbans (cars built by the Niles Car Manufacturing Co.)

4) Chippewa Valley Railway, Light & Power Co. (Wisconsin)
   Knutson retrievers for 2 interurbans (cars built by the St. Louis Car Co.)

5) Milwaukee Northern Railway (Cedarburg, Wisconsin)
   Knutson retrievers on a parlor car (car built by the Niles Car & Manufacturing Co.)

6) Wason Car & Manufacturing Co. (Springfield, Mass.)
   12 Knutson retrievers for unidentified car order(s)

Source: Electric Railway Journal.

Table 3.2. Relative Strength of Steel Gears and Pinions

Gear Material	*Tensile Strength (lbs.)*	*Elastic Limit (lbs.)*
Cast steel	about 60,000	about 25,000
Cast steel (tempered in oil)	about 75,000	about 45,000
Forged steel	about 85,000	about 55,000
Case hardened, hard alloy steel	about 95,000	about 55,000
Forged steel, low carbon	about 55,000	about 25,000
Forged steel, high carbon	about 85,000	about 50,000
Heat-treated forged steel, low carbon	about 100,000	about 55,000
Heat-treated forged steel, high carbon	about 115,000	about 82,000
Case-hardened, special alloy steel	about 95,000	about 55,000

Source: Information obtained from *Proceedings of the American Electric Engineering Association* (1912): 671–75.

Table 4.1. Ohio Brass and Streetcar Building: 1900–14

Partial Summary			
Car Builder	*Total Cars*	*% Total*	*% Ohio*
American Car Company	60	8	—
J. G. Brill Company	52	7	—
Cincinnati Car Company	227	30	45
Danville Car Company	5	1	—
built in-house by railway	18	2	—
Jewett Car Company	28	4	6
G. C. Kuhlman Car Company	207	27	41
Niles Car & Manufacturing Company	40	5	8
Preston Car & Coach Company	6	1	—
St. Louis Car Company	87	11	—
Southern Car Company	5	1	—
unknown	30	4	—
Total cars:	765		
cars equipped with couplers:	553		
cars equipped with sanders:	173		
cars equipped with pole bases:	59		
cars equipped with on-board signals:	10		

Tomlinson Couplers: Non-Car-Builder Orders			
State	*No. of Cars Equipped*	*State*	*No. of Cars Equipped*
Illinois	5	New York	55
Indiana	12	Oklahoma	12
Kentucky	6	Ohio	60
Massachusetts	610	Washington	98
Michigan	100		
Missouri	201	Spain	52

Source: Information obtained from the following (1900–14): *Electric Railway Journal, Electric Traction, Electric Traction Weekly,* and *Street Railway Journal.*

Table 4.2. Ohio Brass Company: Examples of Tomlinson Coupler Orders

1) San Diego Electric Railway Co. (California)
   couplers for 4 interurbans in 1908 (cars built by Niles Car & Manufacturing Co.)

2) Buffalo & Lake Erie Traction Co. (New York)
   couplers for 8 interurbans in 1909 (cars built by Cincinnati Car Co.)

3) Omaha & Council Bluffs Street Railway (Nebraska)
   couplers for 10 interurbans in 1909 (cars built by American Car Co., St. Louis)

4) East St. Louis & Suburban Railway (Illinois)
   couplers for 5 interurbans in 1909 (cars built by American Car Co., St. Louis)

5) Spokane & Inland Empire Railway (Washington)
   86 couplers in 1910

6) Oklahoma Railway (Oklahoma City, Oklahoma)
   12 couplers in 1910

7) Seattle & Everett Interurban Railway (Washington)
   12 couplers in 1910

8) Louisville & Eastern Railroad
   6 couplers in 1910

9) Terre Haute, Indianapolis & Eastern Traction Co. (Indiana)
   12 couplers in 1910

10) Chicago & Joliet Electric Railway (Illinois)
    54 couplers in 1910

11) Puget Sound Electric Railway (Tacoma, Washington)
    couplers on 2 streetcars in 1910 (cars built by Cincinnati Car Co.)

12) Twin City Light & Traction Co. (Canterbury, Washington)
    couplers on 2 suburban cars in 1910 (cars built by Niles Car & Manufacturing Co.)

13) Mahoning & Shenango Railway & Light Co. (Ohio)
    couplers on 6 interurbans in 1910 (cars built by Niles Car & Manufacturing Co.)

14) Chicago & Joliet Electric Railway (Illinois)
    couplers on 2 interurbans in 1910

15) Buffalo & Lake Erie Traction Co. (New York)
    couplers on 5 interurbans in 1911 (cars built by Cincinnati Car Co.)

16) Evansville Railways (Indiana)
    couplers on 1 locomotive in 1911 (locomotive built by Cincinnati Car Co.)

17) Altoona & Logan Valley Electric Railway (Pennsylvania)
    couplers on 5 streetcars in 1911 (cars built by Cincinnati Car Co.)

18) Greenville, Spartanburg & Anderson Railway (South Carolina)
    couplers for 17 interurbans in 1911 (cars built by Jewett Car Co.)

19) Walla Walla Railway (Washington)
    MCB-Tomlinson coupler equipment for 1 interurban in 1911 (car built by the Danville Car Co., Illinois)

(continued)

Table 4.2. *(continued)*

20) Indiana Union Traction Co.
 couplers for 10 interurbans in 1912 (cars built by Cincinnati Car Co.)

21) San Luis Obispo (railway not identified)
 MCB-Tomlinson coupler equipment for 1 interurban in 1913 (car built by Cincinnati Car Co.)

22) New York State Railways
 55 couplers in 1913

23) Detroit United Railways (Michigan)
 100 couplers in 1913

24) Cleveland Railways (Ohio)
 110 couplers in 1913 (50 cars by G. C. Kuhlman Car Co.)

25) United Railways of St. Louis (Missouri)
 210 couplers

26) Cataluña, Spain (railway not identified)
 52 couplers in 1913

27) Tri-City Railway (Davenport, Iowa)
 couplers for 30 streetcars in 1913

28) Montreal Tramways
 couplers for 25 streetcars in 1913 (cars built by J. G. Brill Co.—plant not specified)

29) Union Traction Co. (Indiana)
 MCB-Tomlinson coupler equipment for 10 interurbans in 1913 (cars built by Cincinnati Car Co.)

30) Ohio Electric Railway
 couplers on 6 freight trailers in 1913

31) Ohio Electric Railway
 couplers on 5 interurbans in 1914 (cars built by Cincinnati Car Co.)

32) El Paso Electric Railway (Texas)
 couplers on 6 streetcars in 1914 (cars built by St. Louis Car Co.)

33) Northern Texas Traction Co.
 couplers on 20 streetcars in 1914 (cars built by American Car Co., St. Louis)

34) New York State Railways (Utica, New York)
 MCB-Tomlinson coupler equipment for 1 crane car in 1914 (car built in-house)

35) Philadelphia & Garretford Street Railway (Pennsylvania)
 couplers on 5 interurbans in 1914

Sources: Street Railway Journal and *Electric Railway Journal.*

Table 4.3. Cincinnati Car Company: Car Orders Featuring the Cincinnati Car Company's Couplers

1) Union Traction Co. (Sistersville, West Virginia)
   couplers for 2 streetcars in 1909

2) Hagerstown Railway (Maryland)
   couplers for 1 interurban in 1909

3) Helena Railway, Light & Power Co. (Montana)
   couplers for 2 streetcars in 1909

4 Buffalo & Lackawanna Traction Co. (New York)
   couplers for 10 streetcars in 1909

5) Dallas Consolidated Electric Street Railway
   couplers for 2 streetcars in 1909

6) Northern Texas Traction Co. (Dallas, Texas)
   couplers for 12 streetcars in 1909

7) Cincinnati Traction Co. (Ohio)
   couplers for 50 streetcars in 1910

8) Indianapolis Traction & Terminal Co.
   couplers for 25 interurbans or streetcars in 1910

9) El Paso Electric Co. (Texas)
   couplers for 6 streetcars in 1910

10) Tri-City Railway and Light Co. (Moline, Illinois)
   couplers for 7 streetcars in 1910

11) Jacksonville Electric Co. (Florida)
   couplers for 10 streetcars in 1910

12) Greenville Railway & Light Co. (Texas)
   couplers for 7 streetcars in 1911

13) Pittsburgh, McKeesport & Westmoreland Railway
   couplers for 2 streetcars in 1911

14) Elmira Water, Light & Railroad Co. (New York)
   couplers for 8 streetcars in 1912

15) Seattle Municipal Street Railway Co.
   couplers for 12 streetcars in 1913

16) Manhattan Bridge Three Cent Fare Line
   couplers for 6 streetcars in 1913

Source: Electric Railway Journal.

Table 5.1. Cincinnati Car Company: Car Orders Featuring the Cincinnati Car Company's Sanders (1908–13)

1) Capital Traction Co. (Washington, D.C.)
   sanders for 12 streetcars in 1908

2) Indianapolis Traction & Terminal Co.
   sanders for 44 streetcars in 1909

3) Union Traction Co. (Sistersville, West Virginia)
   sanders for 2 streetcars in 1909

4) Hagerstown Railway (Maryland)
   sanders for 1 interurban in 1909

5) Helena Railway, Light & Power Co. (Montana)
   sanders for 2 streetcars in 1909

6) Indianapolis Traction & Terminal Co.
   sanders on 25 streetcars or interurbans in 1910

7) South Covington & Cincinnati Street Railway (Ohio)
   sanders for 24 streetcars in 1911

8) Greenville Railway & Light Co. (Texas)
   sanders for 7 streetcars in 1911

9) Pittsburgh, McKeesport & Westmoreland Railway
   sanders for 2 streetcars in 1911

10) Seattle Municipal Street Railway Co.
    sanders for 12 streetcars in 1913

11) Manhattan Bridge Three Cent Fare Line
    sanders for 6 streetcars in 1913

Source: Electric Railway Journal.

Table 5.2. Nichols-Lintern Company: Examples of Sander Orders (1905–14)

1) Toledo, Port Clinton & Lake Side Railway
   sanders for 4 interurbans in 1905

2) Lake Shore Electric Railway (Ohio)
   sanders for 10 interurbans in 1906 (cars built by Niles Car & Manufacturing Co.)

3) Cleveland Electric Railway
   sanders for 1 funeral car in 1906

4) Pittsburgh & Butler Street Railway Co. (Pennsylvania)
   sanders for 2 interurbans in 1908

5) Chicago City Railway Co.
   sanders for a 40-ton locomotive in 1908 (locomotive built in-house)

6) Buffalo & Lake Erie Traction Co.
   sanders for 8 interurbans in 1909

7) railway not specified (Ogden, Utah)
   sanders for undisclosed number of cars in 1909 (built by Cincinnati Car Co.)

8) Omaha & Council Bluffs Street Railway
   sanders for 10 streetcars in 1909 (cars built by American Car Co., St. Louis)

9) Milwaukee Northern Railway
   sanders for 2 interurbans in 1909

10) Cleveland City Railway
   sanders for a parlor car in 1909

11) New York & North Shore Traction Co. (Roslyn, New York.)
   sanders for 3 interurbans in 1909 (cars built by G. C. Kuhlman Car Co.)

12) Washington, Baltimore & Annapolis Electric Railway
   sanders for 27 interurbans in 1909

13) Ohio Electric Railway (Cincinnati)
   sanders for 6 express cars in 1909 (cars built by Cincinnati Car Co.)

14) Milwaukee Electric Railway & Light Co.
   sanders for 100 streetcars in 1909 (cars built by St. Louis Car Co.)

15) Shore Line Electric Railway (Connecticut)
   sanders for 12 interurbans in 1909 (cars built by Jewett Car Co.)

16) Wabash & Northern Indiana Traction Co.
   sanders for 7 interurbans in 1910 (cars built by Jewett Car Co.)

17) Cleveland City Railway
   sanders for 100 streetcars in 1910

18) Calumet & South Chicago Railway Co.
   sanders for 2 funeral cars in 1910 (cars built by G. C. Kuhlman Car Co.)

19) Oregon Electric Railway (Portland)
   sanders for 2 parlor interurbans in 1910 (cars built by Niles Car & Manufacturing Co.)

(continued)

Table 5.2. *(continued)*

20) Oklahoma Railway Co. (Oklahoma City)
    sanders for an interurban in 1910 (cars built by Niles Car & Manufacturing Co.)

21) Twin City Light & Traction Co. (Canterbury, Washington)
    sanders for 2 suburban cars in 1910 (cars built by Niles Car & Manufacturing Co.)

22) Gary & Southern Traction Co.
    sanders for 2 interurbans in 1912

23) Cleveland Railways
    sanders for 50 streetcars in 1913 (cars built by G. C. Kuhlman Car Co.)

24) Columbus Railway & Light Co.
    sanders for 376 streetcars in 1914

25) Cleveland Railways
    sanders for 50 streetcars in 1914

26) Scioto Valley Traction Co. (Ohio)
    undisclosed number of sanders in 1914

27) Northern Ohio Traction & Light Co.
    undisclosed number of sanders in 1914

28) Toledo & Western Railway
    undisclosed number of sanders in 1914

Sources: Street Railway Journal and *Electric Railway Journal.*

Table 5.3. Ohio Brass Company: Examples of Sander Orders (1910–14)

1) Buffalo & Lake Erie Traction Co.
    sanders for 5 suburban cars in 1910

2) Winnipeg, Selkirk & Lake Winnipeg Interurban Railway
    sanders adopted as standard on this interurban railway in 1910

3) Mahoning & Shenango Railway & Light Co. (Ohio)
    sanders on 2 interurbans in 1910 (cars built by Niles Car & Manufacturing Co.)

4) City Railway (Dayton, Ohio)
    sanders on 10 streetcars in 1911

5) Northern Ohio Traction & Light Co.
    sanders on undisclosed number of interurbans in 1911 (cars built by G.C. Kuhlman Car Co.)

6) Grand Rapids Railway
    sanders on 10 streetcars in 1912 (cars built by Cincinnati Car Co.)

7) Indianapolis Traction & Terminal Co.
    sanders on 8 streetcars in 1912 (cars built by Cincinnati Car Co.)

8) Lehigh Valley Transit Co. (Pennsylvania)
    sanders on 6 interurbans in 1912 (cars built by Jewett Car Co.)

Table 5.3. *(continued)*

9) Philadelphia Rapid Transit Co.
   sanders on a funeral car in 1912 (car built by J. G. Brill Co., Philadelphia)

10) Lehigh Valley Transit Co. (Pennsylvania)
    sanders on 6 interurbans in 1913 (cars built by Jewett Car Co.)

11) Chicago & Joliet Electric Railway
    sanders on 10 streetcars in 1913 (cars built by St. Louis Car Co.)

12) Puget Sound Traction Co.
    sanders on 10 streetcars in 1913 (cars built by Cincinnati Car Co.)

13) Springfield Railway (Ohio)
    sanders on 10 streetcars in 1913 (cars built by Cincinnati Car Co.)

14) Evanston County Traction Co. (Illinois)
    sanders on 12 streetcars in 1914 (cars built by St. Louis Car Co.)

15) Washington-Virginia Railway (Washington, D.C.)
    sanders on 5 streetcars in 1914 (cars built by Southern Car Co., High Point, North Carolina)

16) Morris County Traction Co. (New Jersey)
    sanders on 10 interurbans in 1914 (cars built by Cincinnati Car Co.)

Source: Electric Railway Journal.

Table 5.4. Eclipse Car Fender Company: Examples of Orders (1911–14)

1) Kansas City (Missouri)
   Eclipse fenders for 25 streetcars in 1911 (cars built by Cincinnati Car Co.)

2) Ontario & San Antonio Heights Railroad (California)
   Eclipse fenders for 3 streetcars in 1912

3) Syracuse Rapid Transit Co.
   Eclipse fenders for 9 streetcars in 1912 (cars built by the St. Louis Car Co.)

4) United Railroads of San Francisco
   Eclipse fenders for 65 streetcars in 1912 (cars built by American Car Co., St. Louis)

5) New Orleans Railway & Light Co.
   undisclosed number of ACME fenders in 1912

6) Los Angeles Railways Co.
   undisclosed number of Eclipse fenders in 1912

7) Geary Street Municipal Railway (San Francisco)
   Eclipse fenders for 86 streetcars in 1912

8) New York State Railways (Rochester)
   Eclipse fenders for 25 interurbans in 1913 (cars built by G.C. Kuhlman Car Co.)

9) Municipal Railways (San Francisco)
   Eclipse fenders for 125 streetcars in 1914 (cars built by Jewett Car Co.)

Source: Electric Railway Journal.

Table 6.1. Ohmer Fare Register Company: Partial Listing of Customers (1902–14)

American Railway Co., Philadelphia
Atlanta-Northern Railway
Canton-Akron Railway
Central Kentucky Traction Co.
Chicago, Ottawa & Peoria Railway
Chippewa Valley Railway, Light & Power Co., Wisconsin (at least 2 cars)
City Railway Co., Dayton
Cleveland, Painesville, & Ashtabula Railroad
Cleveland, Painesville & Eastern Railroad
Cleveland & Southwestern Traction Co.
Conneaut & Erie Traction Co., Ohio
Dayton, Covington & Piqua Traction Co.
Dayton Street Railway
Dayton & Troy Electric Railway
Denver City Tramway
Detroit United Railways
East Saint Louis & Suburban Railway
Eastern Pennsylvania Railway, Pottsville
Evansville & Princeton Traction Co.
Fort Wayne, Van Wert & Lima Traction Co.
Fort Wayne & Wabash Valley Traction Co.
Grand Rapids Street Railroad
Hocking Valley Railroad, Ohio
Illinois Traction System
Indiana, Columbus & Eastern Traction Co., Indiana
Indianapolis, Crawfordsville & Western Traction Co.
Jackson & Battle Creek Traction Co.
Joplin & Pittsburgh Railway Co. (5 interurbans and 1 express car)
Lebanon & Franklin Traction Co., Dayton
Lincoln Traction Co., Nebraska
Los Angeles-Pacific Railway Co.
Mexico Car Co., Mexico City (300 registers)
Muncie, Hartford & Fort Wayne Railway
Northern Ohio Traction & Light Co.
Oakland Traction Co., California
Oakwood Street Railway Co., Dayton
Ogden, Utah (railway not specified)
Ohio Electric Railway, Dayton
Pacific Electric Railway Co., California
People's Railway Co., Dayton
Pittsburgh & Butler Street Railway Co. (2 interurbans only)
Pittsburgh Railways
Portland, Railway, Light & Power Co. (3 fare registers on 38 streetcars)
Providence & Danielson Railway, Rhode Island
Public Service Corp. of New Jersey (300 registers in 1904)
Richmond Passenger & Railway Co. (reputedly two-thirds of its fleet)
Sacramento Electric, Gas & Railway Co. (at least 10 cars)
San Francisco, Vallejo & Napa Valley Railway (at least 2 cars)
Scioto Valley Traction Co., Ohio
Southern Pacific Railroad, California

Table 5.2. *(continued)*

Syracuse Rapid Transit Co., New York
Union Traction Co., Indiana (135 cars)
Washington-Virginia Railway (Washington, D.C.)
Wausau Street Railway, Wisconsin (at least 1 car)

Table 7.1. Dayton Manufacturing Company: Profile of Manufacturing Activity (1901–14)

Car Builder	*# Cars*	*% Cars*	*% Ohio*
American Car Co.	45	4	—
J. G. Brill Co.	139	11	—
Cincinnati Car Co.	823	65	85
Danville Car Co.	14	1	—
Jewett Car Co.	24	2	2
G. C. Kuhlman Car Co.	106	8	11
McGuire-Cummings Mfg. Co.	4	—	—
Niles Car & Manufacturing Co.	19	2	2
Pressed Steel Car Co.	75	6	—
St. Louis Car Co.	17	1	—
Total	1,266		

Components	*# Cars*	*Components*	*# Cars*
baggage racks	21	marker lights	8
gongs	138	sash fixtures	279
handbrakes	104	toilets	31
headlights	691	trimmings	210
lighting	21	ventilators	20

Sources: Street Railway Journal, Electric Railway Journal, Electric Traction Weekly, and *Electric Traction.*

Table 7.2. 1908 Chicago Tests

Test Results System	*Intake (cubic feet/hour)*	*Exhaust (cubic feet/hour)*
Cooke	25,000	47,524
Garland	not given	18,896*
McGerry	22,237	11,582
Perry	28,133	13,043
Taylor	32,062	28,546

* Exhaust was increased to 53,664 cubic feet/hour with rear doors open

Source: Electric Railway Journal 33, no. 19 (8 May 1909): 876–79.

Table 8.1. World War One and Rising Materials Costs

Item	*1915 Price*	*1918 Price*	*Increase*
Boiler tube	$4.60	$18.00	291%
Cement (price per barrel)	$0.80	$2.00	150%
Crushed rock (price per cubic yard)	$0.85	$1.35	59%
Electric switch (track)	$160.00	$180.00	13%
Line pole (5"×30')	$25.75	$41.50	61%
Paint (price per gallon)	$1.15	$2.00	74%
Spikes (price per hundredweight)	$1.58	$3.05	93%
Steel bars (price per hundredweight)	$1.67	$4.50	169%
Steel rails (price per ton)	$36.50	$52.50	44%
Steel sheets (price per hundredweight)	$3.50	$7.00	100%
Streetcar (complete)	$6,000.00	$8,000.00	33%
Ties	$0.53	$0.78	47%
Trolley poles	$1.75	$3.05	74%
Trolley wheels	$1.10	$1.85	68%
Trolley wire (price per foot)	$0.13	$0.38	192%

Source: Electric Railway Journal 51, no. 26 (29 June 1918): 1,260.

Table 9.1. Weight vs. Cost Percentages

Category	*% Weight*	*% Operating Cost*
Carbody	45%	30%
Trucks (without motors)	25%	13%
Two motors	14%	15%
Air brakes and comfort commodities	14%	25%
Controls and cables	2%	12%
Labor	n/a	5%

Source: T. A. Chance, "The Comforts Provided for the Public in Today's Trolley Car," *Electric Traction* 16, no. 1 (Jan. 1920): 34.

Table 9.2. Economic Advantage to Lightweight and Safety Cars

Category	*Older Car Type*	*New, Lightweight Peter Witt*
length	45'3"	50'3"
weight (pounds)	62,000	41,900
height, floor above rail	4'2"	3"
seating capacity	46	54
motor equipment	GE 57	Westinghouse 532-A
motor horsepower	50*	50**
gear ratio	24:63	16:56
wheel size	36"	32"
free running speed, miles per hour at 600v	40	32"
kilowatt hours per car mile, at car	3.26	1.8
kilowatt hours per car mile, at 600v	3.95	1.97
maximum current from trolley, in amperes	540	340

*=at 500v, **=at 600v

Annual Savings in New, Lightweight Peter Witt:	
energy consumption	$18,578
maintenance of car equipment	$19,658
maintenance of roadway	$ 6,770
reduction in crew expense, if cars operated in train	$ 2,422
Total annual saving	$47,428
Investment:	
purchase price, 15 new cars at $15,109 per car	$226,635
annual savings on gross investment	19.36%

Source: P.V.C., "Light Weight Cars for Interurban Service," *Electric Traction* 18, no. 2 (Feb. 1922): 115–16.

Table 11.1. Rail Vehicle Orders

Year	*Streetcars*	*Interurbans*	*Freight*	*Locomotives*	*Total*
1907	3483	1327	1406	*	6216
1908	2208	727	176	*	3111
1909	2537	1245	1175	*	4957
1910	3571	990	820	*	5381
1911	2884	676	505	*	4065
1912	4531	783	687	*	6001
1913	3820	547	1147	*	5514
1914	2147	384	479	*	3010
1915	2072	336	374	*	2782
1916	3046	374	491	31	3942
1917	1998	185	223	49	2455
1918	1842	255	278	44	2419
1919	2129	128	172	18	2447
1920	2889	227	465	17	3598
1921	1059	129	81	7	1276
1922	2910	187	405	34	3536
1923	2915	427	595	92	4029
1924	1985	538	1538	31	4092
1925	1054	320	238	47	1659
1926	1249	309	264	60	1882
1927	824	121	363	40	1348
1928	601	93	171	32	897
1929	963	79	137	77	1256
1930	562	47	18	24	651

* Included in freight.

Source: Electric Railway Journal 75, no. 1 (Jan. 1931): 43.

Table 11.2. Rail Vehicle Orders, 1930s

Year	*Streetcars and Interurbans*	*Rapid Transit Cars*	*Total*
1930	608	0	608
1931	55	500	555
1932	65	0	65
1933	62	0	62
1934	48	0	48
1935	100	651	751
1936	399	176	575
1937	342	300	642
Total	1679	1627	3306

Source: Transit Journal 82, no. 1 (Jan. 1938): 19.

Table 11.3. Vehicles in Service, 1937

	Streetcars and Interurbans	*Rapid Transit*	*Trolley Buses*	*Buses*	*Total*
Cities over 500,000	18418	11283	342	7417	37460
Cities, 100–500,000	10455	0	1184	10041	21680
Cities, 25–100,000	2450	0	136	5331	7917
Under 25,000	320	0	0	979	1299
Interurban areas	2547	0	0	1846	4393
U.S. possessions	60	0	30	61	151
Canada	3614	0	7	550	4171
Total	37864	11283	1699	26225	77071

Source: Transit Journal 82, no. 1 (Jan. 1937): 15.

Table 11.4. Route Miles, 1937

	Streetcars and Interurbans	*Rapid Transit*	*Trolley Buses*	*Buses*	*Total*
Cities over 500,000	6773	1362	148	2987	11270
Cities, 100–500,000	6243	0	895	8687	15825
Cities, 25–100,000	1942	0	141	7275	9358
Under 25,000	314	0	0	2652	2966
Interurban areas	7578	0	0	8554	16132
U.S. possessions	23	0	14	39	76
Canada	1952	0	5	448	2405
Total	24825	1362	1203	30642	58032

Source: Transit Journal 82, no. 1 (Jan. 1937): 17.

Table 11.5. Electric Railway Presidents Conference Committee Membership

Operating Companies

Baltimore Transit Co.	New Orleans Public Services Inc.
Birmingham Electric Co.	Northern Texas Traction Co. (Fort Worth)
Boston Elevated Railway Co.	Omaha & Council Bluffs Street Railway Co.
Brooklyn & Queens Transit Corp.	Pacific Electric Railway Co. (Los Angeles)
Capital Transit Co. (Washington, D.C.)	Railway Equipment & Realty Co.
Chicago Rapid Transit Co.	Key System (Oakland, Calif.)
Chicago Surface Lines	East Bay Street Railways
Cincinnati & Lake Erie Railroad Co.	East Bay Transit Co.
Cleveland Railway Co.	Philadelphia & Western Railway Co.
Honolulu Rapid Transit Co.	Pittsburgh Railways Co.
Houston Electric Co.	Public Service Coordinated Transport (New Jersey)
Los Angeles Railway Corp.	St. Louis Public Service Co.
Louisville Railway Co.	Tennessee Public Service Co.
Memphis Street Railway Co.	Toronto Transportation Commission
Milwaukee Electric Railway & Light Co.	Virginia Electric & Power Co.
Montreal Tramways Co.	

Source: Seymour Kashin and Harre Demoro, *An American Original: The PCC Car* (Glendale, Calif.: Interurban Press, 1986), 43.

Table 11.6. Electric Railway Presidents Conference Committee Membership

Original Manufacturer Members

Aluminum Co. of America	Pantasote Co., Inc.
American Brake Shoe & Foundry Co.	Pullman-Standard Car Manufacturing Co.
Bethlehem Steel Co.	Railway Utility Co.
J. G. Brill Co.	S.K.F. Industries, Inc.
Carnegie-Illinois Co.	Standard Street Works Co.
Cincinnati Car Corp.	St. Louis Car Co.
Consolidated Car-Heating Co., Inc.	Timken-Detroit Axle Corp.
Cummings Car and Coach Co.	Tool Steel Gear & Pinion Co.
Electric Service Supplies Co.	Tuco Products Corp.
General Electric Co.	Twin Coach Corp.
Gold Car Heating & Lighting Co.	Westinghouse Electric & Manufacturing Co.
National Pneumatic Co.	Westinghouse Traction Brake Co.
Ohio Brass Co.	

Source: Seymour Kashin and Harre Demoro, *An American Original: The PCC Car* (Glendale, Calif.: Interurban Press, 1986): 43.

Table 11.7. Reorganization of the J. G. Brill Company, 1931

Eastern Region

District One: New England
District Two: New York (except New York City), New Jersey (except Camden and Newark), Pennsylvania (east of Johnstown, except Philadelphia)
District Three: New York City, Camden (N.J.), Newark (N.J.)
District Four: Kentucky, Michigan (except northwestern), Ohio, Pennsylvania (west of Johnstown), West Virginia, part of Indiana
District Five: Alabama, Delaware, Florida, Georgia, Maryland, Mississippi, North Carolina, South Carolina, Tennessee (except Memphis), Virginia, Washington, D.C., Philadelphia, Chester (Pa.)
Headquarters: District One — J. G. Brill Co. of Massachusetts (formerly Wason)
District Two — J. G. Brill Co. (main plant, Philadelphia)
District Three — J. G. Brill Co. (main plant, Philadelphia)
District Four — J. G. Brill Co. of Ohio (formerly Kuhlman)
District Five — J. G. Brill Co. (main plant, Philadelphia)

Western Region

District One: Illinois (except East St. Louis), Iowa, Michigan (northwestern), Minnesota, Nebraska, North Dakota, South Dakota, Wisconsin, part of Indiana
District Two: Arkansas, Kansas, Louisiana, Missouri, New Mexico, Memphis (Tenn.), East St. Louis (Ill.)
District Three: Arizona, California, Idaho, Montana, Nevada, Oregon, Utah, Washington, Wyoming
Headquarters: District One — Chicago
District Two — J. G. Brill Co. of Missouri (formerly American)
District Three — San Francisco

NOTES

1. An Introduction to the Street Railway Industry

1. The horse to car ratio figures given may be conservative. William Middleton puts the number of horses per car at 5 to 10. The ratio varied from city to city, depending on the terrain and type of service. For instance, Boston's Metropolitan Railroad required 3,600 horses for its fleet of 700 cars. Troy, New York, required 425 horses for 46 cars. Not only were manure and urine a serious public health hazard; urine also was capable of deteriorating lightly built track. George W. Hilton, *The Cable Car in America: A New Treatise upon Cable or Rope Traction as Applied to the Working of Street and Other Railways,* rev. ed. (San Diego: Howell-North Books, 1982), 15; *History of Public Works in the United States, 1776–1976,* ed. Ellis L. Armstrong (Chicago: American Public Works Association, 1976), 164; Clay McShane, *Technology and Reform: Street Railways and the Growth of Milwaukee, 1887–1900* (Madison: Department of History, University of Wisconsin), 6–7; William D. Middleton, *The Time of the Trolley: The Street Railway from Horsecar to Light Rail* (San Marino, Calif.: Golden West Books, 1987), 24; Frank Rowsome Jr., *Trolley Car Treasury: A Century of American Streetcars, Horsecars, Cable Cars, Interurbans, and Trolleys* (New York: Bonanza Books, 1956), 24–25, 26–27.

2. "Report of the Committee on Street-Railway Motors Other Than Animal, Cable, and Electric," *Verbatim Report of the Eighth Annual Meeting of the American Street-Railway Association* (1889): 92–101 (quote taken from 92–93).

3. Eric Schatzberg, "The Mechanization of Urban Transit in the United States," in *Technological Competitiveness: Contemporary and Historical Perspectives on the Electrical, Electronics, and Computer Industries,* ed. William Aspray (Piscataway, N.J.: IEEE Press, 1993), 225–42; Middleton, *Time of the Trolley,* 27, 30, 32, 34; Hilton, *Cable Car in America,* 13; *History of Public Works,* 165; McShane, *Technology and Reform,* 7–8; Rowsome, *Trolley Car Treasury,* 35–36.

4. "Report of the Committee on the Cable System of Motive Power," *Verbatim Report of the Third Annual Meeting of the American Street-Railway Association* (1884), 145–57; "Report of the Committee on the Cable System of Motive Power," *Verbatim Report of the Fourth Annual Meeting of the American Street-Railway Association* (1885), 98–99; Clay McShane, *Down the Asphalt Path: The Automobile and the American City* (New York: Columbia University Press, 1994), 28; Schatzberg, "Mechanization," 231; Middleton, *Time of the Trolley,* 49–51, 54; Hilton, *Cable Car in America,* 6–7, 81–82, 92–96, 100, 103, 106–8, 129, 149–51, 155; *History of Public Works,* 166–67; McShane, *Technology and Reform,* 8–11; Rowsome, *Trolley Car Treasury,* 49–50, 56–57, 60; Harold C. Passer,

Frank Julian Sprague, 1857–1934: Father of Electric Traction (Cambridge: Harvard University Press, 1952), 214–15; Guy Hecker, "The History of Urban Transport," in *Principles of Urban Transportation,* ed. Frank Homer Mossman (Cleveland: Western Reserve University Press, 1951), 1–2; John Anderson Miller, *Fares, Please! From Horsecars to Streamliners* (New York: D. Appleton-Century, 1941), 46.

5. Tarr's investigation of Pittsburgh, however, showed that although the introduction of cable railways resulted in a significant increase in ridership, there was "in most cases . . . no mileage change." Joel A. Tarr, "Transportation Innovation and Changing Spatial Patterns in Pittsburgh, 1850–1934," *Essays in Public Works History* 6 (Apr. 1978): 14, 16.

Another study that addresses the impact of transportation technology on urban growth was conducted by John Adams. Through his study of residential patterns in midwestern cities, Adams argues that it was not until the development of the electric street railway that residential neighborhoods were able to expand outward and away from already developed areas. John S. Adams, "Residential Structure of Midwestern Cities," in *Internal Structure of the City: Readings on Urban Form, Growth, and Policy,* 2d ed., ed. Larry S. Bourne (New York: Oxford University Press, 1982), 173–74, 177.

6. "Report of the Committee on the Progress of Electricity as a Motive Power," *Verbatim Report of the Fourth Annual Meeting of the American Street-Railway Association* (1885), 68–71; "Report of the Committee on the Progress of Electricity as a Motive Power," *Verbatim Report of the Fifth Annual Meeting of the American Street-Railway Association* (1886), 79–80, 82; Frank J. Sprague, "Growth of Electric Railways," *AERA* 5, no. 3 (Oct. 1916): 258–59; J. W. Hammond, "Making the Wheels Turn Faster," *Mass Transportation* 3, no. 7 (July 1935): 211; Middleton, *Time of the Trolley,* 55–57; Miller, *Fares, Please!* 59–80; Victor S. Clark, *History of Manufactures in the United States,* vol. 2, *1860–1893* (New York: Peter Smith, 1949), 379.

7. Report of the Committee on the Progress of Electricity as a Motive Power" (1885), 68–71 and (1886), 79–80, 82; Hammond, "Making the Wheels Turn Faster," 211; Schatzberg, "Mechanization," 235; Middleton, *Time of the Trolley,* 54–55, 57–64; Miller, *Fares, Please!* 55–62; Hecker, "History of Urban Transportation," 2–3; Clark, *History of Manufactures,* 379.

Although no mileage figures were given, the following construction costs were cited at the 1889 ASRA convention: $840,000 for a cable car line and $190,000 for an electric streetcar line (if using overhead power lines, this cost was reduced to $175,000 if storage batteries were used). See "Report on the Conditions Necessary to the Financial Success of Electricity as a Motive Power," *Verbatim Report of the Eighth Annual Meeting of the American Street-Railway Association* (1889), 63–64.

8. It is uncertain whether or not Sprague visited Siemens's early installations. He certainly knew of them and was able to witness the latest developments in European street railway technology at the British Electrical Exhibition held at the Crystal Palace in 1882. Sprague, "Growth of Electric Railways," 260–61; Middleton, *Time of the Trolley,* 66–68; Rowsome, *Trolley Car Treasury,* 81–84; Harold C. Passer, *The Electrical Manufacturers, 1875–1900: A Study in Competition, Entrepreneurship, Technical Change, and Economic Growth* (Cambridge: Harvard University Press, 1953), 237–40; Passer, *Sprague,* 216–22; Miller, *Fares, Please!* 62–63.

9. McShane, *Technology and Reform,* 15. See also Middleton, *Time of the Trolley,* 220; Passer, *Electrical Manufacturers,* 245–46; Passer, *Sprague,* 226.

10. Sprague's initial mounting later became known as "double reduction" gearing. All this meant was that there was one gear between the axle and pinion gears. The gears were of different sizes, reducing the high number of revolutions from the motor to something the car's axles could manage. Sprague favored "single reduction" gearing (where the pinion rested *directly* on the axle gear), but he needed double reduction gearing for his first system in Richmond, Virginia, in order to ascend some difficult hills. Single reduction gearing ultimately became the industry standard. Sprague, "Growth of Electric Railways," 270–74; *Electric Railway Journal* 75, no. 10 (15 Sept.

1931): 520; Middleton, *Time of the Trolley,* 67; Rowsome, *Trolley Car Treasury,* 84, 86; *Public Works,* 169; Passer, *Electrical Manufacturers,* 245–46; Passer, *Sprague,* 226.

11. McShane and Schatzberg point out (correctly) that, although the Richmond system was successful from a *technological* point of view, it was a commercial failure. Sprague built the Richmond system at a tremendous financial loss. Only a few years after it opened, Richmond went back to using horse cars. Apparently the new technology was beyond the capability of the Richmond railway's maintenance men. *Electric Railway Journal* 75, no. 10 (15 Sept. 1931): 520; Schatzberg, "Mechanization," 234–35; McShane, *Technology and Reform,* 15–16, 18–19. Independent confirmation of this can be found in the Frank Seiberling Papers at the Ohio Historical Society. Writing to J. F. Seiberling from Richmond on 14 October 1889, a Mr. Wise wrote the following about the street railway system: "The action in Richmond has no reference whatever to the Sprague system or even to Electric propulsion—the part whose miserable management is to be overhauled has made a worse botch of a street railway propelled *by mules* than of the Electric road. Both lines are in miserable plight." Folder 1, box 1, Frank Seiberling Papers, MSS 347, Ohio Historical Society, Columbus.

12. Hilton cites conservative estimates made in 1889 for a hypothetical 10-mile system-cable would have cost $840,000, whereas electric would have cost $190,000. Even though electric streetcars required heavier rails than either cable or horse cars, the price of heavier rail, roadbed, and all of the overhead wire would only have cost about $100,000, as opposed to $700,000 for a cable conduit. Similar economy could be achieved in powerhouse construction, which cost $30,000 for electric and $125,000 for cable. The only advantage cable railways held over electric was the cars themselves: 15 cable cars would have cost $15,000, while 15 electric cars would have cost four times as much (mostly due to motors and control systems). Hilton, *Cable Car in America,* 153, 155–56.

13. K-controllers included two sets of control positions that enabled motormen to cut resistance out of the motor circuit. The first set of positions connected the motors in *series* (current passed through each motor successively). When all of the resistance was cut out, the motors were said to be in "full series," roughly half the car's total speed. Cars in full series generally traveled between 15 and 25 miles per hour.

The second set of positions reintroduced resistance to the motor circuit but connected the motors in *parallel* (current reached each motor simultaneously). When all of the resistance was cut out, the car was said to be in "full parallel" or "on the post." Cars in full parallel traveled at their top speed, which was usually 25 to 45 miles per hour. Because of their control arrangement, K-controllers were sometimes referred to as "series-parallel" controllers. *Street Railway Journal* 9 no. 7 (July 1893): 480–81; Sydney W. Ashe and J. D. Keiley, "Rolling Stock," in *Electric Railways Theoretically and Practically Treated* (New York: D. Van Nostrand, 1905), 1:102–4, 115–20; Samuel Sheldon and Erich Hausmann, *Dynamo Electric Machinery: Its Construction, Design, and Operation: Direct-Current Machines* (New York: D. Van Nostrand, 1910), 251–56; Daniel H. Braymer, *Armature Winding and Motor Repair* (New York: McGraw-Hill, 1920), 355–58; Frank D. Graham, *Audel's New Electric Library,* vol. 2, *Dynamos, DC Motors: Construction, Installation, Maintenance, and Trouble Shooting* (New York: Theo. Audel, 1929), 826–30; John H. Hanna, "Evolution of Community Transportation," *Electric Railway Journal* 75, no. 10 (15 Sept. 1931): 499; *Electric Railway Journal* 75, no. 10 (15 Sept. 1931): 522; Chester L. Dawes, *Direct Current Machines,* vol. 1 of *A Course in Electrical Engineering* (New York: McGraw-Hill, 1927), 400–402; Rodney Hitt, *Electric Railway Dictionary* (1911; reprint, Novato, Calif.: Newton K. Gregg, 1972), 18, 265–68; Hilton, *Cable Car in America,* 45, 156.

14. McShane, *Technology and Reform,* 4 and *Down the Asphalt Path,* 28.

15. Jon C. Teaford, *The Twentieth-Century American City,* 2nd ed. (Baltimore: Johns Hopkins University Press, 1993), 20–21; Sam Bass Warner Jr., *The Private City: Philadelphia in Three Periods of Its Growth,* 2nd ed. (Philadelphia: University of Pennsylvania Press, 1987), 192–93.

16. George W. Hilton and John F. Due, *The Electric Interurban Railways in America* (Stanford: Stanford University Press, 1960), xvii, 9, 91–92.

17. *Street Railway Journal* 17 (1901): supplement.

18. *Electric Railway Journal* 38, no. 8 (19 Aug. 1911): 311–14.

19. "The History of the American Street-Railway Association," *Street Railway Journal* 24, no. 15 (8 Oct. 1904): 517–42; *Electric Railway Journal* 36, no. 15D (14 Oct. 1910): 806; *Electric Traction* 27, no. 9 (Sept. 1931): 414; *Mass Transportation* 31, no. 10 (Oct. 1935): 334.

20. *Summary Index of Proceedings of the American Transit Association, American Transit Accountants' Association, American Transit Claims Association, American Transit Engineering Association, American Transit Operating Association and Their Predecessors, 1882–1934* (New York: American Transit Association, 1935).

21. *Street Railway Journal* 24, no. 17 (22 Oct. 1904): 771. In his presidential address to the AERA in 1915, C. Loomis Allen remarked on the Manufacturers' Association: "Our own association has never fully recognized the worth, the strength, and the help that [manufacturers] can bring to the industry." C. Loomis Allen, "The Industry and the Association," *Electric Railway Journal* 46, no. 15 (9 Oct. 1915): 704.

22. Hitt, *Electric Railway Dictionary,* 40.

23. The *Electric Railway Journal* defined the Midwest as comprising the following states: Illinois, Indiana, Iowa, Kentucky, Michigan, Minnesota, Montana, Ohio, and Wisconsin.

24. Hilton and Due, *Interurban Railways,* 253, 275, 309.

2. Car Builders of Ohio

1. Walter Licht, *Industrializing America: The Nineteenth Century* (Baltimore: Johns Hopkins University Press, 1995), 63–67, 103, 124–25, 128–29, 133; Jon C. Teaford, *Cities of the Heartland: The Rise and Fall of the Industrial Midwest* (Bloomington: Indiana University Press, 1993), 59.

2. George W. Knepper, *Ohio and Its People,* 2nd ed. (Kent, Ohio: Kent State University Press, 1997), 5–7; Licht, *Industrializing America,* 111–12, 124; Teaford, *Cities of the Heartland,* 2, 4–5, 17–18, 20; Kenneth Warren, *The American Steel Industry, 1850–1970: A Geographical Interpretation* (Oxford: Clarendon Press, 1925), 25; John Merrill Weed, "Business as Usual," in *Ohio in the Twentieth Century, 1900–1938,* comp. Harold Lindley, vol. 6 of *The History of the State of Ohio,* ed. Carl Wittke (Columbus: Ohio State Archaeological and Historical Society, 1941–44), 159.

3. Allan R. Pred, *Urban Growth and the Circulation of Information: The United States System of Cities, 1790–1840* (Cambridge: Harvard University Press, 1973), 132, 137–38; Hugh Allen, *Rubber's Home Town: The Real Life Story of Akron* (New York: Stafford House, 1949), 35–37, 53; David W. Bowman, *Pathway of Progress: A Short History of Ohio* (New York: American Book, 1943); Joseph Butler Jr., *History of Youngstown and the Mahoning Valley,* vol. 1 (Chicago: American Historical Society, 1921), 178, 180–84; H. Roger Grant, *Ohio on the Move: Transportation in the Buckeye State* (Athens: Ohio University Press, 2000), 55–63; Karl H. Grismer, *Akron and Summit County* (Akron: Summit County Historical Society, 1952), 116; Knepper, *Ohio and Its People,* 150–54; and Francis McGovern, *Written on the Hills: The Making of the Akron Landscape* (Akron: University of Akron Press, 1996), 67.

4. Allan R. Pred, *Urban Growth and City Systems in the United States, 1840–1860* (Cambridge: Harvard University Press, 1980), 105, 108; Carl W. Condit, *The Railroad and the City: A Technological and Urbanistic History of Cincinnati* (Columbus: Ohio State University Press, 1977), 3–6, 15, 46–48; Federal Writers' Project, *Cincinnati: A Guide to the Queen City and Its Neighbors* (Cincinnati: Wiesen-Hart Press, 1943), 53–57, 60; Federal Writers' Project, *They Built a City: 150 Years of Industrial Cincinnati* (Cincinnati: Cincinnati Post, 1938), 3, 22, 25–26, 31–32, 34–39.

5. A. W. Drury, *History of the City of Dayton and Montgomery County Ohio* (Chicago: S. J. Clark, 1909), 1:136, 155–56.

6. Carol Poh Miller and Robert A. Wheeler, *Cleveland: A Concise History,* 2nd ed. (Bloomington: Indiana University Press, 1997), 7–9, 15, 17; Licht, *Industrializing America,* 111–14, 124–26; Teaford, *Cities of the Heartland,* 34, 103; Darwin H. Stapleton, "The City Industrious: How Technology Transformed Cleveland," in *Birth of Modern Cleveland,* 74–83; Ronald R. Weiner and Carol A. Beal, "The Sixth City: Cleveland in Three Stages of Urbanization," in *The Birth of Modern Cleveland, 1865–1930,* ed. Thomas F. Campbell and Edward M. Miggins (Cleveland: Western Reserve Historical Society, 1988), 24, 39, 44; David R. Meyer, "Emergence of the American Manufacturing Belt," *Journal of Historical Geography* 9, no. 2 (Apr. 1983): 156; Allen, *Rubber's Home Town,* 10–12; Federal Writers' Project, *Guide to the Queen City,* 70–73; Philip D. Jordan, *Ohio Comes of Age, 1873–1900,* vol. 5 of *The History of the State of Ohio,* ed. Carl Wittke (Columbus: Ohio State University Press, 1943), 222–23; Federal Writers' Project, *They Built a City,* 178–85; William H. Gregory and William B. Guitteau, *History and Geography of Ohio,* rev. ed. (Boston: Ginn, 1935), 274.

7. Gregory and Guitteau, *History and Geography of Ohio,* 243, 257; Vernon Henderson, "Medium Sized Cities," *Regional Science & Urban Economics* 27, no. 6 (Nov. 1997): 592; Jordan, *Ohio Comes of Age,* 226–28, 418–21; Brian Page and Richard Walker, "From Settlement to Fordism: The Agro-Industrial Revolution in the American Midwest," *Economic Geography* 67, no. 4 (Oct. 1991): 303, 307; Warren, *American Steel Industry,* 111, 118, 171, 207, 216; Weed, "Business as Usual," 169.

8. Charlotte Reeve Conover, *The Story of Dayton* (Dayton: Greater Dayton Association, 1917), 146–47, 212–14.

9. *Street Railway Journal* 10, no. 2 (Feb. 1894): 263; no. 8 (Aug. 1894): 518–19; 13, no. 12 (Dec. 1897): 862–63; *Street Railway Review* 3, no. 10 (Oct. 1893): 660; *Williams' Dayton Directory for 1894–95* (Cincinnati: Williams, 1894), 113; Scott D. Trostel, *The Barney & Smith Car Company: Car Builders* (Fletcher, Ohio: Cam-Tech, 1993), 11–14, 22–23, 25–26, 35–37, 87–98, 104–5, 129–34, 148; Michael W. Williams, "Czar of Kossuth: J. D. Moskowitz," *Timeline* 8, no. 5 (Oct./Nov. 1991): 22–25, 31; Carl M. Becker, "A 'Most Complete' Factory: The Barney Car Works," *Cincinnati Historical Society Bulletin* 31, no. 1 (Spring 1973): 49–51, 54–57, 62–64; Warren H. Deem, "The Barney & Smith Car Company: A Study of Business Growth and Decline," paper prepared for course in American economic history, Harvard University (1953), 33 (copy at Ohio Historical Society); Drury, *History of Dayton,* 1:610, 613–15.

10. *Street Railway Journal* 4, no. 9 (Sept. 1888): 243.

11. *Street Railway Journal* Index to Advertisers, 4 (1888): 2–3; no. 1 (Jan. 1888): 54; no. 4 (Apr. 1888): 6, 24–25, 28; no. 9 (Sept. 1888): 243.

12. *Street Railway Journal* 8, no. 6 (June 1892): 384.

13. J. G. Brill Company Records, 1877–1930, series 9, order books 1887–1940, Historical Society of Pennsylvania; *Cleveland City Directories* 1881: 306, 1883: 327–28, 1886: 354, 1897: 594, 1902: 700, 1905: 794; *The Industries of Cleveland: A Resume of the Mercantile and Manufacturing Progress of the Forest City, Together with a Condensed Summary of Her Material Development and History and a Series of Comprehensive Sketches of Her Representative Business Houses* (Cleveland: Elstner, 1888), 85; Sanborn-Perris Map Co., *Cleveland, Ohio* (New York, 1896), 4:480; Sanborn Map Co., *Insurance Map of Cleveland, Ohio* (New York, 1903), 4:551, 524; *Street Railway Journal* 12, no. 1 (Jan. 1896): 74; 16, no. 5 (3 Feb. 1900): 76; no. 17 (5 May 1900): 75; 17, no. 10 (9 March 1901): vi; no. 21 (25 May 1901): viii; no. 26 (29 June 1901): ix; 18, no. 14 (5 Oct. 1901): 520–21; 19, no. 13 (27 Apr. 1902): vi; 24, no. 14 (1 Oct. 1904): 498; 26, no. 1 (1 July 1905): 25–26; no. 4 (22 July 1905): 151; no. 16 (14 Oct. 1905): 742; 27, no. 14 (7 Apr. 1906): 581; *Street Railway Review* 6, no. 1 (15 Jan. 1896): 39; 11, no. 9 (15 Sept. 1901): 599; G. C. Kuhlman Car Co., *The G. C. Kuhlman Car Company: Designers and Builders of Electric Cars* (Collinwood, Ohio, 1903?), 1; *Electric Traction* 3, no. 9 (28 Feb. 1907): 208; *Electric Railway Journal* 46, no. 15 (Oct. 1915): 786; 61, no. 9 (3 March 1923): 392; Betsy Hunter Bradley, *The Works: The Industrial Architecture of the United States* (New York:

Oxford University Press, 1999), 65–66; William D. Middleton, *Time of the Trolley* (San Marino, Calif.: Golden West Books, 1987), 224–25, 227, 229; William Ganson Rose, *Cleveland: The Making of a City* (Cleveland: World, 1950), 438; Weiner and Beal, "The Sixth City," 46.

14. *Street Railway Review* 3, no. 9 (Sept. 1893): 571; *Street Railway Journal* 9, no. 10 (Oct. 1893): 676 (quote taken from the *Journal*).

15. Local historian E. M. P. Brister claimed the Jewett's erecting shop could handle 125 60-foot interurbans. This would have totaled 7,500 feet of track space while the original shop had only 1,600 feet. A later addition had *fewer* tracks than the original shop. E. M. P. Brister, *Centennial History of Newark and Licking County, Ohio* (Columbus: S. J. Clark, 1909; reprint, Defiance, Ohio: Hubbard, 1982), 524.

Folder 4, Jewett Car Company Records, MSS 971, Ohio Historical Society, Columbus; *Street Railway Journal* 9, no. 9 (Sept. 1893): 571; no. 10 (Oct. 1893): 676; 11, no. 5 (May 1895): 345; 13, no. 10 (Oct. 1897): 713; 15, no. 11 (Nov. 1899): 835; 16, no. 13 (4 Aug. 1900): 758–59; 18, no. 14 (5 Oct. 1901): 502; 19, no. 1 (4 Jan. 1902): 60; 22, no. 23 (5 Dec. 1903): 990–91; 24, no. 3 (16 July 1904): 92–93; 25, no. 15 (15 Apr. 1905): 699; *Street Railway Review* 5, no. 4 (15 Apr. 1895): 254; 8, no. 2 (Feb. 1898): 132; 10, no. 2 (3 Feb. 1900): 90; *Electric Railway Journal* 35, no. 1 (1 Jan. 1910): 58; *Electric Traction* 3, no. 28 (11 July 1907): 666; 6, no. 3 (15 Jan. 1910): 72.

16. *Street Railway Review* 5, no. 12 (Dec. 1895): 758; 6, no. 9 (15 Sept. 1895): 526; 13, no. 3 (March 1903): 172; *Street Railway Journal* 14, no. 2 (Feb. 1898): 77–80; 15, no. 5 (May 1899): 309; no. 8 (Aug. 1899): 508–9; 20, no. 19 (8 Nov. 1902): 784–86; 21, no. 5 (31 Jan. 1903): 173; no. 8 (21 Feb. 1903): 289–90; 27, no. 1 (6 Jan. 1906): 65; *McGraw Electrical Directory: Electric Railway Edition* (New York: McGraw, 1910), 106–7; *Electric Traction* 7, no. 8 (25 Feb. 1911): 221; 9, no. 12 (Dec. 1913): 798; *Electric Railway Journal* 41, no. 4 (25 Jan. 1913): 178; *Cincinnati City Directory* (1903), 355; Trostel, *Barney & Smith,* 139; George W. Hilton and John F. Due, *The Electric Interurban Railways in America* (Stanford: Stanford University Press, 1960), 26, 28–33; Federal Writers' Project, *Guide to the Queen City,* 421–23; Federal Writers' Project, *They Built a City,* 40–41; *Cincinnati: The Queen City, 1788–1912* (Cincinnati: S. J. Clark, 1912), 4:656–57.

17. *Warren and Niles City Directory* (1901), 249; *Street Railway Journal* 17, no. 24 (15 June 1901): 708; 19, no. 6 (8 Feb. 1902): xii; 23, no. 4 (23 Jan. 1904): 153; 28, no. 17 (27 Oct. 1906): 847; 30, no. 15 (12 Oct. 1907): 686–87; no. 16 (19 Oct. 1907): 822; *Street Railway Review* 11, no. 6 (15 June 1901): 395; 12, no. 2 (Feb. 1902): 102; *Electric Railway Journal* 32, no. 19A (13 Oct. 1908): 977; Niles Car & Manufacturing Co., *Niles Cars* (Niles, Ohio, n.d.; reprint, Caldwell, Idaho: Caxton Printers, 1982).

18. Charles Henry Davis, "Streetcar Building: III—Material (continued)," *Street Railway Journal* 16, no. 14 (7 Apr. 1900): 364.

19. Rodney Hitt, *Electric Railway Dictionary* (1911; reprint, Novato, Calif.: Newton K. Gregg, 1972), 31.

20. The greatest single source describing car construction during the wooden car period is an 11-part series written by Charles Henry Davis in the *Street Railway Journal* from February 1900 to January 1901. *Street Railway Journal* 11, no. 6 (June 1895): 388–89; W. E. Partridge, "Street Railway Rolling Stock: The Selection of Suitable Woods," 11, no. 10 (Oct. 1895): 636; Charles Henry Davis, "Streetcar Building: II—Material," 16, no. 9 (3 March 1900): 250; "Streetcar Building: III—Material (continued)," no. 14 (7 Apr. 1900): 361–67; John A. Brill, "Twenty Years of Carbuilding," 24, no. 15 (8 Oct. 1904): 562–65; *Electric Railway Journal,* 37, no. 7 (18 Feb. 1911): 303–4; "The Development of the Electric Railway Car," 48, annual convention section (30 Sept. 1916): 561–72; *Electric Traction* 3, no. 11 (14 March 1907): 261; Kuhlman, *Electric Cars,* 2; Middleton, *Time of the Trolley,* 108–9; Andrew D. Young, *St. Louis Car Company Album: A Photographic Record* (Glendale, Calif.: Interurban Press, 1984), 23–26; Alan R. Lind, *From Horsecars to Streamliners: An Illustrated History of the St. Louis Car Company* (Park Forest, Ill.: Transport History Press, 1978), 27; John H. White Jr., *The American Railroad Passenger Car,* part 1 (Baltimore: Johns Hopkins University Press, 1978), 38–43; Hitt, *Electric Railway Dictionary* 15, 20, 26, 33, 52, 58.

21. Folders 2 and 11, "Jewett Car Company Records," MSS 971, Ohio Historical Society, Columbus; "Cars for High Speed Service," *Street Railway Journal* 20, no. 14 (4 Oct. 1902): 538–49; Young, *St. Louis Album,* 10; Lind, *From Horsecars to Streamliners,* 7.

22. Folder 33, "Dayton, Covington & Piqua Traction Company Records, 1897–1927," MSS 346, Ohio Historical Society, Columbus.

23. Folder 2, "Jewett Car Company Records," MSS 971, Ohio Historical Society, Columbus.

24. *Street Railway Journal* 10, no. 11 (Nov. 1894): 741–42; 15, no. 11 (Nov. 1899): 835; 20, no. 16 (18 Oct. 1902): 681–82; 28, no. 17 (27 Oct. 1906): 839, 847; 30, no. 15 (12 Oct. 1907): 686–87; no. 16 (19 Oct. 1907): 822; *Electric Railway Journal* 32, no. 19A (13 Oct. 1908): 977; 41, no. 4 (25 Jan. 1913): 178.

25. There was some debate among street railway men about enclosed platforms. Even though some feared they would limit a motorman's field of vision, most streetcar vestibules were enclosed by 1910. Some railways used removable shells that could protect car crews in winter time but could be removed in warmer weather.

26. Victor S. Clark, *History of Manufacturers in the United States,* vol. 3, *1893–1928* (New York: Peter Smith, 1949), 136–37; Deem, "Barney & Smith," 38–39; White, *Railroad Passenger Car,* 38, 48–49; Seymour Kashin and Harre Demoro, *The PCC Car: An American Original* (Glendale, Calif.: Interurban Press, 1986), 111.

27. Williams, "Czar," 25; Becker, "A 'Most Complete' Factory," 66; Deem, "Barney & Smith," 34–38.

28. Hitt, *Electric Railway Dictionary,* 17, 50; Young, *St. Louis Album,* 23.

29. *Electric Railway Journal* 42, no. 20 (15 Nov. 1913): 1,082; 43, no. 1 (3 Jan. 1914): 64.

30. "Niles Cars: 1914," *Electric Railway Historical Society Bulletin,* no. 30 (n.d.): 13.

31. *Electric Railway Journal* 44, no. 14 (3 Oct. 1914): 600–601.

32. *Street Railway Journal* 11, no. 2 (Feb. 1895): 109–11; Trostel, *Barney & Smith,* 115, 119–25; Williams, "Czar," 25; Deem, "Barney & Smith," 35, 46–47.

3. Making the Cars Go

1. Cincinnati's Bullock Electric Manufacturing Company *did* make motors and control equipment into the early 1900s. However, this was always secondary to Bullock's principal market, electric power-generating stations. Bullock was acquired by Allis Chalmers in 1905, and all electrical manufacturing had been moved from Cincinnati to Milwaukee by 1910. See the following in *Street Railway Journal:* for Bullock's facilities and market, 17, no. 18 (4 May 1901): 77 and 18, no. 11 (14 Sept. 1901): 321; for a glimpse of its products, 22, no. 11 (14 Sept. 1903): 566 and 23, no. 19 (7 May 1904): 718; for early ties to and subsequent merger with Allis Chalmers, 24, no. 23 (3 Dec. 1904): 90 and 25, no. 6 (11 Feb. 1905): 295. See also Walter F. Petersen, *An Industrial Heritage: Allis Chalmers Corporation* (Milwaukee: Milwaukee County Historical Society, 1978), 129–32.

2. There were other methods of current collection used by the street railway industry, although the method described was by far the most common. Pantographs (sometimes spelled "pantagraphs") used wide, flat wire contacts supported by complex, diamond-shaped frameworks of metal tubing. Bow collectors used similar types of contacts, only with a framework that was less complex than the pantograph and behaved in similar fashion to a trolley pole. Cincinnati used two parallel trolley wires (and thus two poles) for current distribution.

Third rail systems placed an electrified rail next to the running rails. Impractical for city streets, third rails were used on some lines with protected rights-of-way and were popular with rapid transit railways and some interurbans. Conduit systems place the conductor in trenches below street level and between the running rails. Contact was made through a device called a "plow" that hung underneath the car and passed through a reinforced metal "slot" in street surface between the rails, in similar fashion to a cable car system.

Rodney Hitt, *Electric Railway Dictionary* (1911; reprint, Novato, Calif.: Newton K. Gregg, 1972), 17, 40, 58–59; William Schaake, "Pantagraph Trolley Development," *Electric Journal* 13, no. 10 (Oct. 1915): 483–88; *Electric Traction* 17, no. 11 (Nov. 1921): 793–804; William D. Middleton, *Time of the Trolley: The Street Railway from Horsecar to Light Rail* (San Marino, Calif.: Golden West Books, 1987), 221.

3. For a brief overview of current collection apparatus, see S. B. Stewart Jr., "Current Collecting Devices for Electric Railways," *General Electric Review* 16, no. 11 (Nov. 1913): 911–20; *Street Railway Journal,* 25, no. 9 (4 March 1905): 432; *Electric Railway Journal* 32, no. 19B (14 Oct. 1908): 1047; 36, no. 20 (12 Nov. 1910): 985; 37, no. 15 (15 Apr. 1911): 674; 39, no. 14 (6 Apr. 1912): 533; 43, no. 4 (24 Jan. 1914): 216; Hitt, *Electric Railway Dictionary,* 59; Middleton, *Time of the Trolley,* 221; Andrew D. Young, *St. Louis Car Company Album: A Photographic Record* (Glendale, Calif.: Interurban Press, 1984), 20, 23, 36, 54, 67; George W. Hilton and John F. Due, *The Electric Interurban Railways in America* (Stanford: Stanford University Press, 1960), 265–66.

4. *Street Railway Journal* 16, no. 41 (13 Oct. 1900): 1,019; 20, no. 3 (19 July 1902): 89–90; 22, no. 9 (29 Aug. 1903): 393; *Electric Railway Journal* 32, no. 19B (14 Oct. 1908): 1,023; 33, no. 25 (19 June 1909): 1,111; 42, no. 4 (26 July 1913): 140.

5. *Street Railway Journal* 22, no. 9 (29 Aug. 1903): 393; 37, no. 2 (14 Jan. 1911) 84.

6. *Street Railway Journal* 24, no. 15 (8 Oct. 1904): 671; 30, no. 21 (23 Nov. 1907): 1,069.

7. *Street Railway Review* 14, no. 4 (Apr. 1904): 269; *Electric Railway Journal* 43, no. 20 (16 May 1914): 1,126.

8. *Electric Railway Journal* 36, no. 15 (8 Oct. 1910): 666; 38, no. 5 (29 July 1911): 214.

9. *Street Railway Journal* 16, no. 41 (13 Oct. 1900): 1,015; *Electric Railway Journal* 38, no. 5 (29 July 1911): 214; 40, no. 14A (8 Oct. 1912): 678.

10. *Street Railway Journal* 30, no. 21 (23 Nov. 1907): 1,069; *Electric Railway Journal* 37, no. 18 (6 May 1911): 804.

11. Hitt, *Electric Railway Dictionary,* 59.

12. *Street Railway Journal* 31, no. 16 (18 Apr. 1908): 646.

13. *Street Railway Journal* 24, no. 5 (30 July 1904): 166; 28, no. 15 (13 Oct. 1906): xxix; 31, no. 14 (4 Apr. 1908): 573.

14. *Electric Railway Journal* 34, no. 14A (5 Oct. 1909): 667.

15. Quote taken from *Electric Traction Weekly* 3, no. 24 (13 June 1907): 571; *Street Railway Journal* 25, no. 9 (4 March 1905): 452; 28, no. 15 (13 Oct. 1906): 696; *Electric Railway Journal* 32, no. 19B (14 Oct. 1908): 1,043; George W. Knepper, *Ohio and Its People,* 2nd ed. (Kent, Ohio: Kent State University Press, 1997), 363; Philip D. Jordan, *Ohio Comes of Age, 1873–1900,* vol. 5 of *The History of the State of Ohio,* ed. Carl Wittke (Columbus: Ohio State University Press, 1943), 228–29; Opha Moore, "Ohio as a Manufacturing State," in Emilius O. Randall and Daniel J. Ryan, *History of Ohio: The Rise and Progress of an American State* (New York: Century History, 1912), 5:316–17; William M. Rockel, *20th Century of Springfield, and Clark County, Ohio and Representative Citizens* (Chicago: Biographical, 1908), 158, 172, 175, 180–82, 185–96, 422–23.

16. *Street Railway Journal* 18, no. 14 (5 Oct. 1901): 503; *Electric Railway Journal* 35, no. 4 (22 Jan. 1910): 158.

17. *Electric Railway Journal* 32, no. 19B (14 Oct. 1908): 994–95.

18. *Electric Railway Journal* 34, no. 22 (4 Dec. 1909): 1,125; 36, no. 27 (31 Dec. 1910): 1,285–86; 43, no. 14 (4 Apr. 1914): 779; Stewart, "Current Collecting Devices"; Hitt, *Electric Railway Dictionary,* 59.

19. *Electric Railway Journal* 36, no. 27 (31 Dec. 1910): 1,285–86; 39, no. 10 (9 March 1912): 399.

20. *Street Railway Journal* 28, no. 15 (13 Oct. 1906): xlviii.

21. *Electric Railway Journal* 43, no. 15 (11 Apr. 1914): 831–32.

22. *Electric Railway Journal* 32, no. 19 (10 Oct. 1908): 969; 33, no. 25 (19 June 1909): 1,110–13.

23. *Electric Railway Journal* 25, no. 14 (28 Jan. 1905): 148; 35, no. 12 (19 March 1910): 514; 40, no. 14A (8 Oct. 1912): 681; *Electric Traction* no. 7 (12 Feb. 1910): 175; Hitt, *Electric Railway Dictionary,* 286.

24. *Street Railway Journal* 21, no. 11 (14 March 1903): 421; 22, no. 11 (12 Sept. 1903): 549–50; no. 13 (26 Sept. 1903): 407; no. 16 (17 Oct. 1903): 742; 23, no. 11 (12 March 1904): 415–16; 25, no. 4 (28 Jan. 1905): 174; 28, no. 17 (27 Oct. 1906): 858; *Electric Railway Journal* 32, no. 19A (13 Oct. 1908): 978; no. 23 (7 Nov. 1908): 1,343; 33, no. 21 (22 May 1909): 957; 34, no. 14A (5 Oct. 1909): 668; 36, no. 15A (11 Oct. 1910): 701; 38, no. 15A (10 Oct. 1911): 716; 40, no. 14A (8 Oct. 1912): 683; *Electric Traction* 6, no. 6 (5 Feb. 1910): 147; 7, no. 39 (30 Sept. 1911): 1,183; 10, no. 3 (March 1914): 199; Knepper, *Ohio and Its People,* 363; Eugene H. Rosenboom and Francis P. Weisenburger, *A History of Ohio,* 2d rev. ed. (Columbus: Ohio Historical Society, 1967), 218; John Merrill Weed, "Business as Usual," in *Ohio in the Twentieth Century, 1900–1938,* comp. Harold Lindley, vol. 6 of *The History of the State of Ohio,* ed. Carl Wittke (Columbus: Ohio State Archaeological and Historical Society, 1941–44), 186; William M. Gregory and William B. Guitteau, *History and Geography of Ohio,* rev. ed. (Boston: Ginn, 1935), 177, 181–83, 255–56; Moore, "Ohio as a Manufacturing State," 5:324–26.

25. *Street Railway Journal* 26, no. 22 (1 June 2907): 981; *Electric Traction Weekly* 3, no. 20 (16 May 1907): 473; no. 26 (27 June 1907): 615; no. 27 (4 July 1907): 641.

26. *Electric Traction Weekly* 3, no. 46 (14 Nov. 1907): 1,187–88; *Electric Railway Journal* 32, no. 2 (13 June 1908): 106; 33, no. 1 (2 Jan. 1909): 39.

27. *Street Railway Journal* 28, no. 15 (13 Oct. 1906): 578; *Electric Railway Journal* 34, no. 14A (5 Oct. 1909): 664; Hitt, *Electric Railway Dictionary,* 286.

28. *Street Railway Journal* 25, no. 24 (14 June 1905): 1,077; 33, no. 26 (29 June 1909): 1,165.

29. *Street Railway Journal* 19, no. 1 (4 Jan. 1902): 57; *Electric Railway* 43, no. 6 (7 Feb. 1914): 322.

30. *Street Railway Journal* 19, no. 1 (4 Jan. 1902): 57.

31. *Street Railway Journal* 21, no. 14 (4 Apr. 1903): 535; *Electric Railway Journal* 43, no. 23 (6 June 1914): 1,289.

32. *Electric Traction Weekly* 3, no. 17 (25 Apr. 1907): 403; *Electric Traction* 7, no. 39 (30 Sept. 1911): 1,183.

33. *Electric Railway Journal* 42, no. 4 (26 July 1913): 148.

34. *Street Railway Journal* 16, no. 1 (6 Jan. 1900): 47; 25, no. 5 (4 Feb. 1905): 197; 27, no. 3 (20 Jan. 1906): 112–13; Frank Rowsome Jr., *Trolley Car Treasury: A Century of American Streetcars, Horsecars, Cable Cars, Interurbans, and Trolleys* (New York: Bonanza Books, 1956), 87.

35. Cincinnati's adoption of the dual overhead wire can be traced to two factors. First, the city was unduly concerned with the possibility of electrolysis (the degradation of underground metal pipes and utility conduit near a streetcar line caused by using the rails as the return circuit for the streetcars). Second, Cincinnati electrified its streetcar lines earlier than most cities and did so using a type of electric street railway system that required two overhead wires. Concern over electrolysis (however unfounded) combined with practical experience with a two-wire system led to Cincinnati Traction's unique method of current collection.

36. *Street Railway Journal* 16, no. 1 (6 Jan. 1900): 47; 20, no. 1 (5 July 1902): 58–59; *Electric Railway Journal* 32, no. 7 (18 July 1908): 331; no. 25 (21 Nov. 1908): 1,427; 40, no. 6 (10 Aug. 1912): 236; no. 8 (24 Aug. 1912): 306.

37. *Street Railway Journal* 27, no. 5 (3 Feb. 1906): 215; *Electric Railway Journal* 35, no. 4 (22 Jan. 1910): 158.

38. *Electric Railway Journal* 32, no. 22 (31 Oct. 1908): 1,292; 33, no. 5 (30 Jan. 1909): 201.

39. *Street Railway Journal* 23, no. 21 (21 May 1904): 759.

40. *Street Railway Journal* 25, no. 8 (25 Feb. 1905): 367–68; *Electric Railway Journal* 33, no. 22 (12 June 1909): 1,073–74; 34, no. 8 (21 Aug. 1909): 278; no. 14D (8 Oct. 1909): 799–800; 35, no. 9 (26 Feb. 1910): 361–63; 39, no. 26 (29 June 1912): 1,111–12; 41, no. 22 (31 May 1913): 955.

41. *Street Railway Journal* 16, no. 35 (1 Sept. 1900): 848–49; 18, no. 10 (15 Oct. 1901): 808; 21, no. 11 (14 March 1903): 444; no. 15 (11 Apr. 1903): 560; 25, no. 1 (7 Jan. 1905): 51–53; no. 10 (11 March 1905): 489; *Electric Railway Journal* 32, no. 19B (14 Oct. 1908): 1,018; 34, no. 14D (8 Oct. 1909): 799–800; 35, no. 9 (26 Feb. 1910): 361–63; 38, no. 15 (7 Oct. 1911): 672; *American Electric Railway Engineering Proceedings* (1912): 671–75; *Electrical Journal* 11, no. 10 (Oct. 1914): 544–47.

42. *Electric Railway Journal* 36, no. 10 (3 Sept. 1910): 357–59.

43. *Street Railway Journal* 24, no. 21 (19 Nov. 1904): 912; 27, no. 18 (5 May 1906): 702; 28, no. 7 (18 Aug. 1906): 261–63; no. 15 (13 Oct. 1906): 639; no. 24 (15 Dec. 1906): 1,115; *Electric Railway Journal* 32, no. 6 (11 July 1908): 269; no. 25 (22 Dec. 1906): 1,153; 35, no. 2 (8 Jan. 1910): 80–81; 38, no. 8 (19 Aug. 1911): 307–8.

44. *Electric Railway Journal* 24, no. 14 (1 Oct. 1904): 661–62, 668.

45. *Electric Railway Journal* 28, no. 25 (22 Dec. 1906): 1,161.

46. *Street Railway Journal* 16, no. 14 (7 Apr. 1900): 76; 20, no. 14 (4 Oct. 1902): 597–99; *Electric Railway Journal* 32, no. 19A (13 Oct. 1908): 978; 36, no. 15A (11 Oct. 1910): 701; 38, no. 8 (19 Aug. 1911): 337; no. 15 (7 Oct. 1911): 653.

47. There were other manufacturers of trucks and truck components in Cincinnati between 1900 and 1914. The Pollak Steel Company made axles, forgings, and heat-treated motor components for street railways, particularly in 1912. However, Pollak Steel was mostly a manufacturer of railroad axles and forgings—especially locomotive crank pins and connecting rods. See *Street Railway Journal* 40, no. 14A (8 Oct. 1912): 682.

The Moffett Electric Railway Bearing Company made roller bearings that could be used on street railway trucks. In 1905 it equipped some cars of the Toledo, Bowling Green & Southern Traction Company with roller bearings instead of conventional journal (friction) axle bearings. See *Street Railway Journal* 26, no. 7 (12 Aug. 1905): 242.

The Buckeye Steel Castings Company made truck frames around 1910, but was mostly known for the manufacture of railroad couplers. See Buckeye Steel Castings Company Records, 1883–1977, MSS 662, Ohio Historical Society, Columbus; Mansel G. Blackford, *A Portrait in Cast Steel: Buckeye International and Columbus, Ohio, 1881–1980* (Westport, Conn.: Greenwood Press, 1982), 25.

48. "Report of the Committee on Equipment," *Proceedings of the American Electric Railway Engineering Association* (1912): 671–75; *Electric Railway Journal* 39, no. 20 (18 May 1912): 860; 40, no. 11 (14 Sept. 1912): 436; 41, no. 1 (4 Jan. 1913): 54; W. L. Allen, "Recent Developments in Railway Motor Gearing," *Electrical Journal* 11, no. 10 (Oct. 1914): 544–47; *Electric Traction* 12, no. 9 (Sept. 1916): 693 (advertisement); 16, no. 11 (Nov. 1920): 917; 17, no. 5 (May 1921): 363; no. 6 (June 1921): 434; no. 7 (July 1921): 498; 18, no. 8 (Aug. 1922): 710; Middleton, *Time of the Trolley*, 122.

49. *Electric Railway Journal* 34, no. 14C (7 Oct. 1909): 766.

50. Samuel Sheldon and Erich Hausmann, *Dynamo Electric Machinery: Its Construction, Design, and Operation: Direct-Current Machines* (New York: D. Van Nostrand, 1910), 81–82; Hitt, *Electric Railway Dictionary*, 10; *Electric Railway Journal* 40, no. 13 (28 Sept. 1912): 491–94; Chester L. Dawes, *Direct Current Machines*, vol. 1 of *A Course in Electrical Engineering* (New York: McGraw-Hill, 1927), 301–2.

51. Cleveland's National Carbon plant is still in use as part of GrafTech International, Ltd., and is in excellent repair. Easily visible from the Red Line on Cleveland's Rapid at West 117th Street, the high, peaked roofs of the plant stand out among the smaller buildings in the neighborhood.

52. *The Industries of Cleveland: A Resume of the Mercantile and Manufacturing Progress of the Forest City, Together with a Condensed Summary of Her Material Development and History and a Series of Comprehensive Sketches of Her Representative Business Houses* (Cleveland: Elstner, 1888), 160; *Cleveland Directory* (1891), 605, (1896), 696, (1901), 836, (1906), 990, (1911), 1,110; *Street Railway Journal* 15, supplement (Oct. 1899): 15; 18, no. 4 (5 Oct. 1901): 522; 20, no. 16 (18 Oct. 1902):

677; 22, no. 1 (12 Sept. 1903): 552; *Electric Railway Journal* 34, no. 14A (5 Oct. 1909): 666–67; 36, no. 15A (11 Oct. 1910): 693, 700; 38, no. 13 (23 Sept. 1911): 518; no. 15 (7 Oct. 1911): 672.

53. *Cleveland City Directory* (1896), 447, (1901): 532, (1906): 627, (1911): 696; *Street Railway Journal* 25, no. 18 (6 May 1905): 839; Sheldon and Hausmann, *Dynamo Electric Machinery*, 46–47; Hitt, *Electric Railway Dictionary*, 16; Dawes, *Direct Current Machines*, 300–30.

4. Couplers

1. *Electric Railway Journal* 41, no. 17 (26 Apr. 1913): 763–64.

2. *Electric Railway Journal* 33, no. 20 (15 May 1909): 927.

3. Rodney Hitt, *Electric Railway Dictionary* (1911; reprint, Novato, Calif.: Newton K. Gregg, 1972), 19–20, 23; George W. Hilton and John F. Due, *The Electric Interurban Railways in America* (Stanford: Stanford University Press, 1960), 76–77.

4. Hitt, *Electric Railway Dictionary*, 23, 38–39, 57. For an overview of multiple unit systems available between 1910 and 1914, see the following: W. B. Kouwenhoven, "The Electrical Controller," *Railway and Locomotive Engineering* 20, no. 12 (Dec. 1907): 545–47 and 21, no. 1 (Jan. 1908): 21–23; "Report of the Committee on Control," *Proceedings of the American Street and Interurban Railway Engineering Association* (1908): 189–221; Karl A. Simmon, "Hand Operated Unit Switch Control," *Electric Journal* 7, no. 10 (Oct. 1910): 802–15; F. E. Case, "Development of the Sprague G-E Type M Multiple Unit Control," *General Electric Review* 16, no. 11 (Nov. 1913): 848–50. For a debate on the merits of trailer versus multiple-unit operation, see Clarence Renshaw, "Trailer Operation versus Multiple-Unit Trams," *Electric Journal* 8, no. 10 (Oct. 1911): 895–904.

5. Hitt, *Electric Railway Dictionary*, 19, 23.

6. *Electric Railway Journal* 25, no. 15 (15 Apr. 1905): 711–12; no. 23 (10 June 1905): 1,040; 26, no. 1 (1 July 1905): 37–38; no. 2 (8 July 1905): 48; 27, no. 5 (3 Feb. 1906): 172; 40, no. 11 (14 Sept. 1912): 405; no. 14B (9 Oct. 1912): 696–97, 709–10; 42, no. 15B (15 Oct. 1913): 775–77.

7. *Street Railway Journal* 28, no. 23 (8 Dec. 1906): 1,107; *Electric Railway Journal* 34, no. 6 (7 Aug. 1909): 213–15; 39, no. 3 (20 Jan. 1912): 88–93, 121–22; no. 22 (1 June 1912): 939; 40, no. 21 (30 Nov. 1912): 1,101–2; 42, no. 10 (6 Sept. 1913): 366; no. 17 (25 Oct. 1913): 935–36; 43, no. 7 (14 Feb. 1914): 367.

8. *Street Railway Journal* 26, no. 22 (25 Nov. 1905): 937; 28, no. 22 (1 Dec. 1906): 1,063–64; *Electric Railway Journal* 32, no. 19D (16 Oct. 1908): 1,138–39.

9. *Street Railway Journal* 26, no. 14 (30 Sept. 1905): 565–68; 30, no. 16 (19 Oct. 1907): 741–42; *Electric Railway Journal* 32, no. 19C (15 Oct. 1908): 1,059–60; 34, no. 1 (3 July 1909): 25–27; no. 14C (7 Oct. 1909): 754–55; 36, no. 1 (2 July 1910): 35–38; *American Street and Interurban Railway Engineering Association Proceedings* (1908): 189–221.

10. Frank J. Sprague, "The Multiple Unit System for Electric Railways," *Cassier's Magazine* 19 (Aug. 1899): 439–60; *Street Railway Journal* 16, no. 44 (3 Nov. 1900): 1,116; 17, no. 18 (4 May 1901): 537–54.

11. *Street Railway Journal* 18, no. 15 (12 Oct. 1901): 489; 19, no. 22 (31 May 1902): 664; 24, no. 14 (1 Oct. 1904): 479–81; *Electric Railway Journal* 40, no. 5 (3 Aug. 1912): 181–82.

12. *Street Railway Journal* 21, no. 20 (16 May 1903): 748–49; no. 26 (27 June 1903): 942; 22, no. 13 (26 Sept. 1903): 616–17; 29, no. 26 (29 June 1907): 1,160–67; *Electric Railway Journal* 34, no. 1 (3 July 1909): 25–27; 36, no. 19 (5 Nov. 1910): 955–57.

13. *Street Railway Journal* 24, no. 24 (10 December 1904): 1,050–51; 27, no. 22 (2 June 1906): 874.

14. *Street Railway Journal* 17, no. 22 (1 June 1901): 671; *Electric Railway Journal* 33, no. 14 (3 Apr. 1909): 638; 18, no. 25 (20 June 1914): 1,372.

15. *Electric Railway Journal* 32, no. 17 (26 Sept. 1908): 711; no. 19C (15 Oct. 1908): 1,056; 37, no. 18 (6 May 12911): 800–803; 38, no. 16 (14 Oct. 1911): 851–58; 39, no. 17 (27 Apr. 1912): 701; 40, no. 14E (12 Oct. 1912): 845–46.

16. The Central Electric Railway Association of the street railway industry should not be confused with the present-day CERA, which is the Central Electric Railfan's Association. The latter organization was founded during the 1940s and deliberately patterned its name after the former.

17. *Street Railway Journal* 28, no. 22 (1 Dec. 1906): 1,063–64; *Electric Railway Journal* 35, no. 23 (4 June 1910): 977–78; 36, no. 9 (27 Aug. 1910): 331; no. 11 (10 Sept. 1910): 403–4; 38, no. 3 (15 July 1911): 115–20; 42, no. 1 (5 July 1913): 29–30.

18. *Street Railway Journal* 28, no. 8 (25 Aug. 1906): 288; 30, no. 15 (12 Oct. 1907): 679, 682; 32, no. 4 (27 June 1908): 178; *Electric Railway Journal* 33, no. 14 (3 Apr. 1909): 638; no. 20 (15 May 1909): 927; 34, no. 14 (2 Oct. 1909): 609–11; 38, no. 5 (29 June 1911): 200.

19. The Chicago-based W. T. Van Dorn Company should not be confused with either the Van Dorn or Van Dorn & Dutton companies of Cleveland. They were not run by the same family.

20. *Street Railway Journal* 16, no. 41 (13 Oct. 1900): 1,020; 23, no. 13 (26 March 1904): 486; 26, no. 3 (15 July 1905): 119; no. 7 (12 Aug. 1905): 251–53; 30, no. 2 (13 July 1907): 74–75; *Electric Railway Journal* 34, no. 8 (21 Aug. 1909): 193–94; no. 9 (28 Aug. 1909): 327.

21. *Street Railway Journal* 30, no. 5 (3 Aug. 1907): 190–91; *Electric Railway Journal* 34, no. 14A (5 Oct. 1909): 664; 37, no. 25 (24 June 1911): 1,135.

22. The potential of the Tomlinson to make automatic electrical connections while coupling was recognized by some railways in advance of Ohio Brass's formal models with this feature. United Railways of St. Louis modified its Tomlinsons with homemade self-connecting electrical contacts in 1913.

23. *Street Railway Journal* 23, no. 20 (14 May 1904): 750–51; 28, no. 22 (1 Dec. 1906): 1,063–64; *Electric Traction Weekly* 3, no. 17 (25 Apr. 1907): 378–79; *Electric Traction* 7, no. 15 (15 Apr. 1911): 429; *Electric Railway Journal* 42, no. 15A (14 Oct. 1913): 739; John Merrill Weed, "The Traveled Ways," in *Ohio in the Twentieth Century, 1900–1938,* comp. by Harold Lindley, vol. 6, *The History of the State of Ohio,* ed. Carl Wittke (Columbus: Ohio State Archaeological and Historical Society, 1944), 146; John Merrill Weed, "Business as Usual," in *Ohio in the Twentieth Century,* 183–84; William M. Gregory and William B. Guitteau, *History and Geography of Ohio,* rev. ed. (Boston: Ginn, 1935), 253–54. Weed observed that "Mansfield was chosen [by Westinghouse] for the factory because it was thought best to locate the plant near the source of parts" (see "The Traveled Ways," 184).

5. Protecting the Public (and Themselves)

1. Randolph E. Bergstrom, *Courting Danger: Injury and Law in New York City, 1870–1910* (Ithaca, N.Y.: Cornell University Press, 1992), 167, 177, 180–81 (quotes from 177).

2. Ibid., 181.

3. For another description of claims departments, along with some key citations from the trade literature, see Barbara Young Welke, *Recasting American Liberty: Gender, Race, Law, and the Railroad Revolution, 1865–1920* (Cambridge: Cambridge University Press, 2001), 105–6. The American Electric Railway Claims Association, as this association was known from 1911 until the AERA reorganization in 1932, began in 1906 as the American Street and Interurban Railway Claims Association.

4. For further information on safety campaigns, see Welke, *Recasting American Liberty,* 35–39.

5. *Street Railway Journal* 17, no. 15 (13 Apr. 1901): 465.

6. *Street Railway Journal* 22, no. 23 (5 Dec. 1903): 994–99.

7. *Street Railway Journal* 32, no. 13 (29 Aug. 1908): 556–57.

8. Rodney Hitt, *Electric Railway Dictionary* (1911; reprint, Novato, Calif.: Newton K. Gregg, 1972), 43; "Dictionary of Electric Railway Material," *Street Railway Journal* 28, no. 15 (13 Oct. 1906):

lii; W. H. Evans, "Sand on Electric Cars," *Electric Railway Journal* 40, no. 2 (21 Sept. 1912): 459–60; *Electric Railway Journal* 1908, 1909, and 1911 directories (the latter included in the 7 Oct. issue).

9. Folder 8, Jewett Car Company Records, MSS 971, Ohio Historical Society, Columbus; *Electric Railway Journal* 32, no. 7 (26 Sept. 1908): 712, 725; 33, no. 8 (20 Feb. 1909): 355; no. 15 (3 Apr. 1909): 711; no. 21 (22 May 1909): 967; 36, no. 6 (6 Aug. 1910): 246; 37, no. 12 (25 March 1911): 544; no. 14 (8 Apr. 1911): 655; no. 18 (3 May 1913): 836; 41, no. 14 (5 Apr. 1913): 633; 42, no. 13 (27 Sept. 1913): 524; Lawrence A. Brough and James H. Graebner, *From Small Town to Downtown: A History of the Jewett Car Company, 1893–1919* (Bloomington: Indiana University Press, 2003), 21.

10. "Dictionary of Electric Railway Materials," lii; *Street Railway Journal* 28, no. 17 (27 Oct. 1906): 857.

11. *Electric Railway Journal* 43, no. 25 (20 June 1914): 1,409.

12. Ohio Brass Company, *Electric Railway and Mine Supplies: Catalogue No. 7* (Mansfield, Ohio, 1907): 261–70; *Street Railway Journal* 27, no. 2 (13 Jan. 1906): 98; 30, no. 16 (19 Oct. 1907): 818; *Electric Railway Journal* 39, no. 1 (6 June 1912): 54; 42, no. 15A (14 Oct. 1913): 747.

13. *Electric Railway Journal* 24, no. 24 (10 Dec. 1904): 1051–52.

14. *Electric Railway Journal* 32, no. 19A (13 Oct. 1908): 977.

15. "Dictionary of Electric Railway Materials," lii; *Street Railway Journal* 28, no. 8 (25 Aug. 1906): 303–4; no. 17 (27 Oct. 1906): 832.

16. *Electric Railway Journal* 35, no. 19 (7 May 1910): 808–9.

17. C. Dorticos, "Headlights for Interurban Service," *Electric Railway Journal* 38, no. 9 (26 Aug. 1911): 349–50.

18. *Street Railway Journal* 16, no. 2 (2 June 1900): 72; 23, no. 14 (2 Apr. 1904): 529.

19. *Street Railway Journal* 22, no. 11 (12 Sept. 1903): 552; 27, no. 5 (3 Feb. 1906): 215.

20. *Street Railway Journal* 28, no. 17 (27 Oct. 1906): 858; 32, no. 1 (6 June 1908): 49; *Electric Railway Journal* 32, no. 19A (13 Oct. 1908): 978; no. 4 (22 July 1911): 166; no. 15A (10 Oct. 1911): 716.

21. *Electric Railway Journal* 33, no. 20 (15 May 1909): 917; 36, no. 15A (11 Oct. 1910): 701.

22. *Street Railway Journal* 26, no. 25 (16 Dec. 1905): 1,072; no. 27 (30 Dec. 1905): 1,136–37.

23. Welke, *Recasting American Liberty,* 30–31.

24. *Electric Railway Journal* 33, no. 22 (29 May 1909): 1,007; Hitt, *Electric Railway Dictionary,* 41.

25. *Street Railway Journal* 6, no. 5 (May 1890): 216–17; *Electric Railway Journal* 33, no. 6 (6 Feb. 1909): 233–36; Hitt, *Electric Railway Dictionary,* 27.

26. *Electric Railway Journal* 33, no. 22 (29 May 1909): 995.

27. *Street Railway Journal* 21, no. 22 (30 May 1903): 814; Hitt, *Electric Railway Dictionary,* 159.

28. *Street Railway Journal* 22, no. 11 (12 Sept. 1903): 551–52; no. 23 (5 Dec. 1903): 1,003; 26, no. 15 (7 Oct. 1905): 714; 30, no. 16 (19 Oct. 1907): 826; *Electric Railway Journal* 32, no. 19A (13 Oct. 1908): 975; 33, no. 6 (6 Feb. 1909): 233–36; 34, no. 14A (5 Oct. 1909): 665; 35, no. 25 (18 June 1910): 1,072; 36, no. 15A (11 Oct. 1910): 698; 37, no. 24 (17 June 1911): 1,074; 38, no. 15A (10 Oct. 1911): 712; 39, no. 9 (2 March 1912): 378; 42, no. 15A (14 Oct. 1913): 567; *Scientific American* 99, no. 12 (19 Sept. 1908): 186; *Engineering News* 61, no. 8 (25 Feb. 1909): 220–21; Hitt, *Electric Railway Dictionary,* 159.

29. Welke, *Recasting American Liberty,* 30–33.

30. *Scientific American* 99, no. 12 (19 Sept. 1908): 186; *Electric Railway Journal* 33, no. 6 (6 Feb. 1909): 233–36; *Engineering News* 61, no. 8 (25 Feb. 1909): 220–22.

31. Welke, *Recasting American Liberty,* 33–34.

32. *Electric Railway Journal* 38, no. 9 (26 August 1911): 341.

33. For a highly detailed look at the design issues surrounding floor heights, see the 1911 report of the Chicago Traction Board of Supervising Engineers. This report explores no fewer than 14 "elements of design" and 4 "elements of variation." For a synopsis, see *Electric Railway Journal* 41, no. 20 (17 May 1913): 893–94.

34. *Electric Railway Journal* 38, no. 9 (26 Aug. 1911): 341.

35. *Electric Railway Journal* 41, no. 20 (17 May 1913): 893–94.

36. *Electric Railway Journal,* 36, no. 24 (10 Dec. 1910): 1155; 39, no. 11 (16 March 1912): 418–22; no. 13 (30 March 1912): 502–3; no. 16 (20 Apr. 1912): 646–52); no. 25 (22 June 1912): 1064, 1066–71; 40, no. 5 (3 Aug. 1912): 154–59; Andrew D. Young, *St. Louis Car Company Album* (Glendale, Calif.: Interurban Press, 1984), 38.

37. *Electric Railway Journal* 43, no. 1 (3 Jan. 1914): 16.

38. Appendix C of "Report of Committee on Equipment," *Proceedings of the American Electric Railway Engineering Association* (1911): 639–41.

39. Hitt, *Electric Railway Dictionary,* 33.

40. Banner Electric Company in Youngstown produced lamps early in the 1900s, including a special "series burning" lamp for street railways. It eventually became the Banner Electric Division of General Electric. *Street Railway Journal* 22, no. 11 (12 Sept. 1903): 556; *Youngstown Official City Directory: 1918* (Akron: Burch Directory, 1918), 241.

41. Federal Writers' Project, *Warren and Trumbull County* (Ohio: Western Reserve Historical Celebration Committee, 1938), 7–10, 21–22, 26–27; Joseph G. Butler Jr., *History of Youngstown and the Mahoning Valley, Ohio,* 3 vols. (Chicago: American Historical Society, 1921), 403–6, 411, 414, 420–22.

42. Federal Writers' Project, *Warren and Trumbull County,* 16–17; Butler, *History of Youngstown,* 680; John Merrill Weed, "Business as Usual," in *Ohio in the Twentieth Century, 1900–1938,* comp. Harold Lindley, vol. 6 of *The History of the State of Ohio,* ed. Carl Wittke (Columbus: Ohio State Archaeological and Historical Society, 1941–44), 163, 165; William M. Gregory and William B. Guitteau, *History and Geography of Ohio,* rev. ed. (Boston: Ginn, 1935), 183, 190–19.

43. Federal Writers' Project, *Warren and Trumbull County,* 29, 42–43; Butler, *History of Youngstown,* 680; *Warren and Niles Official City Directory: 1899* (Akron: Burch Directory, 1898), 129, 132; *Warren and Niles Official City Directory: 1902* (Akron: Burch Directory, 1901), 127, 130; *Warren, Niles and Girard Official City Directory: 1910* (Akron: Burch Directory, 1910), 164; *Warren, Niles and Girard Official City Directory: 1916* (Akron: Burch Directory, 1916), 245.

44. *Electric Railway Journal* 37, no. 11 (18 March 1911): 486.

45. *Electric Railway Journal* 41, no. 20 (17 May 1913): 899; 43, no. 1 (3 Jan. 1914): 64; *Cleveland Directory* (1911), 931, 1,126.

6. Fare Collection and Registration

1. *Electric Railway Journal* 32, no. 19 (10 Oct. 1908): 834–42; Charles S. Sergeant, "Problems Confronting Street Railways," 35, no. 1 (1 Jan. 1910): 6–7, C. L. S. Tingley, "Fares, Taxes, and Regulations," 10–11; W. H. Glenn, "Fares on City Lines," 13–14; L. D. Mathes, "Why Street Railway Fares Should Not Be Lowered," no. 17 (23 Apr. 1910): 750–51; Guy E. Tripp, "Economic Limitations upon the Development of Transportation by Electric Railways," 39, no. 4 (27 Jan. 1912): 135; 43, no. 20 (16 May 1914): 1,080.

2. Henry J. Smith, "The Coupon Transfer," *Electric Railway Journal* 40, no. 23 (14 Dec. 1912): 1,192–94; W. C. Callaghan, "Educational Methods Used in Placing New System of Transfers in Operation," *Electric Railway Journal* 35, no. 10 (5 March 1910): 412–13.

3. *Electric Railway Journal* 32, no. 19 (10 Oct. 1908): 834–42; no. 21 (24 Oct. 1908): 1254; 37, no. 14 (8 Apr. 1911): 624–25.

4. *Electric Railway Journal* 31, no. 2 (13 June 1908): 84–85; 36, no. 20 (12 Nov. 1910): 994–95.

5. *Street Railway Journal* 30, no. 4 (27 July 1907): 138, 140.

6. Theodore Stebbins, "Interurban Fares," *Street Railway Journal* 30, no. 17 (26 Oct. 1907): 866–70; *Electric Railway Journal* 32, no. 19 (10 Oct. 1908): 834–42; 39, no. 2 (13 Jan. 1912): 56.

7. *Street Railway Journal* 27, no. 1 (6 Jan. 1906): 2.

8. *Electric Railway Journal* 32, no. 3 (20 June 1908): 116–17.

9. *Street Railway Journal* 26, no. 16 (14 Oct. 1905): 729; 27, no. 2 (13 Jan. 1906): 89–90; no. 6 (10 Feb. 1906): 253; no. 14 (7 Apr. 1906): 574; "1906 Manufacturers' Directory," 28, no. 15 (13 Oct. 1906): lvii.

10. *Street Railway Journal* 23, no. 8 (20 Feb. 1904): 304; 25, no. 15 (15 Apr. 1905): 719; 27, no. 5 (3 Feb. 1906): 211; 28, no. 15 (13 Oct. 1906): 695; no. 17 (26 Oct. 1906): 839; 30, no. 16 (19 Oct. 1907): 824; *Electric Railway Journal* 32, no. 13 (29 Aug. 1908): 555. MacDonald tickets were popular on many interurbans based in Columbus, Ohio. These included the Columbus, London & Springfield; the Dayton, Springfield, and Urbana; the Columbus, Grove City & Southwestern; the Urbana, Bellefontaine & Northern; the Columbus, Buckeye Lake & Newark; the Columbus, Newark & Zanesville; and the Columbus, Delaware & Marion.

11. *Street Railway Journal* 20, no. 16 (18 Oct. 1902): 677; 22, no. 11 (12 Sept. 1903): 549; 24, no. 11 (10 Sept. 1904): 35; 25, no. 12 (25 March 1905): 568; 28, no. 17 (27 Oct. 1906): 839.

12. *Street Railway Journal* 27, no. 26 (30 June 1906): 1,005, 1,027–29; 29, no. 13 (30 March 1907): 554; *Electric Railway Journal* 32, no. 2 (13 June 1908): 84–85; no. 19 (10 Oct. 1908): 834–35; Howard C. Lake, "What Constitutes Legal Tender for a Fare," 34, no. 7 (14 Aug. 1909): 256–57; 35, no. 8 (15 Feb. 1910): 313; George L. Radcliffe, "The Use of Metal Tickets," 36, no. 15D (14 Oct. 1910): 814–16; 37, no. 18 (6 May 1911): 796.

13. *Electric Railway Journal* 48, convention supplement (30 Sept. 1916): 623–24; Rodney Hitt, *Electric Railway Dictionary* (1911; reprint, Novato, Calif.: Newton K. Gregg, 1972), xxvii; *Electric Traction* 18, no. 11 (Nov. 1922): 988–99. For an example of a specific road, see Alan R. Lind, *Chicago Surface Lines: An Illustrated History* (Park Forest, Ill.: Transport History Press, 1974), 374.

14. *Street Railway Journal* 27, no. 4 (27 Jan. 1906): 136–43; *Electric Railway Journal* 32, no. 1 (6 June 1908): 41; no. 19 (10 Oct. 1908): 834–42; no. 27 (5 Dec. 1908): 1,521; 34, no. 14A (5 Oct. 1909): 661; Debra Brill, *History of the J. G. Brill Company* (Bloomington: Indiana University Press, 2001), 114–16; William D. Middleton, *The Time of the Trolley: The Street Railway from Horsecar to Light Rail* (San Marino, Calif.: Golden West Books, 1987), 217–18.

15. Witt was able to patent his design and thus received royalties on cars built embodying these features. Early users included the Cleveland Railway, which had 150 of these cars in service by May 1916, and the New York State Railways, which had 86. Peter Witt, "The Car Rider's Car," undated pamphlet in container 2, box 5, Peter Witt Papers, MSS 3281 (Cleveland: Western Reserve Historical Society); *Electric Traction* 11, no. 11 (Nov. 1915): 694–96; *Electric Railway Journal* 47, no. 18 (29 Apr. 1916): 845–46; "The Front Entrance, Center-Exit Car, and Higher Schedule Speed," *Electric Railway Journal* 51, no. 3 (9 Jan. 1918): 120–24.

16. A. J. Varrellman, "The Prepayment Car and Its Advantages," *Electric Railway Journal* 35, no. 18 (30 April 1910): 784–85; R. A. Leusser, "Some Experiences with Prepayment Cars," 39, no. 17 (27 Apr. 1912): 698–99; F. T. Wood, "Report of the Committee on Fares and Transfers," 42, no. 15C (16 Oct. 1913): 811.

17. *Street Railway Journal* 17, no. 24 (15 June 1901): 699; 19, no. 15 (12 Apr. 1902): 456; 22, no. 13 (26 Sept. 1903): 605; Albert H. Stanley, "Leaks between Passenger and Treasurer," 28, no. 17 (27 Oct. 1906): 799–802; 31, no. 16 (18 Apr. 1908): 629–30; C. J. Griffith, "The Use and Abuse of Transfers," 33, no. 21 (22 May 1909): 941; 37, no. 25 (24 June 1911): 1,123–24.

18. *Street Railway Journal* 17, no. 24 (15 June 1901): 699; *Electric Railway Journal* 35, no. 8 (19 Feb. 1910): 332–33.

19. *Street Railway Journal* 17, no. 24 (15 June 1901): 699; *Electric Railway Journal* 35, no. 8 (19 Feb. 1910): 332–33.

20. *Street Railway Journal* 30, no. 2 (13 July 1907): 70–72; *Electric Railway Journal* 38, no. 14 (30 Sept. 1911): 550; 41, no. 14 (5 Apr. 1913): 643.

21. *Street Railway Journal* 20, no. 16 (18 Oct. 1902): 680; "Dictionary of Electric Railway Material," *Street Railway Journal* 28, no. 15 (13 Oct. 1906): lxiii; *Electric Railway Journal* 38, no. 15 (7

Oct. 1911): 627–29; no. 15D (13 Oct. 1911): 832–34; 40, no. 23 (14 Dec. 1912): 1197–98; Hitt, *Electric Railway Dictionary,* 27; Arnold Von Schrenk, "Fare Boxes," *Electric Railway Journal* 48, convention supplement (30 Sept. 1916): 624–26.

22. Michael Massouh, "Innovations in Street Railways before Electric Traction: Tom L. Johnson's Contributions," *Technology and Culture* 18, no. 2 (April 1977): 206.

23. *Street Railway Journal* 29, no. 9 (2 March 1907): 392; *Electric Railway Journal* 32, no. 19C (15 Oct. 1908): 1,116; 36, no. 14 (1 Oct. 1909): 523; 37, no. 14 (8 Apr. 1911): 626; 39, no. 1 (6 Jan. 1912): 38; no. 14 (6 Apr. 1912): 590; no. 26 (29 June 1912): 1,135; 43, no. 11 (14 March 1914): 599; Hitt, *Electric Railway Dictionary,* 195–96; Massouh, "Innovations in Street Railways," 202–17.

24. *Electric Railway Journal* 38, no. 15D (13 Oct. 1911): 832–34; see also F. T. Wood, "Report of the Committee on Fares and Transfers," 42, no. 15C (16 Oct. 1913): 811.

25. *Electric Railway Journal* 41, no. 10 (8 March 1913): 442–43, 448.

26. *Electric Railway Journal* 40, no. 11 (14 Sept. 1912): 435–36; no. 12 (30 Sept. 1912): 1,129–30, 1,134.

27. *Electric Railway Journal* 34, no. 22 (2 Dec. 1909): 1,156; 37, no. 10 (11 March 1911): 426–28; 42, no. 15A (14 Oct. 1913): 738.

28. John F. Ohmer, "Fares and Fare Protection," *Street Railway Journal* 22, no. 14 (3 Oct. 1903): 673; *Dayton Directory* (1904–5), 1,017.

29. J. E. Duff, "Operation of Ohmer Fare Registers in City Service," *Electric Railway Journal* 37, no. 12 (25 March 1911): 510–11; Ohmer, "Fares and Fare Protection," 674; A. W. Drury, *History of the City of Dayton and Montgomery County Ohio* (Chicago: S. J. Clark, 1909), 1:628.

30. Ohmer, "Fares and Fare Protection," 675.

31. Duff, "Operation of Ohmer Fare Registers," 510.

32. Ohmer, "Fares and Fare Protection," 675.

33. Folder 86, box 8, Dayton, Covington & Piqua Traction Company Records, 1897–1927, MSS 346, Ohio Historical Society, Columbus.

34. *Street Railway Journal* 20, no. 18 (1 Nov. 1902): 771; 23, no. 14 (2 Apr. 1904): 546.

35. *Street Railway Journal* 26, no. 10 (2 Sept. 1905): 350; 28, no. 17 (27 Oct. 1906): 831–32.

36. *Street Railway Journal* 29, no. 22 (1 June 1907): 979; 30, no. 16 (19 Oct. 1907): 819; *Electric Railway Journal* 36, no. 15A (11 Oct. 1910): 700; no. 25 (17 Dec. 1910): 1,209.

37. *Electric Railway Journal* 32, no. 16 (19 Sept. 1908): 686–87; 37, no. 1 (7 Jan. 1911): 57; 43, no. 14 (4 Apr. 1914): 804.

38. *Electric Railway Journal* 41, no. 24 (14 June 1913): 1,098.

39. *Electric Railway Journal* 34, no. 3 (17 July 1909): 115.

40. *Electric Railway Journal* 33, no. 9 (27 Feb. 1909): 396.

41. *Electric Railway Journal* 33, no. 7 (13 Feb. 1909): 292; 43, no. 16 (18 Apr. 1914): 900.

42. *Street Railway Journal* 16, no. 27 (7 Jan. 1900): 658; 18, no. 24 (14 Dec. 1901): 864; 24, no. 22 (26 Nov. 1904): 960; no. 24 (10 Dec. 1904): 1,052; 30, no. 18 (2 Nov. 1907): 923.

7. Seldom Mentioned

1. *Electric Traction* 15, no. 10 (Oct. 1919): 705; Andrew D. Young, *St. Louis Car Company Album: A Photographic Record* (Glendale, Calif.: Interurban Press, 1984), 27–35.

2. Philip D. Jordan, *Ohio Comes of Age, 1873–1900,* vol. 5 of *The History of the State of Ohio,* ed. Carl Wittke (Columbus: Ohio State University Press, 1943), 226–28.

3. Ibid.; William M. Gregory and William B. Guitteau, *History and Geography of Ohio,* rev. ed. (Boston: Ginn, 1935), 183, 185, 248; A. W. Drury, *History of the City of Dayton and Montgomery County Ohio* (Chicago: S. J. Clark, 1909), 1:539; Opha Moore, "Ohio as a Manufacturing State," in *History of Ohio: The Rise and Progress of an American State,* ed. Emilius O. Randall and Daniel J. Ryan (New York: Century History, 1912), 5:313–15.

4. Post produced everything from track maintenance tools to locomotive gauges and car furnishings. Folder 25, box 3, series 4, Dayton Manufacturing Company, MS-202, Dept. of Archives and Special Collections, Wright State University, Dayton, 120–21; Post & Company, *Catalogue no. 30* (Cincinnati: Post, 1887); Scott D. Trostel, *Barney & Smith Car Company: Car Builders* (Fletcher, Ohio: Cam-Tech, 1993): 79–80; Drury, *History of Dayton,* 1:659–60.

5. Carl M. Becker, "A 'Most Complete' Factory: The Barney & Smith Car Works," *Cincinnati Historical Society Bulletin* 31, no. 1 (Spring 1973): 63. See also "Dayton Manufacturing Company," MS-202, box 3, series 4, folder 25, p. 122, Department of Archives and Special Collections, Wright State University, Dayton.

6. *Dayton Directory* (1889), 97, (1904): 382, (1910): 358.

7. Trostel, *Barney & Smith,* 80–85.

8. Dayton Manufacturing Company, *Dayton Car Lighting Fixtures: Cat. no. 166* (Dayton: 1900): 8.

9. Ibid., 60.

10. *Electric Railway Journal* 47, no. 16 (15 Apr. 1916): 759.

11. The information was obtained by compiling data from each car order reported in *Electric Railway Journal* for 1909–13. While this method is by no means foolproof, it does serve to give a general picture of industry activity during the years in question.

12. *Street Railway Journal* 25, no. 18 (6 May 1905): 829–30; 28, no. 26 (29 Dec. 1906): 1172–73.

13. *Street Railway Journal* 25, no. 18 (6 May 1905): 829–30; American Electric Railway Engineering Association, "Report of the Committee on Equipment, Appendix I—Heating and Ventilation of Cars," *Proceedings of the American Electric Railway Engineering Association* (1911): 639–40; William J. Fleming, "Car Ventilation," *Electric Railway Journal* 40, no. 11 (14 Sept. 1912): 410.

14. "Heating and Ventilation," *Proceedings of the American Electric Railway Engineering Association* (1911): 640–41; Fleming, "Car Ventilation," 410–11; H. E. Lavelle, "Ventilation of Electric Cars," *Electric Railway Journal* 42, no. 1 (5 July 1913): 18.

15. "Heating and Ventilation," *Proceedings of the American Electric Railway Engineering Association* (1911): 640–41.

16. J. F. Biehn and J. B. Gooken, "Results of a Study of Car Ventilation in Chicago," *Electric Railway Journal* 33, no. 19 (8 May 1909): 876–79.

17. *Electric Railway Journal* 32, no. 7 (5 Dec. 1908): 1520–21; 33, no. 3 (16 Jan. 1909): 97; no. 5 (30 Jan. 1909): 201; no. 14 (3 Apr. 1909): 656; no. 19 (8 May 1909): 876–79; 34, no. 14 (2 Oct. 1909): 621–22; H. S. Williams, "Recent Developments in Car Heating and Ventilation," 36, no. 23 (3 Dec. 1910): 1105–6; 37, no. 14 (8 Apr. 1911): 655; 41, no. 26 (28 June 1913): 1187; *Proceedings of the American Electric Railway Engineering Association* (1911): 640.

18. *Electric Railway Journal* 39, no. 25 (22 June 1912): 1066–71.

8. The Decade of Transition, 1910–1919

1. *Electric Railway Journal* 55, no. 1 (3 Jan. 1920): 53.

2. *Electric Railway Journal* 49, no. 23 (9 June 1917): 1,076–77; 50, no. 6 (11 Aug. 1917): 252; 51, no. 11 (6 March 1918): 551.

3. *Electric Railway Journal* 48, no. 14 (30 Sept. 1916): 704.

4. *Electric Railway Journal* 48, no. 16 (14 Oct. 1916): 756.

5. Richard Saunders, *Merging Lines: American Railroads, 1900–1970* (DeKalb: Northern Illinois University Press, 2001): 35.

6. *Electric Railway Journal* 49, no. 3 (20 Jan. 1917): 144–45; 50, no. 8 (25 Aug. 1917): 139; Saunders, *Merging Lines,* 35–36.

7. *Electric Railway Journal* 50, no. 22 (1 Dec. 1917): 1,018–19; 51, no. 17 (27 Apr. 1918): 795.

8. *Electric Railway Journal* 51, no. 7 (16 Feb. 1918): 348; 52, no. 11 (14 Sept. 1918): 471; no. 12 (21 Sept. 1918): 532.

9. *Electric Railway Journal* 50, no. 24 (15 Dec. 1917): 1,103.

10. *Electric Railway Journal* 51, no. 3 (19 Jan. 1918): 137; no. 4 (26 Jan. 1918): 182–83.

11. *Electric Railway Journal* 50, no. 4 (28 July 1917): 170; no. 5 (4 Aug. 1917): 211.

12. *Electric Railway Journal* 51, no. 5 (16 March 1918): 550.

13. *Electric Railway Journal* 50, no. 8 (25 Aug. 1917): 317.

14. *Electric Railway Journal* 50, no. 12 (22 Sept. 1917): 559; no. 13 (29 Sept. 1917): 607.

15. *Electric Railway Journal* 52, no. 26 (14 Dec. 1918): 992.

16. *Electric Railway Journal* 50, no. 8 (25 Aug. 1917): 339; 51, no. 18 (4 May 1918): 887; no. 19 (11 May 1918): 943; Saunders, *Merging Lines,* 37–38.

17. *Electric Railway Journal* 48, no. 13 (23 Sept. 1916): 559; no. 20 (11 Nov. 1916): 1,044; 49, no. 14 (7 Apr. 1917): 673.

18. *Electric Railway Journal* 49, no. 1 (6 Jan. 1917): 12; 50, no. 15 (13 Oct. 1917): 705; 51, no. 26 (22 June 1918): 1,217.

19. *Electric Railway Journal* 50, no. 18 (3 Nov. 1917): 845; 51, no. 1 (5 Jan. 1918): 66–67; no. 24 (15 June 1918): 1,175.

20. *Electric Railway Journal* 50, no. 14 (6 Oct. 1917): 650; 51, no. 16 (20 Apr. 1918): 792; 53, no. 8 (22 Feb. 1919): 391.

21. Debra Brill, *History of the J. G. Brill Company* (Bloomington: Indiana University Press, 2001), 133–39.

22. Folder 3, Jewett Car Company Records, MSS 971, Ohio Historical Society, Columbus; Lawrence A. Brough and James H. Graebner, *From Small Town to Downtown: A History of the Jewett Car Company, 1893–1919* (Bloomington: Indiana University Press, 2004), 63.

23. The orders in question can be found in a surviving account book in the Jewett Car Company Records, MSS 971, Ohio Historical Society, Columbus. For a few examples, see lot numbers 97 (Columbus, Delaware & Marion), 108 (Michigan City, Indiana), 124 (Newark & Granville Street Railway), 172 (University of Illinois), and 195 (Interurban Railway & Terminal Co.).

24. Brough and Graebner, *From Small Town to Downtown,* 72–73.

25. *Electric Railway Journal* 52, no. 25 (12 Dec. 1918): 1,124.

26. *Electric Railway Journal* 53, no. 3 (18 Jan. 1919): 168; no. 8 (22 Feb. 1919): 391; Brough and Graebner, *From Small Town to Downtown,* 69.

27. Scott D. Trostel, *The Barney & Smith Car Company: Car Builders* (Fletcher, Ohio: Cam-Tech, 1993): 115–16, 118.

28. Ibid., 143–44.

29. *Electric Railway Journal* 42, no. 4 (26 July 1913): 165; 46, no. 26 (25 Dec. 1915): 1284; 59, no. 17 (29 Apr. 1922): 785; Trostel, *Barney & Smith,* 161–64; Michael W. Williams, "Czar of Kossuth: J. D. Moskowitz," *Timeline* 8, no. 5 (Oct./Nov. 1991): 25; Warren H. Deem, "The Barney & Smith Car Company: A Study of Business Growth and Decline," paper prepared for course in American economic history, Harvard University (1953), copy at Ohio Historical Society, 45, 47.

30. Deem, "Barney & Smith," 48; *Electric Railway Journal* 46, no. 26 (25 Dec. 1915): 1284; 51, no. 18 (4 May 1918): 887; no. 19 (11 May 1918): 943; 59, no. 17 (29 Apr. 1922): 785; Williams "Czar," 25.

31. *Electric Railway Journal* 57, no. 15 (9 Apr. 1921): 708.

32. *Electric Railway Journal* 57, no. 12 (19 March 1921): 578–79; no. 16 (16 Apr. 1921): 755; 59, no. 17 (29 Apr. 1922): 785; *Electric Traction* 17, no. 3 (March 1921): 223; Deem, "Barney & Smith," 48; Williams, "Czar," 25.

33. *Electric Traction* 17, no. 4 (Apr. 1921): 296.

34. *Electric Railway Journal* 57, no. 26 (25 June 1921): 1,192; 58, no. 13 (24 Sept. 1921): 540; 59, no. 17 (29 Apr. 1922): 785; *Electric Traction* 17, no. 10 (Oct. 1921): 778; Deem, "Barney & Smith," 48.

35. *Electric Railway Journal* 59, no. 17 (29 Apr. 1922): 785; 60, no. 3 (15 July 1922): 106; no. 4 (22 July 1922): 148; 61, no. 14 (7 Apr. 1923): 629; 62, no. 19 (10 Nov. 1923): 843.

36. *Electric Railway Journal* 64, no. 2 (12 July 1924): 73; 66, no. 5 (18 July 1925): 149; Deem, "Barney & Smith," 46–47.

37. *Electric Railway Journal* 48, no. 12 (16 Sept. 1916): 560; Brough and Saunders, *From Small Town to Downtown,* 63–67.

38. *Electric Railway Journal* 47, no. 20 (13 May 1916): 934; no. 21 (20 May 1916): 981; 48, no. 12 (16 Sept. 1916): 520; 49, no. 24 (16 June 1917): 1,123; *Electric Railway Journal* 47, no. 20 (13 May 1916): 934; no. 21 (20 May 1916): 981; 48, no. 12 (16 Sept. 1916): 520; 49, no. 24 (16 June 1917): 1,123

39. *Electric Railway Journal* 44, no. 10 (5 Sept. 1914): 433–35.

40. *Electric Railway Journal* 48, no. 10 (2 Sept. 1916): 425; 49, no. 2 (13 Jan. 1917): 100.

41. *Electric Railway Journal* 48, no. 24 (9 Dec. 1916): 1,229; 49, no. 26 (30 June 1917): 1,211–12; advertisement in *Electric Traction* 14, no. 5 (May 1918): 481. See also *Electric Railway Journal* 49, no. 2 (13 Jan. 1917): 100.

42. *Electric Traction* 13, no. 6 (June 1917): 375.

43. *Electric Railway Journal* 53, no. 20 (17 May 1919): 988.

9. Promise and Stagnation

1. Herbert Hoover, "Relation of the Electric Railway Industry to Industrial Efficiency," *Electric Railway Journal* 58, no. 15 (8 Oct. 1921): 580–81; Steven R. Kale and Richard E. Lonsdale, "Factors Encouraging and Discouraging Plant Location in Nonmetropolitan Areas," in *Nonmetropolitan Industrialization,* ed. Richard E. Lonsdale and H. L. Seyler (Washington, D.C.: V. H. Winston and Sons, 1979), 6.

2. *Electric Railway Journal* 57, no. 1 (1 Jan. 1921): 47; *Transit Journal* 82, no. 1 (Jan. 1938): 15, 17. For a brief summary of street railway decline, see Brian J. Cudahy, *Cash, Tokens, and Transfers* (New York: Fordham University Press, 1990), 151–63.

3. John S. Adams, "Residential Structure of Midwestern Cities," in *Internal Structure of the City: Readings on Urban Form, Growth, and Policy,* 2nd ed., ed. Larry S. Bourne (New York: Oxford University Press, 1982), 173–74.

4. L. F. Eppich, "Necessity of Electric Railways to Communities from the Realtor's Standpoint," *Proceedings of the American Electric Railway Association* (1924): 276–79.

5. Leon Moses and Harold F. Williamson Jr., "The Location of Economic Activity in Cities," *American Economic Review* 57, no. 2 (May 1967): 211–15, and Adams, "Residential Structure," 183.

6. George L. Kippenberger, "Style in Car Design," *Electric Railway Journal* 65, no. 24 (13 June 1925): 938.

7. For an overview of the 1921 standardization effort, as well as an overview of streetcar design and technology, see "Report of the Committee on Unification of Car Design," *Proceedings of the American Electric Railway Engineering Association* (1921): 836–79; "The First Universal Car," *Electric Railway Journal* 52, no. 17 (26 Oct. 1918): 728; Donald Engel, "Lightweight Street and Interurban Cars," *Branford Electric Railway Journal* 37 (1998): 34; Stephen P. Carlson and Fred W. Schneider, *PCC: The Car That Fought Back* (Glendale, Calif.: Interurban Press, 1980): 24–25.

8. Charles Gordon, "Car Architecture—A New Art That Influences Riding," *Electric Railway Journal* 66, no. 13 (26 Sept. 1925): 508.

9. Ibid., 507–10; J. R. Blackhall, "Making Transportation Attractive," *Electric Traction* 23, no. 9 (Sept. 1927): 455–57; "Incorporating Merchandising Features in the Design," *Electric Railway Journal* 62, no. 13 (29 Sept. 1923): 529; Norman Litchfield, "The Car as Transportation Salesman," *Electric Railway Journal* 58, no. 18 (24 Sept. 1921): 491–98; D. W. Pontius, "Improvements in Car Design," *Electric Traction* 21, no. 9 (Sept. 1925): 465–66; H. S. Williams, "What Type of Car Does the Public Want?" *Electric Railway Journal* 67, no. 16 (17 Apr. 1926): 669–72; no. 26 (26 June

1926): 1,086–98; 70, no. 25 (17 Dec. 1927): 1,089; *Electric Traction* 13, no. 12 (Dec. 1917): 808; 21, no. 10 (Oct. 1925): 539–40; 25, no. 10 (Oct. 1929): 527–28.

10. *Electric Railway Journal* 62, no. 13 (29 Sept. 1923): 500–501; 69, no. 15 (9 Apr. 1927): 655–58; Engel, "Lightweight," 10.

11. Despite the "newness" imparted upon one-man operation by the trade press, one-man cars had been common in "small communities" for years. See George Oliver Smith, "Crossing Question with One-Man Cars," *Electric Traction* 12, no. 7 (July 1916): 503–5 (see also table 9.3); "Annual Report of Executive Committee, Appendix H: Report of the Committee on Essential Features of Modern Cars," *Proceedings of the American Electric Railway Engineering Association* (1928): 412–13; C. O. Birney, "The Design and Development for the Safety Car," *Electric Railway Journal* 50, no. 12 (22 Sept. 1917): 478–79; C. O. Birney, "Equipment of the Light-Weight, Safety Cars," *Electric Railway Journal* 50, no. 12 (22 Sept. 1917): 531–32; C. O. Birney, "Present Indications of 'Safety Car' Operation," *Electric Traction* 13, no. 3 (March 1917): 165–66; J. R. Blackhall, "Making Transportation Attractive," *Electric Traction* 23, no. 9 (Sept. 1927): 455–57; W. J. Clardy, "Light-Weight Interurban Cars," *Electric Railway Journal* 65, no. 4 (24 Jan. 1925): 152–53; C. T. Dehore, "Light Weight Interurban Cars," *General Electric Review* 25, no. 6 (June 1922): 352–61; "The Development of the Electric Railway Car," *Electric Railway Journal* 48, convention supplement (30 Sept. 1916): 463–68; Norman Litchfield, "How the Underframe Contributes to the Durability of the Car Body," *Electric Railway Journal* 51, no. 24 (15 June 1918): 1,142–44; Norman Litchfield, "Car Bodies Must Be Designed for Economy as Well as Strength," *Electric Railway Journal* 51, no. 16 (20 April 1918): 755–56; E. A. Palmer, "Service with the Safety Type Car," *Electric Journal* 16, no. 10 (Oct. 1919): 426–28; L. B. Stillwell, "Truss-Side Construction for Railway Cars," *Electric Railway Journal* 48, no. 22 (25 Nov. 1916): 1,112–14; G. M. Woods, "Trends in Electric Railway Equipment," *Electric Railway Journal* 64, no. 1 (5 July 1924): 10.

See also *Electric Railway Journal* 43, no. 22 (30 May 1914): 1,186; 47, no. 12 (18 March 1916): 556–58; 52, no. 18 (2 Nov. 1918): 820; 70, no. 25 (17 Dec. 1927): 1,089; 72, no. 12 (22 Sept. 1928): 519–20; *Electric Traction* 18, no. 1 (Jan. 1922): 32–33; 21, no. 9 (Sept. 1925): 489; Engel, "Lightweight," 6, 17, 19, 27; Seymour Kashin and Harre Demoro, *The PCC Car: An American Original* (Glendale, Calif.: Interurban Press, 1986), 14–16; Carlson and Schneider, *PCC*, 23; Alan R. Lind, *From Horsecars to Streamliners: An Illustrated History of the St. Louis Car Company* (Park Forest, Ill.: Transport History Press, 1978): 55.

12. *Electric Traction* 22, no. 10 (Oct. 1926): 539.

13. W. H. Heulings, "The Light-Weight Safety Car and Car Standardization," *Electric Railway Journal* 50, no. 12 (22 Sept. 1917): 490; Engel, "Lightweight," 27.

14. Horatio Bigelow, "Single-Truck Cars Best Meet the Needs in *Smaller* Communities," *Electric Railway Journal* 70, no. 12 (17 Sept. 1927): 489–92; C. O. Birney, "Design and Development," 478–79; C. O. Birney, "Present Indications," 165–66; *Electric Railway Journal* 42, no. 19 (8 Nov. 1913): 1,025–26; 45, no. 16 (17 Apr. 1915): 765–66; 47, no. 1 (1 Jan. 1916): 21–24; no. 12 (19 March 1916): 556–58; 51, no. 26 (29 June 1918): 1,261; 52, no. 1 (6 July 1918): 39; *Electric Traction* 12, no. 9 (Sept. 1916): 693 (advertisement); 15, no. 5 (May 1919): 300; 61, no. 22 (2 June 1923): 948.

Shoreline Trolley Museum Tripper 15, no. 10 (Oct. 1999): 1–2, 4; Engel, "Lightweight," 6–7, 33–34; Kashin and Demoro, *American Original,* 14–16; Carlson and Schneider, *PCC,* 23; Lind, *From Horsecars to Streamliners,* 55; Frank Rowsome Jr., *Trolley Car Treasury: A Century of American Streetcars, Horsecars, Cable Cars, Interurbans, and Trolleys* (New York: Bonanza Books, 1956), 170–72.

15. *Electric Railway Journal* 57, no. 3 (15 Jan. 1921): 131–34; 62, no. 22 (1 Dec. 1923): 690; 63, no. 7 (16 Feb. 1924): 252–55; A. C. Colby, "Three Car Articulated Train for Detroit," no. 10 (8 March 1924): 357–62; 64, no. 23 (6 Dec. 1924): 963–66; 66, no. 6 (8 Aug. 1925): 191–94; 69, no. 4 (22 Jan. 1927): 165–66; 71, no. 21 (26 May 1928): 883; "Cleveland Articulated Cars Make Uniform Loading

Possible," 72, no. 9 (1 Sept. 1928): 329–31; *Electric Traction* 22, no. 1 (Jan. 1926): 37; Brill, *J. G. Brill Company,* 177–80; Middleton, *Time of the Trolley,* 217–21.

For further reading on articulated streetcars, see "Two-Car Three-Truck Trains," *Electrical News* 30, no. 3 (1 Feb. 1921): 37–40. See also the following in *Electric Railway Journal:* S. E. Emmons, "Articulated Units Are Useful for Mass Transportation," 70, no. 12 (17 Sept. 1927): 482–84, and W. J. Clardy, "Articulated Cars Meet Unusual Requirements," 71, no. 1 (7 Jan. 1928): 17–20.

16. *Electric Traction* 22, no. 4 (Apr. 1926): 186–87; no. 11 (Nov. 1926): 596. Ohio's Lake Shore Electric Railway also operated a fleet of intermodal cars called "Bonner Railwagons." See George W. Hilton and John F. Due, *The Electric Interurban Railways in America* (Stanford: Stanford University Press, 1960), 135.

17. *Electric Railway Journal* 68, no. 20 (13 Nov. 1926): 898.

18. *Electric Railway Journal* 66, no. 15 (10 Oct. 1925): 576.

19. *Electric Railway Journal* 71, no. 7 (18 Feb. 1928): 290.

20. *Electric Railway Journal* 70, no. 27 (31 Dec. 1927): 1,182–84; 71, no. 7 (18 Feb. 1928): 290.

21. *Electric Railway* Journal 69, no. 23 (4 June 1927): 994–95; C. O. Birney, "Construction Details of the New Birney Safety Car," *Electric Railway Journal* 73, no. 3 (19 Jan. 1929): 117–20; Lind, *From Horsecars to Streamliners,* 65.

22. "Fundamental Design Changes Feature Experimental Car at Springfield, Mass.," *Electric Railway Journal* 69, no. 13 (26 March 1927): 562–66; no. 14 (2 Apr. 1927): 602–5; no. 22 (28 May 1927): 954–55; Charles Gordon, "Pittsburgh Seeks More Popular Street Car," 71, no. 22 (2 June 1928): 888–95; 73, no. 16 (July 1929): 712–14; Donald J. Engel, "Lightweight Streetcars and Interurban Cars," *Branford Electric Railway Journal* 37 (1998): 37; Carlson and Schneider, *PCC,* 22–23, 117.

23. "De Luxe Trend Shown in the New Brill Car," *Electric Railway Journal* 70, no. 24 (10 Dec. 1927): 1,055–58; George Frey, "New Group of Standardized Cars Designed," 73, no. 4 (26 Jan. 1929): 157–60; Seymour Kashin and Harre Demoro, *The PCC Car: An American Original* (Glendale, Calif.: Interurban Press, 1986), 24–25, 27.

See also Debra Brill, *J. G. Brill Company,* 173–77. Brill claims that no aluminum Master Units were made, but this is not true. At least one example of a production order of aluminum Master Units has been preserved.

24. "Osgood-Bradley Develops New Model Sample Car," *Electric Railway Journal* 71, no. 25 (23 June 1928): 1,020–25; 73, no. 16 (July 1929): 752. Kashin and Demoro make very brief mention of the Electromobile in *American Original,* 23. Unfortunately, the three illustrations they use are not Electromobiles but two of the three experimental cars that Osgood-Bradley built for Pittsburgh.

25. For some reason, the last third of the article was cropped from the *Electric Railway Journal,* excluding much of the information on the Cincinnati car. "Sample Cars for Louisville Designed to Win Public," *Electric Railway Journal* 73, no. 15 (June 1929): 637–40. See also *Electric Railway Journal* 65, no. 20 (16 May 1925): 769; *Electric Traction* 25, no. (May 1929): 233; Kashin and Demoro, *American Original,* 181–82.

26. *Electric Railway Journal* 46, no. 15 (16 Feb. 1924): 279; 47, no. 25 (17 June 1916): 1,163; 63, no. 7 (16 Feb. 1924): 279. See also *Electric Traction* 12, no. 7 (July 1916): 559; Blaine S. Hays and James A. Toman, *Cleveland Transit through the Years* (Cleveland: Cleveland Landmarks Press, 1999): 15.

27. Statistical information on manufacturing activity in this chapter has been obtained from car orders appearing in *Electric Railway Journal* and *Electric Traction* between 1914 and 1929.

28. *Brill Magazine* 12, no. 2 (July 1923): 59–61; no. 4 (Apr. 1924): 152–54; no. 7 (Nov. 1924): 219–24; no. 9 (May 1925): 283–87; no. 11 (Dec. 1925): 329–33; *Bus Transportation* 1, no. 4 (Apr. 1922):

262; no. 10 (Oct. 1922): 541; Charles Gordon, "Body Construction," *Bus Transportation* 3, no. 1 (Jan. 1924): 21; no. 6 (June 1924): 274; 5, no. 5 (May 1926): 290; *Electric Railway Journal* 58, no. 13 (24 Sept. 1921): 565; 63, no. 14 (Apr. 1924): 555; 64, no. 25 (20 Dec. 1924): 1062; 68, no. 2 (10 July 1926): 83; *Electric Traction* 19, no. 2 (Feb. 1923): 101, 104; 20, no. 10 (Oct. 1924): 96, 98; Donald F. Wood, *American Buses* (Osceola, Wisc.: MBI, 1998), 42.

29. *Electric Traction* 17, no. 12 (Dec. 1921): 988.

30. *Electric Railway Journal* 47, no. 23 (3 June 1916): 1,047.

31. *Electric Railway Journal* 49, no. 6 (10 Feb. 1917): 256–57; no. 16 (21 Apr. 1917): 745–46.

32. *Electric Railway Journal* 57, no. 22 (28 May 1921): 982–83; 59, no. 22 (3 June 1922): 920; 71, no. 14 (7 Apr. 1928): 579 (quote taken from last entry); *Electric Traction* 14, no. 11 (Nov. 1918): 714–19.

33. *Electric Railway Journal* 49, no. 20 (19 May 1917): 938; 57, no. 22 (28 May 1921): 982–83; 59, no. 7 (19 Feb. 1922): 304; no. 10 (11 March 1922): 430; no. 17 (22 Apr. 1922): 735–36; no. 22 (3 June 1922): 920; 60, no. 18 (28 Oct. 1922): 697–99; 62, no. 11 (15 Sept. 1923): 410–11; "New Methods in Body Construction," no. 13 (29 Sept. 1923): 493; Charles Gordon, "Tom Elliott Talks on the Electric Railway Outlook," 67, no. 9 (27 Feb. 1926): 385; 69, no. 23 (4 June 1927): 1,019; *Electric Traction* 14, no. 11 (Nov. 1918): 714–19; W. J. Clardy and N. H. Wilby, "Interesting Features of New Cincinnati Cars," 20, no. 2 (Feb. 1924): 72–74; 23, no. 6 (June 1927): 309–10; 25, no. 10 (Oct. 1929): 569; no. 12 (Dec. 1926): 677; *Williams' Cincinnati Directory* (Cincinnati: Williams Directory, 1905), 338; *Williams' Cincinnati Directory* (Cincinnati: Williams Directory, 1930–31), 391; Engel, "Lightweight," 11; Kashin and Demoro, *American Original,* 17; Richard Wagner and Birdella Wagner, *Curve-Sided Cars Built by Cincinnati Car Company* (Cincinnati: Wagner Car, 1965): 7–8; Hilton and Due, *Interurban Railways,* 26, 28, 80.

Rumors exist of possible legal action between Cincinnati and Brill over similar cars built by Kuhlman, although no hard evidence has been uncovered. Debra Brill discusses this on pages 167–69 of her history of J. G. Brill.

34. *Electric Railway Journal* 71, no. 20 (19 May 1928): 884; 72, no. 26 (29 Dec. 1928): 1,142; *Electric Traction* 25, no. 5 (May 1929): 258.

35. *Bus Transportation* 1, no. 12 (Dec. 1922): 652; 7, no. 12 (Dec. 1928): 70; *Electric Railway Journal* 72, no. 25 (22 Dec. 1928): 1,103; 73, no. 6 (9 Feb. 1929): 235.

10. Parts of the Whole

1. "Equipment of the Light-Weight, Safety Cars," *Electric Railway Journal* 50, no. 12 (22 Sept. 1917): 532–34; G. M. Woods, "Trends in Electric Railway Equipment," *Electric Railway Journal* 64, no. 1 (5 July 1924): 330–32; F. W. McCloskey, "Design of Light Traction Motors," *Electric Traction* 19, no. 7 (July 1923): 330–32; Francis R. Thompson, *Electric Transportation* (Scranton, Pa.: International Textbook, 1940), 111–18, 121–28, 131–37, 142–45; Donald Engel, "Lightweight Street and Interurban Cars," *Branford Electric Railway Journal* 37 (1998): 15–16.

2. Stephen P. Carlson and Fred W. Schneider, *PCC: The Car That Fought Back* (Glendale, Calif.: Interurban Press, 1980), 117.

3. Terrance Scullin, "The 26-Inch Wheel, Low-Floor Cars in Cleveland," *Electric Journal* 13, no. 10 (Oct. 1915): 488–90; *Electric Railway Journal* 45, no. 13 (27 March 1915): 628; "The Development of the Electric Railway Car," 48, convention supplement (30 Sept. 1916): 562–63; "The Car Body from an Operating Standpoint," 48, convention supplement (30 Sept. 1916): 573–82; W. H. Phillips: "Helical Gearing for Railway Motors," 54, no. 20 (29 Oct. 1919): 934–35; "The Helical Gear," 58, no. 1 (2 July 1921): 22–23; "Development of the Modern Gear," 62, no. 12 (15 Sept. 1923): 413–15; Clifford Faust, "Recent Steps in Car Development: Where Are They Heading?" 72, no. 10 (8 Sept. 1928): 398; *Electric Traction* 17, no. 11 (Nov. 1921): 806; Claude L. Van Auken, "Elimination of Noise," 20, no. 9 (Sept. 1924): 405–12; 21, no. 10 (Oct. 1925): 539–40; Timken

Roller Bearing Co., *Wherever Wheels and Shafts Turn* (Canton, Ohio, 1929), 33; Engel, "Lightweight," 23; Carlson and Schneider, *PCC,* 117–19; Thompson, *Electric Transportation,* 146–49.

For additional illustrations and descriptions of gear types, see Herbert L. Nichols Jr., *Moving the Earth: The Workbook of Excavation* (New York: McGraw-Hill, 1976), 12-3–12-4.

4. *Electric Railway Journal* 45, no. 10 (6 March 1915): 491; 48, no. 2 (8 July 1916): 86; 59, no. 24 (17 June 1922): 972; E. S. Sawtelle, "Gear Economies through New Tooth Shapes," 60, no. 3 (15 July 1922): 91–93; 67, no. 3 (16 Jan. 1926): 139; *Electric Traction* 12, no. 3 (March 1915): 265; 18, no. 4 (Apr. 1922): 305; see also the following advertisements: *Electric Traction* 16, no. 10 (Nov. 1920): 917; 17, no. 2 (Feb. 1921): 143; no. 5 (May 1921): 363; no. 7 (July 1921): 498; 18, no. 8 (Aug. 1922): 710.

5. *Electric Railway Journal* 65, no. 7 (14 Feb. 1925): 269; 69, no. 24 (11 June 1927): 1,069; *Electric Traction* 21, no. 3 (March 1925): 154; Timken, *Wherever Wheels and Shafts Turn,* 33; *Timken Trading Post* 6, no. 3 (June 1949): 13–15, 17–18; Bettye H. Pruitt, *Timken: From Missouri to Mars—A Century of Leadership in Manufacturing* (Boston: Harvard Business School Press, 1998), 96–97; Timken Co., *History of the Timken Co.* (Canton, Ohio, 1990), 13.

6. *Electric Traction* 22, no. 1 (Jan. 1926): 33.

7. J. A. Clarke Jr., "Recent Developments in HL Control," *Electric Journal* 13, no. 10 (Oct. 1915): 452–55; *Electric Railway Journal* 48, convention number (30 Sept. 1916): 652; 62, no. 13 (29 Sept. 1923): 529–30; Axtell and R. H. Sjoberg, "Speed, Comfort, and Safety with a New Multipoint Control," *General Electric Review* 33, no. 3 (March 1930): 158–63; Engel, "Lightweight," 19; Thompson, *Electric Transportation,* 155–62, 170–72, 175–79, 192–99.

8. *Electric Railway Journal* 48, convention number (30 Sept. 1916): 660; "Report of Rolling Stock Committee No. 11—Current Collecting Devices," *Proceedings of the American Electric Railway Engineering Association* (1928): 412–13; Thompson, *Electric Transportation,* 200–208.

9. *Electric Railway Journal* 53, no. 18 (3 May 1919): 898; *Electric Traction* 23, no. 8 (Aug. 1927): 44; *Massillon City Directories* 1920, 1925, 1931.

10. Statistical information on manufacturing activity in this chapter has been obtained from car orders appearing in *Electric Railway Journal* and *Electric Traction* between 1914 and 1929.

11. *Electric Railway Journal* 59, no. 2 (14 Jan. 1922): 98; 62, no. 17 (27 Oct. 1923): 748; 70, no. 11 (10 Sept. 1927): 442; 73, no. 22 (Dec. 1929): 1,137; *Electric Traction* 18, no. 12 (Dec. 1922): 1,072; 22, no. 4 (Apr. 1926): 206; 24, no. 12 (Dec. 1928): 645; 25, no. 4 (Apr. 1929): 202.

12. *Electric Railway Journal* 48, no. 25 (16 Dec. 1916): 1,279–80; no. 27 (30 Dec. 1916): 1,358.

13. C. A. Ives, "Self-Lapping Air Brake Valves," *Electric Traction* 26, no. 10 (Oct. 1930): 532.

14. *Electric Railway Journal* 68, no. 3 (17 July 1926): 97–100; Engel, "Lightweight," 28.

15. H. A. Davis, "Air-Magnetic Brakes Makes Quick Stops," *Electric Railway Journal* 74, no. 5 (May 1930): 256.

16. L. M. Aspinwall, "Regenerative Braking Tried on Multiple-Unit Trains," *Electric Railway Journal* 67, no. 11 (13 March 1926): 447–48.

17. *Electric Railway Journal* 69, no. 14 (2 Apr. 1927): 604–5; no. 20 (14 May 1927): 861; 71, no. 25 (23 June 1928): 1,021.

18. *Electric Railway Journal* 48, convention number (30 Sept. 1916): 610–12 (quote from 611); 61, no. 26 (30 June 1923): 1,076–77; 62, no. 13 (29 Sept. 1923): 502–5; L. C. Doane, "An Analysis of Requirements for Modern Street Car Lighting," *Electric Traction* 11, no. 1 (Jan. 1915): 33–34; W. A. Armstrong, "Improvements in Car Lighting" 13, no. 4 (Apr. 1917): 308, 310; 14, no. 4 (Apr. 1918): 224; "Report of Rolling Stock Special Committee no. 4 on Study of Car Lighting (Appendices A–C)," *Proceedings of the American Electric Railway Engineering Association* (1927): 354–73 and lighting glossary appearing in "Report of the Committee on Equipment: Appendix A," *Proceedings of the American Electric Railway Engineering Association* (1926): 532–49.

19. *Electric Railway Journal* 46, no. 21 (20 Nov. 1915): 1,049–50; *Electric Traction* 11, no. 11 (Nov. 1915): 735.

20. *Electric Traction* 10, no. 7 (July 1914): 455.

21. *Electric Railway Journal* 57, no. 8 (19 Feb. 1921): 366–67; *Electric Traction* 10, no. 7 (July 1914): 460; 16, no. 6 (June 1920): 462; 20, no. 5 (May 1924): 238.

22. *Electric Railway Journal* 44, no. 13 (26 Sept. 1914): 583; *Electric Traction* 11, no. 1 (Jan. 1915): 60.

23. *Electric Railway Journal* 72, no. 12 (22 Sept. 1928): 492; Roy G. Benedict and Glenn M. Andersen, "Chicago's Streetcar Fenders and Wheel Guards," *First & Fastest* 16, no. 2 (Summer 2000): 18–21.

24. *Electric Railway Journal* 49, no. 10 (10 March 1917): 468; no. 11 (17 May 1917): 530; 61, no. 2 (13 Jan. 1923): 108. See also *Electric Traction* 19 no. 1 (Jan. 1923): 47.

25. *Electric Railway Journal* 63, no. 11 (15 March 1924): 437; no. 16 (19 Apr. 1924): 642; 64, no. 6 (9 Aug. 1924): 212. See also *Electric Traction* 11, no. 12 (Dec. 1915): 805; 24, no. 5 (May 1928): 219; *Electric Railway Journal* 71, no. 15 (14 Apr. 1928): 644.

26. *Electric Railway Journal* 46, no. 6 (7 Aug. 1915): 256; 60, no. 8 (19 Aug. 1922): 270; *Electric Traction* 18, no. 8 (Aug. 1922): 728; 24, no. 5 (May 1928): 266.

27. *Electric Railway Journal* 45, no. 3 (16 Jan. 1915): 145; no. 23 (5 June 1915): 1,081–82; no. 25 (19 June 1915): 1,173–74; 56, no. 16 (16 Oct. 1923): 854; *Electric Traction* 11, no. 1 (Jan. 1915): 58; no. 3 (March 1915): 186; 12, no. 3 (March 1915): 268; advertisement in no. 7 (July 1916): 559; 13, no. 1 (Jan. 1917): 58–59; 15, no. 4 (Apr. 1919): 264; advertisement in 18, no. 8 (Aug. 1922): 726.

28. *Electric Railway Journal* 65, no. 11 (14 March 1925): 422; 66, no. 25 (19 Dec. 1925): 1,097; 68, no. 24 (11 Dec. 1926): 1,055; *Electric Traction* 13, no. 3 (March 1917): 236; no. 9 (Sept. 1917): 646; 21, no. 12 (Dec. 1925): 676; 22, no. 11 (Nov. 1926): 619–20; 24, no. 9 (Sept. 1928): 483; 25, no. 3 (March 1929): 157.

29. *Electric Railway Journal* 48, no. 26 (23 Dec. 1916): 1,309; *Electric Traction* 13, no. 2 (Feb. 1917): 135–36; advertisement in no. 3 (March 1917): 205; 18, no. 3 (March 1922): 305.

30. *Electric Railway Journal* 45, no. 20 (15 May 1915): 948–49; 49, no. 9 (3 March 1917): 420; 53, no. 12 (22 March 1919): 628; *Electric Traction* 11, no. 9 (Sept. 1915): 550.

31. *Electric Railway Journal* 62, no. 13 (29 Sept. 1923): 506–9; L. P. Hynes, "Getting Best Results with Heater Control Equipment," 63, no. 3 (9 Jan. 1924): 97–100; 72, no. 12 (22 Sept. 1928): 519.

32. "Doors, Seats, and Miscellaneous Interior Equipment," *Electric Railway Journal* 48, convention number (30 Sept. 1916): 613–16; "Equipment of the Light-Weight, Safety Cars," 50, no. 12 (22 Sept. 1917): 531–32; "Facilities for Expediting Passenger Movement," 62, no. 13 (29 Sept. 1923): 486–88; 72, no. 12 (22 Sept. 1928): 519–20.

33. Norman Litchfield, "The Car as Transportation Salesman," *Electric Railway Journal* 58, no. 18 (24 Sept. 1921): 491–98; "Incorporating Merchandising Features in the Design," 62, no. 13 (29 Sept. 1923): 523–28; H. S. Williams, "What Type of Car Does the Public Want?" 67, no. 16 (17 Apr. 1926): 669–72; "New Light-Weight Equipment Pays," 67, no. 26 (26 June 1926): 1,086–98. See also the following in *Electric Traction:* 21, no. 9 (Sept. 1925): 465–66; no. 10 (Oct. 1925): 539–40.

34. Cleveland Tanning Co., untitled booklet (Cleveland, 1925): 20.

35. *Electric Railway Journal* 72, no. 12 (22 Sept. 1928): 527–28; *Electric Traction* 14, no. 4 (Apr. 1918): 224.

36. Cleveland Tanning finished its leathers in three additional fashions. It coated the leather in nitrocellulose solutions and also produced analine-dyed leathers that could be treated with linseed oil in addition to nitrocellulose. The company also embossed its leathers in a number of patterns, including Spanish, seal, walrus, and goat. Cleveland Tanning Company, untitled booklet (Cleveland, 1925): 4, 15, 19; *Cleveland Directory* (1921), 340; (1925), 859; *Electric Railway Journal* 70, no. 18 (29 Oct. 1927): 847.

37. *Electric Railway Journal* 53, no. 3 (18 Jan. 1919): 168; 67, no. 9 (27 Feb. 1926): 388; no. 18 (1 May 1926): 788; no. 23 (5 June 1926): 998; *Electric Traction* 22, no. 7 (July 1926): 383; no. 12 (Dec. 1926): 696.

38. *Electric Railway Journal* 48, no. 16 (14 Oct. 1916): 859; Norman Litchfield, "Paint Used by Electric Railways," *Electric Railway Journal* 56, no. 25 (18 Dec. 1920): 1,232.

39. *Electric Railway Journal* 48, no. 16 (14 Oct. 1916): 859.

40. J. S. Spratt, "Lacquer Painting Reduces Time Needed for Drying," *Electric Railway Journal* 67, no. 8 (20 Feb. 1926): 332–33; D. H. Boll, "Lacquer Finishes for Street Cars," *Electric Railway Journal* 70, no. 7 (13 Aug. 1927): 277–78; *Electric Traction* 23, no. 7 (July 1927): 361–62; W. A. Ernst, "Successful Lacquer Painting," *Electric Traction* 25, no. 1 (Jan. 1929): 27.

41. "Incorporating Merchandising Features in the Design," *Electric Railway Journal* 62, no. 13 (29 Sept. 1923): 523–28.

42. *Electric Railway Journal* 45, no. 6 (6 Feb. 1915): 298–99; *Electric Traction* 11, no. 2 (Feb. 1915): 121; 15, no. 10 (Nov. 1919): 705; Sales and Materials Ledger, 2 Jan. 1919–28 Feb. 1921; Sales and Materials Log, 4 Mar. 1921–Sept. 1925, series 5, Dayton Manufacturing Co., MS-202, Wright State University, Department of Archives and Special Collections, Dayton.

Statistical information on manufacturing activity in this chapter has been obtained from orders appearing in the *Electric Railway Journal* and *Electric Traction* between 1914 and 1929.

11. Streetcar Manufacture during the 1930s

1. Malcolm Muir, "Growing Significance of Obsolescence Problems in American Industry," *Proceedings of the American Transit Association* (1935): 62, 64–65.

2. *Electric Railway Journal* 57, no. 1 (1 Jan. 1921): 47; *Transit Journal* 82, no. 1 (Jan. 1938): 15, 17.

3. J. G. Inglis, "Modern Trends in Interurban Cars," *Electric Traction* 26, no. 2 (Feb. 1930): 85; A. T. Clark, "Interesting Features of Baltimore's New Cars," *Electric Traction* 29, no. 9 (Sept. 1930): 459–63 (quote on 459); Frank Sullivan and Fred Winkowski, *Trolley Cars* (Osceola, Wisc.: Motorbooks International, 1995), 66–67.

4. *Electric Traction and Bus Journal* 30, no. 9 (Sept. 1934): 308; *Mass Transportation* 33, no. 9 (Sept. 1937): 296–97.

5. *Transit Journal* 76, no. 11 (15 Oct. 1932): 436; Sullivan and Winkowski, *Trolley Cars*, 88–89; George W. Hilton and John F. Due, *The Electric Interurban Railways in America* (Stanford: Stanford University Press, 1960), 82.

6. Thomas Conway Jr., "Meeting the Industry's Equipment Problem," *Electric Railway Journal* 74, no. 8 (July 1930): 438–39.

7. *Electric Railway Journal* 74, no. 8 (July 1930): 489; "Cincinnati & Lake Erie's New De Luxe Cars Challenge the Steam Railroad and the Bus," no. 11 (Oct. 1930): 614–21; Conway, "Equipment Problem"; Thomas Conway Jr., "Organization and Purposes of the Electric Railway Presidents Conference Committee," *Proceedings of the American Electric Railway Association* (1931): 81; Kashin and Demoro, *American Original*, 29–30; Stephen P. Carlson and Fred W. Schneider III, *PCC: The Car That Fought Back* (Glendale, Calif.: Interurban Press, 1980): 29–31; Frank Rowsome Jr., *Trolley Car Treasury: A Century of American Streetcars, Horsecars, Cable Cars, Interurbans, and Trolleys* (New York: Bonanza Books, 1956), 181–82.

8. Quote taken from *Transit Journal* 76, no. 11 (15 Oct. 1932): 462. Conway, "Equipment Problem"; Kashin and Demoro, *American Original*, 30.

9. *Mass Transportation* 32, no. 7 (July 1936): 192.

10. W. T. Rossell, "Why Not a Standard Car?" *Electric Traction* 25, no. 9 (Sept. 1929): 454.

11. *Transit Journal* 80, no. 7 (July 1936): 217, 227–29; no. 10 (15 Sept. 1936): 353; Kashin and Demoro, *American Original,* 98–99, 170–71.

12. Clarence Hirshfeld, "PCC Street Car," *AIEE Transactions* 57 (Feb. 1938): 63.

13. Ibid., 63–64; *Mass Transportation* 32, no. 7 (July 1936): 194–206; 34, no. 6 (June 1938): 179–81; C. Bethel, "Economies of the High Speed Street Car Motor and Drive," *Proceedings of the American Electric Railway Engineering Association* (1930): 84; Hirshfeld, "Address to AERA," 160; C. A. Burleson, "Electrical Equipment and Control" (1934): 463–64; C. A. Ives, "Braking Equipment" (1934): 479–84; H. H. Adams, "P.C.C. Car Experience" (1938): 400, 404–6; W. A. Keller, "P.C.C. Car Experience" (1938): 417–18; A. T. Clark, "P.C.C. Car Experience" (1939): 409–10; *Transit Journal* 76, no. 11 (15 Oct. 1932): 465; 80, no. 7 (July 1936): 216, 218–21, 224–26, 230–32; no. 10 (15 Sept. 1936): 352; Kashin and Demoro, *American Original,* 147, 159–63, 165–71; Carlson and Schneider, *PCC,* 117–19, 144–47, 149–51; Francis R. Thompson, *Electric Transportation* (Scranton, Pa.: International Textbook, 1940), 11–12, 17–19.

For a good look at the state of interior car illumination prior to the ERPCC, see *Proceedings of the American Electric Railway Engineering Association* (1931): 649–79. For detailed information of PCC lighting fixtures and lamps, see *Proceedings of the American Transit Engineering Association* (1937): 638–47; *Transit Journal* 76, no. 11 (15 Oct. 1932): 465. See also Hirshfeld, "Address to AERA," 160–66; Hirshfeld, "PCC Street Car," 64.

14. Hirshfeld, "PCC Street Car," 63.

15. *Mass Transportation* 34, no. 6 (June 1938): 179–81; C. A. Burleson, "Electrical Equipment and Control," *Proceedings of the American Transit Engineering Association* (1934): 464–66; C. A. Ives, "Braking Equipment," 479; W. T. Rossell, "Summary of Result of P.C.C. Car Operation" (1938): 371, 373, 377, 379–81, 387; *Transit Journal* 80, no. 7 (July 1936): 216, 218–19, 226–29; no. 10 (15 Sept. 1936): 352; Middleton, *Time of the Trolley,* 156, 168, 218; Kashin and Demoro, *American Original,* 22, 24, 45–55, 58–59, 74–75; Carlson and Schneider, *PCC,* 159.

16. Clarence Hirshfeld, "Production, Maintenance, and Operating Features of the Presidents Conference Committee Car," *Proceedings of the American Transit Engineering Association* (1936): 820–21 (quote from 821).

17. The best summary of Brill, the ERPCC, and the Brilliner can be found in Brill, *J. G. Brill Company,* 191 207; *Mass Transportation* 34, no. 9 (Sept. 1938): advertising pages 103–6 (Brill advertisement), 257–60; W. D. Bearce, "Electrical Equipment for the Brilliner for Atlantic City," 267–68; *Transit Journal* 82, no. 9 (Sept. 1938): 306–7; Kashin and Demoro, *American Original,* 63–66, 191; Carlson and Schneider, *PCC,* 177–87.

18. N. C. Towle, "Keeping Old Cars in Step with New Practice," *Mass Transportation* 33, no. 4 (Apr. 1937): 101–2; Morris Buck, "Revamped Cars Give Higher Speeds," *Transit Journal* 76, no. 4 (Apr. 1932): 170–72; *Transit Journal* 80, no. 7 (July 1936): 233–35; Carlson and Schneider, *PCC,* 188–92.

19. Statistical information on manufacturing activity during the 1930s was obtained for this chapter by compiling data from car orders appearing in *Bus Transportation, Electric Railway Journal, Electric Traction,* and *Transit Journal.*

20. *Electric Railway Journal* 74, no. 3 (March 1930): 152; no. 8 (July 1930): 494–95; *Electric Traction* 26, no. 2 (Feb. 1930): 84; 27, no. 2 (Feb. 1931): 75–76.

21. *Transit Journal* 77, no. 7 (July 1933): 234.

22. The final mention of the Cincinnati Car Company in city directories was in *Williams' Cincinnati Directory: 1936–37* (Cincinnati: Williams Directory, 1936). *Bus Transportation* 9, no. 10 (Oct. 1930): 593; *The Cleveland Directory Co.'s Cleveland City Directory, 1931* 61 (Cleveland: Cleveland Directory, 1931), 475; *Electric Railway Journal* 75, no. 2 (Feb. 1931): 117; *Electric Traction* 27, no. 1 (Jan. 1931): 55–56; 28, no. 4 (Apr. 1932): 185; *Transit Journal* 77, no. 7 (July 1933): 234; Richard J. S. Gutman, *The American Diner Then and Now* (New York: HarperCollins, 1993); *SCA Journal* 18, no. 1 (Spring 2000): 32.

23. *Electric Railway Journal* 75, no. 7 (July 1931): 391; *Electric Traction* 27, no. 7 (July 1931): 362.

24. *Electric Traction* 28, no. 1 (Jan. 1932): 46; *Transit Journal* 76, no. 1 (Jan. 1932): 53. The last mention of J. G. Brill Co. of Ohio is in the *Cleveland City Directory* (1931), 475.

25. Advertisement, *Mass Transportation* 34, no. 9 (Sept. 1938): 102, *Transit Journal* 78, no. 6 (June 1934): 113; 79, no. 4 (Apr. 1935): 113; 80, no. 10 (15 Sept. 1936): 378; 81, no. 10 (15 Sept. 1937): 389; Tool Steel Gear & Pinion Co., *The Tool Steel Gear & Pinion Company, 50th Anniversary* (Cincinnati: Tool Steel Gear & Pinion, 1959), 6.

26. *Electric Railway Journal* 74, no. 2 (Feb. 1930): 109; 75, no. 8 (Aug. 1931): 430; *Electric Traction* 27, no. 3 (March 1931): 164; *Electric Traction and Bus Journal* 29, no. 9 (Sept. 1933): 294–95; *Mass Transportation* 31, no. 9 (Sept. 1935): 291; G. W. Bower, "Selecting Pole Type Current Collection," *Transit Journal* 76, no. 13 (Dec. 1932): 547–48; 80, no. 7 (July 1936): 255; no. 10 (15 Sept. 1936): 378; no. 12 (Nov. 1936): 452; no. 13 (Dec. 1936): 484–85; 81, no. 3 (March 1937): 102.

27. *Bus Transportation* 9, no. 7 (July 1930): 411; 10, no. 12 (Dec. 1931): 669; *Electric Railway Journal* 74, no. 3 (March 1930): 178; 75, no. 3 (March 1931): 172; no. 12 (Nov. 1931): 723; *Electric Traction* 26, no. 1 (Jan. 1930): 43; no. 6 (June 1930): 347; 27, no. 12 (Dec. 1931): 604; *Transit Journal* 80, no. 10 (15 Sept. 1936): 372.

28. *Electric Railway Journal* 75, no. 3 (March 1931): 172.

29. *Electric Traction and Bus Journal* 30, no. 5 (May 1934): 167; no. 8 (Aug. 1934): 263; *Mass Transportation* 33, no. 5 (May 1937): 149; advertisement, no. 9 (Sept. 1937): 38; 34, no. 3 (March 1938): 103; *Transit Journal* 78, no. 6 (June 1934): 181; 79, no. 3 (March 1935): 80; *Polk's Toledo (Lucas County, Ohio) City Directory, 1935* (Toledo: Toledo Directory, 1935), 612; *Polk's Toledo (Lucas County, Ohio) City Directory, 1940* (Toledo: Toledo Directory, 1940), 706.

30. *Bus Transportation* 13, no. 6 (June 1934): 207, 231; 17, no. 8 (Aug. 1938): 400; *Mass Transportation* 31, no. 9 (Sept. 1935): 319; 32, no. 1 (Jan. 1936): 32; no. 9 (Sept. 1936): 296.

31. *Bus Transportation* 16, no. 10 (Oct. 1937): 502; *Mass Transportation* 33, no. 8 (Aug. 1937): 246; *Transit Journal* 80, no. 8 (Aug. 1936): 294; *Cleveland City Directory* (1935), 1,053.

32. *Electric Railway Journal* 75, no. 4 (Apr. 1931): 217.

Afterword

1. It would not be until after World War Two that General Motors would dominate the bus market and force out such firms as Ohio-based Twin Coach and White.

2. *Transit Journal* 82, no. 11 (Oct. 1938): 435.

3. *Transit Journal* 82, no. 9 (Sept. 1938): 328–29.

4. *Bus Transportation* 17, no. 12 (Dec. 1938): 624.

BOOKS IN THE RAILROADS PAST AND PRESENT SERIES

Landmarks on the Iron Railroad: Two Centuries of North American Railroad Engineering by William D. Middleton

South Shore: The Last Interurban (revised second edition) by William D. Middleton

"Yet there isn't a train I wouldn't take": Railroad Journeys by William D. Middleton

The Pennsylvania Railroad in Indiana by William J. Watt

In the Traces: Railroad Paintings of Ted Rose by Ted Rose

A Sampling of Penn Central: Southern Region on Display by Jerry Taylor

Katy Northwest: The Story of a Branch Line Railroad by Donovan L. Hofsommer

The Lake Shore Electric Railway by Herbert H. Harwood, Jr. and Robert S. Korach

The Pennsylvania Railroad at Bay: William Riley McKeen and the Terre Haute and Indianapolis Railroad by Richard T. Wallis

The Bridge at Quebec by William D. Middleton

History of the J. G. Brill Company by Debra Brill

When the Steam Railroads Electrified by William D. Middleton

Uncle Sam's Locomotives: The USRA and the Nation's Railroads by Eugene L. Huddleston

Metropolitan Railways: Rapid Transit in America by William D. Middleton

Limiteds, Locals, and Expresses in Indiana, 1838–1971 by Craig Sanders

Perfecting the American Steam Locomotive by J. Parker Lamb

Invisible Giants: The Empires of Cleveland's Van Sweringen Brothers by Herbert H. Harwood, Jr.

From Small Town to Downtown: A History of the Jewett Car Company, 1893–1919 by Lawrence A. Brough and James H. Graebner

Steel Trails of Hawkeyeland: Iowa's Railroad Experience by Don L. Hofsommer

Still Standing: A Century of Urban Train Station Design by Christopher Brown

The Indiana Rail Road Company: America's New Regional Railroad by Christopher Rund

Amtrak in the Heartland by Craig Sanders

The Men Who Loved Trains: The Story of Men Who Battled Greed to Save an Ailing Industry by Rush Loving, Jr.

The Train Of Tomorrow by Ric Morgan

Evolution of the American Diesel Locomotive by J. Parker Lamb

The Encyclopedia of North American Railroads edited by William D. Middleton, George M. Smerk, and Roberta L. Diehl

INDEX

Page numbers in italics refer to illustrations.

Adams & Westlake (Chicago), 139
Advance Manufacturing & Tool Company (Cleveland), 199
Akerson, D. Hayward, 158–159
Akron Trolley Wheel Company (Akron, Ohio), 51
Allison, W. C., 30–31
American Car & Foundry Company, 160
American Car Company (St. Louis), 16, 173, 179
American Electric Railway Association, 6, 80, 116, 167, 215; Claims Association, 91; conventions, 6, 41–42, 102, 154, 165, 167, 174; Engineering Association, 65, 105–106, 112; Equipment Committee, 168, 171; Fare Research Bureau, 116; predecessor organizations, 6; subsidiary organizations, 6; Traffic and Transportation Association, 126, 129. *See also specific organizations*
American Electric Railway Manufacturers' Association, 6
American Railway Equipment Company, 207
American Railway Supply Company, (N.Y.), 135
American Road Builders' Association, 190
American Society of Mechanical Engineers, 65
American Taximeter Company (N.Y.), 205
Arkwright, P. S., 166
Atha Steel Casting Company (Newark, N.J.), 66
Automatic Car Coupler Company (Los Angeles), 83

Barney, Eliam, 14, 139
Barney & Smith Car Company (Dayton, Ohio), *10–11*, 14–15, 41, 44–45, *140–144*; closure, 160–161; Kossuth, 14; output, 15, 233; physical plant, 15; World War One, 160
Bayonet Trolley Harp Company (Springfield, Ohio), 53–54, 57, 200
Bellamy Vestlette Company (Cleveland), 124
Bendix-Westinghouse, 225
Bergstrom, Randolph, 89–90
Besuden, Edwin, 159, 209
Bolton, C. C., 16
Bonham Fare Recording Company (Hamilton, Ohio), 207
Bonney-Vehslage Tool Company (Newark, N.J.), 129
Boston Elevated Railway, 52, 54, 71
Bloomfield, Russell, 67
brakes, *88*, 200; air *88*, 92, 96, 200; air (storage systems), 93; electrical, 94, 201–202; hand, *88*, *93–94*, 202, 238; PCC, 217; self-lapping, 200
Brill Magazine, 7
Brooklyn Rapid Transit Company, 109–110, 17, 119
Brown, Fayette, 16
bus manufacture, 181, 190, 192, 227–229

C. I. Earll, 57, 223
cable cars. *See also* street railway systems, cable-powered
Canadian Ohmer, 205
Canadian Taximeters, Ltd., 205
Canton, Ohio, 56

car building, *35–39,* 213; body framing, *37, 43;* business, 40; end platforms, 38; finishing, *38–39;* floor framing, *37,* 43; painting, *39,* 209; roofing, *37–38,* 42; transport, *28–29*
Car Lamp Manufacturers Association, 139
car orders, 40
Central Electric Railway Association, 50, 80–81, 134
Chicago, 167
Chicago & Joliet Electric Railway Company, Ill., 176, 199
Chicago City Railway, 145–146
Chicago, North Shore & Milwaukee, Wis., 174, *186*
Chicago Railways, 51
Cincinnati, 13
Cincinnati Car Company, 11, 26, *27–30,* 40, 42–44, 47, 86, 95, *107, 127,* 142, 168, *172,*173–174, 178, 182, 184–185, *186–192, 212,* 214–215, 220, 221, 235; cooperative education, 184; couplers, 243; locomotives, 190; output, 30, 182, 236, 243–244; physical plant, *27–29, 188–190;* sanders, 244; ventilators, 150; World War One, 158–159
Cincinnati Car Corporation, 192
Cincinnati Manufacturing Company, 135
Cincinnati Traction Company (and predecessors), 26–27, 30, 60, 235
Cleveland, 13
Cleveland Fare Box Company (Cleveland), *127*–128, 204–206, 224
Cleveland Railway, *70,* 74, *101, 164,* 169, 179
Cleveland, Southwestern & Columbus Railway, 91
Cleveland Tanning Company (Cleveland), 208–209
Cleveland Trolley Supply Company (Cleveland), 199
Cleveland Trolley Wheel Company (Cleveland), 51
Columbia Machine Works & Malleable Iron Company (Brooklyn, N.Y.), 156–157
Commonwealth Power, Railway & Light Company (Grand Rapids, Mich.), 155
conductors, *114,* 120
Conway, Dr. Thomas, Jr., 212, 214
couplers, 71–72, 78, *79,* 80–81, *82–85,* 86–87; drawbars, 238; knuckle, *79;* link and pin, 79; MCB, *79,* 80–81, 84, 86, *188;* standardization, 78–80; Tomlinson, *83–85,* 86,162, 222, 240–242; Van Dorn, 81, *82*
Covington Brass Manufacturing Company (Covington, Ky.), 139
Cowan, C. E., 16
current collection apparatus, *46,* 47, *48–49,* 50–63, 199–200, 202, 219; catchers and retrievers, *48–49,* 57, *58–58,* 223, *224,* (Earll) 57, 223, (Ideal) 59, (Knutson) 58, *59,* 142, 239, (Ohio Brass) *58, 224,* (Ohio output) 59, (Q-P) 57; failures, 49–50, 56; sleet cutters, *59,* 60, *61–62,* 63, (Bayonet) 63, 200, (Holland) 63, (Ohio Brass) *62;* trolley harp, *48, 51–53,* 54; trolley pole, *46,* 48, 54–55; trolley pole base, *48, 53,* 55–57, 200, 223, (Ohio Brass) *53,* (Peerless) 56, (Simplex) 56, (Star) 56, (Union Standard) 56; trolley shoes, 199; trolley silencer, 200; trolley wheel, *46, 48–49,* 50–51, *52,* 200, (Ideal) 50, (New Haven) 50–51
Curwen, Samuel, 16

Danville Car Company (Danville, Ill.), 16
Dayton, Covington & Piqua Traction Company (Ohio), 131–133
Dayton, Ohio, 13–14, 129, 137–138, 161
Dayton Fare Recorder Company (Dayton, Ohio), 129–130, 134, 135, 205, 206
Dayton Manufacturing Company (Dayton, Ohio), 96, *136,* 137, *138,* 139, *140–144,* 200, 203, 210; output, 139–140, 202, 210, 249; ventilators, 150
Dean, D. B., 16
Denver City Tramway Company, 67
Detroit-Timken Axle Company, 175–176
Doane, W. H., 44
Duplex Headlight Company (Cleveland), 97
Dwight, E. E., 207
Dyer, J. Milton, 16

Ebert, Henry C., 30
Eckert Sanitary Closet, *136, 138*
Eclipse Car Fender Company (Cleveland), *102,* 205, 247
Economic Manufacturing Company (Springfield, Ohio), 54
Electric Railway Journal, 6, 44, 49, 545–56, 80, 91, 104, 154, 157, 174–175, 185, 202; predecessor publications, 6–7, 91, 104; successor publications, 7
Electric Railway Presidents Conference Committee, 215–216, 218, 228–229, 254
Electric Service Supply Company (Philadelphia), 157
Electric Traction, 7
Elliott, Thomas, 184–185, 187, 189, 229–230
Elyria, Ohio, 55
Eureka Trolley Company (Ironton, Ohio), 51
Evans Products Company (Detroit), 225

fare boxes, 124, *125*, 126, *127–128;* registering, 126; Coleman, *125*
fare collection, 120
fare collection and registration, 115–121, *122–123,* 124, *125,* 126, *127–128,* 129–130, *131–133,* 134–135, 205–207; prepayment, 121, 125
fare media, 119–120
fare registers, 129–130, *131–135;* Dayton, 130; industry output, 135; Ohmer, 130–131, *132–133,* 134, *142;* Ohmergraph, 134; turn-in-car, *134*
fares: interurban, 117–118; rapid transit, 118; street railway, 115–117
fares and transfers, 115–118; discounts and free rides, 118; fraud, 122–124
Farmer, Thomas, 16
Fay & Egan (Cincinnati), 44
Franco-American Company, 205
Feist, Charles, 200
Fixler, Edgar R., 56–57
Fixler Trolley Stand Company (Delta, Ohio), 56–57
Fred J. Meyers Company (Hamilton, Ohio), 135

G. C. Kuhlman Car Company, 15–16, *17–21,* *35–39,* 44, 95, *152,* 173–174, 178–179, *180–185,* 255; output, 19, 21, 179, 234; physical plant, 15–16, *35–39, 180;* sale to Brill, 16; World War One, 158–159
Garford Company (Elyria, Ohio), 55
Garland, T. H., 149
gears and pinions, 63, *64,* 65–67, *194,* 196–197, 216, 239
General Electric Company, 4, 53, 56, 64, 75–77, 97, 103, 113, 203, 217, 222
General Railway Supply Company (Chicago), 149
Glidden & Joy (Cleveland), 209
Globe Machinery & Stampings Company (Cleveland), 97
Globe Ticket Company, 119
Gordon, Charles, 168

Harriman, R. A., 16
headlights, 97, *98,* 202–203, 222–223
Herbert, William, 30
Hirschfeld, Dr. Clarence, 215–217
Hitchcock, P. M., 16
Holland Trolley Supply Company (Cleveland), 54, 56
Homer Commutator Company (Cleveland), 68
Hoover, Herbert, 165
horsecars. *See under* street railway systems
Howell, T. P., 16

Illinois Manufacturing Company, 139
Illinois Traction System, 81
Inglis, J. G., 214
International Register Company (Chicago), 129, 135
International Traction Company (Buffalo), 155
Interstate Commerce Commission, 154
interurban cars, *7, 10;* combine, 8, *18, 25–26,* *32,* 186, 237; express, 8, *25, 33;* observation, *34;* passenger, *10, 23–24, 29, 33,* 186, *212,* 214–215
interurban design, curve-sided, 185–188, *189, 191*
interurban railways, 5; Midwest, 8
Ironton, Ohio, 51

J. G. Brill Company, 7–8, *19,* 35, 47, 95, 109, 120, 126, 140, 142, 172–173, 177, 179, 203, 215, 219–221, 255; car builder acquisitions, 16; ventilators, 150; World War One, 158
Jacobs, A. L., 31
James, Lee Warren, 160
James L. Howard & Company (Stafford, Conn.), 139
Jewett Car Company, 11, *21–26,* 40–42, 44, 95, 237; closure, 159, 161; output, 23, 26, 234; physical plant, 22–23; World War One, 159
John Stephenson Works (Elizabeth, N.J.), 16
Johnson, Tom L., 126
Johnson Fare Box Company (Chicago), 126, 224
Jones Taximeter, 205

Kemitex Products (Barberton, Ohio), 208
Kippenberger, George, 167
Kirby, John, Jr., 139
Kittredge, Arthur M.
Kraeuter & Company (Newark, N.J.), 135
Krebs, C. E., 22
Kuhlman, Charles E., 15
Kuhlman, Frederick, 15
Kuhlman, Gustave C., 15–16

L. A. Sayer & Company (Newark, N.J.), 135
lavatory fixtures, *142–144*
Leon L. Wolf Waterproof Fabric Company (Cincinnati), 209
Lev, Benjamin, 102
Libby-Owens-Ford (Toledo), 223–224
Lind, Alan, 45, 120
Lintern, William, 96
Lintern Car Signal Company (Cleveland), 96, 99
Lorentz, W. H., 22
Louisville Railway (Ky.), 178

Ludlow, A. W., 34
Lumen Bearing Company (Buffalo), 50, 52

MacDonald Fare Box Company (Cleveland), *128*, 224
MacDonald Ticket and Ticket Box Company (Cleveland), 119, 134
Marker lights, 98–99, *100;* Nichols-Lintern, *100*, 225; semaphore, 98–99
Master Car Builders' Association, 78
McConway & Torley Company (Pittsburgh), 81
McCorkle, A. G., 30
Midwest United States, 11
Milloy Electric Company (Bucyrus, Ohio), 56
Montreal Tranways, 73
Morgan, Clinton, 189
Morley, I. H., 16
Moskowitz, J. D., 14
motor components, brushes and commutators, 67–68
motors, 47, 94, 107, 195–196
Muir, Malcolm, 213
multiple-unit control systems, 74–75, *76*, 77–78, *85;* General Electric, 75–77; Sprague, 75; Westinghouse, *76*, *77*; wire systems, 77–78
Multiplex Reflector Company (Cleveland), 97

National Carbon Company (Cleveland), 68
National Malleable Castings Company (Cleveland), 84
National Ticket Company (Cleveland), 119
National Tube Company (Pittsburgh), 55
New Haven, Conn., 50
New Haven Trolley Supply Company (Conn.), 205
New York & Ohio Company, 113
New York Railways, 108–109
Newark, N.J., 60
Newark, Ohio, 13, 161
Newark Air Sand Box Company (Newark, Ohio), 96
Nichols-Lintern Company (Cleveland), 96, 203, 208, 224–225; output, 96–97, 204, 208, 224–225, 245–246
Nichols-Lintern Corporation, 225
Niles, Ohio, 13, 161
Niles Car & Manufacturing Company (Niles, Ohio), 11, *30–34*, 41–42, 44, 99, 157, 173, *201;* output, 34, 236; physical plant, 31, 34; ventilators, 150; withdrawal from streetcar market, 161; World War One, 161

O. M. Edwards Company (Syracuse, N.Y.), 142
Ohio: canal systems, 12, 112, 137; industrial development, 12; Ohio Brass Company (Mansfield, Ohio) *49*, *51–53*, 58, *83–84*, 85–86, 96, 162, 199–200, 202–203, 219, 222; output, 86, 97, 199–200, 222, 240–242, 246–247
Ohio Lamp Works (Warren, Ohio), 113
Ohmer, John F., 130–134, 157, 162, 223, 230–231
Ohmer Fare Register Company (Dayton), 130, *132–133*, 134–135, 162, 205, *206*, 207, 223, 225; output, 248–249
Ontario Railway & Municipal Board (Canada), 106
Osgood-Bradley Car Company (Worcester, Mass.), 157, 176
Owosso & Corunna Electric Company (Owosso, Mich.), 237–238

Packard, J. Ward, 113
Packard, William D., 113
Packard Electric Company (Warren, Ohio), 113
Panic of 1893, 14
Parker, Caleb, 14
parlor cars, 8, 41–42, *140*, *143–144*
Partridge Carbon Company (Sandusky, Ohio), 68
Paulson, Neil, 22
Peerless Sectional Gear Company (N. Y.), 66
Perry Ventilator Company (New Bedford, Mass.), 149
Pittsburgh Railways, 73–74, 107, 176
Pneumatic Railway Equipment Company (Cleveland), 96
Post & Company (Cincinnati), 139
Pratt, G. E., 31
Pressed Steel Car Company (Pittsburgh), 43, 160
Public Service Commission of New York, 103, 105–106

Q-P Signal Company (Boston), 57

R. D. Nuttall Company (Pittsburgh), 56, 66
R. Woodman Manufacturing & Supply Company (Boston), 135
rapid transit cars, 8, *28*, *187–188*
Recording & Computing Machines Company, 206
Recording Fare Register Company (New Haven, Conn.), 50, 52, 126, 129, 205
Reymann, Paul, 159
Richmond, Va., 4, 59
Ricks, C. A., 16
Robbins, George B., 30
Rockefeller, Frank, 16

Rooke Automatic Register Company, 124–124
Root Spring Scraper Company (Kalamazoo, Mich.), 60
Root Track Scraper Company (Kalamazoo, Mich.), 60
Rossell, W. T., 216

safety devices, *88*, 89–92, *93–95*, 96–97, *98*, 99, *100–101*, 102–113, *170*, 171, 200, *201*, 202–203, *204*, 205, 214; anti-climbers, 102; bumpers, 101, 238; door-step interlocks, 106, *108–109;* fender and wheelguard testing, 102–104; fenders, *101*, 102, *107*, 204–205, (ACME) 102, (Cincinnati Car) 102, (Eclipse) *101*, 102, 247, (Kuhlman) 102, (Niles Car) 102; gates, 104, (folding) *105–107*, (Minneapolis) 104, (swinging) *105;* gongs and bells, 99; horns, 100; pilots, 100; power off lights, *204*, railings, 110; safety devices, 214, 223–224; sanders, 94, *95*, 96, 203, (Nichols-Lintern) 96, 142, 225, 245–246, (Ohio Brass) 246–247, (Simmons-Moore) 96, (Universal Sander) 96; stanchions, 110; steps, 104–105, 107, *108–109;* 238; straps, 110, *111;* wheel guards, 101, *103*, 104, *107*, 204–205, (Eclipse) 104, (Hudson & Bowring/H-B) 104, (Parmenter) 104, (Watson) 104. *See also* brakes; current collection apparatus, catchers and retrievers; headlights
Sanders, Henry, 189
Sawtelle, Charles, 67, 162
Schall, A. W., 34
Schoepf, W. Kelsey, 30, 189
Sherwin-Williams Company (Cleveland), 209
Sisson, A. H., 22
Smith, Preserved, 14
Snediker, Edward, 160
Soulder, C. P., 30
Sprague, Frank Julian, 4, 59, 75
Springfield, Ohio, 53–54
Springfield Street Railway (Mass.), 175
St. Louis Car Company, 8, 35, 47, 96, 109, 126, 137, 140, 142, 178, 203, 217; ventilators, 150
Standard Brass Foundry Company (Cleveland), 51
Standard Steel Car Company (Pittsburgh), 43, 107
Star Brass Works (Kalamazoo, Mich.), 53
Sterling-Meaker Company (Newark, N. J.), 129, 135
Sterling Fare Register & Fare Box Company (Newark, N. J.), 205
Steubenville & East Liverpool Light Company (Ohio), 142
Stone & Webster, 67
street railway accidents, 89–92; boarding and alighting, 104, 121; fraud, 90
street railway industry, 5, 165–166, 178–179, 213–2145, 227–29, 253; rail and vehicle orders, 252
street railway systems: cable-powered, 2–3; comparative, 5; early electric, 3–4; modern electric, 4; horsecars, 1; steam-powered streetcars, 2
street railways: Chicago, 92; claims departments, 90; safety education, 91; Midwest, 8; New Hampshire, 91. *See also* fare collection and registration; fares and transfers
streetcar design (ca. 1900), 35; articulated, 173–174; Birney,171, *172*, 173; bodies, 167–168, 175–176, *183*, 216; Brilliner, 219; Electromobile,177–178; lightweight, *164*, 169, 176, *180*, *185*, 187; low-floor, 107; Master Unit, 177; nearside, 120; New Birney, 175; 1920s prototypes, 174–178; one-man, 169, *170*, 171, 251; Pay-As-You-Enter, 121; PCC car, 216, 222–223; Peter Witt, 121, *122–123*, 169, *183;* pre-World War One, 35, 42; safety cars, 169, *170*, 171, *181–182*, 251; specialized, 174
streetcar materials: aluminum, 169, 176; composite cars, 43; Duralumin, 169; faced streetcars, 43; plymetl, 169; semi-steel, 43; steel, 36, 42–43, 180; transition from wood to steel, 43–44, 160; wood, 35–38, 42–43, 237
streetcars, basic types: California, 7; center-entrance, 7, *32*, closed, 7, *17*, *27–28*, *39;* convertible, 7, *17–18;* open, 7; semi-convertible, 7, *19*
streetcars, control systems, 47, 177, 198–199; General Electric, 198, 217; K-type, 4; PCC, 217–218; Westinghouse, 198, 217. *See also* multiple-unit control systems
streetcars, heating, *147–149*, 150, 207–208, 217
streetcars, interior lighting, 110–113, 201–203, 217–218; General Electric, 113; Nichols-Lintern controls, *112*, 224; lighting fixtures, 141
streetcars, seating, 208–209, 238
streetcars, trailers, *70*, 72–74
streetcars, trimmings and hardware, *136*, 137, *138*, 139, *140*, 140–142, *143–144*, 208–210
streetcars, ventilation, 142–145, *146–147*, 148–150; ventilation systems, 145–146, 207–208, 217, (Cooke) 146, 150, (Garland) 149, (McGerry) 146, 149, (Nichols-Lintern) 224–225, (Perry) 149–150, (Taylor) 149–150; ventilation systems testing, 145–146, 249

streetcars, weight vs. cost percentages, 250
streetcars, wheels, 217
streetcars, windows, 142, 214, 237–238
suburban cars, 8
Sykes, H. R., 190

Textileather Company (Toledo, Ohio), 209
Thresher, Ebenezer, 14
ticket punches, 135
Timken, Henry, 197
Timken Roller Bearing Company (Canton, Ohio), 197–198
Tomlinson, Charles H., 85, 230
Tool Steel Motor Gear & Pinion Company (Cincinnati), 66–67, 157, 162, 197, 219, 222; output, 67
Trolley Supply Company (Canton, Ohio), 56, 58, 59, 98, 199, 202–204; output, 203, 239
trucks, 47, 108, 188, *191–192*; maximum traction, 108
Trumbull County, Ohio, 112
Twin City Rapid Transit Company (Minneapolis), 104

United Copper Foundry (Boston), 52
United States, industrial development, 12
United States Railroad Administration, 160
Univertsity of Cincinnati, 184

Vacuum Car Ventilating Company (Chicago), 149–150
Van Dorn & Dutton Company (Cleveland), 66
Versare Corporation (Watervliet, N.Y.), 190
von Siemens, Werner, 3

W. H. Killbourn (Greenfield, Mass.), 57
Warren, Ohio, 112
Wason Car & Manufacturing Company (Springfield, Mass.), 16, 175, 179, 255; World War One, 158
Wason Engineering and Supply Company (Milwaukee), 57
Welke, Barbara, 100, 102, 104
Western Electric Company (Chicago), 53
Westinghouse Electrical & Manufacturing Company, 4, 64, 76–77, 103, 107, 176, 217; W-N Drive, 176–177
Westinghouse Traction Air Brake Company, 83, 176
White Motor Company (Cleveland), 181
Whitmore Manufacturing Company (Cleveland), 66
Wilson Trolley Catcher Company (Boston), 57
Wingertner, W. B., 40–41
Winters, Valentine, 160
Witt, Peter, 121
Wood, Clark, 175
World War One: Emergency Fleet Corporation, 158; government regulation of industry, 154–156; labor issues, 156–157; materials costs, 250; materials shortages, 153; U.S. Housing Corporation, 158; War Industries Board, 154
Wright, W. S., 22, 40–41, 159

Craig R. Semsel earned his doctoral degree in history from Case Western Reserve University, where he studied the history of technology and science. He has spent several years operating and helping to restore streetcars in Connecticut. He teaches history at Lakeland Community College and Lorain County Community College in Ohio.